These safety symbols are used in laboratory and field investigations in this book to indicate possible hazards. Learn the meaning of each symbol and refer to this page often. *Remember to wash your hands thoroughly after completing lab procedures.*

PROTECTIVE EQUIPMENT

Do not begin any lab without the proper protection equipment.

GOGGLES 	Proper eye protection must be worn when performing or observing science activities that involve items or conditions as listed below.
APRON 	Wear an approved apron when using substances that could stain, wet, or destroy cloth.
SOAP 	Wash hands with soap and water before removing goggles and after all lab activities.
GLOVES 	Wear gloves when working with biological materials, chemicals, animals, or materials that can stain or irritate hands.

LABORATORY HAZARDS

Symbols	Potential Hazards	Precaution	Response
DISPOSAL	contamination of classroom or environment due to improper disposal of materials such as chemicals and live specimens	• DO NOT dispose of hazardous materials in the sink or trash can. • Dispose of wastes as directed by your teacher.	• If hazardous materials are disposed of improperly, notify your teacher immediately.
EXTREME TEMPERATURE	skin burns due to extremely hot or cold materials such as hot glass, liquids, or metals; liquid nitrogen; dry ice	• Use proper protective equipment, such as hot mitts and/or tongs, when handling objects with extreme temperatures.	• If injury occurs, notify your teacher immediately.
SHARP OBJECTS	punctures or cuts from sharp objects such as razor blades, pins, scalpels, and broken glass	• Handle glassware carefully to avoid breakage. • Walk with sharp objects pointed downward, away from you and others.	• If broken glass or injury occurs, notify your teacher immediately.
ELECTRICAL	electric shock or skin burn due to improper grounding, short circuits, liquid spills, or exposed wires	• Check condition of wires and apparatus for fraying or uninsulated wires, and broken or cracked equipment. • Use only GFCI-protected outlets	• DO NOT attempt to fix electrical problems. Notify your teacher immediately.
CHEMICAL	skin irritation or burns, breathing difficulty, and/or poisoning due to touching, swallowing, or inhalation of chemicals such as acids, bases, bleach, metal compounds, iodine, poinsettias, pollen, ammonia, acetone, nail polish remover, heated chemicals, mothballs, and any other chemicals labeled or known to be dangerous	• Wear proper protective equipment such as goggles, apron, and gloves when using chemicals. • Ensure proper room ventilation or use a fume hood when using materials that produce fumes. • NEVER smell fumes directly. • NEVER taste or eat any material in the laboratory.	• If contact occurs, immediately flush affected area with water and notify your teacher. • If a spill occurs, leave the area immediately and notify your teacher.
FLAMMABLE	unexpected fire due to liquids or gases that ignite easily such as rubbing alcohol	• Avoid open flames, sparks, or heat when flammable liquids are present.	• If a fire occurs, leave the area immediately and notify your teacher.
OPEN FLAME	burns or fire due to open flame from matches, Bunsen burners, or burning materials	• Tie back loose hair and clothing. • Keep flame away from all materials. • Follow teacher instructions when lighting and extinguishing flames. • Use proper protection, such as hot mitts or tongs, when handling hot objects.	• If a fire occurs, leave the area immediately and notify your teacher.
ANIMAL SAFETY	injury to or from laboratory animals	• Wear proper protective equipment such as gloves, apron, and goggles when working with animals. • Wash hands after handling animals.	• If injury occurs, notify your teacher immediately.
BIOLOGICAL	infection or adverse reaction due to contact with organisms such as bacteria, fungi, and biological materials such as blood, animal or plant materials	• Wear proper protective equipment such as gloves, goggles, and apron when working with biological materials. • Avoid skin contact with an organism or any part of the organism. • Wash hands after handling organisms.	• If contact occurs, wash the affected area and notify your teacher immediately.
FUME	breathing difficulties from inhalation of fumes from substances such as ammonia, acetone, nail polish remover, heated chemicals, and mothballs	• Wear goggles, apron, and gloves. • Ensure proper room ventilation or use a fume hood when using substances that produce fumes. • NEVER smell fumes directly.	• If a spill occurs, leave area and notify your teacher immediately.
IRRITANT	irritation of skin, mucous membranes, or respiratory tract due to materials such as acids, bases, bleach, pollen, mothballs, steel wool, and potassium permanganate	• Wear goggles, apron, and gloves. • Wear a dust mask to protect against fine particles.	• If skin contact occurs, immediately flush the affected area with water and notify your teacher.
RADIOACTIVE	excessive exposure from alpha, beta, and gamma particles	• Remove gloves and wash hands with soap and water before removing remainder of protective equipment.	• If cracks or holes are found in the container, notify your teacher immediately.

INTEGRATED iSCIENCE

GLENCOE

COURSE 1

Mc Graw Hill Education

COVER: Dirk Ercken/Shutterstock.com

mheducation.com/prek-12

Send all inquiries to:
McGraw-Hill Education
8787 Orion Place
Columbus, OH 43240

ISBN: 978-0-07-677276-6
MHID: 0-07-677276-4

Printed in the United States of America.

2 3 4 5 6 7 8 9 QVC 22 21 20 19 18 17 16

Contents in Brief

Authors and Contributors

Authors

American Museum of Natural History
New York, NY

Michelle Anderson, MS
Lecturer
The Ohio State University
Columbus, OH

Juli Berwald, PhD
Science Writer
Austin, TX

John F. Bolzan, PhD
Science Writer
Columbus, OH

Rachel Clark, MS
Science Writer
Moscow, ID

Patricia Craig, MS
Science Writer
Bozeman, MT

Randall Frost, PhD
Science Writer
Pleasanton, CA

Lisa S. Gardiner, PhD
Science Writer
Denver, CO

Jennifer Gonya, PhD
The Ohio State University
Columbus, OH

Mary Ann Grobbel, MD
Science Writer
Grand Rapids, MI

Whitney Crispen Hagins, MA, MAT
Biology Teacher
Lexington High School
Lexington, MA

Carole Holmberg, BS
Planetarium Director
Calusa Nature Center and Planetarium, Inc.
Fort Myers, FL

Tina C. Hopper
Science Writer
Rockwall, TX

Jonathan D. W. Kahl, PhD
Professor of Atmospheric Science
University of Wisconsin-Milwaukee
Milwaukee, WI

Nanette Kalis
Science Writer
Athens, OH

S. Page Keeley, MEd
Maine Mathematics and Science Alliance
Augusta, ME

Cindy Klevickis, PhD
Professor of Integrated Science and Technology
James Madison University
Harrisonburg, VA

Kimberly Fekany Lee, PhD
Science Writer
La Grange, IL

Michael Manga, PhD
Professor
University of California, Berkeley
Berkeley, CA

Devi Ried Mathieu
Science Writer
Sebastopol, CA

Elizabeth A. Nagy-Shadman, PhD
Geology Professor
Pasadena City College
Pasadena, CA

William D. Rogers, DA
Professor of Biology
Ball State University
Muncie, IN

Donna L. Ross, PhD
Associate Professor
San Diego State University
San Diego, CA

Marion B. Sewer, PhD
Assistant Professor
School of Biology
Georgia Institute of Technology
Atlanta, GA

Julia Meyer Sheets, PhD
Lecturer
School of Earth Sciences
The Ohio State University
Columbus, OH

Michael J. Singer, PhD
Professor of Soil Science
Department of Land, Air and Water Resources
University of California
Davis, CA

Karen S. Sottosanti, MA
Science Writer
Pickerington, Ohio

Paul K. Strode, PhD
I.B. Biology Teacher
Fairview High School
Boulder, CO

Jan M. Vermilye, PhD
Research Geologist
Seismo-Tectonic Reservoir Monitoring (STRM)
Boulder, CO

Judith A. Yero, MA
Director
Teacher's Mind Resources
Hamilton, MT

Dinah Zike, MEd
Author, Consultant, Inventor of Foldables
Dinah Zike Academy; Dinah-Might Adventures, LP
San Antonio, TX

Margaret Zorn, MS
Science Writer
Yorktown, VA

Consulting Authors

Alton L. Biggs
Biggs Educational Consulting
Commerce, TX

Ralph M. Feather, Jr., PhD
Assistant Professor
Department of Educational Studies and Secondary Education
Bloomsburg University
Bloomsburg, PA

Douglas Fisher, PhD
Professor of Teacher Education
San Diego State University
San Diego, CA

Edward P. Ortleb
Science/Safety Consultant
St. Louis, MO

Series Consultants

Science

Solomon Bililign, PhD
Professor
Department of Physics
North Carolina Agricultural and Technical State University
Greensboro, NC

John Choinski
Professor
Department of Biology
University of Central Arkansas
Conway, AR

Anastasia Chopelas, PhD
Research Professor
Department of Earth and Space Sciences
UCLA
Los Angeles, CA

David T. Crowther, PhD
Professor of Science Education
University of Nevada, Reno
Reno, NV

A. John Gatz
Professor of Zoology
Ohio Wesleyan University
Delaware, OH

Sarah Gille, PhD
Professor
University of California San Diego
La Jolla, CA

David G. Haase, PhD
Professor of Physics
North Carolina State University
Raleigh, NC

Janet S. Herman, PhD
Professor
Department of Environmental Sciences
University of Virginia
Charlottesville, VA

David T. Ho, PhD
Associate Professor
Department of Oceanography
University of Hawaii
Honolulu, HI

Ruth Howes, PhD
Professor of Physics
Marquette University
Milwaukee, WI

Jose Miguel Hurtado, Jr., PhD
Associate Professor
Department of Geological Sciences
University of Texas at El Paso
El Paso, TX

Monika Kress, PhD
Assistant Professor
San Jose State University
San Jose, CA

Mark E. Lee, PhD
Associate Chair & Assistant Professor
Department of Biology
Spelman College
Atlanta, GA

Linda Lundgren
Science writer
Lakewood, CO

Keith O. Mann, PhD
Ohio Wesleyan University
Delaware, OH

Charles W. McLaughlin, PhD
Adjunct Professor of Chemistry
Montana State University
Bozeman, MT

Katharina Pahnke, PhD
Research Professor
Department of Geology and Geophysics
University of Hawaii
Honolulu, HI

Jesús Pando, PhD
Associate Professor
DePaul University
Chicago, IL

Hay-Oak Park, PhD
Associate Professor
Department of Molecular Genetics
Ohio State University
Columbus, OH

David A. Rubin, PhD
Associate Professor of Physiology
School of Biological Sciences
Illinois State University
Normal, IL

Toni D. Sauncy
Assistant Professor of Physics
Department of Physics
Angelo State University
San Angelo, TX

Series Consultants, continued

Malathi Srivatsan, PhD
Associate Professor of Neurobiology
College of Sciences and Mathematics
Arkansas State University
Jonesboro, AR

Cheryl Wistrom, PhD
Associate Professor of Chemistry
Saint Joseph's College
Rensselaer, IN

Reading

ReLeah Cossett Lent
Author/Educational Consultant
Blue Ridge, GA

Math

Vik Hovsepian
Professor of Mathematics
Rio Hondo College
Whittier, CA

Series Reviewers

Thad Boggs
Mandarin High School
Jacksonville, FL

Catherine Butcher
Webster Junior High School
Minden, LA

Erin Darichuk
West Frederick Middle School
Frederick, MD

Joanne Hedrick Davis
Murphy High School
Murphy, NC

Anthony J. DiSipio, Jr.
Octorara Middle School
Atglen, PA

Adrienne Elder
Tulsa Public Schools
Tulsa, OK

Carolyn Elliott
Iredell-Statesville Schools
Statesville, NC

Christine M. Jacobs
Ranger Middle School
Murphy, NC

Jason O. L. Johnson
Thurmont Middle School
Thurmont, MD

Felecia Joiner
Stony Point Ninth Grade Center
Round Rock, TX

Joseph L. Kowalski, MS
Lamar Academy
McAllen, TX

Brian McClain
Amos P. Godby High School
Tallahassee, FL

Von W. Mosser
Thurmont Middle School
Thurmont, MD

Ashlea Peterson
Heritage Intermediate Grade Center
Coweta, OK

Nicole Lenihan Rhoades
Walkersville Middle School
Walkersvillle, MD

Maria A. Rozenberg
Indian Ridge Middle School
Davie, FL

Barb Seymour
Westridge Middle School
Overland Park, KS

Ginger Shirley
Our Lady of Providence Junior-Senior High School
Clarksville, IN

Curtis Smith
Elmwood Middle School
Rogers, AR

Sheila Smith
Jackson Public School
Jackson, MS

Sabra Soileau
Moss Bluff Middle School
Lake Charles, LA

Tony Spoores
Switzerland County Middle School
Vevay, IN

Nancy A. Stearns
Switzerland County Middle School
Vevay, IN

Kari Vogel
Princeton Middle School
Princeton, MN

Alison Welch
Wm. D. Slider Middle School
El Paso, TX

Linda Workman
Parkway Northeast Middle School
Creve Coeur, MO

Teacher Advisory Board

The Teacher Advisory Board gave the authors, editorial staff, and design team feedback on the content and design of the Student Edition. They provided valuable input in the development of *Glencoe Integrated iScience*.

Frances J. Baldridge
Department Chair
Ferguson Middle School
Beavercreek, OH

Jane E. M. Buckingham
Teacher
Crispus Attucks Medical Magnet High School
Indianapolis, IN

Elizabeth Falls
Teacher
Blalack Middle School
Carrollton, TX

Nelson Farrier
Teacher
Hamlin Middle School
Springfield, OR

Michelle R. Foster
Department Chair
Wayland Union Middle School
Wayland, MI

Rebecca Goodell
Teacher
Reedy Creek Middle School
Cary, NC

Mary Gromko
Science Supervisor K–12
Colorado Springs District 11
Colorado Springs, CO

Randy Mousley
Department Chair
Dean Ray Stucky Middle School
Wichita, KS

David Rodriguez
Teacher
Swift Creek Middle School
Tallahassee, FL

Derek Shook
Teacher
Floyd Middle Magnet School
Montgomery, AL

Karen Stratton
Science Coordinator
Lexington School District One
Lexington, SC

Stephanie Wood
Science Curriculum Specialist, K–12
Granite School District
Salt Lake City, UT

Welcome to iSCIENCE

We are your partner in learning by meeting your diverse 21st century needs. Designed for today's tech-savvy middle school students, the Glencoe *iScience* program offers hands-on investigations, rigorous science content, and engaging, real-world applications to make science fun, exciting, and stimulating.

Quick Start Guide
iScience | Student Center

Login information

1. Go to **connected.mcgraw-hill.com.**
2. Enter your registered Username and Password.
3. For **new users** click here to create a new account.
4. Get **ConnectED Help** for creating accounts, verifying master codes, and more.

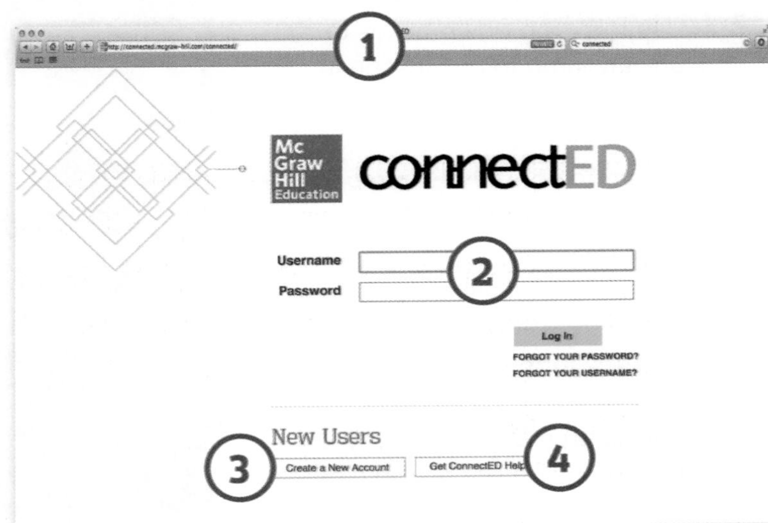

Your ConnectED Center

5. Scroll down to find the program from which you would like to work.

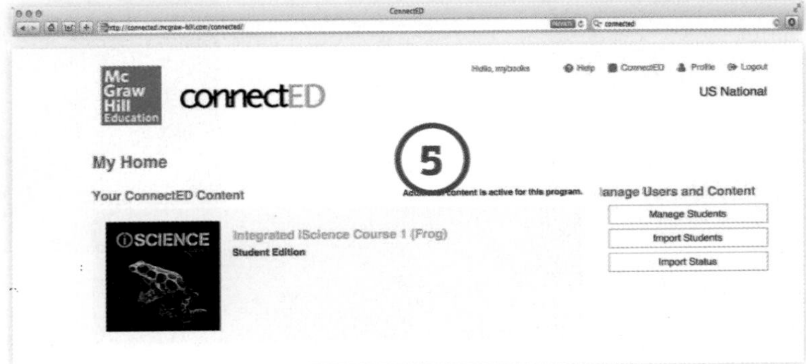

Quick Start Guide
iScience | Student Center

1. The Menu allows you to easily jump to anywhere you need to be.

2. Click the **program icon** at the top left to **return to the main page** from any screen.

3. **Select a Chapter and Lesson** Use the drop down boxes to quickly jump to any lesson in any chapter.

4. Return to your **My Home** page for all your **ConnectED** content.

5. The **Help** icon will guide you to online help. It will also allow for a quick logout.

6. The **Search Bar** allows you to search content by topic or standard.

7. **Access the eBook** **Use** the **Student Edition** to see content.

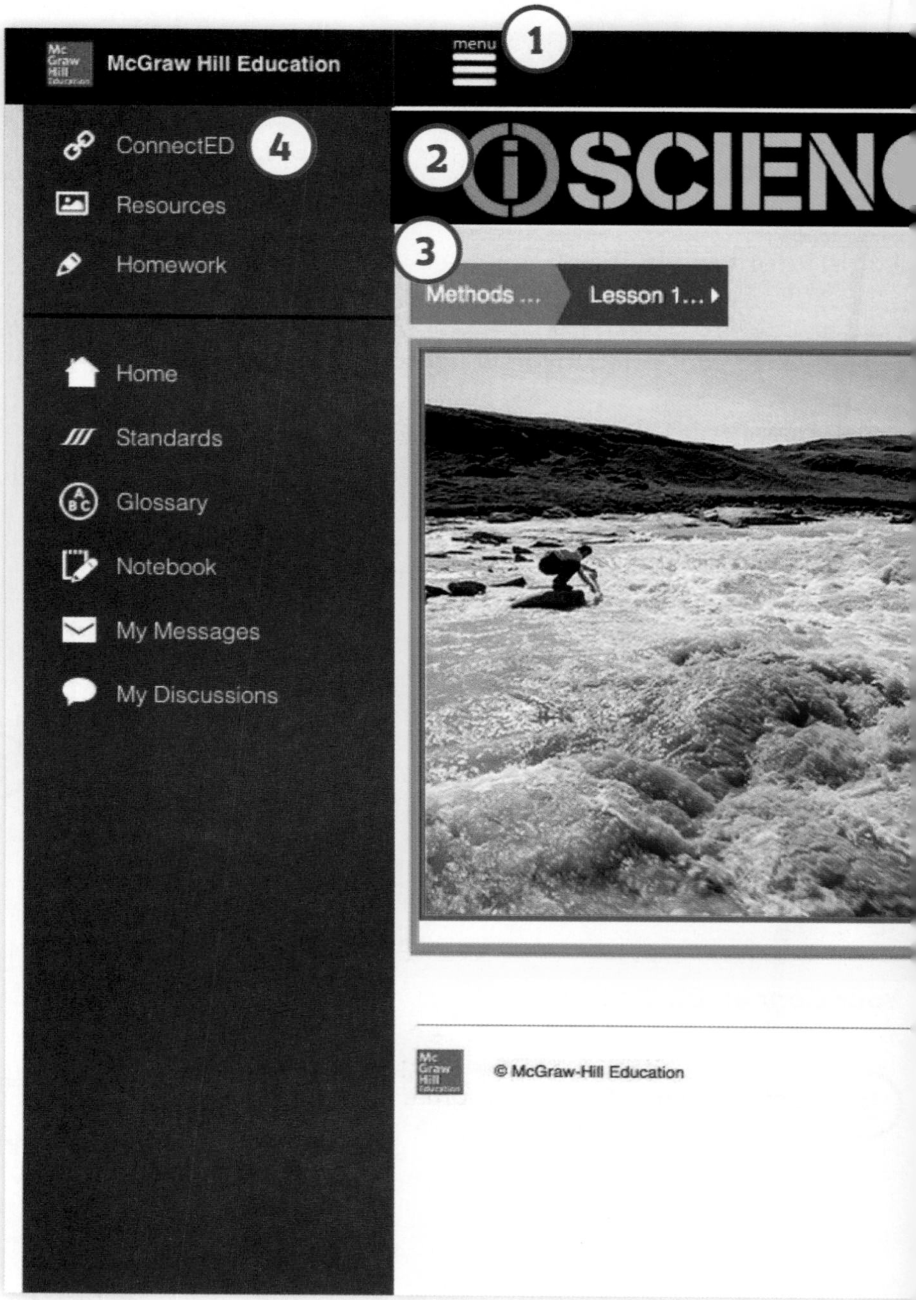

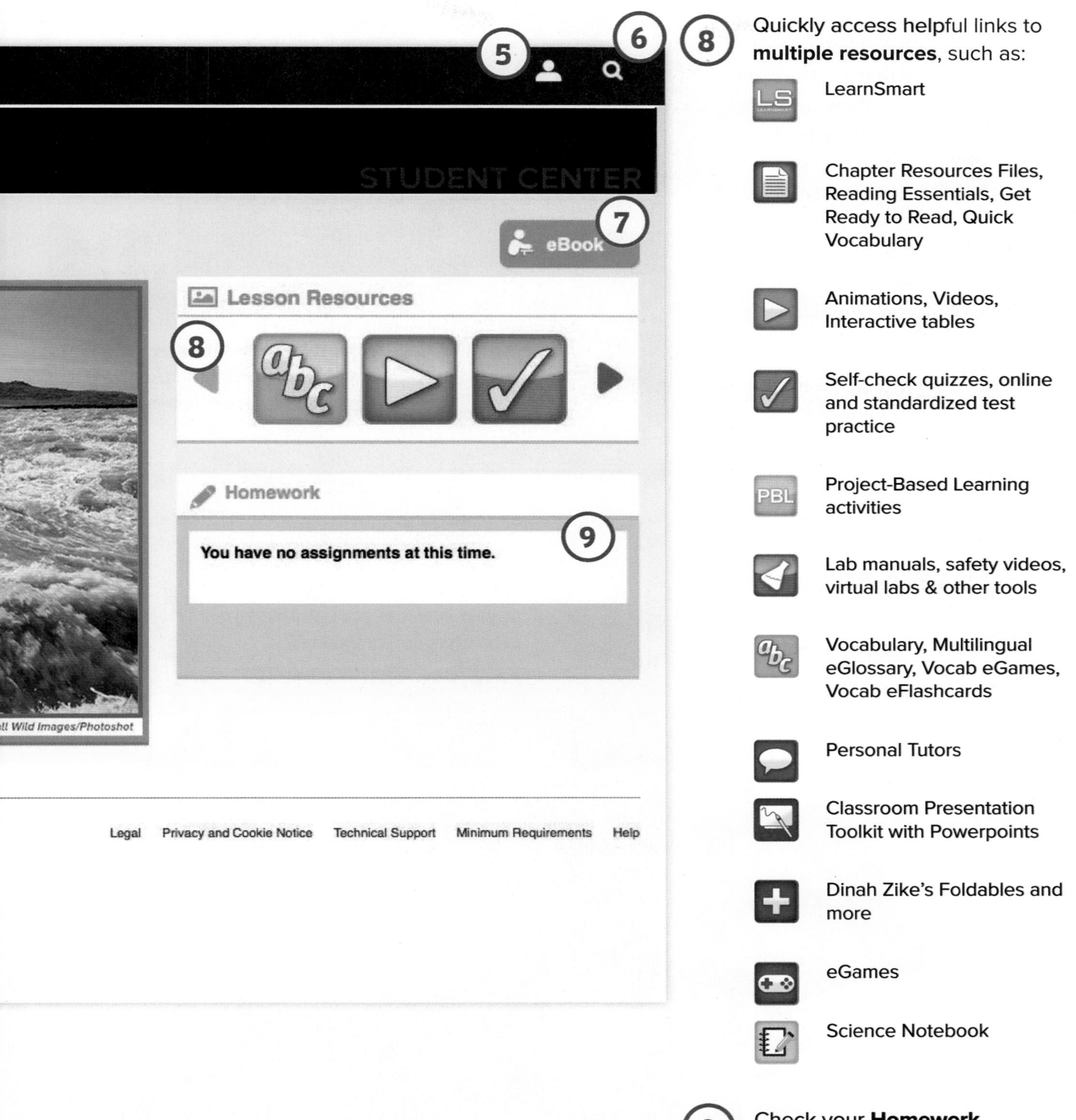

8 Quickly access helpful links to **multiple resources**, such as:

- LearnSmart
- Chapter Resources Files, Reading Essentials, Get Ready to Read, Quick Vocabulary
- Animations, Videos, Interactive tables
- Self-check quizzes, online and standardized test practice
- Project-Based Learning activities
- Lab manuals, safety videos, virtual labs & other tools
- Vocabulary, Multilingual eGlossary, Vocab eGames, Vocab eFlashcards
- Personal Tutors
- Classroom Presentation Toolkit with Powerpoints
- Dinah Zike's Foldables and more
- eGames
- Science Notebook

9 Check your **Homework** assignments here.

Treasure Hunt

Your science book has many features that will aid you in your learning. Some of these features are listed below. You can use the activity at the right to help you find these and other special features in the book.

- THE BIG IDEA can be found at the start of each chapter.
- The Reading Guide at the start of each lesson lists **Key Concepts,** vocabulary terms, and online supplements to the content.
- connectED icons direct you to online resources such as animations, personal tutors, math practices, and quizzes.
- Inquiry Labs and Skill Practices are in each chapter.
- Your FOLDABLES help organize your notes.

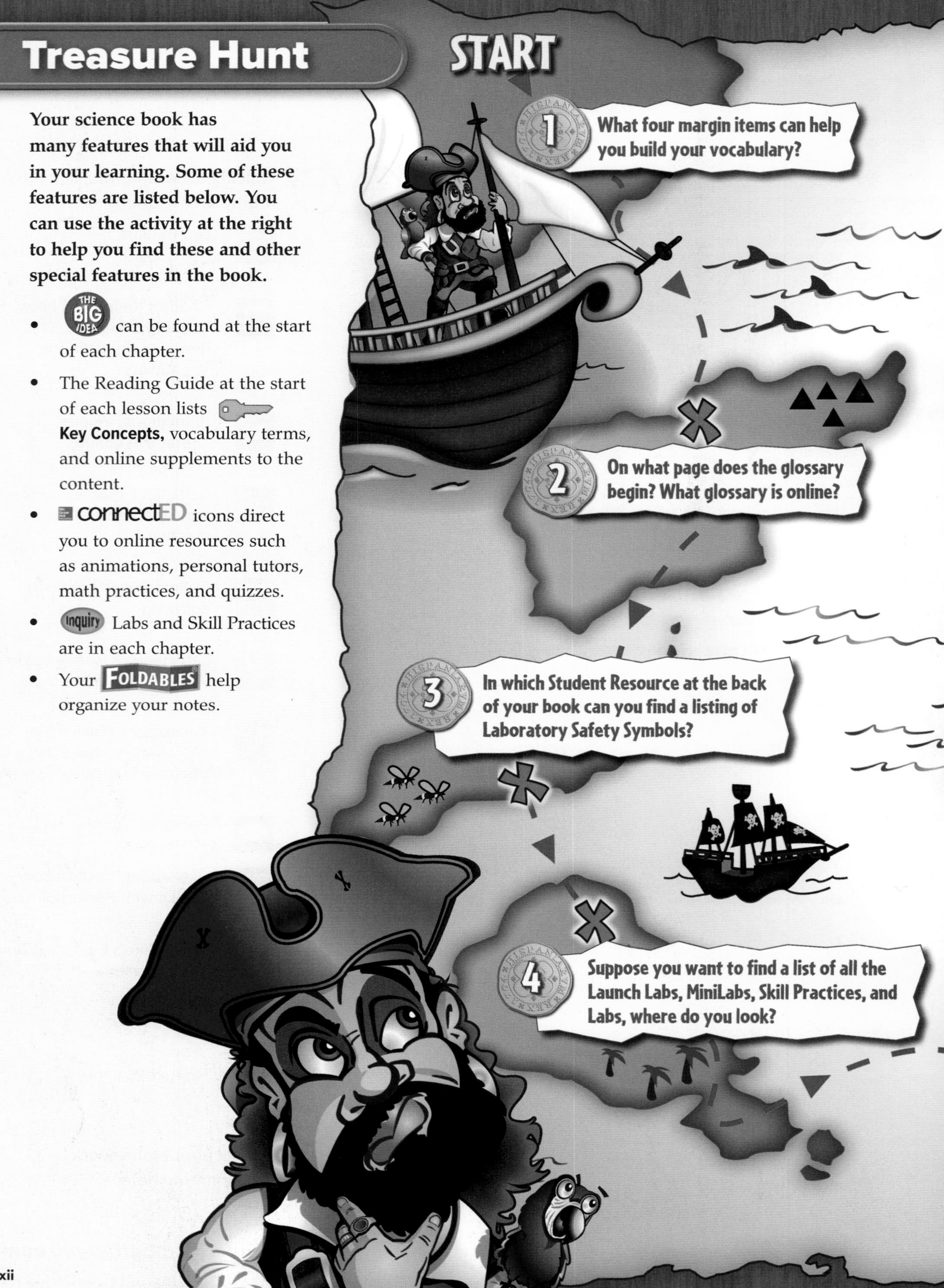

START
1 What four margin items can help you build your vocabulary?
2 On what page does the glossary begin? What glossary is online?
3 In which Student Resource at the back of your book can you find a listing of Laboratory Safety Symbols?
4 Suppose you want to find a list of all the Launch Labs, MiniLabs, Skill Practices, and Labs, where do you look?

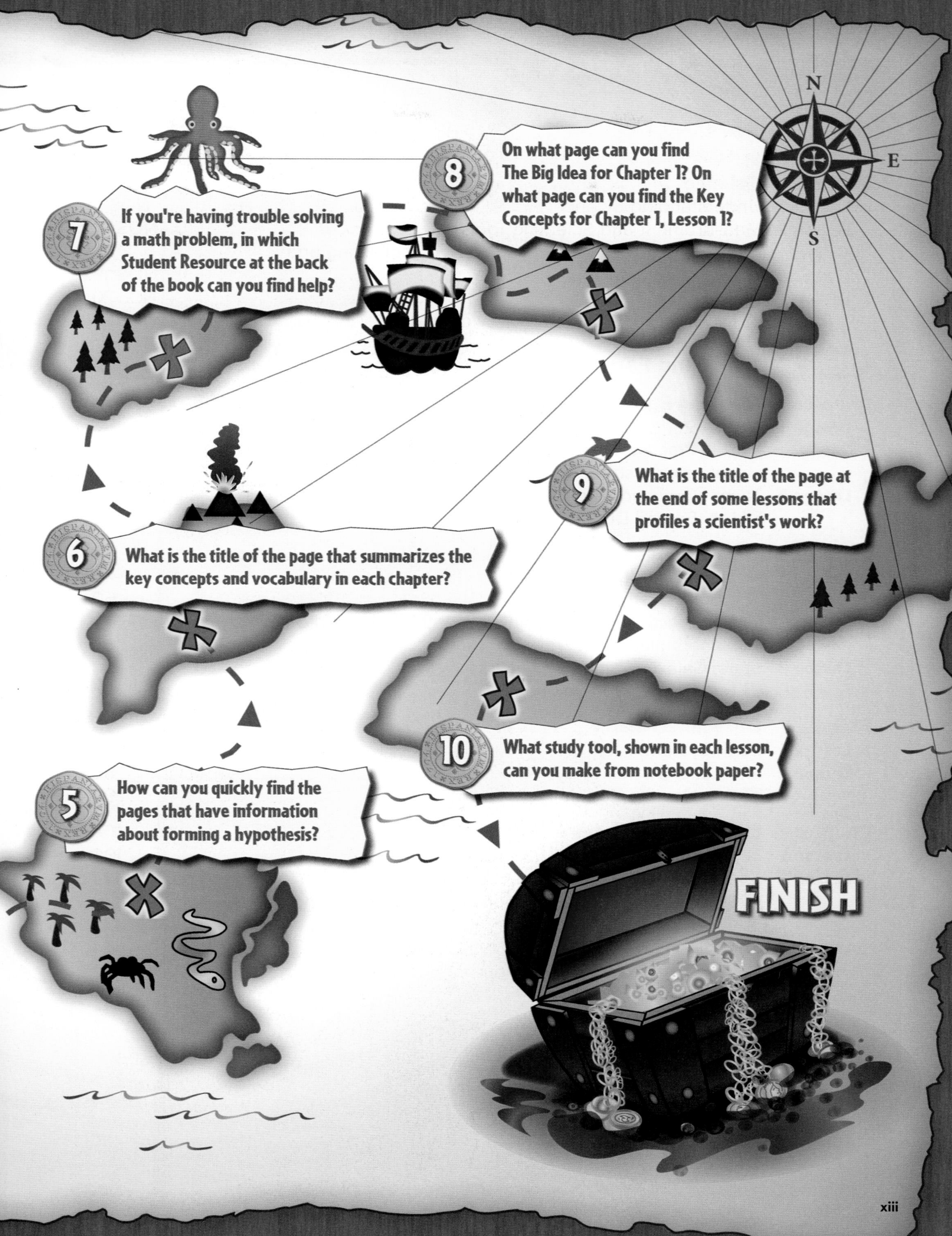
N
E
S
8
On what page can you find The Big Idea for Chapter 1? On what page can you find the Key Concepts for Chapter 1, Lesson 1?
7
If you're having trouble solving a math problem, in which Student Resource at the back of the book can you find help?
9
What is the title of the page at the end of some lessons that profiles a scientist's work?
6
What is the title of the page that summarizes the key concepts and vocabulary in each chapter?
10
What study tool, shown in each lesson, can you make from notebook paper?
5
How can you quickly find the pages that have information about forming a hypothesis?
FINISH

Table of Contents

Table of Contents

Student Resources

Inquiry

TABLE OF CONTENTS

Launch Labs

MiniLabs

Skill Practice

Labs

TABLE OF CONTENTS

GREEN SCIENCE

HOW IT WORKS

SCIENCE & SOCIETY

CAREERS in SCIENCE

Nature of Science

Methods of Science

THE BIG IDEA **What processes do scientists use when they perform scientific investigations?**

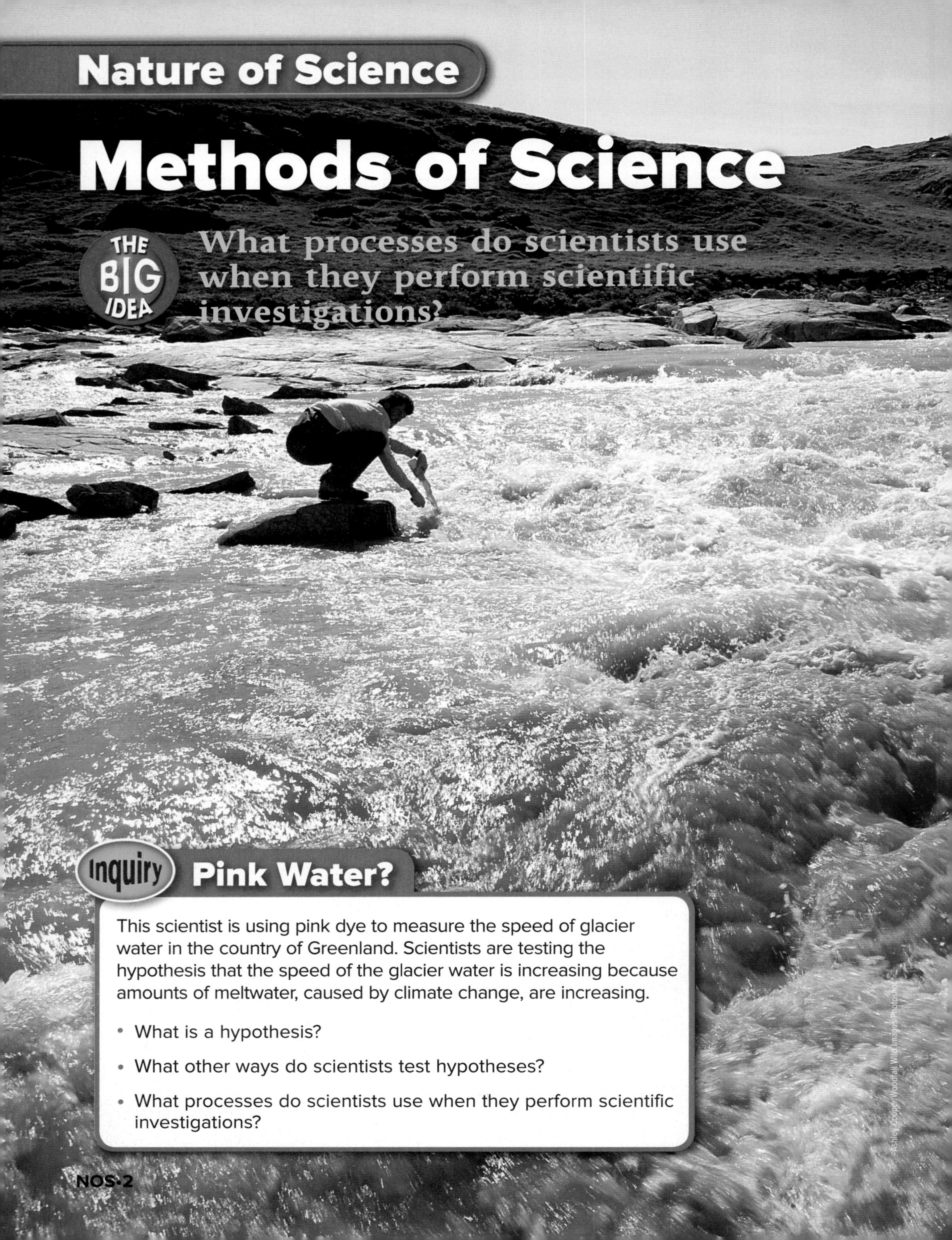

Inquiry Pink Water?

This scientist is using pink dye to measure the speed of glacier water in the country of Greenland. Scientists are testing the hypothesis that the speed of the glacier water is increasing because amounts of meltwater, caused by climate change, are increasing.

- What is a hypothesis?
- What other ways do scientists test hypotheses?
- What processes do scientists use when they perform scientific investigations?

Ashley Cooper/Woodfall Wild Images/Photoshot

Nature of SCIENCE

This chapter begins your study of the nature of science, but there is even more information about the nature of science in this book. Each unit begins by exploring an important topic that is fundamental to scientific study. As you read these topics, you will learn even more about the nature of science.

Systems	Unit 1
Patterns	Unit 2
Technology	Unit 3
Graphs	Unit 4

Your one-stop online resource
connectED.mcgraw-hill.com

LearnSmart®

Chapter Resources Files, Reading Essentials, Get Ready to Read, Quick Vocabulary

Animations, Videos, Interactive Tables

Self-checks, Quizzes, Tests

Project-Based Learning Activities

Lab Manuals, Safety Videos, Virtual Labs & Other Tools

Vocabulary, Multilingual eGlossary, Vocab eGames, Vocab eFlashcards

Personal Tutors

Lesson 1

Reading Guide

Key Concepts
ESSENTIAL QUESTIONS

- What is scientific inquiry?
- How do scientific laws and scientific theories differ?
- What is the difference between a fact and an opinion?

Vocabulary

science p. NOS 4
observation p. NOS 6
inference p. NOS 6
hypothesis p. NOS 6
prediction p. NOS 6
technology p. NOS 8
scientific theory p. NOS 9
scientific law p. NOS 9
critical thinking p. NOS 10

Multilingual eGlossary

BrainPOP® Science Video

Understanding Science

What is science?

Did you ever hear a bird sing and then look in nearby trees to find the singing bird? Have you ever noticed how the Moon changes from a thin crescent to a full moon each month? When you do these things, you are doing science. **Science** *is the investigation and exploration of natural events and of the new information that results from those investigations.*

For thousands of years, men and women of all countries and cultures have studied the natural world and recorded their observations. They have shared their knowledge and findings and have created a vast amount of scientific information. Scientific knowledge has been the result of a great deal of debate and confirmation within the science community.

People use science in their everyday lives and careers. For example, firefighters, as shown in **Figure 1**, wear clothing that has been developed and tested to withstand extreme temperatures and not catch fire. Parents use science when they set up an aquarium for their children's pet fish. Athletes use science when they use high-performance gear or wear high-performance clothing. Without thinking about it, you use science or the results of science in almost everything you do. Your clothing, food, hair products, electronic devices, athletic equipment, and almost everything else you use are results of science.

Figure 1 Firefighters' clothing, oxygen tanks, and equipment are all results of science.

Thomas Del Brase/Getty Images

Branches of Science

There are many different parts of the natural world. Because there is so much to study, scientists often focus their work in one branch of science or on one topic within that branch of science. There are three main branches of science–Earth science, life science, and **physical** science.

WORD ORIGIN

physical
from Latin *physica,* means "study of nature"

Earth Science

The study of Earth, including rocks, soils, oceans, and the atmosphere is Earth science. The Earth scientist to the right is collecting lava samples for research. Earth scientists might ask other questions such as

- How do different shorelines react to tsunamis?
- Why do planets orbit the Sun?
- What is the rate of climate change?

Life Science

The study of living things is life science, or biology. These biologists are attaching a radio collar to a tiger to help track its movements and learn more about its behavior. They are also weighing and measuring the tiger to gain information about this species. Biologists also ask questions such as

- Why do some trees lose their leaves in winter?
- How do birds know which direction they are going?
- How do mammals control their body temperature?

Physical Science

The study of matter and energy is physical science. It includes both physics and chemistry. This research chemist is preparing chemical solutions for analysis. Physicists and chemists ask other questions such as

- What chemical reactions must take place to launch a spaceship into space?
- Is it possible to travel faster than the speed of light?
- What makes up matter?

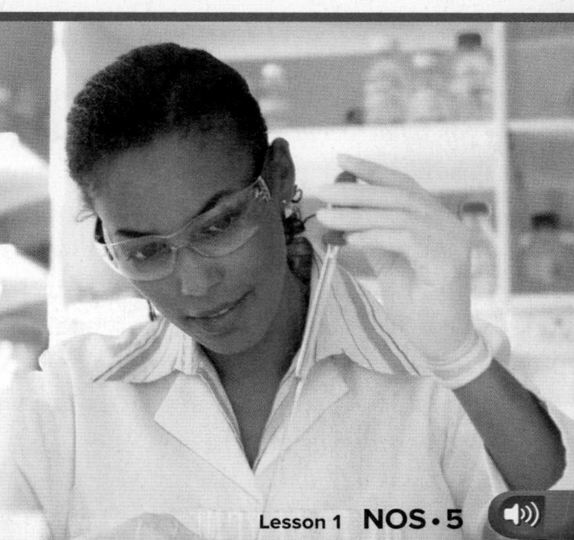

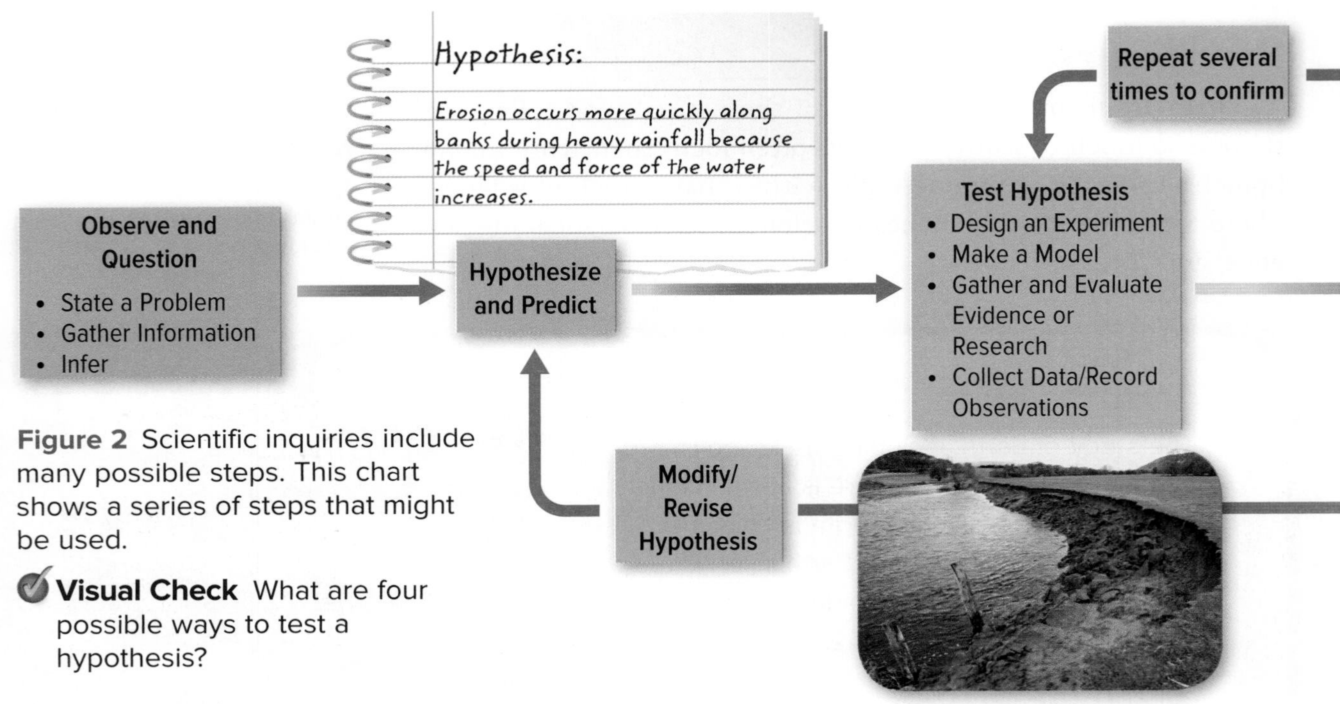

Figure 2 Scientific inquiries include many possible steps. This chart shows a series of steps that might be used.

Visual Check What are four possible ways to test a hypothesis?

Scientific Inquiry

When scientists conduct scientific investigations, they use scientific inquiry. Scientific inquiry is a process that uses a set of skills to answer questions or to test ideas about the natural world. There are many kinds of scientific investigations, and there are many ways to conduct them. The series of steps used in each investigation often varies. The flow chart in **Figure 2** shows an example of the skills used in scientific inquiry.

Key Concept Check What is scientific inquiry?

Ask Questions

One way to begin a scientific inquiry is to observe the natural world and ask questions. **Observation** *is the act of using one or more of your senses to gather information and taking note of what occurs.* Suppose you observe that the banks of a river have eroded more this year than in the previous year, and you want to know why. You also note that there was an increase in rainfall this year. After these observations, you make an inference based on these observations. *An* **inference** *is a logical explanation of an observation that is drawn from prior knowledge or experience.*

You infer that the increase in rainfall caused the increase in erosion. You decide to investigate further. You develop a hypothesis and a method to test it.

Hypothesize and Predict

A **hypothesis** *is a possible explanation for an observation that can be tested by scientific investigations.* A hypothesis states an observation and provides an explanation. For example, you might make the following hypothesis: More of the riverbank eroded this year because the amount, the speed, and the force of the river water increased.

When scientists state a hypothesis, they often use it to make predictions to help test their hypothesis. *A* **prediction** *is a statement of what will happen next in a sequence of events.* Scientists make predictions based on what information they think they will find when testing their hypothesis. For example, predictions for the hypothesis above could be: If rainfall increases, then the amount, the speed, and the force of river water will increase. If the amount, the speed, and the force of river water increase, then there will be more erosion.

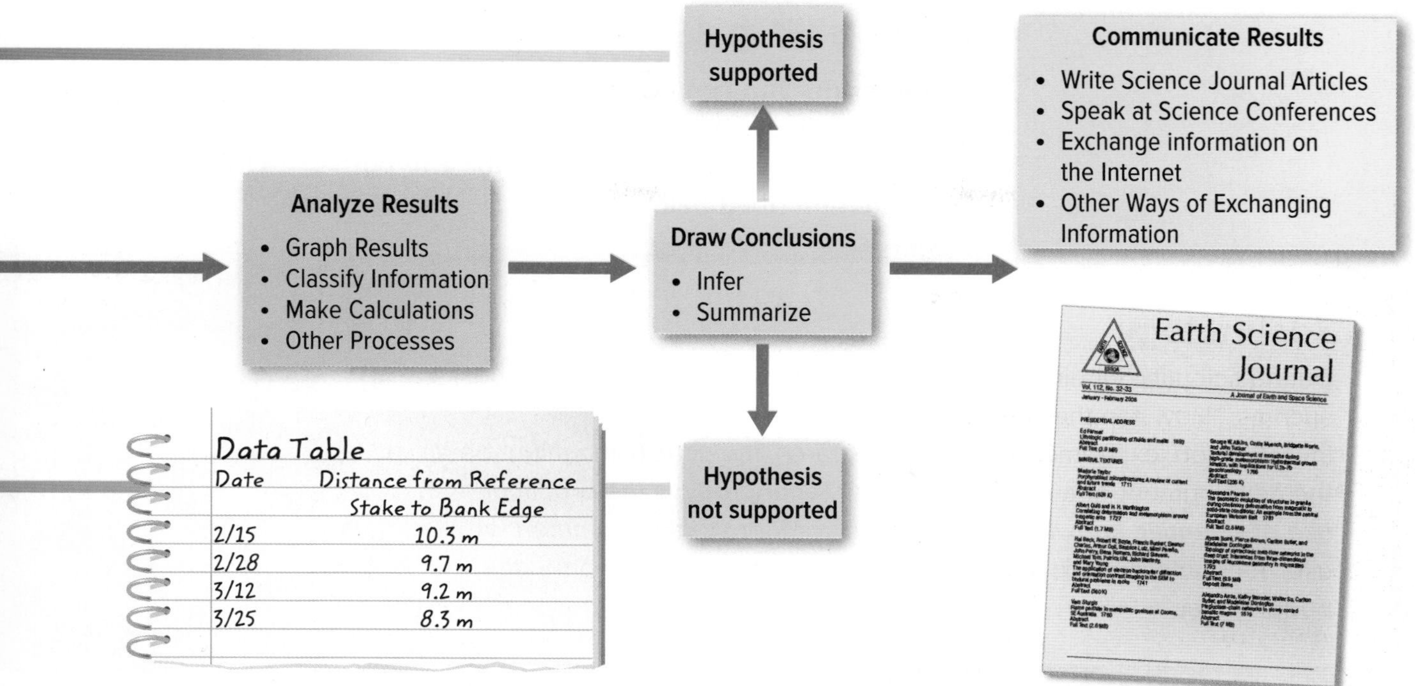

Test Hypothesis

When you test a hypothesis, you often test whether your predictions are true. If a prediction is confirmed, then it supports your hypothesis. If your prediction is not confirmed, you might need to modify your hypothesis and retest it.

There are several ways to test a hypothesis when performing a scientific investigation. Four possible ways are shown in **Figure 2.** For example, you might make a model of a riverbank in which you change the speed and the amount of water and record results and observations.

Analyze Results

After testing your hypothesis, you analyze your results using various methods, as shown in **Figure 2.** Often, it is hard to see trends or relationships in data while collecting it. Data should be sorted, graphed, or classified in some way. After analyzing the data, additional inferences can be made.

Draw Conclusions

Once you find the relationships among data and make several inferences, you can draw conclusions.

A conclusion is a summary of the information gained from testing a hypothesis. Scientists study the available information and draw conclusions based on that information.

Communicate Results

An important part of the scientific inquiry process is communicating results. Several ways to communicate results are listed in **Figure 2.** Scientists might share their information in other ways, too. Scientists communicate results of investigations to inform other scientists about their research and their conclusions. When a scientist uses that information to repeat another scientist's experiment, he or she is replicating the experiment to confirm results.

Further Scientific Inquiry

After finishing an experiment, a scientist must verify his or her results. If the hypothesis is supported, the scientist will repeat the experiment several times to make sure the conclusions are the same–this is called experimental repetition. If the hypothesis is not supported, any new information gained can be used to revise the hypothesis. Hypotheses can be revised and tested many times.

Results of Science

The results and conclusions from an investigation can lead to many outcomes, such as the answers to a question, more information on a specific topic, or support for a hypothesis. Other outcomes are described below.

Technology

During scientific inquiry, scientists often look for answers to questions such as, "How can the hearing impaired hear better?" After investigation, experimentation, and research, the conclusion might be the development of a new technology. **Technology** *is the practical use of scientific knowledge, especially for industrial or commercial use.* Technology, such as the cochlear implant, can help some deaf people hear.

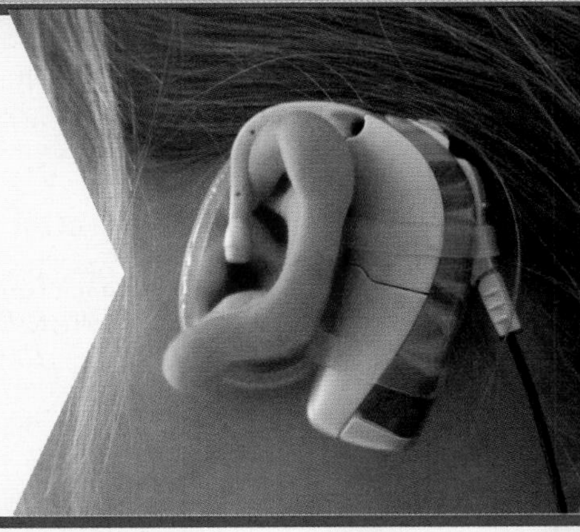

New Materials

Space travel has unique challenges. Astronauts must carry oxygen to breathe. They also must be protected against temperature and pressure extremes, as well as small, high-speed flying objects. Today's spacesuit, a result of research, testing, and design changes, consists of layers of material. The outer layer is made of a blend of materials. One material is waterproof and another material is heat and fire-resistant.

Possible Explanations

Scientists often perform investigations to find explanations as to why or how something happens. NASA's *Spitzer Space Telescope,* which has aided in our understanding of star formation, shows a cloud of gas and dust with newly formed stars.

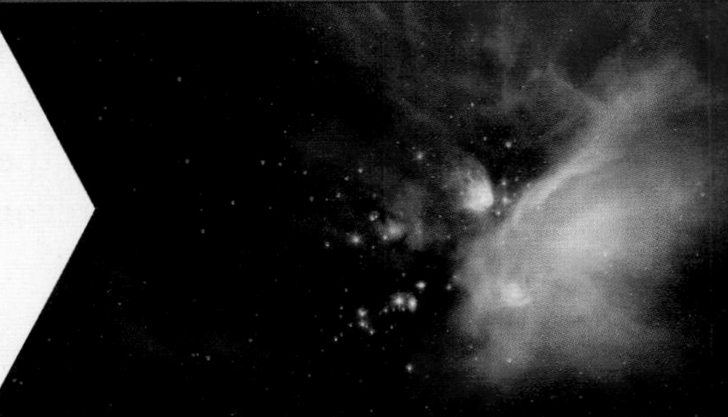

 **Reading Check** What are some results of science?

Scientific Theory and Scientific Law

Another outcome of science is the development of scientific theories and laws. Recall that a hypothesis is a possible explanation about an observation that can be tested by scientific investigations. What happens when a hypothesis or a group of hypotheses has been tested many times and has been supported by the repeated scientific investigations? The hypothesis can become a scientific theory.

(t) Hannah Gal/Science Source, (c)john angerson/Alamy, (b)NASA/JPL-Caltech/Harvard-Smithsonian CfA

Scientific Theory

Often, the word *theory* is used in casual conversations to mean an untested idea or an opinion. However, scientists use *theory* differently. *A* **scientific theory** *is an explanation of observations or events that is based on knowledge gained from many observations and investigations.*

Scientists regularly question scientific theories and test them for validity. A scientific theory generally is accepted as true until it is disproved. An example of a scientific theory is the theory of plate tectonics. The theory of plate tectonics explains how Earth's crust moves and why earthquakes and volcanoes occur. Another example of a scientific theory is discussed in **Figure 3.**

▲ **Figure 3** Scientists once believed Earth was the center of the solar system. In the 16th century, Nicolaus Copernicus hypothesized that Earth and the other planets actually revolve around the Sun.

Scientific Law

A scientific law is different from a social law, which is an agreement among people concerning a behavior. *A* **scientific law** *is a rule that describes a pattern in nature.* Unlike a scientific theory that explains why an event occurs, a scientific law only states that an event will occur under certain circumstances. For example, Newton's law of gravitational force implies that if you drop an object, it will fall toward Earth. Newton's law does not explain why the object moves toward Earth when dropped, only that it will.

Key Concept Check How do scientific laws and theories differ?

New Information

Scientific information constantly changes as new information is discovered or as previous hypotheses are retested. New information can lead to changes in scientific theories, as explained in **Figure 4.** When new facts are revealed, a current scientific theory might be revised to include the new facts, or it might be disproved and rejected.

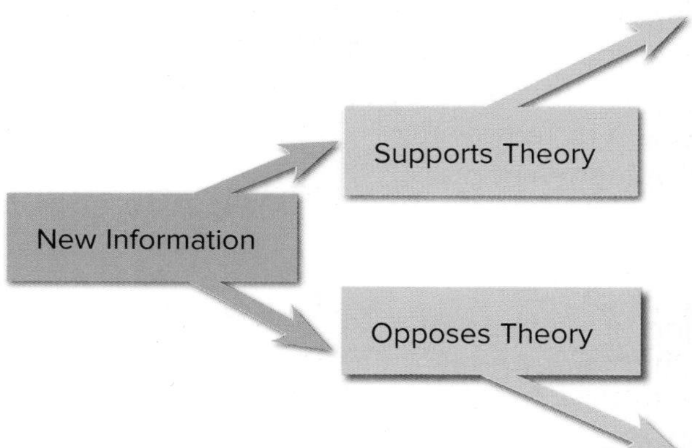

If new information supports a current scientific theory, then the theory is not changed. The information might be published in a scientific journal to show further support of the theory. The new information might also lead to advancements in technology or spark new questions that lead to new scientific investigations.

If new information opposes, or does not support a current scientific theory, the theory might be modified or rejected altogether. Often, new information will lead scientists to look at the original observations in a new way. This can lead to new investigations with new hypotheses. These investigations can lead to new theories.

▲ **Figure 4** New information can lead to changes in scientific theories.

NASA/JPL

FOLDABLES®

Make a six-tab book, and label it as shown. Use it to organize your notes about a scientific inquiry.

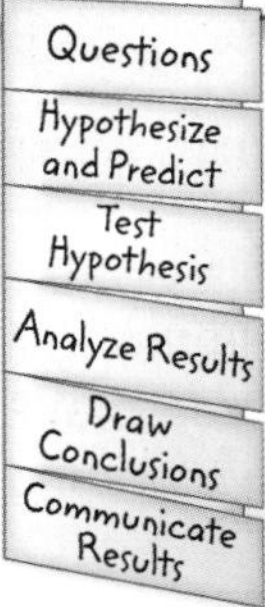

Evaluating Scientific Evidence

Did you ever read an advertisement, such as the one below, that made extraordinary claims? If so, you have practiced **critical thinking**–*comparing what you already know with the information you are given in order to decide whether you agree with it.* To determine whether information is true and scientific or pseudoscience (information incorrectly represented as scientific), you should be skeptical and identify facts and opinions. This helps you evaluate the strengths and weaknesses of information and make informed decisions. Critical thinking is important in all decision making–from everyday decisions to community, national, and international decisions.

Key Concept Check How do a fact and an opinion differ?

Learn Algebra *While You Sleep!*

Have you struggled to learn algebra? Struggle no more.

Math-er-ific's new algebra pillow is scientifically proven to transfer math skills from the pillow to your brain while you sleep. This revolutionary scientific design improved the algebra test scores of laboratory mice by 150 percent.

Dr. Tom Equation says, "I have never seen students or mice learn algebra so easily. This pillow is truly amazing."

For only $19.95, those boring hours spent studying are a thing of the past. So act fast! If you order today, you can get the algebra pillow and the equally amazing geometry pillow for only $29.95. That is a $10 savings!

Skepticism

To be skeptical is to doubt the truthfulness or accuracy of something. Because of skepticism, science can be self-correcting. If someone publishes results or if an investigation gives results that don't seem accurate, a skeptical scientist usually will challenge the information and test the results for accuracy.

Identifying Facts

The prices of the pillows and the savings are facts. A fact is a measurement, observation, or statement that can be strictly defined. Many scientific facts can be evaluated for their validity through investigations.

Identifying Opinions

An opinion is a personal view, feeling, or claim about a topic. Opinions are neither true nor false.

Mixing Facts and Opinions

Sometimes people mix facts and opinions. You must read carefully to determine which information is fact and which is opinion.

©Sigrid Olsson/PhotoAlto/Corbis

Science cannot answer all questions.

Scientists recognize that some questions cannot be studied using scientific inquiry. Questions that deal with opinions, beliefs, values, and feelings cannot be answered through scientific investigation. For example, questions that cannot be answered through scientific investigation might include

- Are comedies the best kinds of movies?
- Is it ever okay to lie?
- Which food tastes best?

The answers to all of these questions are based on opinions, not facts.

Safety in Science

It is very important for anyone performing scientific investigations to use safe practices, such as the student shown in **Figure 5.** You should always follow your teacher's instructions. If you have questions about potential hazards, use of equipment, or the meaning of safety symbols, ask your teacher. Always wear protective clothing and equipment while performing scientific investigations. If you are using live animals in your investigations, provide appropriate care and ethical treatment to them. For more information on practicing safe and ethical science, consult the Science Safety Skill Handbook in the back of this book.

Figure 5 Always use safe lab practices when doing scientific investigations.

ACADEMIC VOCABULARY

potential
(adjective) possible, likely, or probable

Lesson 1 Review

Online Quiz Virtual Lab

Use Vocabulary

1. The practical use of science, especially for industrial or commercial use, is ________.
2. **Distinguish** between a hypothesis and a prediction.
3. **Define** *observation* in your own words.

Understand Key Concepts

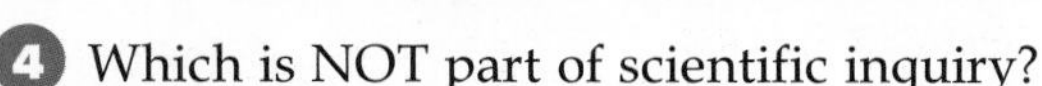

4. Which is NOT part of scientific inquiry?
   A. analyze results **C.** make a hypothesis
   B. falsify results **D.** make observations
5. **Explain** the difference between a scientific theory and a scientific law. Give an example of each.
6. **Write** an example of a fact and an example of an opinion.

Interpret Graphics

7. **Organize** Draw a graphic organizer similar to the one below. List four ways a scientist can communicate results.

Critical Thinking

8. **Identify** a real-world problem related to your home, your community, or your school that could be investigated scientifically.
9. **Design** a scientific investigation to test one possible solution to the problem you identified in the previous question.

Hutchings Photography/Digital Light Source

Lesson 2

Reading Guide

Key Concepts
ESSENTIAL QUESTIONS

- Why is it important for scientists to use the International System of Units?
- What causes measurement uncertainty?
- What are mean, median, mode, and range?

Vocabulary

description p. NOS 12

explanation p. NOS 12

International System of Units (SI) p. NOS 12

significant digits p. NOS 14

Multilingual eGlossary

Measurement and Scientific Tools

Description and Explanation

The scientist in **Figure 6** is observing a volcano. He describes in his journal that the flowing lava is bright red with a black crust, and it has a temperature of about 630°C. *A* **description** *is a spoken or written summary of observations.* There are two types of descriptions. When making a qualitative description, such as *bright red,* you use your senses (sight, sound, smell, touch, taste) to describe an observation. When making a quantitative description, such as *630°C,* you use numbers and measurements to describe an observation. Later, the scientist might explain his observations. *An* **explanation** *is an interpretation of observations.* Because the lava was bright red and about 630°C, the scientist might explain that these conditions indicate the lava is cooling and the volcano did not recently erupt.

The International System of Units

At one time, scientists in different parts of the world used different units of measurement. Imagine the confusion when a British scientist measured weight in pounds-force, a French scientist measured weight in Newtons, and a Japanese scientist measured weight in momme (MOM ee). Sharing scientific information was difficult, if not impossible.

In 1960, scientists adopted a new system of measurement to eliminate this confusion. *The* **International System of Units (SI)** *is the internationally accepted system for measurement.* SI uses standards of measurement, called base units, which are shown in **Table 1** on the next page. A base unit is the most common unit used in the SI system for a given measurement.

Figure 6 Scientists use descriptions and explanations when observing natural events.

Table 1 SI Base Units		
Quantity Measured	**Unit**	**Symbol**
Length	meter	m
Mass	kilogram	kg
Time	second	s
Electric current	ampere	A
Temperature	Kelvin	K
Amount of substance	mole	mol
Intensity of light	candela	cd

◄ **Table 1** You can use SI units to measure the physical properties of objects.

SI Unit Prefixes

In addition to base units, SI uses prefixes to identify the size of the unit, as shown in **Table 2.** Prefixes are used to indicate a fraction of ten or a multiple of ten. In other words, each unit is either ten times smaller than the next larger unit or ten times larger than the next smaller unit. For example, the prefix *deci-* means 10^{-1}, or 1/10. A decimeter is 1/10 of a meter. The prefix *kilo-* means 10^3, or 1,000. A kilometer is 1,000 m.

Converting Between SI Units

Because SI is based on ten, it is easy to convert from one SI unit to another. To convert SI units, you must multiply or divide by a factor of ten. You also can use proportions as shown below in the Math Skills activity.

Key Concept Check Why is it important for scientists to use the International System of Units (SI)?

Table 2 Prefixes are used in SI to indicate the size of the unit. ▼

Table 2 Prefixes	
Prefix	**Meaning**
Mega- (M)	1,000,000 (10^6)
Kilo- (k)	1,000 (10^3)
Hecto- (h)	100 (10^2)
Deka- (da)	10 (10^1)
Deci- (d)	0.1 (10^{-1})
Centi- (c)	0.01 (10^{-2})
Milli- (m)	0.001 (10^{-3})
Micro- (μ)	0.000 001 (10^{-6})

Math Skills ×÷ **Use Proportions**

Math Practice

Personal Tutor

A book has a mass of **1.1 kg**. Using a proportion, find the mass of the book in grams.

1. Use the table to determine the correct relationship between the units. One kg is 1,000 times greater than 1 g. So, there are 1,000 g in 1 kg.
2. Then set up a proportion.

$$\left(\frac{x}{1.1\text{ kg}}\right) = \left(\frac{1{,}000\text{ g}}{1\text{ kg}}\right)$$

$$x = \left(\frac{(1{,}000\text{ g})(1.1\,\cancel{\text{kg}})}{1\,\cancel{\text{kg}}}\right) = 1{,}100\text{ g}$$

3. Check your units. The answer is 1, 100 g.

Practice

1. Two towns are separated by 15,328 m. What is the distance in kilometers?
2. A dosage of medicine is 325 mg. What is the dosage in grams?

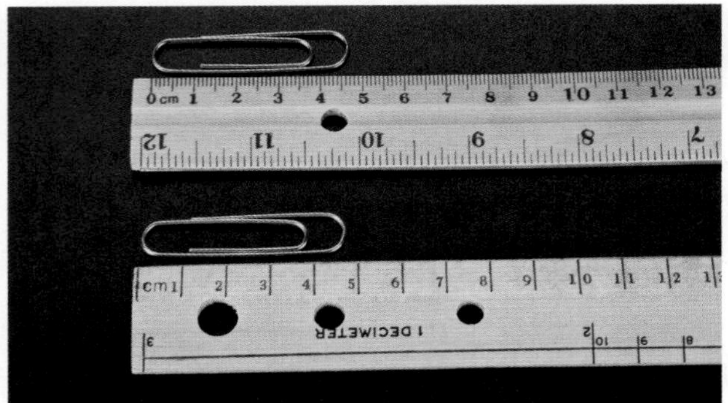

Figure 7 All measurements have some uncertainty.

FOLDABLES®

Make a vertical two-tab book using the labels shown. Use it to organize your notes about SI conversions and rounding significant digits.

Conversions Between SI Units

Rounding Significant Digits

Personal Tutor

Table 3 Significant Digits Rules

1. All nonzero numbers are significant.
2. Zeros between significant digits are significant.
3. All final zeros to the right of the decimal point are significant.
4. Zeros used solely for spacing the decimal point are NOT significant. The zeros only indicate the position of the decimal point.

* The blue numbers in the examples are the significant digits.

Number	Significant Digits	Applied Rules
1.234	4	1
1.02	3	1, 2
0.023	2	1, 4
0.200	3	1, 3
1,002	4	1, 2
3.07	3	1, 2
0.001	1	1, 4
0.012	2	1, 4
50,600	3	1, 2, 4

Measurement and Uncertainty

Have you ever measured an object, such as a paper clip? The tools used to take measurements can limit the accuracy of the measurements. Look at the bottom ruler in **Figure 7.** Its measurements are divided into centimeters. The paper clip is between 4 cm and 5 cm. You might guess that it is 4.7 cm long. Now, look at the top ruler. Its measurements are divided into millimeters. You can say with more precision that the paper clip is about 4.75 cm long. This measurement is more precise than the first measurement.

Key Concept Check What causes measurement uncertainty?

Significant Digits and Rounding

Because scientists duplicate each other's work, they must record numbers with the same degree of precision as the original data. Significant digits allow scientists to do this. **Significant digits** *are the number of digits in a measurement that you know with a certain degree of reliability.* **Table 3** lists the rules for expressing and determining significant digits.

In order to achieve the same degree of precision as a previous measurement, it often is necessary to round a measurement to a certain number of significant digits. Suppose you have the number below, and you need to round it to four significant digits.

1,348.527 g

To round to four significant digits, you need to round the 8. If the digit to the right of the 8 is 0, 1, 2, 3, or 4, the digit being rounded (8) remains the same. If the digit to the right of the 8 is 5, 6, 7, 8, or 9, the digit being rounded (8) increases by one. The rounded number is 1,349 g.

What if you need to round 1,348.527 g to two significant digits? You would look at the number to the right of the 3 to determine how to round. 1,348.527 rounded to two significant digits would be 1,300 g. The 4 and 8 become zeros.

Matt Meadows

Mean, Median, Mode, and Range

A rain gauge measures the amount of rain that falls on a location over a period of time. A rain gauge can be used to collect data in scientific investigations, such as data shown in **Table 4a.** Scientists often need to analyze their data to obtain information. Four values often used when analyzing numbers are median, mean, mode, and range.

 **Key Concept Check** What are mean, median, and mode?

Median

The median is the middle number in a data set when the data are arranged in numerical order. The rainfall data are listed in numerical order in Table 4b. If you have an even number of data items, add the two middle numbers together and divide by two to find the median.

$$\text{median} = \frac{8.18 \text{ cm} + 8.84 \text{ cm}}{2}$$

$$= 8.51 \text{ cm}$$

Table 4a Rainfall Data

Month	Rainfall
January	7.11 cm
February	11.89 cm
March	9.58 cm
April	8.18 cm
May	7.11 cm
June	1.47 cm
July	18.21 cm
August	8.84 cm

Mean

The mean, or average, of a data set is the sum of the numbers in a data set divided by the number of entries in the set. To find the mean, add the numbers in your data set and then divide the total by the number of items in your data set.

$$\text{mean} = \frac{\text{(sum of numbers)}}{\text{(number of items)}}$$

$$= \frac{72.39 \text{ cm}}{8 \text{ months}}$$

$$= \frac{9.05 \text{ cm}}{\text{month}}$$

Mode

The mode of a data set is the number or item that appears most often. The number in blue in Table 4b appears twice. All other numbers appear only once.

mode = 7.11 cm

Table 4b Rainfall Data (numerical order)

Rainfall
1.47 cm
7.11 cm
7.11 cm
8.18 cm
8.84 cm
9.58 cm
11.89 cm
18.21 cm

Range

The range is the difference between the greatest number and the least number in the data set.

$$\text{range} = 18.21 \text{ cm} - 1.47 \text{ cm}$$

$$= 16.74 \text{ cm}$$

Scientific Tools

As you engage in scientific inquiry, you will need tools to help you take quantitative measurements. Always follow appropriate safety procedures when using scientific tools. For more information about the proper use of these tools, see the Science Skill Handbook at the back of this book.

◄ Science Journal

Use a science journal to record observations, questions, hypotheses, data, and conclusions from your scientific investigations. A science journal is any notebook that you use to take notes or record information and data while you conduct a scientific investigation. Keep it organized so you can find information easily. Write down the date whenever you record new information in the journal. Make sure you are recording your data honestly and accurately.

Rulers and Metersticks ►

Use rulers and metersticks to measure lengths and distances. The SI unit of measurement for length is the meter (m). For small objects, such as pebbles or seeds, use a metric ruler with centimeter and millimeter markings. To measure larger objects, such as the length of your bedroom, use a meterstick. To measure long distances, such as the distance between cities, use an instrument that measures in kilometers. Be careful when carrying rulers and metersticks, and never point them at anyone.

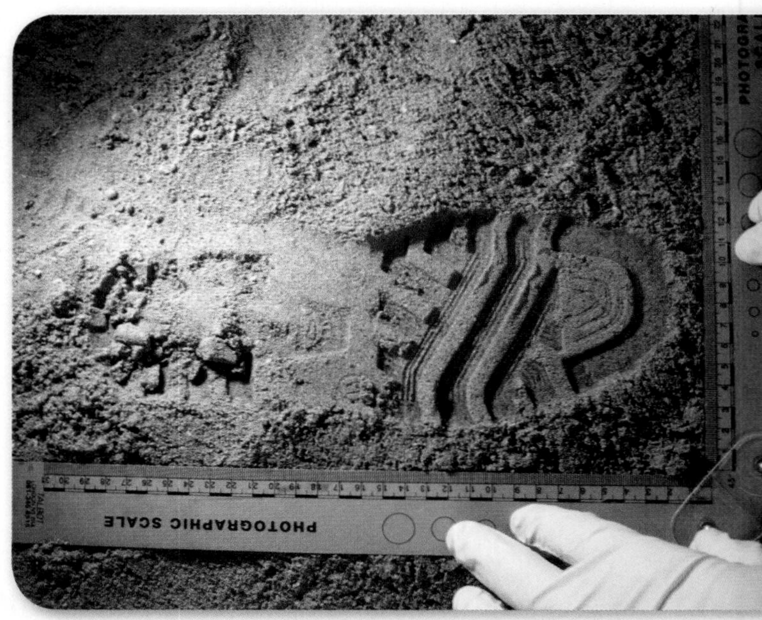

◄ Glassware

Use beakers to hold and pour liquids. The lines on a beaker do not provide accurate measurements. Use a graduated cylinder to measure the volume of a liquid. Volume is typically measured in liters (L) or milliliters (mL).

(t)Matt Meadows, (c)By Ian Miles-Flashpoint Pictures/Alamy, (b) Joseph Clark/Photodisc/Getty Images

Triple-Beam Balance ►

Use a triple-beam balance to measure the mass of an object. The mass of a small object is measured in grams. The mass of large object is usually measured in kilograms. Triple-beam balances are instruments that require some care when using. Follow your teacher's instructions so that you do not damage the instrument. Digital balances also might be used.

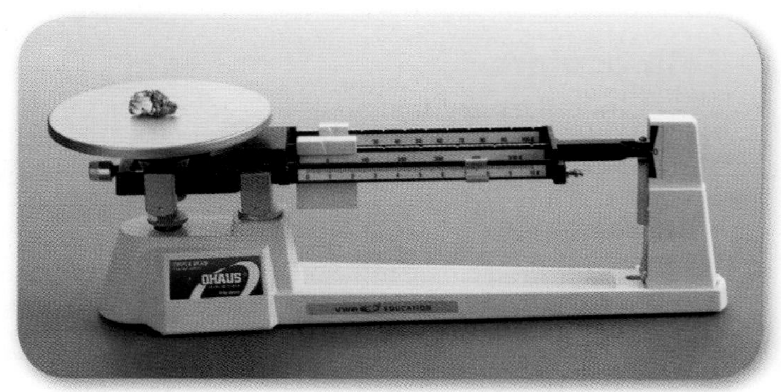

◄ Thermometer

Use a thermometer to measure the temperature of a substance. Kelvin is the SI unit for temperature, but you will use a thermometer to measure temperature in degrees Celsius (°C). To use a thermometer, place a room-temperature thermometer into the substance for which you want to measure temperature. Do not let the thermometer touch the bottom of the container that holds the substance or you will get an inaccurate reading. When you finish, remember to place your thermometer in a secure place. Do not lay it on a table, because it can roll off the table. Never use a thermometer as a stirring rod.

Computers and the Internet ►

Use a computer to collect, organize, and store information about a research topic or scientific investigation. Computers are useful tools to scientists for several reasons. Scientists use computers to record and analyze data, to research new information, and to quickly share their results with others worldwide over the Internet.

(t) Hutchings Photography/Digital Light Source, (c) McGraw-Hill Education, (b) Steve Cole/Getty Images

Tools Used by Earth Scientists

Binoculars

Binoculars are instruments that enable people to view faraway objects more clearly. Earth scientists use them to view distant landforms, animals, or even incoming weather.

Compass

A compass is an instrument that shows magnetic north. Earth scientists use compasses to navigate when they are in the field and to determine the direction of distant landforms or other natural objects.

Wind Vane and Anemometer

A wind vane is a device, often attached to the roofs of buildings, that rotates to show the direction of the wind. An anemometer, or wind-speed gauge, is used to measure the speed and the force of wind.

Streak Plate

A streak plate is a piece of hard, unglazed porcelain that helps you identify minerals. When you scrape a mineral along a streak plate, the mineral leaves behind powdery marks. The color of the mark is the mineral's streak.

Lesson 2 Review

Online Quiz

Use Vocabulary

1. **Distinguish** between description and explanation.
2. **Define** *significant digits* in your own words.

Understand Key Concepts

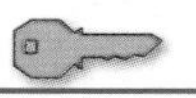

3. Which base unit is NOT part of the International System of Units?
   A. ampere
   B. meter
   C. pound
   D. second
4. **Give an example** of how scientific tools cause measurement uncertainty.
5. **Differentiate** among mean, median, mode, and range.

Interpret Graphics

6. **Change** Copy the graphic organizer below, and change the number shown to have the correct number of significant digits indicated.

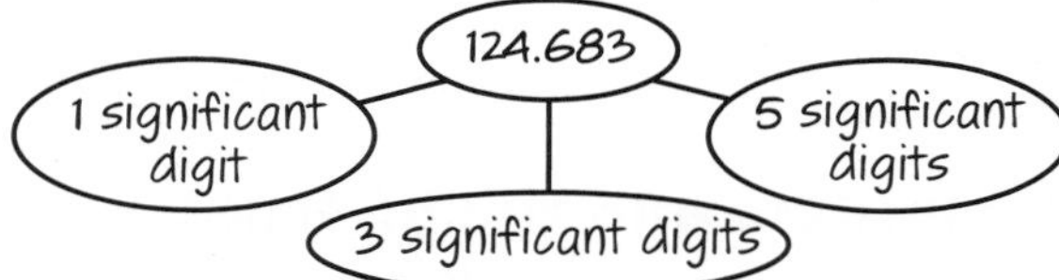

Critical Thinking

7. **Write** a short essay explaining why the United States should consider adopting SI as the measurement system used by supermarkets and other businesses.

Math Skills

Math Practice

8. **Convert** 52 m to kilometers. Explain how you got your answer.

(tl) ©Lawrence Manning/Corbis, (tr) Paul Rapson/Science Source, (bl) Jacques Cornell/McGraw-Hill Education, (br) ©Doug Sherman/Geofile

Skill Practice

Collect and Analyze Data

45 minutes

What can you learn by collecting and analyzing data?

People who study ancient cultures often collect and analyze data from soil samples. Soil samples contain bits of pottery, bones, seeds, and other clues to how ancient people lived and what they ate. In this activity, you will separate and analyze a simulated soil sample from an ancient civilization.

Materials

250-mL beaker

large piece of newsprint

1-L containers

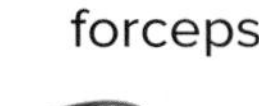

forceps

strainer

probe

Also needed: soil mixture, balance, plastic containers

Safety

Learn It

Data includes observations you can make with your senses and observations based on measurements of some kind. **Collecting and analyzing data** includes collecting, classifying, comparing and contrasting, and interpreting (looking for meaning in the data).

Try It

1. Read and complete a lab safety form.
2. Obtain a 200-mL sample of "soil."
3. Spread the newsprint over your workspace. Slowly pour the soil through a strainer over a plastic container. Shake the strainer gently so that all of the soil enters the container.
4. Pour the remaining portion of the soil sample onto the newsprint. Use a probe and forceps to separate objects. Classify different types of objects, and place them into the other plastic containers.
5. Copy the data tables from the board into your Science Journal.
6. Use the balance to measure and record the masses of each group of objects found in your soil sample. Write your group's data in the data table on the board.
7. When all teams have finished, use the class data from the board to find the mean, the median, the mode, and the range for each type of object.

Apply It

8. **Make Inferences** Assuming that the plastic objects represented animal bones, how many different types of animals were indicated by your analysis? Explain.
9. **Evaluate** Archaeologists often include information about the depth at which soil samples are taken. If you received a soil sample that kept the soil and other objects in their original layers, what more might you discover?
10. **Key Concept** Why didn't everyone in the class get the same data? What were some possible sources of uncertainty in your measurements?

Lesson 3 Case Study

Reading Guide

Key Concepts

ESSENTIAL QUESTIONS

- How are independent variables and dependent variables related?
- How is scientific inquiry used in a real-life scientific investigation?

Vocabulary

variable p. NOS 21

independent variable p. NOS 21

dependent variable p. NOS 21

Multilingual eGlossary

Go to the resource tab in ConnectED to find the PBL *Solutions for Pollution.*

The Iceman's Last Journey

The Tyrolean Alps border western Austria, northern Italy, and eastern Switzerland, as shown in **Figure 8**. They are popular with tourists, hikers, mountain climbers, and skiers. In 1991, two hikers discovered the remains of a man, also shown in **Figure 8**, in a melting glacier on the border between Austria and Italy. They thought the man had died in a hiking accident. They reported their discovery to the authorities.

Initially authorities thought the man was a music professor who disappeared in 1938. However, they soon learned that the music professor was buried in a nearby town. Artifacts near the frozen corpse indicated that the man died long before 1938. The artifacts, as shown in **Figure 9**, were unusual. The man, nicknamed the Iceman, was dressed in leggings, a loincloth, and a goatskin jacket. A bearskin cap lay nearby. He wore shoes made of red deerskin with thick bearskin soles. The shoes were stuffed with grass for insulation. In addition, investigators found a copper ax, a partially constructed longbow, a quiver containing 14 arrows, a wooden backpack frame, and a dagger at the site.

Figure 8 Excavators used jackhammers to free the man's body from the ice, which caused serious damage to his hip. Part of a longbow also was found nearby.

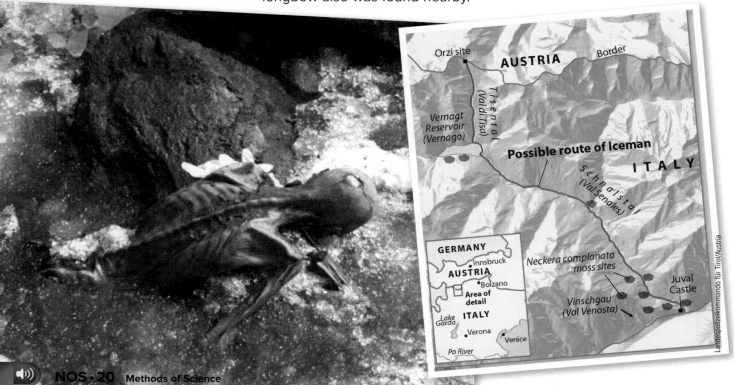

Landespolizeikommando für Tirol/Austria

A Controlled Experiment

The identity of the corpse was a mystery. Several people hypothesized about his identity, but controlled experiments were needed to unravel the mystery of who the Iceman was. Scientists and the public wanted to know the identity of the man, why he had died, and when he had died.

Figure 9 These models show what the Iceman and the artifacts found with him might have looked like.

Identifying Variables and Constants

When scientists design a controlled experiment, they have to identify factors that might affect the outcome of an experiment. *A* **variable** *is any factor that can have more than one value.* In controlled experiments, there are two kinds of variables. *The* **independent variable** *is the factor that you want to test. It is changed by the investigator to observe how it affects a dependent variable. The* **dependent variable** *is the factor you observe or measure during an experiment.* When the independent variable is changed, it causes the dependent variable to change.

A controlled experiment has two groups–an experimental group and a control group. The experimental group is used to study how a change in the independent variable changes the dependent variable. The control group contains the same factors as the experimental group, but the independent variable is not changed. Without a control, it is difficult to know if your experimental observations result from the variable you are testing or from another factor.

Scientists used inquiry to investigate the mystery of the Iceman. As you read the rest of the story, notice how scientific inquiry was used throughout the investigation. The blue boxes in the margins point out examples of the scientific inquiry process. The notebooks in the margin identify what a scientist might have written in a journal.

Scientific investigations often begin when someone asks a question about something observed in nature.

Observation: A corpse was found buried in ice in the Tyrolean Alps.
Hypothesis: The corpse found in the Tyrolean Alps is the body of a missing music professor because he disappeared in 1938, and had not been found.
Observation: Artifacts near the body suggested that the body was much older than the music professor would have been.
Revised Hypothesis: The corpse found was dead long before 1938 because the artifacts found near him appear to date before the 1930s.
Prediction: If the artifacts belong to the corpse, and date back before 1930, then the corpse is not the music professor.

An inference is a logical explanation of observations based on past experiences.

Inference: Based on its construction, the ax is at least 4,000 years old.
Prediction: If the ax is at least 4,000 years old, then the body found near it is also at least 4,000 years old.
Test Results: Radiocarbon dating showed the man to be 5,300 years old.

After many observations, revised hypotheses, and tests, conclusions often can be made.

Conclusion: The Iceman is about 5,300 years old. He was a seasonal visitor to the high mountains. He died in autumn. When winter came the Iceman's body became buried and frozen in the snow, which preserved his body.

Figure 10 This ax, bow and quiver, and dagger and sheath were found with the Iceman's body.

South Tyrol Museum of Archaeo... (www.iceman.it)

An Early Conclusion

Konrad Spindler was a professor of archeology at the University of Innsbruck in Austria when the Iceman was discovered. Spindler estimated that the ax, shown in **Figure 10**, was at least 4,000 years old based on its construction. If the ax was that old, then the Iceman was also at least 4,000 years old. Later, radiocarbon dating showed that the Iceman actually lived about 5,300 years ago.

The Iceman's body was in a mountain glacier 3,210 m above sea level. What was this man doing so high in the snow- and ice-covered mountains? Was he hunting for food, shepherding his animals, or looking for metal ore?

Spindler noted that some of the wood used in the artifacts was from trees that grew at lower elevations. He concluded that the Iceman was probably a seasonal visitor to the high mountains.

Spindler also hypothesized that shortly before the Iceman's death, the Iceman had driven his herds from their summer high mountain pastures to the lowland valleys. However, the Iceman soon returned to the mountains where he died of exposure to the cold weather.

The Iceman's body was extremely well preserved. Spindler inferred that ice and snow covered the Iceman's body shortly after he died. Spindler concluded that the Iceman died in autumn and was quickly buried and frozen, which preserved his body and all his possessions.

More Observations and Revised Hypotheses

When the Iceman's body was discovered, Klaus Oeggl was an assistant professor of botany at the University of Innsbruck. His area of study was plant life during prehistoric times in the Alps. He was invited to join the research team studying the Iceman.

Upon close examination of the Iceman and his belongings, Professor Oeggl found three plant materials–grass from the Iceman's shoe, as shown in **Figure 11**, a splinter of wood from his longbow, and a tiny fruit called a sloe berry.

Over the next year, Professor Oeggl examined bits of charcoal wrapped in maple leaves that had been found at the discovery site. Examination of the samples revealed the charcoal was from the wood of eight different types of trees. All but one of the trees grew only at lower elevations than where the Iceman's body was found. Like Spindler, Professor Oeggl suspected that the Iceman had been at a lower elevation shortly before he died. From Oeggl's observations, he formed a hypothesis and made some predictions.

Oeggl realized that he would need more data to support his hypothesis. He requested that he be allowed to examine the contents of the Iceman's digestive tract. If all went well, the study would show what the Iceman had swallowed just hours before his death.

Figure 11 Professor Oeggl examined the Iceman's belongings along with the leaves and grass that were stuck to his shoe.

South Tyrol Museum of Archaeology, Italy (www.iceman.it)

Scientific investigations often lead to new questions.

Observations: Plant matter near body to study–grass on shoe, splinter from longbow, sloe berry fruit, charcoal wrapped in maple leaves, wood in charcoal from 8 different trees–
7 of 8 types of wood in charcoal grow at lower elevations
Hypothesis: The Iceman had recently been at lower elevations before he died because the plants identified near him grow only at lower elevations.
Prediction: If the identified plants are found in the digestive tract of the corpse, then the man actually was at lower elevations just before he died.
Question: What did the Iceman eat the day before he died?

Experiment to Test Hypothesis

The research teams provided Professor Oeggl with a tiny sample from the Iceman's digestive tract. He was determined to study it carefully to obtain as much information as possible. Oeggl carefully planned his scientific inquiry. He knew that he had to work quickly to avoid the decomposition of the sample and to reduce the chances of contaminating the samples.

There is more than one way to test a hypothesis. Scientists might gather and evaluate evidence, collect data and record their observations, create a model, or design and perform an experiment. They also might perform a combination of these skills.

Test Plan:

- Divide a sample of the Iceman's digestive tract into four sections.
- Examine the pieces under microscopes.
- Gather data from observations of the pieces and record observations.

His plan was to divide the material from the digestive tract into four samples. Each sample would undergo several chemical tests. Then, the samples would be examined under an electron microscope to see as many details as possible.

Professor Oeggl began by adding a saline solution to the first sample. This caused it to swell slightly, making it easier to identify particles using the microscope at a relatively low magnification. He saw particles of a wheat grain known as einkorn, which was a common type of wheat grown in the region during prehistoric times. He also found other edible plant material in the sample.

Oeggl noticed that the sample also contained pollen grains in the digestive tract of the Iceman, who is shown in **Figure 12**. To see the pollen grains more clearly, he used a chemical that separated unwanted substances from the pollen grains. He washed the sample a few times with alcohol. After each wash, he examined the sample under a microscope at a high magnification. The pollen grains became more visible. Many more microscopic pollen grains could now be seen. Professor Oeggl identified these pollen grains as those from a hop-hornbeam tree.

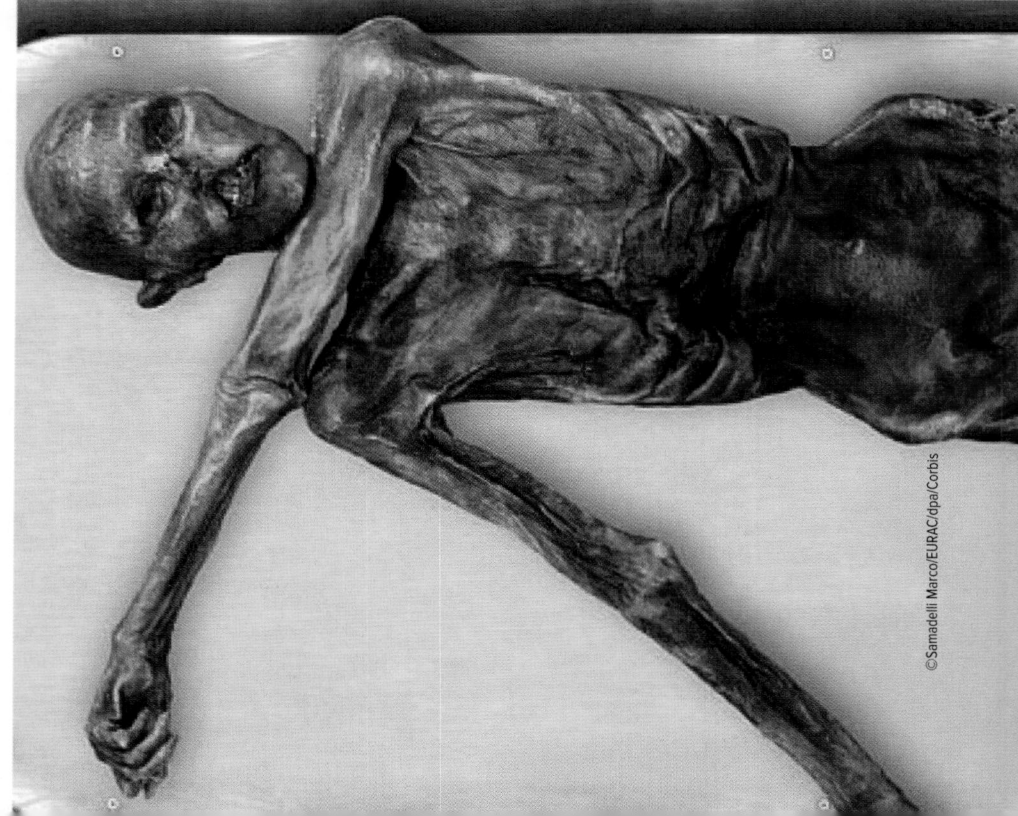

Figure 12 The Iceman, shown here, had pollen grains from hop hornbeam trees in his digestive tract.

Analyzing Results

Professor Oeggl observed that the hop-hornbeam pollen grains had not been digested. Therefore, the Iceman must have swallowed them within hours before his death. But, hop-hornbeam trees only grow in lower valleys. Oeggl was confused. How could pollen grains from trees at low elevations be ingested within a few hours of this man dying in high, snow-covered mountains? Perhaps the samples from the Iceman's digestive tract had been contaminated. Oeggl knew he needed to investigate further.

Further Experimentation

Oeggl realized that the most likely source of contamination would be Oeggl's own laboratory. He decided to test whether his lab equipment or saline solution contained hop-hornbeam pollen grains. To do this, he prepared two identical, sterile slides with saline solution. Then, on one slide, he placed a sample from the Iceman's digestive tract. The slide with the sample was the experimental group. The slide without the sample was the control group.

The independent variable, or the variable that Oeggl changed, was the presence of the sample on the slide. The dependent variable, or the variable Oeggl measured, was whether hop-hornbeam pollen grains showed up on the slides. Oeggl examined the slides carefully.

Analyzing Additional Results

The experiment showed that the control group (the slide without the digestive tract sample) contained no hop-hornbeam pollen grains. Therefore, the pollen grains had not come from his lab equipment or solutions. Each sample from the Iceman's digestive tract was closely re-examined. All of the samples contained the same hop-hornbeam pollen grains. The Iceman had indeed swallowed the hop-hornbeam pollen grains.

Error is unavoidable in scientific research. Scientists are careful to document procedures and any unanticipated factors or accidents. They also are careful to document possible sources of error in their measurements.

Procedure:
- Sterilize laboratory equipment.
- Prepare saline slides.
- View saline slides under electron microscope. Results: no hop-hornbeam pollen grains
- Add digestive tract sample to one slide.
- View this slide under electron microscope. Result: hop-hornbeam pollen grains present

Controlled experiments contain two types of variables.

Dependent Variables: amount of hop-hornbeam pollen grains found on slide
Independent Variable: digestive tract sample on slide

Without a control group, it is difficult to determine the origin of some observations.

Control Group: sterilized slide
Experimental Group: sterilized slide with digestive tract sample

An inference is a logical explanation of an observation that is drawn from prior knowledge or experience. Inferences can lead to predictions, hypotheses, or conclusions.

Observation: The Iceman's digestive tract contains pollen grains from the hop-hornbeam tree and other plants that bloom in spring.
Inference: Knowing the rate at which food and pollen decompose after swallowed, it can be inferred that the Iceman ate three times on the day that he died.
Prediction: The Iceman died in the spring within hours of digesting the hop-hornbeam pollen grains.

Mapping the Iceman's Journey

The hop-hornbeam pollen grains were helpful in determining the season the Iceman died. Because the pollen grains were whole, Professor Oeggl inferred that the Iceman swallowed the pollen grains during their blooming season. Therefore, the Iceman must have died between March and June.

After additional investigation, Professor Oeggl was ready to map the Iceman's final trek up the mountain. Because Oeggl knew the rate at which food travels through the digestive system, he inferred that the Iceman had eaten three times in the final day and a half of his life. From the digestive tract samples, Oeggl estimated where the Iceman was located when he ate.

First, the Iceman ingested pollen grains native to higher mountain regions. Then he swallowed hop-hornbeam pollen grains from the lower mountain regions several hours later. Last, the Iceman swallowed other pollen grains from trees of higher mountain areas again. Oeggl proposed the Iceman traveled from the southern region of the Italian Alps to the higher, northern region as shown in **Figure 13**, where he died suddenly. He did this all in a period of about 33 hours.

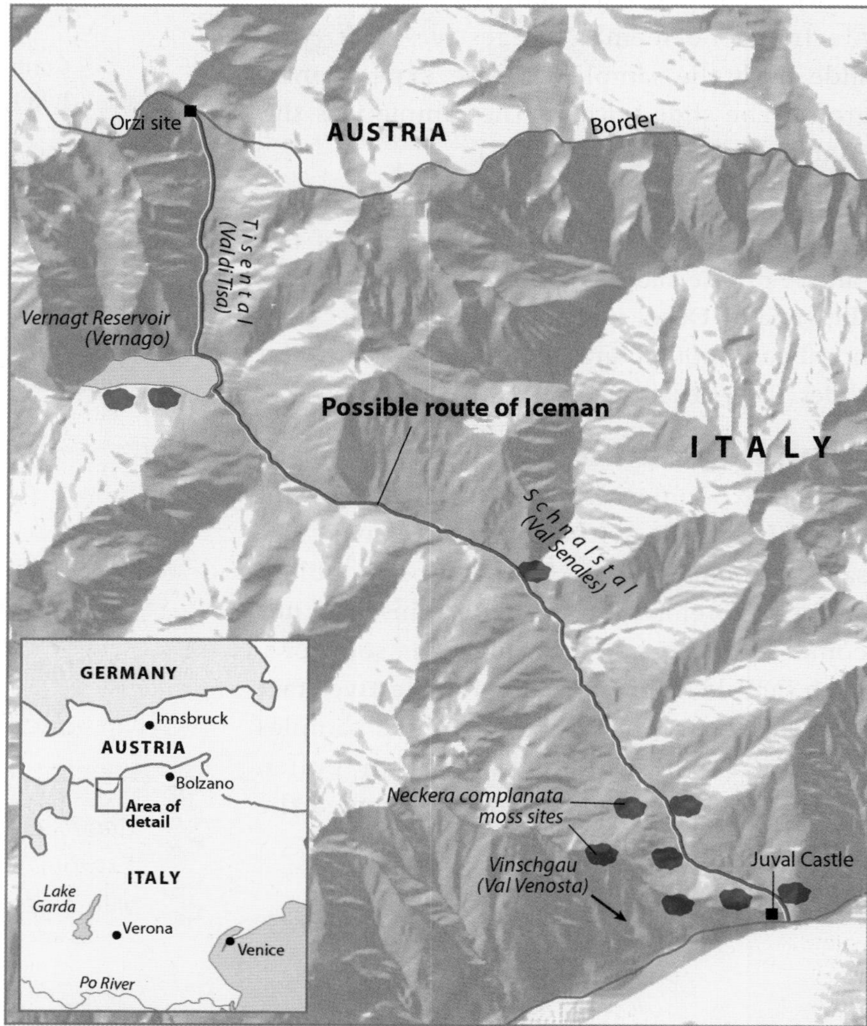

Figure 13 By examining the contents of the Iceman's digestive tract, Professor Oeggl was able to reconstruct the Iceman's last journey.

Conclusion

Researchers from around the world worked on different parts of the Iceman mystery and shared their results. Analysis of the Iceman's hair revealed his diet usually contained vegetables and meat. Examining the Iceman's one remaining fingernail, scientists determined that he had been sick three times within the last six months of his life. X-rays revealed an arrowhead under the Iceman's left shoulder. This suggested that he died from that serious injury rather than from exposure.

Finally, scientists concluded that the Iceman traveled from the high alpine region in spring to his native village in the lowland valleys. There, during a conflict, the Iceman sustained a fatal injury. He retreated back to the higher elevations, where he died. Scientists recognize their hypotheses can never be proved, only supported or not supported. However, with advances in technology, scientists are able to more thoroughly investigate mysteries of nature.

Scientific investigations may disprove early hypotheses or conclusions. However, new information can cause a hypothesis or conclusion to be revised many times.

Revised Conclusion:
In spring, the Iceman traveled from the high country to the valleys. After he was involved in a violent confrontation, he climbed the mountain into a region of permanent ice where he died of his wounds.

Lesson 3 Review

Use Vocabulary

1. A factor that can have more than one value is a(n) __________.
2. **Differentiate** between independent and dependent variables.

Understand Key Concepts 

3. Which part of scientific inquiry was NOT used in this case study?
   A. Draw conclusions.
   B. Make observations.
   C. Hypothesize and predict.
   D. Make a computer model.
4. **Determine** which is the control group and which is the experimental group in the following scenario: Scientists are testing a new kind of aspirin to see whether it will relieve headaches. They give one group of volunteers the aspirin. They give another group of volunteers pills that look like aspirin but are actually sugar pills.

Interpret Graphics

5. **Summarize** Copy and fill in the flow chart below summarizing the sequence of scientific inquiry steps that was used in one part of the case study. Draw the number of boxes needed for your sequence.

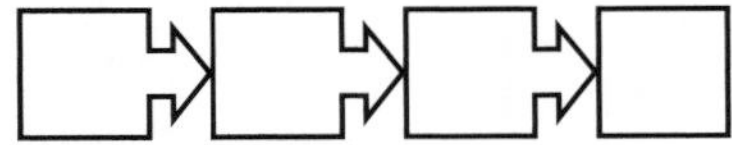

6. **Explain** What is the significance of the hop-hornbeam pollen found in the Iceman's digestive tract?

Critical Thinking

7. **Formulate** more questions about the Iceman. What would you want to know next?
8. **Evaluate** the hypotheses and conclusions made during the study of the Iceman. Do you see anything that might be an assumption? Are there holes in the research?

Lab

1-2 class periods

Inferring from Indirect Evidence

Materials

owl pellet

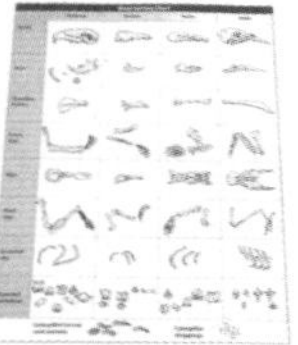

bone identification chart

probe

forceps

magnifying lens

Also needed: toothpicks, small brush, paper plate, ruler

Safety

In the case study about the Iceman, you learned how scientists used evidence found in or near the body to learn how the Iceman might have lived and what he ate. In this investigation, you will use similar indirect evidence to learn more about an owl.

An owl pellet is a ball of fur and feathers that contains bones, teeth, and other undigested parts of animals eaten by the owl. Owls and other birds, such as hawks and eagles, swallow their prey whole. Stomach acids digest the soft parts of the food. Skeletons and body coverings are not digested and form a ball. When the owl coughs up the ball, it might fall to the ground. Feathers, straw, or leaves often stick to the moist ball when it strikes the ground.

Ask a Question

What kinds of information can I learn about an owl by analyzing an owl pellet?

Make Observations

1. Read and complete a lab safety form.
2. Carefully measure the length, the width, and the mass of your pellet. Write the data in your Science Journal.
3. Gently examine the outside of the pellet using a magnifying lens. Do you see any sign of fur or feathers? What other substances can you identify? Record your observations.
4. Use a probe, toothpicks, and forceps to gently pull apart the pellet. Try to avoid breaking any of the tiny bones. Spread out the parts on a paper plate.
5. Copy the table into your Science Journal. Use the bone identification chart to identify each of the bones and other materials found in your pellet. Make a mark in the table for each part you identify.

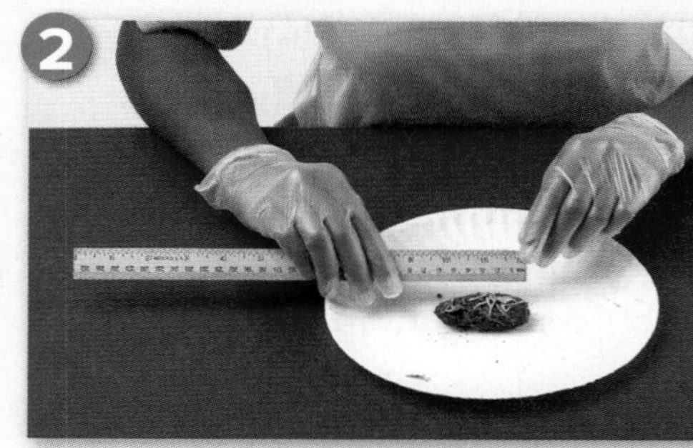

Bone Identification Chart

Bone	Animal	Number
Skull		
Jaw		
Shoulder blade		
Forelimb		
Hind limb		
Hip/pelvis		
Rib		
Vertebrae		
Insect parts		

(t to b) Ken Karp/McGraw-Hill Education, McGraw-Hill Education, (3,4,5,6) Hutchings Photography/Digital Light Source

Analyze and Conclude

6. **Assemble** the bones you find into a skeleton. You may need to locate pictures of rodents, shrews, moles, and birds.
7. **Discuss** with your teammates why parts of an animal skeleton might be missing.
8. **Write** a report that includes your data and conclusions about the owl's diet.
9. **Identify Cause and Effect** Is every bone you found in the pellet necessarily from the owl's prey? Why or why not?
10. **Analyze** What conclusions can you reach about the diet of the particular owl from which your pellet came? Can you extend this conclusion to the diets of all owls? Why or why not?
11. THE BIG IDEA **The Big Idea** How did the scientific inquiry you used in the investigation compare to those used by the scientists studying the Iceman? In what ways were they the same? In what ways were they different?

Communicate Your Results

Compare your results with those of several other teams. Discuss any evidence to support that the owl pellets did or did not come from the same area.

Inquiry Extension

Put your data on the board. Use the class data to determine a mean, median, mode, and range for each type of bone.

Lab Tips

- ☑ When using your forceps, squeeze the sides very lightly so that you don't crush fragile bones.
- ☑ Use the brush to clean each bone. Try rotating the bones as you match them to the chart.
- ☑ Lay the bones on the matching box on the chart as you separate them. Then count them when you are finished.

4

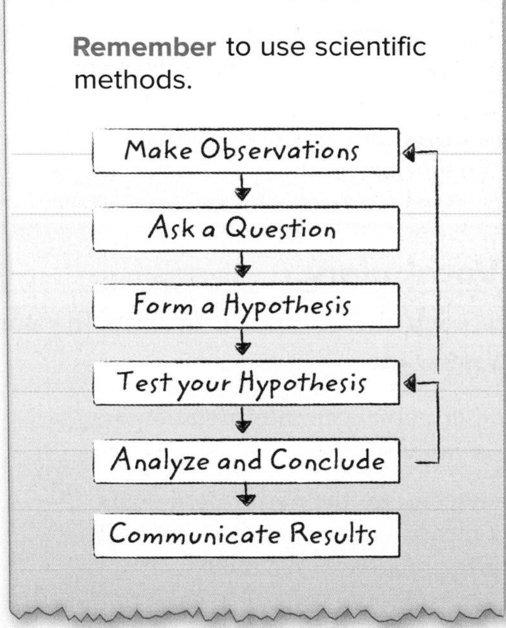

Chapter Study Guide and Review

Scientists use the process of scientific inquiry to perform scientific investigations.

Key Concepts Summary	Vocabulary
Lesson 1: Understanding Science • Scientific inquiry is a process that uses a set of skills to answer questions or to test ideas about the natural world. • A **scientific law** is a rule that describes a pattern in nature. A **scientific theory** is an explanation of things or events that is based on knowledge gained from many **observations** and investigations. • Facts are measurements, observations, and theories that can be evaluated for their validity through objective investigation. Opinions are personal views, feelings, or claims about a topic that cannot be proven true or false.	**science** p. NOS 4 **observation** p. NOS 6 **inference** p. NOS 6 **hypothesis** p. NOS 6 **prediction** p. NOS 6 **technology** p. NOS 8 **scientific theory** p. NOS 9 **scientific law** p. NOS 9 **critical thinking** p. NOS 10
Lesson 2: Measurement and Scientific Tools • Scientists worldwide use the **International System of Units (SI)** because their work is easier to confirm and repeat by their peers. • Measurement uncertainty occurs because no scientific tool can provide a perfect measurement. • Mean, median, mode, and range are statistical calculations that are used to evaluate sets of data.	**description** p. NOS 12 **explanation** p. NOS 12 **International System of Units (SI)** p. NOS 12 **significant digits** p. NOS 14
Lesson 3: Case Study: The Iceman's Last Journey • The **independent variable** is the factor a scientist changes to observe how it affects a **dependent variable.** A dependent variable is the factor a scientist measures or observes during an experiment. • Scientific inquiry was used throughout the investigation of the Iceman when hypotheses, predictions, tests, analysis, and conclusions were developed.	**variable** p. NOS 21 **independent variable** p. NOS 21 **dependent variable** p. NOS 21

Use Vocabulary

Replace each underlined term with the correct vocabulary word.

1. A <u>description</u> is an interpretation of observations.
2. The <u>means</u> are the numbers of digits in a measurement that you know with a certain degree of reliability.
3. The act of watching something and taking note of what occurs is a(n) <u>inference</u>.
4. A <u>scientific theory</u> is a rule that describes a pattern in nature.

Understand Key Concepts

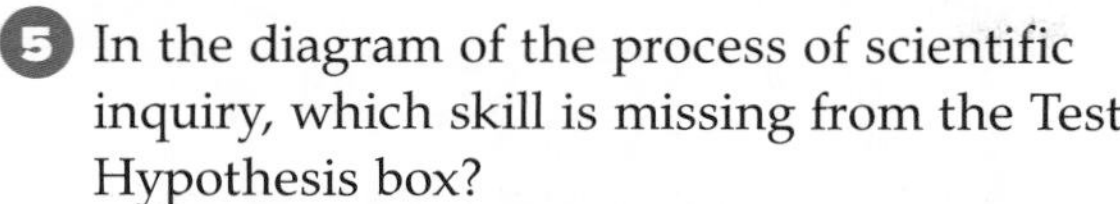

5 In the diagram of the process of scientific inquiry, which skill is missing from the Test Hypothesis box?

Test Hypothesis
- Design an Experiment
- Gather and Evaluate Evidence
- Collect Data/Record Observations

A. Analyze results.

B. Communicate results.

C. Make a model.

D. Make observations.

6 You have the following data set: 2, 3, 4, 4, 5, 7, and 8. Is 6 the mean, the median, the mode, or the range of the data set?

A. mean

B. median

C. mode

D. range

7 Which best describes an independent variable?

A. It is a factor that is not in every test.

B. It is a factor the investigator changes.

C. It is a factor you measure during a test.

D. It is a factor that stays the same in every test.

Critical Thinking

8 **Predict** what would happen if every scientist tried to use all the skills of scientific inquiry in the same order in every investigation.

9 **Assess** the role of measurement uncertainty in scientific investigations.

10 **Evaluate** the importance of having a control group in a scientific investigation.

Writing in Science

11 **Write** a five-sentence paragraph explaining why the International System of Units (SI) is an easier system to use than the English system of measurement. Be sure to include a topic sentence and a concluding sentence in your paragraph.

REVIEW THE BIG IDEA

12 What process do scientists use to perform scientific investigations? List and explain three of the skills involved.

13 Infer the purpose of the pink dye in the scientific investigation shown in the photo.

Math Skills

Math Practice

Use Numbers

14 Convert 162.5 hg to grams.

15 Convert 89.7 cm to millimeters.

Unit 1

Exploring Earth

SIR, we're lost and our GPS unit is dead.

LOST?! Nonsense, I have just what we need.

Or NOT. On second thought, I guess we already knew we were in Mexico, so this globe won't help.

And we certainly don't need a road map. We're nowhere near a road.

What to do? What to do?

Ahem

10000 B.C. | 1500 | 1900

12000 B.C.
A map scratched into a mammoth jawbone, the oldest surviving map, depicts a group of settlements and the surrounding countryside in what is now Mehirich, Ukraine.

2300 B.C.
The oldest surviving city map, a map of the Mesopotamian city of Lagash that includes the layout of the city, is created.

150 A.D.
Ptolemy illustrates a world map with Earth as a sphere from 60°N to 30°S latitudes.

1506
Francesco Rosselli produces the first map to show the "New World."

1930s
Maps become increasingly accurate and factual due to the widespread use of aerial photography after World War I.

1960s
Geographic Information Systems (GIS) are developed. GIS displays large amounts of information and includes computer software and hardware as well as digital data and storage.

1993
The space-based Global Positioning System achieves initial operational capability.

2005
An internet mapping tool displays satellite images of Earth's surface.

Visit ConnectED for this unit's **STEM** activity.

Unit 1

Nature of SCIENCE

Systems

Input—things, such as milk, lids, labels, energy, and information, that enter a system to achieve a goal

Processing—the changes that the system makes to the inputs, such as fastening lids to milk jugs

Output—material, information, or energy that leaves the system, such as sealed and labeled milk jugs

Feedback—information a system uses to regulate the input, process, and output.

▲ **Figure 1** Many systems are designed to achieve a goal.

You have probably heard about a computer's operating system, a weather system, and the system of government in the United States. What exactly is a system? A **system** is a collection of parts that influence or interact with one another. Systems often are used to achieve a goal or are developed with a specific purpose in mind. Like most systems, a milk-bottling manufacturing system is described in terms of the system's input, processing, output, and feedback, as shown in **Figure 1.**

Some systems, such as ecosystems and solar systems, are natural systems. Political, educational, and health-care systems are social systems that involve interactions among people. Transportation, communication, and manufacturing systems are designed systems that provide services or products.

Subsystems and Their Interactions

Large systems often are made of groups of smaller subsystems. Subsystems within a large system interact. **Figure 2** shows specialized transportation subsystems that are a part of an overall transportation system. These subsystems interact with one another, moving people and goods from place to place.

Figure 2 The subsystems within a transportation system interact with one another. ▼

Waterways transport ships and boats on rivers, lakes, and oceans.

Roadways conveniently transport cars, buses, and trucks on highways, streets, and roads.

Airways quickly transport people and materials long distances.

Pipelines transport large amounts of fluids, such as oil, through a network of pipes.

Railways transport trains and subways on rail lines between stations, at relatively low cost.

Bike paths and sidewalks transport people and packages over short distances at low cost.

(t) Jason Lugo/Getty Images, (cl)Cathlyn Melloan/Getty Images, (c)Alan SCHEIN/Alamy, (cr)Bill Barley/SuperStock, (bl)Digital Vision/Getty Images, (bc) Slow Images/Getty Images, (br) Andrew Resek/McGraw-Hill Education

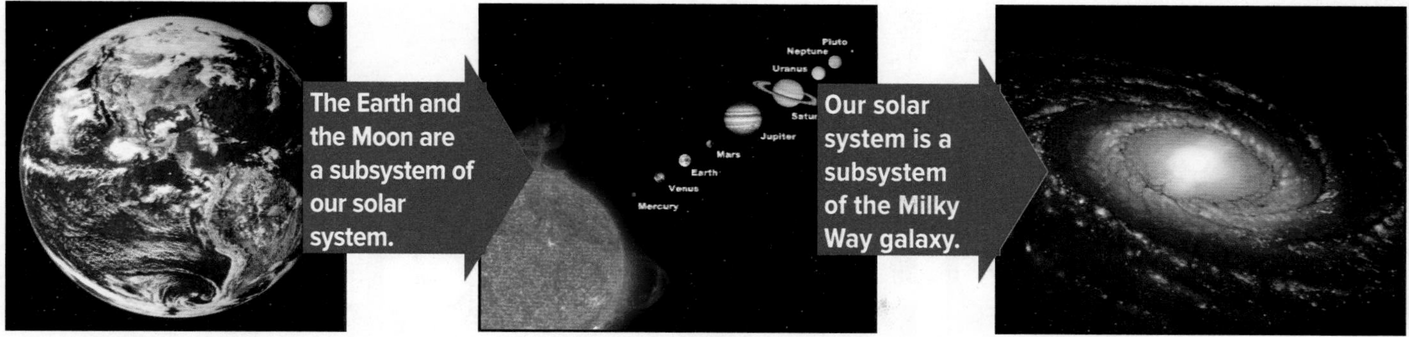

The Earth and the Moon are a subsystem of our solar system.
Pluto
Neptune
Uranus
Jupiter
Mars
Earth
Venus
Mercury
Our solar system is a subsystem of the Milky Way galaxy.

▲ **Figure 3** Natural systems, such as the Milky Way galaxy, consist of subsystems.

Natural Systems

Like a designed system, natural systems have interacting subsystems. As shown in **Figure 3,** Earth and the Moon are a subsystem of our solar system. Our solar system is a subsystem of the Milky Way galaxy. The interactions among these subsystems depend on the gravitational forces that affect the motions of planets, stars, moons, and other objects in space.

Most natural systems constantly change. For example, gravity between Earth and the Moon causes tides. The movements of water against seafloors is altered by friction. This friction affects the tilt of Earth on its axis, if only just a little. The tilt of the Earth is important because it causes the seasons and influences weather patterns. As shown in **Figure 4,** the northern hemisphere is warmer during summer because of the tilt of Earth.

Thinking in Terms of Systems

If you think of the world as one system made up of interacting subsystems, you will better understand the effects of your choices. For example, when people choose to disrupt an ecosystem in one part of the world by clearing huge areas of forests, the resulting climate change affects the weather system in another part of the world. This in turn affects the agricultural system and, therefore, the cost of food. Actions you take locally affect everyone globally.

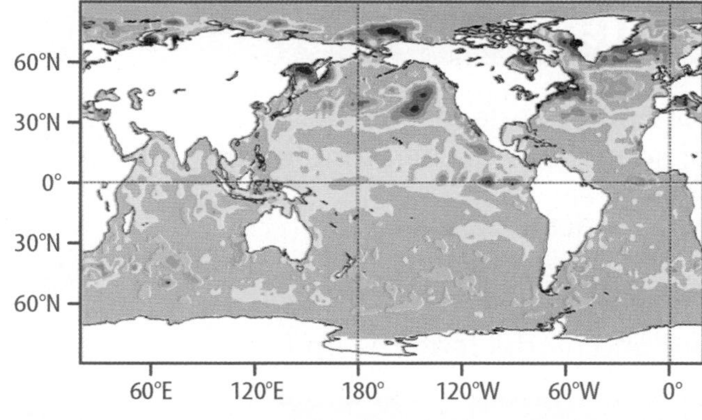

Figure 4 The darker reds show that the temperatures are warmer in the northern hemisphere in summer. ►

MiniLab
25 minutes

How can you get a package to Mars?

Suppose humans colonize Mars. How could materials be transported from Earth to Martian space colonies?

1. Plan a transportation system that could deliver a product made in Montana to colonies on Mars.
2. Draw a diagram of your system, labeling subsystems and explaining how the parts of the system interact.

Analyze and Conclude

1. **Analyze** Identify the input, processing, output, and feedback in your system.
2. **Explain** How does your system include or interact with designed systems, social systems, and natural systems?

Chapter 1

Mapping Earth

THE BIG IDEA

How are Earth's surface features measured and modeled?

Inquiry Do these maps show the same area?

Maps show the features of Earth's surface, such as mountains, roads, or different rock types. Notice that these maps show different features of the same area.

- What features are shown in each map?
- How were these maps made?
- How would you measure and model features on Earth's surface?

U.S. Geological Survey (USGS); (inset)Courtesy of HISTOPO

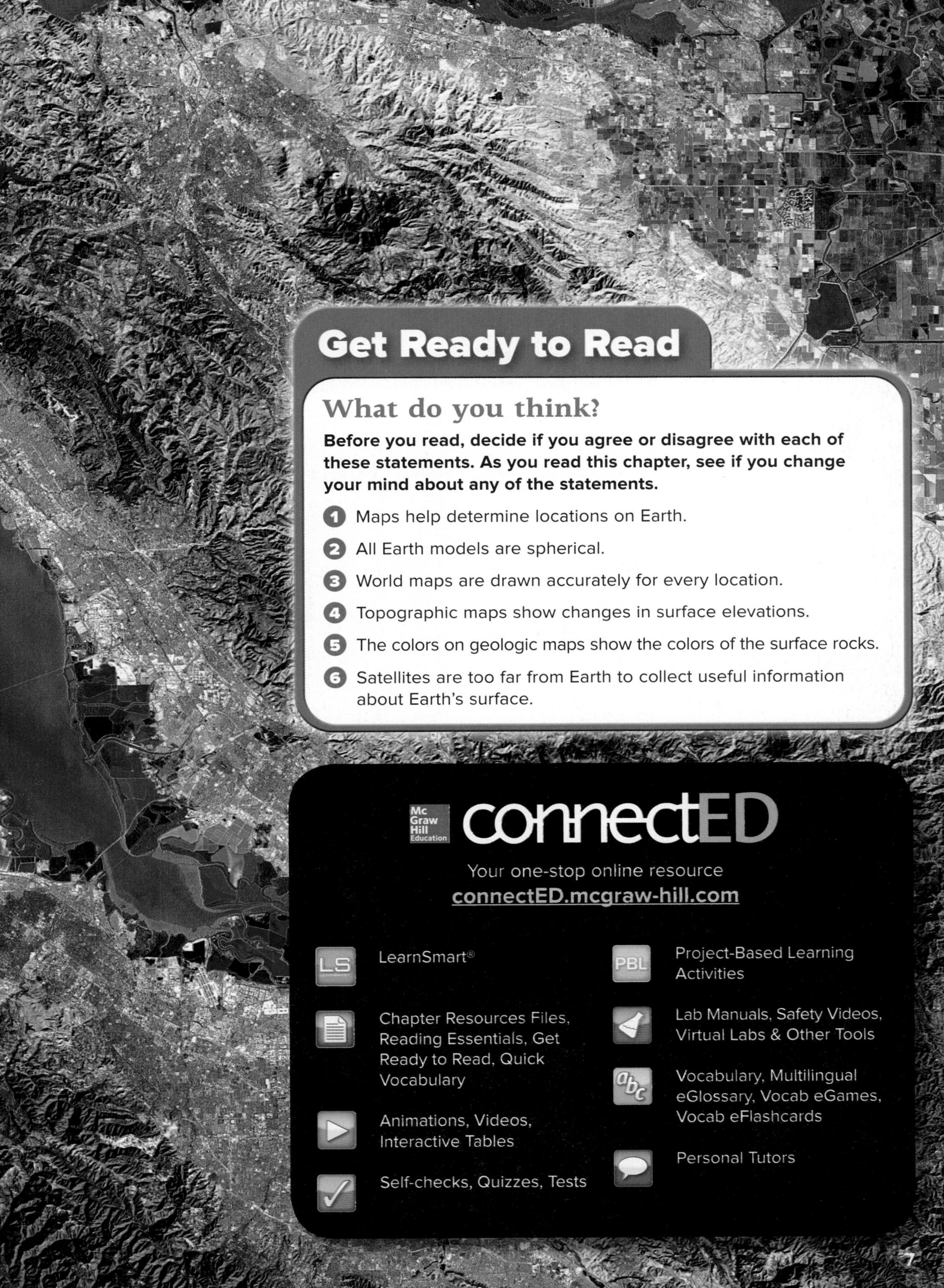

Get Ready to Read

What do you think?

Before you read, decide if you agree or disagree with each of these statements. As you read this chapter, see if you change your mind about any of the statements.

1. Maps help determine locations on Earth.
2. All Earth models are spherical.
3. World maps are drawn accurately for every location.
4. Topographic maps show changes in surface elevations.
5. The colors on geologic maps show the colors of the surface rocks.
6. Satellites are too far from Earth to collect useful information about Earth's surface.

Lesson 1

Reading Guide

Key Concepts
ESSENTIAL QUESTIONS

- How can a map help determine a location?
- Why are there different map projections for representing Earth's surface?

Vocabulary

map view p. 9
profile view p. 9
map legend p. 10
map scale p. 11
longitude p. 12
latitude p. 12
time zone p. 14
International Date Line p. 14

Multilingual eGlossary

Maps

Inquiry Where are they?

Look at the horizon—the place where the blue sky and the blue water come together. What do you see? Have you ever had to figure out where you were without using any landmarks? Suppose you are sailing in the South Pacific. How would you navigate without landmarks to use as reference points?

Hemera Technologies/age fotostock

Launch Lab

15 minutes

How will you get from here to there?

When you need to get to a place you have never visited, you might use a map to help you find your way. Maps help people get where they are going without getting lost.

1. Suppose it is a new student's first day at your school. Write directions for the student to get from the science classroom to the cafeteria.
2. Now draw a map for the student to get from the science classroom to the cafeteria.

Think About This

1. How were the written instructions different from the map?
2. **Key Concept** How are maps useful?

Understanding Maps

When was the last time you used a map to find information? Maybe you looked at a map of your school to find all your classrooms. Or, maybe you reviewed the map of the school to practice for a fire drill or a disaster drill. A map might show all the exits or the safest room to go to if there were a tornado. There are many kinds of maps, such as road maps, trail maps, and weather maps. Each type of map contains different information and serves a different purpose.

A map can be used to model Earth. In order to model Earth's surface, you can make a flat representation of an area of Earth on a piece of paper. In order to model the entire planet and its shape, you can make a globe.

Map Views

Most maps are drawn in **map view**–*drawn as if you were looking down on an area from above Earth's surface.* Map view also can be referred to as plan view, and is shown in **Figure 1.**

A **profile view** *is a drawing that shows an object as though you were looking at it from the side.* A profile view is like a side view of a house. To help you visualize this concept, a map view and a profile view of a house are shown in **Figure 1.** Map views and profile views will be used to describe topographic maps and geologic maps at the end of this chapter. Also, you will use profile views when you study cross sections, or models of the inner structures of Earth.

Map Views

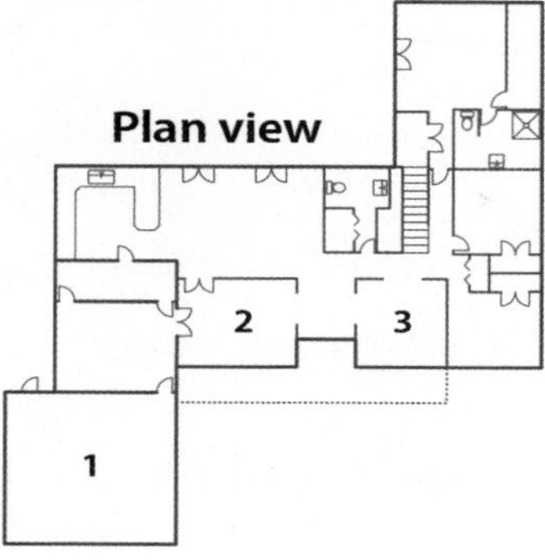

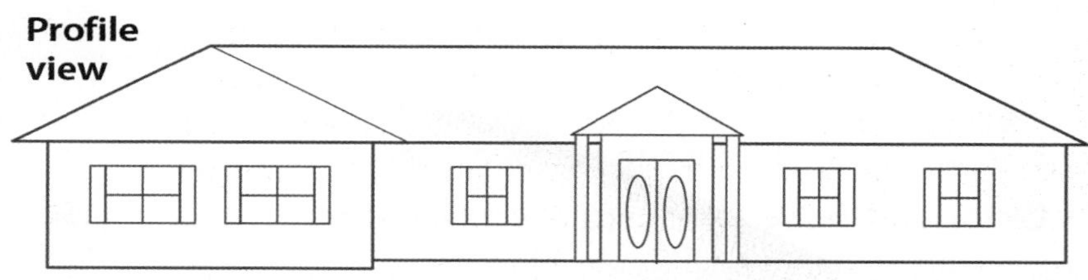

Figure 1 A map view, or plan view, looks down on an object, while a profile view looks from the side.

Hutchings Photography/ Digital Light Source

Figure 2 The legend on this map explains what the symbols mean.

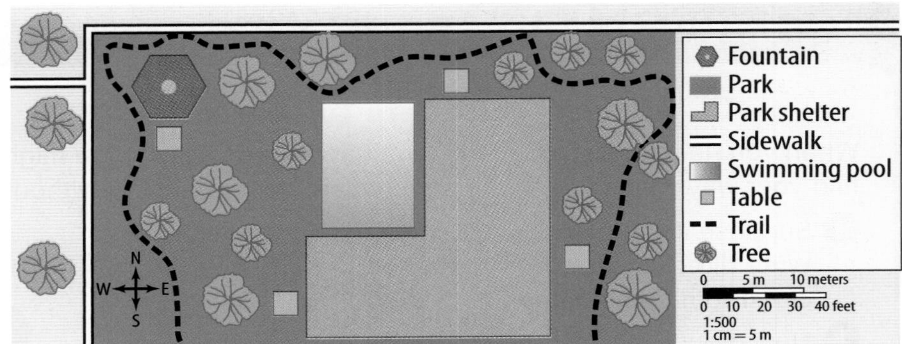

SCIENCE USE V. COMMON USE

legend

Science Use part of a map that explains the map symbols

Common Use a story coming down from the past

Map Legends and Scales

Maps have two features to help you read and understand the map. One feature is a series of symbols called a map legend. The other is a ratio, which establishes the map scale.

Map Legends Maps use specific symbols to represent certain features on Earth's surface, such as roads in a city or restrooms in a park. These symbols allow mapmakers to fit many details on a map without making it too cluttered. All maps include a **map legend**–*a key that lists all the symbols used on the map*–so you can interpret the symbols. It also explains what each symbol means. For example, in the map legend shown in **Figure 2**, a dashed line represents a trail.

Reading Check What is the purpose of a map legend?

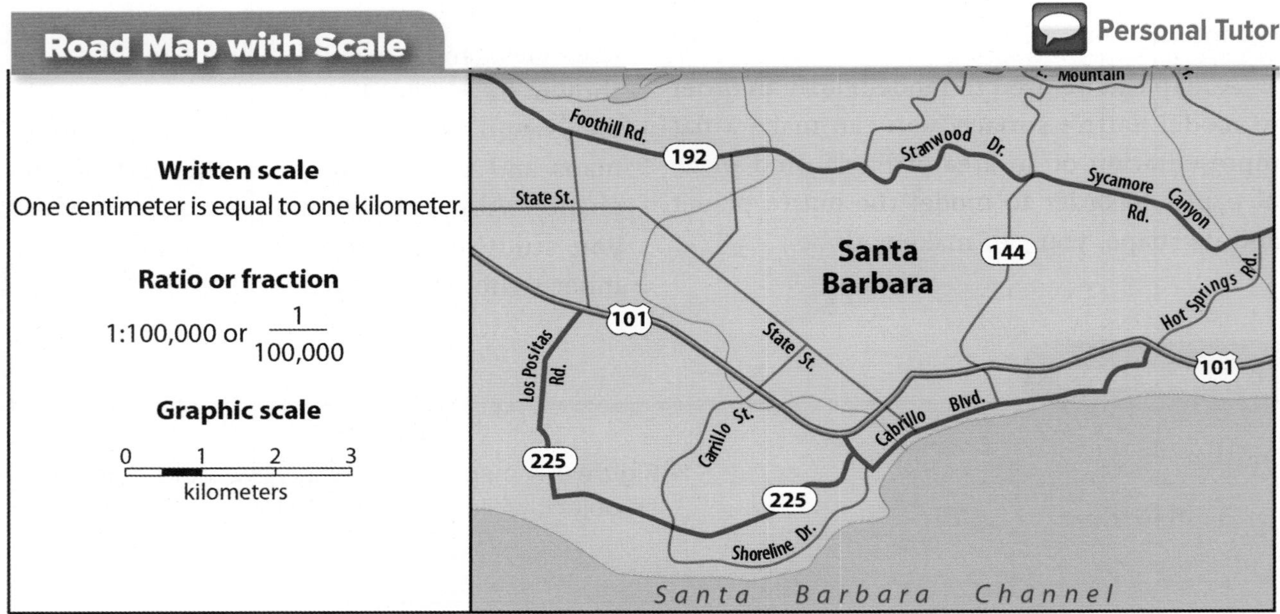

Figure 3 Different types of scales can be used with maps. For example, the graphic scale compares map distance to actual distance.

Visual Check Which scale would you use to measure the distance between the two rivers that intersect Route 192?

Map Scales When mapmakers draw a map, they need to decide how big or small to make the map. They need to decide on the map's scale. **Map scale** *is the relationship between a distance on the map and the actual distance on the ground.* The scale can be verbally written such as "one centimeter is equal to 1 kilometer." The scale also can be written as a ratio, such as 1:100. Because this is a ratio, there are no units. Verbally, you would say, "every unit on the map is equal to 100 units on the ground." If your unit were 1 cm on the map, it would be equal to 100 cm on the ground. If you drew a map of your school at a scale of 1:1, your map would be as large as your school! **Figure 3** gives you a written scale, a ratio scale, and a graphic scale in the map legend. Each one can be useful in different ways. For example, the graphic scale, or scale bar, would be useful in measuring distances on the map. You would have to measure it, however, to find that 1 cm is equal to 1 km.

Figure 4 shows another way in which scales are useful. Models are built with scaled measurements that can be increased or decreased relative to the measurements of real objects. Models have the same relative proportions as the objects they represent, similar to a map scale.

Math Skills 

Ratio Scale

A ratio is a comparison of two quantities by division. For example, a map scale is the ratio of the distance on the map to the actual distance. A map might use a scale in which **1 cm** on the map represents **5 km** of actual distance. This may be written as a ratio:

1 cm to **5 km** or

1 cm:**5 km** or

$\frac{1\text{ cm}}{5\text{ km}}$

This ratio is the map scale.

Practice

A map uses a scale of **1 cm**: **1 km.** If the distance between two points on the map is **3 cm**, what is the actual distance between the points?

 **Math Practice**

 **Personal Tutor**

Figure 4 These images have different scales. In the large photo, the scale is 1:25. In the smaller photo to the right, the scale is 12:1.

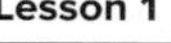

Reading Maps

To find your way to a specific place, you need a way to determine where you are on Earth. Imagine telling someone your exact position on the snow-covered continent of Antarctica. It would be difficult to describe. Ship captains and airplane pilots experience the same problems as they plot their courses across the oceans or above a cloud-covered Earth.

A Grid System for Plotting Locations

Have you ever played a game of chess? If you have, you know that the board is set up with grid lines to help you choose your moves and the position of the pieces on the board. Long ago mapmakers created a system for identifying locations on Earth that uses a similar grid system. This system uses two sets of imaginary lines that encircle Earth. The two sets of lines are called latitude and longitude. The intersection of a line of latitude and a line of longitude can pinpoint a location on a map or a globe.

WORD ORIGIN
longitude
from Latin *longitudo,* means "length"

Longitude Mapmakers started the grid system with a line that circled Earth and passed through the North Pole and the South Pole. The half of the circle from the North Pole to the South Pole passes through Greenwich, England, and is known as the prime meridian. The prime meridian is shown in **Figure 5.** The other half of the circle is the 180° meridian. Similar circles are drawn at every degree east and west of the prime meridian. **Longitude** *is the distance in degrees east or west of the prime meridian.* The prime meridian and the 180° meridian combine to divide Earth into east and west halves, or hemispheres–the eastern hemisphere and the western hemisphere. East of the prime meridian, longitude is measured in degrees east, and west of the prime meridian, longitude is measured in degrees west. They both meet at the 180° meridian. All the meridians pass through the North Pole and the South Pole.

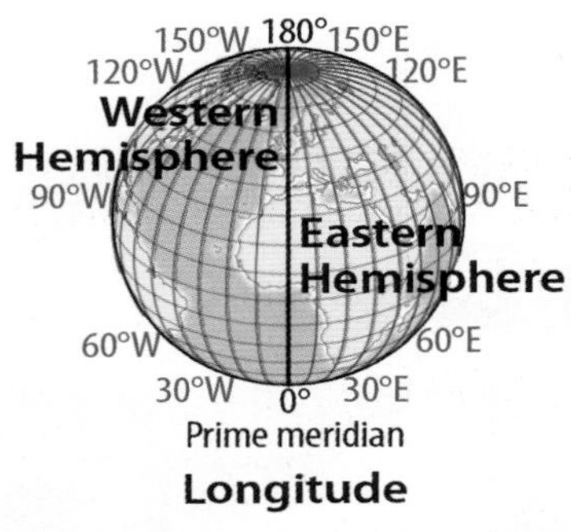

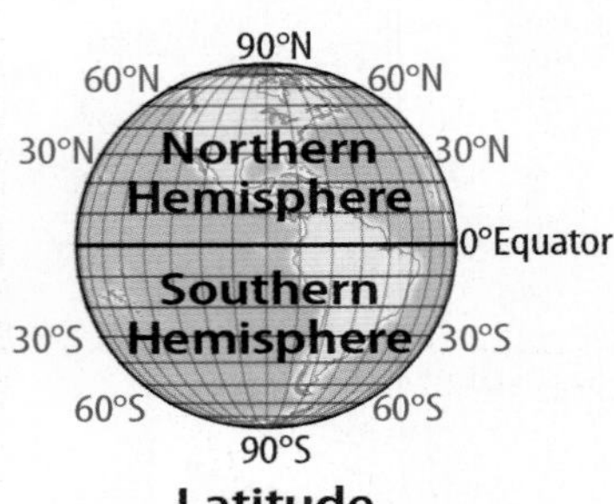

Figure 5 Longitude and latitude are imaginary lines used to pinpoint places on Earth.

Latitude Mapmakers also drew lines east to west around Earth. These lines, called lines of latitude, are somewhat perpendicular to lines of longitude. The center line, called the equator, divides Earth into the northern hemisphere and the southern hemisphere. **Latitude** *is the distance in degrees north or south of the equator.* Unlike lines of longitude, lines of latitude are parallel, as shown in **Figure 5.** The equator is the largest circle. All the other circles become smaller and smaller the closer they are to Earth's poles.

Key Concept Check What relationship do lines of longitude and lines of latitude have?

Plotting Locations

How can you use Earth's grid system to plot locations? First, think about why longitude and latitude are measured in degrees. Earth is a sphere–a ball-shaped object. If you look straight down on a sphere, it looks like a circle. Like a circle, a sphere can be divided into 360 degrees. Look back at **Figure 5.** The latitude at the equator is 0°. All other lines of latitude are measured in degrees north and south of the equator. The North Pole is located at 90 degrees north latitude (90°N), and the South Pole is located at 90 degrees south latitude (90°S).

Lines of longitude are measured in degrees east or west of the prime meridian. There are 180 degrees of east longitude and 180 degrees of west longitude. Any location on Earth can be described by the intersection of the closest line of latitude and the closest line of longitude. Latitude is always stated before longitude.

Minutes and Seconds Longitude and latitude lines are far apart. To better help determine locations, each degree is divided into 60 parts. These parts are called minutes ('). Each minute is divided into 60 parts. These parts are called seconds ("). Degrees, minutes, and seconds allow you to accurately locate places on Earth.

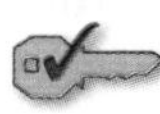

Key Concept Check How do latitude and longitude describe a location on Earth?

MiniLab — 20 minutes

Can you find latitude and longitude?

Use the diagram on the right to answer the following questions.

1. Which city is located at 20°N, 155°W?
2. Which city is located at 40°N, 75°W?
3. What is the latitude and longitude of Seward, Alaska; Memphis, Tennessee; and Denver, Colorado?

- Hilo, HI
- Denver, CO
- Philadelphia, PA
- Memphis, TN
- Seward, AK

Analyze and Conclude

1. **Explain** why the latitude is °N and the longitude is °W for questions 1 and 2 above.
2. **Estimate** What are the latitude and longitude for the city closest to you?
3. **Key Concept** How do latitude and longitude help people locate cities on a map?

Time Zones When it is high noon at your location, the Sun is directly overhead. But as Earth rotates, the Sun is directly above different locations at different times. Businesses in cities many miles apart would have a hard time doing business with each other if every city had its own time. Time zones were created to make travel, communicating, and doing business easier for everyone. *A* **time zone** *is the area on Earth's surface between two meridians where people use the same time.* The reference or starting point for time zones is the **prime** meridian. Earth is divided into 24 time zones, and the width of a time zone is 15° longitude. But, as shown in **Figure 6,** the time zones do not follow the meridians exactly. Their locations are sometimes altered at political boundaries. Notice how the time changes by one hour at the boundary of each time zone. What happens then when you go halfway around the globe?

ACADEMIC VOCABULARY

prime
(adjective) first in rank

Reading Check Why do we need a starting point for time zones?

International Date Line *The line of longitude 180° east or west of the prime meridian is called the* **International Date Line.** Recall that there are 24 time zones, and 24 hours in a day. Because one location can't have two different times on the same day, the day changes as you cross the date line. If you cross from east to west, it is the next day in the west. And if you cross from west to east, it is the day before in the east.

Notice that the International Date Line does not follow the 180° meridian exactly. This is because some island nations would be divided by the line. It would be one day on one island and a different day for another island in the same nation. To avoid this, the International Date Line goes around them.

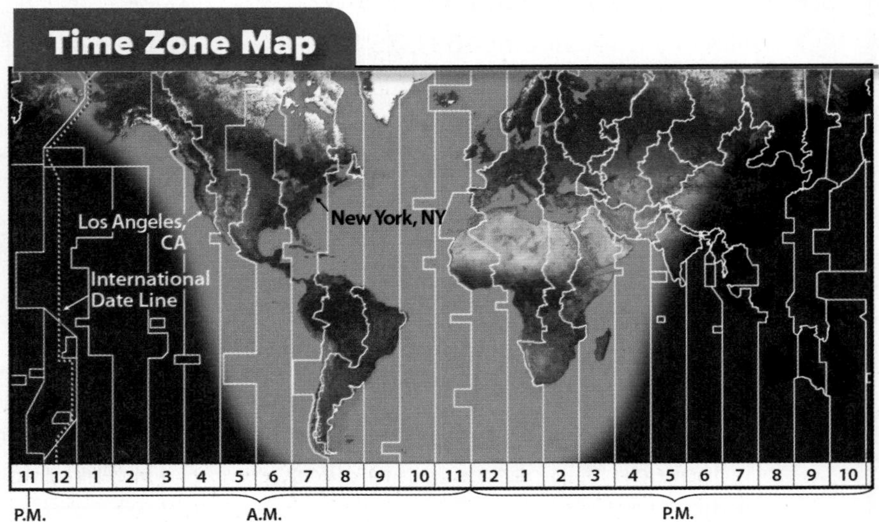

Figure 6 There are 24 time zones around the world.

Visual Check If it is 2:00 P.M. in New York City, what time is it in Los Angeles, California?

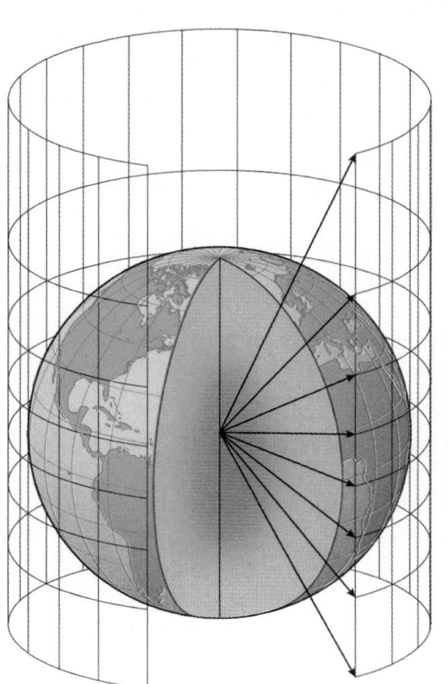

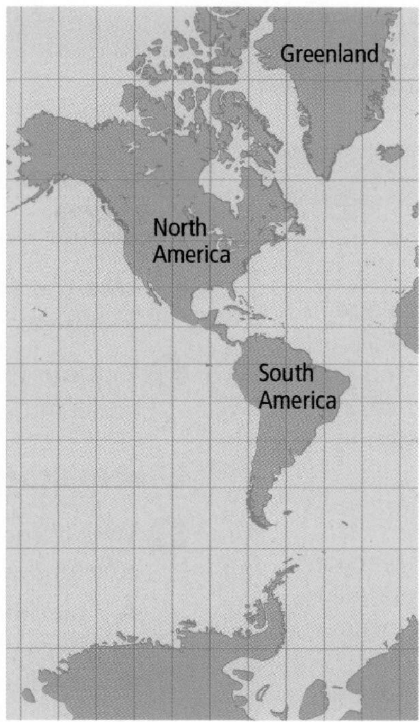

Cylindrical Projection

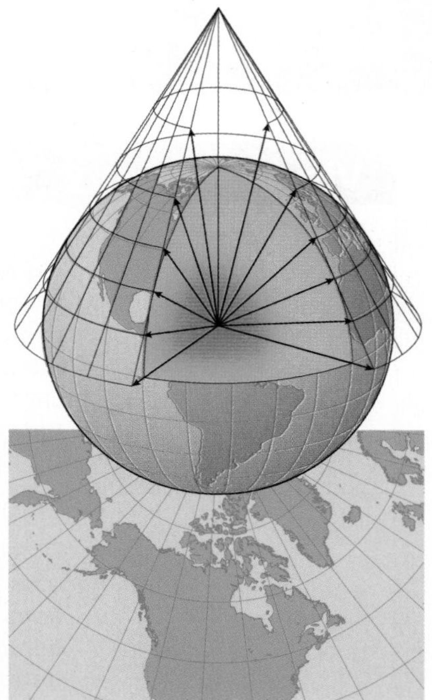

Conical Projection

Figure 7 The cylindrical and conical projections transfer features located on a sphere to a flat map. This process always results in some distortion.

Animation

Map Projections

Since a globe is spherical like Earth, Earth's features are not distorted on a globe. Maps, however, are flat. How can a flat map be made from a sphere? One way to transfer features from a globe to a flat map is to make a projection.

Cylindrical Projections Imagine a light at the center of a globe. It would throw shadows of the continents and the latitude and longitude lines onto a sheet of paper if it were wrapped around the globe. Because the paper is shaped like a cylinder, as shown in **Figure 7**, this is called a cylindrical projection. The resulting map represents shapes near the equator very well. However, shapes near the poles are enlarged. Notice that in the cylindrical projection, Greenland appears to be larger than South America. However, Greenland is about one-eighth the size of South America.

Conical Projections Wrapping a cone around the globe makes a conical projection. It has little distortion near the line of latitude where the cone touches the globe, but it is distorted elsewhere. All types of projections distort the shapes observed on a sphere. In other projections, the continents are represented accurately only because the other areas, such as the oceans, are distorted or cut away.

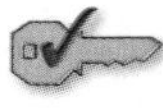

Key Concept Check What are the advantages and disadvantages of cylindrical projections and conical projections?

FOLDABLES®

Make a folded book from a sheet of paper. Label it as shown. Use it to record information about map projections. Label the outside of the book Map Projections.

Conical Projections | Cylindrical Projections

Lesson 1 Review

Visual Summary

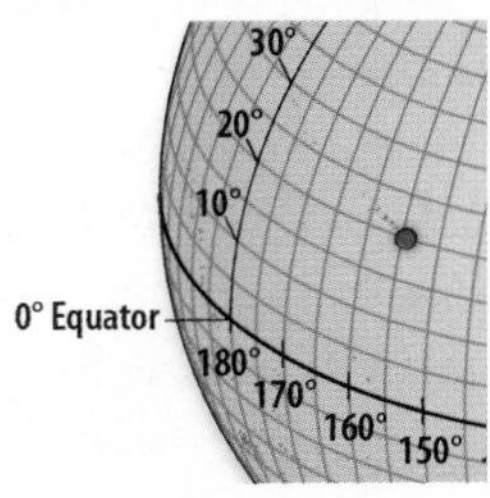

Finding locations on a map or a globe can be done accurately by using grid lines called longitude and latitude.

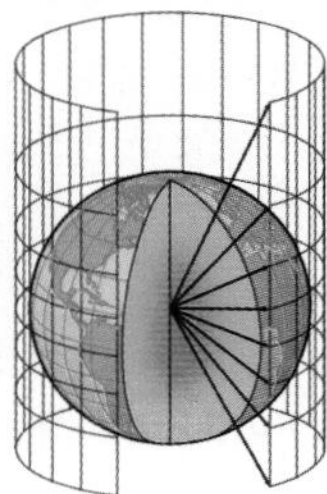

Different projections offer different solutions to the distortion problem of transferring three dimensions into two dimensions.

FOLDABLES®

Use your lesson Foldable to review the lesson. Save your Foldable for the project at the end of the chapter.

What do you think NOW?

You first read the statements below at the beginning of the chapter.

1. Maps help determine locations on Earth.
2. All Earth models are spherical.
3. World maps are drawn accurately for every location.

Did you change your mind about whether you agree or disagree with the statements? Rewrite any false statements to make them true.

Use Vocabulary

1. **Define** *profile view* in your own words.
2. **Use the terms** *latitude* and *longitude* in a sentence.
3. **Explain** the difference between a map scale and a map legend.

Understand Key Concepts 

4. Which lines are used to measure the distance south of the equator?
   - **A.** meridians
   - **B.** lines of latitude
   - **C.** the International Date Line
   - **D.** lines of longitude
5. **Compare** a globe to a map. Explain why distortions occur.
6. **Explain** why the International Date Line does not match the 180° meridian exactly.

Interpret Graphics

7. **Identify** Copy and fill in the graphic organizer below to identify the three units used to measure latitude and longitude.

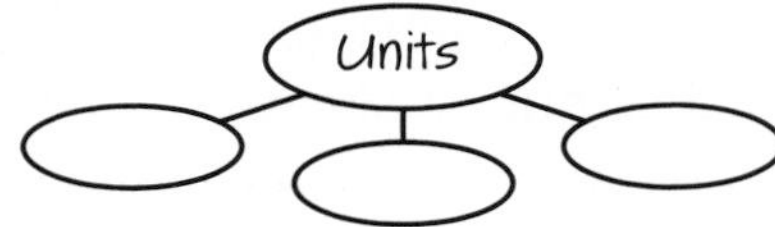

Critical Thinking

8. **Suggest** a reason that the time zones do not exactly follow meridians.
9. **Evaluate** Which type of projection—conical or cylindrical—would show less distortion of central Africa? Explain your choice.

Math Skills

10. The distance between two towns on a map is 7 cm. The map scale is 1 cm:100 km. What is the actual distance between the two towns?

Skill Practice Compare and Contrast 25 minutes

How can you fit your entire classroom on a single sheet of paper?

Mapmakers must measure objects and distances very carefully to produce accurate maps. Without detailed and accurate measurements, maps would not be useful. Most maps are scaled down. This means that the map and details in it are smaller than what they represent. Sizes and distances on a scaled map are proportions of the actual values. For example, if a map has a 1 cm to 1 m scale, 5 cm on the map represents 5 m.

Materials

metric ruler

meterstick

graph paper

Learn It

When you look for similarities between two things, you **compare** them. When you find differences between two things, you **contrast** them. Creating a ratio in order to scale down the dimensions of a room to make a map compares the room's actual dimensions to the map's scale dimensions. The difference between the map and the room are the units of measurement (cm:m).

Try It

1. Read and complete a lab safety form.
2. On a blank sheet of graph paper, sketch your classroom as if you were looking down on it. Do not worry about accuracy right now.
3. Select several objects or structures lining the classroom, such as windows or doors. Measure how far each is from the corners of the walls. Record your data in your Science Journal.

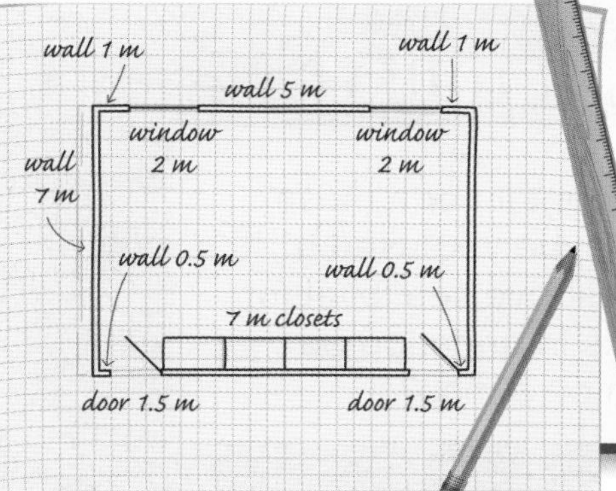

4. Your teacher will tell you the dimensions of the classroom. Choose a scale for a map of the room. Use the dimensions and your scale to draw a map of the classroom on a single piece of graph paper.
5. Make sure to include all the features from your sketch. Also, include a scale, legend, and the total area.

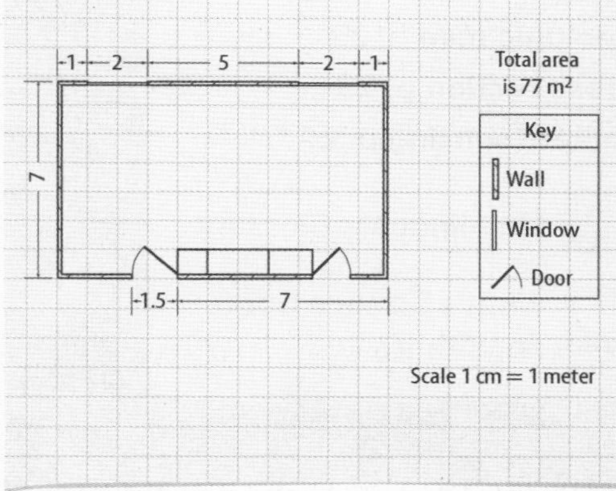

Apply It

6. What scale did you use in your map? Explain why you chose that scale.
7. How is your map similar to a map of Earth? How is it different?
8. **Key Concept** Would the sketch or the map you made be more useful to help someone locate an object in the room? Support your reasoning.

Lesson 2

Reading Guide

Key Concepts
ESSENTIAL QUESTIONS

- What can a topographic map tell you about the shape of Earth's surface?
- What can you learn from geologic maps about the rocks near Earth's surface?
- How can modern technology be used in mapmaking?

Vocabulary

topographic map p. 20
elevation p. 20
relief p. 20
contour line p. 20
contour interval p. 21
slope p. 21
geologic map p. 23
cross section p. 23
remote sensing p. 27

Multilingual eGlossary

What's Science Got to do With It?

Technology and Mapmaking

Inquiry Mountains or Molehills?

Have you ever hiked to the top of a mountain or a mesa? How did you know how high it was? Maybe you used a map with information about the height above sea level for locations along the trail. Why would this information be helpful? How is this information shown on a map?

Launch Lab

20 minutes

Will this be an easy hike or a challenging hike?

If you were going for a hike, you would probably want to know if it would be easy or hard. Would you have to climb a steep hill or is the area flat? How could you find this information?

1. Obtain a map with elevation information on it.
2. Plan two hikes that cover the same distance on the map. Plan one easy hike over flat terrain and one challenging hike in which a hill will be climbed.
3. Share with a partner how both hikes would be different. How are the elevations of locations on your map shown?

Think About This

1. What are the benefits of knowing where there are steep and gentle slopes on a map?
2. **Key Concept** How would you describe elevation information on a map?

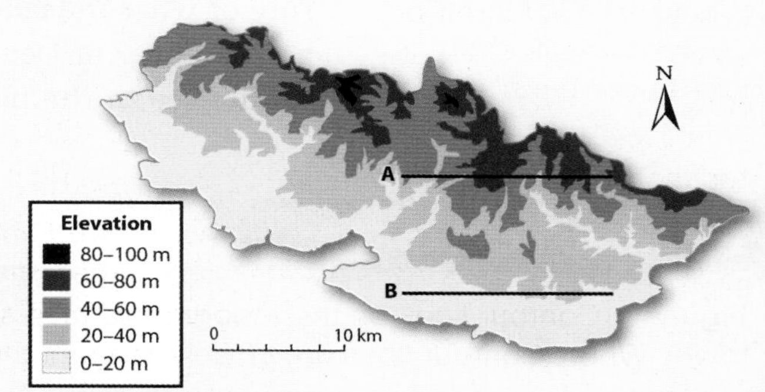

Types of Maps

If you were going to join two pieces of wood together, you might use a hammer and nails. To scramble eggs you could use a whisk and a skillet. Just as there are tools for doing different jobs, there are maps for different purposes.

General-Use Maps

The first maps were hand-drawn by explorers and sailors to record their trading routes. Today we use maps in a variety of situations. You might use a map to help a friend find your house or the quickest route to the mall. If you go to a park there might be a trail map outlining the route you will hike. Some everyday maps you might use include:

- **Physical maps** use lines, shading, and color to indicate features such as mountains, lakes, and streams.
- **Relief maps** use shading and shadows to identify mountains and flat areas.
- **Political maps** show the boundaries between countries, states, counties, or townships. The boundaries can be shown as a variety of solid or dashed lines. Different colors might be used to indicate areas within the boundaries.
- **Road maps,** as shown in **Figure 8,** can show interstates or a range of roads from four-lane expressways to gravel roads. Maps all are useful in helping you find your way.

Figure 8 Road maps of counties or cities can be very detailed, but an atlas of maps of the 50 states might show only the important or main roads.

Hutchings Photography/Digital Light Source

Topographic Maps

If you were hiking across the United States, you might want to follow level terrain. If you were piloting an airplane across the United States, you would definitely want to fly higher than a mountain. Showing you how high or low land features are is a feature of one kind of specialty map.

WORD ORIGIN

topography
from Greek *topos*, means "place"; and *graphein*, means "to write"

The shape of the land surface is called topography. *A* **topographic map** *shows the detailed shapes of Earth's surface, along with its natural and human-made features.* It helps give you a picture of what the landscape looks like without seeing it. The topographic map of Devil's Tower in **Figure 9** shows the details you cannot see in the photo.

Topographic Map

Figure 9 Contour lines on the topographic map show differences in elevation on this volcanic tower. Where contour lines are closely spaced, the topography is much steeper.

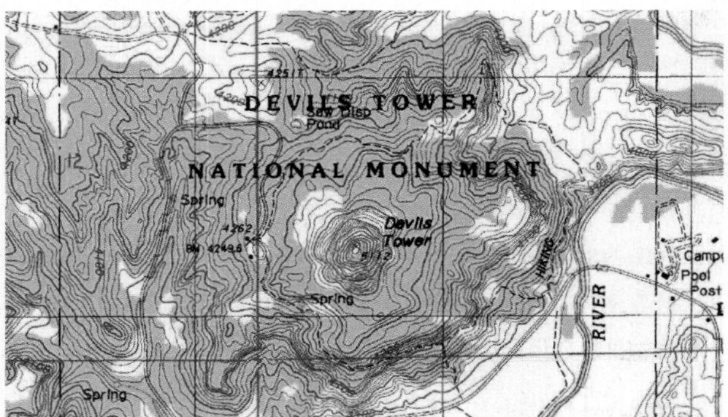

Elevation and Relief *The height above sea level of any point on Earth's surface is its* **elevation.** For example, Mt. Rainier in Washington is 4,392 m above sea level. The city of Olympia, Washington, is about 43 m above sea level. *The difference in elevation between the highest and lowest point in an area is called* **relief.** For example, the relief between Mt. Rainier and Olympia is calculated 4,392 m − 43 m = 4,349 m.

Contour lines *are lines on a topographic map that connect points of equal elevation.* Similar to lines of latitude and longitude, contour lines do not really exist on Earth's surface. Using contour lines, you can measure both elevation and relief on a topographic map. If the top of Devil's Tower is 5,112 ft and the base is 4,400 ft, what is the relief?

Reading Check How are contour lines similar to lines of latitude and longitude?

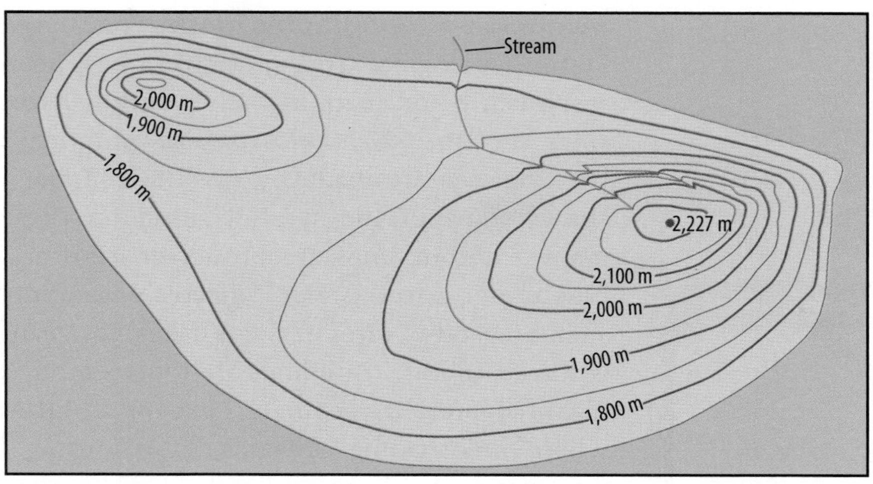

Figure 10 Contour lines connect points of equal elevation.

Visual Check Where is the slope gentle or steep?

Interpreting Contours Contour lines that represent a mountain are shown in **Figure 10.** Notice that the elevation is not written on every contour line. The darker contour lines, called index contours, are labeled with the elevation. How do you find the elevation of contours other than by using index contours? You need to know the elevation difference between the lines.

The elevation difference between contours that are next to each other is called the **contour interval.** The map in **Figure 10** has a contour interval of 50 m. You can find the elevation of an unlabeled contour by using the numbered index contours. First, find the closest index contour below the contour you are identifying. Then, count up to it by 50s from the index contour.

Study **Figure 10** again. Notice that a contour line at the top of the mountain forms a small enclosed loop with a dot in the middle of it. This dot represents the highest point on the mountain–2,227 m. The V-shaped contours pointing downhill indicate ridges. A small V pointing uphill indicates a stream valley or drainage.

The spacing of contours indicates slope. **Slope** *is a measure of the steepness of the land.* If the contours are spaced far apart, the slope is gradual or flat. If the contours are close together, the slope is steep.

Key Concept Check What can you learn about the features at Earth's surface from studying contour lines?

Topographic Profiles You read about map views and profile views in Lesson 1. The information contour lines provided on a topographic map can be used to draw an accurate profile of the topography. Making profiles like this can help you determine the easiest path to take when crossing the land.

FOLDABLES®

Make a vertical trifold book. Label the front *Topographic Maps,* and label the inside of the book as shown. Use it to collect information about what a topographic map can show you.

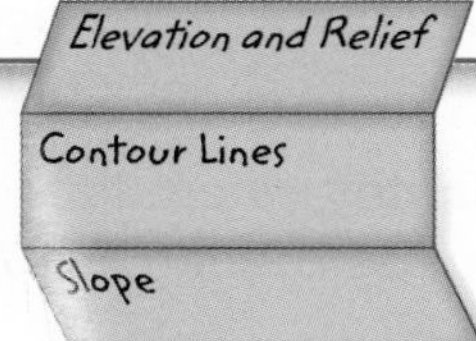

MiniLab 20 minutes

Can you construct a topographic profile?

A topographic profile of line AB helps you identify geological features of a contour map.

1. Use a piece of **graph paper** to set up your topographic profile graph. Label the x-axis *Distance Between A and B.* Label the y-axis *Elevation (m).*
2. Measure the length of line AB on the contour map below. Use a **ruler** to measure the distance from point A to the intersection of the first contour line. Plot the point on your graph.
3. Plot distance and elevation pairs for each contour line where it intersects line AB.
4. Connect the points on your graph and observe the topographic profile.

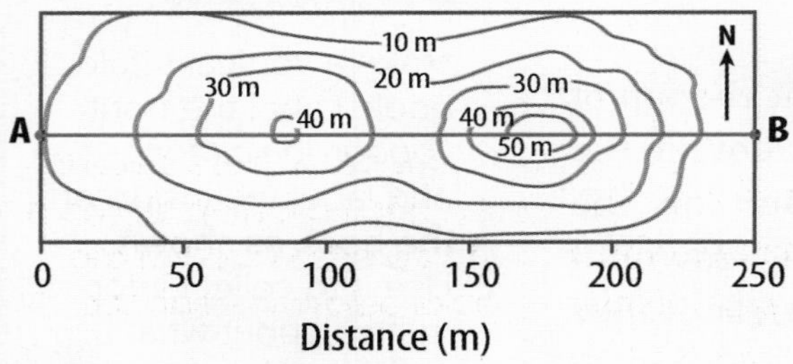

Analyze and Conclude

1. **Analyze** At what distance from point A is the highest point on line AB? The lowest?
2. **Identify** where the topography is the steepest along line AB. Explain how you know this.
3. **Predict** how a contour map and topographic profile would be useful as you design a skateboard park.
4. **Key Concept** Describe three topographic features depicted in your topographic profile.

Symbols on Topographic Maps The United States Geological Survey (USGS) has been responsible for mapping the United States since the late 1800s. Most topographic maps that you see are made by the USGS. **Table 1** shows some of the symbols used on these maps. Contour lines are brown on land and blue under water. Green indicates vegetation, such as woods. Water in rivers, lakes, or oceans is shown in blue. Buildings are represented as black squares or rectangles, except in cities where pink shading indicates dense housing.If information has been updated since the original map was made, it is added in purple.

Reading Check Why is it important for a topographic map to have a legend?

Interactive Table

Table 1 USGS Topographic Map Symbols

Description	Symbol
Primary highway	
Secondary highway	
Unimproved road	
Railroad	
Buildings	
Urban area	
Index contour	100
Intermediate contour	
Perennial streams	
Intermittent streams	
Wooded marsh	
Woods or brushwood	

Visual Check How are primary and secondary highways differentiated on a USGS topographic map?

Geologic Maps

Another kind of specialty map is a geologic map. **Geologic maps** *show the surface geology of the mapped area.* This can include the rock types, their ages, and locations of faults. The geologic map in **Figure 11** shows the geology of the Grand Canyon.

Geologic Map

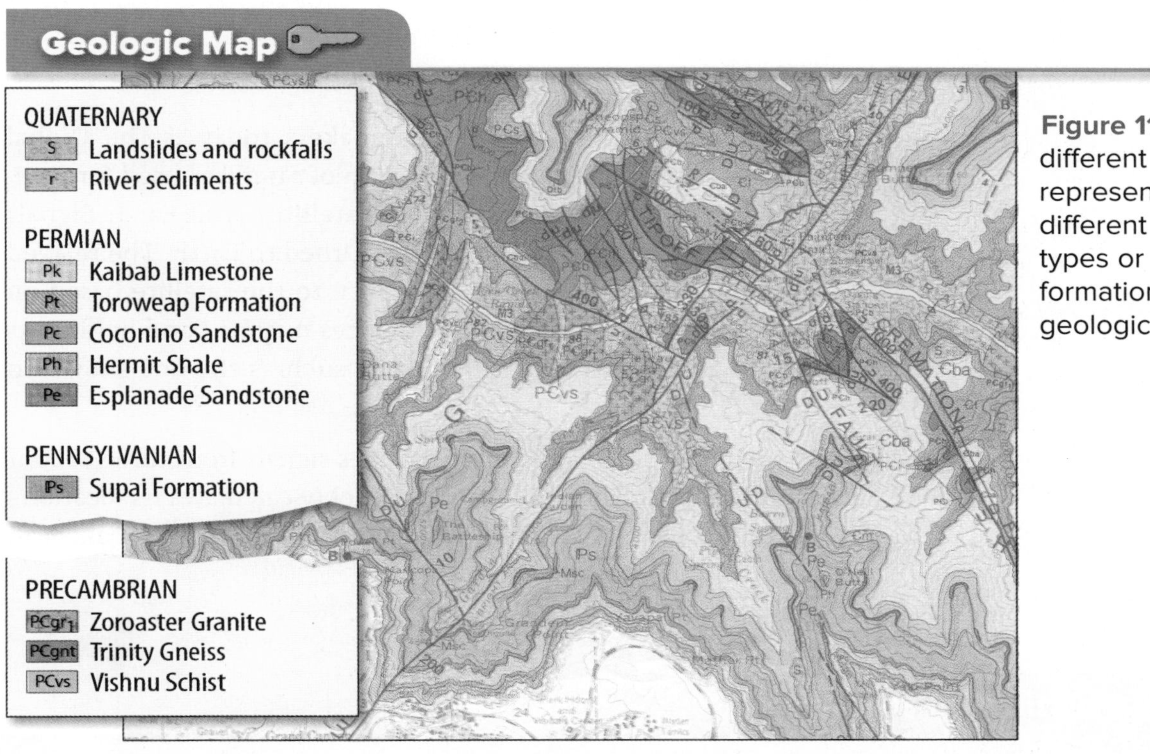

Figure 11 The different colors represent different rock types or formations on a geologic map.

Geologic Formations On a geologic map, different colors and symbols represent different geologic formations. A geologic formation is a rock unit with similar origins, rock type, and age. The map legend lists the colors and symbols along with the age of the rock formation. The colors do not indicate the rock's true colors; they show the many formations on the map. Find the Kaibab formation in the map legend of **Figure 11.** It tells you this limestone rock was made during the Permian period.

Geologic Cross Sections Sometimes geologists need to know what the rocks are like underground as well as on the surface. Information can be gathered by drilling for samples, studying earthquake waves, or looking at cliffs. A cliff face is like a profile view of the ground. Geologists use this information to produce a profile view of the rocks below the ground. The resulting diagram is called a **cross section**–*a profile view that shows a vertical slice through rocks below the surface.* **Figure 12** shows a cross section of a geologic map.

 **Key Concept Check** How is color used in a geologic map?

Figure 12 A cross section of a geologic map shows a vertical slice through the rocks below the surface. ▼

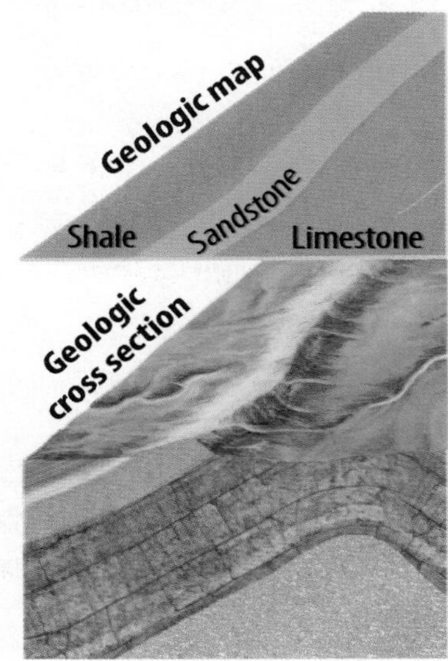

Making Maps Today

For centuries mapmakers made observations of Earth and gathered information from explorers. First, mapmakers and explorers used instruments such as a compass, a telescope, and a sextant, which is used to find latitude, to make and record measurements. Then, mapmakers carefully drew new maps by hand. Today, mapmakers use computers and data from satellites to make maps.

Global Positioning System

One important resource for mapmakers today is the Global Positioning System (GPS). GPS is a group of satellites used for navigation. As shown in **Figure 13,** 24 GPS satellites orbit Earth. Signals sent from devices on the surface are returned to Earth. The relayed signals are used to calculate the distance to the satellite based on the average time of the signal. The devices may be hand-held units the size of a cell phone, or larger units such as the one shown in **Figure 14.**

At any given time, a GPS unit receives signals from three or four different satellites. Then the receiver quickly calculates its location—its latitude, longitude, and altitude. GPS is used by mapmakers to accurately locate reference points.

Global Positioning System

Figure 13 GPS receivers detect signals from the 24 GPS satellites orbiting Earth. Using signals from at least three satellites, the receiver can calculate its location within 10 m.

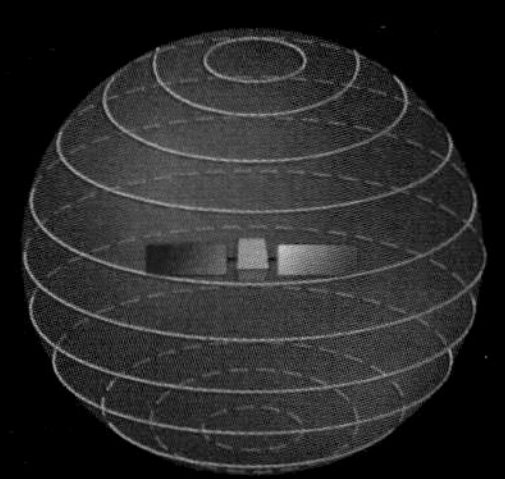

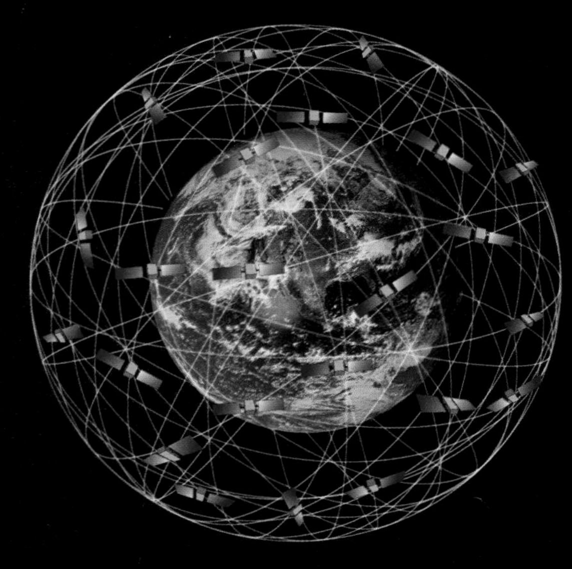

1 The information from one satellite tells the GPS receiver that it is somewhere on a sphere surrounding that satellite. Suppose the distance to the satellite is 20,000 km. This limits the possible location of the receiver to a spherical radius of 20,000 km from the satellite. If the receiver is on Earth, that limits the location to a large circle somewhere on Earth.

Originally designed for military purposes, it is now a continuously available service for everyone worldwide. Airplanes, ships, and cars have navigational systems that use GPS technology. People can find their way to restaurants, hotels, or sporting events and home again. Other uses include tracking wildlife for scientific data collection, detecting earthquakes, hiking, biking, and land surveying. **Figure 14** shows a portable GPS receiver you might use while traveling in a car.

Reading Check What are some common uses of GPS?

GPS technology continues to improve. Land-based units used in combination with satellites are already being used to pinpoint people and places to the centimeter. Future improvement will include guiding self-driven automobiles and additional civil channels projected to improve safety and rescue operations.

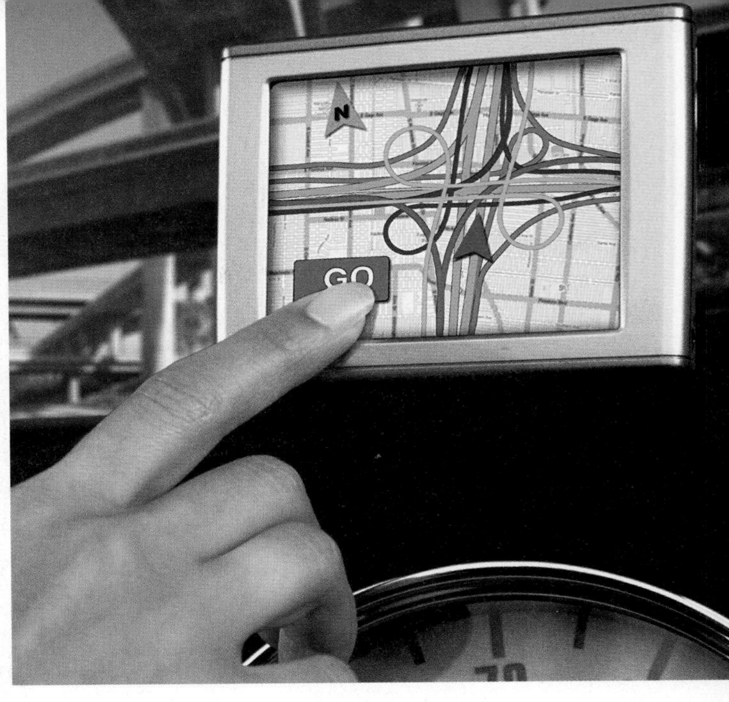

▲ **Figure 14** Portable GPS receivers like this one can be used to help find your way.

Key Concept Check How can GPS technology be used in mapmaking?

Animation

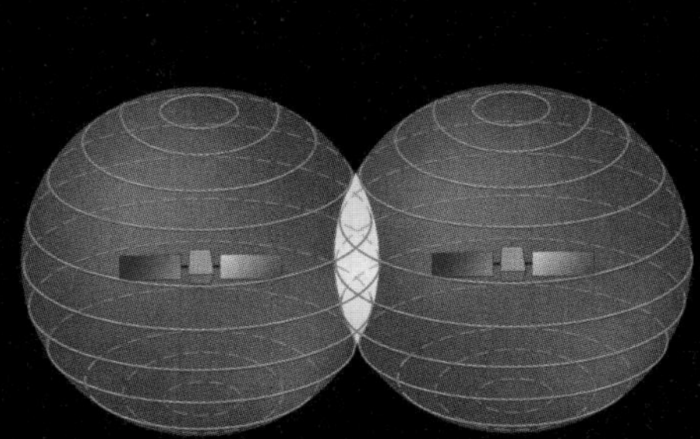

2 Next, the receiver measures the distance to a second satellite. Suppose this distance is calculated to be 21,000 km away. The location of the receiver has to be somewhere within the area where the two spheres intersect, shown here in yellow.

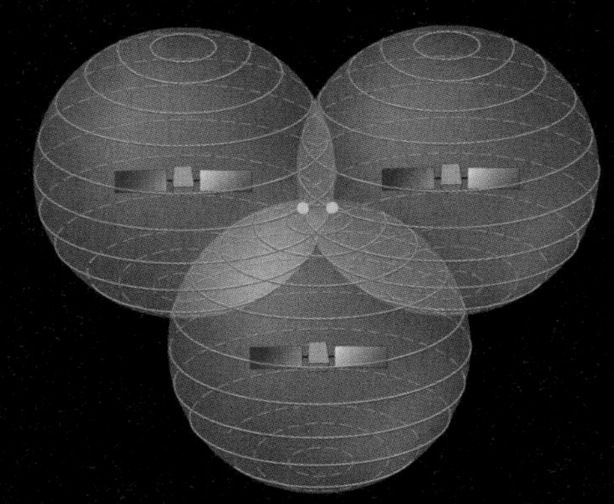

3 Finally, the distance to a third satellite is calculated. Using this information, the location of the receiver can be narrowed even further. By adding a third sphere, the location can be calculated to be one of two points as shown. Often one of these points can be rejected as an improbable or impossible location. Information from a fourth satellite can be used to tell elevation above Earth's surface. A pilot or a climber might find this useful.

Geographic Information Systems

Geographic Information Systems (GIS) are computerized information systems used to store and analyze map data. GIS combine data collected from many different sources, including satellites, scanners, and **aerial** photographs. Aerial photographs are taken from above the ground. Data collection that at one time took months now takes hours or minutes.

ACADEMIC VOCABULARY
aerial
(adjective) operating or occurring overhead

Mapmakers use GIS to analyze and organize those data and then create digital maps. One of the features of GIS is that it creates different map layers of the same location. As shown in **Figure 15**, the map layers are like the layers of a cake. However, when the map layers are placed on top of each other, you can see through to the lower layers. Different layers can show land use, elevation, roads, streams and lakes, or the type of soil on the ground.

Three Views Imagine setting up a model for an airplane landing under certain weather conditions using GIS.

- Database view begins the process by assembling information from existing databases on winds, airplane flight, landing procedures, and airport layouts.
- Map view would draw from a set of interactive, digital maps that show features and their relationship to Earth's surface.
- Model view then would pull all the information together so you could run simulations under changing weather conditions.

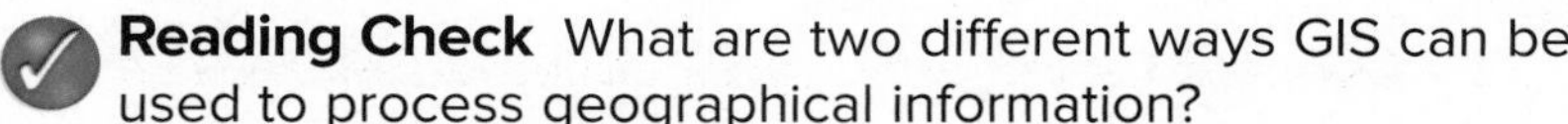

Reading Check What are two different ways GIS can be used to process geographical information?

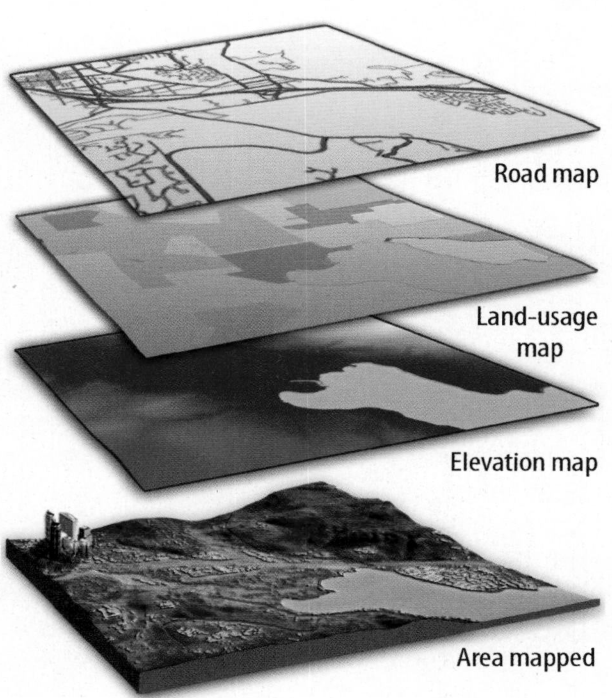

Figure 15 GIS combines the data from many maps to give detailed information about a mapped area.

Remote Sensing

A cup of hot chocolate looks very hot. You place your hand over it and feel the heat from the liquid. Without even touching it, you know it's still too hot to drink. You have just avoided burning your mouth by using remote sensing.

Remote sensing *is the process of collecting information about an area without coming into physical contact with it.* There are many applications for remote sensing. This process produces maps that show detailed information about agriculture, forestry, geology, land use, and many other subjects. Often these maps cover huge areas.

Mapmaking was transformed when it became possible to take aerial photographs from airplanes. Now an even more powerful type of remote sensing is being used. Since the 1970s, satellites orbiting thousands of kilometers above Earth's surface have been used to collect data.

Monitoring Change with Remote Sensing Satellites orbit Earth repeatedly. This means images of a location made at different times can be used to study change. For example, a 3-year drought in northen California dropped water levels in many local reservoirs. Before and after images of the Shasta Lake reservoir are shown in **Figure 16.** Images like these help mapmakers to quickly make maps of areas affected by natural disasters. The maps are then used to monitor damage and help organize rescue efforts.

Key Concept Check How can remote sensing be an advantage to mapmakers?

Figure 16 These satellite images show the changing shoreline of the Shasta Lake reservoir after a 3-year drought. The after image shows the increase in shoreline that occurred due to periods of extreme drought.

Before

After

Landsat One series of satellites used to collect data about Earth's surface is the Landsat group. *Landsat 7,* launched in 1999, and *Landsat 8,* launched in 2013, scan Earth's entire surface. Comparing today's data to similar data collected years ago, scientists can recognize global changes in agriculture, forestry, geology, and changes caused by natural disasters such as hurricanes, flooding, and forest fires. Images obtained from *Landsat 7* and *Landsat 8* can be used for emergency response and disaster relief. Landsat has been used to contribute to the GIS database as well.

OSTM/*Jason-2* and *Jason-3* A series of satellite missions, including the OSTM/*Jason-2* and *Jason-3,* have been used to determine ocean topography and circulation, sea level, tides, and climate changes. **Figure 17** shows ocean surface changes due to El Niño and La Niña as captured by *Jason-2.*

SeaBeam A device that uses sonar to map the bottom of the ocean is SeaBeam. SeaBeam is mounted onboard a ship. Computers calculate the time a sound wave takes to bounce off the ocean floor and return to the ship. This gives the operators an accurate image of the seafloor and the depth of the ocean at that point. SeaBeam is used by fishing fleets, drilling operations, and various scientists.

Reading Check What are some methods used to collect remote sensing data?

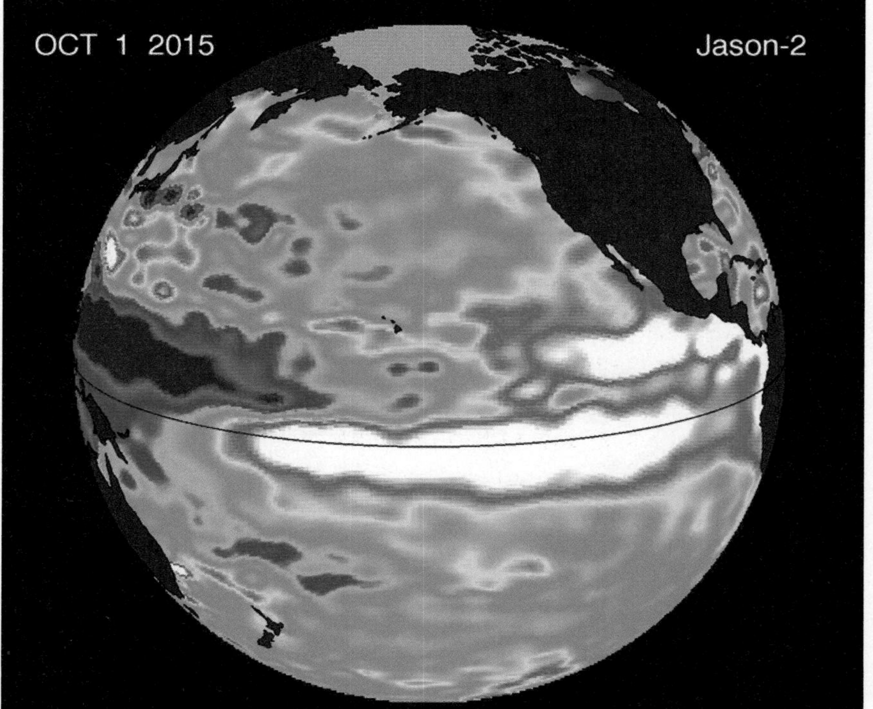

Figure 17 This image of the Pacific was captured by *Jason-2* during an El Niño event. The white area shows an increase in the ocean surface temperature compared to normal.

NASA/JPL Ocean Surface Topography Team

Lesson 2 Review

Visual Summary

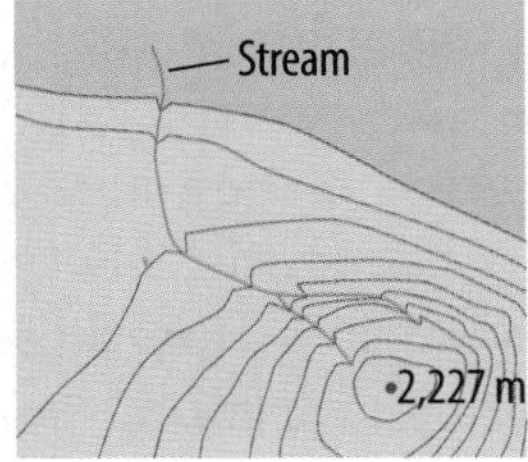

Topographic maps have contour lines that help describe the elevation and relief of the surface of Earth at a particular location.

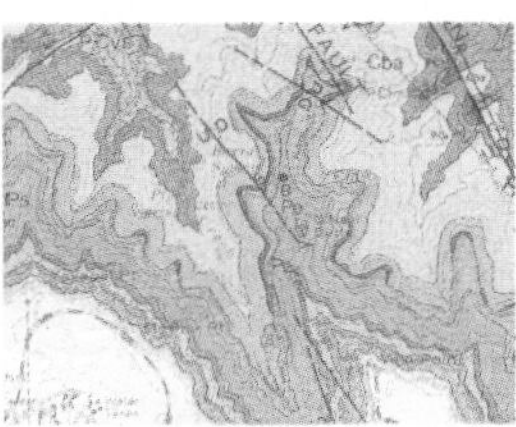

Geologic maps are useful in determining rock type, rock age, and the rock formations in an area.

FOLDABLES®

Use your lesson Foldable to review the lesson. Save your Foldable for the project at the end of the chapter.

What do you think NOW?

You first read the statements below at the beginning of the chapter.

4. Topographic maps show changes in surface elevations.

5. The colors on geologic maps show the colors of the surface rocks.

6. Satellites are too far from Earth to collect useful information about Earth's surface.

Did you change your mind about whether you agree or disagree with the statements? Rewrite any false statements to make them true.

Use Vocabulary

1. **Define** *cross section* in your own words.
2. Change in elevation is called ________.
3. **Use the terms** *contour lines* and *contour interval* in a complete sentence.

Understand Key Concepts 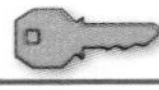

4. Which type of map would be more useful for determining the quickest route by car?
   A. geologic map
   B. political map
   C. road map
   D. topographic map
5. **Illustrate** a mountaintop that is 850 m high using a contour interval of 50 m.
6. **Separate** the symbols shown on a topographic map legend into groups of natural and cultural features.

Interpret Graphics

7. **Summarize** Copy and fill in the graphic organizer below to identify three things you can learn about the shape of Earth's surface from contour lines.

8. **Determine** the contour interval for the contour map below.

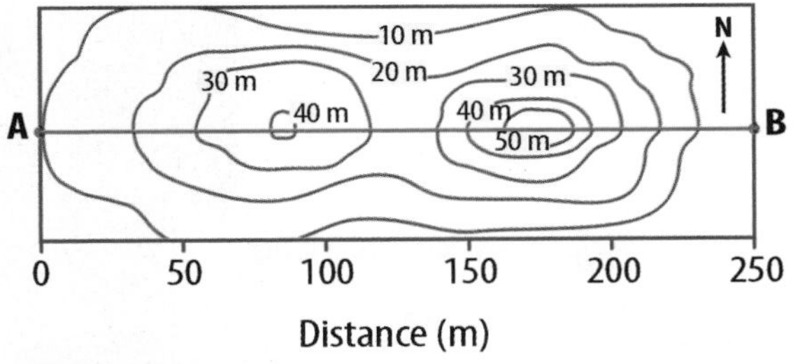

Critical Thinking

9. **Suggest** how a photograph of the Grand Canyon could be used to make a geologic map of Arizona.

Lab

40 minutes

Materials

metric ruler

Seeing Double?

The satellite images shown here are called stereo photographs. At first glance, the photos in each set might appear the same. How-ever, if you look very closely, you should see slight differences between the pictures in each set. These differences will allow your eyes to change the 2-D images into 3-D views of the land. Let's see if your eyes can create these 3-D optical illusions so that you can study these features of Earth's surface.

Question

How can stereo photographs be used to study Earth's surface features?

Procedure

1. Look at the images on this page. Find the thin, gray lines in the image. They are rivers that were filled with ash from a volcanic eruption.
2. Now locate the black shapes near the center of each image. These are lakes.
3. Much of the white and gray area in the photo is one landform made of a volcano and its ash and debris deposits.
4. Now study the images on the next page. The bright blue areas on the images are lakes.
5. The brown and reddish-brown colored areas on this set of images are rocks and soil.
6. Now locate the different green-colored areas on this set of images. These are trees and other vegetation.

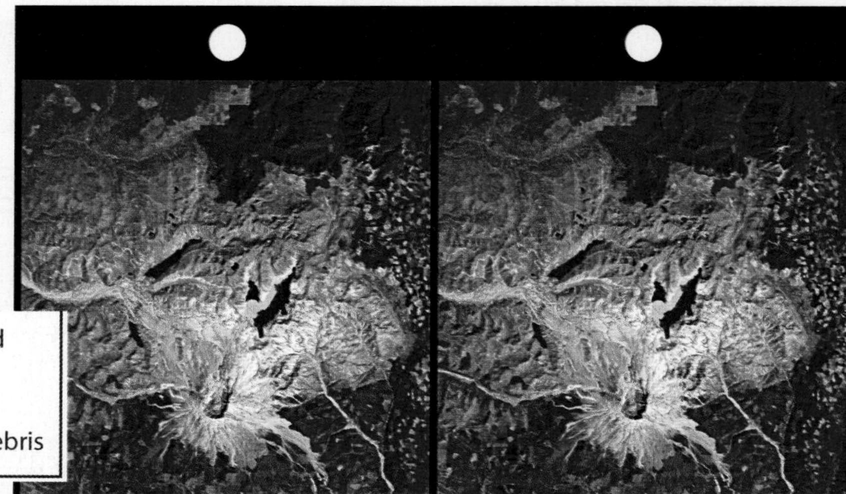

Forest-covered land

Lake water

Volcanic ash and debris

(t)Michael Scott/McGraw-Hill Education; (b)NASA/JPL

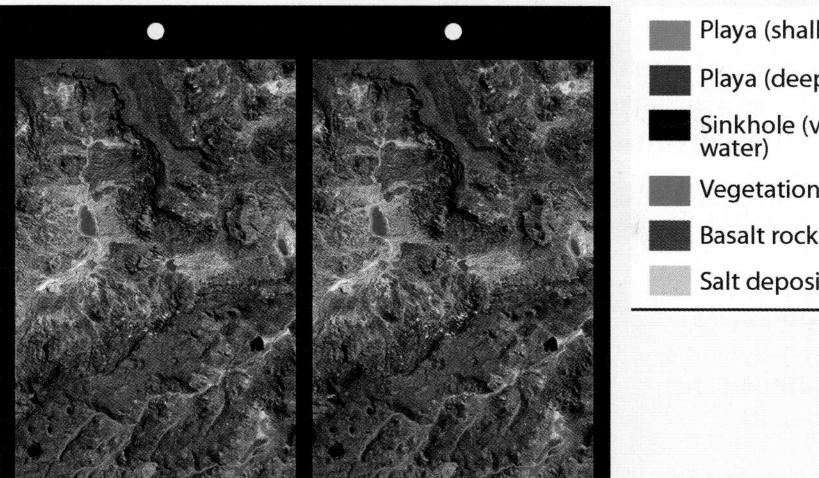

7 View each landscape in 3-D. To do this, slightly cross your eyes while looking at the two white dots above the images. A third white dot will appear between the two dots. At this point, you should see a 3-D view of the landscape. Study the landscape. Then repeat this procedure for the other set of images.

Lab Tips

- ☑ Cross your eyes only slightly to help form the 3-D view.
- ☑ Relax! Sometimes, if you try too hard to see in 3-D, your eyes will not be able to form the optical illusion.

Analyze and Conclude

8 **Explain** What features changed when you viewed the photos in 3-D?

9 **Measure** The scale for the image above is 1 cm = 2.39 km. In the left-hand image, what is the actual distance from the western edge of the largest lake to the eastern tip of the large mesa at the eastern edge?

10 **The Big Idea** What types of models are satellite images, and why are they sometimes used to study Earth's features?

Communicate Your Results

Write several sentences to describe the general topography of each landscape.

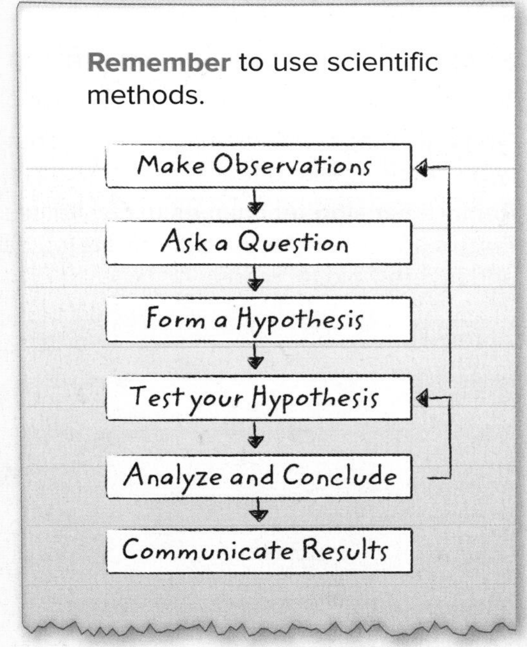

Which landscape is best for your favorite outdoor activity? Justify your answer.

NASA/JPL/NIMA

Chapter 1 Study Guide

Geologists model Earth's features using map projections, topographic maps, and geologic maps. They measure Earth's features using remote sensing, primarily from satellites.

Key Concepts Summary 

Lesson 1: Maps

- Maps represent the features of Earth's surface and have symbols, scales, and a grid of **latitude** and **longitude** lines that identify locations.
- Distortion occurs because maps are flat representations of Earth. Mapmakers use different map projections to reduce distortion in certain areas.

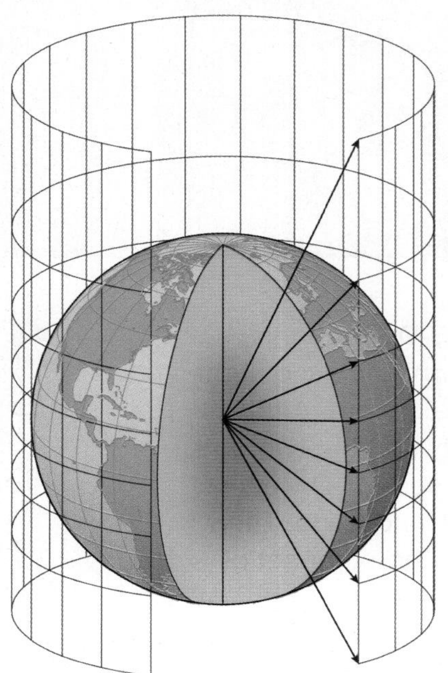

Vocabulary

map view p. 9
profile view p. 9
map legend p. 10
map scale p. 11
longitude p. 12
latitude p. 12
time zone p. 14
International Date Line p. 14

Lesson 2: Technology and Map Making

- Topographic maps show **elevation** through **contour lines.**
- **Geologic maps** contain information about rocks, such as rock types, rock age, and faults.
- **Remote sensing** techniques use satellites and create maps of Earth's surface features. **GIS** integrates data and creates detailed and layered digital maps.

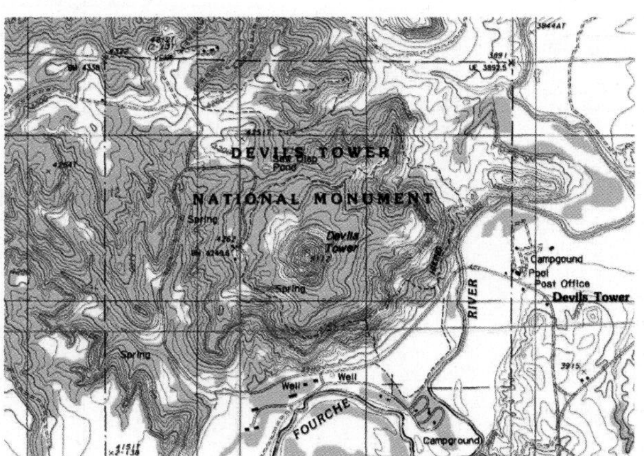

Vocabulary

topographic map p. 20
elevation p. 20
relief p. 20
contour line p. 20
contour interval p. 21
slope p. 21
geologic map p. 23
cross section p. 23
remote sensing p. 27

USGS

Vocabulary eFlashcards
Vocabulary eGames
Personal Tutor

FOLDABLES® Chapter Project

Assemble your lesson Foldables as shown to make a Chapter Project. Use the project to review what you have learned in this chapter.

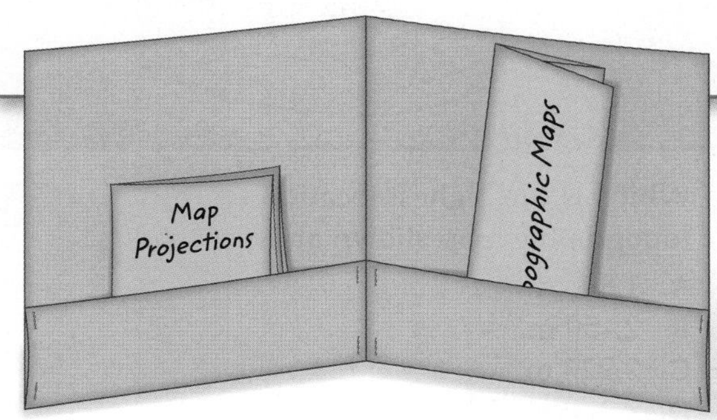

Use Vocabulary

1. Both a map scale and a ________ are added to a map to help interpret the features shown on the map.
2. In order to simplify timekeeping in locations that are close to each other, Earth is divided into 24 ________.
3. In order to learn about geologic formations underground, you need to look at a ________.
4. The ________ indicates the difference in elevation between two adjacent contour lines.
5. Aerial photographs are part of the map-making technology known as ________.
6. A map that shows the shape and features of Earth's surface is a(n) ________.

Link Vocabulary and Key Concepts

Interactive Concept Map

Copy this concept map, and then use vocabulary terms from the previous page to complete the concept map.

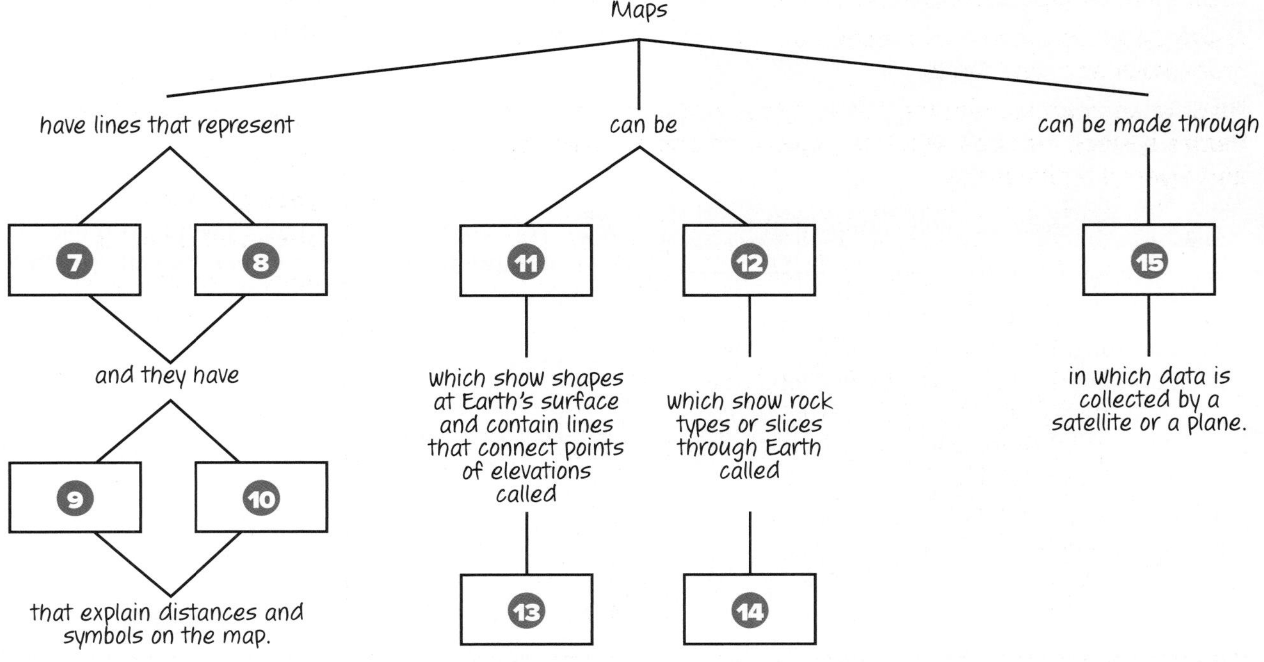

Chapter 1 Review

Understand Key Concepts 

1. Which group of terms describes lines on a map?
   A. latitude, meridians, International Date Line
   B. parallels, profiles, time zones
   C. legends, meridians, International Date Line
   D. longitude, latitude, legends

2. If you traveled west from one time zone to another, what would the time difference be?
   A. one minute
   B. two minutes
   C. one hour
   D. two hours

3. Which model of Earth does not distort surface features?
   A. a conical projection
   B. a cylindrical projection
   C. a globe
   D. a map

4. What do the white lines represent in the figure below?

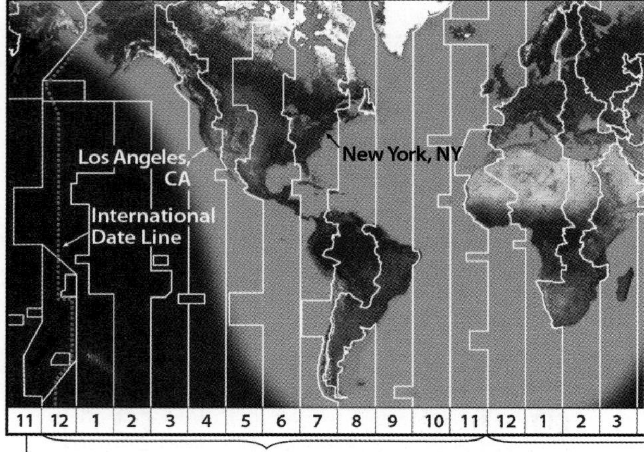

   A. time zones
   B. meridians
   C. lines of latitude
   D. lines of longitude

5. A diagram that represents a slice of Earth is called
   A. a cross section.
   B. topography.
   C. relief.
   D. an elevation.

6. Study the topographic map below.

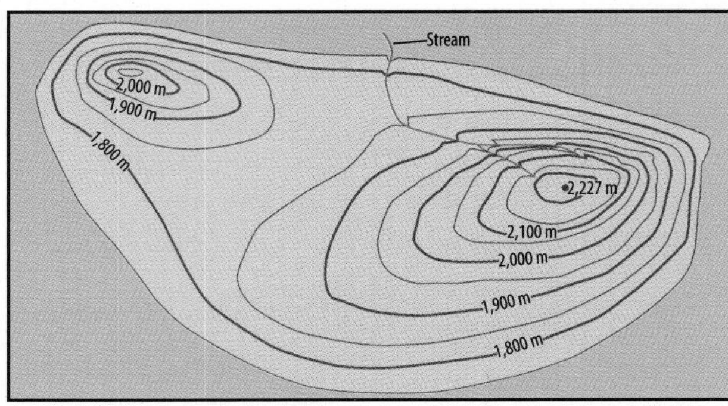

   Which is the highest location on the topographic map shown above?
   A. 1,800 m
   B. 2,150 m
   C. 2,227 m
   D. 2,300 m

7. What information on a topographic map would be helpful if you were making a geologic cross section?
   A. formations
   B. streams
   C. mountaintops
   D. cliffs

8. What is the minimum number of satellites needed to find your exact location using GPS?
   A. 1
   B. 3
   C. 12
   D. 24

9. Which of these legend items would you use to measure the distance between two cities on a map?
   A. contour interval
   B. graphic scale
   C. map projection
   D. road symbol

Critical Thinking

10 **Compare** GIS to GPS.

11 **Construct** a map of an imaginary city. Include a scale and a legend.

12 **Distinguish** between political maps and geologic maps.

13 **Suggest** a reason that the contour interval used on a map of a mountain might be different from the contour interval used on a map of a plain.

14 **Evaluate** the benefit of creating a topographic profile.

15 **Analyze** Which projection in the image below has less distortion? What explanation can you give for this difference?

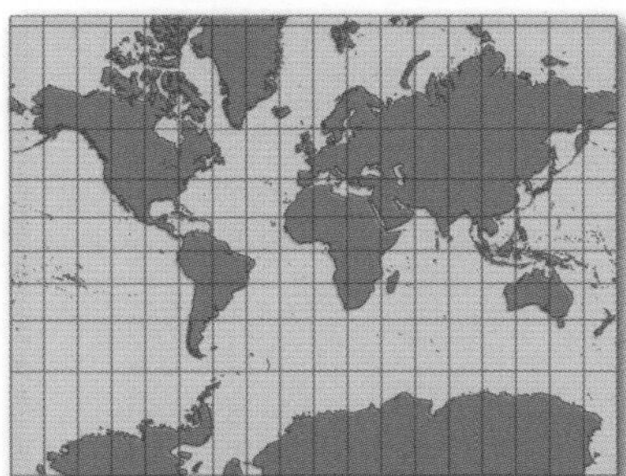

Mercator

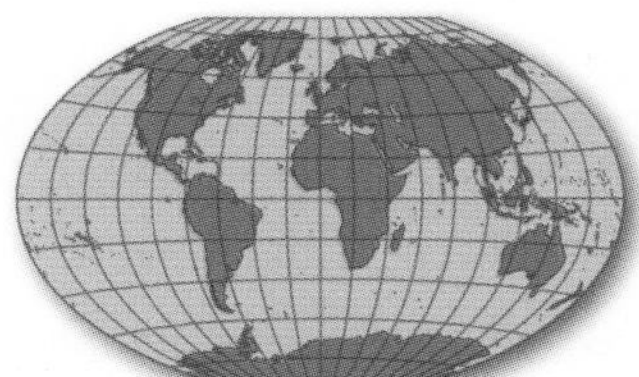

Winkel

16 **Justify** using remote sensing technology during a natural disaster.

Writing in Science

17 **Write** a paragraph that justifies the expense of making and placing satellites in orbit for observing Earth.

REVIEW THE BIG IDEA

18 How do different types of maps model the features of Earth? Explain how the method of data collection or the way in which a map is constructed affects the information on the map.

19 What is the value of having different types of maps of the same area?

20 How can you determine the scale of the chair in the photo below?

Math Skills

Math Practice

Ratio Scale

21 When making a map of the school, you decide to let 1 cm on your map represent 10 m of actual distance. Write this ratio in three different ways.

22 A large wall map of a museum uses a scale of 1 cm:2 m. If the length of one room on the map measures 25 cm, what is the actual length of the room?

23 You are making a map of your city with a scale of 1 cm:4 km. If two buildings in the city are 6 km apart, how far apart will they be on your map?

Standardized Test Practice

Record your answers on the answer sheet provided by your teacher or on a sheet of paper.

Multiple Choice

1 What is the relationship between map distance and actual distance?

A location latitude

B location longitude

C map legend

D map scale

Use the diagram below to answer questions 2 and 3.

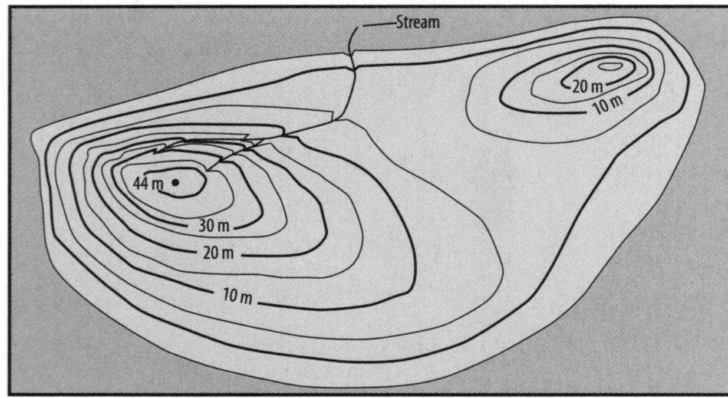

2 What is the approximate height of the lower peak in the diagram above?

A 20 m

B 27 m

C 30 m

D 32 m

3 What is the relief on the map?

A 27 m

B 30 m

C 32 m

D 44 m

4 How many minutes comprise each degree of longitude and latitude?

A 60

B 90

C 180

D 360

Use the diagram below to answer question 5.

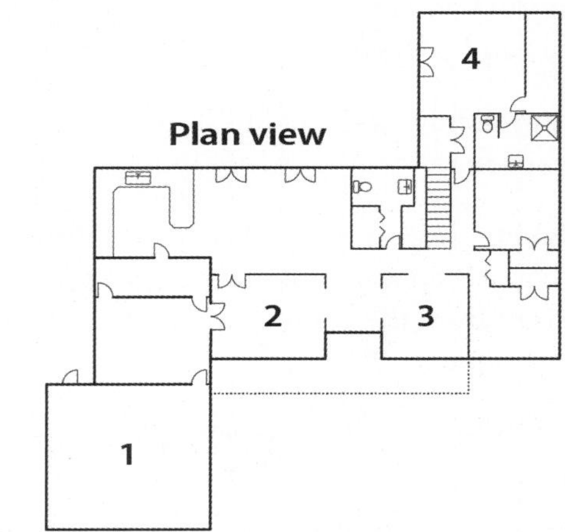

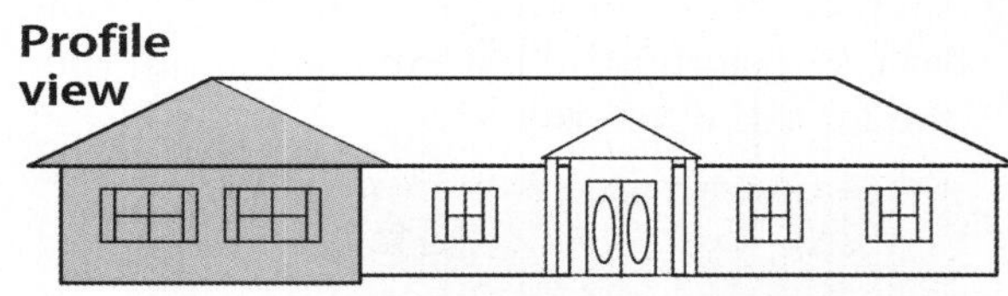

5 In the diagram above, which number on the plan view corresponds to the shaded area on the profile view?

A 1

B 2

C 3

D 4

6 Which is a feature of GIS?

A different map layers of the same location

B navigational satellites

C rock types and locations

D surface geology of a particular area

7 Which map mainly shows boundaries between states and counties?

A physical

B political

C relief

D road

Use the diagram below to answer question 8.

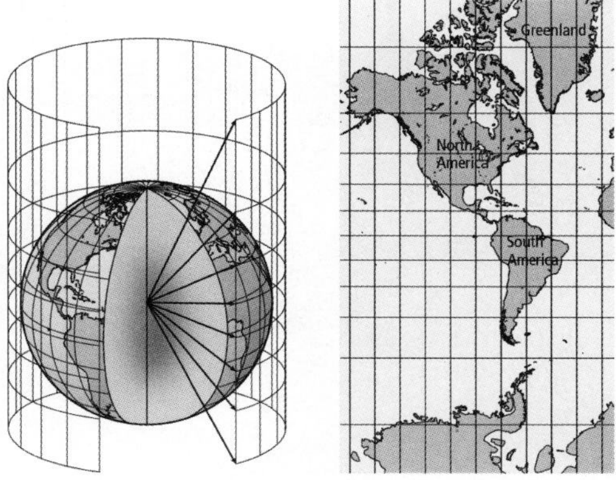

8 Where is map distortion greatest in the type of projection shown above?

A along the mid-latitudes

B along the prime meridian

C at the equator

D at the poles

9 What do contour lines on a topographic map connect?

A areas with similar climates

B highest and lowest points

C points of equal elevation

D regions under the same rule

10 If Earth science students want to find a sandstone rock layer beneath the surface layer on which they stand, which map should they study?

A geologic

B physical

C relief

D road

Constructed Response

Use the diagram below to answer questions 11 and 12.

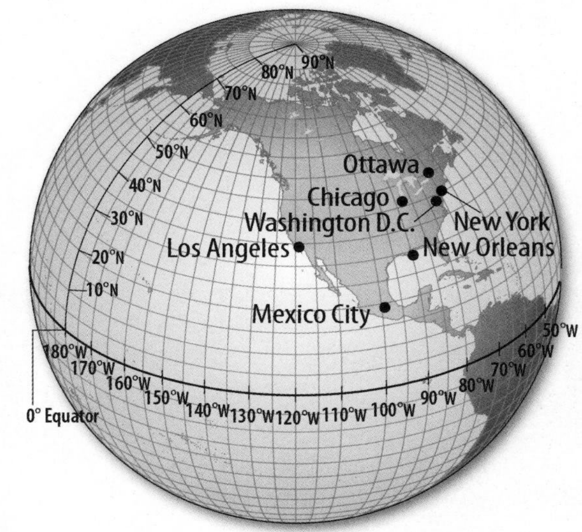

11 In the diagram above, what are the coordinates of Mexico City and New Orleans to the nearest 10 degrees? What is the difference in latitude between the two cities?

12 What are the coordinates of New York and Los Angeles to the nearest 5 degrees? What is the difference in longitude between the two cities?

13 How do modern technologies, such as remote sensing and global positioning systems, help mapmakers?

14 Why do hikers often use topographic maps? Describe map symbols, colors, and features that hikers might find helpful.

NEED EXTRA HELP?														
If You Missed Question...	1	2	3	4	5	6	7	8	9	10	11	12	13	14
Go to Lesson...	1	2	2	1	1	2	2	1	2	2	1	1	2	2

Chapter 2

Earth in Space

THE BIG IDEA **Where is Earth in the universe, and how is Earth related to other objects in the universe?**

Inquiry How Big?

When you see a photograph of Earth taken from space, Earth seems enormous. Although Earth is large compared to the Moon, it is tiny compared to the solar system and barely a speck compared to the universe.

- How do the motions of Earth and the Moon affect Earth?
- How does Earth compare with other objects in the solar system?
- What is Earth's location in the universe?

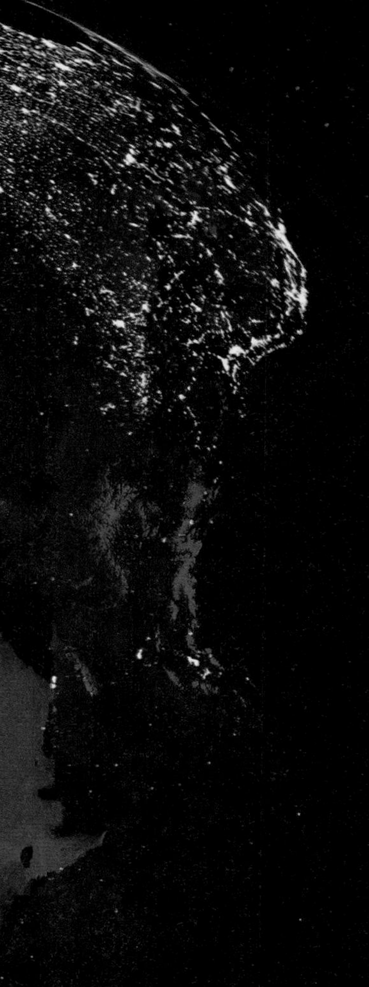

Get Ready to Read

What do you think?

Before you read, decide if you agree or disagree with each of these statements. As you read this chapter, see if you change your mind about any of the statements.

1. Seasons are caused by the changing distance between Earth and the Sun.
2. The Moon has a dark side upon which the Sun never shines.
3. The solar system contains nine planets.
4. Earth is the only planet that has a moon.
5. The Sun is more massive than 90 percent of other stars.
6. The solar system is at the center of the Milky Way.

Your one-stop online resource
connectED.mcgraw-hill.com

LearnSmart®

Chapter Resources Files, Reading Essentials, Get Ready to Read, Quick Vocabulary

Animations, Videos, Interactive Tables

Self-checks, Quizzes, Tests

Project-Based Learning Activities

Lab Manuals, Safety Videos, Virtual Labs & Other Tools

Vocabulary, Multilingual eGlossary, Vocab eGames, Vocab eFlashcards

Personal Tutors

Lesson 1

The Sun-Earth-Moon System

Reading Guide

Key Concepts

ESSENTIAL QUESTIONS

- What causes seasons on Earth?
- How does the Moon affect Earth?
- How do solar and lunar eclipses differ?

Vocabulary

Multilingual eGlossary

What's Science Got to do With It?

Inquiry Crescent Sun?

This is an eclipse. The Moon is moving in front of the Sun, blocking a part of it. Do you know what kind of eclipse this is?

Launch Lab

20 minutes

What is the center of the solar system?

For thousands of years, people believed Earth was the center of the solar system. Then, in the 1600s, Galileo viewed Venus with a telescope and discovered that Venus shows phases similar to the phases of the Moon. This discovery helped prove that the Sun is the center of the solar system.

1. Read and complete a lab safety form.
2. In groups of three, assume roles as the Sun, Venus, or Earth. The Sun holds a **battery-powered lamp.** Venus holds a **plastic foam ball** on a **pencil.** Line up so Venus is between the Sun and Earth.
3. To model an Earth-centered system: In a darkened room, Earth stands still while the Sun and Venus orbit it, staying close together. Earth watches the ball held by Venus and draws diagrams in his or her Science Journal of what the ball looks like at four locations in its orbit.
4. To model the Sun-centered system: The Sun stands still while Earth and Venus orbit it. Earth takes small steps. Venus takes large steps. Earth watches the ball held by Venus and draws diagrams of what the ball looks like at four locations in its orbit.

Think About This

1. How did the appearance of Venus differ in the Earth-centered and Sun-centered systems?
2. **Key Concept** How do you think Earth and other objects in the solar system move?

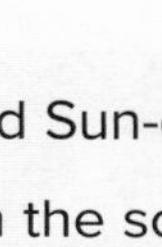

Earth and the Universe

Have you ever noticed how the Moon is in a different place each night, or wondered why summer days seem longer than winter days? Long ago, people carefully studied the positions and motions of the Sun, the Moon, and other objects in the sky. They noticed patterns in the motions. Using the patterns, they were able to predict future positions of objects in the sky. But they did not understand how the objects were related.

People today know that Earth is not the center of the universe. The Moon moves around, or orbits, Earth. Earth is just one of eight planets that orbit the Sun. The Sun is one of billions of stars that make up the Milky Way galaxy. And the Milky Way is one of billions of galaxies in the universe.

Reading Check How many planets orbit the Sun?

Objects in the solar sytem orbit the Sun because it has a huge gravitational pull. The Sun contains more than 99 percent of the solar system's mass. It is also the biggest object in the solar system. As shown in **Figure 1**, the Sun's diameter is 100 times greater than Earth's and 10 times greater than Jupiter's.

Personal Tutor

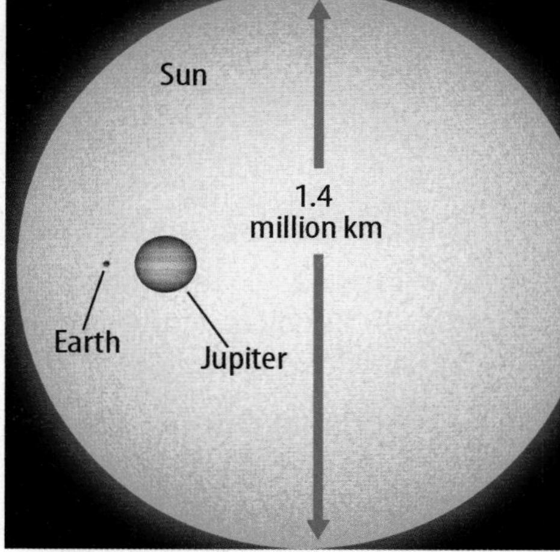

Figure 1 The Sun is 100 times wider than Earth and 10 times wider than Jupiter, the second largest object in the solar system.

FOLDABLES®

Make an envelope book. Label it as shown. Draw an image of the Sun on the inside center. On the inside tabs, draw the position of Earth for each season.

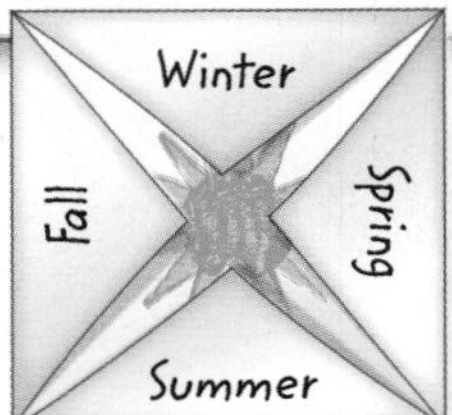

Motions of Earth

Have you ever flown in an airplane? Some airplanes can travel over 900 km/h. Yet, as you sit in one, you hardly feel that you are moving. Living on Earth is like traveling in an airplane. It seems as if Earth is still and the Sun and stars move around it. But Earth is not still. It moves in space.

Earth's Orbit

As you read this, Earth is moving around the Sun because of the Sun's huge gravitational pull. Without the Sun's pull, Earth would move off into space in a straight line, as shown in **Figure 2.** Earth's orbit is elliptical, or nearly round. *The orbit of an object around another object is called* **revolution.** It takes Earth 365.25 days, or one year, to revolve around the Sun once.

As shown in **Figure 2,** the distance between Earth and the Sun is not always the same. An astronomical unit (AU) is the average distance between Earth and the Sun. One AU is nearly 150 million km. Scientists often use AUs to measure distances to planets and other objects within the solar system.

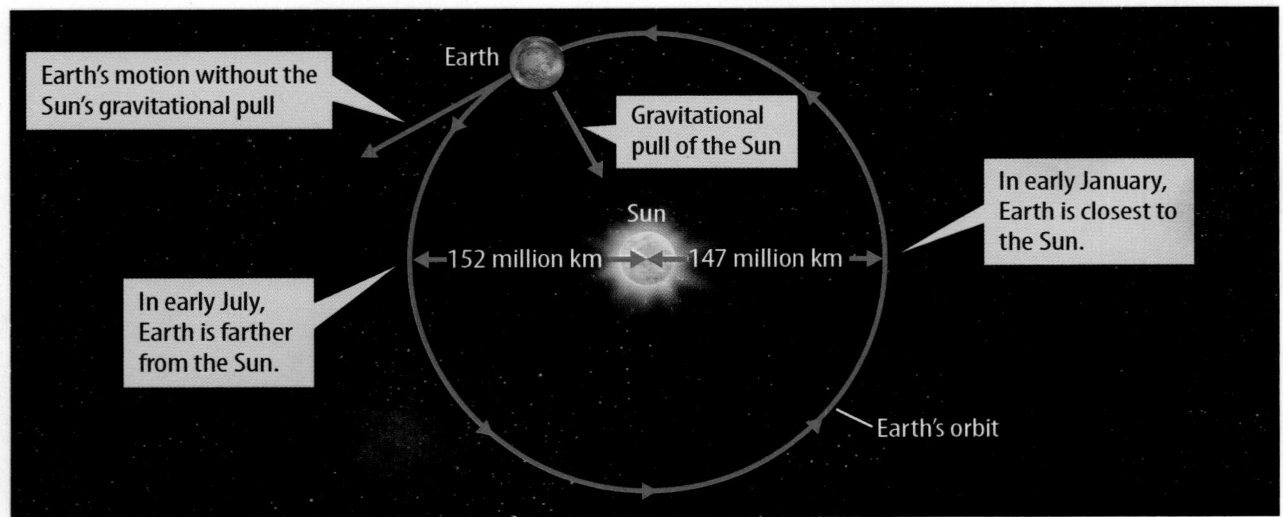

Figure 2 Earth orbits the Sun because of the Sun's gravitational pull.

Visual Check When is Earth closest to the Sun?

SCIENCE USE V. COMMON USE

revolution

Science Use orbit of one object around another object

Common Use a complete change, often in government or technology

Earth's Rotation

Imagine a rod pushed through the center of Earth, from the North Pole to the South Pole, as shown in the images of Earth in **Figure 3.** The rod represents Earth's axis. Earth spins, or rotates, on its axis like a top. **Rotation** *is the spin of an object around its axis.* Rotation is what causes day and night. The side of Earth facing the Sun is in daylight, and the side of Earth away from the Sun is in darkness. Earth makes one full rotation every 24 hours.

Reading Check How long does it take Earth to rotate once?

Earth's Tilt and Seasons

You might think that summer occurs when Earth is closest to the Sun. However, Earth is actually closest to the Sun in January, when it is winter in the northern hemisphere. As shown in **Figure 3**, seasons occur because Earth's tilt does not change as Earth orbits the Sun. This alternates the amount of direct sunlight that each hemisphere receives.

If you drew a line perpendicular to Earth's orbital path, the angle of tilt between Earth's axis and that line would be 23.5°. As Earth moves, this angle of tilt remains the same. The North Pole and the South Pole always point in the same directions. However, as shown in **Figure 4**, the position of Earth's tilt as it relates to the Sun does change.

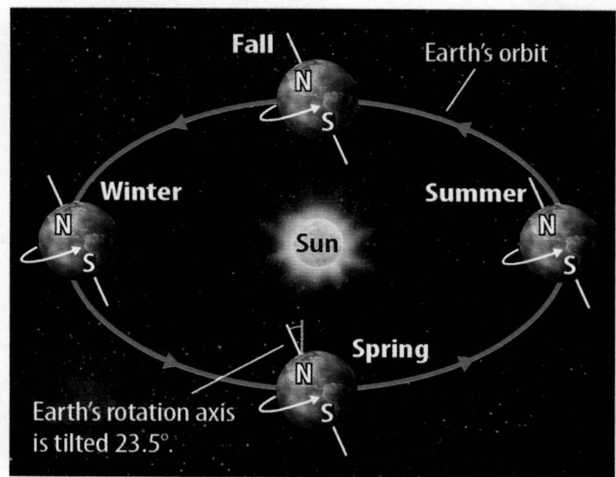

▲ **Figure 3** The tilt of Earth's axis does not change as Earth moves around the Sun.

 **Personal Tutor**

Spring and Fall

An **equinox** (EE kwuh nahks) *occurs when Earth's rotation axis is tilted neither toward nor away from the Sun. Equinox* means "equal night." Hours of daylight equal hours of darkness during an equinox. An equinox occurs two days of the year, one in March and one in September. These days are used to signify the beginning of spring or fall.

Summer and Winter

When the Earth's rotation axis is tilted directly toward or away from the Sun a **solstice** (SAHL stuhs) *occurs, as shown in the bottom of* **Figure 4**. Solstices happen in June and December. When the North Pole is toward the Sun, the northern hemisphere experiences summer. The northern hemisphere receives more direct sunlight, and there are more hours of sunlight during the day. At the same time, the South Pole is tilted away from the Sun, and the southern hemisphere experiences winter. This hemisphere receives less direct sunlight, and there are fewer hours of sunlight. Six months later, the seasons are reversed.

 **Key Concept Check** What causes seasons?

Figure 4 The position of Earth's tilt relative to the Sun causes the seasons. Each season starts at an equinox or a solstice.

Visual Check How are the beginnings of spring and fall similar?

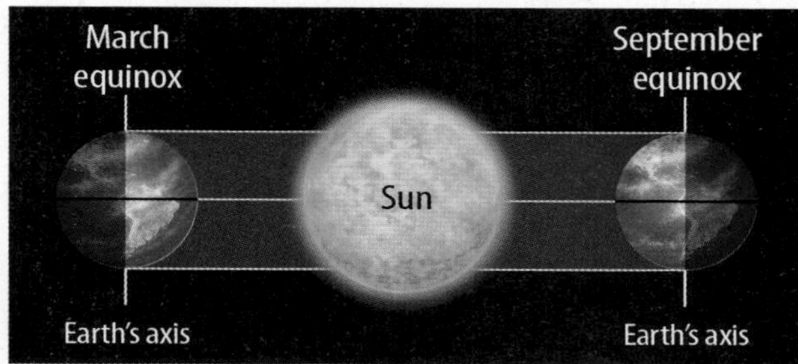

At two points in Earth's orbit—the March and September equinoxes— Earth's axis does not point either toward or away from the Sun. Light is distributed equally in the northern and southern hemispheres.

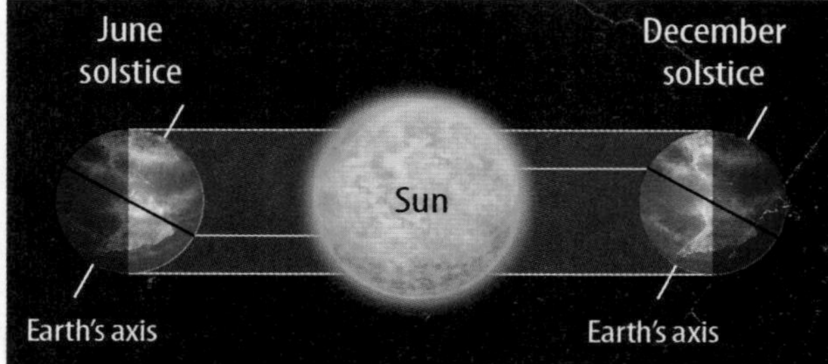

At two points in Earth's orbit—the June and December solstices—Earth's axis points the most toward or away from the Sun. Light is not distributed equally in the northern and southern hemispheres.

▲ **Figure 5** Early in the Moon's history, collisions with asteroids and comets created huge impact craters. The large crater above is about 93 km in diameter and is on the far side of the Moon.

REVIEW VOCABULARY

lava
molten volcanic material

Figure 6 Because the Moon rotates once as it makes one orbit around Earth, the same side of the Moon always faces Earth. ▼

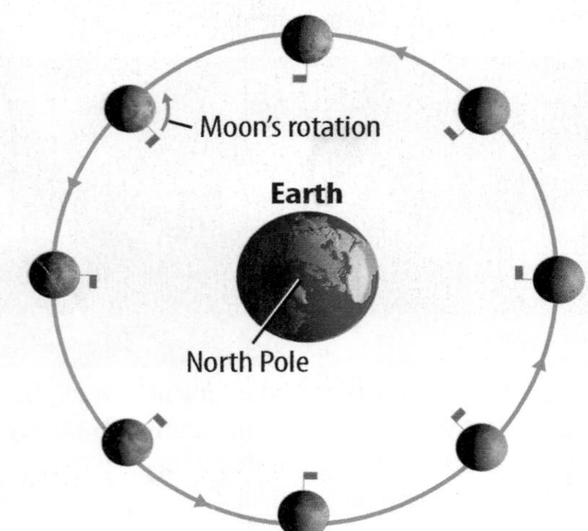

Earth's Moon

You can probably guess what force holds the Moon in orbit around Earth. It's the same force that holds Earth in orbit around the Sun–gravity! The Moon is about one-fourth the size of Earth. It is a dry, airless object made mostly of rock. Early in the Moon's history, many asteroids and comets crashed into it, leaving huge craters on its surface, such as those shown in **Figure 5.** The Moon also has mountains and smooth, dark lava plains from ancient volcanoes.

Reading Check What created the Moon's craters?

Formation of the Moon

Scientists propose that the Moon formed when a Mars-sized object collided with Earth soon after Earth formed. This collision threw debris into orbit around Earth. Gravity pulled the debris together forming the Moon.

Motions of the Moon

Like Earth, the Moon moves in different ways. It rotates on its axis, and it revolves around Earth. It orbits Earth once every 27.3 days. That is also how long it takes the Moon to rotate once. Because the Moon revolves and rotates in the same amount of time, the same side of the Moon always faces Earth, as shown in **Figure 6.** The side of the Moon that does not face Earth is called the far side. You cannot see the Moon's far side from Earth.

Phases of the Moon

The Moon does not create its own light. The Moon is visible only because it reflects sunlight. As the Moon orbits Earth, the half of the Moon facing the Sun is in sunlight, and the half facing away is in shadow, as shown in **Figure 7.** However, as the Moon orbits Earth, the visible part of the Moon seems to change shape. These shapes are the phases of the Moon. The Moon completes a cycle of phases every 29.5 days.

Kevin Kelley/Getty Images

New Moon and Waxing Phases When the Moon is between Earth and the Sun, the sunlit half of the Moon faces away from Earth. The half facing Earth is dark because it is in shadow, as shown in **Figure 7.** This phase is called a new moon. During the two weeks following a new moon, more of the Moon becomes visible. *As the lit portion of the Moon becomes larger, the Moon is* **waxing.** The waxing phases are waxing crescent, first quarter, and waxing gibbous.

Full Moon and Waning Phases When Earth is between the Moon and the Sun, the entire sunlit half of the Moon faces Earth. This phase, represented by the image of the Moon at the top of **Figure 7,** is called a full moon. During the two weeks following a full moon, less of the sunlit side of the Moon is visible. *As the lit portion of the Moon becomes smaller, the Moon is* **waning.** The waning phases are waning gibbous, third quarter, and waning crescent.

Phases of the Moon

Animation

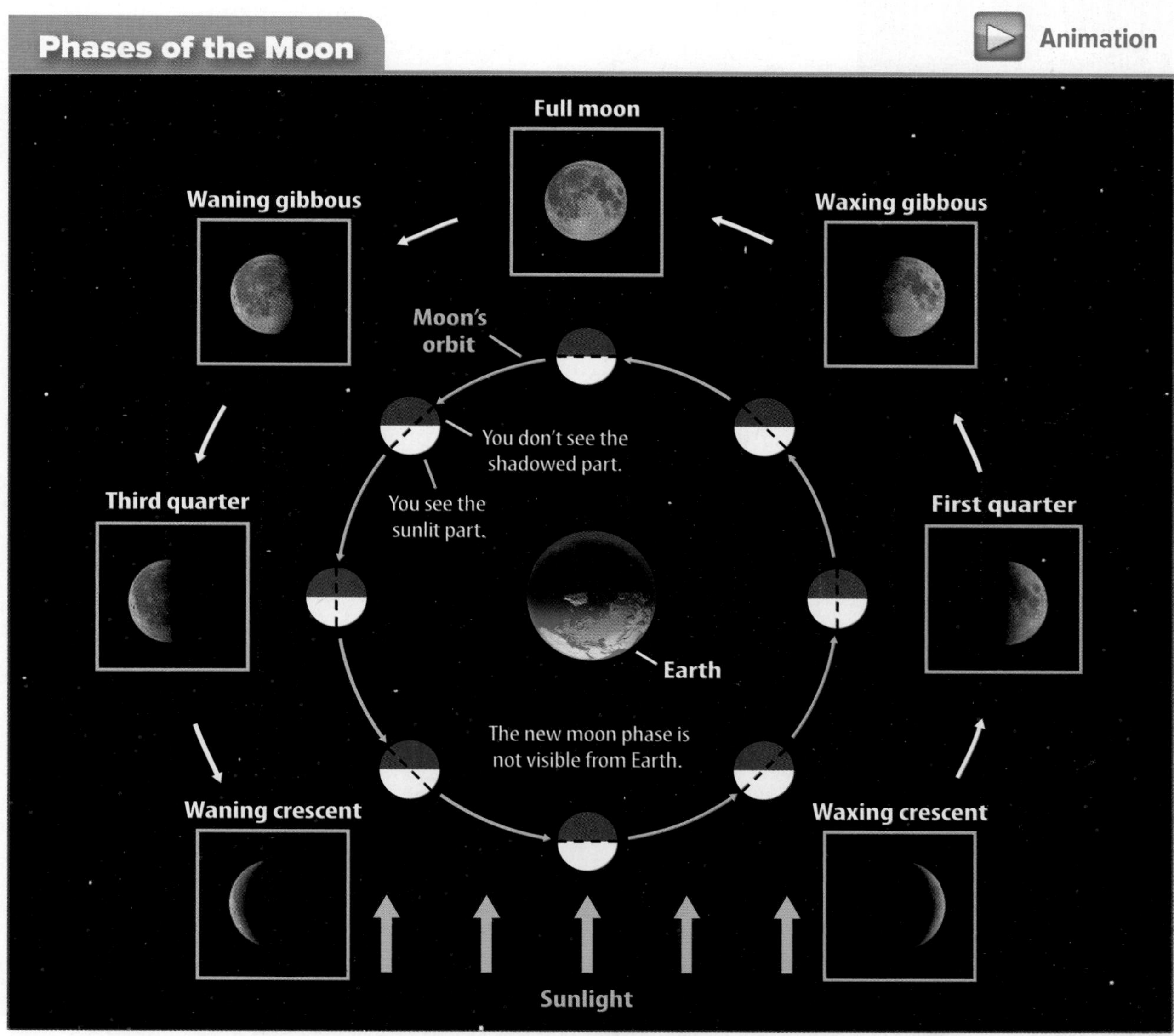

Figure 7 The Sun always shines on half of the Moon as the Moon orbits Earth. But only part of the Moon's sunlit half is visible from Earth.

Visual Check When does the Moon appear to get larger? When does is appear to get smaller?

▲ **Figure 8** These pictures were taken at the same place but at different times of day. The top photograph shows high tide, and the bottom photograph shows low tide.

Tides

Water levels of the ocean change, as shown in **Figure 8.** **Tides** *are the periodic rise and fall of the oceans' surfaces caused by the gravitational force between Earth and the Moon and the Sun.* The Moon has about twice as much influence on tides because it is so much closer to Earth than the Sun.

Effect of the Moon

Locations on Earth closest to and farthest from the Moon undergo the largest tidal effect. Water on Earth bulges slightly at these locations, as illustrated in **Figure 9,** and high tides occur. Places on Earth halfway between the two high-tide regions have low tides. As Earth rotates, the locations of high tide and low tide change in predictable ways. Most coastlines have two high tides and two low tides each day. But tides also are affected by water depth, coastline shape, and weather.

 **Key Concept Check** How does the Moon cause tides on Earth?

Effect of the Sun

When Earth and the Moon are in line with the Sun, the Sun's gravitational pull adds to the Moon's gravitational pull. As a result, high tides are higher than usual, as shown in the top of **Figure 9.** Tides at this time are called spring tides. Spring tides occur during full moon and new moon phases.

During the first and third quarter moons, the gravitational pull of the Moon is perpendicular to the gravitational pull of the Sun, as shown in the bottom of **Figure 9.** High tides are lower than usual. Tides at these times are called neap tides.

Figure 9 The Sun's gravity affects tides more when the Moon is in line with, not perpendicular to, the Sun. ▼

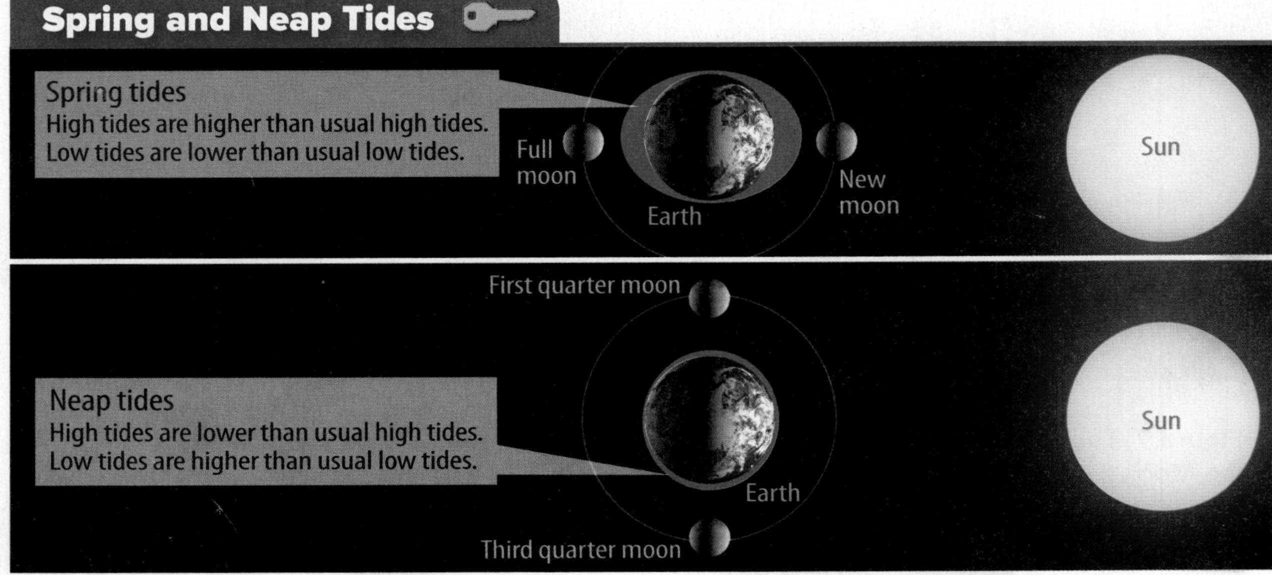
Spring and Neap Tides
Spring tides
High tides are higher than usual high tides.
Low tides are lower than usual low tides.
Full moon
Earth
New moon
Sun
First quarter moon
Neap tides
High tides are lower than usual high tides.
Low tides are higher than usual low tides.
Earth
Third quarter moon
Sun

MiniLab

10 minutes

What causes eclipses?

Throughout human history, people have interpreted eclipses as signs of war or disaster. But there is nothing superstitious about eclipses. They are natural events.

1. Read and complete a lab safety form.
2. Hold a **plastic foam ball** between your face and a **lightbulb.** The ball represents the Moon, the lightbulb represents the Sun, and your head represents Earth. Sit or stand so the Moon covers the Sun. What is this phase of the Moon? Have a partner observe where the Moon's shadow falls. Record your observations in your Science Journal.
3. Turn so your head is directly between the Sun and the Moon. Hold the ball in front of your face. What is this phase of the Moon? Record your observations.

Analyze and Conclude

1. **Identify** Which type of eclipse did you model each time?
2. **Key Concept** How do the positions of Earth and the Moon cause eclipses?

Eclipses

An **eclipse** *is the movement of one solar system object into the shadow of another object.* You can view solar and lunar eclipses from Earth.

Solar Eclipses

A solar eclipse can only occur during a new moon, as shown in the top of **Figure 10.** During a solar eclipse, a small part of Earth is in the Moon's shadow. The Moon appears to completely or partially cover the Sun.

Lunar Eclipses

A lunar eclipse only can occur during a full moon, as shown in the bottom of **Figure 10.** During a lunar eclipse, Earth's shadow completely or partially covers the Moon. The Moon is visible during a total lunar eclipse because light changes direction as it passes through Earth's atmosphere. The light that reaches the Moon appears red.

Key Concept Check How do solar and lunar eclipses differ?

Figure 10 The type of eclipse depends on the positions of the Moon, Earth, and the Sun.

Visual Check Where would you have to be on Earth to see this total solar eclipse?

Animation

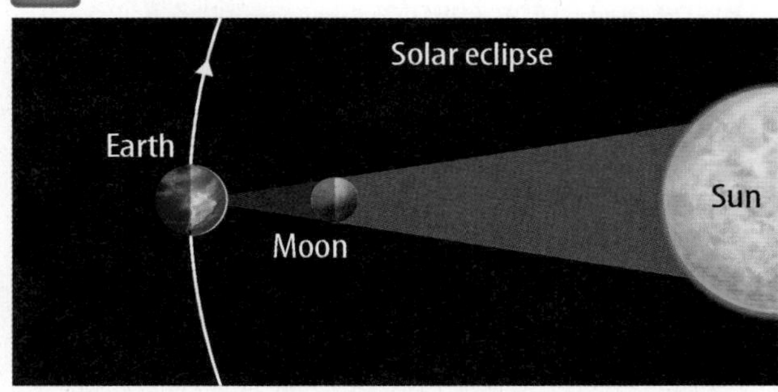

During a total solar eclipse, only a small part of Earth is covered by the Moon's shadow.

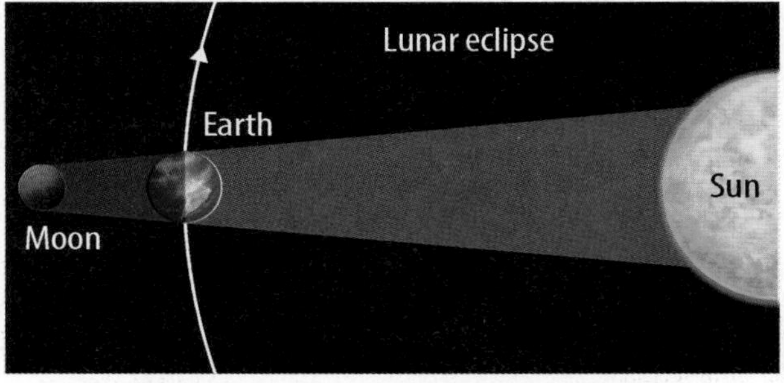

During a total lunar eclipse, the Moon is completely covered by Earth's shadow.

Lesson 1 Review

Visual Summary

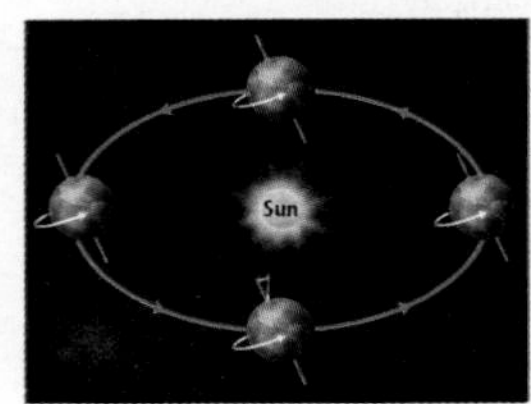

Earth's tilt does not change as Earth orbits the Sun.

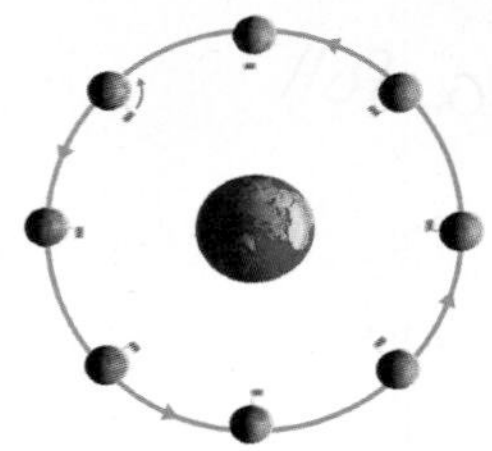

The same side of the Moon always faces Earth.

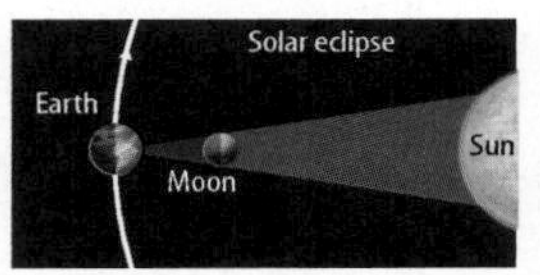

During a solar eclipse, only a small part of Earth is covered by the Moon's shadow.

FOLDABLES®

Use your lesson Foldable to review the lesson. Save your Foldable for the project at the end of the chapter.

What do you think NOW?

You first read the statements below at the beginning of the chapter.

1. Seasons are caused by the changing distance between Earth and the Sun.
2. The Moon has a dark side upon which the Sun never shines.

Did you change your mind about whether you agree or disagree with the statements? Rewrite any false statements to make them true.

Use Vocabulary

1. When the Sun, the Moon, and Earth are in a direct line, a(n) ________ can occur.
2. **Define** *equinox* and *solstice* in your own words.
3. **Distinguish** between a waxing moon and a waning moon.

Understand Key Concepts 

4. **List** the phases of the Moon, starting and ending with a new moon.
5. Which does Earth's rotation affect?
   - **A.** change of seasons
   - **B.** Earth-Sun distance
   - **C.** hours of daylight
   - **D.** length of month
6. **Explain** the influence of the Sun and the Moon on Earth's tides.

Interpret Graphics

7. **Identify** the season in the southern hemisphere shown in the image of Earth and the Sun to the right. Explain your reasoning.

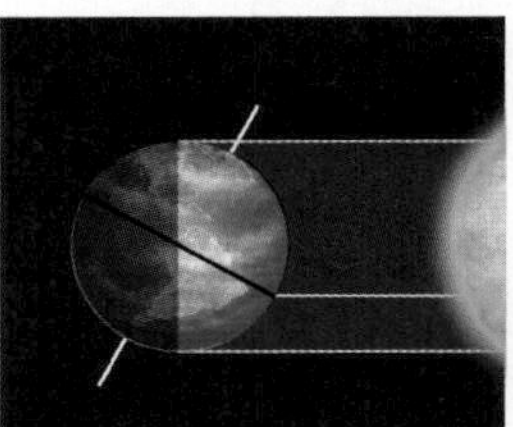

8. **Organize Information** Copy and fill in the graphic organizer below to list three effects of the Moon's motions.

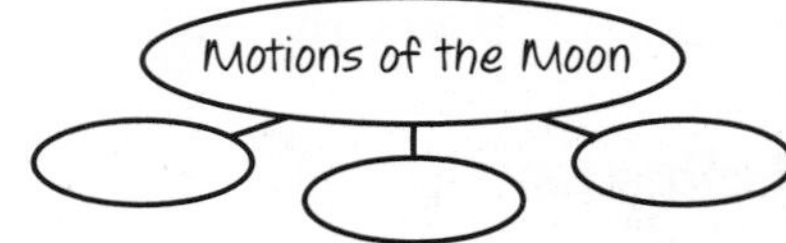

Critical Thinking

9. **Explain** why the Moon has a dark side.
10. **Conclude** People often collect the best seashells when tides are low. During which phases of the Moon would people find the best shells?

Tidal Energy

GREEN SCIENCE

Harnessing the Power of Tides

Have you ever built a sand castle on a beach far from the water, only to see the ocean wash it away later in the day? If so, you've seen the effects of tides—the rise and fall of water levels caused by the gravitational pulls of the Moon and the Sun.

Tides can do more than wash away sand castles. The energy of flowing tidal water can be converted to electric energy and used to heat and cool buildings and to power appliances.

The power of tides can be harnessed in several ways. In an area where the difference in water level between high tide and low tide is great, a damlike structure called a tidal barrage (BAHR ij) can be built. A barrage temporarily holds water at high tide. When enough water has accumulated, the water is released through turbines. The turbines convert the mechanical energy of moving water into electrical energy.

Tidal barrages hold water like dams do. But instead of stopping the flow of a river, a tidal barrage collects ocean water at high tide and releases it at low tide. ▼

◀ Tidal barrages work only in places where the water level at high tide is at least 5–7 m higher than the water level at low tide.

Tides also generate currents, or streams of moving water, in the ocean. Turbines placed in these currents harness the energy of the moving water, which turns the turbines. The turbines convert the mechanical energy of the moving water into electric energy.

Tidal power is an inexhaustible resource. It is also a predictable source of energy. So why doesn't tidal power meet more of our energy needs? First, most tidal regions are not well-suited to tidal power production. Second, the structures needed to capture tidal power can be expensive to build. And third, constructing tidal power structures in the ocean or along ocean shorelines can impact the environment and natural habitats. However, as costs become lower and environmental concerns are addressed, tidal power could meet more of our energy needs in the future.

Turbines that capture tidal energy are similar to turbines on land that capture wind energy. ▼

It's Your Turn

RESEARCH AND REPORT If you were going to build a tidal barrage, where would you build it? What location would enable your barrage to generate the most electricity? How would the Sun and the Moon affect your barrage's energy production?

Lesson 2

Reading Guide

Key Concepts

ESSENTIAL QUESTIONS

- How does gravity influence the shape and the motion of objects in the solar system?
- What objects are in the solar system?
- How does Earth compare with other objects in the solar system?

Vocabulary

planet p. 51
dwarf planet p. 52
moon p. 53
asteroid p. 53
comet p. 53
meteoroid p. 53
meteor p. 53

Multilingual eGlossary

BrainPOP®

Believe it or not, this is a moon! It is Saturn's moon Hyperion. Most moons are spherical in shape. But some small moons, such as Hyperion, have irregular shapes. What other objects does the solar system contain?

NASA/JPL/Space Science Institute

Launch Lab

20 minutes

How does rotation affect shape?

When the solar system formed, it was an enormous ball of gas and dust spinning slowly in space. As gravity pulled it closer together, the solar system spun faster. What happened to the shape of the solar system as it spun faster?

1. Read and complete a lab safety form.
2. Make a round ball about the size of your fist from a piece of **salt dough.**
3. Place the dough in a **small bucket.** Attach 1 m of sturdy **string** to the bucket's handle. ⚠ Be sure the string is securely attached.
4. ⚠ Stand away from all furniture and people. Whirl the bucket around your head for 1 min.
5. Lower the bucket. Observe the salt dough and record your observations in your Science Journal.

Think About This

1. What happened to the salt dough? What other objects change shape as they spin?
2. **Key Concept** How do you think gravity influenced the shape of the early solar system?

The Solar System

The Sun and everything that orbits it make up the solar system. The solar system formed 4.6 billion years ago from a cloud of gas and dust. As gravity pulled the cloud together, it became smaller and hotter and began to spin. In the center of the cloud, where the gas was hottest and densest, a star formed–the Sun.

At first, the solar system was shaped like a ball. As it rotated, gravity caused it to flatten, and it became a disk. Gravity also caused leftover gas and dust from the solar system's formation to clump together and form small, rocky or icy bodies. These bodies merged and formed planets and other objects.

Key Concept Check What role did gravity play in the formation of the solar system?

Other than the Sun, planets are the largest objects in the solar system. A **planet** *orbits the Sun, is massive enough to be nearly spherical in shape, and has no other large object in its orbital path.* All eight planets revolve in the same direction. The closer a planet is to the Sun, the faster it revolves. Mercury orbits the Sun once every 88 Earth days. The planet farthest from the Sun, Neptune, orbits once every 165 Earth years.

Recall that Earth orbits the Sun at a distance of 1 AU. Neptune is 30 times farther from the Sun. But the Sun's gravitational pull extends far beyond Neptune. Billions of small, icy objects orbit the Sun at a distance of 50,000 AU.

FOLDABLES®

Make a horizontal tri-fold Venn book. Label it as shown. Use it to compare and contrast characteristics of the inner and outer planets.

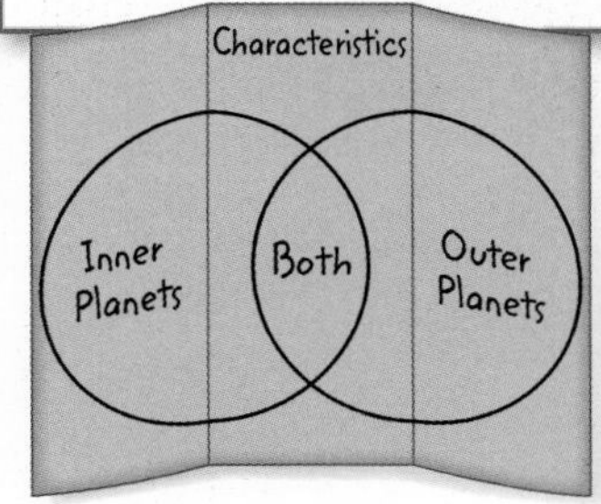

Hutchings Photography/Digital Light Source

Objects in the Solar System

As shown in **Figure 11**, the solar system contains many different objects. These objects include planets as well as objects too small to be classified as planets.

Planets and Dwarf Planets Recall that planets are massive objects that do not have other objects of similar size in their orbital paths around the Sun. Some spherical objects that orbit the Sun are similar to planets but are not massive enough to be planets. Some of these are dwarf planets. **Dwarf planets** *orbit the Sun and are nearly spherical in shape, but they share their orbital paths with other objects of similar size.* Pluto was once considered a planet but is now classified as a dwarf planet.

Figure 11 The solar system includes the Sun, planets, and many other objects.

Visual Check What percentage of the mass in the solar system exists outside the Sun?

The Solar System

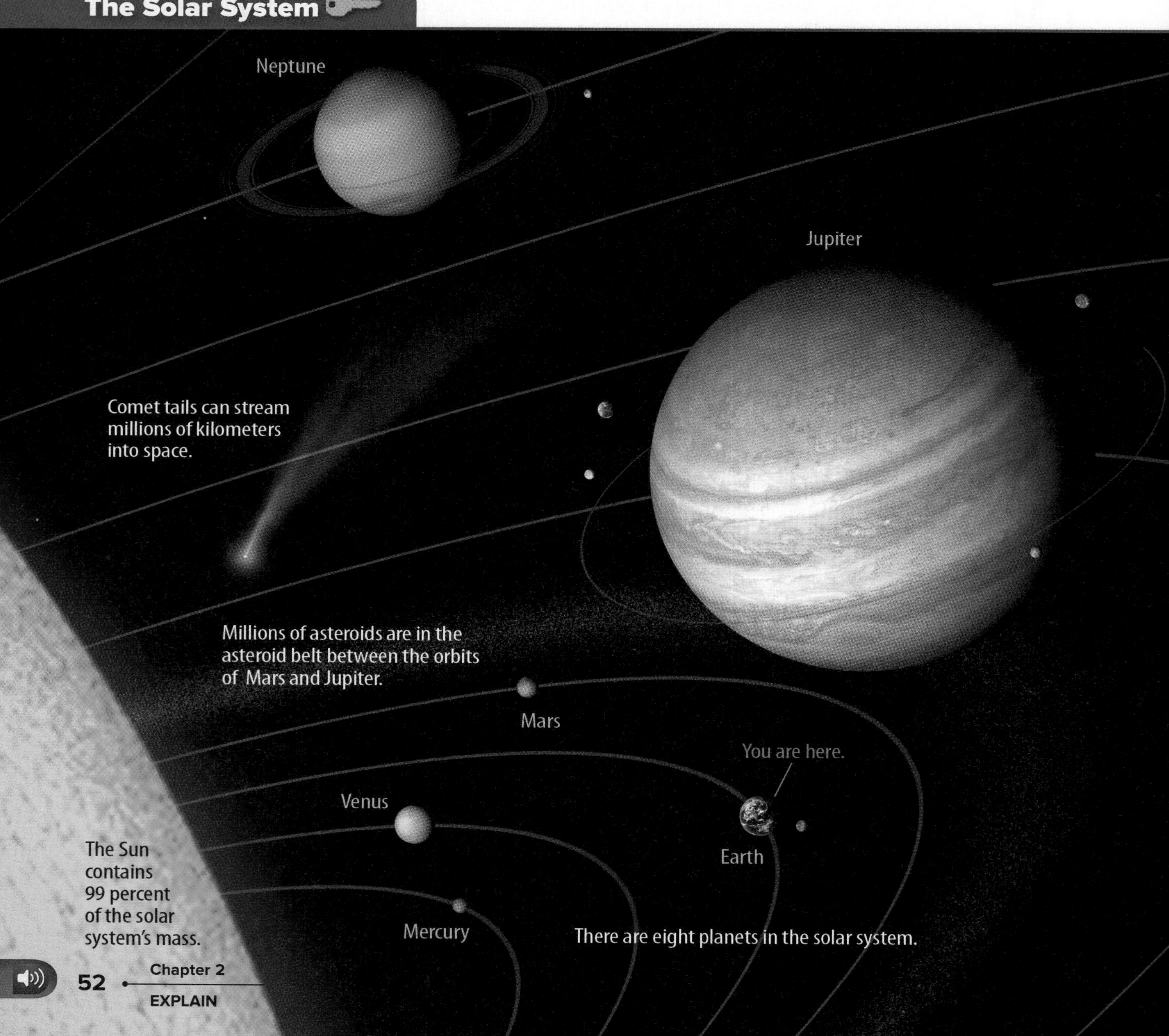
Neptune
Jupiter
Comet tails can stream millions of kilometers into space.
Millions of asteroids are in the asteroid belt between the orbits of Mars and Jupiter.
Mars
You are here.
Venus
Earth
The Sun contains 99 percent of the solar system's mass.
Mercury
There are eight planets in the solar system.

Other Solar System Bodies Not all spherical bodies in the solar system are planets. Many moons are massive enough to be spherical. *A* **moon** *is a natural satellite that orbits an object other than a star.* Some asteroids also are spherical. **Asteroids** *are small, rocky objects that orbit the Sun.* Most known asteroids are in the asteroid belt located between the orbits of Mars and Jupiter. **Comets** *are small, rocky, icy objects that orbit the Sun.* As comets near the Sun, the ice melts and the water forms a "tail" behind the comet. The orbital paths of comets extend to the outer solar system, beyond Neptune. **Meteoroids** *are small, rocky particles that move through space. When a meteoroid enters Earth's atmosphere, it produces a streak of light called a* **meteor.** A meteoroid becomes a meteorite only if it impacts Earth.

WORD ORIGIN
comet
from Greek *komētēs*, means "long-haired"

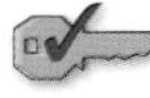 **Key Concept Check** What objects are in the solar system?

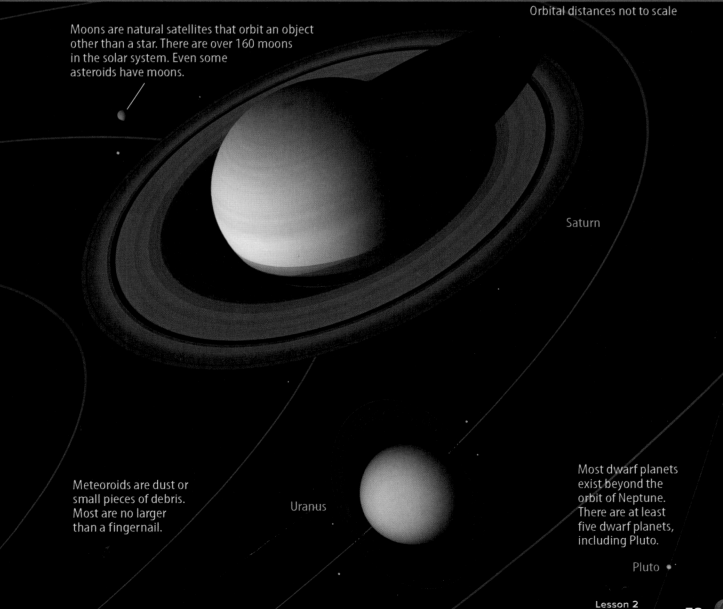

Orbital distances not to scale

Moons are natural satellites that orbit an object other than a star. There are over 160 moons in the solar system. Even some asteroids have moons.

Saturn

Meteoroids are dust or small pieces of debris. Most are no larger than a fingernail.

Uranus

Most dwarf planets exist beyond the orbit of Neptune. There are at least five dwarf planets, including Pluto.

Pluto

Inner Planets

The center of the solar system was extremely hot when it formed. Gases and materials with low boiling points escaped from the region closest to the Sun. The four inner planets, also called the rocky planets, formed from the rocks and the heavy elements, including metals, left behind. The cores of the inner planets are mostly iron. The inner planets are the smallest planets. They have few or no moons, no rings, and they rotate more slowly than the outer planets. The inner planets are shown and described in **Table 1.**

Table 1 The inner planets are mostly rock and metal.

Table 1 The Inner Planets

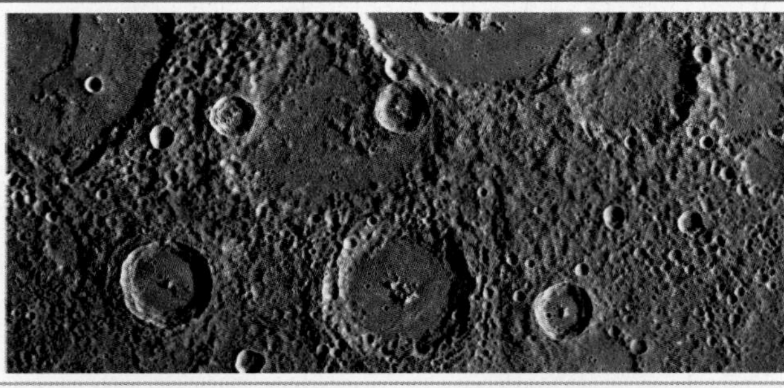	**Mercury** At 0.39 AU from the Sun, Mercury is the planet closest to the Sun. It is also the smallest planet, only about one-third the diameter of Earth. Mercury rotates slowly. As its surface heats and cools during its long day, temperatures can vary by as much as 500°C. Mercury has almost no atmosphere. Its gray surface has many impact craters and resembles the Moon.
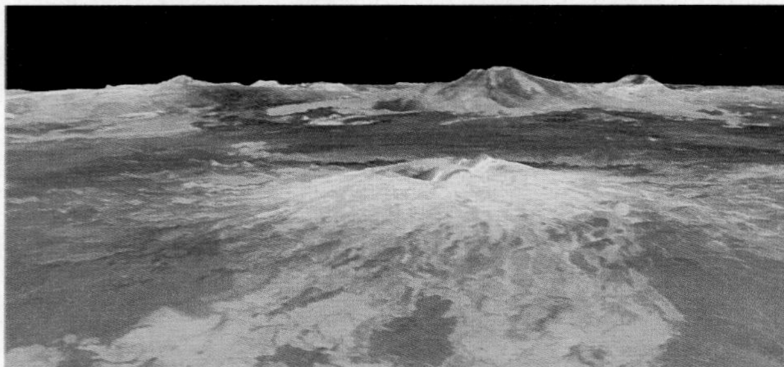	**Venus** Venus is 0.72 AU from the Sun. It is almost the same size and has nearly the same makeup as Earth. It has the slowest rotation of any planet. One day on Venus is equal to 244 Earth days. Its heavy layer of clouds and thick carbon-dioxide atmosphere traps energy from the Sun. This makes Venus the hottest planet. Scientists think some volcanoes on its surface might be active.
	**Earth** Earth is 1 AU from the Sun. The largest and densest of the inner planets, Earth is the only planet where life is known to exist. It is also the only planet with large amounts of liquid water on its surface. Earth's water and water vapor appear blue and white when viewed from space. Earth's atmosphere is 78 percent nitrogen and 21 percent oxygen.
	**Mars** Mars is half the size of Earth and orbits at 1.5 AU from the Sun. Scientists have found evidence that water flows periodically on the Martian surface. Rocks on Mars contain iron oxides, which give Mars a reddish color. Mars has some of the largest volcanoes in the solar system, including Olympus Mons.

 **Key Concept Check** How does Earth differ from other inner planets?

Outer Planets

The four outermost planets, shown in **Table 2**, formed from gases and other materials that escaped from the region closest to the Sun. They are often called the gas giants. They are larger than the inner planets, they rotate more quickly, and they each have rings. Except for Saturn's rings, the rings are barely visible. Each outer planet also has many moons. Scientists suspect each has a small, rocky core. These planets do not have solid surfaces. They have thick atmospheres of hydrogen and helium.

Table 2 The outer planets are made almost entirely of gas and ice.

Visual Check What makes Uranus and Neptune appear blue?

Table 2 The Outer Planets

Jupiter Though it is made mostly of hydrogen and helium, Jupiter contains more mass than the rest of the planets combined. Jupiter revolves around the Sun at a distance of 5 AU. It has the fastest rotation of any planet—a day lasts just 10 Earth hours. Jupiter's clouds swirl with various colors because they contain small amounts of sulfur and phosphorus. Jupiter has strong weather systems.	
Saturn Saturn is the second-largest planet. At 9.5 AU from the Sun, it is nearly twice as far from the Sun as Jupiter, but its makeup is similar. Saturn has thousands of thin rings made of billions of pieces of ice ranging in size from pebbles to boulders. Saturn's clouds form bands and spots, but it is hard to see them. Saturn's hazy upper atmosphere hides its colorful lower layers.	
Uranus Uranus orbits the Sun at a distance of nearly 20 AU. Uranus is tilted so much that its axis sometimes points directly toward the Sun. The planet is a bluish-green color because of a small amount of methane in its atmosphere. Scientists suspect that a layer of icy liquid water, ammonia, and other compounds lies deep below Uranus's thick atmosphere.	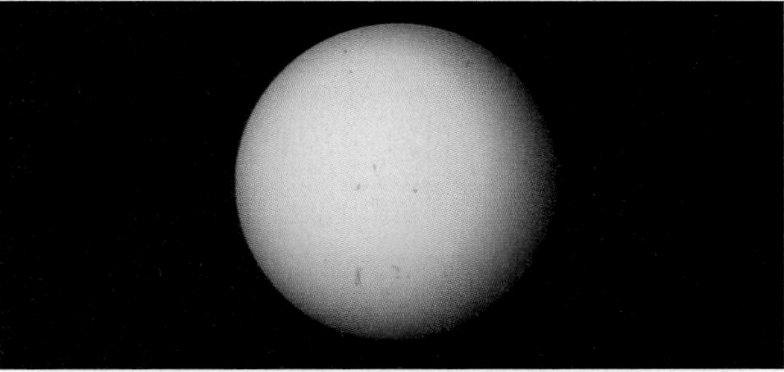
Neptune At 30 AU, Neptune is so far away that it cannot be seen from Earth without a telescope. Neptune's makeup is similar to that of Uranus, although it has more methane in its atmosphere and is deeper blue in color. Neptune has the fastest winds of any planet, recorded at over 1,100 km/h. The spots on its surface are hurricane-like storms, which do not last long.	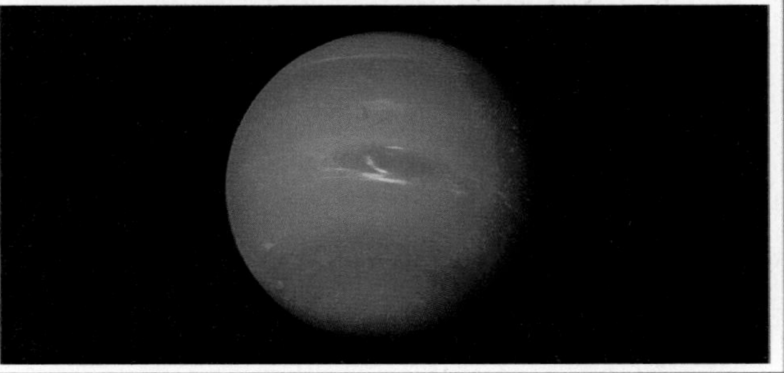

Key Concept Check How do the inner and outer planets differ?

Lesson 2 Review

Visual Summary

The solar system includes planets, moons, asteroids, comets, and many other objects.

Jupiter contains more mass than the rest of the planets combined.

Saturn's rings are made of billions of pieces of ice.

FOLDABLES®

Use your lesson Foldable to review the lesson. Save your Foldable for the project at the end of the chapter.

What do you think NOW?

You first read the statements below at the beginning of the chapter.

3. The solar system contains nine planets.

4. Earth is the only planet that has a moon.

Did you change your mind about whether you agree or disagree with the statements? Rewrite any false statements to make them true.

Use Vocabulary

1. **Define** *comet* in your own words.
2. **Distinguish** between a meteor and a meteoroid.
3. Pluto is classified as a(n) ________.

Understand Key Concepts 

4. Between which planets is the asteroid belt located?
   - **A.** Earth and Mars
   - **B.** Mars and Jupiter
   - **C.** Saturn and Uranus
   - **D.** Uranus and Neptune
5. **Discuss** gravity's role in the formation of the solar system.
6. **Compare and Contrast** Venus is often called Earth's twin. In what ways is this true? In what ways is it not true?

Interpret Graphics

7. **Explain** why each of the planets below has a blue color.

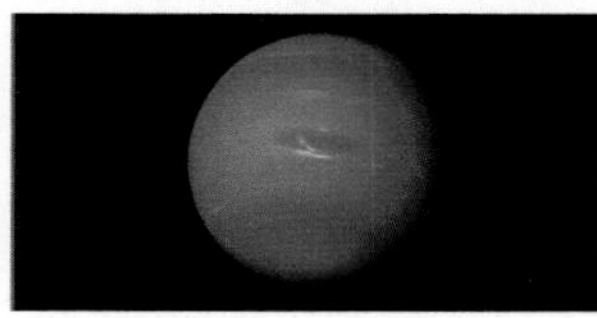

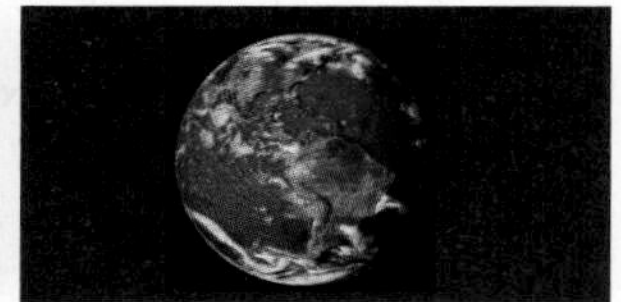

8. **Organize Information** Copy and fill in the graphic organizers below. List these objects in the solar system, from smallest to largest: the Sun, meteoroids, planets, asteroids, comets, dwarf planets, satellites.

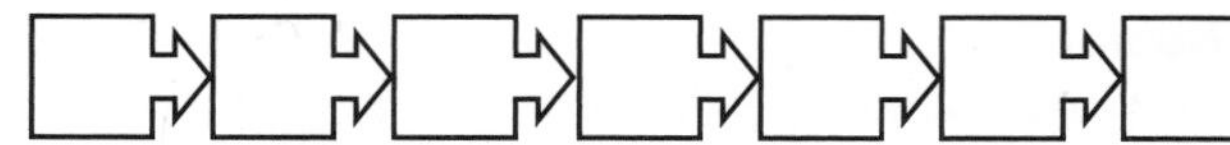

Critical Thinking

9. **Conclude** why comets become smaller each time they near the Sun.
10. **Defend** Ganymede is an object that orbits Jupiter. It is larger than Mercury, and it is round in shape. Should Ganymede be classified as a planet, a dwarf planet, a moon, or an asteroid? Defend your choice.

(tl)NASA/JPL/University of Arizona, (bl)NASA/JPL/Space Science Institute, (bc)NASA/JPL, (br)NASA;

Skill Practice Measure

40 minutes

How do densities of the inner and outer planets differ?

The inner and outer planets differ in makeup. What properties can you measure in simple objects to help you determine how densities of the inner and outer planets differ?

Materials

metal ball

marble encased in an ice cube

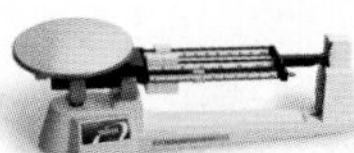
ruler

balance

Safety

Learn It

You use measurements every day. You might **measure** your height or your weight. Scientists also take measurements; it is one method of collecting data. When taking measurements, it is important to know how to use all of the instruments properly so your measurements are accurate.

3

Try It

1. Read and complete a lab safety form.
2. Copy the table below into your Science Journal.
3. Measure the diameter and the mass of a metal ball. Record the data in your table.
4. Measure the sides and the mass of an ice cube. Record the data.
5. Calculate the volume and the density of both objects.

Apply It

6. **Compare** the ball and the ice to the inner and outer planets. How do you think the makeup and the densities are similar? How do you think they are different?
7. **Key Concept** Explain how densities of the inner and outer planets differ.

To calculate volume (v), use these formulas.

Sphere: $v = \frac{4}{3}\pi r^3$

Cube: $v = \ell \times w \times h$

$\pi = 3.14$

	Inner Planet Model (metal ball)	Outer Planet Model (ice cube)
Sketch of object with measurements		
Mass (g)		
Volume (cm^3)		
Density (g/cm^3) $D = m \div v$		

(t to b, 2,3) McGraw-Hill Education (4, r)Hutchings Photography/Digital Light Source

Lesson 3

Reading Guide

Key Concepts 

ESSENTIAL QUESTIONS

- What are stars?
- How does the Sun compare to other stars?
- Where is Earth located in the universe?
- How is the universe structured?

Vocabulary

star p. 59

light-year p. 59

galaxy p. 61

Big Bang theory p. 62

Multilingual eGlossary

Stars, Galaxies, and the Universe

Inquiry Explosion in Space?

Yes, this is the remnant of a star explosion. When massive stars run out of fuel, they explode and release gas and other material into space. Do you think you have anything in common with an exploding star?

Launch Lab

20 minutes

Where does a star's energy come from?

The inside of a star is so hot that light elements combine, or fuse, and make heavier elements. This reaction is called nuclear fusion. It occurs in a sequence of steps.

1. Read and complete a lab safety form.
2. Obtain a **cup of chocolate puffs and corn puffs.** A chocolate puff represents the one proton of hydrogen. A corn puff represents a neutron.
3. Bring two protons together. One proton decays into a neutron and gives off energy. This forms deuterium. To model this reaction, crush one proton—to represent the release of energy—and replace it with a neutron.
4. Combine the deuterium (the proton and neutron) with a proton to make helium-3.
5. Repeat steps 3 and 4 to make two helium-3s.
6. Combine two helium-3s and make beryllium-6.
7. Beryllium-6 becomes one helium-4 (two protons and two neutrons) and two protons. The helium-4 is stable. The two protons start the process over again.

Think About This

1. Draw a picture showing how nuclear fusion in the cores of stars makes energy.
2. When hydrogen is gone, what will be left?
3. **Key Concept** How do you think stars shine?

Stars

Do you know the song "Twinkle, twinkle, little star"? Have you ever wondered what stars really are or why they twinkle? *A* **star** *is a large sphere of hydrogen gas hot enough for nuclear reactions to occur in its core.* A star's core heats as gravity pulls gas inward. Once the gas becomes hot enough, nuclear reactions begin and energy begins to travel outward. When the energy reaches the star's surface, the star shines. A star appears to twinkle because its light passes through Earth's atmosphere before reaching your eyes. As particles in the atmosphere move, the star's light slightly changes directions.

Key Concept Check What is a star?

Light from Stars

When measuring distances to stars, astronomers often use a unit based on the speed of light rather than astronomical units. *A* **light-year** *is the distance light travels in one year.* Light travels 300,000 km/s. One light-year equals 9.46 trillion km. Because it takes time for light to travel, stars are not seen as they are now, but as they were in the past. Proxima Centauri, the star nearest the Sun, is 4.2 light-years away. The light from Proxima Centauri we see today left the star 4.2 years ago.

FOLDABLES®

Make a vertical three-tab Venn book. Label it as shown. Use it to compare the Sun to other stars.

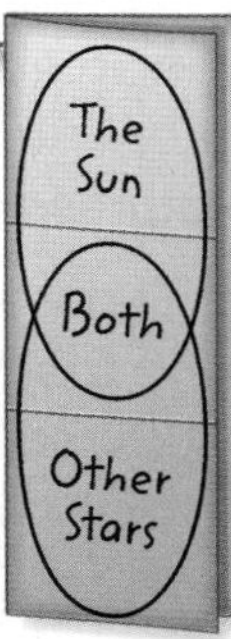

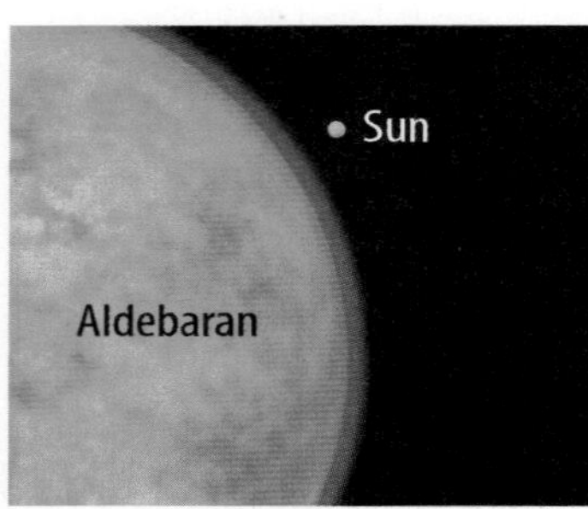

Figure 12  The star Aldebaran is 44 times wider than the Sun. The largest known star is 1,000 times wider than the Sun.

Types of Stars

At first glance, all stars appear white. But if you look closely at the brightest stars in the night sky, you will see that some stars are red, some are orange, and some even appear blue. The color of a star indicates its temperature. Blue stars are the hottest stars, followed by blue-white, white, yellow, and orange stars. Red stars are the coolest stars. The Sun is a yellow star.

When you look at stars, they appear to be the same size. But stars vary in size. The Sun is larger and more massive than 90 percent of other stars. But the Sun is tiny compared to the giant star shown in **Figure 12.**

Key Concept Check How does the Sun compare in size to other stars?

The Sun is a solitary star. Many other stars are members of binary star systems or multiple-star systems. In a binary star system, two stars orbit each other's center of mass. In a multiple-star system, two or more stars orbit the entire system's center of mass. Stars differ in other ways, too. For example, stars called variable stars change in brightness over time.

ACADEMIC VOCABULARY

estimate
(verb) to determine roughly the value, size, or extent of something

Earth's Star—the Sun

The Sun is the closest star to Earth. It has been shining for nearly 5 billion years. Scientists **estimate** that it has a lifetime of approximately 10 billion years, so it will continue to shine for 5 billion more years. When it stops shining, it will become a small, dense star that emits little light called a white dwarf star.

MiniLab

20 minutes

How does mass affect a star?

Stars vary in size and mass. Mass affects many properties of most stars, including their temperatures and their expected life spans.

1. The Sun is one solar mass. Find the Sun on the table. How does the Sun compare to other stars in the table? Record your observations in your Science Journal.
2. Use the data to make a graph plotting mass v. temperature and another graph plotting mass v. expected life span.

Mass (in solar masses)	Temperature (K)	Expected life span (millions of years)
40	38,000	1
18	30,000	7
6.5	16,400	93
3.2	10,800	550
2.1	8,620	1,560
1.7	7,240	2,650
1.3	6,540	5,190
1	5,920	10,000
0.78	5,150	18,600
0.47	3,920	66,000

Analyze and Conclude

Key Concept How is the mass of a star related to the star's temperature and expected life span?

Figure 13 There are three main types of galaxies in the universe.

Visual Check Which galaxy type has a well-defined center?

Types of Galaxies

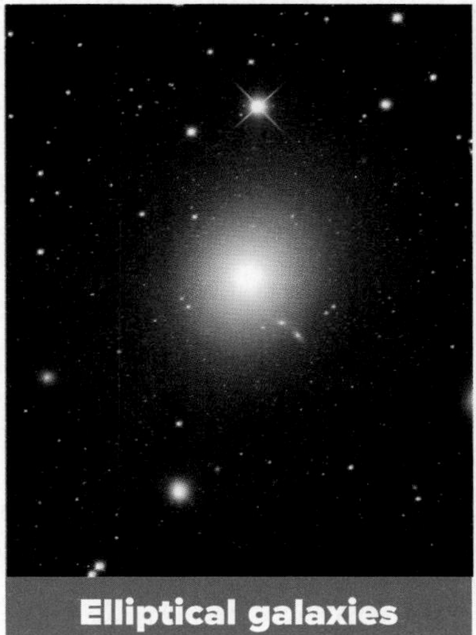

Elliptical galaxies

Shaped like basketballs or footballs, elliptical galaxies contain older, redder stars and have little gas or dust. Because stars form from gas and dust, elliptical galaxies contain few young stars.

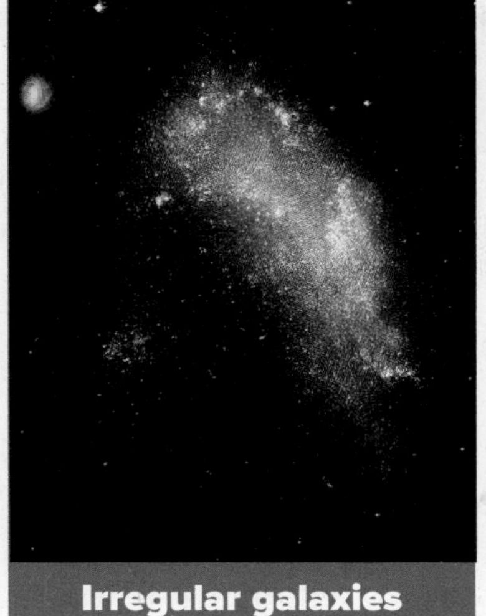

Irregular galaxies

These oddly shaped galaxies contain large amounts of gas and dust. They exhibit the highest rate of star formation of any galaxy type. Irregular galaxies have many young stars. These galaxies do not have bright centers.

Spiral galaxies

These galaxies are shaped like disks. They contain dust, gas, and young stars in their bluish arms. Older, redder stars are in their central bulges. Spiral galaxies are surrounded by spherical halos containing older stars.

Galaxies

Stars are not randomly scattered throughout the universe. Most stars are bound by gravity into galaxies. *A* **galaxy** *is a huge collection of stars, gas, and dust.* Astronomers classify galaxies by their shapes. Examples of the three main types of galaxies–elliptical, irregular, and spiral–are shown in **Figure 13.**

The universe contains hundreds of billions of galaxies. Each galaxy can contain hundreds of billions of stars. The solar system where you live is part of the Milky Way, a spiral galaxy. The Milky Way is larger than most galaxies in the universe. It contains over 100 billion stars.

Because Earth is inside the Milky Way, scientists cannot see the Milky Way from the outside as they can see other galaxies. Even though they cannot see all of the Milky Way, scientists have determined that the Milky Way has at least two major spiral arms. The Sun is near one of the arms a little more than halfway from the Milky Way's center.

Key Concept Check In which galaxy is Earth located?

WORD ORIGIN

galaxy
from Latin *galactos,* means "milk"

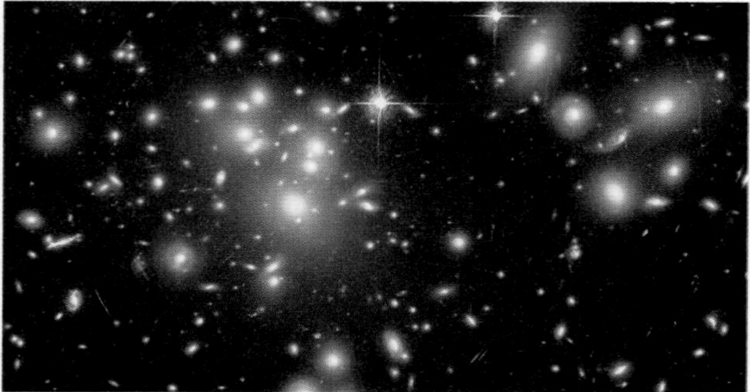

Figure 14 Gravity causes galaxies to gather in clusters, where they interact and sometimes merge with one another.

 Animation

Math Skills

Use Dimensional Analysis

Light-years (ly) describe distances to nearby stars. Astronomers often use parsecs (pc) to describe greater distances in space.

1 pc = 3.26 ly
1 ly = 9.46 trillion km

The star Proxima Centauri is 4.2 ly from Earth. What is that distance in parsecs?

1. Select a conversion factor with the unit you want in the numerator and the given unit in the denominator.

$$\frac{1\text{ pc}}{3.26\text{ ly}}$$

2. Multiply the starting quantity and units by the conversion factor.

$$\frac{4.2\text{ ly} \times 1\text{ pc}}{3.26\text{ ly}}$$

3. Complete the calculation.

$$\frac{4.2\text{ pc}}{3.26} = 1.3\text{ pc}$$

Practice

The nearest galaxy to the Milky Way is the Andromeda galaxy. It is approximately 2.5 million ly from Earth. What is that distance in parsecs?

 Math Practice

 Personal Tutor

The Universe

Most galaxies, such as those shown in **Figure 14**, are pulled by gravity into clusters of galaxies. The Milky Way is part of a cluster called the Local Group, which contains about 30 galaxies. The Local Group, in turn, is part of a supercluster of galaxies called the Local Supercluster. Superclusters are some of the largest structures in the universe. Some superclusters contain thousands of galaxies. But even superclusters are parts of larger structures. Superclusters form enormous sheetlike walls in space.

 **Key Concept Check** How is the universe structured?

By studying the rotations and the interactions of galaxies in clusters, astronomers can determine how much mass the galaxies contain. Astronomers have discovered that only 5–10 percent of the mass in galaxies emits light. They hypothesize that the rest of the mass in galaxies–and in the universe–is invisible dark matter or dark energy.

Recycled Matter

Most of the elements in your body were originally made in stars. Hydrogen is combined into more-complex elements during nuclear reactions in stars. When a massive star explodes, such as the star in the photo at the beginning of this lesson, it releases those elements into space. This material can then form new stars and planets. In this way, matter in the universe is recycled.

Big Bang Theory

Most scientists agree that the universe formed 13–14 billion years ago and that it had a hot and dense beginning. *The* **Big Bang theory** *states that the universe began from one point and has been expanding and cooling ever since.* Will the universe expand forever, or will gravity cause it eventually to contract? These questions remain unanswered. Scientists have not yet been able to determine the fate of the universe.

Lesson 3 Review

Visual Summary

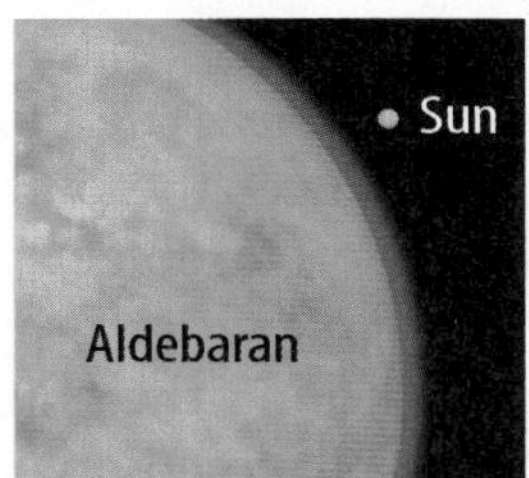

The largest stars are much larger than the Sun.

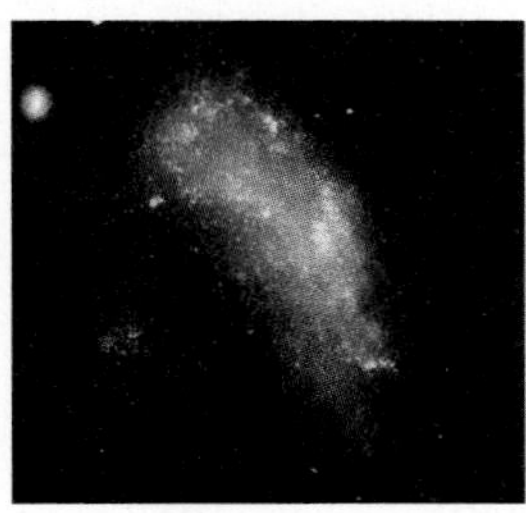

Irregular galaxies have many young stars.

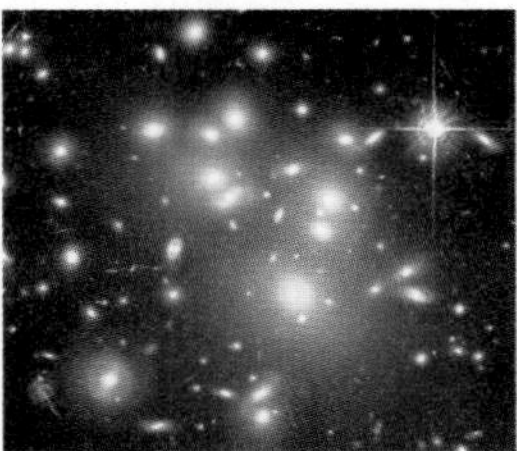

Gravity causes galaxies to gather in clusters.

FOLDABLES®

Use your lesson Foldable to review the lesson. Save your Foldable for the project at the end of the chapter.

What do you think NOW?

You first read the statements below at the beginning of the chapter.

5. The Sun is more massive than 90 percent of other stars.

6. The solar system is at the center of the Milky Way.

Did you change your mind about whether you agree or disagree with the statements? Rewrite any false statements to make them true.

Use Vocabulary

1. The __________ states that the universe expanded from one point.
2. **Define** *light-year* in your own words.
3. **Use the term** *galaxy* in a sentence.

Understand Key Concepts 

4. **Describe** the location of Earth within the universe.
5. What percentage of stars are larger and more massive than the Sun?
   A. 10 percent
   B. 30 percent
   C. 50 percent
   D. 90 percent
6. **Discuss** the importance of gravity for stars, galaxies, and the universe.

Interpret Graphics

7. **Classify** the galaxy below.

8. **Organize Information** Copy and fill in the graphic organizer below to list structures in the universe larger than the Sun, in order of their size.

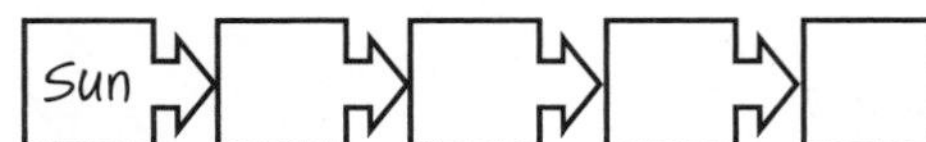

Critical Thinking

9. **Infer** why astronomers study far-distant galaxies to learn about the early universe.

Math Skills

 Math Practice

10. The Milky Way is about 100,000 ly across. What is that distance in kilometers?

Lab

2 class periods

Planetary Revolutions

Materials

ball of string

masking tape

measuring tape

Safety

Gravity continuously pulls objects in the universe closer together. The more massive the objects and the closer they are together, the stronger the pull of gravity. The solar system is made up of eight planets and other objects, such as asteroids and comets, that orbit the Sun. The Sun's gravitational pull influences the speed planets and other space objects orbit around the Sun. Can you model gravity's effect on the time it takes for a planet to complete one orbit around the Sun?

Question

Do the planets take the same amount of time to complete their orbits?

Procedure

1. Read and complete a lab safety form.
2. Working as a group in an open area, use string to make a circle 1 m across. Use masking tape to secure this circle to the ground. Think of this circle as a bull's-eye.
3. As shown below, make seven concentric circles around the 1-m circle. Make these circles 3 m, 5 m, 7 m, 10 m, 12 m, 14 m, and 16 m across.
4. Label the 1-m circle Mercury. Then label the others Venus, Earth, Mars, Jupiter, Saturn, Uranus, and Neptune in that order. These circles represent the orbital paths of the planets.
5. Have a student stand in the middle of the 1-m circle and hold an 18-m length of string . Have another student hold the other end of the string and stand just beyond the outermost circle. Stretch the string tight to mark a straight line from the center of the 1-m circle to beyond the outermost circle. Place the string on the ground, and use masking tape to secure it. This line is your reference line and will help you to plot the planets' orbits.

6. Choose eight students to represent the eight planets. Have each student stand on a circle where the reference line crosses it. Eight other students will record data for the planets.
7. When your teacher gives the signal, students should begin walking their orbits, moving in the same direction at the same rate of speed. When "Earth" completes one orbit, all planets should stop and stand in place.
8. Student recorders should use masking tape to mark the position of his or her assigned planet. Label this position as *Trial 1.* A recorder also should record how many revolutions the "planet" completed.
9. Starting from the ending position in step 8, repeat steps 7 and 8 three times, trials 2, 3, and 4.

Analyze and Conclude

10. The orbital period of a planet is the time it takes to complete one orbit. Which planet has the shortest orbital period?
11. Which planet has the longest orbital period?
12. **The Big Idea** In the time it took for the student representing Mars to orbit the Sun once, how many times did the student representing Earth orbit the Sun?

Communicate Your Results

Create a graphic novel or picture book that contains photos or drawings of a variety of possible views of planets crossing the night skies from different planets in our solar system. Explain how the views would differ from views on Earth.

Inquiry Extension

Research the orbital paths and periods of asteroids, comets, meteoroids, and other space objects, such as telescopes and space probes. Compare and contrast these orbits to the planets. Are their orbital patterns consistent or do they vary?

Lab Tips

- ☑ Remember that these circles are scale models of the orbits of the planets and not correct representations of the distances between the planets.
- ☑ Carefully consider how far your planet of choice is from the Sun when creating your graphic novel or picture book.

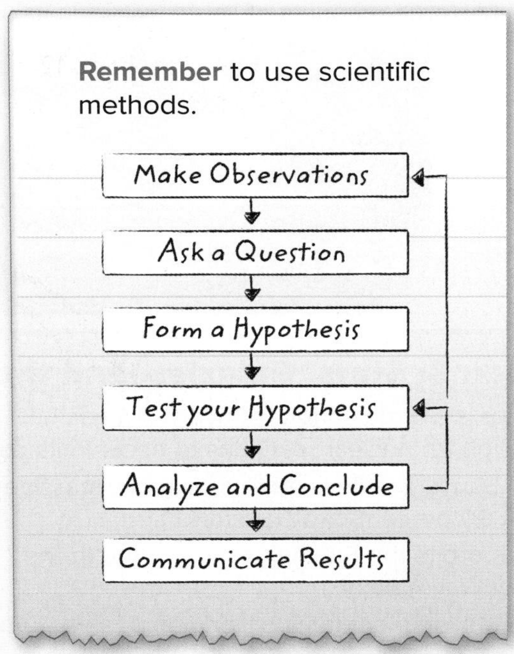

Chapter 2 Study Guide

The Moon orbits Earth as Earth orbits the Sun. The Sun is at the center of the solar system. The solar system is part of the Milky Way, one of billions of galaxies in the universe.

Key Concepts Summary

Lesson 1: The Sun-Earth-Moon System

- Earth has seasons because it is tilted as it revolves around the Sun.
- The Moon's position in relation to Earth and the Sun causes **waxing** and **waning** moon phases. The Moon's gravitational pull is largely responsible for **tides.** As the Moon orbits Earth, it causes **eclipses.**
- A solar eclipse occurs when the Moon moves between Earth and the Sun and the Moon's shadow covers part of Earth. A lunar eclipse occurs as the Moon passes through Earth's shadow.

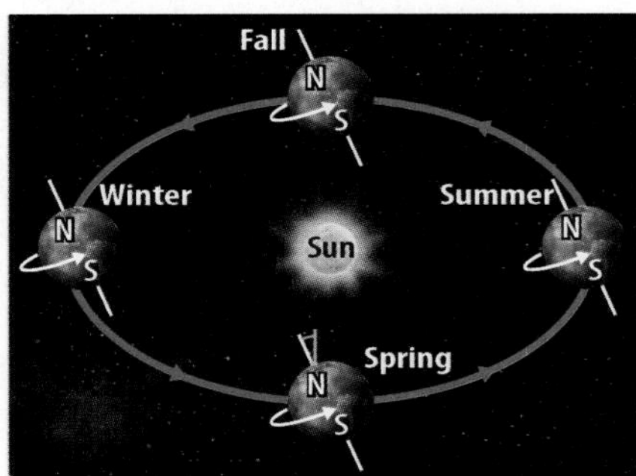

Vocabulary

revolution p. 42
rotation p. 42
equinox p. 43
solstice p. 43
waxing p. 45
waning p. 45
tide p. 46
eclipse p. 47

Lesson 2: The Solar System

- Gravity affected how the solar system and objects within it formed, and it continues to affect how solar system objects orbit the Sun.
- The solar system contains the Sun, **planets, dwarf planets, asteroids, comets, meteoroids, moons,** and other objects.
- Earth is the third planet from the Sun and the only planet known to have large amounts of liquid water on its surface.

Vocabulary

planet p. 51
dwarf planet p. 52
moon p. 53
asteroid p. 53
comet p. 53
meteoroid p. 53
meteor p. 53

Lesson 3: Stars, Galaxies, and the Universe

- A **star** is a large sphere of hydrogen gas hot enough for nuclear reactions to occur in its core.
- The Sun is a yellow star and is more massive than 90 percent of other stars.
- Earth orbits the Sun, which is located in the Milky Way **galaxy.**
- The universe contains billions of galaxies, which are grouped into clusters and superclusters.

Vocabulary

star p. 59
light-year p. 59
galaxy p. 61
Big Bang theory p. 62

Vocabulary eFlashcards
Vocabulary eGames

Personal Tutor

FOLDABLES® Chapter Project

Assemble your lesson Foldables as shown to make a Chapter Project. Use the project to review what you have learned in this chapter.

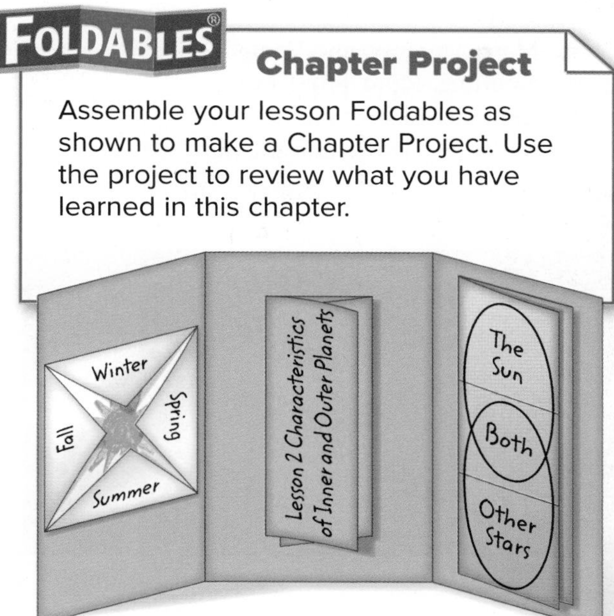

Use Vocabulary

1. Compare Earth's rotation and revolution.
2. As the Moon appears to get smaller, its phases are ________.
3. As the Moon appears to get larger, its phases are ________.
4. Distinguish between a solstice and an equinox.
5. Define the terms *meteoroid* and *meteor.*
6. A collection of stars, gas, and dust is a(n) ________.
7. When the Moon moves between Earth and the Sun, a(n) ________ can occur.
8. Define the word *tides.*

Interactive Concept Map

Link Vocabulary and Key Concepts

Copy this concept map, and then use vocabulary terms from the previous page to complete the concept map.

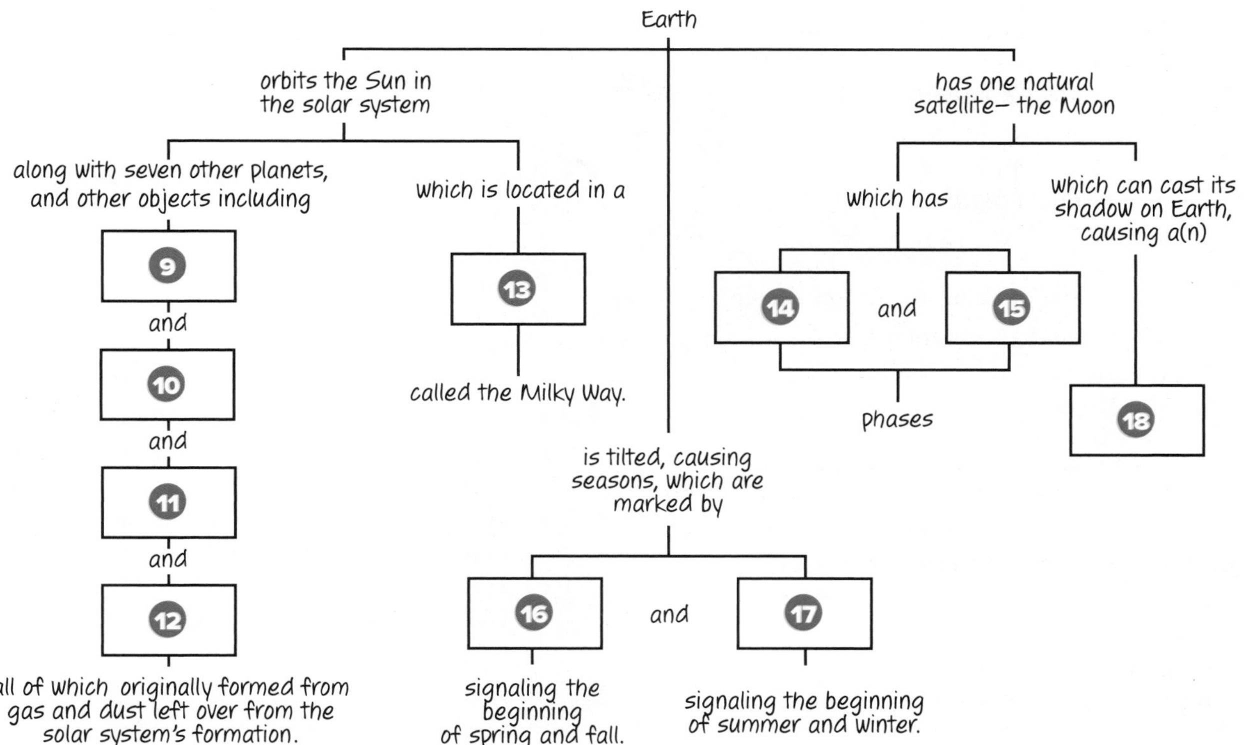

Chapter 2 Review

Understand Key Concepts

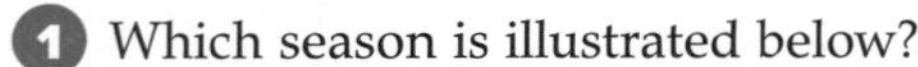

1. Which season is illustrated below?

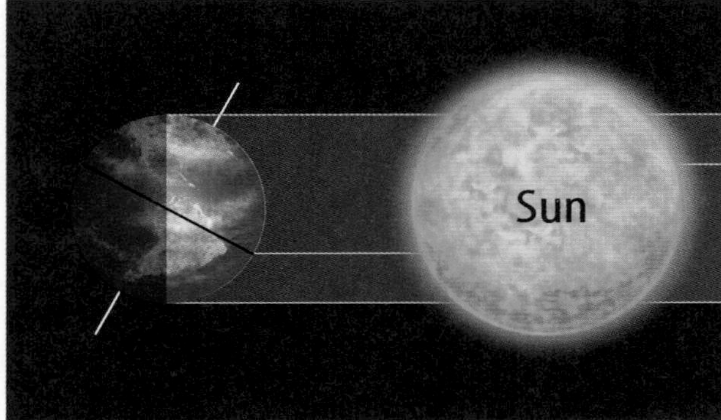

   A. fall in the northern hemisphere; spring in the southern hemisphere
   B. spring in the northern hemisphere; fall in the southern hemisphere
   C. summer in the northern hemisphere; winter in the southern hemisphere
   D. winter in the northern hemisphere; summer in the southern hemisphere

2. When is a solar eclipse visible?
   A. only when the Moon is full
   B. only when the Moon is new
   C. only when the Moon is waning
   D. only when the Moon is waxing

3. Where is the solar system located?
   A. in the center of the Milky Way
   B. in the halo of the Milky Way
   C. near a spiral arm of the Milky Way
   D. outside the Milky Way

4. Which statement about the Moon is true?
   A. The Moon does not rotate.
   B. The Moon orbits the Sun.
   C. The Moon has one side that never faces the Sun.
   D. The Moon has one side that never faces Earth.

5. Which is a property of the outer planets?
   A. few moons
   B. ring systems
   C. rocky surfaces
   D. short orbits

6. What objects in the solar system are larger than Earth?
   A. Mars, Mercury, and Venus
   B. Neptune, Pluto, and Uranus
   C. the inner planets and the Sun
   D. the outer planets and the Sun

7. Which star below is the coolest?

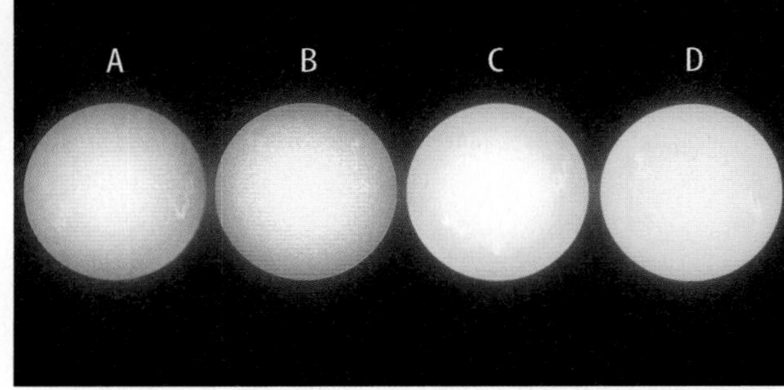

   A. A
   B. B
   C. C
   D. D

8. Which solar system objects have orbits that take them farthest from the Sun?
   A. asteroids
   B. comets
   C. meteoroids
   D. planets

9. Which planet is most like Earth in size and makeup?
   A. Mars
   B. Mercury
   C. Saturn
   D. Venus

10. Which statement about tides is true?
   A. High tides occur twice each day in all coastal areas on Earth.
   B. Low tides and high tides never vary in height.
   C. Tides can be predicted.
   D. Weather has no effect on tides.

Critical Thinking

11 **Hypothesize** The first photo of the far side of the Moon was taken in 1959. Why were none taken earlier?

12 **Deduce** The ancient Greeks used lunar eclipses as evidence that Earth is round and not flat. Why?

13 **Decide** Astronomers prefer to observe the sky during moonless nights. Which phase of the Moon would be best for their observations?

14 **Support** the statement, "The universe has structure."

15 **Imagine** that Earth had no moon. What would be different?

16 **Interpret Graphics** The planets shown below are in the wrong order according to distance from the Sun. List them in the correct order. Assume the Sun is at the left.

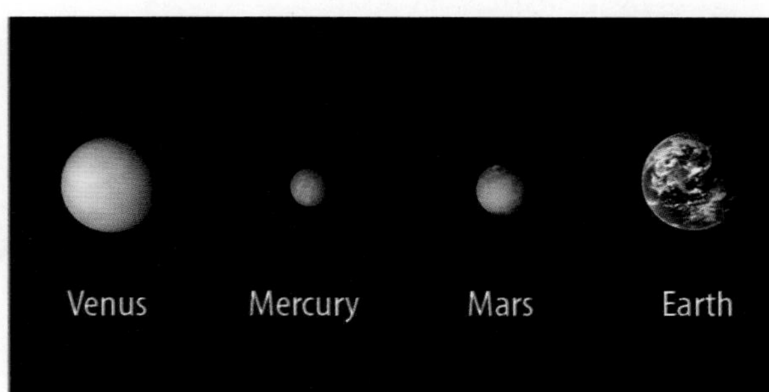

Writing in Science

17 **Write a letter** Pluto was classified as a planet after its discovery in 1930. In 2006, the International Astronomical Union (IAU) reclassified Pluto as a dwarf planet. Write a letter to the IAU, at least five sentences long, either agreeing or disagreeing with the IAU's decision. Include the definitions of *planet* and *dwarf planet* in your letter. If you disagree with the IAU, suggest how the definitions might be changed to support your argument.

REVIEW THE BIG IDEA

18 Draw a diagram that shows at least 10 objects correctly positioned within 50,000 AU of Earth. How is Earth related to these objects?

19 The photo below shows a view of Earth in space. Where is Earth's place in the universe? How is Earth related to other objects in the universe?

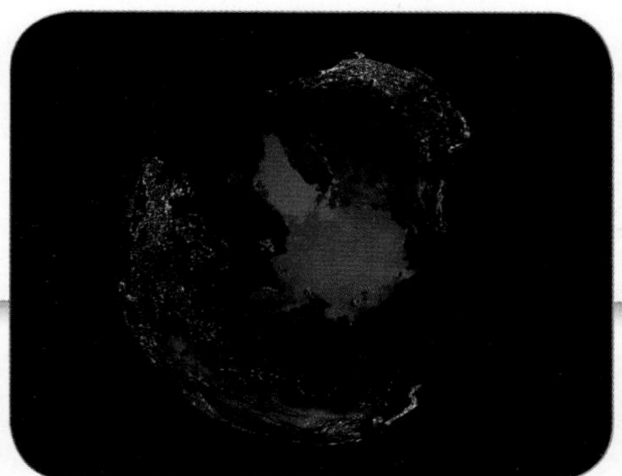

Math Skills

 Math Practice

Use Dimensional Analysis

20 The *Hubble Space Telescope* has taken photographs of a galaxy that is 123,000,000,000,000,000,000,000 km away from Earth.

a. How far is that distance in light-years?

b. How far is that distance in parsecs?

21 Polaris, the North Star, is about 430 ly from Earth.

a. What is that distance in parsecs?

b. How long does it take light from Polaris to reach Earth?

Standardized Test Practice

Record your answers on the answer sheet provided by your teacher or on a sheet of paper.

Multiple Choice

1 What time of year is Earth's northern hemisphere closest to the Sun?

A in January, during winter

B in July, during summer

C in April, during spring

D in October, during fall

2 Which is the main component of stars?

A dust

B hydrogen

C nitrogen

D rock

Use the figure below to answer question 3.

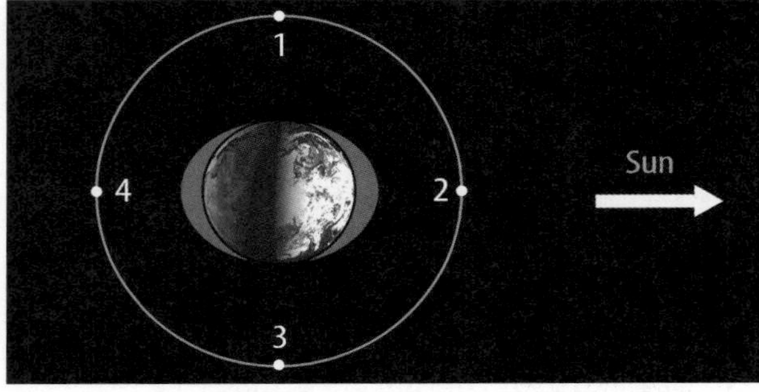

3 The figure shows a model for spring tides, which are characterized by high tides that are higher and low tides that are lower than usual. What numbers show the positions of the Moon that would cause spring tides?

A 1 and 2

B 1 and 3

C 2 and 4

D 3 and 4

4 Pluto is an example of what kind of object?

A asteroid

B comet

C dwarf planet

D meteoroid

5 How does the Sun compare to other stars in the universe?

A It is farther away than most other stars.

B It is hotter than most other stars.

C It is more massive than most other stars.

D It is whiter than most other stars.

Use the figure below to answer question 6.

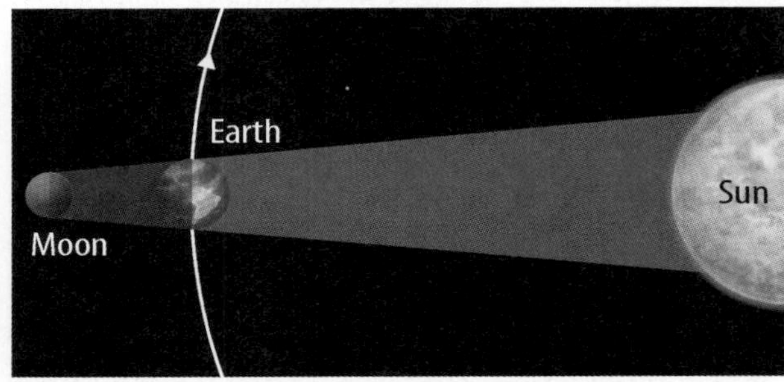

6 What event occurs when Earth, the Moon, and the Sun are in the positions shown?

A lunar eclipse

B neap tide

C new moon

D solar eclipse

7 Galaxies are classified by shape. What shape is the galaxy that includes the Sun, Earth, and the rest of the solar system?

A elliptical

B irregular

C regular

D spiral

8 Which describes the organization of the universe from the smallest unit to the largest unit?

A cluster, supercluster, galaxy, star

B galaxy, star, supercluster, cluster

C star, cluster, supercluster, galaxy

D star, galaxy, cluster, supercluster

Use the figure below to answer question 9.

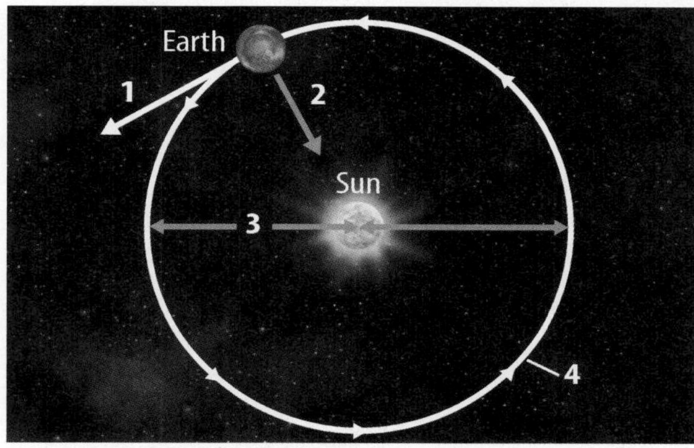

9 Which arrow shows the motion of Earth if there were no gravity between Earth and the Sun?

A 1

B 2

C 3

D 4

10 Which objects in the solar system develop long tails during part of their orbits?

A asteroids

B comets

C meteoroids

D moons

Constructed Response

Use the figure below to answer questions 11 and 12.

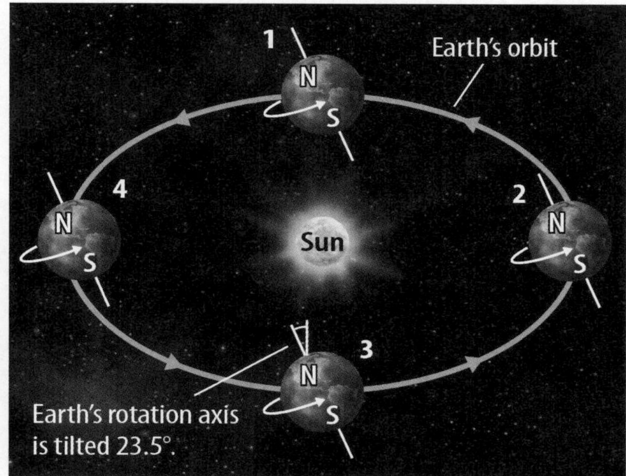

11 Suppose you are planning a trip to Australia. Your friend tells you that it is best to go there during Australia's winter. In which position should Earth be when you visit? Explain your answer.

12 Sketch what the figure above would look like if Earth were not tilted on its axis. What would a year on Earth be like?

13 Both asteroids and planets orbit the Sun. All planets are spherical, but most asteroids are not. Explain why.

14 Suppose a person on Earth sees a star twinkling in the night sky. At the same moment, an astronaut in a space shuttle looks at the same star. Explain why the astronaut does not see the star twinkle.

NEED EXTRA HELP?														
If You Missed Question...	1	2	3	4	5	6	7	8	9	10	11	12	13	14
Go to Lesson...	1	3	1	2	3	1	3	3	1	2	1	1	2	3

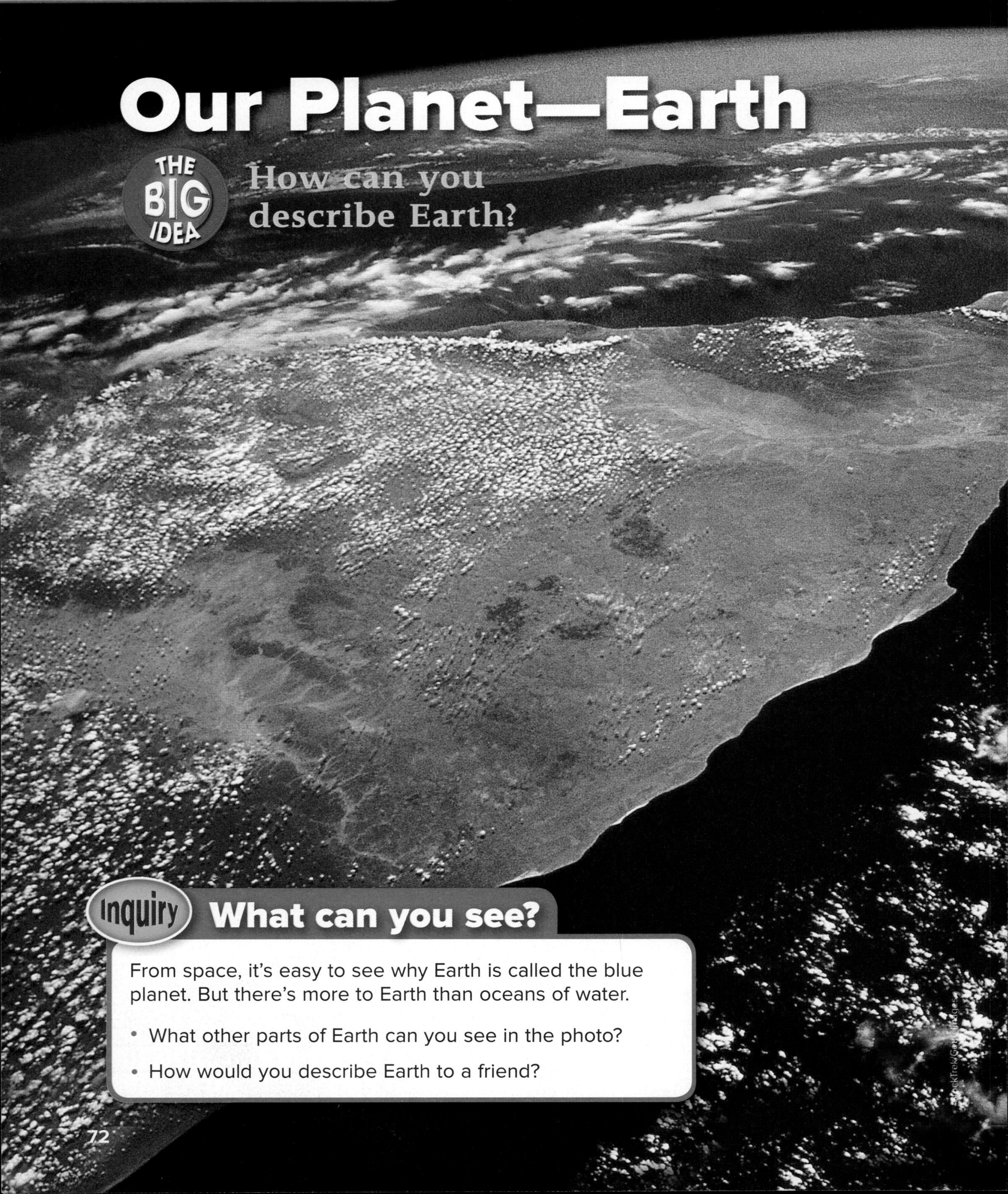

Chapter 3

Our Planet—Earth

THE BIG IDEA How can you describe Earth?

Inquiry What can you see?

From space, it's easy to see why Earth is called the blue planet. But there's more to Earth than oceans of water.

- What other parts of Earth can you see in the photo?
- How would you describe Earth to a friend?

StockTrek/Getty Images

Get Ready to Read

What do you think?

Before you read, decide if you agree or disagree with each of these statements. As you read this chapter, see if you change your mind about any of the statements.

1. Earth is a simple system made of rocks.
2. Most of Earth is covered by one large ocean.
3. Earth's interior is made of distinct layers.
4. The water cycle begins in the ocean.
5. Earth's air contains solids, liquids, and gases.
6. Rocks are made of minerals.

Your one-stop online resource
connectED.mcgraw-hill.com

LearnSmart®

Chapter Resources Files, Reading Essentials, Get Ready to Read, Quick Vocabulary

Animations, Videos, Interactive Tables

Self-checks, Quizzes, Tests

Project-Based Learning Activities

Lab Manuals, Safety Videos, Virtual Labs & Other Tools

Vocabulary, Multilingual eGlossary, Vocab eGames, Vocab eFlashcards

Personal Tutors

Lesson 1

Earth Systems

Reading Guide

Key Concepts
ESSENTIAL QUESTIONS

- What are the composition and the structure of the atmosphere?
- How is water distributed in the hydrosphere?
- What are Earth's systems?
- What are the composition and the structure of the geosphere?

Vocabulary

biosphere p. 76
atmosphere p. 77
hydrosphere p. 79
groundwater p. 80
geosphere p. 81
mineral p. 81
rock p. 82

Multilingual eGlossary

BrainPOP®
Science Video
What's Science Got to do With It?

Inquiry A Hot Mix?

Earth is made of more than soil, minerals, and melted rocks flowing out of volcanoes. What other parts of Earth do you see in the photo? How do these parts interact?

Launch Lab

20 minutes

How can you describe Earth?

When you look out the window, you might see wispy white clouds, birds in the trees, and rolling hills in the distance. All these things are part of Earth. What else makes up Earth?

1. Read and complete a lab safety form.
2. With your partner, brainstorm a list of words that describe Earth. Limit the list to 20 words. Be creative! Record the list in your Science Journal.
3. Use **markers** to rewrite your list of words using different colors and letter shapes. Use **scissors** to cut out each word.
4. Group the words that you think relate to each other. Use a **glue stick** to fix the words to a piece of **colored paper.**

Think About This

1. What words did you use to describe Earth?
2. How did your list compare to those of other students?
3. **Key Concept** What things do you think make up Earth?

What is Earth?

The puffy, white clouds over your head and the hard ground under your feet are both parts of Earth. The water in the oceans and the fish that live there are also parts of Earth. The planet Earth is more than a solid ball in space. It includes air molecules that float near the boundaries of outer space and molten rock that churns deep below Earth's surface.

Earth is a complex place. People often divide complex things into smaller parts in order to study them. Scientists divide Earth into four systems to help better understand the planet. The systems contain different materials and work in different ways, but they all interact. What happens in one system affects the others.

Earth's Air

The outermost Earth system is an invisible layer of gases that surrounds the planet. Even though you cannot see air, you can feel it when the wind blows. Moving air is blowing the tree in **Figure 1**.

Figure 1 Even though you cannot see air, you can see its power when it makes objects move.

Earth's Water

Below the layer of air is the system that contains Earth's water. Like air, water can move from place to place. Some of the water is salty, and some is fresh. Fresh river water flows into the salty Pacific Ocean in Hawaii, as shown in **Figure 2.**

The Solid Earth

The next system is the solid part of Earth. It contains a thin layer of soil covering a rocky center. It is by far the largest Earth system. Because it is solid, materials in this system move more slowly than air or water. But they do move, and over time, landforms rise up and then wear away. It took millions of years for the canyon shown in **Figure 2** to form.

Life on Earth

The Earth system that contains all living things is the **biosphere.** Living things are found in air, water, and soil. So, the biosphere has no distinct boundaries; it is found within the other Earth systems. The living things shown in **Figure 2** are part of the biosphere. You will learn more about the biosphere when you study life science, or biology. The rest of this chapter will describe the three Earth systems made of nonliving things.

Reading Check Why doesn't the biosphere have distinct boundaries?

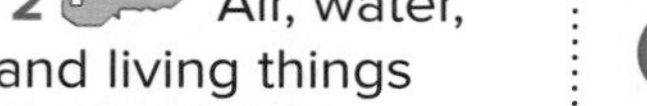

Figure 2 Air, water, rocks, and living things are all part of Earth.

The Atmosphere

The force of Earth's gravity pulls molecules of gases into a layer surrounding the planet. *This mixture of gases forms a layer around Earth called the* **atmosphere.** The atmosphere is denser near Earth's surface and becomes less dense farther from Earth. It keeps Earth warm by trapping thermal energy from the Sun that bounces back from Earth's surface. If the atmosphere did not regulate temperature, life as it is on Earth could not exist.

WORD ORIGIN

atmosphere
from Greek *atmos-*, means "vapor"; and Greek *spharia*, means "sphere"

What makes up the atmosphere?

The atmosphere contains a mixture of nitrogen, oxygen, and smaller amounts of other gases. The graph in **Figure 3** shows the percentages of these gases. The most common gas is nitrogen, which makes up 78 percent of the atmosphere. Most of the remaining gas is oxygen.

The other gases are called trace gases because they make up only 1 percent, or a trace, of the atmosphere. Nonetheless, trace gases are important. Carbon dioxide, methane, and water vapor help regulate Earth's temperature. Note that **Figure 3** shows the percentages of gases in dry air. The atmosphere also contains water vapor. The amount of water vapor in the atmosphere generally ranges from 0 to 4 percent.

Along with gases and water vapor, the atmosphere contains small amounts of solids. Particles of dust float along with the gases and water vapor. Sometimes you can see these tiny specks as sunlight reflects off them as it shines through a window.

 **Key Concept Check** What is the composition of the atmosphere?

Figure 3 Dry air contains a mixture of gases. Though the atmosphere is made mainly of nitrogen and oxygen, trace gases are also important.

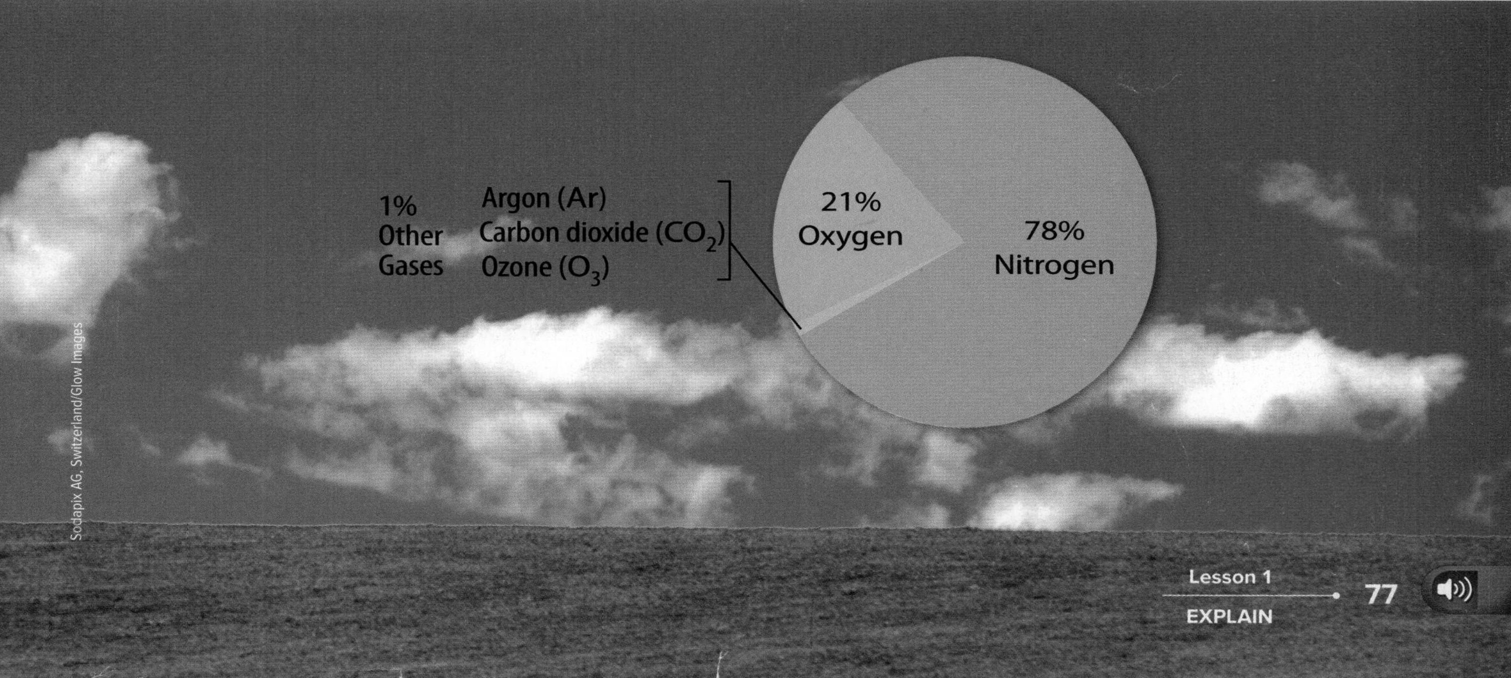

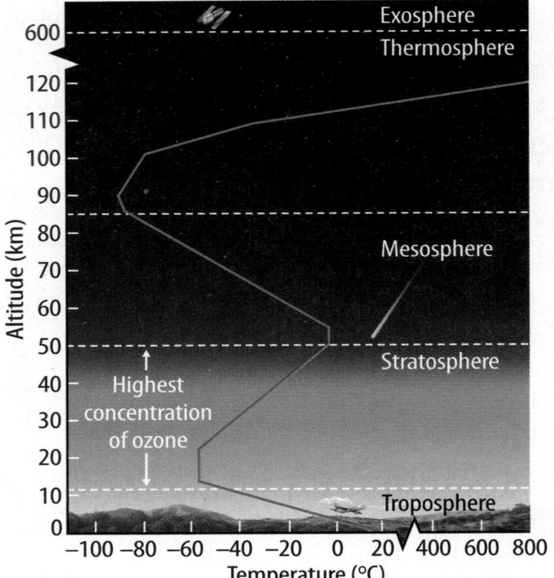

Figure 4 The atmosphere is divided into layers according to differences in temperature.

Visual Check Summarize how temperature changes as altitude increases.

Animation

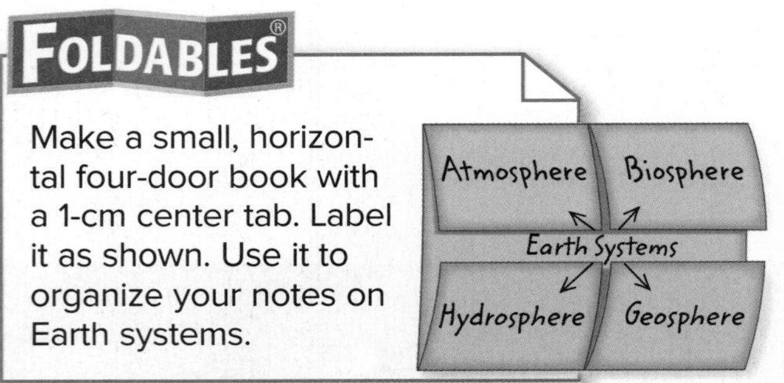

Make a small, horizontal four-door book with a 1-cm center tab. Label it as shown. Use it to organize your notes on Earth systems.

Layers of the Atmosphere

The composition of the atmosphere does not change much over time. However, the temperature of the atmosphere does change. Radiant energy from the Sun heats Earth's atmosphere; however, different parts of the atmosphere absorb or reflect the Sun's energy in different ways. The red line in **Figure 4** shows changes in temperature as altitude increases. These temperature changes are used to distinguish layers in the atmosphere.

The Troposphere If you have ever hiked up a mountain, you might have noticed that the temperature decreases as you climb higher. In the bottom layer of the atmosphere, called the troposphere, temperature decreases as you move upward from Earth's surface. Gases flow and swirl in the troposphere, causing weather. Although the troposphere does not extend very far upward, it contains most of the mass in the atmosphere.

The Stratosphere Above the troposphere is the stratosphere. Unlike gases in the troposphere, gases in the stratosphere do not swirl around. They are more stable and form flat layers. Within the stratosphere is a layer of ozone, a form of oxygen. This ozone layer protects Earth's surface from harmful radiation from the Sun. It acts like a layer of sunscreen, protecting the biosphere. Because ozone absorbs solar radiation, temperatures increase in the stratosphere.

Upper Layers Above the stratosphere is the mesosphere. Temperature decreases in this layer, then increases again in the next layer, the thermosphere. The last layer of Earth's atmosphere is the exosphere. The lowest density of gas molecules is in this layer. Beyond the exosphere is outer space.

Key Concept Check What are the layers of the atmosphere?

The Hydrosphere

Water is one of the most common and important substances on Earth. *The system containing all Earth's water is called the* **hydrosphere.** Most water is stored on Earth's surface, but some is located below the surface or within the atmosphere and biosphere. The hydrosphere contains more than 1.3 billion km^3 of water. The amount of water does not change. But like the gases in the atmosphere, water in the hydrosphere flows. It moves from one location to another over time. Water also changes state. It is found as a liquid, a solid, and a gas on Earth.

 **Reading Check** How much water is in the hydrosphere?

Ocean

Scientists call the natural locations where water is stored reservoirs (REH zuh vworz). The largest reservoir on Earth is the world ocean. Though the oceans have separate names, they are all connected, making one large ocean. Water flows freely throughout the world ocean. About 97 percent of Earth's water is in the ocean, as shown in **Figure 5.**

Many minerals dissolve easily in water. As water in rivers and underground reservoirs flows toward the ocean, it dissolves materials from the solid Earth. These dissolved minerals make ocean water salty. Most plants and animals that live on land, including humans, cannot use salt water. They need **freshwater** to survive.

REVIEW VOCABULARY

freshwater
water that contains less than 0.2 percent dissolved salts

Distribution of Earth's Water

Figure 5 Water in the hydrosphere is found in several different reservoirs.

Visual Check Where is most water on Earth located?

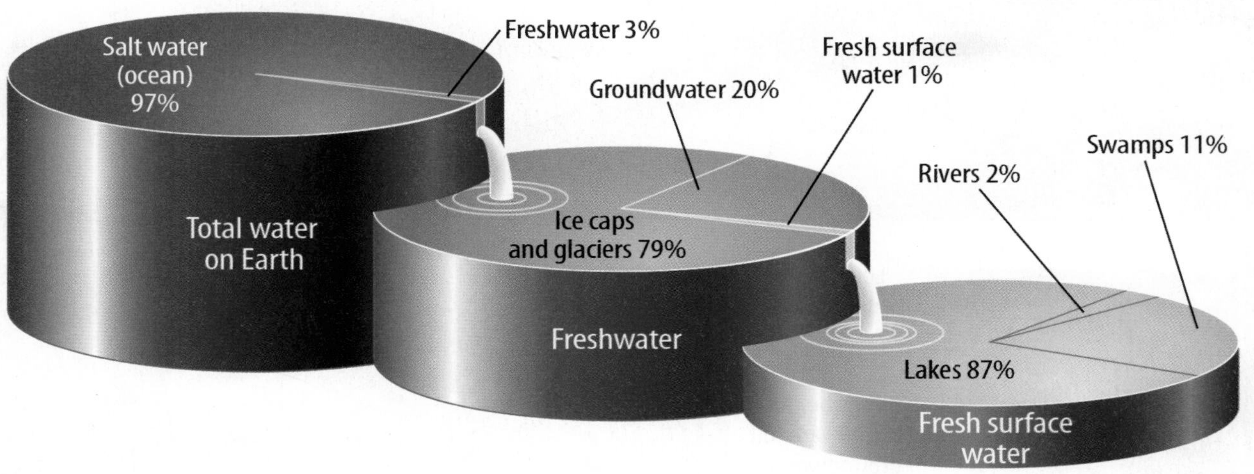

Lakes and Rivers

Less than 1 percent of freshwater is easily accessible on Earth's surface. This small percentage of Earth's total water must meet the needs of people and other organisms that require freshwater. Rain and snow supply water to the surface reservoirs–lakes and rivers. Water in these reservoirs moves through the water cycle much faster than water frozen in glaciers and ice caps.

Groundwater

Ice, lakes, and rivers hold about 80 percent of Earth's freshwater. Where is the remaining 20 percent? It is beneath the ground. Some rain and snow seep into the ground and collect in small cracks and open spaces called pores. **Groundwater** *is water that is stored in cracks and pores beneath Earth's surface.* As shown in **Figure 6**, groundwater collects in layers. Many people get their water by drilling wells down into these layers of groundwater.

 **Key Concept Check** How is water distributed in the hydrosphere?

The Cryosphere

Did you know that most of Earth's freshwater is frozen? The frozen portion of water on Earth's surface is called the cryosphere. About 79 percent of the planet's freshwater is in the cryosphere. The cryosphere consists of snow, glaciers, and icebergs. Water can be stored as ice for thousands of years before melting and becoming liquid water in other reservoirs.

Figure 6 Freshwater in lakes, rivers, and glaciers is visible on Earth's surface, but large amounts of groundwater are hidden below the surface.

The Geosphere

The last nonliving Earth system is the geosphere. *The* **geosphere** *is the solid part of Earth.* It includes a thin layer of soil and broken rock material along with the underlying layers of rock. The rocks and soil on land and beneath the oceans are part of the geosphere.

Key Concept Check What are Earth systems?

Materials in the Geosphere

The geosphere is made of soil, rock, and metal. All of these materials are composed of smaller particles.

Minerals Have you ever seen a sparkling diamond ring? Diamond is a mineral that is mined and then later cut and polished. **Minerals** *are naturally occurring, inorganic solids that have crystal structures and definite chemical compositions.*

To be considered a mineral, a material must have all five characteristics listed above. For example, materials that are made by people are not minerals because they did not form naturally. Materials that were once alive are organic and cannot be minerals. A mineral must be solid, so liquids and gases are not minerals. The atoms in minerals must be arranged in an orderly, repeating pattern. Finally, each mineral has a unique composition made of specific elements.

Minerals are identified by their physical properties, which include color, streak, hardness, luster, and crystal shape. Streak is the color of a mineral's powder. Even though some minerals have different colors, the color of the streak is the same. Hardness describes how easily a mineral can be scratched. Luster describes how a mineral reflects light. Usually, you must test several properties to identify a mineral. Examples of minerals with different properties are shown in **Figure 7**.

Figure 7 Minerals have different properties. The quartz shown on the left has a visible crystal structure. The olivine shown on the right has a striking color.

Personal Tutor

Figure 8 Diorite (top) is an igneous rock. Gneiss (center) is metamorphic. The conglomerate (bottom) is sedimentary.

Rocks Minerals are the building blocks of rocks. *A* **rock** *is a naturally occurring solid composed of minerals and sometimes other materials such as organic matter.* Scientists classify rocks according to how they formed. As shown in **Figure 8,** there are three major rock types: igneous, sedimentary, and metamorphic.

Igneous rocks form when molten material cools and hardens. Magma is molten material below Earth's surface. When it erupts from volcanoes and flows onto Earth's surface, it is called lava. So, igneous rocks can form inside Earth or on Earth's surface.

Sedimentary rocks form when forces such as water, wind, and ice break down rocks into small pieces called sediment. These same forces carry and deposit the sediment in layers. The bottom layers of sediment are compressed and then cemented together by natural substances to form rocks. Sedimentary rocks also form when minerals crystallize directly out of water.

Metamorphic rocks form when extreme temperatures and pressure within Earth change existing rocks into new rocks. The rocks do not melt. Instead, their compositions or their structures change.

MiniLab

20 minutes

What makes the geosphere unique?

Rocks and minerals, minerals and rocks—they always seem to go together. Can you have one without the other?

1. Read and complete a lab safety form.
2. Select a set of samples. You should have eight **minerals** and one **rock.** Identify which samples are minerals and which is the rock. Check with your teacher before moving on to step 3.
3. Use a **magnifying lens** to examine each mineral carefully. Note its color and other properties. Record your observations in your Science Journal.
4. Now, examine the rock. Do you recognize any of the minerals in the rock? Make two sets of samples—minerals that are present in the rock and minerals that are not.

Analyze and Conclude

1. **Summarize** the mineral and rock properties you observed.
2. **Generalize** What general statement can you make about the differences between rocks and minerals?
3. **Key Concept** Based on your observations, what kinds of materials make up the geosphere?

Structure of the Geosphere

Figure 9 Earth's major layers include the crust, the mantle, and the core.

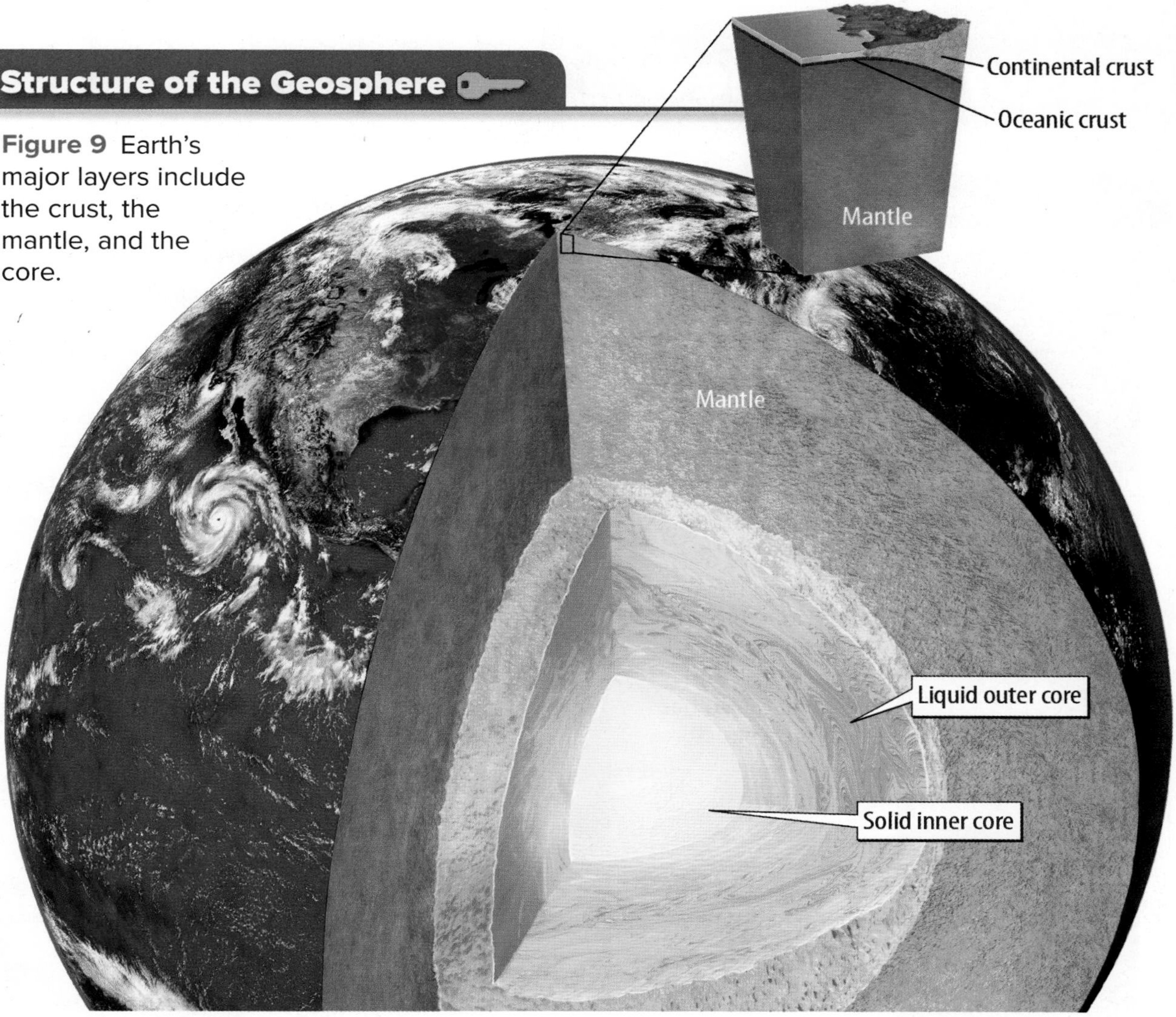

Structure

Earth's internal structure is layered like the layers of a hard-cooked egg. The three basic layers of the geosphere are shown in **Figure 9**. Similar to an egg, each layer of the geosphere has a different composition.

Crust The brittle outer layer of the geosphere is much thinner than the inner layers, like the shell on a hard-cooked egg. This thin layer of rock is called the crust. The crust is found under the soil on continents and under the ocean. Oceanic crust is thinner and denser than continental crust. This is due to their different compositions. Continental crust is made of igneous, sedimentary, and metamorphic rocks. Oceanic crust is made of only igneous rock.

Mantle The middle and largest layer of the geosphere is the mantle. Like the crust, the mantle is made of rock; however, mantle rocks are hotter and denser than those in the crust. In parts of the mantle, temperatures are so high that rocks flow, a bit like partially melted plastic.

Core The center of Earth is the core. If you use a hard-cooked egg as a model of Earth, then the yolk would be the core. Unlike the crust and the mantle, the core is not made of rock. Instead, it is made mostly of the metal iron and small amounts of nickel. The core is divided into two parts. The outer core is liquid. The inner core is a dense ball of solid iron.

Key Concept Check What are the composition and the structure of the geosphere?

Lesson 1 Review

Visual Summary

Earth is made of interacting systems: the atmosphere, the hydrosphere, the cryosphere, the geosphere, and the biosphere.

The atmosphere is made mainly of gases and has a layered structure. The geosphere is made of rock, soil, and metal and also has a layered structure.

Most water in the hydrosphere is in the world ocean.

FOLDABLES®

Use your lesson Foldable to review the lesson. Save your Foldable for the project at the end of the chapter.

What do you think NOW?

You first read the statements below at the beginning of the chapter.

1. Earth is a simple system made of rocks.
2. Most of Earth is covered by one large ocean.
3. Earth's interior is made of distinct layers.

Did you change your mind about whether you agree or disagree with the statements? Rewrite any false statements to make them true.

Use Vocabulary

1. **Use the term** *atmosphere* in a sentence.
2. **Distinguish** between the geosphere and the hydrosphere.
3. **Define** *mineral* in your own words.

Understand Key Concepts

4. Which Earth system contains living things?
   - **A.** atmosphere
   - **B.** biosphere
   - **C.** geosphere
   - **D.** hydrosphere
5. **Compare** the structure of the geosphere to that of a hard-cooked egg.
6. **Organize** the reservoirs in the hydrosphere according to how much water they hold. Begin with the reservoir that holds the most water.
7. **Distinguish** among Earth systems based on the states of matter found in each system.

Interpret Graphics

8. **Describe** How are Earth systems interacting in the photo shown here?

9. **Summarize** Copy and fill in the graphic organizer below to identify Earth systems.

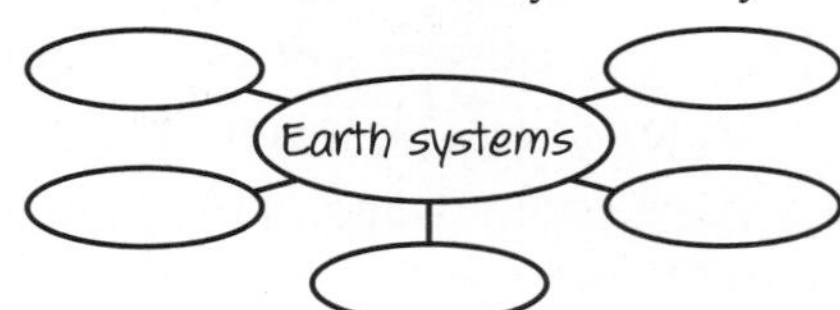

Critical Thinking

10. **Hypothesize** Earth systems interact with and affect one another. What might happen to your local hydrosphere and geosphere if conditions in the troposphere caused rain for several weeks?

Desalination

Taking the Salt out of Salt Water

HOW IT WORKS

Anyone who's been toppled by a big ocean wave knows salt water doesn't taste like the water we drink. People can't drink salt water. It's about 200 times more salty than freshwater. About 97 percent of Earth's water is salty. Most freshwater is frozen in glaciers and ice caps, leaving less than 1 percent of the planet's water available for 7 billion people and countless other organisms that require freshwater to live.

The need for freshwater has scientists searching for efficient ways to take the salt out of salt water. One solution is a desalination plant, where dissolved salts are separated from seawater through a process called reverse osmosis. This is how it works:

▲ **Desalination plants are found all over the world, including the United States.**

1 Salt water is pumped from the ocean.

2 High pressure forces water through a semipermeable membrane.

3 The semipermeable membrane acts as a filter, allowing the water, but not the salt, to pass through.

4 Clean freshwater is collected in a separate tank.

5 Water containing the waste salts flows out of the tank.

Because it takes a lot of energy to change salt water into freshwater, desalination plants are expensive to operate. But desalination is used in places such as Saudi Arabia and Japan, where millions of people have few freshwater resources.

It's Your Turn

RESEARCH What is the cost of desalinated water for households? How does it compare to the cost of water for households in your area? Present your findings to the class.

AMERICAN MUSEUM of NATURAL HISTORY

Lesson 2

Reading Guide

Key Concepts

ESSENTIAL QUESTIONS

- How does the water cycle show interactions of Earth systems?
- How does weather show interactions of Earth systems?
- How does the rock cycle show interactions of Earth systems?

Vocabulary

water cycle p. 87
evaporation p. 88
transpiration p. 88
condensation p. 89
precipitation p. 89
weather p. 90
climate p. 91
rock cycle p. 92
uplift p. 92

Multilingual eGlossary

BrainPOP®

Go to the resource tab in ConnectED to find the PBL *Campers in the Mist.*

Interactions of Earth Systems

All Systems Go?

A storm is moving from over the ocean toward land. Waves are crashing against the shore. All Earth systems are affected by the storm. How does water in clouds enter the atmosphere? How are Earth systems interacting in this storm?

Launch Lab

20 minutes

How do some Earth systems interact?

Earth's systems constantly interact with each other. In this activity, you'll model some common interactions.

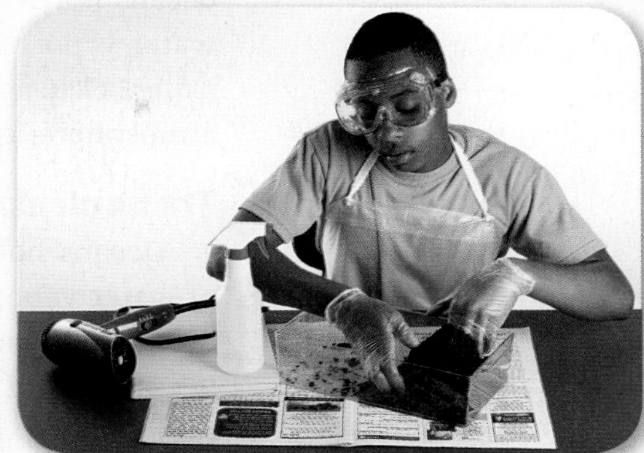

1. Read and complete a lab safety form.
2. Place a **plastic container** on a sheet of **newspaper.** In one end of the container, mold about 5 cups of **soil** into a landform of your choice.
3. Hold a **hair dryer** about 20 cm from the model landform. Using the hair dryer set on low, blow air across the model landscape for 1 min. Be careful not to blow the soil out of the container. Record your observations in your Science Journal.
4. Using a **spray bottle,** spray water onto your landform. Record your observations.

Think About This

1. How did you use the materials in this activity to model Earth's systems?
2. How could you improve your model? What changes would you make?
3. **Key Concept** Describe how Earth systems interacted in your model.

The Water Cycle

You read that the amount of water on Earth does not change. The water that you drink has been on Earth for a long time. Millions of years ago, a dinosaur might have swallowed the same water that you are drinking today. Or, that water might have raged down a river, flooding an ancient city. How does water move from place to place as time passes?

The **water cycle** *is the continuous movement of water on, above, and below Earth's surface.* The Sun provides the energy that drives the water cycle and moves water from place to place. As this occurs, liquid water can change state to a gas or a solid and then back again to a liquid. The change of state requires either an input or an output of thermal energy. **Figure 10** illustrates how energy is absorbed during evaporation and released during condensation.

Because the water cycle is continuous, there is no beginning or end. You will start your investigation of the water cycle in the hydrosphere's largest reservoir, the world ocean.

Reading Check What is the source of energy for the water cycle?

Personal Tutor

Figure 10 When water changes state from a gas to a liquid, thermal energy is released. Thermal energy is absorbed when liquid water changes into water vapor.

Hutchings Photography/Digital Light Source

Evaporation

When the Sun shines on an ocean, water near the surface absorbs energy and becomes warmer. As a molecule of water absorbs energy, it begins to vibrate faster. When it has enough energy, it breaks away from the other water molecules in the ocean. It rises into the atmosphere as a molecule of gas called water vapor. **Evaporation** *is the process by which a liquid, such as water, changes into a gas.* Water vapor, like other gases in the atmosphere, is invisible.

Transpiration and Respiration

Oceans hold most of Earth's water, so they are major sources of water vapor. But, water also evaporates from rivers, lakes, puddles, and even soil. These sources, along with oceans, account for 90 percent of the water that enters the atmosphere. Most of the remaining 10 percent is produced by transpiration. **Transpiration** *is the process by which plants release water vapor through their leaves.*

Some water vapor also comes from organisms through cellular respiration. Cellular respiration takes place in many cells. Water and carbon dioxide are produced during cellular respiration. When animals breathe, they release carbon dioxide and water vapor from their lungs into the atmosphere. The blue arrows in **Figure 11** show how water vapor enters the atmosphere.

Reading Check How are transpiration and respiration similar? How are they different?

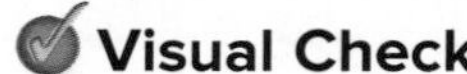

Animation

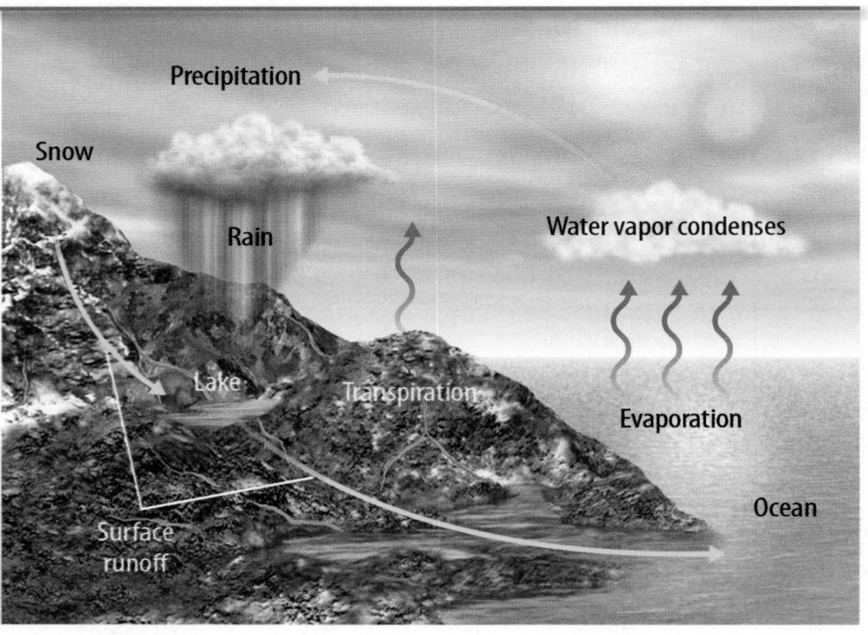

Figure 11 In the water cycle, water moves through the hydrosphere, the atmosphere, the geosphere, and the biosphere.

Visual Check Through which processes does water vapor enter the atmosphere?

Condensation

Recall that the temperatures of the troposphere decrease with increasing altitude. So, as water vapor rises through the troposphere, it becomes cooler. Eventually it loses so much thermal energy that it returns to the liquid state. *The process by which a gas changes to a liquid is* **condensation.** Tiny droplets of liquid water join to form larger drops. When millions of water droplets come together, a cloud forms.

Precipitation

Eventually, drops of water in the clouds become so large and heavy that they fall to Earth's surface. *Moisture that falls from clouds to Earth's surface is* **precipitation.** Rain and snow are forms of precipitation.

WORD ORIGIN

precipitation
from Latin *praecipitationem,* means "act or fact of falling headlong"

More than 75 percent of precipitation falls into the ocean, and the rest falls onto land. Some of this water evaporates and goes back into the atmosphere. Some flows into lakes or rivers, and the rest seeps into soil and rocks.

In the water cycle, water continually moves between the hydrosphere, the atmosphere, the biosphere, and the geosphere. As water flows across the land, it interacts with soil and rocks in the geosphere. You will learn more about these interactions when you read about the rock cycle.

Key Concept Check How do Earth systems interact in the water cycle?

MiniLab

20 min

How do plants contribute to the water cycle?

You have learned how water moves through Earth systems. How does the biosphere contribute to the water cycle?

1. Read and complete a lab safety form.
2. Choose a **potted plant.**
3. Carefully slide the plant into a **self-sealing plastic bag.** Close the bag tightly.
4. Place your bag on a sunny windowsill and leave it undisturbed overnight.
5. Observe the plant and the bag. Record your observations in your Science Journal.

Analyze and Conclude

1. **Recognize** Where did the moisture in the bag come from?
2. **Identify** What process of the water cycle did you model?
3. **Key Concept** How does your model show interactions among Earth systems?

Hutchings Photography/Digital Light Source

Math Skills 

Use a Formula

The amount of water vapor in air is called vapor density. Relative humidity (RH) compares the actual vapor density in air to the amount of water vapor the air could contain at that temperature. For example, at 15°C, air can contain a maximum of **12.8 g/m³** of water vapor. If the air contains 10.0 g/m³ of water vapor, what is the RH?

1. Use the formula:

$$RH = \left(\frac{\text{actual vapor density}}{\text{maximum vapor density}}\right) \times 100$$

2. Work out the equation.

$$RH = \left(\frac{10.0\ g/m^3}{12.8\ g/m^3}\right) \times 100$$

$$RH = 0.781 \times 100 = 78.1\%$$

Practice

At 0°C, air can contain 4.85 g/m³ of water vapor. Assume the actual water vapor content is 0.970 g/m³. What is the RH?

 Math Practice

 Personal Tutor

Changes in the Atmosphere

The atmosphere is continually changing. These changes take place mainly within the troposphere, which contains most of the gases in the atmosphere. Some changes occur within hours or days. Others can take decades or even centuries.

Weather

When you wake up in the morning and get ready for school, you might look outside to check the weather. **Weather** *is the state of the atmosphere at a certain time and place.* In most places, the weather changes to some degree every day. How do scientists describe weather and its changes?

Describing Weather Scientists use several factors to describe weather, as shown in **Figure 12.** Air temperature is a measure of the average amount of energy produced by the motion of air molecules. Air process is the force exerted by air molecules in all directions. Wind is the movement of air caused by differences in air pressure. Humidity is the amount of water vapor in a given volume of air. High humidity makes it more likely that clouds will form and precipitation will fall.

Interactions Weather is influenced by conditions in the atmosphere, the geosphere, and the hydrosphere. For example, air masses take on the characteristics of the area over which they form. So, an air mass that forms over a cool ocean will bring cool, moist air. In addition to these interactions, the hydrosphere provides much of the water for cloud formation and precipitation. Warm tropical waters provide the thermal energy that produces hurricanes.

 **Key Concept Check** How does weather show interactions of Earth systems?

Figure 12 Scientists describe weather using air temperature and pressure, wind speed and direction, and humidity.

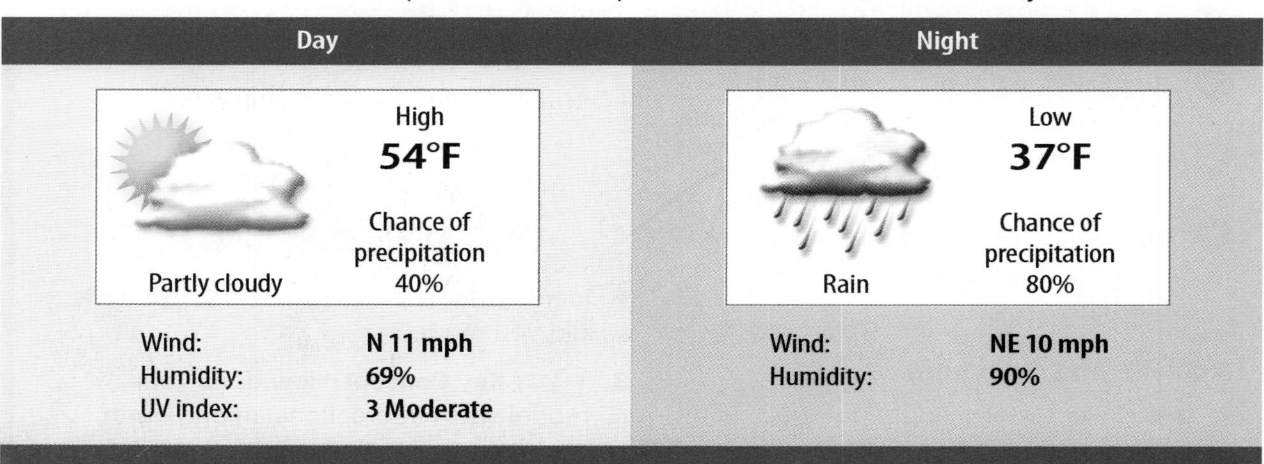

Day
High
54°F
Chance of precipitation 40%
Partly cloudy
Wind: N 11 mph
Humidity: 69%
UV index: 3 Moderate
Night
Low
37°F
Chance of precipitation 80%
Rain
Wind: NE 10 mph
Humidity: 90%

Climate

The weather in the area where you live might change each day, but weather patterns can remain nearly the same from season to season. For example, the weather might differ each day in the summer. But overall, summer is warm. These weather patterns are called climate. **Climate** *is the average weather pattern for a region over a long period of time.* Earth has many climates. Climates differ in part because of interactions between the atmosphere and other Earth systems.

 **Reading Check** How does weather differ from climate?

Mountains Recall that air temperature decreases with altitude. So the climate near the top of a mountain often is cooler than the climate near the mountain's base. Mountains also can affect the amount of precipitation an area receives–a phenomenon known as the rain-shadow effect. As shown in **Figure 13**, warm, wet air rises and cools as it moves up the windward side of a mountain. Clouds form and precipitation falls, giving this side of the mountain a wet climate. The air, now dry, continues to move over the mountain's peak and down the leeward side of the mountain. This side of the mountain often has a dry climate.

Ocean Currents As wind blows over an ocean, it creates surface currents in the water. Surface currents are like rivers in an ocean–the water flows in a predictable pattern. These currents transport thermal energy in water from place to place. For example, the Gulf Stream carries warm waters from tropical regions to northern Europe, making the climate of northern Europe warmer than it would be without these warm water currents.

SCIENCE USE V. COMMON USE

pressure

Science Use the force exerted over an area

Common Use the burden of physical or mental distress

Rain-Shadow Effect

Figure 13 Moist air on the windward side of mountains cools as it rises. Rain falls on this side of the mountain, resulting in a wet climate. This leaves little precipitation for the leeward side of the mountain, resulting in a dry climate.

Visual Check How can mountains affect the amount of precipitation an area receives?

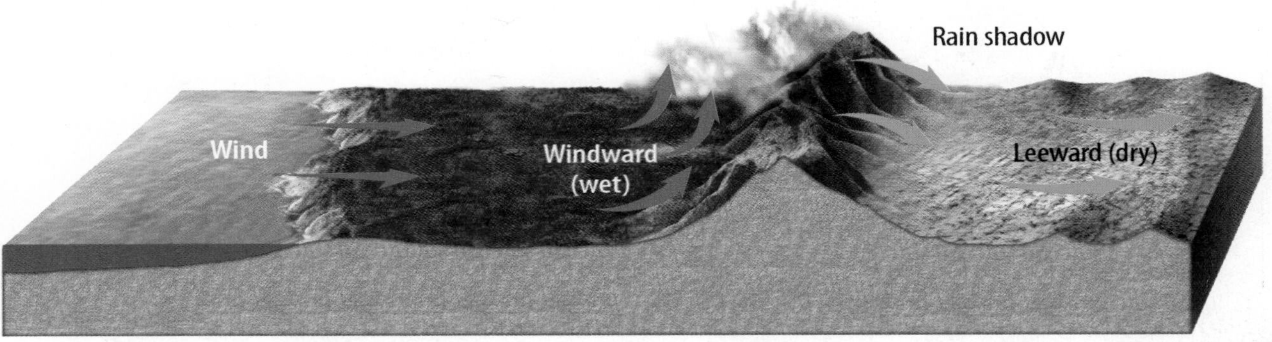

The Rock Cycle

In the water cycle, water moves throughout the hydrosphere, the atmosphere, the biosphere, and the geosphere. Another natural cycle is the rock cycle. *The* **rock cycle** *is the series of processes that transport and continually change rocks into different forms.* This cycle, shown in **Figure 14**, takes place in the geosphere, but it is affected by interactions with the other Earth systems.

As rocks move through the rock cycle, they might become igneous rocks, sedimentary rocks, or metamorphic rocks. At times they might not be rocks at all. Instead, they might take the form of sediments or hot, flowing magma. Like the water cycle, the rock cycle has no beginning or end. Some processes in this cycle take place on Earth's surface, and others take place deep within the geosphere.

Cooling and Crystallization

As shown in **Figure 14**, magma is located inside the geosphere. When magma flows out onto Earth's surface, it is called lava. Mineral crystals form as magma cools below the surface or as lava cools on the surface. This crystallization changes the molten material into igneous rock.

Uplift

Even rocks formed deep within Earth can eventually be exposed at the surface. **Uplift** *is the process that moves large bodies of Earth materials to higher elevations.* Uplift is often associated with mountain building. After millions of years of uplift, rocks that formed deep below Earth's surface could move up to the surface.

Reading Check How can a rock buried deep within Earth eventually reach the surface?

The Rock Cycle

Figure 14 As rocks move slowly through the rock cycle, they change from one form to another.

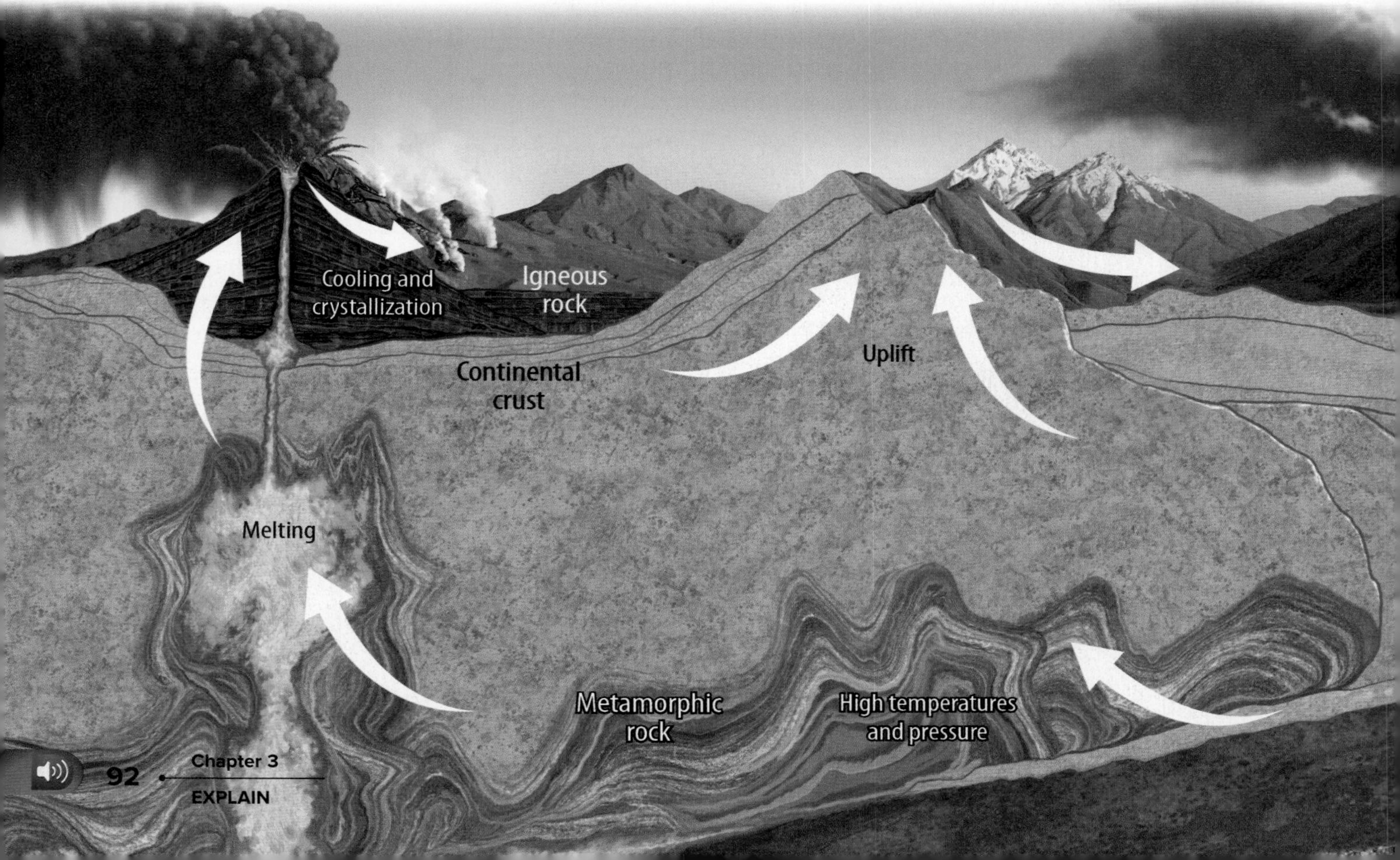

Weathering and Erosion

Rocks on Earth's surface are exposed to the atmosphere, the hydrosphere, and the biosphere. Glaciers, wind, and rain, along with the activities of some organisms, break down rocks into sediment. This process is called weathering. In **Figure 14,** weathering is shown in the mountains, where uplift has exposed rocks. Weathering of rocks into sediments is often accompanied by erosion. Erosion occurs when the sediments are carried by agents of erosion–water, wind, or glaciers–to new locations.

ACADEMIC VOCABULARY

process
(noun) a natural phenomenon marked by gradual changes that lead toward a particular result

Deposition

Eventually, agents of erosion lose their energy and slow down or stop. When this occurs, eroded sediments are deposited, or laid down, in new places. Deposition forms layers of sediment. Over time, more and more layers are deposited.

Compaction and Cementation

As more layers of sediment are deposited, their weight pushes down on underlying layers. The deeper layers are compacted. Minerals dissolved in surrounding water crystallize between grains of sediment and cement the sediments together. Compaction and cementation produce sedimentary rocks.

Visual Check How do weathering and erosion change rocks?

FOLDABLES®

Make a pyramid book, and label it as shown. Use it to organize your notes on Earth-system interactions including the water cycle, the rock cycle, and weather.

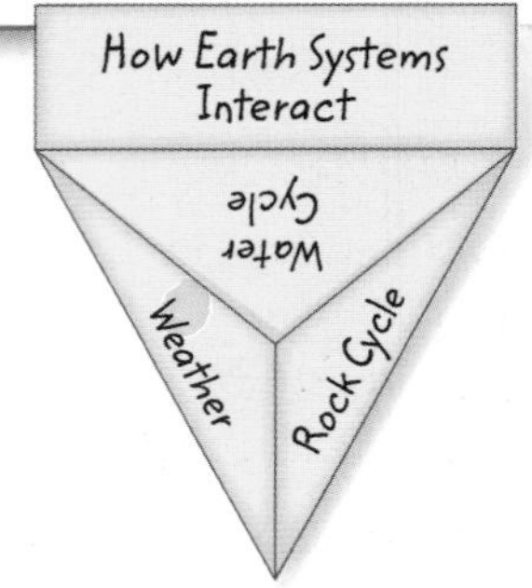

High Temperatures and Pressure

Metamorphic rocks form when rocks are subjected to high temperatures and pressure. This usually occurs far beneath Earth's surface. Igneous, sedimentary, and even metamorphic rocks can become new metamorphic rocks. Then, uplift can bring the rocks to the surface. There, the rocks are broken down and continue moving through the rock cycle.

Most interactions between the geosphere, the hydrosphere, and the atmosphere occur on Earth's surface. The atmosphere and hydrosphere alter rocks in the geosphere, and the geosphere in turn alters the other Earth systems. For example, energy from the Sun reaches Earth. The energy is reflected by Earth's surface and heats the atmosphere.

These are just a few examples of different interactions among Earth's systems. You have read about four different Earth systems in this chapter. But as **Figure 15** shows, the systems interact and function together as one unified system–planet Earth.

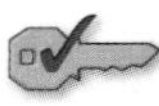

Key Concept Check How do Earth systems interact in the rock cycle?

Figure 15 Earth is a unified system made of four interacting subsystems.

Earth Systems

Personal Tutor

Atmosphere:
layer of gases surrounding Earth

Hydrosphere:
water found on Earth

Geosphere:
Earth's entire solid body

Biosphere:
all living organisms on Earth

(tl)Brand X Pictures/PunchStock, (tr)Gary Vestal/Getty Images, (c)NASA, (bl)Corbis/SuperStock, (br)age fotostock/SuperStock

Lesson 2 Review

Visual Summary

In the water cycle, water continually moves through the hydrosphere, the atmosphere, the geosphere, and the biosphere.

Weather and climate are influenced by interactions between the atmosphere and the other Earth systems.

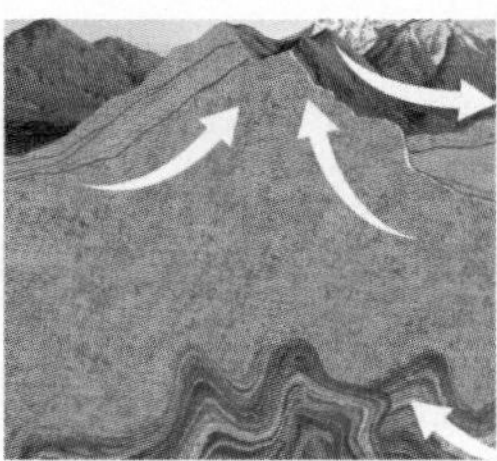

In the rock cycle, rocks continually change from one form to another.

FOLDABLES®

Use your lesson Foldable to review the lesson. Save your Foldable for the project at the end of the chapter.

What do you think NOW?

You first read the statements below at the beginning of the chapter.

4. The water cycle begins in the ocean.

5. Earth's air contains solids, liquids, and gases.

6. Rocks are made of minerals.

Did you change your mind about whether you agree or disagree with the statements? Rewrite any false statements to make them true.

Use Vocabulary

1. **Distinguish** between weather and climate.
2. **Define** the *water cycle* in your own words.
3. The process that changes liquid water to water vapor is ________.

Understand Key Concepts 

4. Which is an example of an interaction between the atmosphere and the geosphere?
   A. breathing
   B. ocean currents
   C. storms
   D. weathering
5. **Outline** Make an outline about the rock cycle. Include information about processes, rock types, and interactions with Earth systems.
6. **Compare** how the hydrosphere affects weather and how it affects climate.

Interpret Graphics

7. **Organize Information** Copy and fill in the graphic organizer below. Identify the processes of the water cycle.

Critical Thinking

8. **Design** a model that shows an interaction between two Earth systems.
9. **Assess** Some gasoline was spilled in a driveway. Could the pollutant pose a problem for the hydrosphere? Why or why not?

Math Skills

Math Practice

10. Air at 20°C has a vapor density of 8.65 g/m^3. The maximum amount of vapor density at that temperature is 17.3 g/m^3. What is the relative humidity?

 Lab

1 hour

How do Earth's systems interact?

Materials

water

lamp

sand

table fan

Safety

You've learned about the rock cycle and the water cycle. These are just two examples of how Earth systems work together. Each system interacts with the others to help maintain an ecological balance on Earth. What happens if one system is disrupted?

Ask a Question

How does a change in one system affect other systems? How can you model interactions among Earth systems?

Make Observations

1. Read and complete a lab safety form.
2. Think about Earth's four systems and how they interact with each other. In your Science Journal, describe a real-world scenario that shows these interactions. The photos on the next page show examples of real-world scenarios.
3. Use the materials shown here, or make a list of your own materials. Then, design a model of your scenario. Think about the following as you plan your model:
   - How can you represent each of Earth's systems?
   - How will you show the systems interacting?
   - Will your model be self-contained or open to the air?
4. After your teacher approves your design, build your model according to your design plans.

(t to b)(1) McGraw-Hill Education, (2,3) Ken Cavanagh/McGraw-Hill Education, (4)Jules Frazier/Getty Images, (5)Hutchings Photography/Digital Light Source

Form a Hypothesis

5 After building your model, formulate a hypothesis on how a change in one system might affect the other systems.

Test Your Hypothesis

6 Add or take away something in your model to cause one system to change. Is the change realistic? Could this happen in real life?

7 Observe and record the results immediately after the change occurs. Examine your model again on the following day. Be sure to record the results.

Analyze and Conclude

8 **Identify** Which parts of your model represent each system?

9 **Summarize** how the change you made to one system affected the others.

10 **Interpret** Was the change you modeled helpful or harmful? Was it caused by human activities or natural events? Explain.

11 **The Big Idea** Earth is sometimes described as a rocky planet. Based on what you observed in this lab, does that statement accurately describe Earth? Why or why not?

Communicate Your Results

Take your classmates on a "tour" of your model. Point out each Earth system, explain your hypothesis, recreate the change you introduced, and describe your results. Invite your classmates to ask questions and offer suggestions about improving your model.

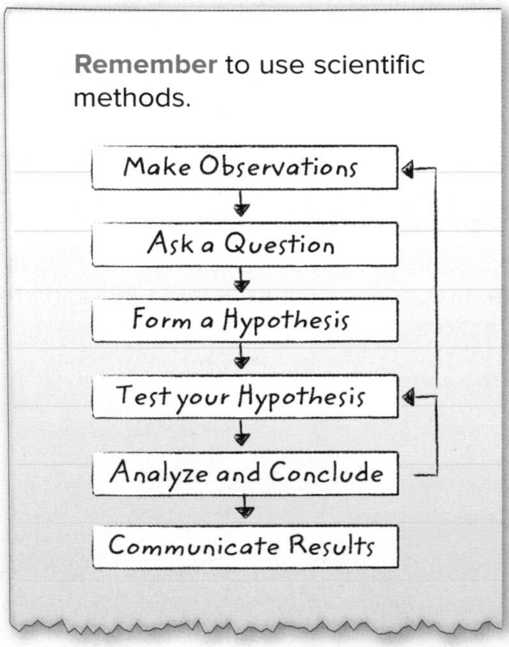

Inquiry Extension

Conduct research to locate a place where the change you observed in your model has occurred. Find out what impact it had on the living things in the area. Determine if the change is still impacting life in the area.

(t)DEA/C. SAPPA/Getty Images, (c) Maria Stenzel/Getty Images, (b) D.W. Peterson/USGS

Chapter 3 Study Guide

Earth is a unified system that can be modeled by dividing it into four interacting subsystems: the biosphere, the atmosphere, the hydrosphere, and the geosphere.

Key Concepts Summary

Lesson 1: Earth Systems

- Earth is made of the **biosphere,** the **atmosphere,** the **hydrosphere,** and the **geosphere.**
- The atmosphere has a layered structure that includes the troposphere, the stratosphere, the mesosphere, the thermosphere, and the exosphere. It is made of nitrogen, oxygen, and trace gases.
- Water is found on Earth in oceans, lakes, rivers, and as ice and **groundwater.** Small amounts of water are also found within the atmosphere and the biosphere.
- The geosphere is made of soil, metal, and **rock.** It has a layered structure that includes the crust, the mantle, and the core.

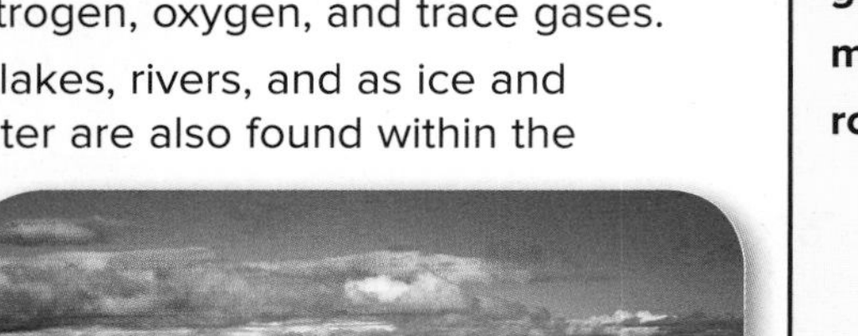

Vocabulary

biosphere p. 76
atmosphere p. 77
hydrosphere p. 79
groundwater p. 80
geosphere p. 81
mineral p. 81
rock p. 82

Lesson 2: Interactions of Earth Systems

- The **water cycle** shows how water moves between reservoirs of the hydrosphere, the atmosphere, the geosphere, and the biosphere.
- **Weather** and **climate** are influenced by transfers of water and energy among the atmosphere, the geosphere, and the hydrosphere.
- Rocks continually change form as they move through the **rock cycle.** Processes such as weathering and erosion are examples of interactions among Earth systems.

Vocabulary

water cycle p. 87
evaporation p. 88
transpiration p. 88
condensation p. 89
precipitation p. 89
weather p. 90
climate p. 91
rock cycle p. 92
uplift p. 92

Vocabulary eFlashcards
Vocabulary eGames

Personal Tutor

FOLDABLES® Chapter Project

Assemble your lesson Foldables as shown to make a Chapter Project. Use the project to review what you have learned in this chapter.

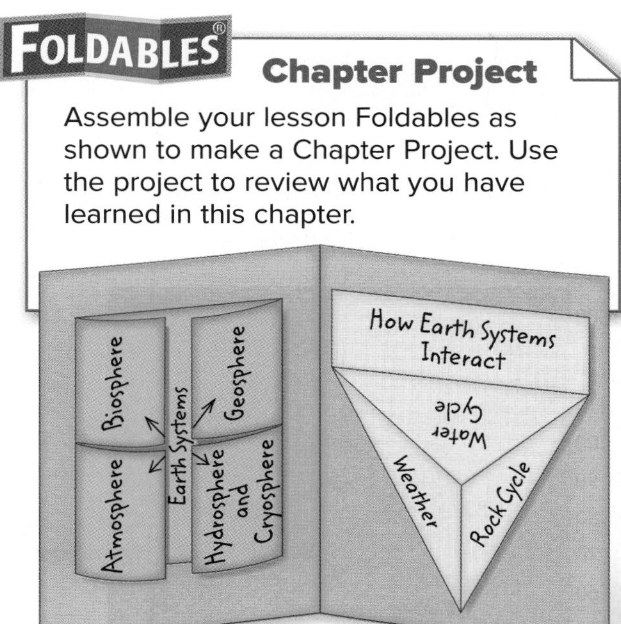

Use Vocabulary

1. The Earth system containing all living things is the ________.
2. Use the term *mineral* in a sentence.
3. Distinguish between rocks and minerals.
4. Conditions in the atmosphere at a given time and place are called ________.
5. Define the word *uplift* in your own words.
6. Distinguish between condensation and precipitation.

Link Vocabulary and Key Concepts

Interactive Concept Map

Copy this concept map, and then use vocabulary terms from the previous page to complete the concept map.

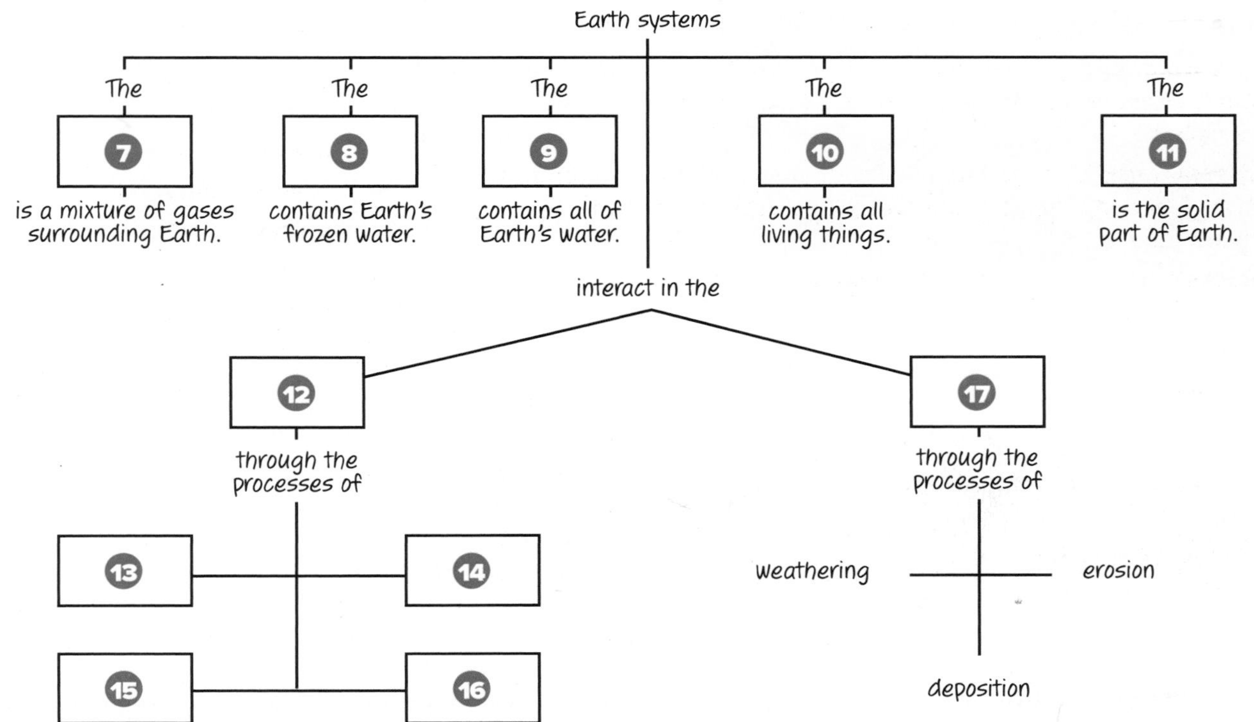

Chapter 3 Review

Understand Key Concepts

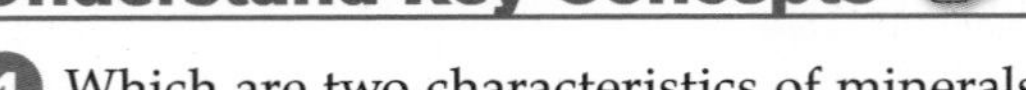

1. Which are two characteristics of minerals?
   A. artificial and organic
   B. liquid and gas
   C. living and inorganic
   D. solid and natural

2. What are the major gases of the atmosphere?
   A. carbon dioxide and water vapor
   B. nitrogen and carbon dioxide
   C. nitrogen and oxygen
   D. oxygen and water vapor

3. Which reservoir holds the largest amount of freshwater?
   A. groundwater
   B. ice
   C. lakes
   D. rivers

4. The diagram below shows the water cycle. Which number represents precipitation?

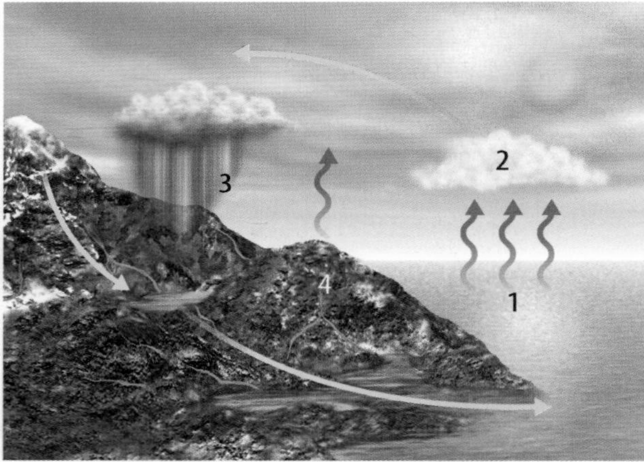

   A. 1
   B. 2
   C. 3
   D. 4

5. In which layer of the atmosphere does weather occur?
   A. hydrosphere
   B. mesosphere
   C. stratosphere
   D. troposphere

6. What does the hydrosphere contain?
   A. air
   B. plants
   C. soil
   D. water

7. The diagram below shows the layers of the atmosphere. The arrow is pointing to which layer?

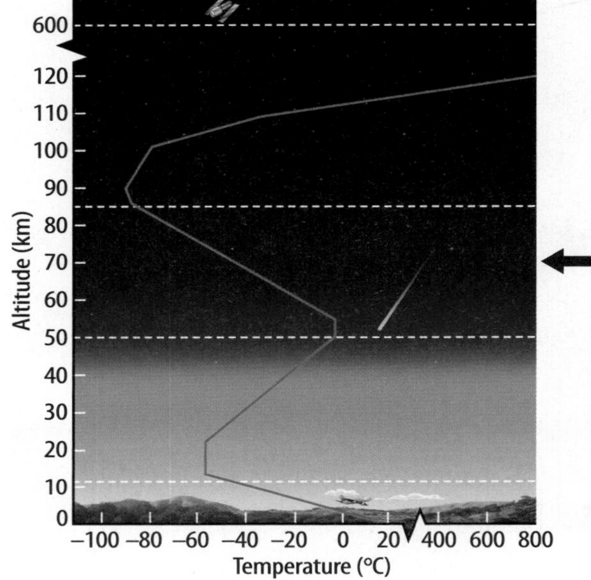

   A. troposphere
   B. mesosphere
   C. stratosphere
   D. exosphere

8. What is the middle layer of the geosphere?
   A. inner core
   B. crust
   C. mantle
   D. core

9. Rocks are classified according to
   A. color.
   B. formation.
   C. size.
   D. structure.

Critical Thinking

10 **Give** an example of how the water cycle impacts the rock cycle.

11 **Construct** Describe how you might construct a terrarium that models Earth systems.

12 **Design** Based on what you have learned about the water cycle, design a device for turning salt water into freshwater.

13 **Assess** How does the geosphere affect organisms that live in an ocean?

14 **Infer** How might the distribution of freshwater on Earth change if surface temperatures decreased?

15 **Evaluate** the relationship between weathering and erosion. How do the processes work together to change Earth's surface? How might the surface be different if only one of these processes occurred?

16 **Simplify** The diagram below shows the path of one rock through the rock cycle. What terms are missing from the diagram? Use the terms to describe how the rock changed.

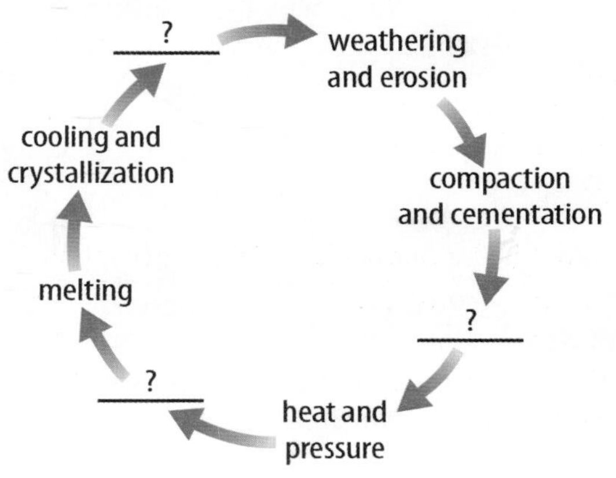

Writing in Science

17 **Create** A haiku is a poem with three lines. The lines contain five, seven, and five syllables respectively. Create a haiku that describes interactions among Earth systems.

REVIEW THE BIG IDEA

18 How would you describe Earth to a younger student?

19 What Earth systems do you see in the photo? What does each system include?

StockTrek/Getty Images

Math Skills

Math Practice

Use a Formula

Use the data in the table below to answer questions 20–22.

Temperature (°C)	Maximum Vapor Density (g/m^3)
10	9.4
24	23.0
30	30.4

20 The current temperature is 24°C. The water vapor in the air has a density of 5.75 g/m^3. What is the relative humidity?

21 At a temperature of 30°C, the air contains 22.8 g/m^3 of water vapor. What is the relative humidity?

22 Based on the data in the table, what is the relationship between the temperature and the amount of water vapor air can contain?

Standardized Test Practice

Record your answers on the answer sheet provided by your teacher or on a sheet of paper.

Multiple Choice

1 Which of Earth's systems includes the crust, the mantle, and the core?

A atmosphere

B biosphere

C geosphere

D hydrosphere

2 How much of Earth's water is freshwater?

A 1 percent

B 3 percent

C 79 percent

D 97 percent

Use the diagram below to answer question 3.

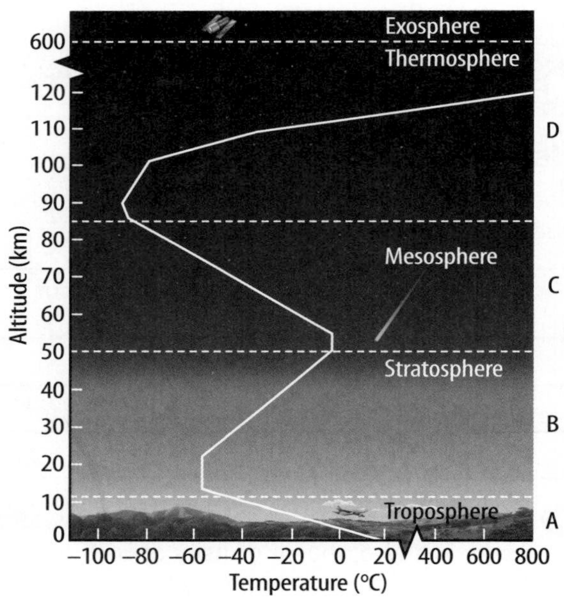

3 Earth's ozone layer absorbs solar radiation, protecting the biosphere. Which atmospheric layer includes the ozone layer?

A A

B B

C C

D D

4 Through which process does water leave the hydrosphere and enter the atmosphere?

A condensation

B deposition

C evaporation

D precipitation

5 Though the geosphere is described as the solid part of the Earth, which part is liquid?

A crust

B inner core

C mantle

D outer core

Use the image below to answer question 6.

6 Which process is occurring in the area circled in the figure?

A condensation

B deposition

C precipitation

D transpiration

7 Which process recycles water from the biosphere to the atmosphere?

A condensation

B deposition

C precipitation

D transpiration

8 Which sequence accurately shows the events that form sedimentary rock?

A compaction → cementation → melting

B erosion → volcanic eruption → weathering

C volcanic eruption → cooling → crystallization

D weathering → erosion → deposition

Use the diagram below to answer question 9.

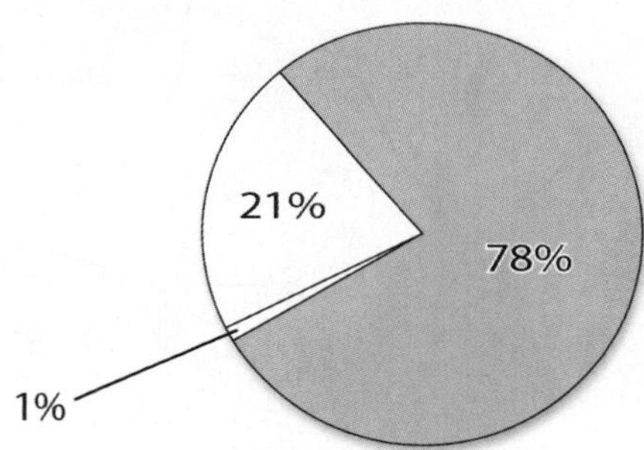

9 Which gas is represented by the shaded portion on the pie chart?

A carbon dioxide

B nitrogen

C oxygen

D water vapor

Constructed Response

10 Describe the path an igneous rock could take through the rock cycle. Begin and end with an igneous rock.

Use the figure below to answer questions 11 and 12.

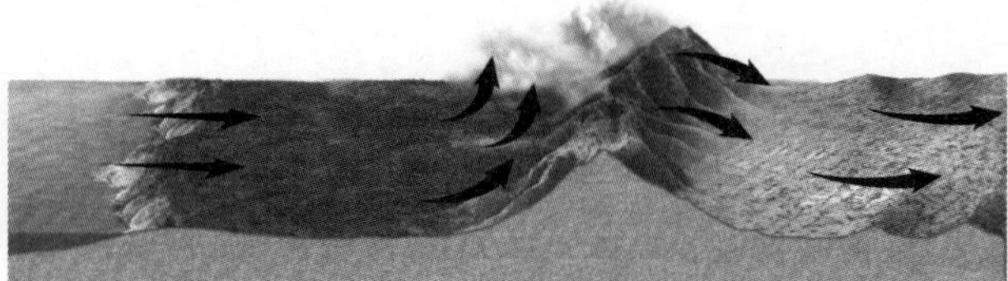

11 Use the figure to describe why the weather differs on the left and right sides of the mountains.

12 Describe how the hydrosphere, the atmosphere, and the geosphere interact to produce the rain-shadow effect in the figure.

13 Millions of years ago, a dinosaur might have drunk the same water that you drink today. Explain how this is possible.

14 Describe how the hydrosphere, the atmosphere, the biosphere, and the geosphere interact in the rock cycle to form sedimentary rock.

NEED EXTRA HELP?														
If You Missed Question...	1	2	3	4	5	6	7	8	9	10	11	12	13	14
Go to Lesson...	1	1	1	2	1	2	2	2	1	2	2	2	2	2

Chapter 4

Earth's Dynamic Surface

THE BIG IDEA What processes change Earth's surface?

Inquiry Can you break down a mountain?

Few people can break rocks apart with their hands. Usually they need hammers or other tools because rocks and mountains are solid and strong. However, something has affected the strength of this mountain.

- How does an entire mountain break apart?
- Can you see this as it happens?
- What processes change Earth's surface?

Get Ready to Read

What do you think?

Before you read, decide if you agree or disagree with each of these statements. As you read this chapter, see if you change your mind about any of the statements.

1. Earth's surface is made up of tectonic plates.
2. Tectonic plate motion is too slow to measure.
3. Most earthquakes occur near tectonic plate boundaries.
4. Volcanoes can erupt anywhere.
5. Wind erosion only occurs in the desert.
6. Rivers are the only cause of erosion.

Your one-stop online resource
connectED.mcgraw-hill.com

LearnSmart®

Chapter Resources Files, Reading Essentials, Get Ready to Read, Quick Vocabulary

Animations, Videos, Interactive Tables

Self-checks, Quizzes, Tests

Project-Based Learning Activities

Lab Manuals, Safety Videos, Virtual Labs & Other Tools

Vocabulary, Multilingual eGlossary, Vocab eGames, Vocab eFlashcards

Personal Tutors

Lesson 1

Earth's Moving Surface

Reading Guide

Key Concepts
ESSENTIAL QUESTIONS

- What is the theory of plate tectonics?
- What are the differences between divergent, convergent, and transform plate boundaries?
- What causes tectonic plates to move on Earth's surface?

Vocabulary

plate tectonics p. 107
lithosphere p. 108
asthenosphere p. 108
divergent boundary p. 109
convergent boundary p. 109
subduction p. 109
transform boundary p. 109
convection p. 111

Multilingual eGlossary

What's Science Got to do With It?

Why are these mountains here?

The Sierra Nevada, the Cascade Range, and the Andes are mountain ranges along the west coasts of North America and South America. The mountains shown here are the Southern Alps on South Island, New Zealand. They are in the middle of the island. Why are some mountains on coasts and others in the middle of a continent?

Launch Lab

15 minutes

How can movement deep within Earth change its surface?

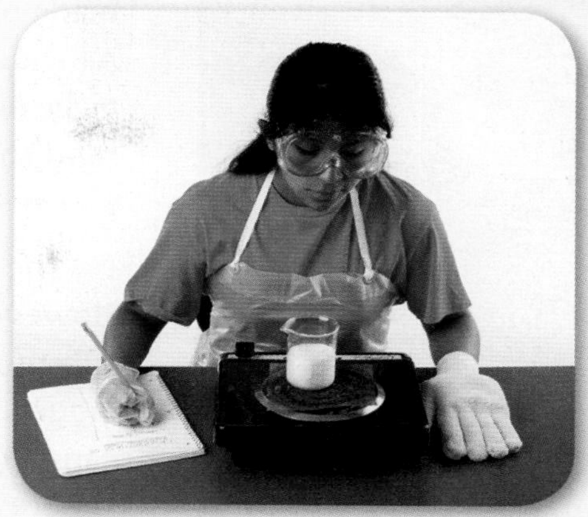

Have you ever warmed a pot of soup on a stove ? If so, you probably noticed that the noodles and vegetables moved around the pot. How is this movement similar to movement within Earth?

1. Read and complete a lab safety form.
2. Pour the **fluid** into a **250-mL glass beaker** until the beaker is about half full.
3. Place the beaker in the center of a **hot plate.**
4. Turn the hot plate on low, and slowly warm the fluid.
5. When the fluid starts to move, observe the beaker from the top and the side. Record your observations in your Science Journal.

Think About This

1. Describe, in detail, how the fluid moves as it warms.
2. **Key Concept** Based on your observations, how do you think this relates to movement within Earth?

Plate Tectonics

Have you ever looked at a map like the one shown in **Figure 1**? Did you notice that the surface is not the same everywhere? Some regions have tall, rugged mountains. Some regions have flat plains. What causes different landforms? What processes shape and change Earth's surface?

During the 1960s, scientists developed a theory to explain many of the features on Earth's surface. The theory is called **plate tectonics** (tek TAH nihks) and states that *Earth's surface is broken into large, rigid pieces that move with respect to each other.* These pieces, or tectonic plates, move slowly over Earth's surface. You will read how tectonic plate motion forms volcanoes and mountains and causes earthquakes. The theory of plate tectonics was revolutionary because it explained much about how Earth changes.

Key Concept Check What is the theory of plate tectonics?

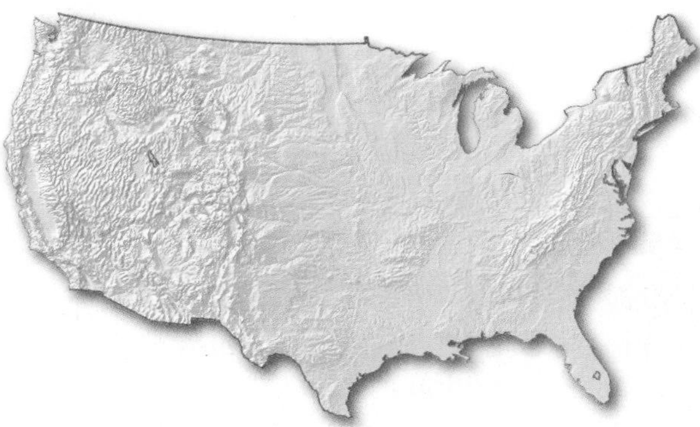

Figure 1 Tectonic plate motion and other processes have affected North America. The western United States has tall mountains, while the central region is flat.

Word Origin

tectonic
from Greek *tekton*, means "builder"

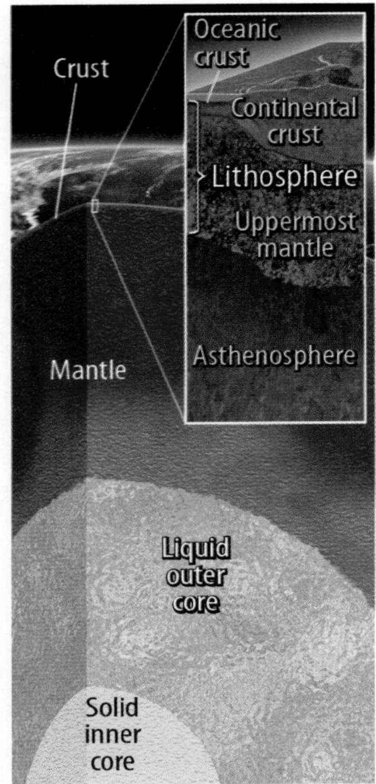

▲ **Figure 2** The lithosphere consists of the crust and the upper part of the mantle. The lithosphere is above the hotter asthenosphere.

What is a tectonic plate?

You might know that Earth is not a solid ball of rock but is made of layers of material. The outermost layer is called the crust. As shown in **Figure 2**, *the crust and uppermost part of the mantle, make up the* **lithosphere** (LIH thuh sfihr). The lithosphere forms a rigid shell on the outside of Earth. However, it is broken into large pieces–tectonic plates.

The rocks in the lithosphere are strong and do not bend easily. However, *the partially melted portion of the mantle below the lithosphere is the* **asthenosphere** (as THEN uh sfihr), also shown in **Figure 2**. The asthenosphere is hotter than the lithosphere and can bend more easily. As you will read, the ability of the asthenosphere to bend is related to tectonic plate movement.

Major Tectonic Plates

Scientists have identified 15 large tectonic plates, as shown in **Figure 3**. Some plates are so large they support entire continents. Other plates are so small that they cannot be represented on a map of this scale.

Many of you live on the North American Plate. To the east of it is the Eurasian Plate. To the west are two plates–a small plate called the Juan de Fuca Plate and the largest plate, the Pacific Plate. Oceans completely cover some plates, such as the Juan de Fuca Plate. Other plates, such as the North American Plate, are made of both oceanic crust and continental crust.

Reading Check What is a tectonic plate?

Earth's Tectonic Plates

Figure 3 Earth's tectonic plates fit together like puzzle pieces. They are in constant motion across the surface.

Visual Check How many large plates have continents on them?

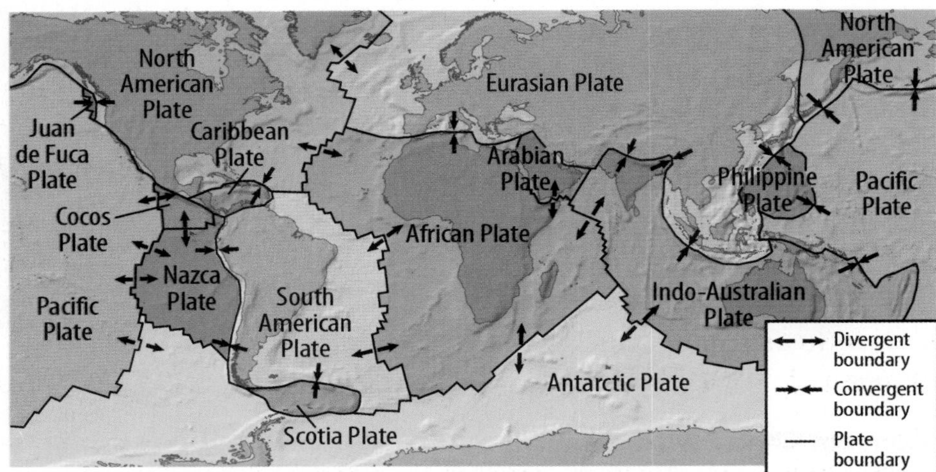

Divergent

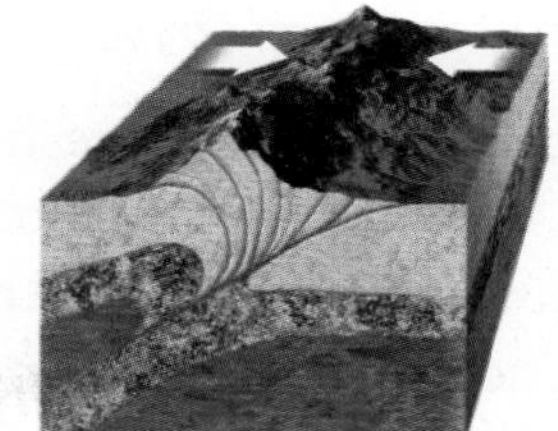
Convergent

Transform

▲ **Figure 4** The relative movement of tectonic plates creates three types of plate boundaries.

Animation

Plate Boundaries

How do scientists describe the movement of a tectonic plate? They describe a plate's relative motion–how it moves in relation to another plate. For example, the North American Plate is moving away from the Eurasian Plate, but it is also moving toward the Pacific Plate.

Place two pieces of paper side by side on your desk. How can you make one sheet move relative to the other? You can push the sheets together so one goes over or under the other. Or you might move them apart or slide them by each other. Tectonic plates move in a similar way, as illustrated in **Figure 4.** As plates move relative to each other, they form different types of boundaries. The type of boundary depends on the relative motion of the plates.

FOLDABLES®

Make a horizontal two-tab book with an extended tab. Label it as shown. Use it to summarize the causes and effects of tectonic plate movement.

Divergent Boundaries

A boundary where two plates move away from each other is called a **divergent boundary.** The boundary between the North American Plate and the Eurasian Plate is a divergent boundary. As plates move apart, new crust forms between them.

Convergent Boundaries

A boundary where two plates move toward each other is a **convergent boundary.** In some locations, one plate is pushed under the other plate and down into the mantle. That plate is eventually recycled into the mantle. **Subduction** *is the process that occurs when one tectonic plate moves under another tectonic plate,* as shown in **Figure 5.** The Pacific Plate is being subducted under the North American Plate at the convergent plate boundary adjacent to Alaska.

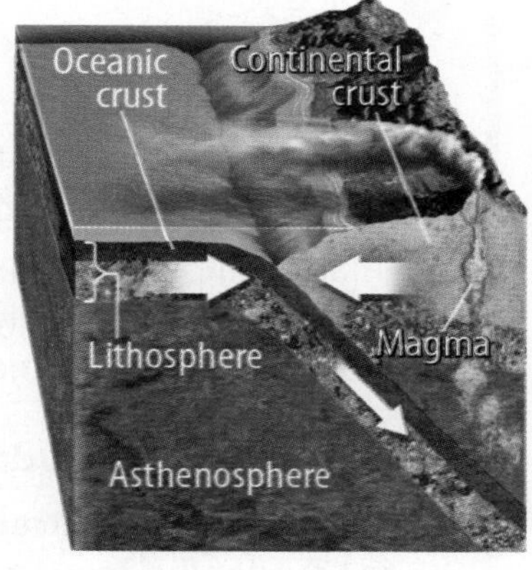

▲ **Figure 5** At a convergent boundary, the process of subduction forces one plate under the other.

Transform Boundaries

Two plates slide past each other at a **transform boundary.** The boundary between the Pacific Plate and the North American Plate in California is an example of a transform boundary.

Key Concept Check What are the differences between divergent, convergent, and transform plate boundaries?

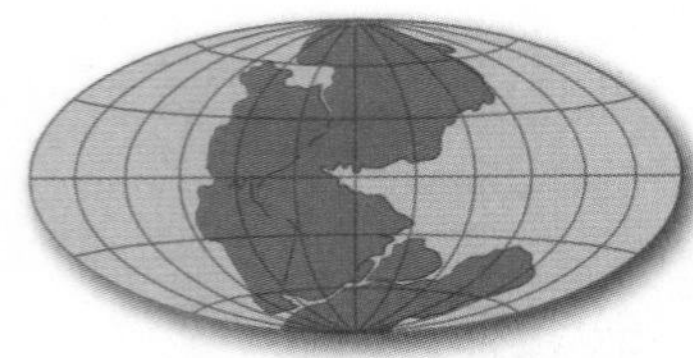

Figure 6 Over time, tectonic plate motion broke apart the supercontinent Pangaea.

**Visual Check** Where were North America and Europe joined as part of Pangaea?

Animation

Measuring Plate Movement

Tectonic plates move horizontally over Earth's surface. They move so slowly that before the mid-twentieth century, geologists could not measure their movement. However, during the 1970s, scientists and engineers developed new technologies that enabled them to measure how fast tectonic plates move. This technology has determined that North America is separating from Europe at an average rate of just 2.5 cm/y.

The position of any point on Earth's surface can be accurately measured using the network of satellites known as the Global Positioning System (GPS). GPS is a set of 24 satellites in orbit around Earth that send signals to help locate and track various moving objects. By tracking tectonic plate positions over several years, scientists can measure the speed and the direction of plate movement.

Even though plate movement is slow, dramatic changes occur over long periods of time. North America and Europe once were part of a large continent called Pangaea (pan GEE uh), illustrated in **Figure 6.** A divergent boundary formed between North America and Europe about 200 million years ago. The plates moved apart, and the Atlantic Ocean formed.

MiniLab

15 minutes

What happens at plate boundaries?

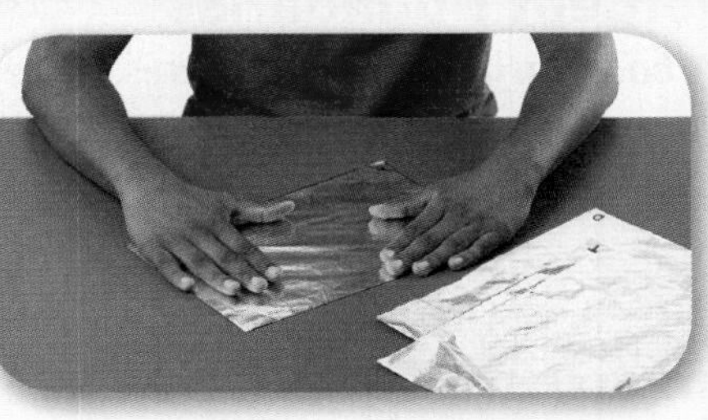

Most rocks seem hard and unchangeable. But, under the right conditions, they can bend or break. Can you model how rocks change along plate boundaries?

1. Read and complete a lab safety form.
2. Obtain three pieces of **foil,** each 20 cm long. Use a **marker** to mark the pieces with a *C,* a *D,* and a *T.*
3. Place foil C flat on your desktop. Place the palms of your hands on the foil as shown. Slowly move your hands toward each other. Observe and record the results in your Science Journal.
4. Place foil piece D flat on your desktop. Place your palms along the edges of the foil. Slowly pull outward on the foil. Observe and then record the results.
5. Place foil piece T on your desktop. Place your palms flat against each end of the foil. At the same time, slowly move your right hand toward you and your left hand away from you. Observe and then record the results.

Analyze and Conclude

1. **Compare and Contrast** How did each piece of foil change? How does the motion of the foil model the movement of plate boundaries?
2. **Key Concept** What are the differences between divergent, convergent, and transform plate boundaries?

Why do tectonic plates move?

You have read that tectonic plates move over Earth's surface. As the plates move and interact with each other, they form convergent, divergent, and transform boundaries. But what causes plates to move?

Convection

Recall that density is the amount of matter per unit of volume. When a fluid is heated, its molecules spread out. It has less matter in the same amount of volume. So, it becomes less dense. However, fluids do not heat evenly. Some of a fluid can be warmer and less dense, while some is cooler and more dense. The warmer, less dense fluid rises, and the cooler, denser fluid sinks, as shown in **Figure 7**. *The circulation within fluids caused by differences in density and thermal energy is called* **convection.**

Convection also occurs in Earth's asthenosphere, just below the lithosphere. Recall that rocks in the mantle are hot enough to bend easily. They can flow in a way similar to fluids. Convection in the mantle can drag plates over Earth's surface, as illustrated in **Figure 7**.

Two other forces also contribute to plate motion. Rising mantle material at mid-ocean ridges helps push plates away from the ridge with a force called ridge push. When two plates collide, one can sink under the other. When this happens, it pulls on the rest of the plate with a force called slab pull.

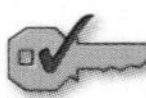 **Key Concept Check** What causes tectonic plates to move?

Convection Currents

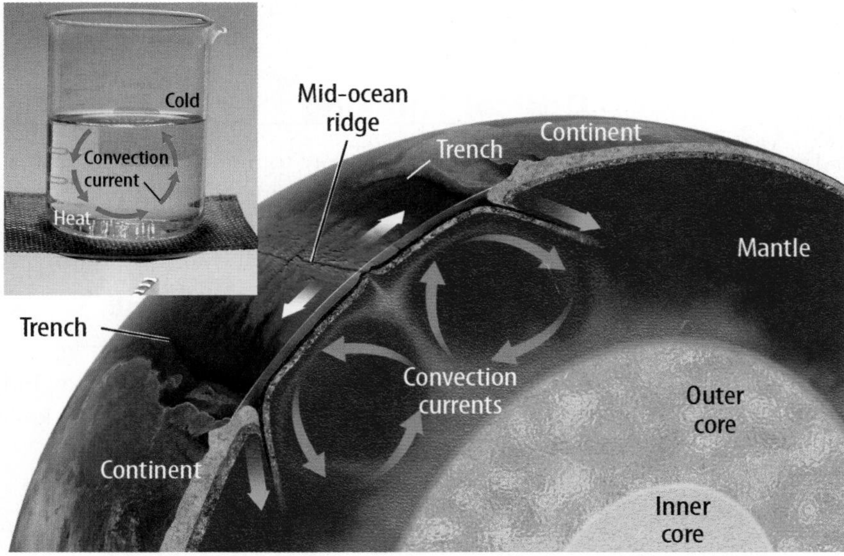

Figure 7 The left image shows convection in water—heated water rises to the surface, cools, and sinks back down. As shown on the right, heated rock in the mantle does the same. As the mantle convects, it pulls and pushes the tectonic plates above.

Math Skills 

Use Proportions

The plates along the Mid-Atlantic Ridge spread at an average rate of 2.5 cm/y. How long will it take the plates to spread 100 m? Use proportions to find the answer.

1. Convert the distances to the same unit.
   100 cm = 1 m so
   2.5 cm = 0.025 m
2. Set up a proportion.
   $\frac{0.05 \text{ m}}{1 \text{ y}} = \frac{100 \text{ m}}{x \text{ y}}$
3. Cross multiply and solve for x.
   $0.025\text{m} \times x \text{ y} = 100\text{m} \times 1 \text{ y}$
4. Divide both sides by 0.025 m.
   $x \text{ y} = \frac{100 \text{ m/y}}{0.025 \text{ m}}$
   $x = 4{,}000 \text{ y}$

Practice

The Eurasian Plate travels at about 0.7 cm/y. How long would it take the plate to travel 1 km?

(1 km = 100,000 cm)

 **Math Practice**

 **Personal Tutor**

Lesson 1 Review

Visual Summary

Earth's lithosphere is broken into pieces called plates. These plates move on Earth's surface and interact in three different ways.

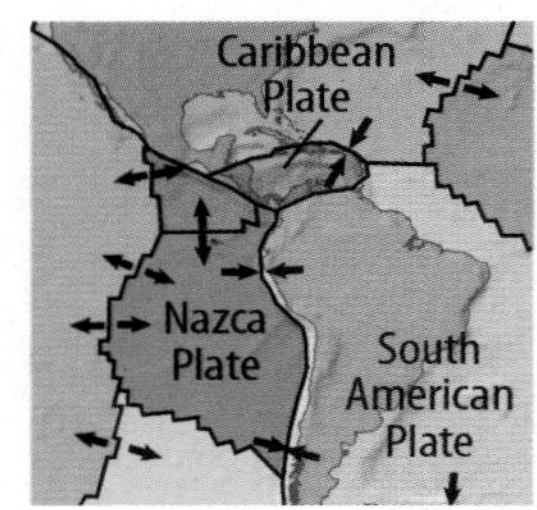

GPS can be used to measure and track plate movement.

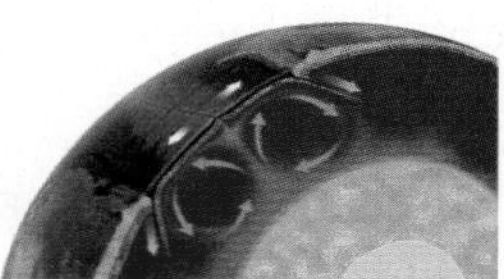

Convection in the mantle causes tectonic plates to move across Earth's surface.

FOLDABLES®

Use your lesson Foldable to review the lesson. Save your Foldable for the project at the end of the chapter.

What do you think NOW?

You first read the statements below at the beginning of the chapter.

1. Earth's surface is made up of tectonic plates.
2. Tectonic plate motion is too slow to measure.

Did you change your mind about whether you agree or disagree with the statements? Rewrite any false statements to make them true.

Use Vocabulary

1. **Describe** the asthenosphere.
2. **Identify** the type of movement that occurs at convergent boundaries.
3. **Define the term** *convection* in your own words.

Understand Key Concepts 

4. **Distinguish** between the mantle and the lithosphere.
5. **Contrast** the motions of plates at divergent and transform boundaries.
6. Which is not part of the lithosphere?
   - **A.** asthenosphere
   - **B.** continental crust
   - **C.** oceanic crust
   - **D.** upper mantle

Interpret Graphics

7. **Identify** What type of plate boundary is shown in the figure to the right?

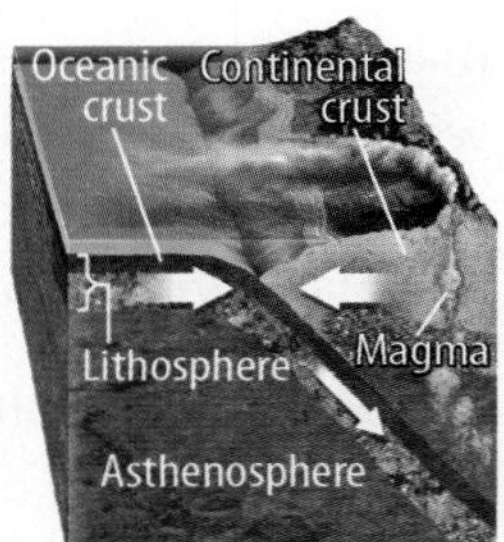

8. **Sequence** Use a graphic organizer to place these words in the correct sequence: *plate motion, convection, subduction, convergent boundary.*

Critical Thinking

9. **Recommend** a way to determine how fast North America is moving away from Europe without using GPS.

Math Skills

10. Mountains at a convergent boundary may build at a rate of 3 mm/y. How long would it take for a mountain to build to a height of 3,000 m? (1 m = 1,000 mm)

GPS and Plate Tectonics

HOW IT WORKS

Measuring Earth's Movement from Space

A satellite sends a radio signal to a GPS receiver.
The signal sends information about the satellite's location and current time.
An exact position is determined by comparing the information from 3 or more satellites.
San Andreas Fault
Direction of plate movement
N
Stationary receiver
Moving receiver
North American Plate
Pacific Plate

The shape of Earth's crust is always changing because tectonic plates are always moving. The Global Positioning System, or GPS, can be used to measure the distance, the speed, and the direction a tectonic plate moves.

How GPS Works GPS is made up of 24 satellites that orbit about 11,000 km above Earth's surface. The position of an object is determined by a GPS receiver's distance to a particular satellite. Distance is measured by determining the time it takes a one-way radio signal sent from the satellite to reach the receiver. To calculate an accurate position, receivers use measurements from three separate satellites using a process called triangulation. Position is described by latitude, longitude, and height.

Differential GPS, or dGPS, uses two receivers—one is stationary and another records ground movement. They are cemented into the ground, so they don't move unless the ground moves. dGPS receivers record east-west movement, north-south movement, and up-and-down movement. The stationary receiver calculates differences between measurements taken at the stationary received itself. Then it compares them to the measurements taken by the moving receiver to determine how much the moving receiver, and the tectonic plate, has moved.

A GPS receiver like this one is used to measure tectonic plate movement on top of Mount St. Helens. Measurements are taken as often as every 30 s. ▶

It's Your Turn

INVESTIGATE Is there tectonic activity in your area? Find GPS or dGPS data that might indicate how the land you live on is moving. Share what you learn with your class.

Lesson 2

Reading Guide

Key Concepts
ESSENTIAL QUESTIONS

- Where do most earthquakes occur?
- How are landforms related to plate tectonics?
- Where do most volcanoes form?
- How does plate movement form mountains?

Vocabulary
earthquake p. 115
fault p. 115
magma p. 118
lava p. 118
volcano p. 118
mid-ocean ridge p. 120

Multilingual eGlossary

Science Video
What's Science Got to do With It?

Shaping Earth's Surface

Inquiry Is the water on fire?

What type of liquid looks as if it is on fire? This is melted rock flowing out of a volcano on the island of Iceland. The position of Iceland makes it possible for volcanic activity such as this to occur. What do you think is unusual about Iceland that makes these events common?

Launch Lab

15 minutes

What happens when a volcano erupts?

Some volcanic eruptions are quiet. The lava flows out of the volcano and over Earth's surface. Other eruptions are explosive and send gases, lava, and pieces of rock high into the air.

1. Read and complete a lab safety form.
2. Cover your work area with **newspaper.** Place a **small plastic cylinder** in the center of the paper.
3. Use a **funnel** to pour one heaping spoonful of **baking soda** into the cylinder. Then add one spoonful of **small plastic beads** and five of the **larger plastic beads** to the cylinder.
4. Pour about 50 mL of **white vinegar** into a small **beaker.**
5. Pour the vinegar into the cylinder. Record your observations in your Science Journal.

Think About This

1. **Describe** What happens when you add the vinegar to the baking soda? How are the different-sized beads erupted?
2. **Key Concept** How do you think a volcano can change Earth's surface?

Earthquakes

An **earthquake** *is the vibrations caused by the rupture and sudden movement of rocks along a break or a crack in Earth's crust.* Earthquakes occur every day on Earth. The strong shaking of Earth's surface can damage both natural features and human-made structures.

Fault

Earthquakes can occur at faults. A **fault** *is a crack or a fracture in Earth's crust along which movement occurs.* One place where a fault can exist is at a plate boundary. Tectonic plates do not continually slide past each other along faults. But, because of the convection currents beneath the tectonic plates, forces build up along faults. Eventually, these forces become so great that the rocks on either side of the fault move and slide along the fault, as shown in Figure 8. When this happens, the fault is said to rupture, and Earth's crust moves along the fault, causing an earthquake.

Reading Check What is the relationship between faults and plate boundaries?

Figure 8 Blocks of crust move along the surface at a fault. This is an aerial view of a transform fault in Iran.

Visual Check In which direction are the plates moving?

Interactive Table

Table 1 Earthquake Magnitudes

Magnitude	Average Number per Year	Typical Fault Length on Surface	Typical Movement on Fault
3	>100,000	15 m	1 mm
4	15,000	100 m	5 mm
5	3,000	800 m	3 cm
6	100	6 km	20 cm
7	20	40 km	1 m
8	2	300 km	6 m

▲ **Table 1** Earthquakes occur every day, and there are more than a few major earthquakes each year.

WORD ORIGIN

magnitude
from Latin *magnitudo,* means "great bulk or size"

SCIENCE USE V. COMMON USE

fault
Science Use a fracture in the crust of a planet

Common Use responsibility for wrongdoing or failure

Where Earthquakes Occur

Most earthquakes occur at plate boundaries. Plate boundaries are long and do not rupture all at once. Instead, usually only small segments rupture. As shown in **Table 1,** small earthquakes occur more frequently than large ones. The size of an earthquake is determined by how much energy is released during the earthquake. This is called **magnitude** and can range from less than one to at least 9.9.

A plate boundary is made up of more than one **fault.** The boundary covers a large region, and many smaller faults can branch out from the main fault. Faults can be many kilometers from the plate boundary. Earthquakes can occur on these remote faults, just as they do on faults at plate boundaries.

Faults are largest where one plate subducts into the mantle. The strongest and most damaging earthquakes occur at these locations. Higher magnitude earthquakes occur when movement along faults covers large distances. For example, in 2004 the boundary between two plates, shown in **Figure 9,** ruptured. The earthquake had a magnitude greater than 9 and devastated the country of Sumatra.

 **Key Concept Check** Where do most earthquakes occur?

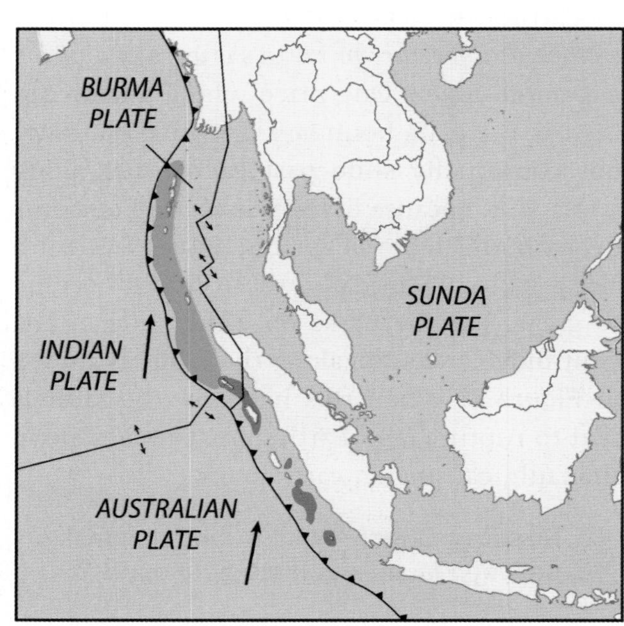

Figure 9 The convergent boundary between the Indian Plate and the Burma Plate ruptured the greatest fault length ever recorded, spanning 1,500 km. The area of the rupture is indicated in green on the map. Smaller earthquakes have occurred in the purple areas. ▶

Visual Check How does the size of the 2004 earthquake compare to the other earthquakes?

How Earthquakes Change Earth's Surface

The movement of crust along faults can make mountains, valleys, and other landforms. Different types of movement occur at the three types of plate boundaries. These are illustrated and described in **Table 2**.

Key Concept Check How are landforms related to plate tectonics?

Table 2 Changes at Plate Boundaries

Plate Boundary	Relative Motion	Example
Transform boundary At a transform boundary, blocks of crust move horizontally past each other. Features that cross the fault, such as streams, are shifted both by plate movement and by earthquakes. Transform faults also are called strike-slip faults.	Transform fault Continental crust Continental crust Lithosphere	
Divergent boundary At divergent boundaries between oceanic plates, mid-ocean ridges form, as illustrated to the right. Between continental plates, one side of the fault moves down relative to the other side of the fault. Normal faults form valleys at these boundaries, as shown in the photograph to the right.	Mid-ocean ridge Rift valley Oceanic crust Lithosphere Asthenosphere	
Convergent boundary—subduction zone At these types of boundaries, the plate that does not subduct deforms and crumples as the two plates push toward each other. As the mantle near the subducted plate melts, magma rises and forms a volcanic arc on the plate that does not subduct.	Deep ocean trench Volcanoes Oceanic crust Lithosphere Continental crust Asthenosphere	
Convergent boundary—no subduction At these types of boundaries, the edges of both tectonic plates become crumpled and deformed. Because neither plate subducts, blocks of crust slide upward along a complex series of faults called reverse faults. This results in the formation of tall mountains.	Mountains Continental crust Lithosphere Asthenosphere	

MiniLab

15 minutes

What are three types of fault motion?

A fault is a fracture in Earth's crust along which movement has occurred. There are three main types of faults. Can you model these faults and relate them to the movement at plate boundaries?

1. Read and complete a lab safety form.
2. Obtain a **pair of blocks.** What do the different colors represent? Where is the fault?
3. Line up the rock beds. Model a fault by moving the block above the fault plane down about 5 cm. In your Science Journal, use **colored pencils** to draw and label the positions of the two blocks.
4. Line up the rock beds. Model another fault by moving the block above the fault plane up about 5 cm. Draw and label the positions of the two blocks.

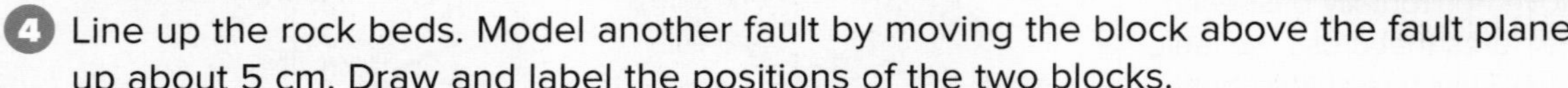

5. Model a third fault by lining up the blocks, then moving one block to the side. Do not move the block vertically. Draw and label the positions of the blocks.

Analyze and Conclude

1. **Cause and Effect** Identify the three types of faults modeled. Along which type of plate boundary does each of these faults occur?
2. **Key Concept** How are faults related to plate tectonics?

FOLDABLES®

Make a vertical three-tab Venn book. Label it as shown. Use it to compare and contrast earthquakes and faults.

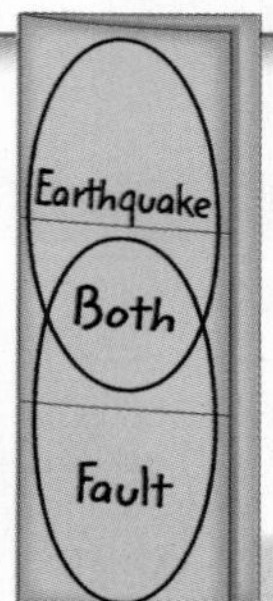

Volcanoes

The temperature inside Earth is hot enough to melt rock. Geologists call *molten rock stored beneath Earth's surface* **magma.** **Lava** *is magma that erupts onto Earth's surface.* **Volcanoes** *are vents in Earth's crust through which molten rock flows.* Volcanoes are common on Earth. During the last 10,000 years, more than 1,500 different volcanoes have erupted. Although they are common, volcanoes do not form everywhere.

Where Volcanoes Occur

Most volcanoes form at convergent plate boundaries. Recall that at some convergent boundaries, one plate subducts under another plate. Some rocks contain water within their structure. As the rocks subduct, heat and pressure drive the water out. This water can lower the melting temperature of the mantle above the subducting plate. Magma then rises toward the surface and forms volcanoes on the plate that does not subduct. A line of volcanoes forms parallel to the plate boundary directly above the plate that subducted. The volcanoes in Washington and Oregon, such as Mount Rainier, Mount St. Helens, and Mount Hood, formed above the subducting Juan de Fuca Plate.

Key Concept Check Where do most volcanoes form?

Hutchings Photography/Digital Light Source

How Volcanoes Change Earth's Surface

Volcanoes are some of Earth's most distinctive landforms. Compared to other mountains, volcanoes can form quickly. Mountains can form over millions of years, but volcanoes can form in hundreds to thousands of years. Sometimes it happens even more quickly. Paricutín volcano, shown in **Figure 10**, formed within one year.

▲ **Figure 10** Within one year, Paricutín volcano in Mexico grew 365 m above its surroundings.

Volcanoes erupt in two ways. Sometimes, lava can flow over Earth's surface before cooling, hardening, and becoming solid rock, as pictured in **Figure 11**. This is called a lava flow. Lava flows can be more than 10 km long and, over time, can cover large areas.

Figure 11 Lava flows can slowly cover the region surrounding a volcano. ▼

At other times, volcanoes can erupt explosively. Much of Mount St. Helens, shown in **Figure 12**, was destroyed during an eruption in 1980. This kind of eruption can produce tiny pieces of glass made from solidified lava. These pieces are called ash and can be blown high into the atmosphere. When the ash falls back to Earth's surface, it can cover vast areas. Ash from Mount St. Helens in Washington fell as far away as Minnesota and Oklahoma.

Reading Check How can volcanoes change Earth's surface?

Figure 12 The image on the left shows Mount St. Helens before it erupted. In 1980, the volcano's explosive eruption destroyed part of the mountain, as shown in the image on the right. ▼

Formation of Oceanic Crust 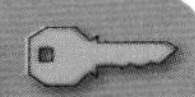

Figure 13 At a divergent plate boundary, magma rises between two plates and forms new crust. A long mountain range called a mid-ocean ridge forms at a divergent boundary beneath the ocean.

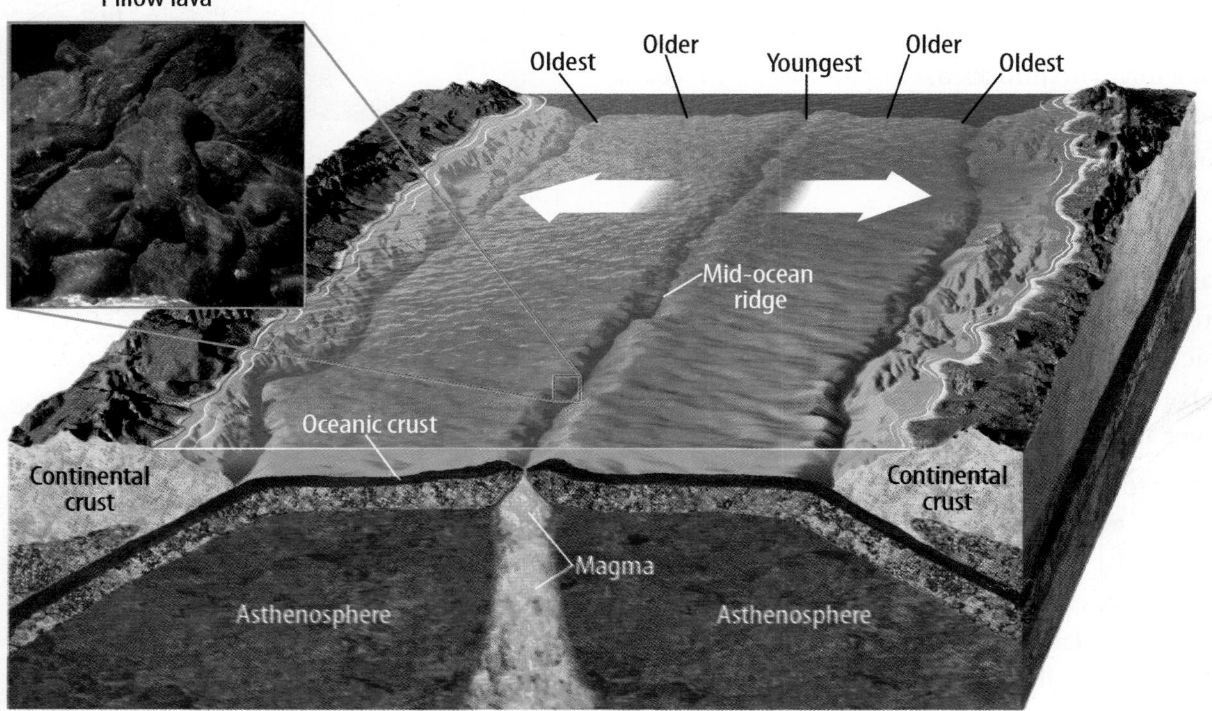

Ocean Basins

Lava does not erupt only from volcanoes on land. Recall that the land masses that make up North America and Europe separated 200 million years ago when a divergent plate boundary formed between them. What happened in the area between these land masses?

Lava on the Ocean Floor

Lava erupts at both convergent plate boundaries and divergent plate boundaries, as shown in **Figure 13.** This lava hardens and forms new crust. At an oceanic divergent plate boundary, the newly formed crust is added to the edges of the plates as new ocean crust. As the plates move apart, more lava fills in the space and forms more ocean crust. The seafloor between North America and Europe is made of ocean crust that formed after the continents began to spread apart.

Mountains on the Ocean Floor

The ocean crust made at divergent plate boundaries is not flat. **Mid-ocean ridges** *are long, narrow mountains formed by magma at divergent boundaries.* The mid-ocean ridge in the Atlantic Ocean begins near the North Pole and continues down the middle of the Atlantic Ocean, nearly all the way to the South Pole. Mid-ocean ridges usually have gentle slopes and are about 2 km high.

Even though explosive volcanic eruptions usually occur near convergent plate boundaries, more lava erupts at divergent plate boundaries. Three-quarters of all lava erupts at mid-ocean ridges. As lava erupts under water, it hardens into flows and unique shapes such as pillow lava, shown in **Figure 13.**

 **Reading Check** What is a mid-ocean ridge?

Mountains at Convergent Boundaries

Mountains form when Earth's crust folds and crumples. Where do you think this happens? Recall that tectonic plates are rigid pieces of lithosphere. The center of these rigid plates usually does not fold as a result of collisions. Folding and crumpling usually occur at the edges of plates. This is why most mountains form near plate boundaries.

Recall that volcanoes form at convergent plate boundaries where one plate subducts under the other. These volcanoes form volcanic mountain chains along the plate boundaries. The Andes in South America and the Cascade Range in North America formed this way.

When two continents collide at a convergent plate boundary, large mountain ranges form. The tectonic plates are under extreme pressure and fold or crumple upward. The Himalayas, shown in **Figure 14**, formed as the Indian Plate converged with the Eurasian Plate. The Himalayas are the largest and highest mountain range in the world, and they are still growing!

The Appalachian Mountains in the eastern United States and the Caledonian mountains in Scotland and Scandinavia formed at convergent plate boundaries. However, over millions of years, tectonic motion has moved these mountain chains. **Figure 15** shows two maps comparing the locations of these mountain chains on Pangaea and on present-day landmasses.

 **Key Concept Check** How does plate movement form mountains?

Figure 14 Earth's tallest mountains are in the Himalayas. The Himalayas formed at the convergent plate boundary between the Indian Plate and the Eurasian Plate.

Mountains and Plate Movement 

Animation

Figure 15 The Appalachian Mountains in North America and the Caledonian mountains in Eurasia formed at the same convergent boundary. Over time, plate motion separated the mountains.

Visual Check What mountain chains once lined up across land masses?

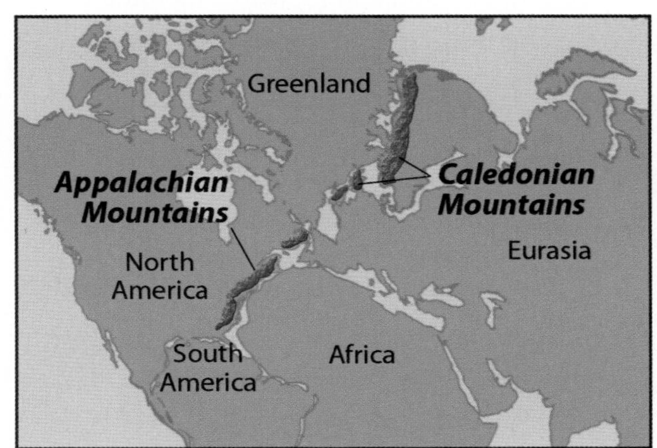

Lesson 2 Review

Visual Summary

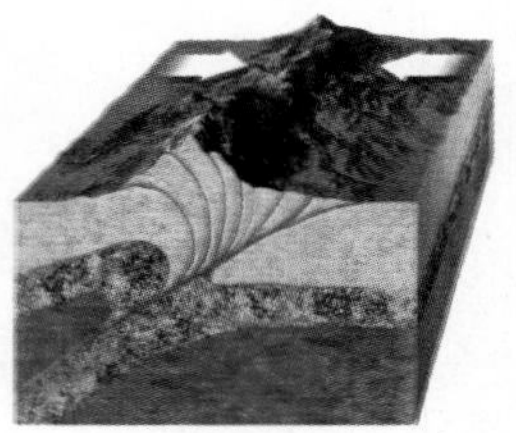

Interactions between plates can cause the crust to spread apart and form valleys or to deform and form mountains.

Earthquakes occur when stresses or forces on a fault cause the crust on either side of the fault to move.

Volcanic mountains form at subduction zones, where the mantle above the plate melts and magma erupts through Earth's crust.

FOLDABLES®

Use your lesson Foldable to review the lesson. Save your Foldable for the project at the end of the chapter.

What do you think NOW?

You first read the statements below at the beginning of the chapter.

3. Most earthquakes occur near tectonic plate boundaries.

4. Volcanoes can erupt anywhere.

Did you change your mind about whether you agree or disagree with the statements? Rewrite any false statements to make them true.

Use Vocabulary

1. **Identify** What is the name given to the underwater mountains in the middle of oceans?
2. **Use the terms** *volcano, lava,* and *magma* in a sentence.
3. **Identify** What geological feature is a fracture in Earth's crust?

Understand Key Concepts 

4. Which is evidence that a convergent boundary once existed?
   - **A.** matching mountains
   - **B.** transform faults
   - **C.** volcanoes
   - **D.** wide oceans
5. **Contrast** the eruption of magma at divergent and convergent plate boundaries.
6. **Distinguish** between a fault and an earthquake.

Interpret Graphics

7. **Illustrate** Copy the diagram to the right. Draw arrows to indicate the direction of plate movement. Indicate the location of the fault, and name the type of boundary.

8. **Sequence** Copy the graphic organizer below. Fill in the steps in forming a mountain at a convergent plate boundary, beginning with the convergence and ending with a mountain.

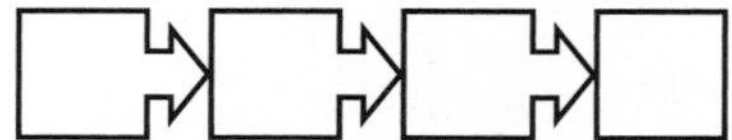

Critical Thinking

9. **Justify** investing money to make buildings more earthquake-resistant in the states of Washington and Oregon but not in Florida.
10. **Infer** Where is the youngest part of the North American Plate?

Skill Practice **Interpret Scientific Illustrations** **30 minutes**

Do the locations of earthquakes and volcanoes form a pattern?

Materials

transparencies

masking tape

fine-tipped markers

Also needed: earthquake data map, volcano data map

Safety

You've learned that many earthquakes occur on Earth every day and that hundreds of volcanoes are active. How are they related?

Learn It

Scientists sometimes use maps to organize and display data. Maps are **scientific illustrations** that show information about places on Earth. Data on maps are often represented by symbols and colors.

Try It

1. Obtain the volcano and earthquake location maps.
2. Tape the earthquake data map to your desktop. Tape a transparency over the map. Use a fine-tipped marker to transfer the earthquake data onto the transparency.
3. Repeat step 2 using the volcano data map, a different colored marker, and another transparency.
4. When you are finished transferring the data, place one of the transparencies over the map on this page. Record your observations in your Science Journal.
5. Repeat step 4 with the other transparency. Record your observations.
6. Place both transparencies over the map on this page. Use your observations to answer the questions.

Apply It

7. **Interpret Scientific Illustrations** What patterns did you observe between the two sets of data? Infer why these patterns exist.
8. **Key Concept** Where do most earthquakes and volcanoes occur? How are they related to each other?

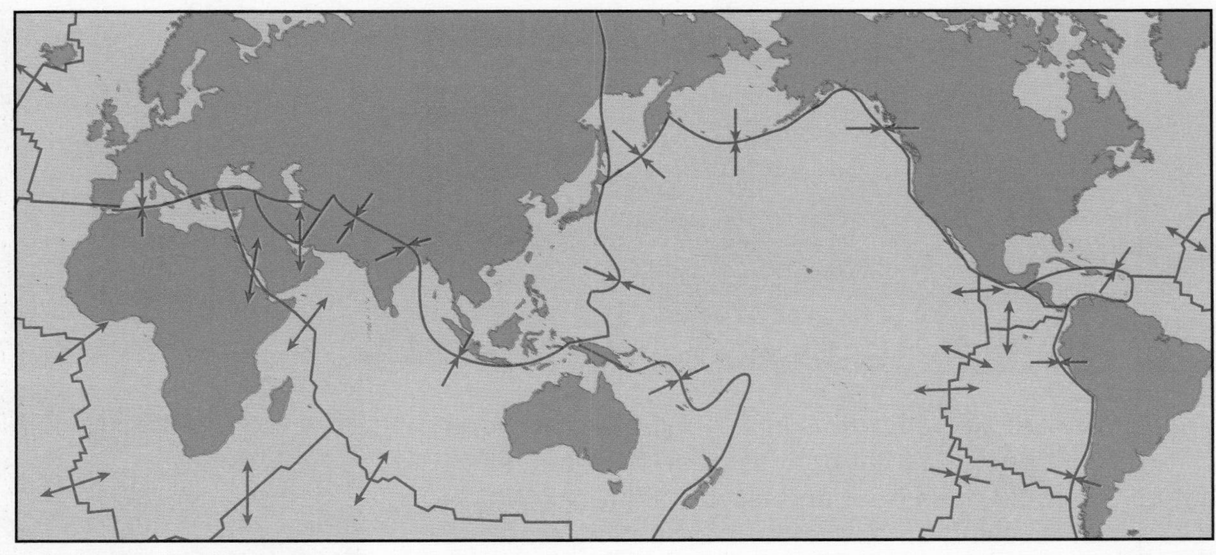

Lesson 3

Reading Guide

Key Concepts
ESSENTIAL QUESTIONS

- What is the difference between physical and chemical weathering?
- How do water, ice, and wind change Earth's surface?

Vocabulary

weathering p. 125
sediment p. 125
physical weathering p. 126
chemical weathering p. 127
erosion p. 128
deposition p. 128
mass wasting p. 128
glacier p. 130

Multilingual eGlossary

BrainPOP®

Changing Earth's Surface

Inquiry Why is this rock all alone?

Uluru, also called Ayers Rock, is in Australia. It is part of a rock layer that once was flat but has been bent and tilted. What forces could have bent and tilted solid rock? Why is this rock now sticking up above the surrounding land?

Yann Arthus-Bertrand/Corbis

Launch Lab

15 minutes

How do rocks change?

Weathering is any natural process that changes a rock. The processes can be physical or chemical. How can you tell the difference between physical and chemical weathering?

1. Read and complete a lab safety form.
2. Use a **graduated cylinder** to pour 100 mL of water into a **beaker.** Use a **wax pencil** to mark this beaker with a *W.*
3. Pour 100 mL of **vinegar** into a second beaker. Mark this beaker with a *V.*
4. Break a piece of **chalk** into two equal pieces. Use a **mortar and pestle** to crush one half. Put the crushed chalk on a piece of **paper.** Repeat for the other piece of chalk.
5. At the same time, add the crushed chalk to each beaker. In your Science Journal, describe what happens.

Think About This

1. When did physical weathering take place? When did chemical weathering take place? Explain.
2. **Key Concept** What do you think is the difference between physical and chemical weathering?

Breaking Down Earth Materials

Tall mountains form as a result of movement along faults near plate boundaries. But mountains don't get taller forever. Other processes wear away and break down mountains. These processes often are so slow that it is difficult to see changes in the mountains during a human's lifetime.

Think of an old castle made of stone, such as the one shown in **Figure 16.** The stones might be round or broken. Some walls might have collapsed. Over time, many processes acted together and broke down the stones. The effects of rain and wind have gradually made the stones more rounded. Perhaps an earthquake knocked down some of the castle. Mountains are similar. Over time, the same processes that changed the stones of the castle can change the rocks that make up mountains.

Weathering *refers to the mechanical and chemical processes that change Earth's surface over time.* Weathering can affect rocks in different ways. The processes of weathering can break, scrape, smooth, or chemically change rock. **Sediment** *is the material formed from rocks broken down by weathering.* Weathering can produce sediment of different sizes. Sediment can be rock fragments, sand, silt, or clay.

Reading Check How is weathering related to sediment?

Figure 16 Over time, some of the same processes that changed the stone walls of this building also can wear away and break down mountains.

Visual Check What changes have occurred to the stone walls of this castle?

(t)Hutchings Photography/Digital Light Source, (b)David Toase/Getty Images

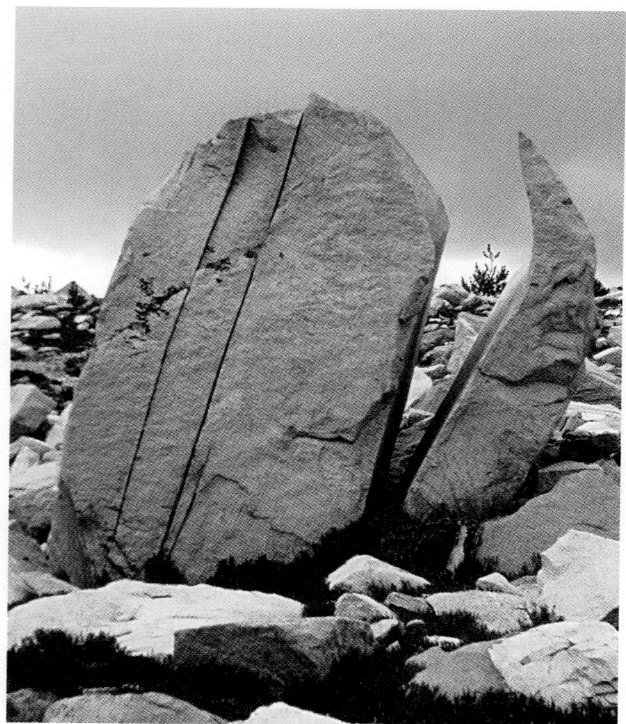

▲ **Figure 17** The repeated freezing and thawing of water broke this rock.

Physical Weathering

The first step in making sediment is to make smaller pieces of rock from larger ones. **Physical weathering** *is the process of breaking down rock without changing the composition of the rock.* Several natural processes cause physical weathering. For example, if a boulder rolls off a cliff and breaks apart, it experiences physical weathering. Forces from plate motion, such as when faults rupture, also can cause rock to break.

Physical weathering also can occur because of changes in weather. Water can seep into rocks. If the temperature is cold enough, the water can freeze. Unlike most liquids, water expands when it freezes. The force from the expanding ice pushes outward and, over time, can shatter rocks, as shown in **Figure 17.**

Plants and animals also can break rocks. The roots of plants can grow into cracks in rocks. The force from the growing roots can pry the rock open, as shown in **Figure 18.** Even the roots of lichen, the small flat plants that grow on rock surfaces, can cause small cracks to form.

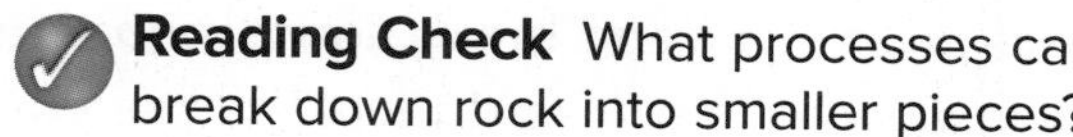

Reading Check What processes can break down rock into smaller pieces?

Figure 18 As roots grow, they push pieces of rock apart. ▼

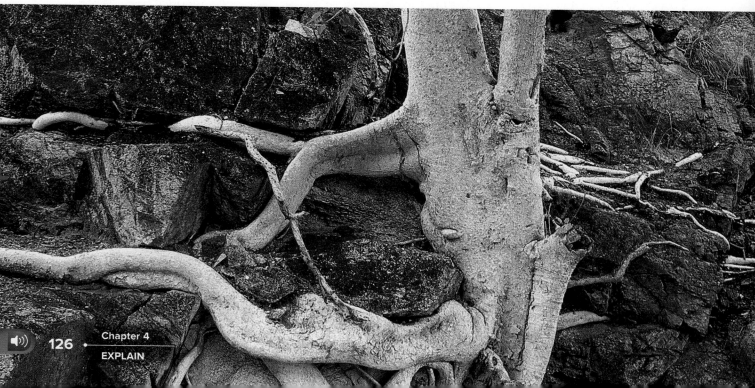

Chemical Weathering

Some minerals can react with water, air, or substances in water and air, such as carbon dioxide (CO_2). **Chemical weathering** *is the process that changes the composition of rocks.*

Some minerals can dissolve in slightly acidic water, such as rainwater. Limestone contains calcite, a mineral that dissolves in slightly acidic water, as shown in **Figure 19.**

Other minerals react with air and water to form new minerals. CO_2 in the atmosphere and in water can react with minerals, such as feldspar, to form clay. Some minerals contain iron, which can react with oxygen in the atmosphere to form iron oxide, or rust. The rust color of many rocks is caused by iron oxide, as shown in **Figure 19.**

Chemical weathering happens faster where water is abundant. It also happens faster where it is warm because chemical reactions happen faster at higher temperatures.

Chemical weathering and physical weathering affect each other. For example, physical weathering makes smaller pieces of rock. When rocks break, chemical weathering can occur on the newly exposed surfaces. Chemical weathering can make rocks weaker, making the rocks break more easily. The rounded hill pictured at the beginning of this chapter formed from the interaction between physical and chemical weathering.

Key Concept Check What is the difference between physical and chemical weathering?

FOLDABLES®

Make a vertical three-column chart book. Label it as shown. Use it to organize your notes on weathering and erosion.

Figure 19 Chemical weathering can change rocks. Water and oxygen weathered the iron in this rock on the left, forming rust-colored iron oxide. Limestone, shown on the right, can dissolve easily in rainwater.

Figure 20 Mass wasting can cause quick and drastic changes to landscapes. This landslide, in Emerald Bay, California, occurred in weathered granite.

Moving Earth Materials

Mountains wear away for many reasons. Weathering produces smaller rocks, which can be moved more easily. Chemical weathering dissolves minerals that make up these smaller rocks. But this slow weathering is not the only way mountains wear down. Rock also can be removed from the tops of mountains. Geologists use the term **erosion** to describe *the moving of weathered material, or sediment, from one location to another.* **Deposition** *is the laying down or settling of eroded material.* Together, erosion and deposition change the surface of Earth.

Gravity's Influence

Gravity causes material to move downhill. **Mass wasting** *is the downhill movement of a large mass of rocks or soil due to gravity.* If mountains are tall enough or slopes are too steep, the force of gravity can create landslides, a type of mass wasting. In just a few moments, large amounts of rock and soil can come crashing downhill. Some landslides start from the tops of mountains and end at valley floors, as shown in **Figure 20.**

Erosion requires energy. During a landslide, gravity provides this energy. But, flowing water, wind, and moving ice also can have enough energy to move rocks and soil.

Reading Check What provides the forces that can cause rock to move downhill?

MiniLab

15 minutes

How can wind change Earth's surface?

Wind changes Earth's surface in different ways. How can you demonstrate how wind can change the land?

1. Read and complete a lab safety form.
2. Pour dry **sand** into a **clear plastic box.** Place two or three **large pebbles** and **small clumps of grass** on the surface of the sand.
3. Place the box into a **cardboard box** that is open on one side.
4. Plug in the **hair dryer.** Hold the hair dryer about 10 cm from the edge of the box at a 45° angle relative to the plastic box.
5. Turn the dryer on low. Hold it in the same position for 2–3 min. Record your observations in your Science Journal.

Analyze and Conclude

1. **Observe** How did the hair dryer—the wind model—change the sand? Be specific in your answer.
2. **Key Concept** How does wind change Earth's surface?

(t) John A. Karachewski, (b) Hutchings Photography/Digital Light Source

Water

Most erosion occurs by flowing water. Water flows fastest where the land is steep. Water also flows faster in larger rivers than in smaller ones. Large rivers cause the most erosion. The faster water flows, the larger the pieces of sediment it can carry. Rivers can even wear away solid rock, such as Niagara Falls, shown in **Figure 21.**

Water often slows as it flows downstream. Slowly flowing water has less energy and can carry less sediment. As water slows, the sediment in the water is deposited on the sides of the river. Sediment also is deposited when rivers enter oceans or lakes, creating land features called deltas.

REVIEW VOCABULARY

delta
triangular deposit of sediment that forms where a stream enters a large body of water.

Wind

Sometimes, wind is strong enough to cause erosion. In deserts, erosion by wind can be the most important process that changes landforms. Wind can slowly weather and erode solid rock, as pictured in **Figure 22.** Wind also can pick up sand grains and carry them from one place to another. Sand dunes and ripples, such as those shown in **Figure 23,** are examples of landforms made by wind.

 **Reading Check** What causes most erosion on Earth?

Figure 22 Strong wind can carve rocks into unusual shapes that rise above the surroundings. ▼

 **Personal Tutor**

▲ **Figure 21** Flowing water can erode solid rock and make land features such as waterfalls.

▲ **Figure 23** Blowing wind can carry sand, forming ripples on sandy surfaces.

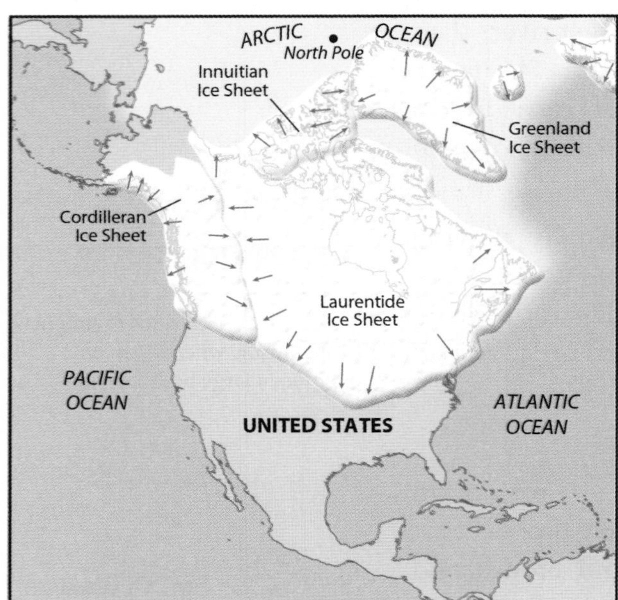

Figure 24 Large glaciers flowed from Northern Canada into the United States 20,000 years ago. These ice sheets brought sediment with them. When the ice melted, sediment, such as these large boulders, was left behind.

Ice

Large masses of ice, formed by snow accumulation on land, that move slowly across Earth's surface are called **glaciers.** Glaciers form in cold climates, such as high mountains or near the North Pole and the South Pole. The force of gravity causes this ice to flow downhill.

WORD ORIGIN

glacier
from Greek *glacies,* means "ice"

Sliding and flowing ice can weather the rocks over which the ice moves. This process creates sediment that glaciers carry away. Over time, glaciers can carve deep valleys.

When a glacier melts, it deposits the sediment it carried. Much of North America was covered by ice 20,000 years ago, as shown in **Figure 24.** When this ice sheet melted, it left behind sediment. Rocks and smaller sediment were carried from northern Canada and deposited in the United States.

Earth's Changing Surface

Many of Earth's surface features and the processes that occur on it can be related and explained by plate tectonics. The processes that move Earth material depend on climate, or the average weather in a region over a long period of time. Temperature, amount of precipitation, the pattern of winds, and circulation of the ocean affect climate. The location of continents affects ocean circulation. The locations of mountains affect wind patterns and precipitation. The processes that change the features made by plate movement are affected by plate movement itself.

ACADEMIC VOCABULARY

processes
(noun) a series of actions or operations that lead to an end result

Key Concept Check How do water, ice, and wind change Earth's surface?

Lesson 3 Review

Visual Summary

Physical weathering breaks down rocks into smaller pieces but does not change the mineral composition.

Chemical weathering changes a rock's mineral composition. This makes rocks weaker and helps break down rocks into smaller pieces.

Gravity, wind, water, and ice can cause physical and chemical weathering. They break down, erode, transport, and deposit rocks and sediment.

FOLDABLES®

Use your lesson Foldable to review the lesson. Save your Foldable for the project at the end of the chapter.

What do you think NOW?

You first read the statements below at the beginning of the chapter.

5. Wind erosion only occurs in the desert.

6. Rivers are the only cause of erosion.

Did you change your mind about whether you agree or disagree with the statements? Rewrite any false statements to make them true.

Use Vocabulary

1. **Define** *weathering* in your own words.
2. Material transported by water, wind, and ice is called ________.
3. The opposite of deposition is ________.

Understand Key Concepts 

4. **Explain** why large boulders are present in steep mountain streams.
5. On steep slopes, gravity causes
   - **A.** chemical weathering.
   - **B.** mass wasting.
   - **C.** physical weathering.
   - **D.** strong winds.
6. **Contrast** chemical and physical weathering.

Interpret Graphics

7. **Analyze** What happened to the speed of the water as it rounded this bend in the river shown above? Explain.
8. **Sequence** Put these terms in the correct sequence as they relate to the breakdown and movement of Earth materials: *rock, sediment, deposition, erosion, weathering.*

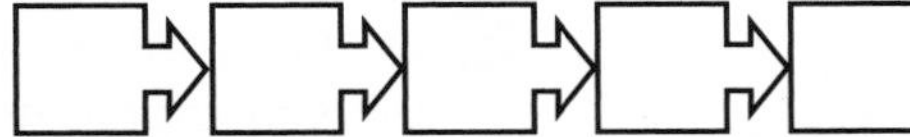

Critical Thinking

9. **Predict** how increased temperature and precipitation would affect a mountain.
10. **Reconstruct** the ways physical and chemical weathering rounded the rocks in the lesson opener photo. Explain how the two types of weathering interact.

(t) Susan Rayfield/Science Source, (c) ©Magdalena Rehova/Alamy, (b) Vladimir Melnik/Getty Images

Earth's Changing Surface

Materials

glue stick

colored pencils

scissors

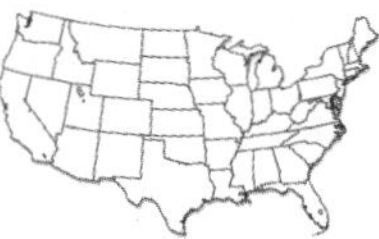

outline map of the United States

physiograpic map of United States

Also needed: large paper 11 × 18

Safety

Your teacher just opened a travel agency and has hired all of the students in your class as tour guides. Unlike a typical travel agency, Teacher's Trusty Travel Tours offers tours to any part of the lower 48 states to explain how the region formed and continues to change. You will be a Trusty Travel Guide for one of these regions. As the expert, you will

- develop a travel slogan for your region;
- explain how the region formed and continues to change;
- make a travel brochure to persuade tourists that your region is the best place in the lower 48 states to visit.

Question

Find out which region will be your tour-guide territory. What Earth processes do you think might have produced this region? How do you think Earth processes continue to change the region? What information about your region will be most interesting and persuade tourists to choose your area to visit?

Procedure

1. A physiographic region is an area with certain relief, or elevation; specific rock types; and specific geologic structures. You will work with an online, interactive physiographic map of the United States. Obtain the Web address for the map from your teacher.

2. Locate your region on the physiographic map. Use a colored pencil to mark its location on the outline map of the lower 48 states. You may use the outline map in your travel brochure.
3. Research to find photographs of your region. Study the photographs carefully, keeping in mind what you have learned in this chapter about the processes that change Earth's surface. Based on your observations, explain how you think your region formed.

4 Research your region and record your findings in your Science Journal. Use at least two references to obtain data about your region. Your data can include the geologic history of the region, landforms characteristic of the region, diagrams showing how the landforms formed, and any other information you find interesting.

5 Use the information you have gathered to make your travel brochure. Include a map of your region, the region's name, your travel slogan, two or more paragraphs describing the processes that formed and that continue to change the region, and several photographs and/or drawings of the region.

Analyze and Conclude

6 **Research Information** How did your research compare to how you thought your region formed?

7 **Classify** the major Earth processes that produced your region as well as the processes that continue to change it.

8 **The Big Idea** How do the motion of tectonic plates and the processes of weathering, erosion, and deposition change Earth's surface?

Communicate Your Results

Use your travel brochure to put together a 1-minute speech about your region and why people should come to visit it. Practice your speech. Be prepared to answer any questions potential tourists might have!

Extension

Find the tour guides for two other regions that are close to yours. Talk about any similarities and differences among the regions and how they formed.

Lab Tips

- ☑ Remember to rewrite any information you find in books or on the Internet in your own words.
- ☑ When using the Internet, make sure the information is from a reliable source, such as a college or a university, an online encyclopedia, or a government agency.
- ☑ Your brochure must serve two purposes. It must be informative, yet easy to understand. It also must persuade people to visit that region of the country.

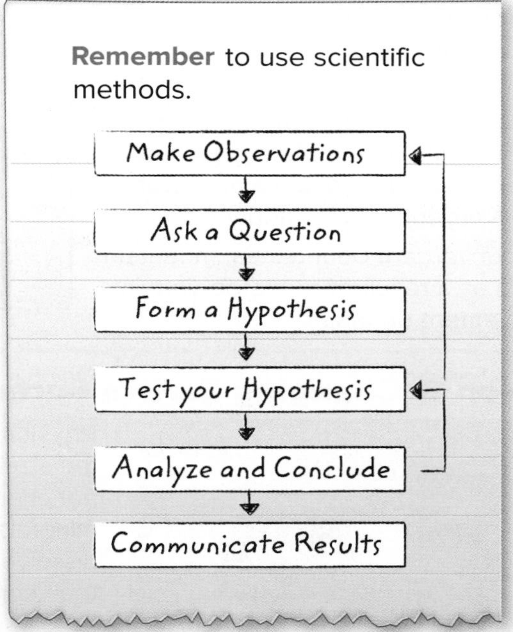

Chapter 4 Study Guide

Interaction of tectonic plates at plate boundaries forms earthquakes, volcanoes, and mountains. Erosion and deposition by water, wind, and ice, along with mass wasting, can change the landforms made by plate motion.

Key Concepts Summary

Lesson 1: Earth's Moving Surface

- The theory of **plate tectonics** states that Earth's surface is broken into rigid plates that move with respect to each other. Tectonic plates are pieces of the **lithosphere.**

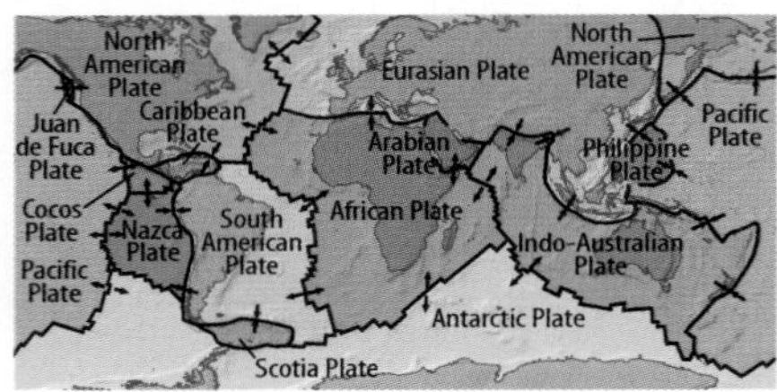

- At a **convergent boundary,** plates come together. At a **divergent boundary,** they move apart. At a **transform boundary,** plates slide past each other.
- **Convection** in Earth's mantle causes tectonic plates to move.

Lesson 2: Shaping Earth's Surface

- Earthquakes occur at **faults.** Most faults are near plate boundaries.
- Deformation near plate boundaries forms **volcanoes** and mountain ranges and causes **earthquakes.**
- Volcanoes form at convergent plate boundaries on the plate that does not get subducted. **Lava** also erupts at divergent plate boundaries.
- At convergent boundaries, mountains form as a result of repeated vertical movement on faults.

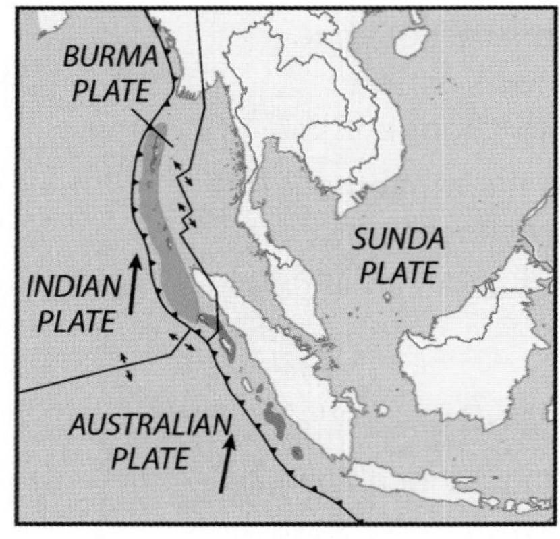

Lesson 3: Changing Earth's Surface

- **Physical weathering** changes the size of sediment but does not change the composition of minerals. **Chemical weathering** dissolves minerals or changes their composition.
- Gravity, water, ice, and wind erode landscapes and move materials from one place and deposit them in others.

Vocabulary

Lesson 1

plate tectonics p. 107
lithosphere p. 108
asthenosphere p. 108
divergent boundary p. 109
convergent boundary p. 109
subduction p. 109
transform boundary p. 109
convection p. 111

Lesson 2

earthquake p. 115
fault p. 115
magma p. 118
lava p. 118
volcano p. 118
mid-ocean ridge p. 120

Lesson 3

weathering p. 125
sediment p. 125
physical weathering p. 126
chemical weathering p. 127
erosion p. 128
deposition p. 128
mass wasting p. 128
glacier p. 130

Study Guide

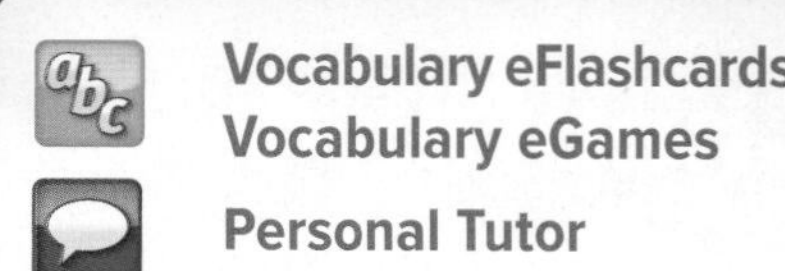

Foldables Chapter Project

Assemble your lesson Foldables as shown to make a Chapter Project. Use the project to review what you have learned in this chapter.

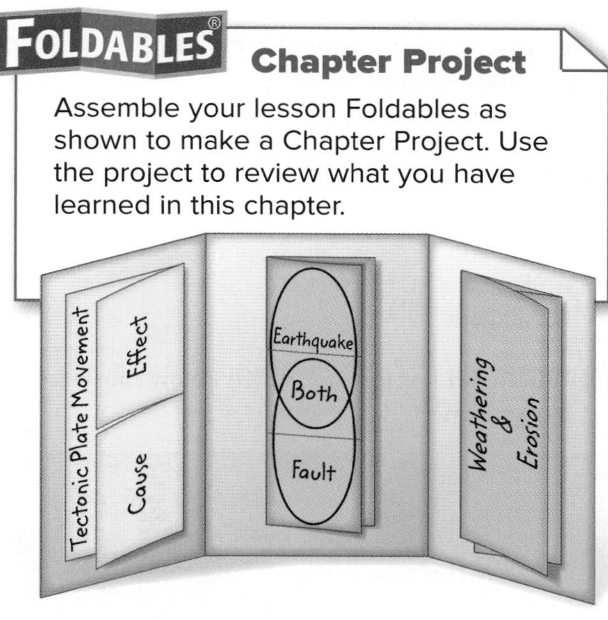

Use Vocabulary

1. Use the words *asthenosphere* and *lithosphere* in a sentence.
2. What is the difference between magma and lava?
3. The removal and transport of sediment is called ________.
4. What is the name given to the mountain range that formed in oceans at divergent plate boundaries?
5. Sheets of flowing ice are called ________.
6. One tectonic plate slides under another plate in a process called ________.

Link Vocabulary and Key Concepts

Copy this concept map, and then use vocabulary terms from the previous page to complete the concept map.

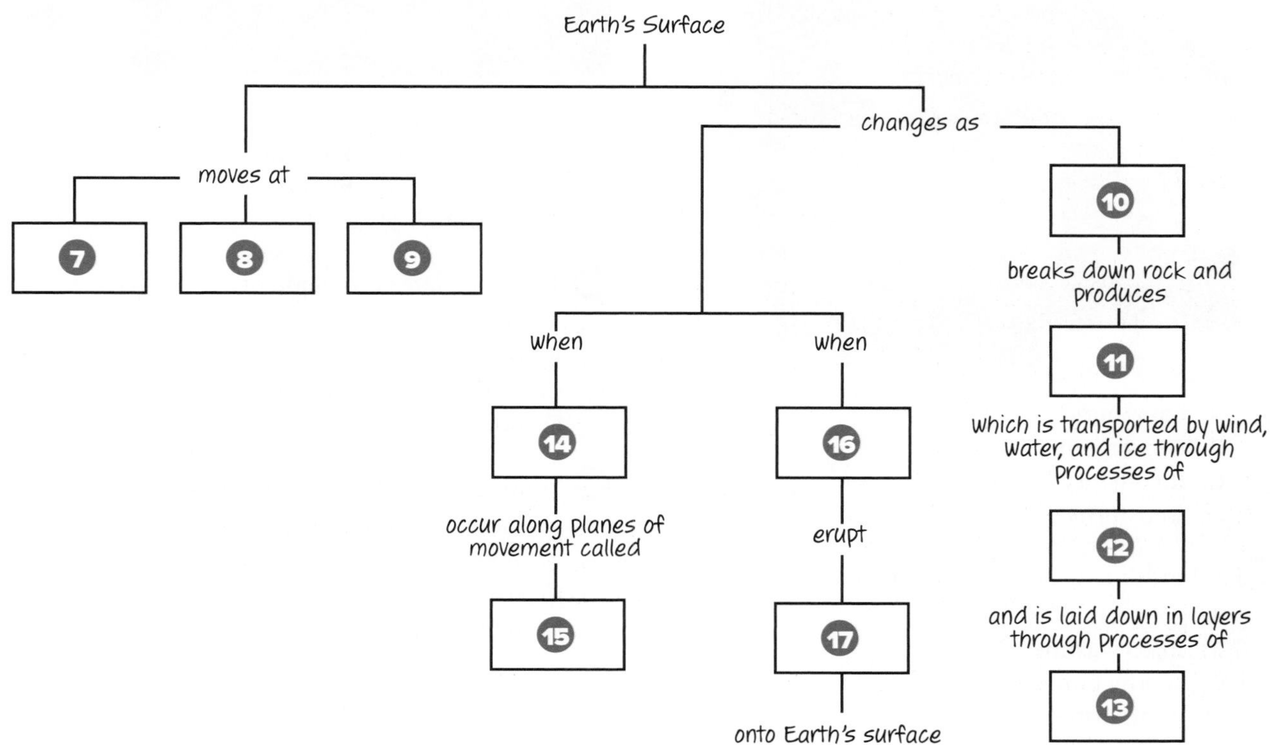

Chapter 4 Review

Understand Key Concepts 

1. The largest earthquakes occur
   A. at convergent plate boundaries.
   B. at divergent plate boundaries.
   C. at transform plate boundaries.
   D. far from plate boundaries.

2. At which type of boundary does one plate subduct beneath another plate?
   A. compound
   B. convergent
   C. divergent
   D. transform

3. The process of loosening and transporting rock and sediment is called
   A. convection.
   B. deposition.
   C. erosion.
   D. weathering.

4. The plate boundary shown below is an example of a ________ boundary.

   A. compound
   B. convergent
   C. divergent
   D. transform

5. A landslide is an example of what process?
   A. convection
   B. mass wasting
   C. subduction
   D. weathering

6. A rusty nail is altered by which process?
   A. chemical weathering
   B. erosion
   C. mass wasting
   D. physical weathering

7. Melting rock produces
   A. ash.
   B. erosion.
   C. faults.
   D. magma.

8. Where did the feature shown below most likely form?

   A. at a convergent boundary
   B. at a divergent boundary
   C. at a transform boundary
   D. far from a plate boundary

9. Tectonic plates are made of
   A. continental crust only.
   B. oceanic crust only.
   C. crust and flowing mantle.
   D. crust and rigid mantle.

10. Sand dunes in deserts are made by
   A. blowing wind.
   B. flowing water.
   C. mass wasting.
   D. moving glaciers.

Patrick Shyu/Getty Images

Online Test Practice

Critical Thinking

11 **Interpret** Discuss the tectonic activity that occurs at the plate boundary pictured in the center of the image above. Include the type of boundary, the name of the landform, and the relative ages of the crust.

12 **Summarize** some of the observations that support the claim that most geological activity happens near plate boundaries.

13 **Critique** the hypothesis that tectonic plates are perfectly rigid.

14 **Infer** what type of weathering is most important in Hawaii where it is hot and wet. Explain.

15 **Criticize** the claim that river erosion has always been the most important type of erosion in North America.

16 **Justify** the assumption that Earth's mantle under tectonic plates is hot.

17 **Rank** the relative importance of different ways of transporting sediment in your region.

18 **Predict** where on Earth most landslides occur and where they are largest.

Writing in Science

19 **Report** Write a short news story about a recent volcanic eruption or large earthquake. Explain how plate tectonics affected where it occurred and how it occurred.

REVIEW THE BIG IDEA

20 Explain why the mountains in the western United States are so much taller and more jagged than the mountains in the eastern United States. Write about how plate tectonics and changes at Earth's surface make these mountains look the way they do.

21 The photo below shows rock layers in Capitol Reef National Park. What processes have changed the surface of this landform?

Math Skills

Math Practice

Use Proportions

22 Two plates in the South Pacific separate at an average rate of 15 cm/y. How long would it take them to move 10 km apart? (1 km = 100,000 cm)

23 During part of Earth's history, the Pacific Plate moved along the San Andreas Fault zone past the North American Plate. It covered a distance of 110 km in 4 million years.

a. How many kilometers per year did the plates move during that period?

b. What was the rate of movement in centimeters per year during that period? (1 km = 100,000 cm)

Standardized Test Practice

Record your answers on the answer sheet provided by your teacher or on a sheet of paper.

Multiple Choice

1 What is the term for the large, rigid pieces of Earth's crust?

A faults

B sediment

C mountain

D tectonic plates

2 Which location generally experiences the strongest earthquakes?

A the top of a mountain

B the center of a tectonic plate

C near a subduction zone

D next to a mid-ocean ridge

Use the figure below to answer question 3.

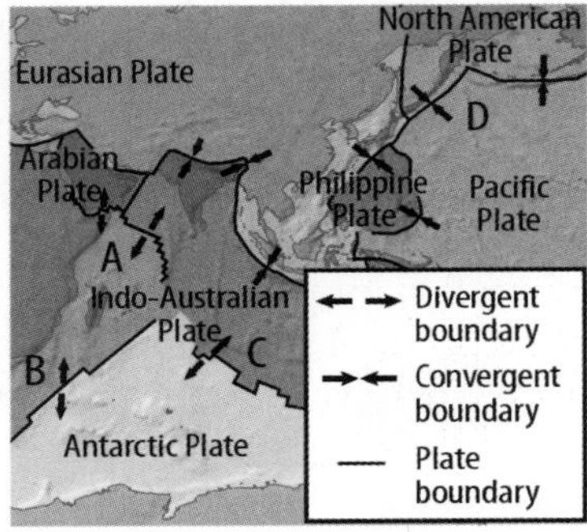

3 Near which plate boundary would you most likely find volcanoes?

A A

B B

C C

D D

4 Volcanoes form on land at which type of boundary?

A divergent

B transform

C convergent with no subduction

D convergent with subduction

5 How do scientists measure the yearly movement of tectonic plates?

A They count the number of volcanic eruptions.

B They measure the strength and the number of earthquakes.

C They record the change in mountain height.

D They analyze data from satellites orbiting Earth.

Use the figure below to answer question 6.

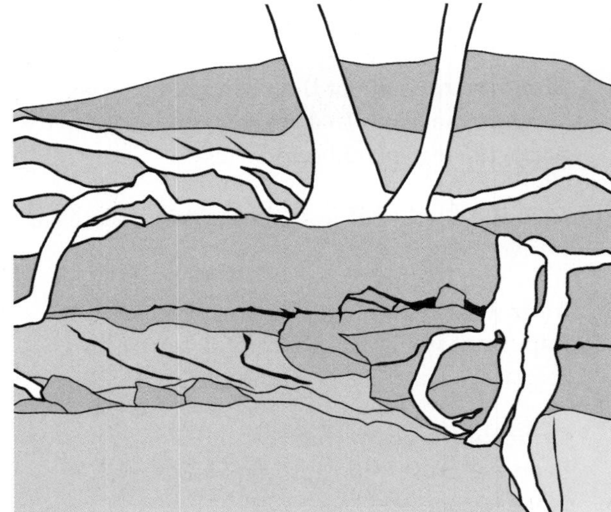

6 Which process is illustrated in the diagram?

A deposition

B erosion

C chemical weathering

D physical weathering

7 Which weathering agent causes the most erosion and deposition of sediment on Earth?

A flowing water

B gravity

C moving wind

D sliding ice

8 Which is an example of chemical weathering?

A ice cracking rocks

B landslides moving downhill

C water dissolving limestone

D ripples forming on sand dunes

Use the figure below to answer question 9.

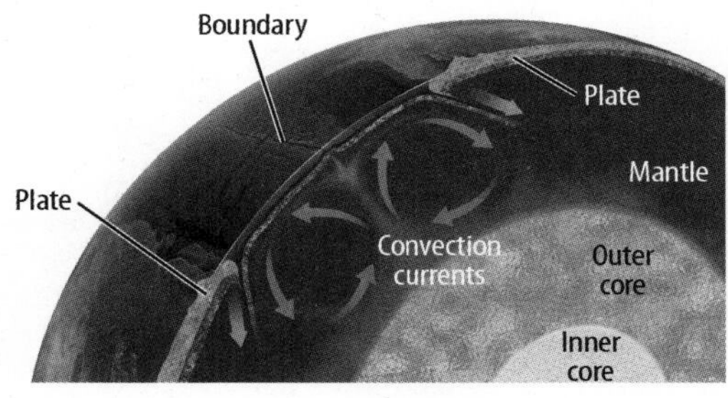

9 Convection currents cause tectonic plates above them to move. In which direction are the two oceanic plates on either side of the labeled boundary moving?

A Both plates are moving away from the boundary.

B Both plates are moving toward the boundary.

C The plate on the left is moving under the plate on the right.

D The plate on the right is moving under the plate on the left.

10 Other than convection in the mantle, which process can cause tectonic plate movement?

A erosion

B deposition

C mass wasting

D subduction

Constructed Response

Use the figure below to answer questions 11 and 12.

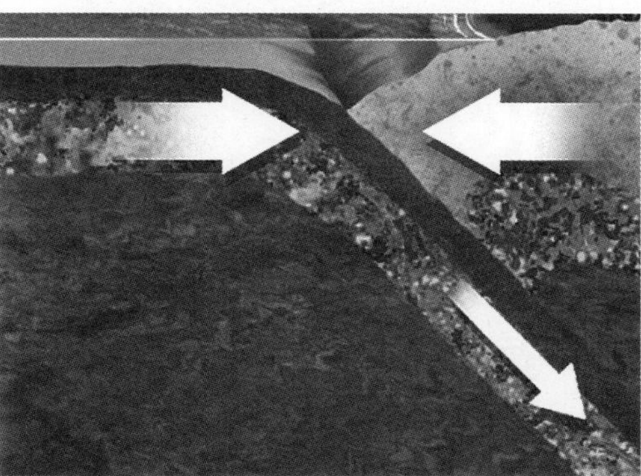

11 What kind of boundary is shown in the figure? How do you know?

12 What landforms would you expect to see around the boundary shown? How are these landforms formed?

13 Some boulders in Iowa are the same type of rock as rock formations in Minnesota. Some boulders in Illinois are the same type of rock as rock formations in Michigan's Upper Peninsula. What agent of erosion could have carried boulders this far? What other changes would this agent cause to Earth's surface?

14 Many earthquakes occur every day in California. Why don't you hear about all these earthquakes on the news? Why do so many occur in California?

NEED EXTRA HELP?														
If You Missed Question...	1	2	3	4	5	6	7	8	9	10	11	12	13	14
Go to Lesson...	1	2	2	2	1	3	3	3	1	1	1	2	3	2

Chapter 5

Natural Resources

THE BIG IDEA

Why is it important to manage natural resources wisely?

Inquiry What do these colors mean?

This image shows where thermal energy escapes from the inside of a house. Red and yellow areas represent the greatest loss. Blue areas represent low or no loss.

- Which energy resources are used to heat this house?
- Why is it important to reduce thermal energy loss from houses, cars, or electrical appliances?
- Why is it important to manage natural resources wisely?

Tyrone Turner/National Geographic Stock

Get Ready to Read

What do you think?

Before you read, decide if you agree or disagree with each of these statements. As you read this chapter, see if you change your mind about any of the statements.

1. Nonrenewable energy resources include fossil fuels and uranium.
2. Energy use in the United States is lower than in other countries.
3. Renewable energy resources do not pollute the environment.
4. Burning organic material can produce electricity.
5. Cities cover most of the land in the United States.
6. Minerals form over millions of years.
7. Humans need oxygen and water to survive.
8. About 10 percent of Earth's total water can be used by humans.

Your one-stop online resource
connectED.mcgraw-hill.com

LearnSmart®

Chapter Resources Files, Reading Essentials, Get Ready to Read, Quick Vocabulary

Animations, Videos, Interactive Tables

Self-checks, Quizzes, Tests

Project-Based Learning Activities

Lab Manuals, Safety Videos, Virtual Labs & Other Tools

Vocabulary, Multilingual eGlossary, Vocab eGames, Vocab eFlashcards

Personal Tutors

Lesson 1

Reading Guide

Key Concepts
ESSENTIAL QUESTIONS

- What are the main sources of nonrenewable energy?
- What are the advantages and disadvantages of using nonrenewable energy resources?
- How can individuals help manage nonrenewable resources wisely?

Vocabulary

nonrenewable resource p. 143
renewable resource p. 143
nuclear energy p. 147
reclamation p. 149

Multilingual eGlossary

Energy Resources

Inquiry What's in the pipeline?

The Trans-Alaska Pipeline System carries oil more than 1,200 km from beneath Prudhoe Bay, Alaska, to the port city of Valdez, Alaska. How might the pipeline's construction and operation affect the habitats and the organisms living along it? How do getting and using fossil fuels impact the environment?

Launch Lab

20 minutes

How do you use energy resources?

In the United States today, the energy used for most daily activities is easily available at the flip of a switch or the push of a button. How do you use energy in your daily activities?

1. Design a three-column data chart in your Science Journal. Title the columns *Activity, Type of Energy Used,* and *Amount of Time.*
2. Record every instance that you use energy during a 24-hr period.
3. Total your usage of the different forms of energy, and record them in your Science Journal.

Think About This

1. How many times did you use each type of energy?
2. Compare and contrast your usage with that of other members of your class.
3. **Key Concept** Are there instances of energy use when you could have conserved energy? Explain how you would do it.

Sources of Energy

Think about all the times you use energy in one day. Are you surprised by how much you depend on energy? You use it for electricity, transportation, and other needs. That is one reason it is important to know where energy comes from and how much is available for humans to use.

Table 1 lists different energy sources. Most energy in the United States comes from nonrenewable resources. **Nonrenewable resources** *are resources that are used faster than they can be replaced by natural processes.* Fossil fuels, such as coal and oil, and uranium, which is used in nuclear reactions, are both nonrenewable energy resources.

Renewable resources *are resources that can be replaced by natural processes in a relatively short amount of time.* The Sun's energy, also called solar energy, is a renewable energy resource. You will read more about renewable energy resources in Lesson 2.

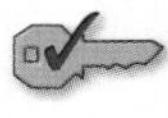

Key Concept Check What are the main nonrenewable energy resources?

Table 1 Energy resources can be nonrenewable or renewable.

Table 1 Energy Sources

Nonrenewable Energy Resources	Renewable Energy Resources
fossil fuels uranium	solar wind water geothermal biomass

WORD ORIGIN

resource
from Latin *resurgere,* means "to rise again"

Nonrenewable Energy Resources

You might turn on a lamp to read, turn on a heater to stay warm, or ride the bus to school. In the United States, the energy to power lamps, heat houses, and run vehicles probably comes from nonrenewable energy resources, such as fossil fuels.

Fossil Fuels

Coal, oil, also called petroleum, and natural gas are fossil fuels. They are nonrenewable because they form over millions of years. The fossil fuels used today formed from the remains of prehistoric organisms. The decayed remains of these organisms were buried by layers of sediment and changed chemically by extreme temperatures and pressure. The type of fossil fuel that formed depended on three factors:

- the type of organic matter
- the temperature and pressure
- the length of time that the organic matter was buried

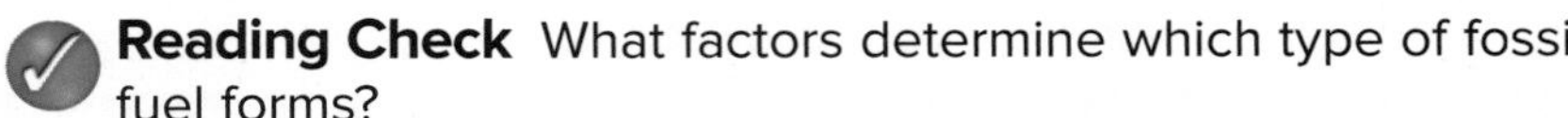

Reading Check What factors determine which type of fossil fuel forms?

Coal Earth was very different 350 million years ago, when the coal used today began forming. Plants, such as ferns and trees, grew in prehistoric swamps. As shown in **Figure 1**, the first step of coal formation occurred when those plants died.

Bacteria, extreme temperatures, and pressure acted on the plant remains over time. Eventually a brownish material, called peat, formed. Peat can be used as a fuel. However, peat contains moisture and produces a lot of smoke when it burns. As shown in **Figure 1**, peat eventually can change into harder and harder types of coal. The hardest coal, anthracite, contains the most carbon per unit of volume and burns most efficiently.

Figure 1 Much of the coal used today began forming more than 300 million years ago from the remains of prehistoric plants.

Animation

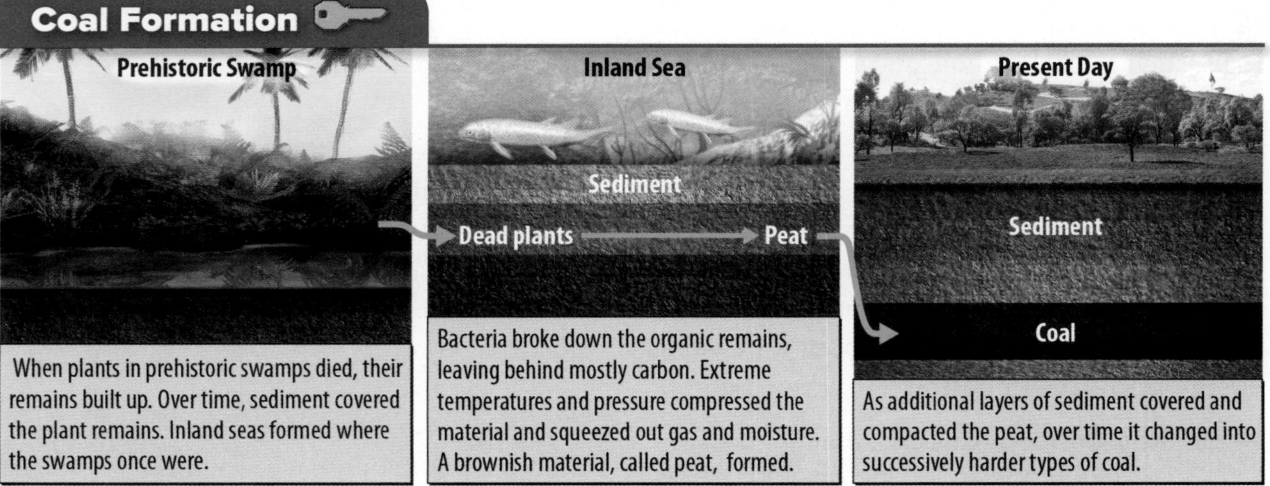

Coal Formation
Prehistoric Swamp
Inland Sea
Present Day
Sediment
Dead plants
Peat
Sediment
Coal
When plants in prehistoric swamps died, their remains built up. Over time, sediment covered the plant remains. Inland seas formed where the swamps once were.
Bacteria broke down the organic remains, leaving behind mostly carbon. Extreme temperatures and pressure compressed the material and squeezed out gas and moisture. A brownish material, called peat, formed.
As additional layers of sediment covered and compacted the peat, over time it changed into successively harder types of coal.

Figure 2 Reservoirs of oil and natural gas often are under layers of impermeable rock.

Visual Check What prevents oil and natural gas from rising to the surface?

Oil and Natural Gas Like coal, the oil and natural gas used today formed millions of years ago. The process that formed oil and natural gas is similar to the process that formed coal. However, oil and natural gas formation involves different types of organisms. Scientists theorize that oil and natural gas formed from the remains of microscopic marine organisms called plankton. The plankton died and fell to the ocean floor. There, layers of sediment buried their remains. Bacteria decomposed the organic matter, and then pressure and extreme temperatures acted on the sediments. During this process, thick, liquid oil formed first. If the temperature and pressure were great enough, natural gas formed.

Most of the oil and natural gas used today formed where forces within Earth folded and tilted thick rock layers. Often hundreds of meters of sediments and rock layers covered oil and natural gas. However, oil and natural gas were less dense than the surrounding sediments and rock. As a result, oil and natural gas began to rise to the surface by passing through the pores, or small holes, in rocks. As shown in **Figure 2**, oil and natural gas eventually reached layers of rock through which they could not pass, or impermeable rock layers. Deposits of oil and natural gas formed under these impermeable rocks. The less-dense natural gas settled on top of the denser oil.

Reading Check How is coal formation different from oil formation?

Advantages of Fossil Fuels

Do you know that fossil fuels store chemical energy? Burning fossil fuels transforms this energy. The steps involved in changing chemical energy in fossil fuels into electric energy are fairly easy and direct. This process is one advantage of using these nonrenewable resources. Also, fossil fuels are relatively inexpensive and easy to transport. Coal is often transported by trains, and oil is transported by pipelines or large ships called tankers.

Disadvantages of Fossil Fuels

Although fossil fuels provide energy, there are disadvantages to using them.

Limited Supply One disadvantage of fossil fuels is that they are nonrenewable. No one knows for sure when supplies will be gone. Scientists estimate that, at current rates of consumption, known reserves of oil will last only another 50 years.

Habitat Disruption In addition to being nonrenewable, the process of obtaining fossil fuels disturbs environments. Coal comes from underground mines or strip mines, such as the one shown in **Figure 3.** Oil and natural gas come from wells drilled into Earth. Mines in particular disturb habitats. Forests might be fragmented, or broken into areas of trees that are no longer connected. Fragmentation can negatively affect birds and other organisms that live in forests.

Reading Check How much longer are known oil reserves predicted to last?

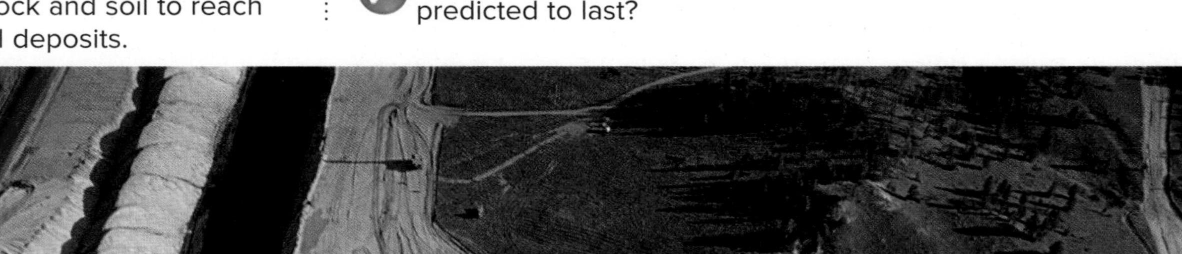

Figure 3 Strip-mining involves removing layers of rock and soil to reach coal deposits.

Larry Mayer/Creatas/SuperStock

Pollution Another disadvantage of fossil fuels as an energy resource is pollution. For example, runoff from coal mines can pollute soil and water. Oil spills from tankers can harm living things, such as the bird shown in **Figure 4.**

Pollution also occurs when fossil fuels are used. Burning fossil fuels releases chemicals into the atmosphere. These chemicals react in the presence of sunlight and produce a brownish haze. This haze can cause respiratory problems, particularly in young children. The chemicals also can react with water in the atmosphere and make rain and snow more acidic. The acidic precipitation can change the chemistry of soil and water and harm living things.

Figure 4 One disadvantage of fossil fuels is pollution, which can harm living things. This bird was covered with oil after an oil spill.

Key Concept Check What is one advantage and one disadvantage of using fossil fuels?

Nuclear Energy

Atoms are too small to be seen with the unaided eye. Even though they are small, atoms can release large amounts of energy. *Energy released from atomic reactions is called* **nuclear energy.** Stars release nuclear energy by fusing atoms. The type of nuclear energy used on Earth involves a different process.

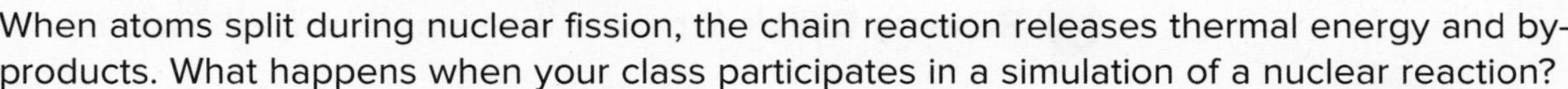

What is your reaction?

When atoms split during nuclear fission, the chain reaction releases thermal energy and by-products. What happens when your class participates in a simulation of a nuclear reaction?

1. Read and complete a lab safety form.
2. Use a **marker** to label three **sticky notes.** Label one note *U-235.* Label two notes *Neutron.* Stick the U-235 note on your **apron.** Hold the Neutron notes in one hand and a **Thermal Energy Card** in the other. You now represent a uranium-235 atom.
3. When you are tagged with a Neutron label from another student, tag two other student U-235 atoms with your Neutron labels. Drop your Thermal Energy Card into the **Energy Box.**
4. Observe as the remainder of the U-235 atoms are split, and imagine this happening extremely fast at the atomic level.

Analyze and Conclude

1. **Describe** what the simulation illustrated about nuclear fission.
2. **Predict** what would happen if, in the simulation, your classroom was filled wall-to-wall with U-235 atoms and the chain reaction got out of control.
3. **Key Concept** Identify one advantage and one disadvantage of nuclear energy.

Nuclear Energy

Animation

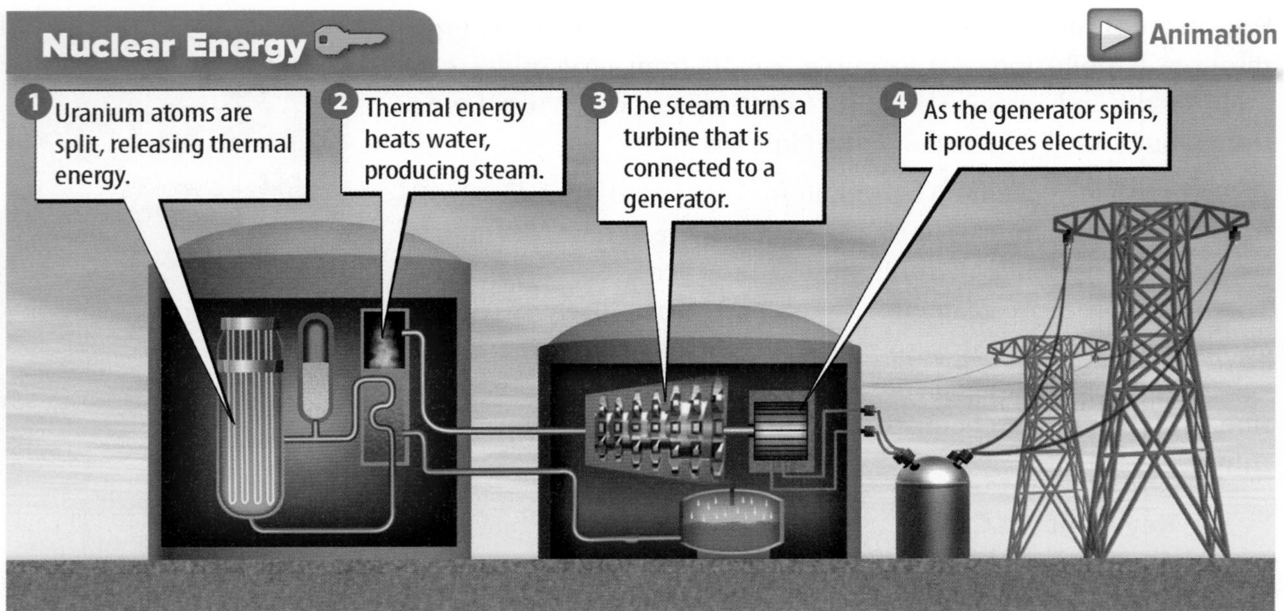

1 Uranium atoms are split, releasing thermal energy.
2 Thermal energy heats water, producing steam.
3 The steam turns a turbine that is connected to a generator.
4 As the generator spins, it produces electricity.

Figure 5 In a nuclear power plant, thermal energy released from splitting uranium atoms is transformed into electrical energy.

Nuclear Fission Nuclear power plants, such as the one shown in **Figure 5**, produce electricity using nuclear fission. This process splits atoms. Uranium atoms are placed into fuel rods. Neutrons are aimed at the rods and hit the uranium atoms. Each atom splits and releases two to three neutrons and thermal energy. The released neutrons hit other atoms, causing a chain reaction of splitting atoms. Countless atoms split and release large amounts of thermal energy. This energy heats water and changes it to steam. The steam turns a turbine connected to a generator, which produces electricity.

 **Reading Check** What are the steps in nuclear fission?

Advantages and Disadvantages of Nuclear Energy

One advantage of using nuclear energy is that a relatively small amount of uranium produces a large amount of energy. In addition, a well-run nuclear power plant does not pollute the air, the soil, or the water.

However, using nuclear energy has disadvantages. Nuclear power plants use a nonrenewable resource–uranium–for fuel. In addition, the chain reaction in the nuclear reactor must be carefully monitored. If it gets out of control, it can lead to a release of harmful radioactive substances into the environment.

The waste materials from nuclear power plants are highly radioactive and dangerous to living things. The waste materials remain dangerous for thousands of years. Storing them safely is important for both the environment and public health.

 **Reading Check** Why is it important to control a chain reaction?

Managing Nonrenewable Energy Resources

As shown in **Figure 6**, fossil fuels and nuclear energy provide about 87 percent of U.S. energy. Because these sources eventually will be gone, we must understand how to manage and conserve them. This is particularly important because energy use in the United States is higher than in other countries. Although only about 4.5 percent of the world's population lives in the United States, it uses more than 18 percent of the world's total energy.

Management Solutions

Mined land must be reclaimed. **Reclamation** *is a process in which mined land must be recovered with soil and replanted with vegetation.* Laws also help ensure that mining and drilling take place in an environmentally safe manner. In the United States, the Clean Air Act limits the amount of pollutants that can be released into the air. In addition, the U.S. Atomic Energy Act and the Energy Policy Act include regulations that protect people from nuclear emissions.

What You Can Do

Have you ever heard of vampire energy? Vampire energy is the energy used by appliances and other electronic equipment, such as microwave ovens, washing machines, televisions, and computers, that are plugged in 24 h a day. Even when turned off, they still consume energy. These appliances consume about 5 percent of the energy used each year. You can conserve energy by unplugging DVD players, printers, and other appliances when they are not in use.

You also can walk or ride your bike to help conserve energy. And, you can use renewable energy resources, which you will read about in the next lesson.

Key Concept Check How can you help manage nonrenewable resources wisely?

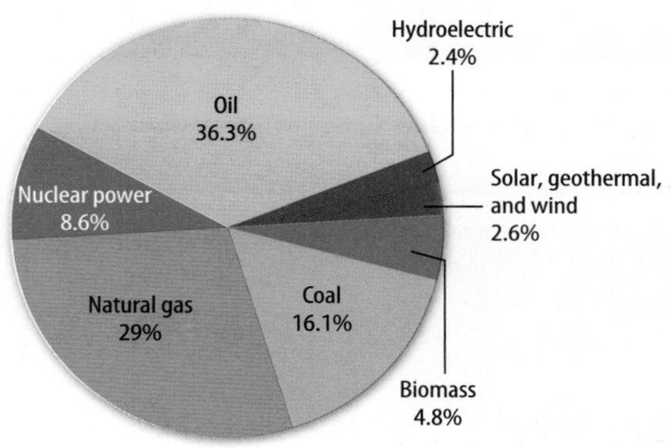

Figure 6 About 87 percent of the energy used in the United States comes from nonrenewable resources.

Visual Check Which energy source is used most in the United States?

ACADEMIC VOCABULARY

regulation
(noun) a rule dealing with procedures, such as safety

FOLDABLES®

Make a three-tab book. Before cutting the tabs, draw a Venn diagram and label as illustrated. Compare and contrast the use of fossil fuels and nuclear energy.

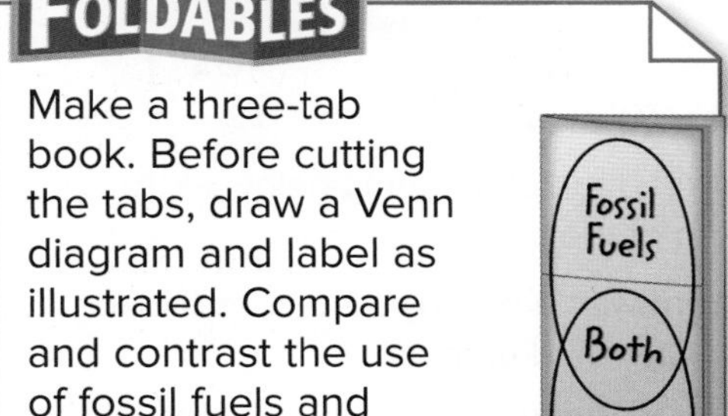

Lesson 1 Review

Visual Summary

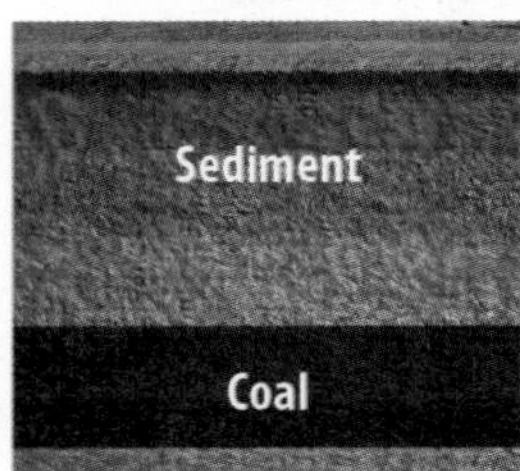

Fossil fuels include coal, oil, and natural gas. Fossil fuels take millions of years to form. Humans use fossil fuels at a much faster rate.

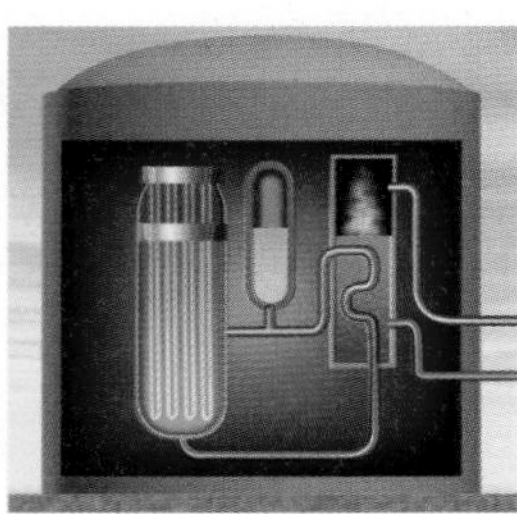

Nuclear energy comes from splitting atoms, or fission. Nuclear power plants must be monitored for safety, and nuclear waste must be stored properly.

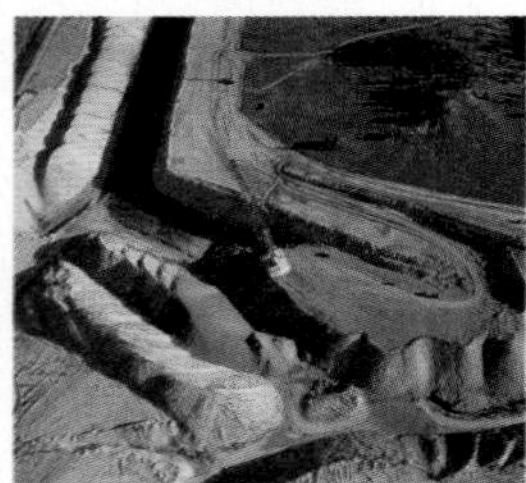

It is important to manage nonrenewable energy resources wisely. This includes mine reclamation, limiting air pollutants, and conserving energy.

Use your lesson Foldable to review the lesson. Save your Foldable for the project at the end of the chapter.

What do you think NOW?

You first read the statements below at the beginning of the chapter.

1. Nonrenewable energy resources include fossil fuels and uranium.
2. Energy use in the United States is lower than in other countries.

Did you change your mind about whether you agree or disagree with the statements? Rewrite any false statements to make them true.

Use Vocabulary

1. Energy produced from atomic reactions is called ________.
2. **Distinguish** between renewable and nonrenewable resources.
3. **Use the term** *reclamation* in a sentence.

Understand Key Concepts 

4. What is the source of most energy in the United States?
   - **A.** coal
   - **B.** oil
   - **C.** natural gas
   - **D.** nuclear energy
5. **Summarize** the advantages and disadvantages of using nuclear energy.
6. **Illustrate** Make a poster showing how you can conserve energy.

Interpret Graphics

7. **Sequence** Draw a graphic organizer like the one below to sequence the events in the formation of oil.

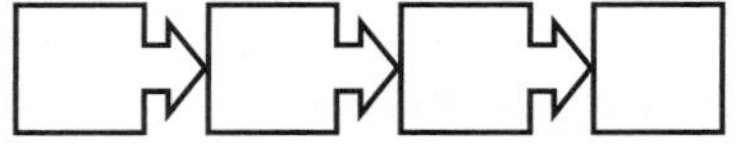

8. **Describe** Use the diagram below to describe the energy conversions that take place in a nuclear power plant.

Critical Thinking

9. **Suppose** that a nuclear power plant will be built near your town. Would you support the plan? Why or why not?
10. **Consider** Do the advantages of using fossil fuels outweigh the disadvantages? Explain your answer.

How can you identify bias and its source?

Whenever an author is trying to persuade, or convince, readers to share a particular opinion, you must read and evaluate carefully for bias. Bias is a way of thinking that tells only one side of a story, sometimes with inaccurate information.

Learn It

Sometimes a scientific investigation involves making judgments. When you make a judgment, you form an opinion. It is important to be honest and not allow any expectations of results to **bias** your judgments.

Try It

1. Read the passage to the right for sources of bias, such as

- claims not supported by evidence;
- persuasive statements;
- the author wanting to believe what he or she is saying, whether or not it is true.

Apply It

2. Analyze the passage, and identify two instances of bias and the source of each. Record this information in your Science Journal.
3. If you were the moderator at an EPA hearing about the issue, what would you do to solve the problem of bias?
4. **Key Concept** In your own words, explain how you would formulate an argument about the wise management of air resources while avoiding bias.

Factory Spews Out Air Pollution

Environmental organizations claim a coal-burning factory is polluting the air with toxic particulate matter in violation of U.S. Environmental Protection Agency (EPA) standards. Particulate matter is a mix of both solid and liquid particles in the air.

Citizens of the town collected air samples for a period of six months. An independent laboratory analysis of the samples showed a dangerously high level of the toxic particulate materials. Levels this high have been cited in medical journals as contributing to illness and death from asthma, respiratory disease, and lung cancer.

The factory has not updated its pollution-control equipment, claiming that it cannot afford the cost. At a town meeting, a company spokesperson claimed that the particulate matter is not harmful to human health. The state environmental director, who previously worked at the factory, stated that the jobs provided by the factory are more important to the state than environmental concerns.

Lesson 2

Reading Guide

Key Concepts
ESSENTIAL QUESTIONS

- What are the main sources of renewable energy?
- What are the advantages and disadvantages of using renewable energy resources?
- What can individuals do to encourage the use of renewable energy resources?

Vocabulary

solar energy p. 153
wind farm p. 154
hydroelectric power p. 154
geothermal energy p. 155
biomass energy p. 155

Multilingual eGlossary

Renewable Energy Resources

Inquiry What do these panels do?

These solar panels convert energy from the Sun into electrical energy. This solar power plant, at Nellis Air Force Base in Nevada, produces up to 42 percent of the electricity used on the base. What are some of the advantages of using energy from the Sun? What are some of the disadvantages?

Launch Lab

20 minutes

How can renewable energy sources generate energy in your home?

Renewable energy technologies can contribute to reducing our dependence on fossil fuels.

1. Review the table below. It shows how much energy, in Watt-hours, it takes to run certain appliances.
2. In one hour, a typical bicycle generator generates 200 W-h of electric energy; a small solar panel generates 150 W-h; and small wind turbines typically generate 100 W-h. Complete the table by calculating the time it would take for each alternative form of energy to generate the electricity needed to run each appliance for 1 h.

Hint: Use the following equation to solve for the time used by each energy source:

$$\begin{pmatrix}\text{Time used by}\\\text{energy source}\end{pmatrix} = \frac{\begin{pmatrix}\text{Time to use}\\\text{appliance}\end{pmatrix} \times \begin{pmatrix}\text{Energy used per hour}\\\text{by appliance}\end{pmatrix}}{\begin{pmatrix}\text{Energy produced per hour}\\\text{by energy source}\end{pmatrix}}$$

Think About This

1. Which appliance required the longest energy-generating time from the alternative energy sources? Why?
2. **Key Concept** What issues would you have to consider when using solar or wind energy to generate electricity in your home?

Appliance	Energy Used Per Hour	Time on Bike	Time for Solar Panel	Time for Wind Turbine
Desktop computer	75 W-h			
Hair dryer	1000 W-h			
Television	200 W-h			

Renewable Energy Resources

Could you stop the Sun from shining or the wind from blowing? These might seem like silly questions, but they help make an important point about renewable resources. Renewable resources come from natural processes that have been happening for billions of years and will continue to happen.

Solar Energy

Solar energy *is energy from the Sun.* Solar cells, such as those in watches and calculators, capture light energy and transform it to electrical energy. Solar power plants can generate electricity for large areas. They transform energy in sunlight, which then turns turbines connected to generators.

Some people use solar energy in their homes, as shown in **Figure 7.** Active solar energy uses technology, such as solar panels, that gathers and stores solar energy that heats water and homes. Passive solar energy uses design elements that capture energy in sunlight. For example, windows on the south side of a house can let in sunlight that helps heat a room.

Figure 7 People can use solar energy to provide electricity for their homes.

Russel Illig/Photodisc/Getty Images

▲ **Figure 8** Offshore wind farms are called wind parks. This wind park is in Denmark.

Wind Energy

Have you ever dropped your school papers outside and had them scattered by the wind? If so, you experienced wind energy. This renewable resource has been used since ancient times to sail boats and to turn windmills. Today, wind turbines, such as the ones shown in **Figure 8**, can produce electricity on a large scale. *A group of wind turbines that produce electricity is called a* **wind farm.**

Reading Check How is wind energy a renewable resource?

Water Energy

Like wind energy, flowing water has been used as an energy source since ancient times. Today, water energy produces electricity using different methods, such as hydroelectric power and tidal power.

Hydroelectric Power *Electricity produced by flowing water is called* **hydroelectric power.** To produce hydroelectric power, humans build a dam across a powerful river. **Figure 9** shows how flowing water is used to produce electricity.

Tidal Power Coastal areas that have great differences between high and low tides can be a source of tidal power. Water flows across turbines as the tide comes in during high tides and as it goes out during low tides. The flowing water turns turbines connected to generators that produce electricity.

Figure 9 In a hydroelectric power plant, energy from flowing water produces electricity. ▼

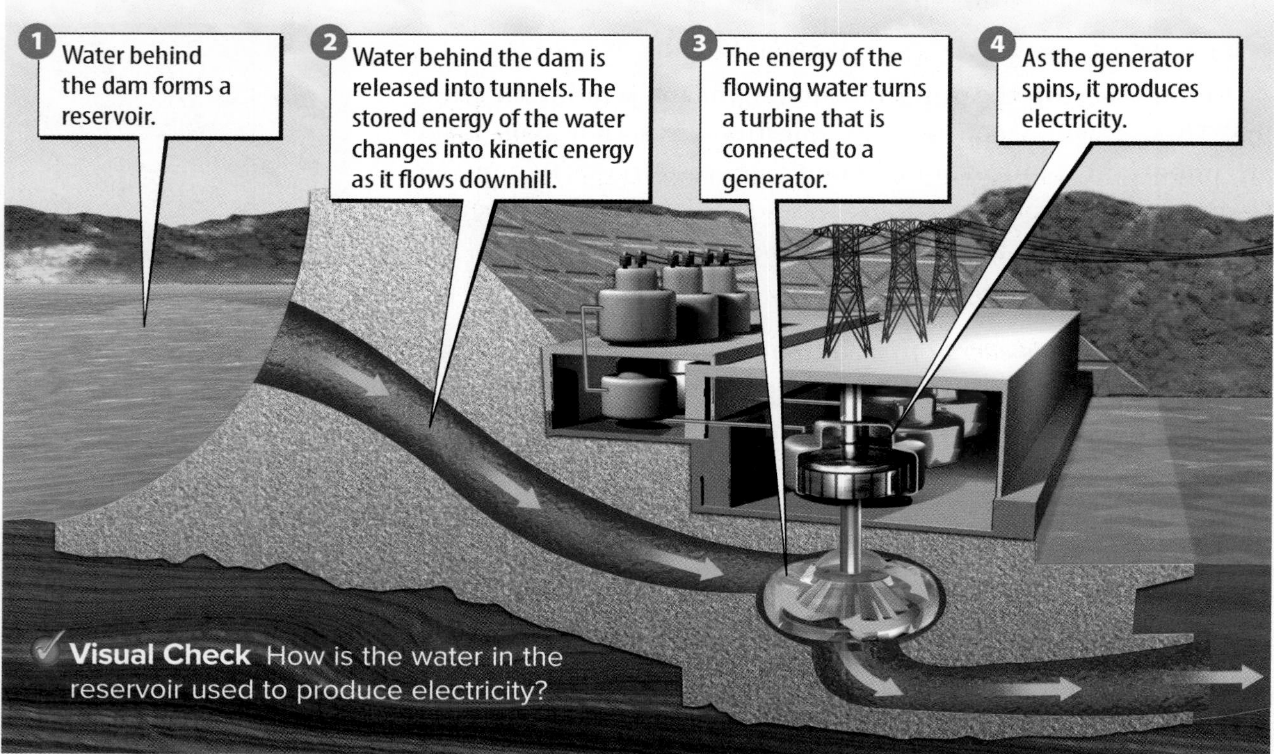

1 Water behind the dam forms a reservoir.
2 Water behind the dam is released into tunnels. The stored energy of the water changes into kinetic energy as it flows downhill.
3 The energy of the flowing water turns a turbine that is connected to a generator.
4 As the generator spins, it produces electricity.
Visual Check How is the water in the reservoir used to produce electricity?

Clynt Garnham Renewable Energy/Alamy

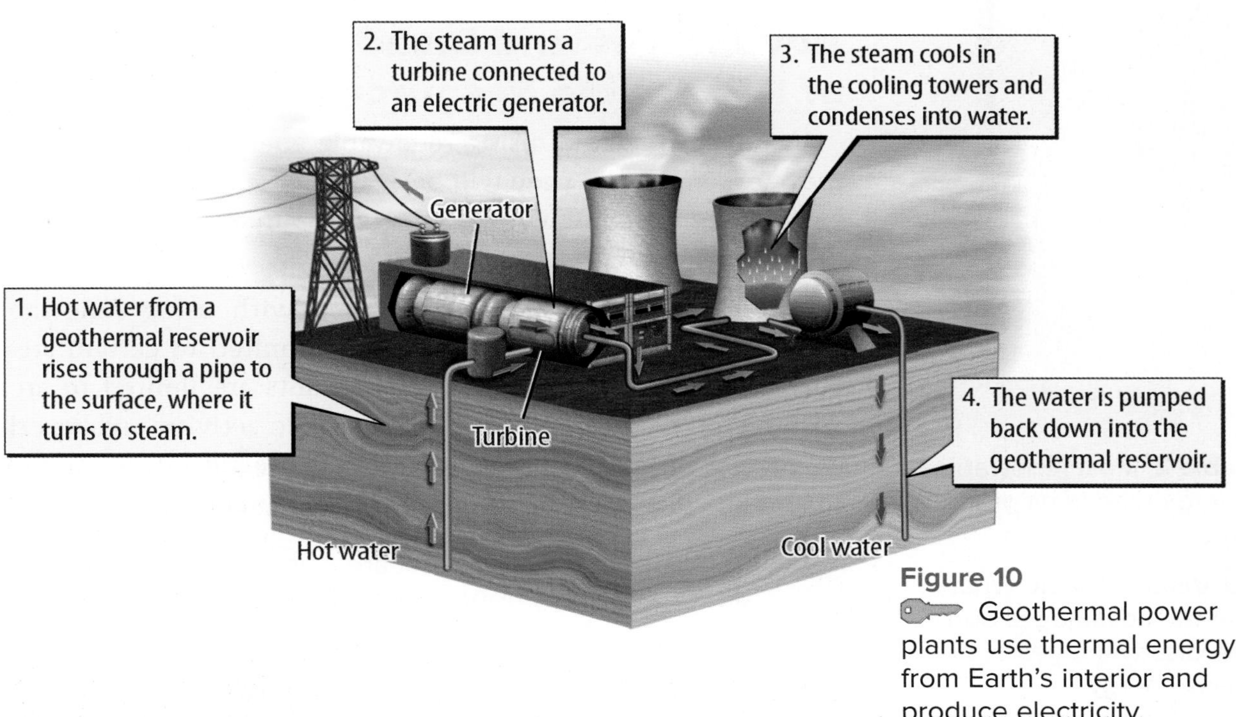

2. The steam turns a turbine connected to an electric generator.
3. The steam cools in the cooling towers and condenses into water.
Generator
1. Hot water from a geothermal reservoir rises through a pipe to the surface, where it turns to steam.
Turbine
4. The water is pumped back down into the geothermal reservoir.
Hot water
Cool water

Figure 10 Geothermal power plants use thermal energy from Earth's interior and produce electricity.

Animation

Geothermal Energy

Earth's core is nearly as hot as the Sun's surface. This thermal energy flows outward to Earth's surface. *Thermal energy from Earth's interior is called* **geothermal energy.** It can be used to heat homes and generate electricity in power plants, such as the one shown in **Figure 10.** People drill wells to reach hot, dry rocks or bodies of magma. The thermal energy from the hot rocks or magma heats water that makes steam. The steam turns turbines connected to generators that produce electricity.

WORD ORIGIN

geothermal
from Greek *ge-*, means "Earth"; and Greek *therme*, means "heat"

Biomass Energy

Since humans first lit fires for warmth and cooking, biomass has been an energy source. **Biomass energy** *is energy produced by burning organic matter, such as wood, food scraps, and alcohol.* Wood is the most widely used biomass. Industrial wood scraps and organic materials, such as grass clippings and food scraps, are burned to generate electricity on a large scale.

Biomass also can be converted into fuels for vehicles. Ethanol is made from sugars in plants, such as corn. Ethanol often is blended with gasoline. This reduces the amount of oil used to make the gasoline. Adding ethanol to gasoline also reduces the amount of carbon monoxide and other pollutants released by vehicles. Another renewable fuel, biodiesel, is made from vegetable oils and fats. It emits few pollutants and is the fastest-growing renewable fuel in the United States.

FOLDABLES®

Make a vertical five-tab Foldable. Label the tabs as illustrated. Identify the advantages and disadvantages of alternative fuels.

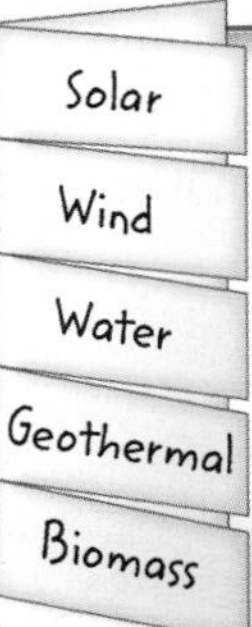

 **Key Concept Check** What are the main sources of renewable energy?

Advantages and Disadvantages of Renewable Resources

A big advantage of using renewable energy resources is that they are renewable. They will be available for millions of years to come. In addition, renewable energy resources produce less pollution than fossil fuels.

There are disadvantages associated with using renewable resources, however. Some are costly or limited to certain areas. For example, large-scale geothermal plants are limited to areas with tectonic activity. Recall that tectonic activity involves the movement of Earth's plates. **Table 2** lists the advantages and disadvantages of using renewable energy resources.

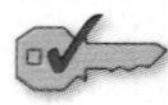

Key Concept Check What are some advantages and disadvantages of using renewable energy resources?

Table 2 Most renewable energy resources produce little or no pollution.

Visual Check What are the advantages and the disadvantages of biomass energy?

Interactive Table

Table 2 Renewable Resources—Advantages and Disadvantages

Renewable Resource	Advantages	Disadvantages
Solar energy	• nonpolluting • available in the United States	• less energy produced on cloudy days • no energy produced at night • high cost of solar cells • requires a large surface area to collect and produce energy on a large scale
Wind energy	• nonpolluting • relatively inexpensive • available in the United States	• large-scale use limited to areas with strong, steady winds • best sites for wind farms are far from urban areas and transmission lines • potential impact on bird populations
Water energy	• nonpolluting • available in the United States	• large-scale use limited to areas with fast-flowing rivers or great tidal differences • negative impact on aquatic ecosystems • production of electricity affected by long periods of little or no rainfall
Geothermal energy	• produces little pollution • available in the United States	• large-scale use limited to tectonically active areas • habitat disruption from drilling to build a power plant
Biomass energy	• reduces amount of organic material discarded in landfills • available in the United States	• air pollution results from burning some forms of biomass • less energy efficient than fossil fuels, costly to transport

Managing Renewable Energy Resources

Renewable energy currently meets only 7 percent of U.S. energy needs. As shown in **Figure 11**, most renewable energy comes from biomass. Solar energy, wind energy, and geothermal energy meet only a small percentage of U.S. energy needs. However, some states are passing laws that require the state's power companies to produce a percentage of electricity using renewable resources. Management of renewable resources often focuses on encouraging their use.

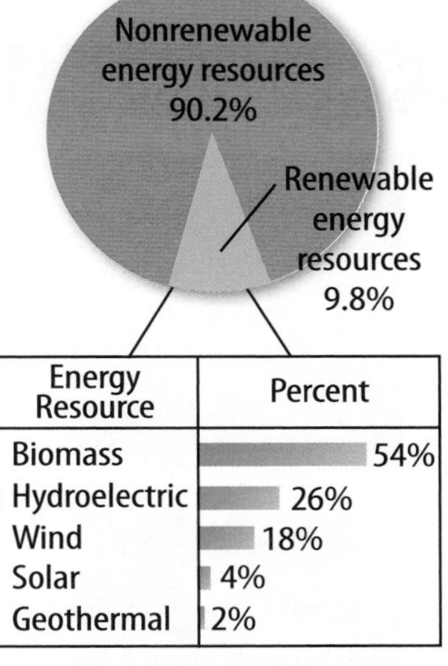

Figure 11 The renewable energy resource used most in the United States is biomass energy.

Management Solutions

The U.S. government has begun programs to encourage use of renewable resources. In 2009, billions of dollars were granted to the U.S. Department of Energy's Office of Energy Efficiency and Renewable Energy for renewable energy research and programs that reduce the use of fossil fuels.

What You Can Do

You might be too young to own a house or a car, but you can help educate others about renewable energy resources. You can talk with your family about ways to use renewable energy at home. You can participate in a renewable energy fair at school. As a consumer, you also can make a difference by buying products that are made using renewable energy resources.

Key Concept Check What can you do to encourage the use of renewable energy resources?

MiniLab

20 minutes

How are renewable energy resources used at your school?

Complete a survey about the use of renewable resources in your school.

1. Prepare interview questions about the use of renewable energy resources at your school. Each group member should come up with at least two questions.
2. Choose one group member to interview a school staff member.
3. Copy the table at the right into your Science Journal, and fill in the interview data.

Renewable Energy Source	Yes/No	Where is it used?	Why is it used? or Why isn't it used?
Sun			
Wind			
Water			
Geothermal			
Biomass			

Analyze and Conclude

1. **Explain** Which renewable energy resources are and are not being used? Why or why not?
2. **Key Concept** Choose one "why not" reason and describe how it could be addressed by communication with school planners.

Lesson 2 Review

Visual Summary

Renewable energy resources can be used to heat homes, produce electricity, and power vehicles.

Advantages of renewable energy resources include little or no pollution and availability.

Management of renewable energy resources includes encouraging their use and continuing to research more about their use.

FOLDABLES®

Use your lesson Foldable to review the lesson. Save your Foldable for the project at the end of the chapter.

What do you think NOW?

You first read the statements below at the beginning of the chapter.

3. Renewable energy resources do not pollute the environment.

4. Burning organic material can produce electricity.

Did you change your mind about whether you agree or disagree with the statements? Rewrite any false statements to make them true.

Use Vocabulary

1. **Define** *hydroelectric power* in your own words.
2. Burning wood is an example of _________ energy.

Understand Key Concepts 

3. Which can reduce the amount of organic material discarded in landfills?
   - **A.** biomass energy
   - **B.** solar energy
   - **C.** water energy
   - **D.** wind energy
4. **Compare and contrast** solar energy and wind energy.
5. **Determine** Your family wants to use renewable energy to heat your home. Which renewable energy resource is best suited to your area? Explain your answer.

Interpret Graphics

6. **Organize** Copy and fill in the graphic organizer below. In each oval, list a type of renewable energy resource.

7. **Compare** the use of renewable resources and nonrenewable resources in the production of electricity in the United States, based on the table below.

Sources of Electricity Generation, 2015

Energy Source	Percent
Fossil fuels	67%
Nuclear power	20%
Solar, wind, geothermal, biomass	7%
Hydroelectric	6%

Critical Thinking

8. **Design** and explain a model that shows how a renewable resource produces energy.

Skill Practice Analyze Data

40 minutes

How can you analyze energy-use data for information to help conserve energy?

As a student, you are not making large governmental policy decisions about uses of resources. As an individual, however, you can analyze data about energy use. You can use your analysis to determine some personal actions that can be taken to conserve energy resources.

Learn It

To **analyze the data** of fuel usage, you will need to look for patterns in the data, compare and categorize them, and determine cause and effect.

Try It

1. Study the fuel usage graph shown below. The data were collected from a house that uses natural gas as a source of energy to heat it.
2. Identify the time period that is covered by the graph.
3. Explain what is represented by the values on the vertical axis of the graph.
4. Describe the range of monthly gas usage over the 12-month period.
5. Group the monthly gas usage into three levels. Give each level a title. Enter these in your Science Journal.

Apply It

6. Categorize the three levels based on the amount of natural gas use.
7. Identify the three highest and four lowest months of gas usage. What might explain the usage patterns during these months?
8. Suppose the house from which the data came was heated with an electric furnace, instead of a furnace that used natural gas. What would you expect a usage graph for an electric furnace to look like?
9. **Key Concept** Formulate a list of heat conservation practices for homes.

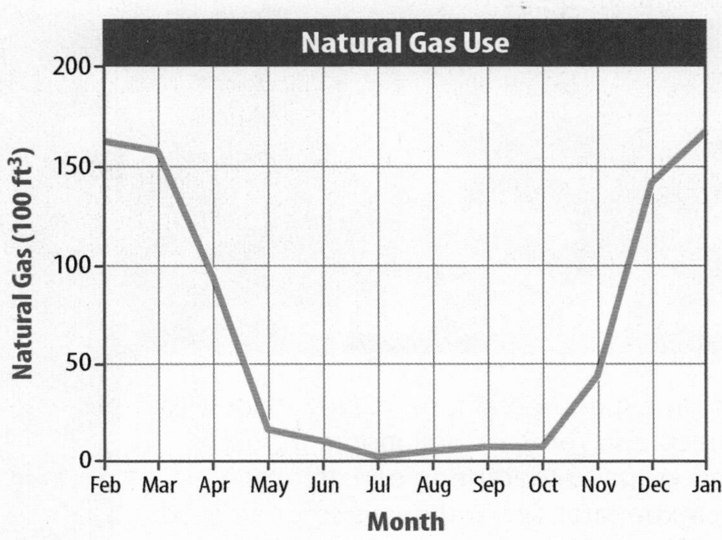

Lesson 3

Land Resources

Reading Guide

Key Concepts

ESSENTIAL QUESTIONS

- Why is land considered a resource?
- What are the advantages and disadvantages of using land as a resource?
- How can individuals help manage land resources wisely?

Vocabulary

ore p. 163

deforestation p. 164

Multilingual eGlossary

What's Science Got to do With It?

Inquiry A Garden on the Water?

The Science Barge is an experimental farm in New York City, New York. It saves space and reduces pollution and fossil fuel use while growing crops to feed people in an urban area. Why are people experimenting with ways to grow food that have fewer environmental impacts? Why is it important for humans to use land resources wisely?

Launch Lab

20 minutes

What resources from the land do you use every day?

The land on which humans live is part of Earth's crust. It provides resources that enable humans and other organisms to survive.

1. Make a list of every item you use in a 24-h period as you carry out your daily activities.
2. Combine your list with your group members' lists and decide which items contain resources from the land. Design a graphic organizer to group the materials into categories.
3. Fill in the graphic organizer on **chart paper.** Use a **highlighter** or **colored markers** to show which resources are renewable and which are nonrenewable.
4. Post your chart and compare it with the others in your class.

Think About This

1. Are there any times in your day when you do not use a resource from the land? Provide an example.
2. Describe the major categories that you used to organize your list of resources.
3. **Key Concept** Why do you think land is considered a resource?

Land as a Resource

A natural resource is something from Earth that living things use to meet their needs. People use soil for growing crops and forests to harvest wood for making furniture, houses, and paper products. They mine minerals from the land and clear large areas for roads and buildings. In each of these cases, people use land as a natural resource to meet their needs.

Key Concept Check Why is land considered a resource?

Living Space

No matter where you live, you and all living things use land for living space. Living space includes natural habitats, as well as the land on which buildings, sidewalks, parking lots, and streets are built. As shown in **Figure 12,** cities make up only a small percentage of land use in the United States. Most land is used for agriculture, grasslands, and forests.

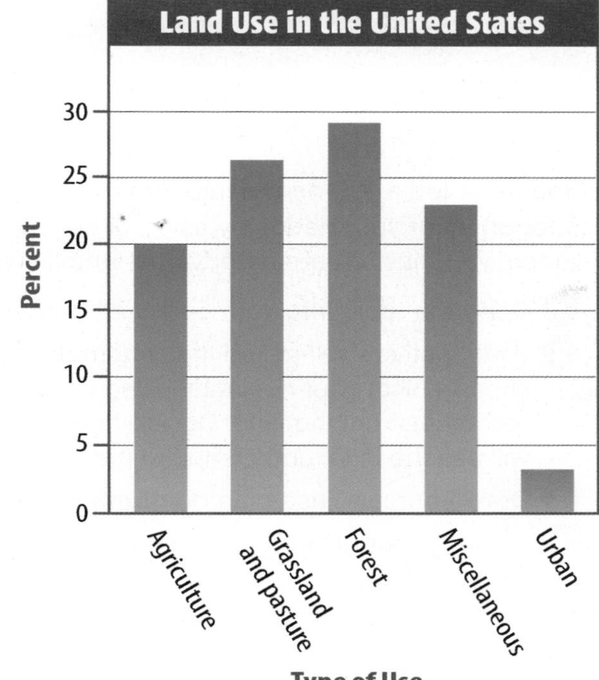

Figure 12 Forests and grasslands make up the largest categories of U.S. land use.

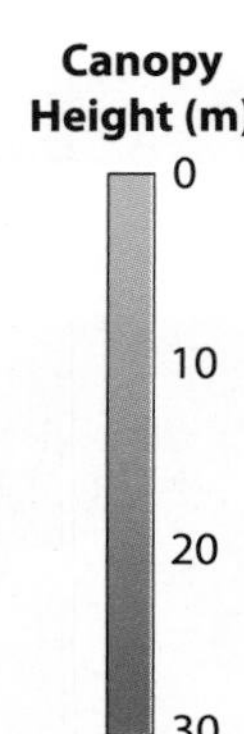

Figure 13 Much of the U.S. eastern forest has been replaced by cities, farms, and other types of development.

Visual Check Compare forest cover in the eastern United States in 1650 and 1920.

Forests and Agriculture

As shown in **Figure 13**, forests covered much of the eastern United States in 1650. By 1920, the forests had nearly disappeared. Although some of the trees have grown back, they are not as tall and the forests are not as complex as they were originally.

Forests were cut down for the same reasons that forests are cut down today: for fuel, paper products, and wood products. People also cleared land for development and agriculture. Today, about one-fifth of U.S. land is used for growing crops and about one-fourth is used for grazing livestock.

Reading Check Why are forests cut down?

MiniLab

20 minutes

How can you manage land resource use with environmental responsibility?

You inherited a 100-acre parcel of forested land. Your relative's will stated that you must support all of your needs by using or selling resources from the land. To receive the inheritance, you must create an environmentally responsible land-use plan.

1. Copy the table into your Science Journal.
2. Your relative's will stated that you must support all of your needs by using or selling resources from the land. Decide how you will use the land, and complete the table.
3. Draw your land use plan on **graph paper.**
4. Present your group's land use plan to the class, and explain your reasoning.

Land Use	Percent of Total Area	Reasoning
Forest		
House and yard		
Garden		
Mineral mine		
Other		

Analyze and Conclude

1. **Compare and contrast** the design and reasoning of your plan with another group's plan.
2. **Identify** additional information about the land parcel that you would need to refine your plan.
3. **Key Concept** Summarize two environmentally responsible practices that were used by more than one group in their plan.

Mineral Resources

Recall that coal, an energy resource, is mined from the land. Certain minerals also are mined to make products you use every day. These minerals often are called ores. **Ores** *are deposits of minerals that are large enough to be mined for a profit.*

The house in **Figure 14** contains many examples of common items made from mineral resources. Some of these come from metallic mineral resources. Ores such as bauxite and hematite are metallic mineral resources. They are used to make metal products. The aluminum in automobiles and refrigerators comes from bauxite. The iron in nails and faucets comes from hematite. Some mineral resources come from nonmetallic mineral resources, such as gypsum, halite, and minerals found in sand and gravel. Nonmetallic mineral resources also are mined from the land. The sulfur used in paints and rubber and the fluorite used in paint pigments are other examples of nonmetal mineral resources.

WORD ORIGIN

ore
from Old English *ora*, means "unworked metal"

Figure 14 Many common products are made from mineral resources.

Visual Check Identify two products made from nonmetallic mineral resources.

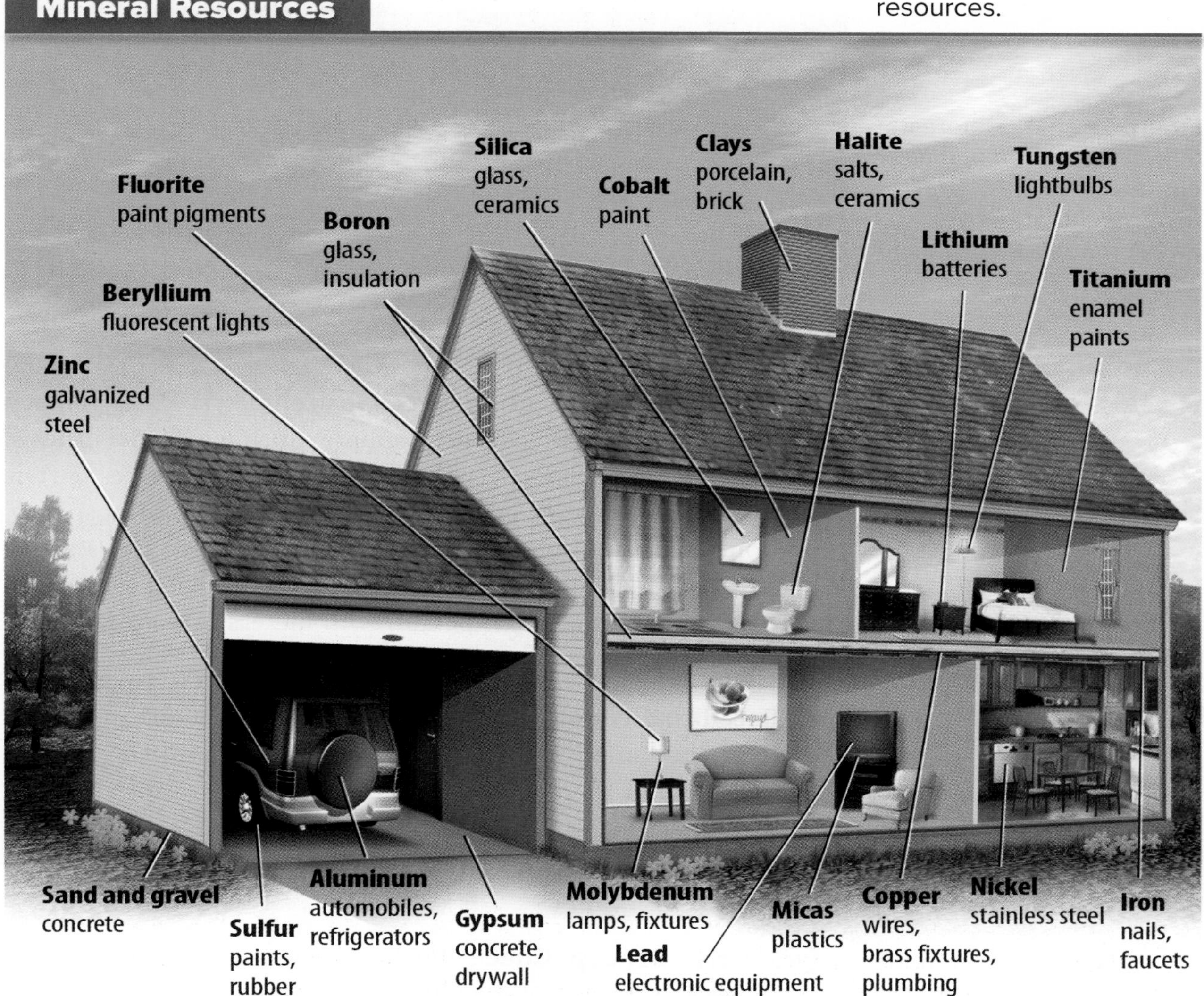

FOLDABLES®

Make a horizontal two-tab book. Label the tabs as illustrated. Use the Foldable to record your notes about renewable and nonrenewable land resources.

Renewable Land Resources | Nonrenewable Land Resources

Advantages and Disadvantages of Using Land Resources

Land resources such as soil and forests are widely available and easy to access. In addition, crops and trees are renewable—they can be replanted and grown in a relatively short amount of time. These are all advantages of using land resources.

Some land resources, however, are nonrenewable. It can take millions of years for minerals to form. This is one disadvantage of using land resources. Other disadvantages include deforestation and pollution.

Deforestation

As shown in **Figure 15**, humans sometimes cut forests to clear land for grazing, farming, and other uses. **Deforestation** *is the cutting of large areas of forests for human activities.* It leads to soil erosion and loss of animal habitats. In tropical rain forests—complex ecosystems that can take hundreds of years to replace—deforestation is a serious problem.

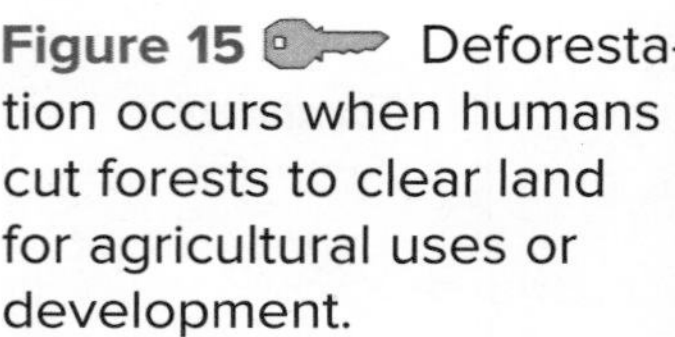

Figure 15 Deforestation occurs when humans cut forests to clear land for agricultural uses or development.

Karen Huntt/Getty Images

Deforestation also can affect global climates. Trees remove carbon dioxide from the atmosphere during photosynthesis. Rates of photosynthesis decrease when large areas of trees are cut down, and more carbon dioxide remains in the atmosphere. Carbon dioxide helps trap thermal energy within Earth's atmosphere. Increased concentrations of carbon dioxide can cause Earth's average surface temperatures to increase.

Pollution

REVIEW VOCABULARY

runoff
rainwater that does not soak into the ground and flows over Earth's surface

Recall that runoff from coal mines can affect soil and water quality. The same is true of mineral mines. Runoff that contains chemicals from these mines can pollute soil and water. In addition, many farmers use chemical fertilizers to help grow crops. Runoff containing fertilizers can pollute rivers, soil, and underground water supplies.

Key Concept Check What are some advantages and disadvantages of using land resources?

Managing Land Resources

Because some land uses involve renewable resources while others do not, managing land resources is complex. For example, a tree is renewable. But forests can be nonrenewable because some can take hundreds of years to fully regrow. In addition, the amount of land is limited, so there is competition for space. Those who manage land resources must balance all of these issues.

Management Solutions

One way governments can manage forests and other unique ecosystems is by preserving them. On preserved land, logging and development is either banned or strictly controlled. Large areas of forests cannot be cut. Instead, loggers cut selected trees and then plant new trees to replace ones they cut.

Science Use v. Common Use

preserve

Science Use to keep safe from injury, harm, or destruction

Common Use to can, pickle, or save something for future use

Land mined for mineral resources also must be preserved. On both public and private lands, mined land must be restored according to government regulations.

Land used for farming and grazing can be managed to conserve soil and improve crop yield. Farmers can leave crop stalks after harvesting to protect soil from erosion. They also can use organic farming techniques that do not use synthetic fertilizers.

What You Can Do

You can help conserve land resources by recycling products made from land resources. You can use yard waste and vegetable scraps to make rich compost for gardening, reducing the need to use synthetic fertilizers. Compost is a mix of decayed organic material, bacteria and other organisms, and small amounts of water. **Figure 16** shows another way you can help manage land resources wisely.

Key Concept Check What can you do to help manage land resources wisely?

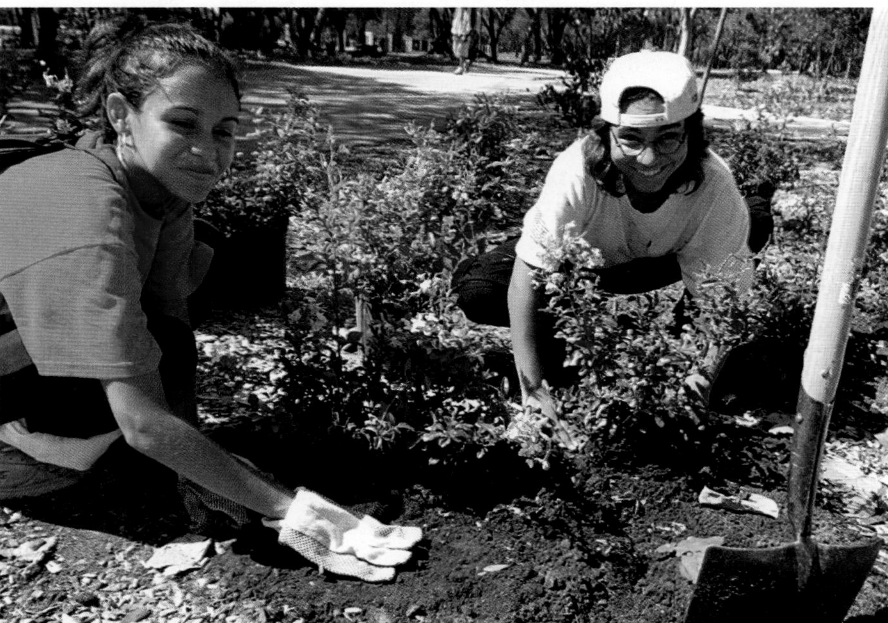

Jeff Greenberg/Alamy

Figure 16 A community garden is one way to help manage land resources wisely.

Lesson 3 Review

Visual Summary

Land is a natural resource that humans use to meet their needs.

Disadvantages of using land as a resource include deforestation, which leads to increased erosion and increased carbon dioxide in the atmosphere.

Individuals can help manage land resources wisely by recycling, composting, and growing food in community gardens.

FOLDABLES®

Use your lesson Foldable to review the lesson. Save your Foldable for the project at the end of the chapter.

What do you think NOW?

You first read the statements below at the beginning of the chapter.

5. Cities cover most of the land in the United States.

6. Minerals form over millions of years.

Did you change your mind about whether you agree or disagree with the statements? Rewrite any false statements to make them true.

Use Vocabulary

1. Cutting down forests for human activities is called ________.
2. **Use the word** *ore* in a sentence.

Understand Key Concepts 

3. One disadvantage of using metallic mineral resources is that these resources are
   - **A.** easy to mine.
   - **B.** inexpensive.
   - **C.** nonrenewable.
   - **D.** renewable.
4. **Give an example** of how people use land as a resource.
5. **Compare** the methods used by governments and individuals to manage land resources wisely.

Interpret Graphics

6. **Take Notes** Copy the graphic organizer below, and list at least two land resources mentioned in this lesson. Describe how using each affects the environment.

Land Resource	How Use Affects Environment

7. **Identify** whether the resources shown here are from metallic or nonmetallic mineral resources.

Critical Thinking

8. **Design** a way to manage land resources wisely. Use a method that is not discussed in this lesson.
9. **Decide** Land is a limited resource. There often is pressure to develop preserved land. Do you think this should happen? Why or why not?

(t)Karen Hunt/Getty Images (b)Jeff Greenberg/Alamy

AMERICAN MUSEUM OF NATURAL HISTORY

GREEN SCIENCE

A Greener Greensburg

A town struck by disaster makes the world a greener place.

In May 2007, a powerful tornado struck the small Kansas town of Greensburg. The tornado destroyed almost every home, school, and business. Six months later, the town's officials and residents decided to rebuild Greensburg as a model green community.

The town's residents pledged to use fewer natural resources; to produce clean, renewable energy; and to reuse and recycle waste. As part of this effort, every new home and building would be designed for energy efficiency. The homes also would be constructed of materials that are healthful for the people who live and work in them.

What is a model green town? Here are some ways Greensburg will help the environment, save money, and make life better for its residents.

▲ Rain gardens help improve water quality by filtering pollutants from runoff.

USE RENEWABLE ENERGY

- **Produce clean energy** with renewable energy sources such as wind and sunlight. Wind turbines capture the abundant wind power of the Kansas plains.
- **Cut back on greenhouse gas emissions** with electric or hybrid city vehicles.

BUILD GREEN BUILDINGS

- **Design every home, school, and office** to use less energy and promote better health.
- **Make the most of natural daylight** for indoor lighting with many windows, which also can be opened for fresh air.
- **Use green materials** that are nontoxic and locally grown or made from recycled materials.

CONSERVE WATER

- **Capture runoff and rainwater** with landscape features such as rain gardens, bowl-shaped gardens designed to collect and absorb excess rainwater.
- **Use low-flow** faucets, shower heads, and toilets.

CREATE A HEALTHY ENVIRONMENT

- **Provide parks and green spaces** filled with native plants that need little water or care.
- **Create a "walkable community"** to encourage people to drive less and be more active, with a town center connected to neighborhoods by sidewalks and trails.

PROBLEM SOLVING With your group, choose one of Greensburg's projects. Make a plan describing how it could be implemented in your community and what its benefits would be.

Lesson 4

Reading Guide

Key Concepts

ESSENTIAL QUESTIONS

- Why is it important to manage air and water resources wisely?
- How can individuals help manage air and water resources wisely?

Vocabulary

photochemical smog p. 170

acid precipitation p. 170

Multilingual eGlossary

BrainPOP®

Go to the resource tab in ConnectED to find the PBLs *Where in the World...?* and *7 Billion and Counting.*

Air and Water Resources

Inquiry Are these crop circles?

No, this dotted landscape in Colorado is the result of circle irrigation. The fields are round because the irrigation equipment pivots from the center of the field and moves in a circle to water the crops. Crop irrigation accounts for about 34 percent of water used in the United States.

Launch Lab

20 minutes

How often do you use water each day?

In most places in the United States, people are fortunate to have an adequate supply of clean water. When you turn on the faucet, do you think about the value of water as a resource?

1. Prepare a two-column table to collect data on the number of times you use water in one day. Title the first column *Purpose* and the second column *Times Used.*
2. In the *Purpose* column, describe how you used the water, such as *Faucet, Toilet, Shower/Bath, Dishwasher, Laundry, Leaks,* and *Other.*
3. In the *Times Used* column, record and tally the total number of times you used water.
4. Calculate the percent that you use water for each category. Construct a circle graph showing the percentages of use in a day.

Think About This

1. For which purpose did you use water the most? The least?
2. **Key Concept** In which category, or categories, could you conserve water? How?

Importance of Air and Water

Using some natural resources, such as fossil fuels and minerals, makes life easier. You would miss them if they were gone, but you would still survive.

Air and water, on the other hand, are resources that you cannot live without. Most living things can survive only a few minutes without air. Oxygen from air helps your body provide energy for your cells.

Water also is needed for many life functions. As shown in **Figure 17**, water is the main component of blood. Water helps protect body tissues, helps maintain body temperature, and has a role in many chemical reactions, such as the digestion of food. In addition to drinking water, people use water for other purposes that you will learn about later in this lesson, including agriculture, transportation, and recreation.

Reading Check What are the functions of water in the human body?

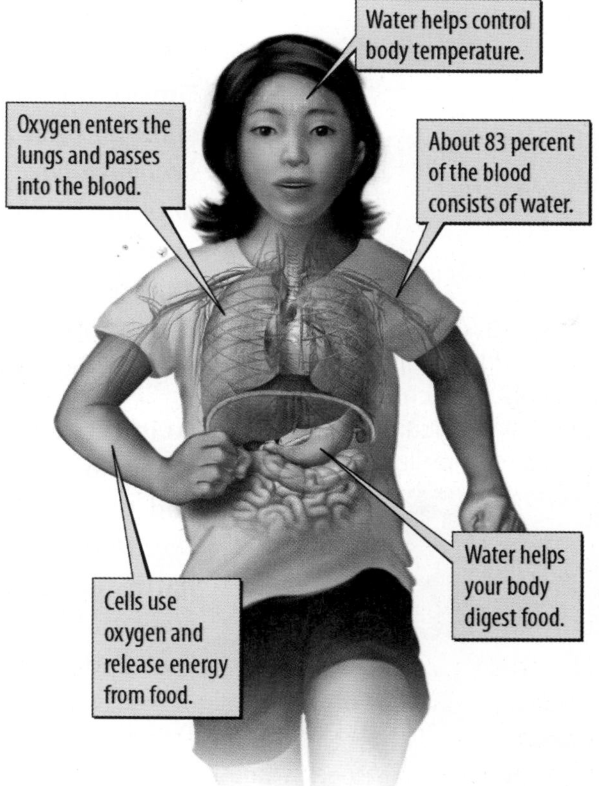

Water helps control body temperature.
Oxygen enters the lungs and passes into the blood.
About 83 percent of the blood consists of water.
Cells use oxygen and release energy from food.
Water helps your body digest food.

Figure 17 Your body needs oxygen and water to carry out its life-sustaining functions.

Michelle D. Bridwell/PhotoEdit

Figure 18 Sometimes a layer of warm air can trap smog in the cooler air close to Earth's surface. The smog can cover an area for days.

Visual Check Where does the pollution that forms smog come from?

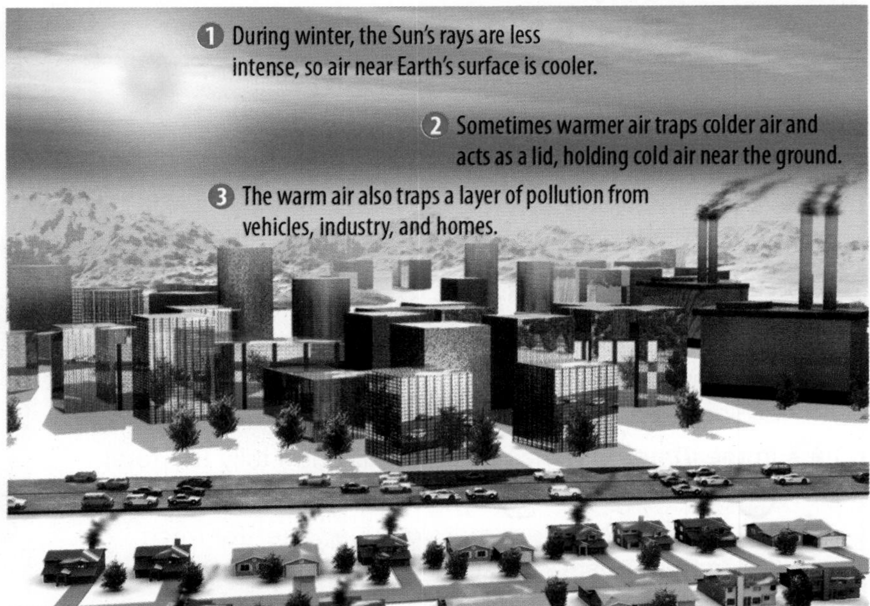

1 During winter, the Sun's rays are less intense, so air near Earth's surface is cooler.
2 Sometimes warmer air traps colder air and acts as a lid, holding cold air near the ground.
3 The warm air also traps a layer of pollution from vehicles, industry, and homes.

Air

Most living things need air to survive. However, polluted air, such as the air in **Figure 18**, can actually harm humans and other living things. Air pollution is produced when fossil fuels burn in homes, vehicles, and power plants. It also can be caused by natural events, such as volcanic eruptions or forest fires.

Reading Check What activities can cause air pollution?

Smog Burning fossil fuels releases not only energy, but also substances, such as nitrogen compounds. **Photochemical smog** *is a brownish haze produced when nitrogen compounds and other pollutants in the air react in the presence of sunlight.* Smog can irritate your respiratory system. In some individuals, it can increase the chance of asthma attacks. Smog can be particularly harmful when it is trapped under a layer of warm air and remains in an area for several days, also shown in **Figure 18.**

Acid Precipitation Nitrogen and sulfur compounds released when fossil fuels burn can react with water in the atmosphere and produce acid precipitation. **Acid precipitation** *is precipitation that has a pH less than 5.6.* When it falls into lakes, it can harm fish and other organisms. It also can pollute soil and kill trees and other plants. Acid precipitation can even damage buildings and statues made of some types of rocks.

Natural Events Forest fires and volcanic eruptions, such as the one shown in **Figure 19**, release gases, ash, and dust into the air. Dust and ash from one volcanic eruption can spread around the world. Materials from forest fires and volcanic eruptions can cause health problems similar to those caused by smog.

Figure 19 Gas and dust released by erupting volcanoes, such as Karymsky Volcano in Russia, can pollute the air.

Water

Suppose you saved $100, but you were only allowed to spend 90 cents. You might be very frustrated! If all of the water on Earth were your $100, freshwater that we can use is like that 90 cents you can spend. As shown in **Figure 20**, most water on Earth is salt water. Only 3 percent is freshwater, and most of that is frozen in glaciers. That leaves just a small part, 0.9 percent, of the total amount of water on Earth for humans to use.

This relatively small supply of freshwater must meet many needs. In addition to drinking water, people use water for farming, industry, electricity production, household activities, transportation, and recreation. Each of these uses can affect water quality. For example, water used to irrigate fields can mix with fertilizers. This polluted water then can run off into rivers and groundwater, reducing the quality of these water supplies. Water used in industry often is heated to high temperatures. The hot water can harm aquatic organisms when it is returned to the environment.

Reading Check How can farming affect water quality?

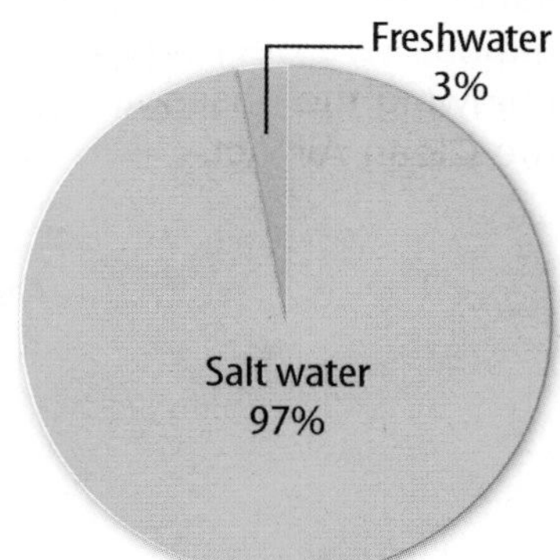

Figure 20 Freshwater makes up only 3 percent of Earth's water.

MiniLab

20 minutes

How much water can a leaky faucet waste?

You are competing for the job of environmental consultant at your school. One of the competition requirements is to complete an analysis of water waste from existing faucets.

1. Read and complete a lab safety form.
2. Catch the water from a **leaking faucet** in a **beaker.** Time the collection for 1 min with a **stopwatch.**
3. Use a **50-mL graduated cylinder** to measure the amount of water lost. Record the amount of water, in milliliters per minute, that leaked from the faucet in your Science Journal.
4. Make a table to show the amount of water that would leak from the faucet in 1 hour, 1 day, 1 week, 1 month, and 1 year.

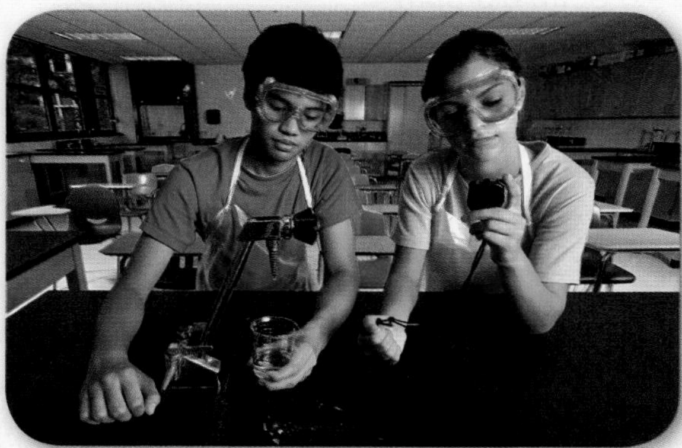

Hutchings Photography/Digital Light Source

Analyze and Conclude

1. **Construct** a graph of your data. Label the axes and title your graph. Explain what the graph illustrates.
2. **Describe** how many liters of water would be wasted by the leak over a period of one year. Explain how you arrived at that figure.
3. **Key Concept** As an environmental consultant, what information and recommendations would your report contain about water waste in the school?

Figure 21 The amount of sulfur compounds in the atmosphere decreased following the passage of the Clean Air Act.

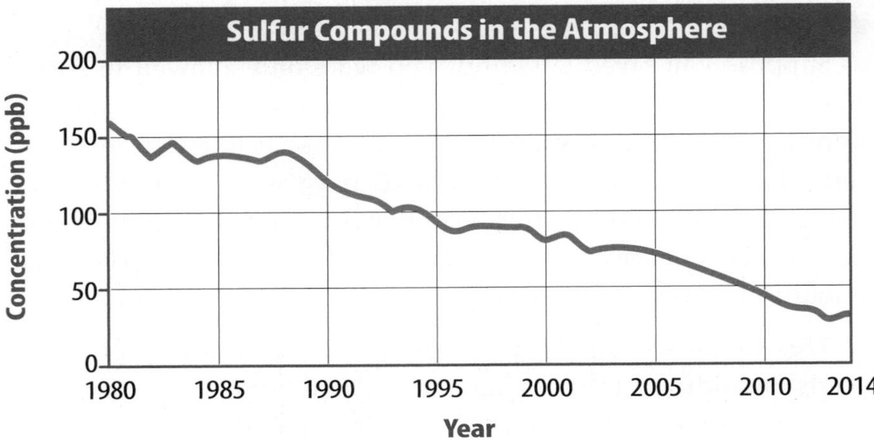

Math Skills 

Use Percentages

The carbon monoxide (CO) level in Portland air went from 6.3 parts per million (ppm) in 1990 to 1.3 ppm in 2014. What was the percent change in CO levels?

1. Subtract the starting value from the final value.

   1.3 ppm − 6.3 ppm = −5.0 ppm

2. Divide the difference by the starting value.

   −5.0 ppm/6.3 ppm = −0.794

3. Multiply by 100 and add a % sign.

   −0.794 × 100 = −79.4%

It decreased by 79.4%.

Practice

Between 1990 and 2014, the ozone (O_3) levels in Richmond went from 0.084 ppm to 0.062 ppm. What was the percent change in ozone levels?

 Math Practice

 Personal Tutor

Managing Air and Water Resources

Animals and plants do not use natural resources to produce electricity or to raise crops. But they do use air and water. Management of these important resources must consider both human needs and the needs of other living things.

 **Key Concept Check** Why is it important to manage air and water resources wisely?

Management Solutions

Legislation is an effective way to reduce air and water pollution. The regulations of the U.S. Clean Air Act, passed in 1970, limit the amount of certain pollutants that can be released into the air. The graph in **Figure 21** shows how levels of sulfur compounds have decreased since the act became law.

Similar laws are now in place to maintain water quality. The U.S. Clean Water Act legislates the reduction of water pollution. The Safe Drinking Water Act legislates the protection of drinking water supplies. By reducing pollution, these laws help ensure that all living things have access to clean air and water.

What You Can Do

You have learned that reducing fossil fuel use and improving energy efficiency can reduce air pollution. You can make sure your home is energy efficient by cleaning air-conditioning or heating filters and using energy-saving lightbulbs.

You can help reduce water pollution by properly disposing of harmful chemicals so that less pollution runs off into rivers and streams. You can volunteer to help clean up litter from a local stream. You also can conserve water so there is enough of this resource for you and other living things in the future.

 **Key Concept Check** How can individuals help manage air and water resources wisely?

Lesson 4 Review

Visual Summary

Sources of air pollution include the burning of fossil fuels in vehicles and power plants, and natural events such as volcanic eruptions and forest fires.

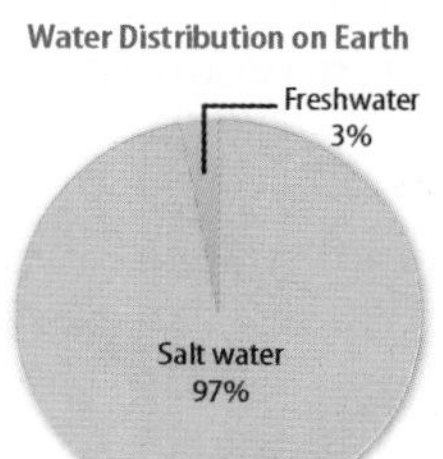

Only a small percentage of Earth's water is available for humans to use. Humans use water for agriculture, industry, recreation, and cleaning.

Management of air and water resources includes passing laws that regulate sources of air and water pollution. Individuals can reduce energy use and dispose of chemicals properly to help keep air and water clean.

FOLDABLES®

Use your lesson Foldable to review the lesson. Save your Foldable for the project at the end of the chapter.

What do you think NOW?

You first read the statements below at the beginning of the chapter.

7. Humans need oxygen and water to survive.

8. About 10 percent of Earth's total water can be used by humans.

Did you change your mind about whether you agree or disagree with the statements? Rewrite any false statements to make them true.

Use Vocabulary

1. **Define** *acid precipitation* in your own words.
2. Air pollution caused by the reaction of nitrogen compounds and other pollutants in the presence of sunlight is ________.

Understand Key Concepts

3. About how much of Earth's water is available for humans to use?
   A. 0.01 percent
   B. 0.90 percent
   C. 3.0 percent
   D. 97.0 percent
4. **Relate** In terms of human health, why is it important to manage air resources wisely?
5. **List** ways your classroom could improve its energy efficiency.

Interpret Graphics

6. **Determine Cause and Effect** Copy and fill in the graphic organizer below to describe three effects of acid precipitation.

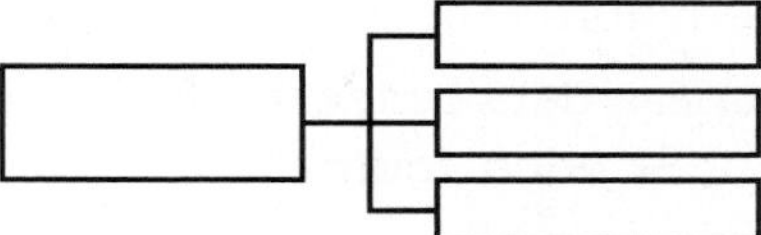

Critical Thinking

7. **Evaluate** The top three categories of household water use in the United States are flushing the toilet, washing clothes, and taking showers. Evaluate your water use, and list one thing you could do to reduce your use in each category.

Math Skills

Math Practice

8. Between 1990 and 2014, the amount of sulfur dioxide (SO_2) in Miami's air went from 14 ppb to 1 ppb. What was the percent change of SO_2?

Research Efficient Energy and Resource Use

A community organization is encouraging your school's board of education to participate in the "Green Schools" program. Your class has been nominated to research and report on the present status of energy efficiency and resource use in the school. The results of the report will be used as information for the presentation. Your task is to choose a natural resource and collect data about how it is presently used in the school. Your group will then recommend environmentally responsible management practices.

Question

How can a natural resource be used more wisely at school?

Procedure

1. Read and complete a lab safety form.
2. With your group, choose one of these resources to research its use in your school: water, land, air, or an energy resource.
3. For your chosen resource, plan how you will research resource use. What questions will you ask? How much of the resource is used by the school? Is it used efficiently? How could it be used more efficiently, or how could it be conserved? Have your teacher approve your plan.
4. Prepare data collection forms like the one below to record the results of your research in your Science Journal.
5. Conduct your research, and enter the data on the forms.

Sample Data Table				
Resource: Water Areas of Research: Water Loss Through Leaks and Recycling System				
Location	Faucets	Water Fountains	Toilets	"Gray" Water Recycling System
Washroom	6 good 2 poor 2 leaking		4 good 1 leaking	no
Hallway		3 good 1 leaking		no
Classroom 101	1 good			no
Classroom 102	1 good			no
Classroom 103	1 leaking			no

6. Review and summarize the data. Perform any necessary calculations to convert values to annual usage.
7. Conduct interviews, or collect more data about areas of research for which you need additional information.
8. After analyzing your data, write a proposal suggesting how the resource can be wisely managed in your school.
9. Compare the elements you addressed in your research with those recommended by a state or a national environmental organization. Did your research include everything?
10. Modify your proposal, if necessary. Record your revisions in your Science Journal.

Analyze and Conclude

11. **Graph** and explain the results of your data analysis.
12. **Predict** one impact on the environment of the existing management practices of the resource that you audited.
13. **The Big Idea** Describe two recommendations that you would make to the school's board of education about changes in resource management practices.

Communicate Your Results

Present the results of your research and your proposal to the class. Use appropriate visual aids to help make your points.

Extension

Combine information and reports from groups that investigated other resources from the list in step 2 so that all four resources are represented. Make a final report that includes recommendations for efficient use of each resource at your school.

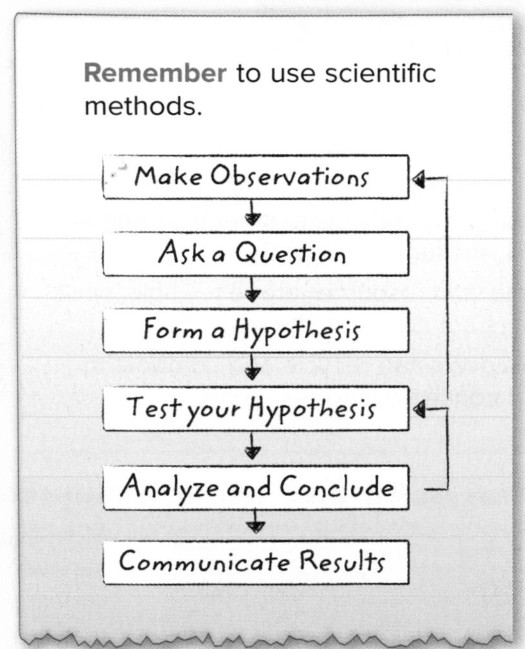

Chapter 5 Study Guide

Wise management of natural resources helps extend the supply of nonrenewable resources, reduce pollution, and improve soil, air, and water quality.

Key Concepts Summary	Vocabulary
Lesson 1: Energy Resources  • **Nonrenewable resources** include fossil fuels and uranium, which is used for **nuclear energy.** • Nonrenewable energy resources are widely available and easy to convert to energy. However, using these resources can cause pollution and habitat disruption. Safety concerns also are an issue. • People can conserve energy to help manage these resources.	**nonrenewable resource** p. 143 **renewable resource** p. 143 **nuclear energy** p. 147 **reclamation** p. 149
Lesson 2: Renewable Energy Resources  • Renewable energy resources include **solar energy,** wind energy, water energy, **geothermal energy,** and **biomass energy.** • Renewable resources cause little to no pollution. However, some types of renewable energy are costly or limited to certain areas. • Individuals can help educate others about renewable resources.	**solar energy** p. 153 **wind farm** p. 154 **hydroelectric power** p. 154 **geothermal energy** p. 155 **biomass energy** p. 155
Lesson 3: Land Resources  • Land is considered a resource because it is used by living things to meet their needs for food, shelter, and other things. • Some land resources are renewable, while others are not. • Individuals can recycle and compost to help conserve land resources.	**ore** p. 163 **deforestation** p. 164
Lesson 4: Air and Water Resources  • Most living things cannot survive without clean air and water. • Individuals can make their homes and schools more energy efficient.	**photochemical smog** p. 170 **acid precipitation** p. 170

Study Guide

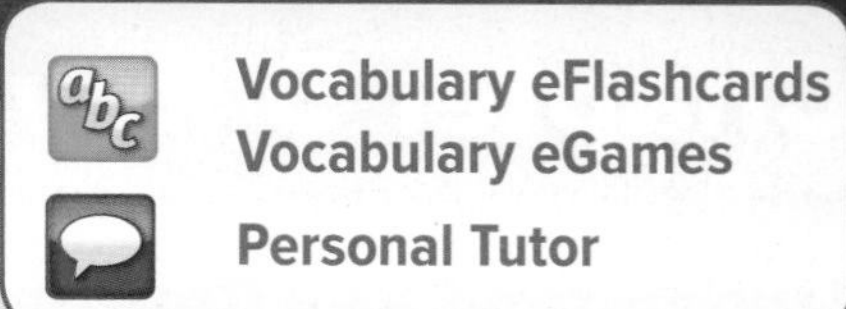

FOLDABLES® Chapter Project

Assemble your lesson Foldables as shown to make a Chapter Project. Use the project to review what you have learned in this chapter.

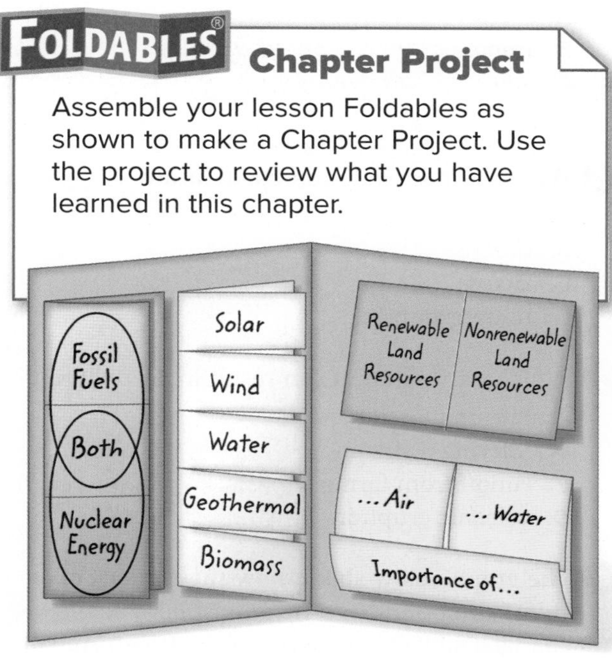

Use Vocabulary

1. Distinguish between renewable resources and nonrenewable resources.
2. Replace the underlined words with the correct vocabulary word: <u>Energy produced from atomic reactions</u> can be used to generate electricity.
3. How does biomass energy differ from geothermal energy?
4. Energy from the Sun is ________.
5. Define the term *ore* in your own words.
6. Distinguish between photochemical smog and acid precipitation.

Link Vocabulary and Key Concepts

Copy these concept maps, and then use vocabulary terms from the previous page and other terms from the chapter to complete the concept maps.

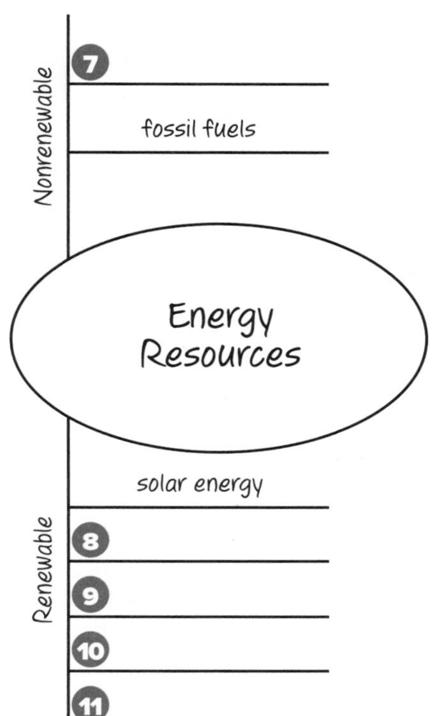

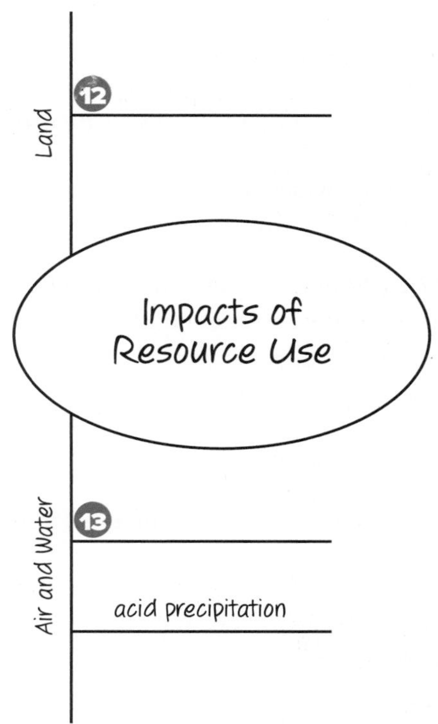

Chapter 5 Review

Understand Key Concepts 

1. Which energy source produces radioactive waste?
   A. biomass
   B. geothermal
   C. hydroelectric power
   D. nuclear power

2. The table below shows the energy sources used to produce electricity in the United States. What can you infer from the table?

Electricity Production	
Energy Source	**Percent**
Coal	33
Natural gas	33
Nuclear power	20
Solar, wind, geothermal, biomass	7
Hydroelectric power	6
Oil	1
Other	<1

   A. About 19 percent of U.S. electricity comes from renewable sources.
   B. Hydroelectric power is more widely used for electricity than nuclear power.
   C. About 87 percent of U.S. electricity comes from nonrenewable sources.
   D. Oil is more widely used for electricity than hydroelectric power.

3. Which factor would best determine whether a home is suitable for solar energy?
   A. difference in tidal heights
   B. strength of daily winds
   C. nearness to tectonically active areas
   D. number of sunny days per year

4. Which product comes from a metallic mineral resource?
   A. aluminum
   B. drywall
   C. gravel
   D. table salt

5. Which is a renewable land resource?
   A. forests
   B. minerals
   C. soil
   D. trees

6. Where is most water on Earth located?
   A. lakes
   B. oceans
   C. rivers
   D. underground

7. Which natural event can result in air pollution?
   A. burning fossil fuels
   B. littering a stream
   C. runoff from farms
   D. volcanic eruption

8. The graph below shows how the amount of sulfur compounds in the atmosphere has changed since the passage of the Clean Air Act. Based on the data in the graph, what can you infer about the act?

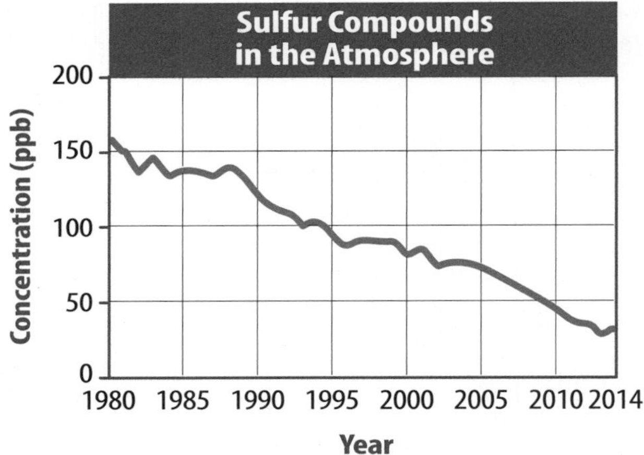

   A. The act has helped decrease pollutants in the atmosphere.
   B. The act has helped increase pollutants in the atmosphere.
   C. The act has incentives for use of renewable resources.
   D. The act has not impacted the amount of pollutants in the atmosphere.

 Online Test Practice

Critical Thinking

9 **Organize** the list of energy sources into renewable and nonrenewable energy resources.

• coal	• nuclear energy
• solar energy	• wind energy
• oil	• natural gas
• geothermal energy	• tidal power
• hydroelectric power	• biomass

10 **Create** a cartoon showing a chain reaction in a nuclear power plant.

11 **Compare** hydroelectric and tidal power.

12 **Design** a way to use passive solar energy in your classroom.

13 **Distinguish** between geothermal energy and solar energy.

14 **Consider** What factors must governments consider when managing land resources?

15 **Evaluate** the use of forests as natural resources. Do the advantages outweigh the disadvantages? Explain.

16 **Infer** When would you expect more smog to form—on cloudy days or on sunny days? Explain.

17 **Design** a way to remove salt from salt water. Then evaluate your plan. Could it be used to produce freshwater on a large scale? Why or why not?

18 **Formulate** a way to demonstrate the importance of air and water resources to younger students.

Writing in Science

19 **Compose** a song about vampire energy. The lyrics should describe vampire energy and explain how it can be reduced.

REVIEW THE BIG IDEA

20 Select a natural resource and explain why it is important to manage the resource wisely.

21 Suppose the house below is heated by electricity produced from burning coal. Which areas of the house have the greatest loss of thermal energy? Why is it important for this house to reduce thermal energy loss?

Math Skills

Math Practice

Use Percentages

22 Between 2012 and 2013, the carbon monoxide level in the air in Denver, Colorado, went from 2 ppm to 2.5 ppm. What was the percent change in CO?

23 There often is a considerable difference between pollutants in surface water and pollutants in groundwater in the same area. For example, in Portland, Oregon, there were 4.6 ppm of sulfates in the groundwater and 0.9 ppm in the surface water. What was the percent difference? (Hint: Use 4.6 ppm as the starting value.)

Tyrone Turner/National Geographic Stock

Standardized Test Practice

Record your answers on the answer sheet provided by your teacher or on a sheet of paper.

Multiple Choice

1 Which activity does NOT reduce the use of fossil fuels?

A riding a bicycle to school

B unplugging DVD players

C walking to the store

D watering plants less often

Use the graph below to answer questions 2 and 3.

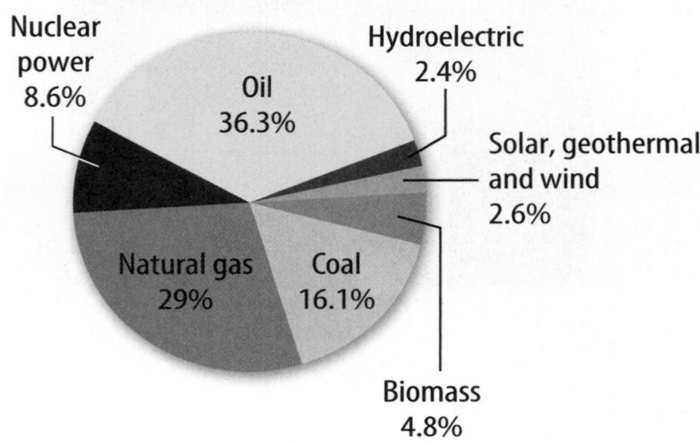

2 Which is the most-used renewable energy resource in the United States?

A biomass

B hydroelectric

C natural gas

D nuclear energy

3 What percentage of the energy used in the United States comes from burning fossil fuels?

A About 40 percent

B About 45 percent

C About 80 percent

D About 93 percent

4 Which practice emphasizes the use of renewable energy resources?

A buying battery-operated electronics

B installing solar panels on buildings

C replacing sprinklers with watering cans

D teaching others about vampire energy

5 Which is a nonrenewable land resource?

A crops

B minerals

C streams

D trees

Use the figure below to answer question 6.

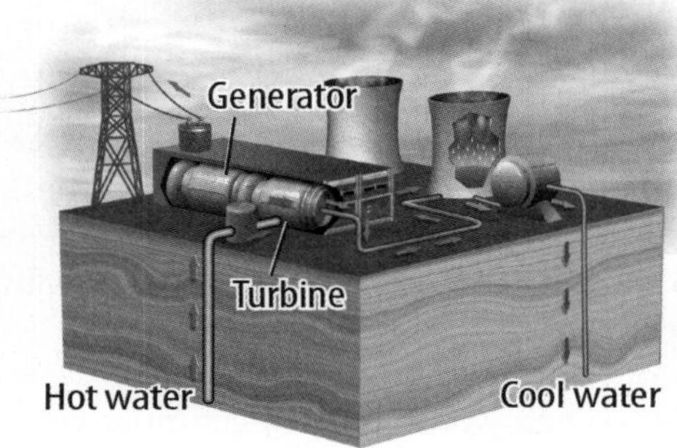

6 Which alternative energy resource is used to make electricity in the figure?

A solar energy

B tidal power

C geothermal energy

D hydroelectric power

7 Which practice is a wise use of land resources?

A composting

B conserving water

C deforestation

D strip mining

Use the figure below to answer question 8.

8 Which type of air pollution is labeled *A* in the figure?

- **A** acid precipitation
- **B** fertilizer runoff
- **C** nuclear waste
- **D** photochemical smog

9 Approximately how much water on the Earth is in oceans?

- **A** 1 percent
- **B** 3 percent
- **C** 75 percent
- **D** 97 percent

10 Which is a source of biomass energy?

- **A** sunlight
- **B** uranium
- **C** wind
- **D** wood

Constructed Response

Use the figure below to answer questions 11 and 12.

11 Which resource powers the turbine in the figure? Describe what happens at steps A–D to produce electricity.

12 What are two advantages and two disadvantages of producing electricity in the way shown in the figure?

13 Describe an example of how forests are used as a resource. What is one advantage of using the resource in this way? What is a disadvantage?

14 Agree or disagree with the following statement: "Known oil reserves will last only another 50 years. Thus, the United States should build more nuclear power plants to deal with the upcoming energy shortage." Support your answer with at least two advantages or two disadvantages of using nuclear energy.

NEED EXTRA HELP?														
If You Missed Question...	1	2	3	4	5	6	7	8	9	10	11	12	13	14
Go to Lesson...	1	2	1	2	3	2	3	4	4	2	2	2	3	1

Unit 2

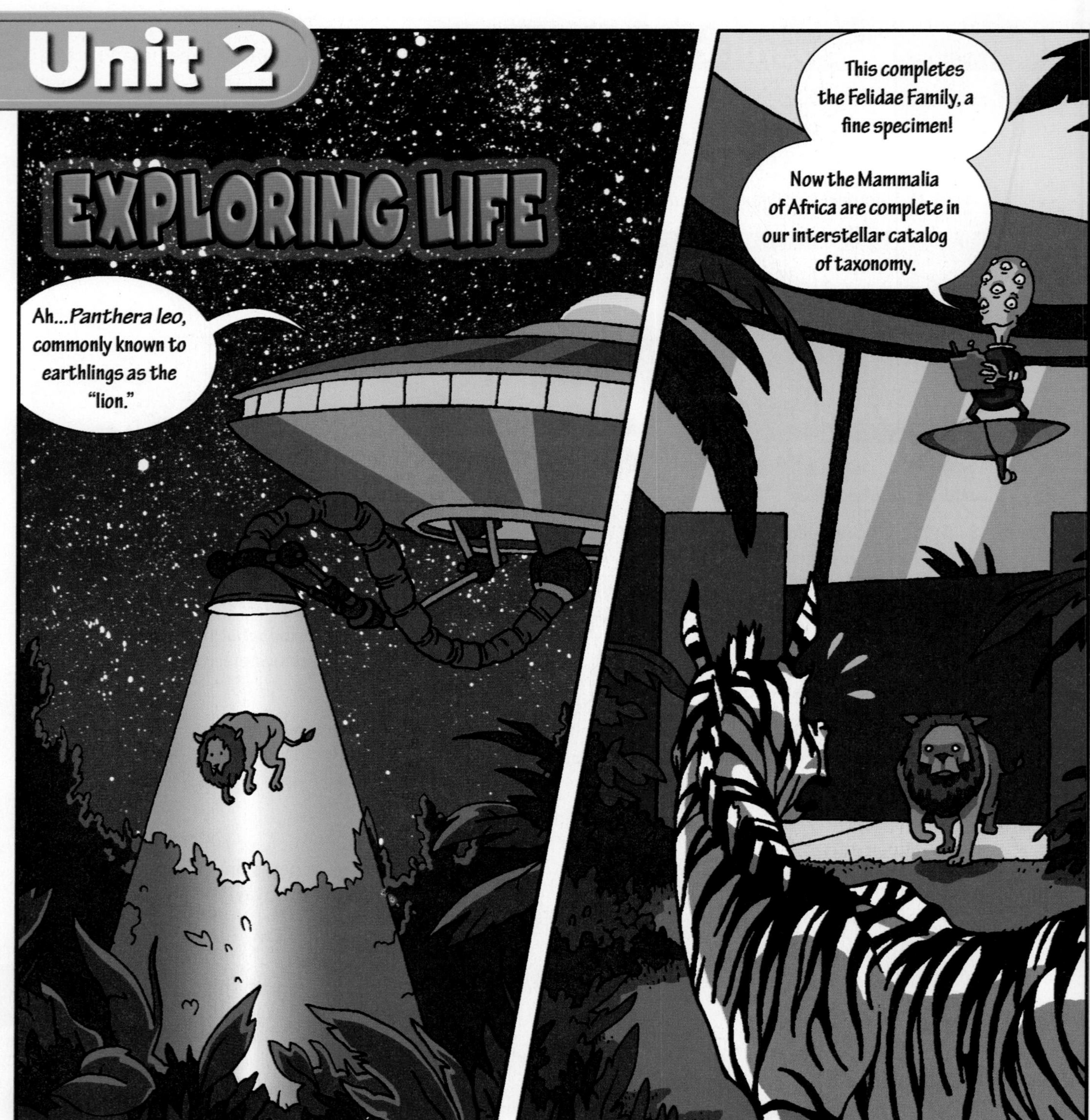

100 B.C. · 1700 · 1875 · 1900

350–341 B.C.
Greek philosopher Aristotle classifies organisms by grouping 500 species of animals into eight classes.

1735
Carl Linnaeus classifies nature within a hierarchy and divides life into three kingdoms: mineral, vegetable, and animal. He uses five ranks: class, order, genus, species and variety. Linnaeus's classification is the basis of modern biological classification.

1859
Charles Darwin publishes *On the Origin of Species,* in which he explains his theory of natural selection.

1866
German biologist Ernst Haeckel coins the term *ecology.*

1950

2000

1969
American ecologist Robert Whittaker is the first to propose a five-kingdom taxonomic classification of the world's biota. The five kingdoms are Animalia, Plantae, Fungi, Protista and Monera.

1973
Konrad Lorenz, Niko Tinbergen, and Karl von Frisch are jointly awarded the Nobel Prize for their studies in animal behavior.

1990
Carl Woese introduces the three-domain system that groups cellular life-forms into Archaea, Bacteria, and Eukaryote domains.

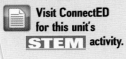

Visit ConnectED for this unit's STEM activity.

Unit 2

Nature of SCIENCE

Patterns

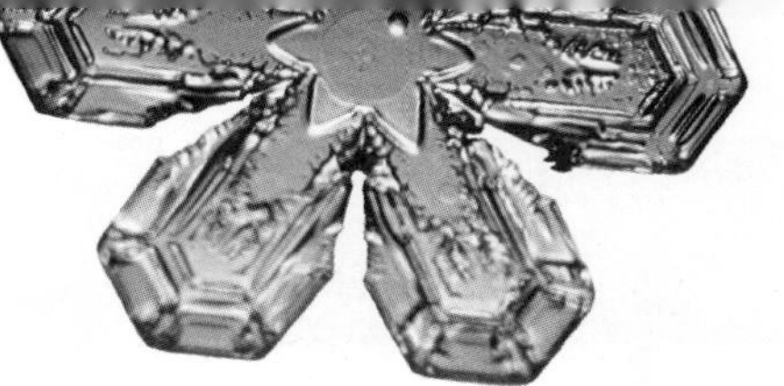

Have you ever caught a snowflake in your hand or seen one close-up in a book or on TV? You might have heard someone say that no two snowflakes are alike. While this is true, it is also true that all snowflakes have similar patterns. A **pattern** is a consistent plan or model used as a guide for understanding or predicting things. Patterns can be created or occur naturally. The formation of snowflakes is an example of a repeating pattern. They form piece by piece, as water drops in the air freeze into a six-sided crystal.

How Scientists Use Patterns

Studying and using patterns is useful to scientists because it can help explain the natural world or predict future events. A biologist might study patterns in DNA to predict what organisms will look like. A meteorologist might study cloud formation patterns to predict the weather. When doing research, scientists also try to match patterns found in their data with patterns that occur in nature. This helps to determine whether data are accurate and helps to predict outcomes.

Types of Patterns

Cyclic Patterns

A cycle, or repeated series of events, is a form of pattern. An organism's life cycle typically follows the pattern of birth, growth, and death. Scientists study an organism's life cycle to predict the life of its offspring.

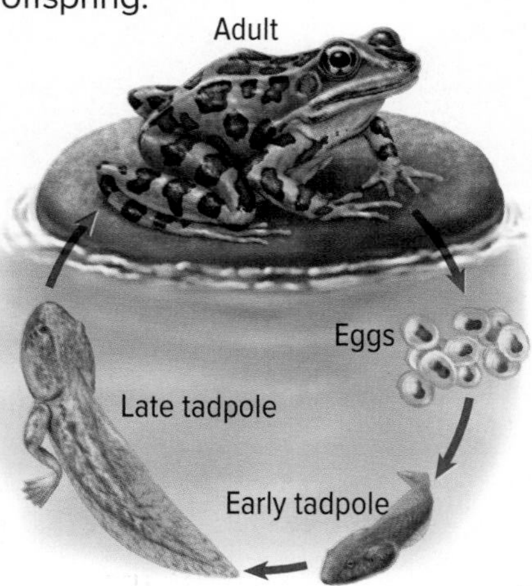

Physical Patterns

Physical patterns have an artistic or decorative design. Physical patterns can occur naturally, such as the patterns in the colors on butterfly wings or flower petals, or they can be created intentionally, such as a design in a brick wall.

Patterns in Life Science

Why do police detectives or forensic scientists take fingerprints at a crime scene? Forensic scientists know that every fingerprint is unique. Fingerprints contain patterns that can help detectives narrow a list of suspects. The patterns on the fingerprints can then be examined more closely to identify an exact individual. This is because no two humans have the same fingerprint, just as no two zebras have the same stripe pattern.

Patterns are an important key to understanding life science. They are found across all classifications of life and are studied by scientists. Patterns help scientists understand the genetic makeup, lifestyle, and similarities of various species of plants and animals. Zoologists might study the migration patterns of animals to determine the effects climate has on different species. Botanists might study patterns in the leaves of flowering plants to classify the species of the plant and predict the characteristics of the offspring.

Mathematical Patterns

Patterns are applied in mathematics all the time. Whenever you read a number, perform a mathematical operation, or describe a shape or graph, you are using patterns.

2, 5, 8, 11, ___, ___, ___

What numbers come next in this number pattern?

What will the next shape look like according to the pattern?

MiniLab — 15 minutes

Leaf Patterns

Each species of flowering plant has leaves with unique patterns.

Leaf Venation

Pinnate
one main vein with smaller branching veins

Palmate
several main veins that branch from one point

Parallel
veins that do not branch

1. Obtain a collection of leaves.
2. Use the leaves above to identify the venation, or vein patterns, of each leaf.

Analyze and Conclude

1. **Describe** the physical pattern you see in each leaf.
2. **Choose** Besides venation, what other patterns can you use to group the leaves?
3. **Identify** What types of patterns can be used to classify other organisms?

Chapter 6

Life's Classification and Structure

How is the classification of living things related to the structure of their cells?

Inquiry Why All the Hooks?

This color-enhanced scanning electron micrograph shows the hooked fruit of the goosegrass plant. The hooks attach to the fur of passing animals. This enables the plant's seeds, which are in the fruit, to spread.

- What characteristics would you use to classify this plant?
- How is the classification of living things related to the structure of their cells?

Andrew Syred/Photo Researchers

Get Ready to Read

What do you think?

Before you read, decide if you agree or disagree with each of these statements. As you read this chapter, see if you change your mind about any of the statements.

1. All living things are made of cells.
2. A group of organs that work together and perform a function is called a tissue.
3. Living things are classified based on similar characteristics.
4. *Cell wall* is a term used to describe the cell membrane.
5. Prokaryotic cells contain a nucleus.
6. Plants use chloroplasts to process energy.

Your one-stop online resource
connectED.mcgraw-hill.com

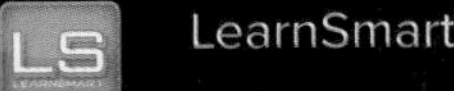

- LearnSmart®
- Chapter Resources Files, Reading Essentials, Get Ready to Read, Quick Vocabulary
- Animations, Videos, Interactive Tables
- Self-checks, Quizzes, Tests
- Project-Based Learning Activities
- Lab Manuals, Safety Videos, Virtual Labs & Other Tools
- Vocabulary, Multilingual eGlossary, Vocab eGames, Vocab eFlashcards
- Personal Tutors

Lesson 1

Reading Guide

Key Concepts

ESSENTIAL QUESTIONS

- What are living things?
- What do living things need?
- How are living things classified?

Vocabulary

autotroph p. 192

heterotroph p. 192

habitat p. 193

binomial nomenclature p. 194

taxon p. 195

Multilingual eGlossary

Classifying Living Things

Inquiry Living or Not?

This tide pool contains sea anemones, barnacles, and sea stars that are living and rocks that are not living. How can you tell whether something is alive? Do all living things move? All living things have certain characteristics that you will read about in this lesson.

Launch Lab

10 minutes

How can you tell whether it is alive?

Living things share several basic characteristics. Think about what you have in common with other living things such as a bug or a tree. Do other things have some of those same characteristics?

1. Read and complete a lab safety form.
2. Observe a **lit candle** for 1–2 min. Pay attention to both the candle and the flame.
3. Write what you observe in your Science Journal.
4. Write what you think you would observe if you were to observe the candle for several hours.

Think About This

1. What characteristics does the flame have that would lead some people to think the flame is alive?
2. What qualities did you think of earlier (that you share with other living things) that the candle does not possess?
3. **Key Concept** What characteristics do you think something must have to be considered alive?

What are living things?

It might be easy to tell whether a bird, a tree, or a person is alive. But for some organisms, it is harder to tell whether they are living things. Look at the moldy bread shown in **Figure 1**. Is the bread a living thing? What about the green mold and white mold on the bread? All living things have six characteristics in common:

- Living things are made of cells.
- Living things are organized.
- Living things grow and develop.
- Living things respond to their environment.
- Living things reproduce.
- Living things use energy.

The bread shown in **Figure 1** is not living, but the molds growing on the bread are living things. Mold is a type of fungus. If you looked at the mold using a microscope, you would see that it is made of cells. Mold cells respond to their environment by growing and reproducing. The molds obtain energy, which they need to grow, from the bread.

 **Key Concept Check** What are living things?

Figure 1 Mold is a living thing.

Living things are organized.

Marching bands are made up of rows of people playing different instruments. Some rows are made up of people playing flutes, and other rows are filled with drummers. Although marching bands are organized into different rows, all band members work together to play a song. Like marching bands, living things also are organized. Some living things are more complex than others, but all organisms are made of cells. In all cells, **macromolecules** are organized into different structures that help cells function. You might recall that there are four macromolecules in cells–nucleic acids, lipids, proteins, and carbohydrates. Nucleic acids, such as DNA, store information. Lipids are the main component of cell membranes and provide structure. Some proteins are enzymes, and others provide structure. Carbohydrates are used for energy.

REVIEW VOCABULARY

macromolecule
substance in a cell that forms by joining many small molecules together

Reading Check Name the four macromolecules in cells.

Unicellular Organisms Some living things are unicellular, which means they are made up of only one cell. In fact, most living things on Earth are unicellular organisms. Unicellular organisms are the oldest forms of life. There are many groups of unicellular organisms, each with **unique** characteristics. Bacteria, amoebas (uh MEE buhz), and paramecia (per uh MEE see ah) are examples of unicellular organisms. Unicellular organisms have everything needed to obtain and use energy, reproduce, and grow inside one cell. Some unicellular organisms are tiny and cannot be seen without a microscope. Other unicellular organisms, such as the plasmodial (plaz MOH dee ul) slime mold shown in **Figure 2**, can be large.

ACADEMIC VOCABULARY

unique
(adjective) without an equal, distinctive

Figure 2 A plasmodial slime mold is a huge cell formed by many cells that join together and form one cell.

Multicellular Organisms Soccer teams are made up of many types of players, including goalkeepers, forwards, and fullbacks. Each team member has a specific job, but they all work together when playing a game. Many living things are made of more than one cell and are called multicellular organisms. Like the different types of players on a soccer team, multicellular organisms have different types of cells that carry out specialized functions. The ladybug shown in **Figure 3** has cells that form wings and other cells that form eyes.

Multicellular organisms have different levels of organization. Groups of cells that work together and perform a specific function are called tissues. Tissues that work together and carry out a specific function are called organs. Organs that work together and perform a specific function are called organ systems. Organ systems work together and perform all the functions an organism needs to survive.

Figure 3 Multicellular organisms, such as this ladybug, contain groups of cells that carry out special functions.

Visual Check What structures can you identify in the ladybug?

Living things grow, develop, and reproduce.

During their lifetimes, living things grow, or increase in size. For a unicellular organism, the size of its cell increases. For a multicellular organism, the number of its cells increases. Living things also develop, or change, during their lifetimes. For some organisms, it is easy to see the changes that happen as they grow and develop. As shown in **Figure 4**, ladybug larva grow into pupae (PYEW pee; singular, pupa), an intermediate stage, before developing into adults.

Once an organism is an adult, it can reproduce either asexually or sexually and form new organisms. Unicellular organisms, such as bacteria, reproduce asexually when one cell divides and forms two new organisms. Some multicellular organisms also can reproduce asexually; one parent organism produces offspring when body cells replicate and divide. Sexual reproduction occurs when the reproductive cells of one or two parent organisms join and form a new organism. Multicellular organisms such as humans and other mammals reproduce sexually. Some organisms such as yeast can reproduce both asexually and sexually.

Figure 4 A ladybug grows and develops from a larva to a pupa.

Visual Check What differences do you see between the two stages?

▲ **Figure 5** Algae are autotrophic because they use sunlight to produce energy.

Living things use energy.

All living things need energy to survive. Some organisms are able to convert light energy to chemical energy that is used for many cellular processes. *Organisms that convert energy from light or inorganic substances to usable energy are called* **autotrophs** (AW tuh trohfs).

Many autotrophs use energy from light and convert carbon dioxide and water into carbohydrates, or sugars. Autotrophs use the carbohydrates for energy. Plants and the algae shown growing on the pond in **Figure 5** are autotrophs.

Other autotrophs, called chemoautotrophs (kee moh AW tuh trohfs), grow on energy released by chemical reactions of inorganic substances such as sulfur and ammonia. Many chemoautotrophs are bacteria that live in extreme environments such as deep in the ocean or in hot sulfur springs.

Reading Check How do some autotrophs use energy from sunlight?

Heterotrophs (HE tuh roh trohfs) *are organisms that obtain energy from other organisms.* Heterotrophs eat autotrophs or other heterotrophs to obtain energy. Animals and fungi are examples of heterotrophs.

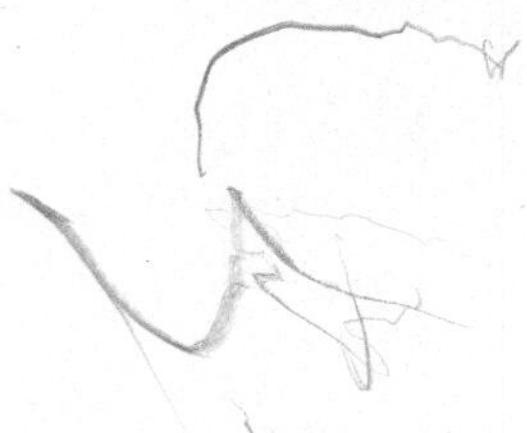

Living things respond to stimuli.

All living things sense their environments. If an organism detects a change in its external environment, it will respond to that change. A change in an organism's environment is called a stimulus (STIHM yuh lus; plural, stimuli). Responding to a stimulus might help an organism protect itself. For example, the octopus in **Figure 6** responds to predators by releasing ink, a black liquid. In many organisms, nerve cells detect the environment, process the information, and coordinate a response.

Figure 6 An octopus responds to potential harm by secreting ink. The ink hides the octopus while it escapes. ▼

What do living things need?

You just read that all living things need energy in order to survive. Some organisms obtain energy from food. What else do living things need to survive? Living things also need water and a place to live. Organisms live in environments specific to their needs where they are protected, can obtain food and water, and can get shelter.

A Place to Live

Living things are everywhere. Organisms live in the soil, in lakes, and in caves. Some living things live on or in other organisms. For example, bacteria live in your intestines and on other body surfaces. *A specific environment where an organism lives is its* **habitat.** Most organisms can survive in only a few habitats. The land iguana shown in **Figure 7** lives in warm, tropical environments and would not survive in cold places such as the Arctic.

Food and Water

Living things also need food and water. Food is used for energy. Water is essential for survival. You will read about how water is in all cells and helps them function in Lesson 2. The type of food that an organism eats depends on the habitat in which it lives. Marine iguanas live near the ocean and eat algae. Land iguanas, such as the one in **Figure 7,** live in hot, dry areas and eat cactus fruits and leaves. The food is processed to obtain energy. Plants and some bacteria use energy from sunlight and produce chemical energy for use in cells.

 **Key Concept Check** What do living things need?

WORD ORIGIN

habitat
from Latin *habitare,* means "to live or dwell"

FOLDABLES®

Make a vertical three-column chart book. Label it as shown. Use it to organize your notes about living things, their needs, and classification criteria.

Definition of a Living Thing	Survival Requirements	Classification Criteria

Needs of Living Things

Figure 7 This Galápagos land iguana is eating the fruit of a prickly pear cactus.

Carolyn Jenkins/Alamy

Math Skills 

Use Ratios

A ratio expresses the relationship between two or more things. Ratios can be written

3 to 5, 3:5, or $\frac{3}{5}$.

Reduce ratios to their simplest form. For example, of about 3 million species in the animal kingdom, about 50,000 are mammals. What is the ratio of mammals to animals?

Write the ratio as a fraction.

$$\frac{50{,}000}{3{,}000{,}000}$$

Reduce the fraction to the simplest form.

$$\frac{50{,}000}{3{,}000{,}000} = \frac{5}{300} = \frac{1}{60}$$

(or 1:60 or 1 to 60)

Practice

Of the 5,000 species of mammals, 250 species are carnivores. What is the ratio of carnivores to mammals? Write the ratio in all three ways.

 Math Practice

 Personal Tutor

How are living things classified?

You might have a notebook with different sections. Each section might contain notes from a different class. This organizes information and makes it easy to find notes on different subjects. Scientists use a classification system to group organisms with similar traits. Classifying living things makes it easier to organize organisms and to recall how they are similar and how they differ.

Naming Living Things

Scientists name living things using a system called binomial nomenclature (bi NOH mee ul • NOH mun klay chur). **Binomial nomenclature** *is a naming system that gives each living thing a two-word scientific name.*

More than 300 years ago a scientist named Carolus Linnaeus created the binomial nomenclature system. All scientific names are in Latin. *Homo sapiens* is the scientific name for humans. As shown in **Table 1**, the scientific name for an Eastern chipmunk is *Tamias striatus*.

Table 1 Classification of the Eastern Chipmunk

Taxonomic Group	Number of Species	Examples
Domain Eukarya	about 4–10 million	
Kingdom Animalia	about 2 million	
Phylum Chordata	about 50,000	
Class Mammalia	about 5,000	
Order Rodentia	about 2,300	
Family Sciuridae	299	
Genus *Tamias*	25	
Species *Tamias striatus*	1	

Frank Cezus/Getty Images

Classification Systems

Linnaeus also classified organisms based on their behavior and appearance. Today, the branch of science that classifies living things is called taxonomy. *A group of organisms is called a* **taxon** (plural, taxa). There are many taxa, as shown in **Table 1.** Recall that all living things share similar traits. However, not all living things are exactly the same.

Taxonomy

Using taxonomy, scientists divide all living things on Earth into three groups called domains. Domains are divided into kingdoms, and then phyla (FI luh; singular, phylum), classes, orders, families, genera (singular, genus), and species. A species is made of all organisms that can mate with one another and produce offspring that can reproduce. The first word in an organism's scientific name is the organism's genus (JEE nus), and the second word might describe a distinguishing characteristic of the organism. For example, dogs belong to the genus *Canis*. The *Canis* genus also includes wolves, coyotes, and jackals.

Recall that Linnaeus used similar physical traits to group organisms. Today, scientists also look for other similarities, such as how an organism reproduces, how it processes energy, and the types of genes it has.

Dichotomous Keys

A dichotomous (di KAH tah mus) key is a tool used to identify an organism based on its characteristics. Dichotomous keys contain descriptions of traits that are compared when classifying an organism. Dichotomous keys are organized in steps. Each step might ask a yes or a no question and have two answer choices. Which question is answered next depends on the answer to the previous question. Based on the features, a choice is made that best describes the organism.

Key Concept Check How are living things classified?

MiniLab 20 minutes

Whose shoe is it?

A dichotomous key is a tool to help identify an unknown object or organism.

1. Read and complete a lab safety form.
2. Have each person in your group place one of his or her **shoes** in a pile.
3. Observe the shoes, looking for similarities and differences among them.
4. In your Science Journal, write a question that can be used to separate the shoes into two groups.
5. Divide the shoes into the two groups.
6. Continue asking questions for each subgroup until all of the shoes are identified.
7. Number the questions from the top of the key down, and create your key this way:

   1. Question 1?
   Yes . . . . go to question ________
   No . . . . go to question ________

Analyze and Conclude

1. **Classify** What characteristics probably should not be used when creating a dichotomous key?
2. **Key Concept** Describe how a doctor and a pest exterminator could use a dichotomous key.

SCIENCE USE V. COMMON USE

key

Science Use an aid to identification

Common Use a device to open a lock

(t)Ryan McVay/Getty Images, (c)Stockbyte/Getty Images, (b)Photodisc/Getty Images

Lesson 1 Review

Visual Summary

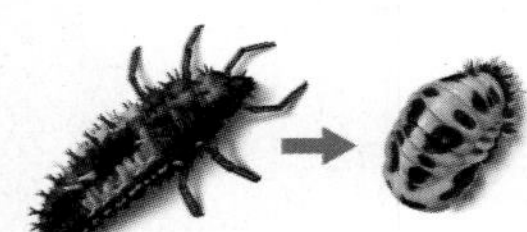

All living things grow, develop, and reproduce.

All living things are organized, respond to their environment, and use energy.

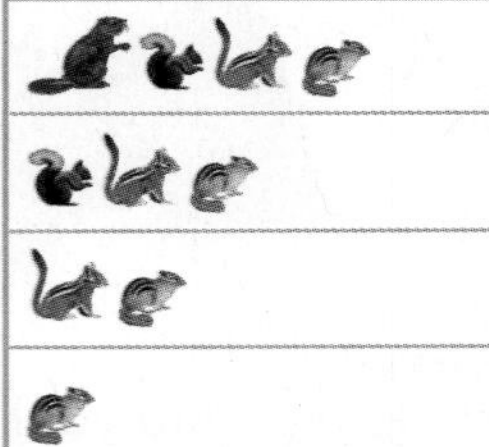

Scientists use a classification system to group organisms with similar traits and genetic makeup.

FOLDABLES®

Use your lesson Foldable to review the lesson. Save your Foldable for the project at the end of the chapter.

What do you think NOW?

You first read the statements below at the beginning of the chapter.

1. All living things are made of cells.
2. A group of organs that work together and perform a function is called a tissue.
3. Living things are classified based on similar characteristics.

Did you change your mind about whether you agree or disagree with the statements? Rewrite any false statements to make them true.

Use Vocabulary

1. **Use the term** *taxon* in a sentence.
2. **Distinguish** between the terms *autotroph* and *heterotroph.*
3. Linnaeus created a two-word naming system for organisms called ________.

Understand Key Concepts 

4. An environment where specific organisms live is called a(n)
   - **A.** autotroph.
   - **B.** habitat.
   - **C.** heterotrophy.
   - **D.** taxon.
5. **Explain** how binomial nomenclature helps scientists classify organisms.
6. **Relate** the number of cells an organism has to the way it reproduces.

Interpret Graphics

7. **Summarize** Copy and fill in the graphic organizer below to summarize the characteristics of living things.

Critical Thinking

8. **Differentiate** between living and nonliving things in the picture at right.

Math Skills 

Math Practice

9. There are 3 million species in the animal kingdom. Of those, about 270 species are carnivores. What is the ratio of carnivores to animals? Write the ratio all three ways.

(l) Hutchings Photography/Digital Light Source, (r) James Zipp/Science Source

CAREERS in SCIENCE

On a Quest for Leeches

How do you catch a leech? Let it bite you!

Mark Siddall travels the world searching for creatures that make most people cringe—leeches. He collects leeches to understand how they are related and how they evolved. This is a huge job since there are more than 600 known species of leeches!

Siddall is a scientist at the American Museum of Natural History. He travels to remote places, such as the jungles of Rwanda and the swamps of Argentina, to collect leeches. Once he's there, he lets the leeches find him. Barefoot, he treks through damp forests or wades in streams. Leeches attach to his skin, draw blood until they're full, and then fall off. That's when Mark adds them to the museum's collection.

Siddall identifies the leeches by their body parts. Some have jaws and teeth. Others have thin tubes for sucking in liquid. They use these parts to draw blood or fluids from animals, such as frogs, humans, snails, and other worms. Some even swallow their prey whole!

Through his research, Siddall is helping build a family tree for leeches to learn how they evolved. For example, leeches today live on land, in freshwater, and in the ocean. Siddall's research shows that leeches first appeared in freshwater and then moved onto land and into the ocean. Many species of leeches are being threatened by habitat destruction. Siddall hopes his research will help protect leeches and their habitats.

▲ **Mark Siddall uses himself as bait to catch leeches. When a leech is done feeding, it falls off into a collection bag.**

In just 30 minutes, a leech can swallow more than five times its weight in blood. It might not need to feed again for a few months.

It's Your Turn

DIAGRAM Leeches are classified according to how they feed. Choose a jawed leech *(Hirudinidae)*, a jawless leech *(Erpobdellidae)*, or a leech that uses a proboscis *(Glossiphoniidae)*. Research how it feeds. Draw a diagram and present your findings.

Lesson 2

Reading Guide

Key Concepts
ESSENTIAL QUESTIONS

- What is a cell made of?
- How do the parts of a cell enable it to survive?

Vocabulary

prokaryotic cell p. 200
eukaryotic cell p. 200
cytoplasm p. 202
mitochondrion p. 203

Multilingual eGlossary

BrainPOP®

Cells

Inquiry Weird Web?

This isn't a spider's strange web. These are nerve cells shown in a color-enhanced electron micrograph. The larger green parts are the cell bodies. The threadlike parts carry electrical signals from one nerve cell to another. How do these parts help the cells?

SPL/Science Source

Launch Lab

15 minutes

Are all cells alive?

There are many bacteria that live on and in people. These unicellular organisms have all the characteristics of life and are alive. Are human cells, which the bacteria live on and in, also alive?

1. In your Science Journal, draw a circle that takes up half of the page. The circle represents a human cell.
2. Draw and label the following things in your cell:

   A power plant to represent the need for and use of energy; label it *energy production.*

   A garbage truck to represent waste removal; label it *waste removal.*

   A city hall with a mayor to represent the organization and processes of the cell; label it *organization.*

   A road system to represent the transportation that occurs in the cell; label it *transportation.*

   A cement truck to represent the construction of new structures in the cell; label it *growth.*

   A fire truck to represent a cell's ability to respond to changes in its surroundings; label it *response to environment.*

   A copy machine in city hall to represent the cell's ability to follow instructions and make more cells; label it *reproduction.*

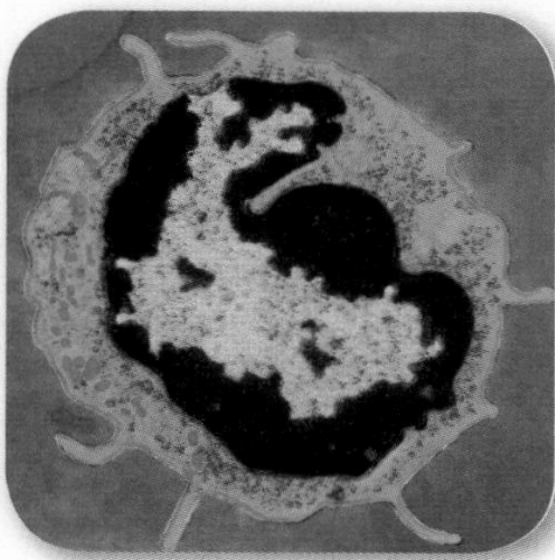

Think About This

1. Does the human cell you drew have all the characteristics of life? Explain your answer.
2. **Key Concept** Do you think each of the trillions of cells that are part of you are either alive or once-living? Why?

What are cells?

What is one thing all living things have in common? All living things have cells, the basic unit of an organism. As you read in Lesson 1, most organisms have only one cell. Other organisms have many cells. Humans have about 100 trillion cells! Most cells are so small that they cannot be seen without a microscope. Microscopes, such as the one shown in **Figure 8,** are used to view details of small objects or to view things that are too small to be seen by the unaided eye.

Scientists first used microscopes to look at cells over 300 years ago. Cells come in different shapes and sizes. Nerve cells are long and slender. Many female reproductive cells, or eggs, are large and round.

Reading Check Why is a microscope needed to view most cells?

LM Magnification: 21×

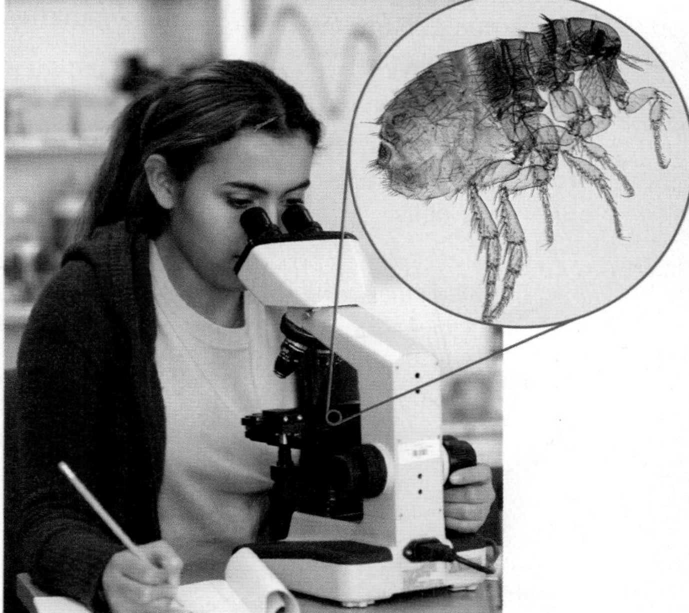

Figure 8 Microscopes increase the size of an image so that a small thing, such as the flea shown here, can be observed.

What are cells made of?

Recall that all cells are made of four macromolecules–nucleic acids, lipids, proteins, and carbohydrates. Cells also have many other characteristics. For example, all cells are surrounded by an outer structure called a cell membrane. The cell membrane keeps substances such as macromolecules inside the cell. It also helps protect cells by keeping harmful substances from entering. About 70 percent of the inside of a cell is water. Many of the substances inside cells are dissolved in water so they can move easily about the cell.

 **Key Concept Check** What is a cell made of?

Types of Cells

There are two main types of cells, as shown in **Figure 9.** **Prokaryotic** (pro kayr ee AH tihk) **cells** *do not have a nucleus or other membrane-bound organelles.* Organelles are structures in cells that carry out specific functions. The few organelles in prokaryotic cells are not surrounded by membranes. Organisms with prokaryotic cells are called prokaryotes. Most prokaryotes are unicellular organisms, such as bacteria.

Eukaryotic (yew ker ee AH tihk) **cells** *have a nucleus and other membrane-bound organelles.* Most multicellular organisms and some unicellular organisms are eukaryotes. The eukaryotic cell shown in **Figure 9** contains many structures that are not in a prokaryotic cell. In eukaryotes, membranes surround most of the organelles, including the nucleus.

Figure 9 Prokaryotic cells do not have a nucleus. Eukaryotic cells have a nucleus and many other organelles.

Visual Check What structures are in both prokaryotic cells and eukaryotic cells?

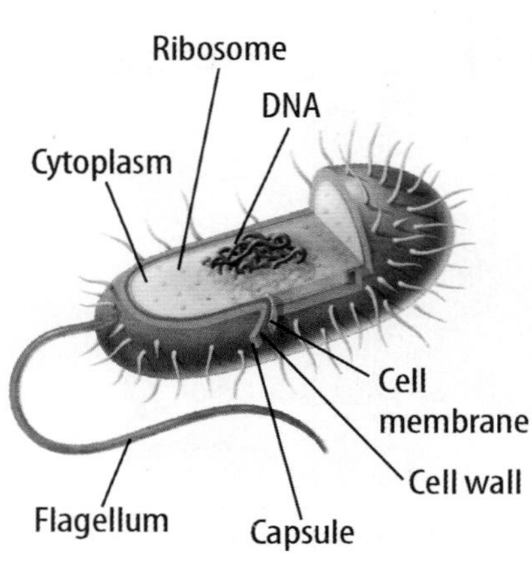

Prokaryotic Cell

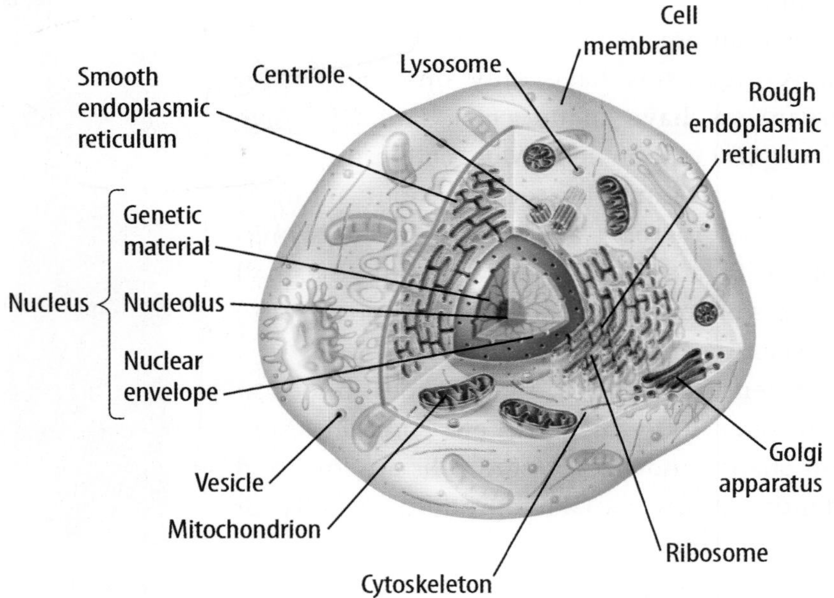

Eukaryotic Cell

The Outside of a Cell

As you have just read, the cell membrane surrounds a cell. Much like a fence surrounds a school, the cell membrane helps keep the substances inside a cell separate from the substances outside a cell. Some cells also are surrounded by a more rigid layer called a cell wall.

Cell Membrane

The cell membrane is made of lipids and proteins. Recall that lipids and proteins are macromolecules that help cells function. Lipids in the cell membrane protect the inside of a cell from the external environment. Proteins in the cell membrane transport substances between a cell's environment and the inside of the cell. Proteins in the cell membrane also communicate with other cells and organisms and sense changes in the cell's environment.

Reading Check Summarize the major components of cell membranes.

Cell Wall

In addition to a cell membrane, some cells also have a cell wall, as shown in **Figure 10.** The cell wall is a strong, rigid layer outside the cell membrane. Cells in plants, fungi, and many types of bacteria have cell walls. Cell walls provide structure and help protect the cell from the outside environment. Most cell walls are made from different types of carbohydrates.

Animation

Figure 10  This plant cell has a cell membrane and a cell wall.

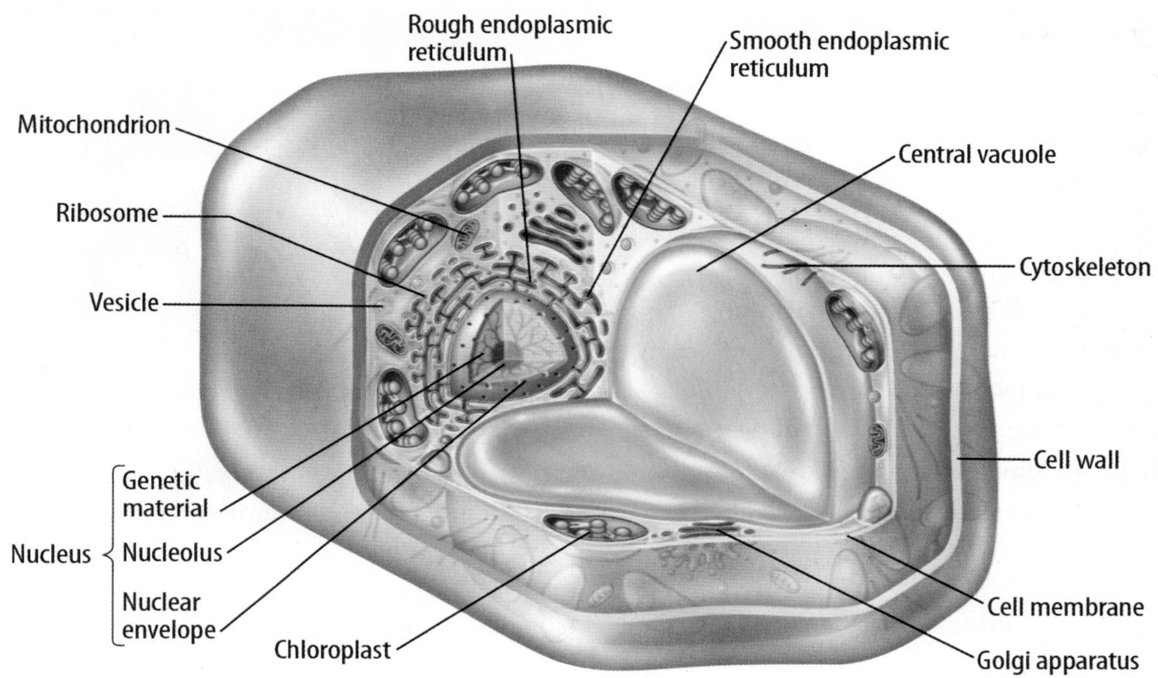

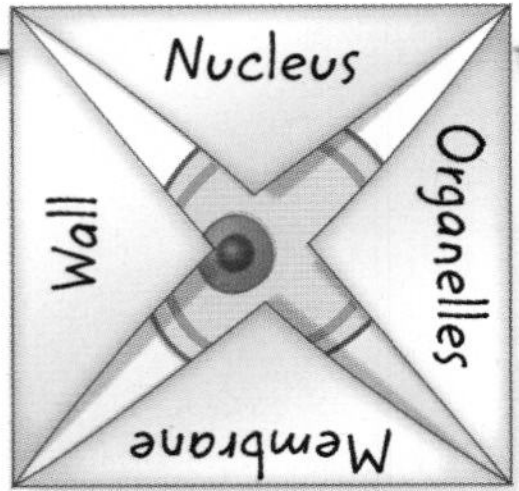

FOLDABLES®

Make an envelope book. Label it as shown. Use the center inside tab to illustrate a cell. Use the inside tabs to describe the cellular structures.

The Inside of a Cell

Recall that the inside of a cell is mainly water. Many substances used for communication, energy, and growth dissolve in water. This makes it easier for the substances to move around inside a cell. Water also gives cells their shapes and helps keep the structures inside a cell organized. The organelles inside a cell perform specific functions. They control cell activities, provide energy, transport materials, and store materials.

Cytoplasm

The liquid part of a cell inside the cell membrane is called the **cytoplasm.** It contains water, macromolecules, and other substances. The organelles in eukaryotic cells are located in the cytoplasm. Proteins in the cytoplasm provide structure and help organelles and other substances move around.

Controlling Cell Activities

The information that controls all of a cell's activities is stored in its genetic material, called DNA. DNA is a type of macromolecule called a nucleic acid. The information in DNA is transferred to another nucleic acid called RNA. RNA gives cells instructions about which proteins need to be made. In prokaryotic cells, DNA is in the cytoplasm. In eukaryotic cells, DNA is stored in an organelle called the nucleus. A membrane, called the nuclear membrane, surrounds the nucleus. Tiny holes in the nuclear membrane let certain substances move between the nucleus and the cytoplasm.

MiniLab

20–30 minutes

What can you see in a cell?

When people developed microscopes, they were able to see things that they could not see with their eyes alone.

1. Read and complete a lab safety form.
2. Carefully remove a thin layer of membrane from a piece of **onion.**
3. Place the membrane on the center of a dry **microscope slide.**
4. Add a drop of **iodine** on top of the sample.
5. Place a **cover slip** on top of the sample.
6. Use a **microscope** to focus on the slide using low power. Sketch what you see in your Science Journal.
7. View and sketch the sample on medium and high powers.

Analyze and Conclude

1. **Observe** What structures did you see at low, medium, and high powers?
2. **Infer** How might your view of the cells change if you view them at an even higher power?
3. **Key Concept** How does a microscope help you learn more about the onion plant?

Hutchings Photography/Digital Light Source

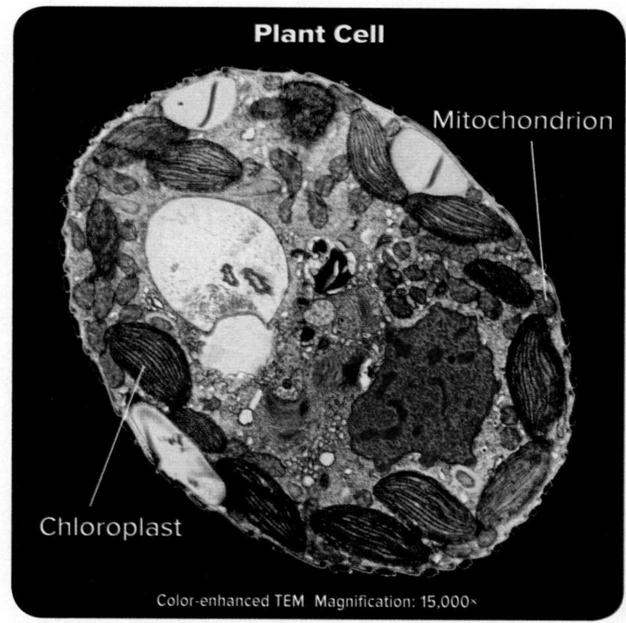

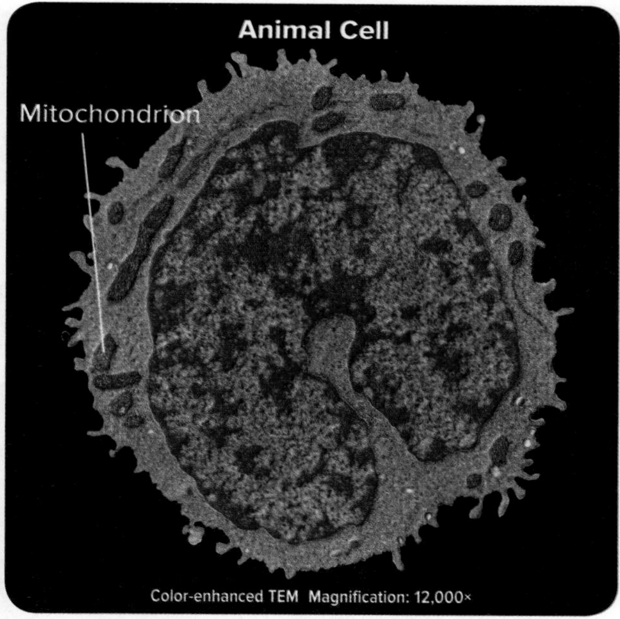

Figure 11 Plant cells have mitochondria and chloroplasts. Animal cells only contain mitochondria.

Visual Check Where are the mitochondria located in a cell?

Energy for the Cell

You read in Lesson 1 that all living things use energy. Proteins in the cytoplasm process energy in prokaryotes. Eukaryotes have special organelles, the chloroplasts and mitochondria (mi tuh KAHN dree uh; singular, mitochondrion) shown in **Figure 11,** that process energy.

Mitochondria Most eukaryotes contain hundreds of mitochondria. **Mitochondria** *are organelles that break down food and release energy.* This energy is stored in molecules called ATP–adenosine triphosphate (uh DEN uh seen • tri FAHS fayt). ATP provides a cell with energy to perform many functions, such as making proteins, storing information, and communicating with other cells.

Reading Check What energy molecule is made in a mitochondrion?

Chloroplasts Energy also can be processed in organelles called chloroplasts, shown in **Figure 11.** Plants and many other autotrophs have chloroplasts and mitochondria. Chloroplasts capture light energy and convert it into chemical energy in a process called photosynthesis. Chloroplasts contain many structures that capture light energy. Like the reactions that occur in mitochondria, ATP molecules are produced during photosynthesis. However, photosynthesis also produces carbohydrates such as glucose that also are used to store energy.

WORD ORIGIN

mitochondrion
from Greek *mitos,* means "thread"; and *khondrion,* means "little granule"

Protein Production

You just read that cells use protein for many functions. These proteins are made on the surface of ribosomes that are in the cytoplasm of both prokaryotic and eukaryotic cells. In eukaryotic cells, some ribosomes are attached to an organelle called the endoplasmic reticulum (en duh PLAZ mihk • rih TIHK yuh lum), as shown in **Figure 12.** It is made of folded membranes. The proteins can be processed and can move inside the cell through the endoplasmic reticulum.

Color-enhanced TEM Magnification: Unavailable

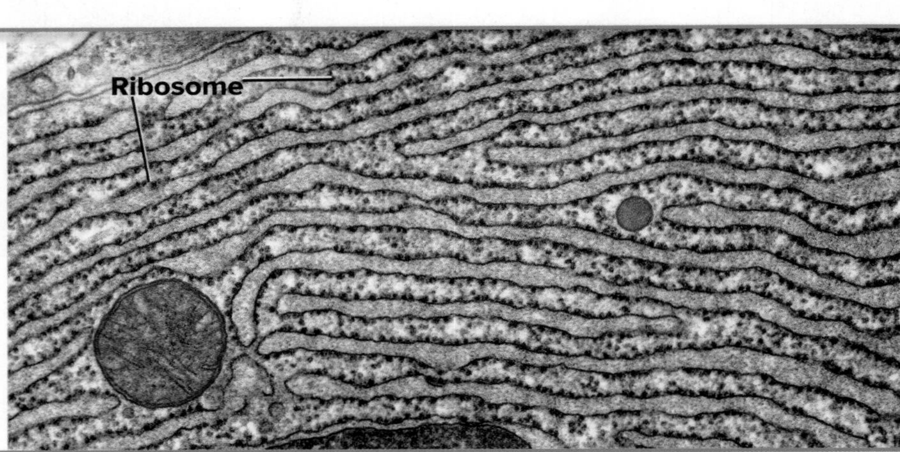

Figure 12 Ribosomes are attached to the rough endoplasmic reticulum. ▶

Cell Storage

What happens to the molecules that are made in a cell? An organelle called the Golgi (GAWL jee) apparatus packages proteins into tiny organelles called vesicles. Vesicles transport proteins around a cell. Other molecules are stored in organelles called vacuoles. A vacuole is usually the largest organelle in a plant cell, as shown in **Figure 13.** In plant cells, vacuoles store water and provide support. In contrast to all plant cells, only some animal and bacterial cells contain vacuoles. The vacuoles in animal and bacterial cells are smaller than the ones in plant cells.

Key Concept Check How do the parts of a cell enable it to survive?

Color-enhanced TEM Magnification: 2,000×

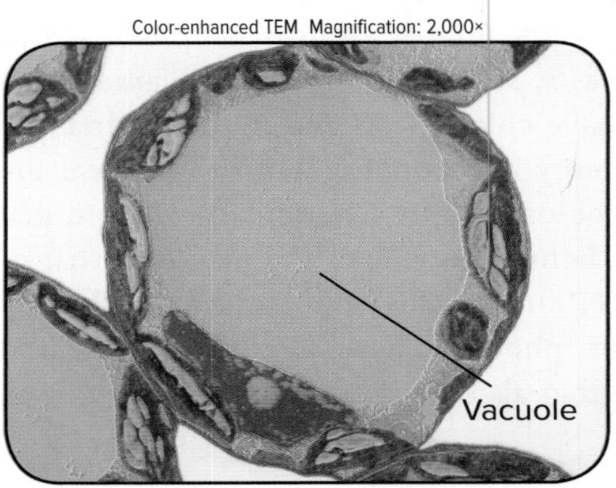

Figure 13 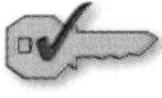 Vacuoles are used by plant cells for storage and to provide structure. ▶

(t) MedImage/Science Source, (b) Dr. Jeremy Burgess/Science Source

Lesson 2 Review

Visual Summary

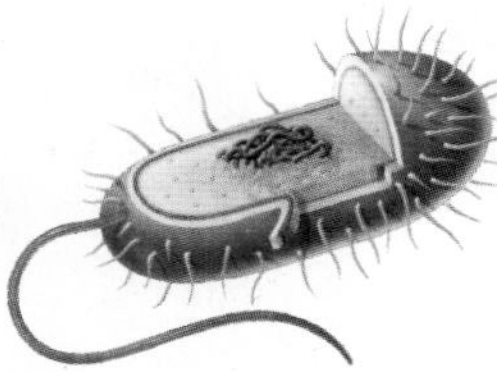

Prokaryotic cells are surrounded by a cell membrane but have no internal organelles with membranes.

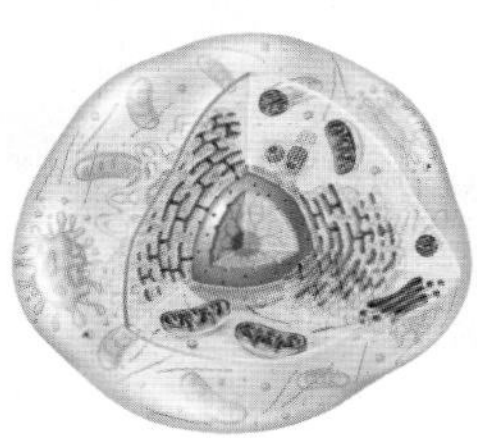

Eukaryotic cells contain a nucleus and many other organelles.

Plant cells have cell walls, chloroplasts, and a large vacuole.

FOLDABLES®

Use your lesson Foldable to review the lesson. Save your Foldable for the project at the end of the chapter.

What do you think NOW?

You first read the statements below at the beginning of the chapter.

4. *Cell wall* is a term used to describe the cell membrane.

5. Prokaryotic cells contain a nucleus.

6. Plants use chloroplasts to process energy.

Did you change your mind about whether you agree or disagree with the statements? Rewrite any false statements to make them true.

Use Vocabulary

1. **Distinguish** between prokaryotic cells and eukaryotic cells.
2. Water, proteins, and other substances are found in the _________ of a cell.
3. **Define** *mitochondrion* in your own words.

Understand Key Concepts 

4. Which organelles store water, carbohydrates, and wastes in plants?
   A. chloroplasts
   B. mitochondria
   C. nuclei
   D. vacuoles
5. **Compare** how energy is processed in animal and plant cells.
6. **Distinguish** between a cell membrane and a cell wall.

Interpret Graphics

7. **Summarize** Use the table below to identify organelles and their functions.

Organelle	Function
Nucleus	
	energy processing
Vacuole	

8. **Compare and contrast** the structures of the two cells shown below.

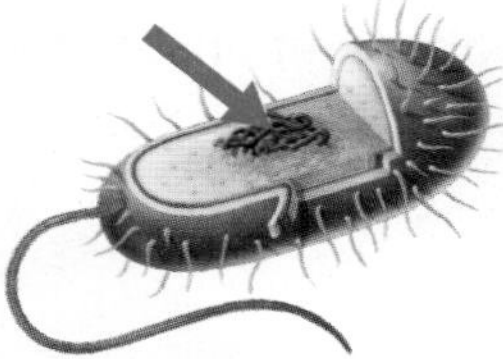

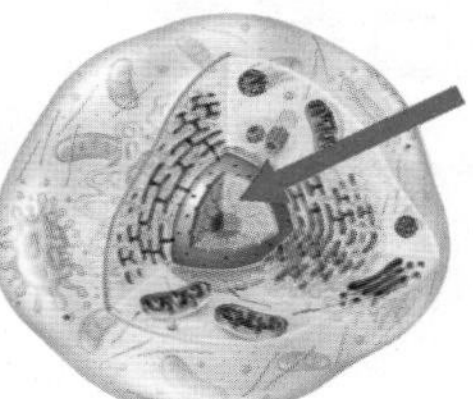

Critical Thinking

9. **Assess** the role of water in cell function.
10. **Relate** the cell wall to protection in bacteria.

Dr. Jeremy Burgess/Science Source

 Lab

2 class periods

How can living things be classified?

Materials

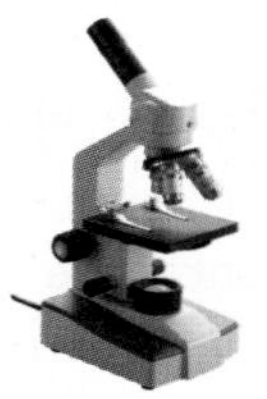

compound microscope

dissecting microscope

magnifying lens

ruler

Also needed: specimens

Safety

Thousands of new organisms are discovered each year. Today, an organism's DNA can be used to determine how closely a newly discovered organism is related to living things that are already known. A long time ago, taxonomists had to rely on what they could observe with their senses to determine the relationships between organisms. They looked at characteristics such as an organism's parts, behaviors, or the environments in which they lived to help them determine relationships among organisms. The father of taxonomy, Carolus Linnaeus, developed a system in the 1700s by which he classified over 9,000 organisms, primarily based on their external features.

Question

What characteristics can be used to distinguish among different types of organisms?

Procedure

1. Read and complete a lab safety form.
2. Use your background knowledge of the specimens provided and the available tools to identify distinguishing characteristics of the specimens. Be sure to observe each of the specimens thoroughly and completely.
3. In your Science Journal, record as much information as possible about each of the organisms. This information can include your observations and any knowledge you have of the organism.

4. Using the information you have recorded, create a dichotomous key that can be used to identify all of the specimens.
5. Trade your key with another student.
6. Verify that the key you received works by trying to identify all ten of the organisms, one at a time.
7. If your key did not work, repeat steps 2–5. If your key did work, move on to the **Analyze and Conclude** section.

Analyze and Conclude

8. **Compare and Contrast** How are the questions in the key you made similar to and different from the questions in the key that you checked?
9. **Classify** How would an elephant, bread mold, and a rose be identified if the key you created was used to identify them? Are these identifications accurate? Why did this happen?
10. **The Big Idea** How would the questions in your key be different if all ten organisms were more closely related, such as ten different plants?

Communicate Your Results

Share your key and questions with a small group of students. After everyone shares, make a group key that combines the most objective questions that were asked among the various keys.

Inquiry Extension

Research the scientific names of the specimens that you observed, and find the meanings of the species name of each organism. Research the characteristics by which bacteria are classified. Design a key to be used by younger students to help identify different polygons (triangles, pentagons, octagons, and so on) using correct mathematical terms.

Lab Tips

☑ Recall from earlier in the chapter the types of characteristics that are good to use in creating a dichotomous key.

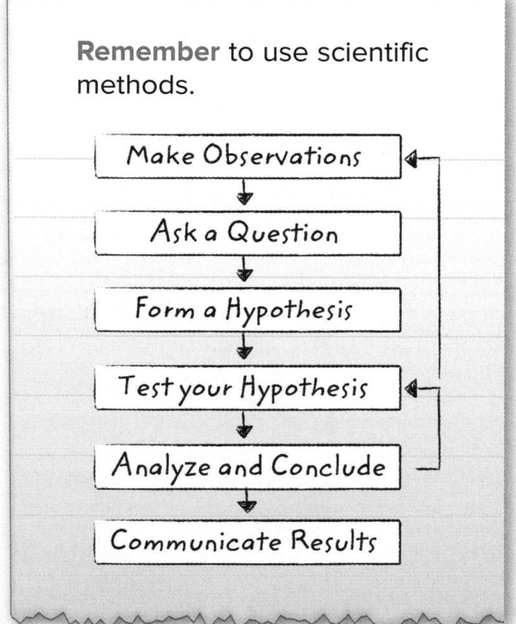

Chapter 6 Study Guide

Organisms are classified based on similar characteristics, including cell structure and function.

Key Concepts Summary

Lesson 1: Classifying Living Things

- Living things are organized, process energy, grow, reproduce, respond to stimuli, and contain cells.
- Living things need food, water, and a **habitat.**
- Organisms are classified based on similar characteristics.

Vocabulary

autotroph p. 192
heterotroph p. 192
habitat p. 193
binomial nomenclature p. 194
taxon p. 195

Lesson 2: Cells

- Cells are made of water and macromolecules.
- Different parts of a cell enable it to perform special functions.

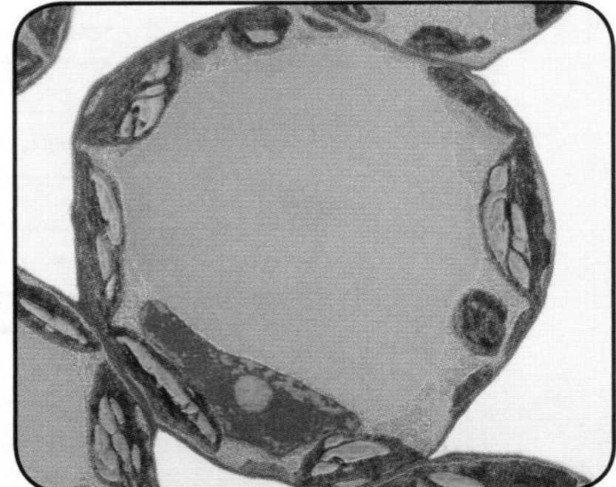

Vocabulary

prokaryotic cell p. 200
eukaryotic cell p. 200
cytoplasm p. 202
mitochondrion p. 203

(t) Jeff Rotman/Science Source, (b) Dr. Jeremy Burgess/Science Source

Vocabulary eFlashcards
Vocabulary eGames

Personal Tutor

FOLDABLES® **Chapter Project**

Assemble your lesson Foldables as shown to make a Chapter Project. Use the project to review what you have learned in this chapter.

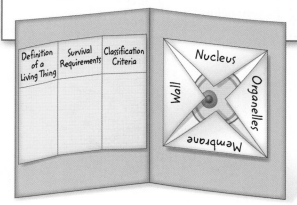

Use Vocabulary

1. The Latin term *Homo sapiens* is an example of ________.
2. Organisms that obtain energy by eating other organisms are called ________.
3. Use the term *habitat* in a sentence.
4. Define the term *cytoplasm* in your own words.
5. Animal cells obtain energy by breaking down food in ________.
6. Use the term *prokaryotic cell* in a sentence.

Link Vocabulary and Key Concepts

Interactive Concept Map

Copy this concept map, and then use vocabulary terms from the previous page to complete the concept map.

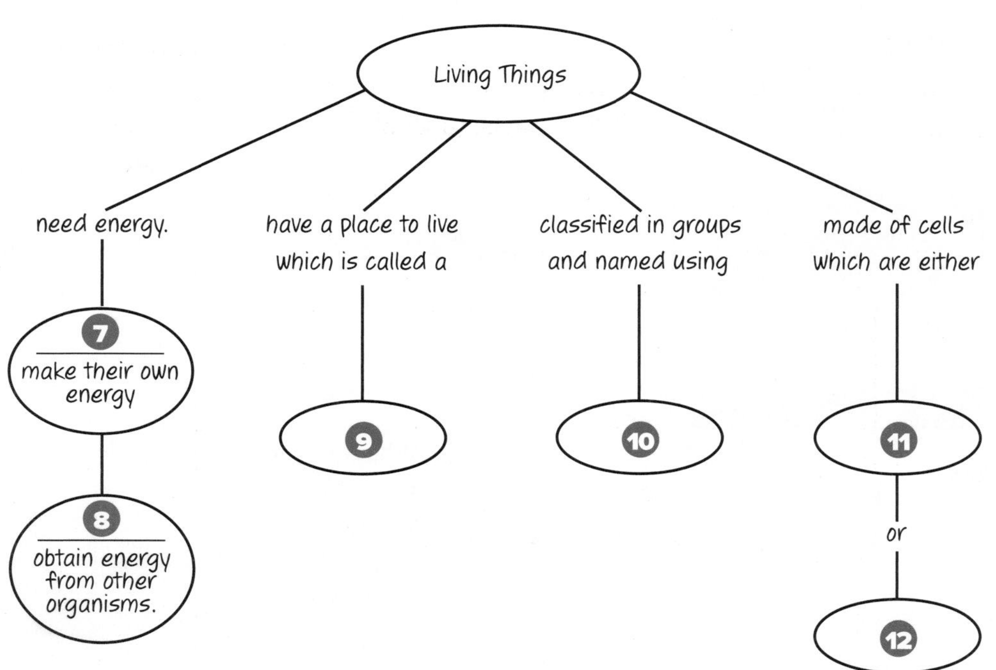

Chapter 6 Review

Understand Key Concepts 

1. What is a rigid structure that provides support and protection to plants and some types of bacteria?
   - **A.** chloroplast
   - **B.** nucleus
   - **C.** cell membrane
   - **D.** cell wall

2. What type of reproduction occurs when a cell divides to form two new cells?
   - **A.** autotrophic
   - **B.** heterotrophic
   - **C.** asexual reproduction
   - **D.** sexual reproduction

3. Which is the binomial nomenclature for humans?
   - **A.** *Canis lupos*
   - **B.** *Felis catus*
   - **C.** *Homo sapiens*
   - **D.** *Tamias striatus*

4. What is a group of organisms called?
   - **A.** taxon
   - **B.** tissue
   - **C.** dichotomous key
   - **D.** organ system

5. Which organelle is the arrow pointing to in the picture below?

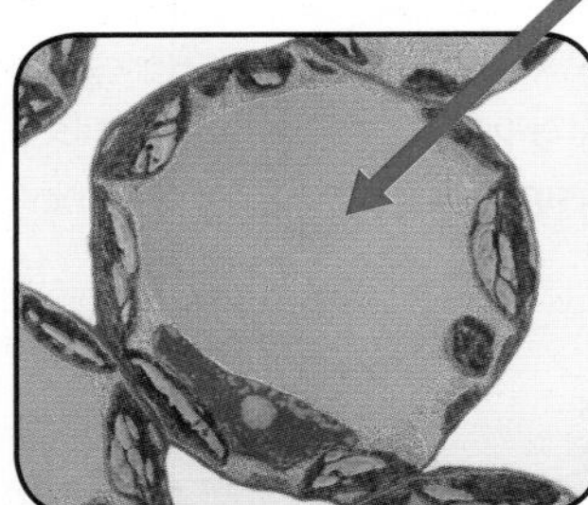

   - **A.** chloroplast
   - **B.** cytoplasm
   - **C.** mitochondrion
   - **D.** vacuole

6. Which is NOT a characteristic of all living things?
   - **A.** grow
   - **B.** reproduce
   - **C.** have organelles
   - **D.** use energy

7. Which organelle is the arrow pointing to in the picture below?

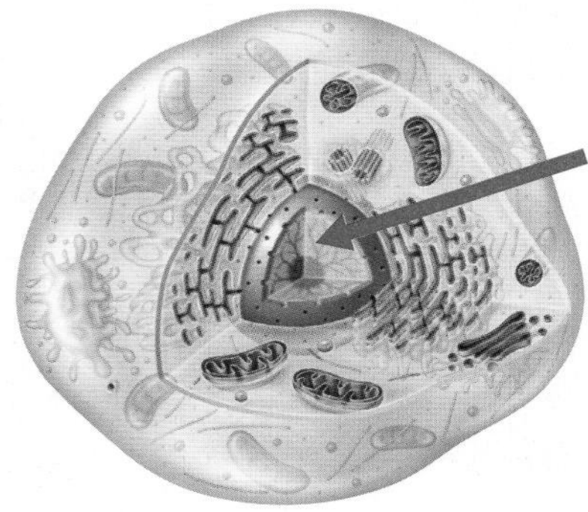

   - **A.** chloroplast
   - **B.** cytoplasm
   - **C.** mitochondrion
   - **D.** nucleus

8. What is the name used to describe the specific place where an organism lives?
   - **A.** autotroph
   - **B.** habitat
   - **C.** heterotroph
   - **D.** taxon

9. What is the smallest unit of all living things?
   - **A.** cell
   - **B.** organ
   - **C.** organelle
   - **D.** tissue

10. What are cells mostly made of?
    - **A.** DNA
    - **B.** lipids
    - **C.** proteins
    - **D.** water

Dr. Jeremy Burgess/Science Source

Online Test Practice

Critical Thinking

11 **Summarize** the characteristics of all living things.

12 **Describe** how the organization of a multicellular organism helps it function. Diagram the relationships.

13 **Assess** how taxonomy relates to the diversity of species.

14 **Explain** why different organisms live in different habitats.

15 **Assess** the role of organelles in the functions of eukaryotic cells.

16 **Relate** the structure in the plant cell shown at the pointer in the picture below to how it obtains energy.

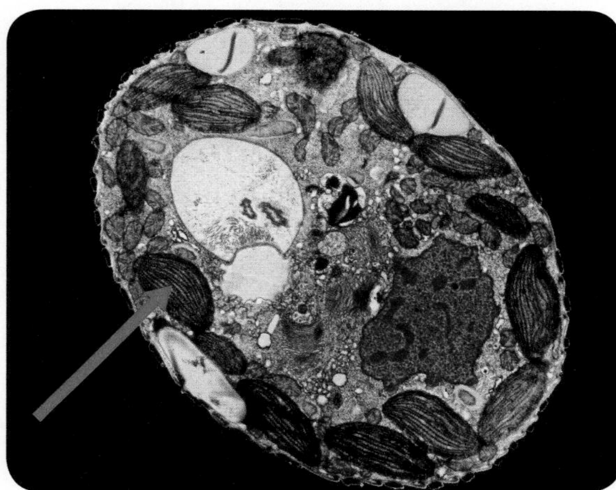

17 **Summarize** the role of nucleic acids in controlling cell functions.

18 **Discuss** how heterotrophs process energy.

Writing in Science

19 **Write** a five-sentence paragraph that describes the characteristics that all living things share.

REVIEW THE BIG IDEA

20 **Assess** how the classification of prokaryotes and eukaryotes relates to the structure of their cells.

21 How is the classification of living things related to the structure of their cells? Use the plant in the photo below as an example.

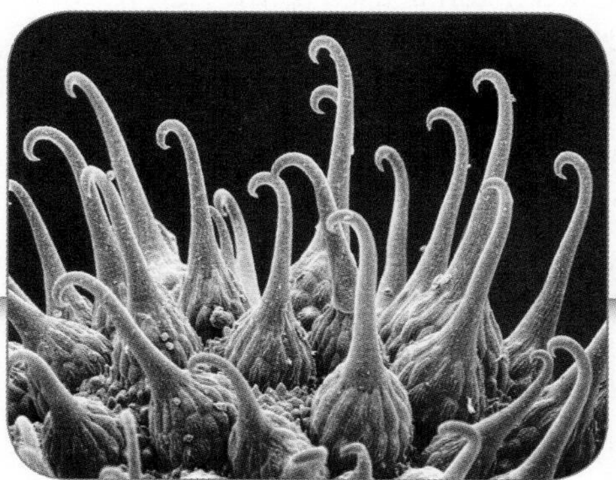

Math Skills

Math Practice

Use Ratios

22 There are about 300,000 species of plants. Of those, 12,000 are mosses. What is the ratio of mosses to plants? Express the answer all three ways.

23 Out of 300,000 plant species, 260,000 are flowering plants. What is the ratio of flowering plants to all plant species? Express the ratio in all three ways.

24 Out of 12,000 species of mosses, only about 400 are club mosses. What is the ratio of club mosses to all mosses? Express the ratio in all three ways.

Standardized Test Practice

Record your answers on the answer sheet provided by your teacher or on a sheet of paper.

Multiple Choice

1 Which would a chemoautotroph use to produce energy?

A sulfur

B sunlight

C carbon dioxide

D other organisms

2 Which taxon is used as the first word in an organism's scientific name?

A class

B genus

C kingdom

D order

Use the diagram below to answer question 3.

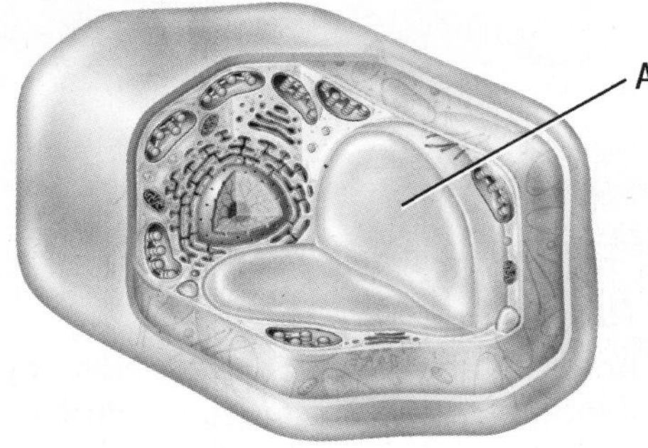

3 The diagram shows the parts of a plant cell. What is the name and function of structure A?

A chloroplast, making carbohydrates

B chloroplast, producing energy

C vacuole, storing water

D vacuole, transporting proteins

4 Which molecule stores energy for cells?

A ATP

B DNA

C proteins

D ribosomes

5 What do scientists call the largest taxonomic level of organization for organisms?

A domains

B genera

C kingdoms

D phyla

Use the image below to answer question 6.

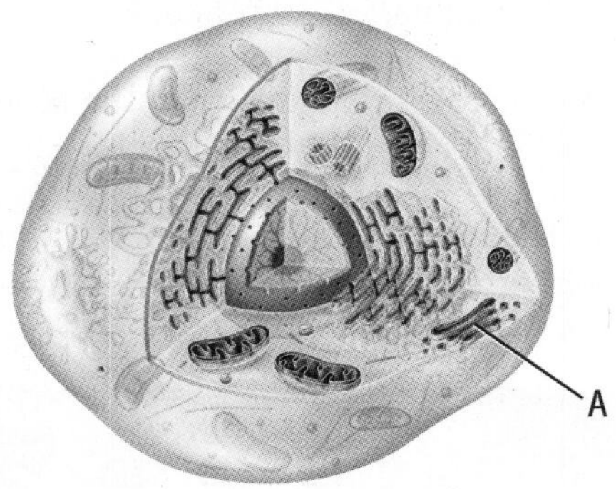

6 In the diagram, the organelle labeled *A* packages proteins into vesicles. What is this organelle called?

A central vacuole

B endoplasmic reticulum

C Golgi apparatus

D nuclear envelope

7 Which cell structures break down food and release energy?

A chloroplasts

B mitochrondria

C ribosomes

D vacuoles

8 Carl Linnaeus grouped organisms into categories based on which characteristic?

A energy production
B gene type
C physical traits
D reproduction habits

9 Which term defines a group of cells that work together and perform a function?

A organ
B taxon
C tissue
D phylum

Use the diagram to answer question 10.

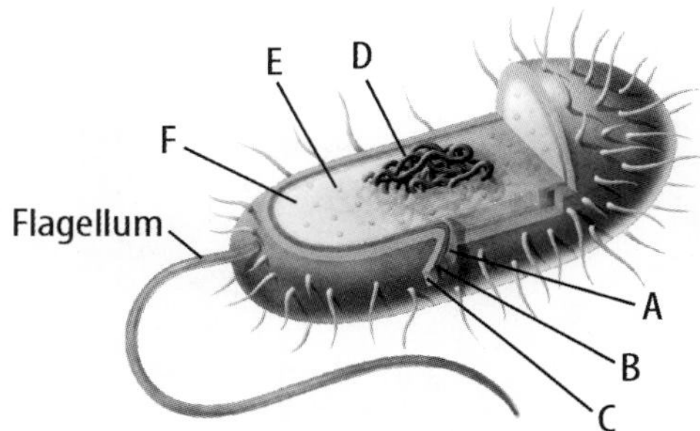

10 In the cell shown, what is the letter for the structure that provides much of the cell's support and helps protect it from the outside environment?

A A
B B
C C
D D

Constructed Response

Use the figure to answer questions 11 and 12.

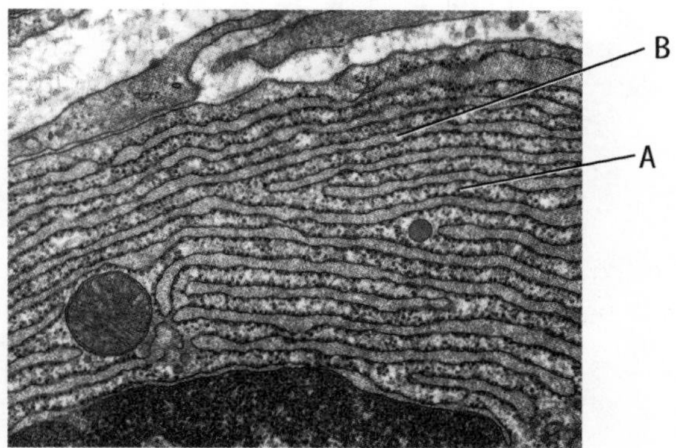

11 Identify the structure labeled *A* in the diagram. What is its function?

12 How are the organelles labeled *A* and *B* related? Are they found in prokaryotic cells, eukaryotic cells, or both?

13 Explain the relationship between cells, tissues, organs, and organ systems in a multicellular organism.

14 Cell membranes are made up mainly of proteins and carbohydrates. How do these molecules function in the cell membrane?

NEED EXTRA HELP?														
If You Missed Question...	1	2	3	4	5	6	7	8	9	10	11	12	13	14
Go to Lesson...	1	1	2	2	1	2	2	1	1	2	2	2	1	2

Chapter 7

Inheritance and Adaptations

How do inherited traits become adaptations?

Inquiry Follow me?

The strong legs and hooves of these mountain goats enable them to climb up and down mountains with relative ease. Their thick, white fur helps keep them warm and blend in against a snowy background.

- How are offspring similar to but different from their parents?
- How do adaptations help species survive?
- How do inherited traits become adaptations?

FRANK HILDEBRAND/Getty Images

Get Ready to Read

What do you think?

Before you read, decide if you agree or disagree with each of these statements. As you read this chapter, see if you change your mind about any of the statements.

1. Genes are made of chromosomes.
2. A mutation is a permanent change in a gene.
3. The environment cannot affect an inherited trait.
4. Mutations are a source of variation.
5. All species on Earth are uniquely adapted to their environments.
6. Plants have adaptations for movement.

Your one-stop online resource
connectED.mcgraw-hill.com

LearnSmart®

Chapter Resources Files, Reading Essentials, Get Ready to Read, Quick Vocabulary

Animations, Videos, Interactive Tables

Self-checks, Quizzes, Tests

Project-Based Learning Activities

Lab Manuals, Safety Videos, Virtual Labs & Other Tools

Vocabulary, Multilingual eGlossary, Vocab eGames, Vocab eFlashcards

Personal Tutors

Lesson 1

Inheritance and Traits

Reading Guide

Key Concepts
ESSENTIAL QUESTIONS

- What is inheritance?
- What is the role of genes in inheritance?
- How do environmental factors influence traits?
- How do mutations influence traits?

Vocabulary

trait p. 217
inheritance p. 217
gene p. 218
genotype p. 220
phenotype p. 220
mutation p. 222

Multilingual eGlossary

BrainPOP®

Go to the resource tab in ConnectED to find the PBL *Ready, Set, Grow!*

No; due to a genetic mutation, about 1 in 5 million lobsters are naturally blue. What is a mutation? How do you think mutations affect traits?

IMNATURE/Getty Images

Launch Lab

10 minutes

What role does chance play in inheritance?

You probably look like your parents in many ways, but you are not identical to them. For instance, you might have blue eyes like your father, but brown hair like your mother. Inheriting traits is a matter of chance.

1. Obtain two **dice** of different colors. With a partner, roll the dice 10 times. Make a data table in your Science Journal to record the number of dots on each die for each roll.
2. Discuss with your partner how this activity might model reproduction. What do the colors represent? What do the dots represent?

Think About This

1. Did you get the same combination for any of your rolls?
2. What if each die had 12 faces, or 100 faces? How do you think these changes would affect your chances of getting the same combination?
3. **Key Concept** In what ways do you think rolling dice models how traits are inherited? What role does chance play in inheritance?

What is inheritance?

You probably resemble your parents or grandparents. If you have brothers or sisters, they probably resemble your parents and grandparents, too. You all might have some of the same characteristics, such as being tall or having brown eyes. *A distinguishing characteristic of an organism is a* **trait.** During reproduction, many traits are passed from one generation to the next. *The passing of traits from generation to generation is* **inheritance.** Inheritance is the reason offspring resemble their parents, their grandparents, and even their distant ancestors.

Every organism has a range of inherited traits. The parrot shown in **Figure 1** has green feathers, wings, and a hooked beak. All of these traits can be passed to its offspring.

Not all traits are inherited. If the parrot in **Figure 1** lost a claw in an accident, its offspring would not be born missing a claw. Similarly, the parrot's offspring would not be born knowing how to put a ball into a basket. Losing a claw and learning tricks are examples of acquired traits. An acquired trait is a trait that an organism acquires or develops during its lifetime.

Key Concept Check What is inheritance?

Figure 1 This bird's color, shape, and body structure are inherited traits. However, the trick it has learned—putting a ball in a basket—is an acquired trait.

(t)Hutchings Photography/Digital Light Source, (b)Carolyn A. McKeone, courtesy of Tracey Charleson/Photo Researchers

Figure 2 DNA is a molecule that contains genes. It is coiled inside a cell's nucleus and forms a chromosome. ▶

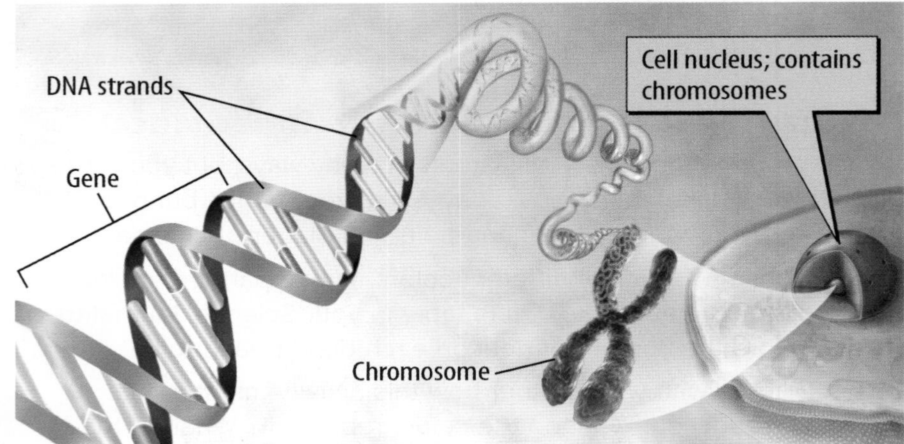

Inheritance and DNA

Organisms pass inherited traits to their offspring in one of two ways, depending on whether they reproduce asexually or sexually. Some organisms, such as amoebas, bacteria, and some plants, pass traits to their offspring by cell division and mitosis. This process is called asexual reproduction. It produces offspring that are identical to the original organism. Many other organisms, including humans, reproduce sexually. This process produces offspring that are similar–but not identical–to the parent or parents.

DNA and Genes

WORD ORIGIN
gene
from Greek *genea,* means "generation"

Sexual reproduction requires DNA from a sperm cell and an egg cell. DNA, shown in **Figure 2,** is a molecule inside a cell's nucleus that looks like a twisted zipper. Genes are distinct segments of DNA. *A* **gene** *is a section of DNA that has genetic information for one trait.* Genes carry this information in a unique sequence within DNA, much as words convey information by the unique sequence of their letters.

DNA is long. If you stretched out the DNA in one of your cells, it would be almost 2 m long. DNA fits into a cell's nucleus because it is tightly coiled with proteins to form chromosomes. A chromosome is a structure made of long chains of DNA.

Key Concept Check How are traits and genes related?

Chromosomes

Figure 3 Each of your body cells contains 23 pairs of chromosomes. ▼

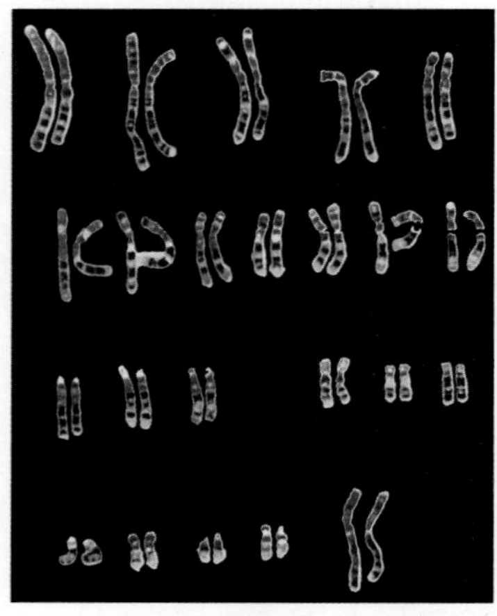

Color-enhanced LM Magnification: 2,000×

The number of chromosomes in a cell differs depending on the species. In most species, chromosomes come in pairs. Humans have 23 pairs of chromosomes in each body cell, as shown in **Figure 3.** Each pair contains one chromosome from the father and one chromosome from the mother. Human reproductive cells–called sperm and eggs–each contain only 23 single chromosomes. Along each of these chromosomes lies hundreds or thousands of genes.

L. Willatt, East Anglian Regional Genetics Service / Science Source

Meiosis

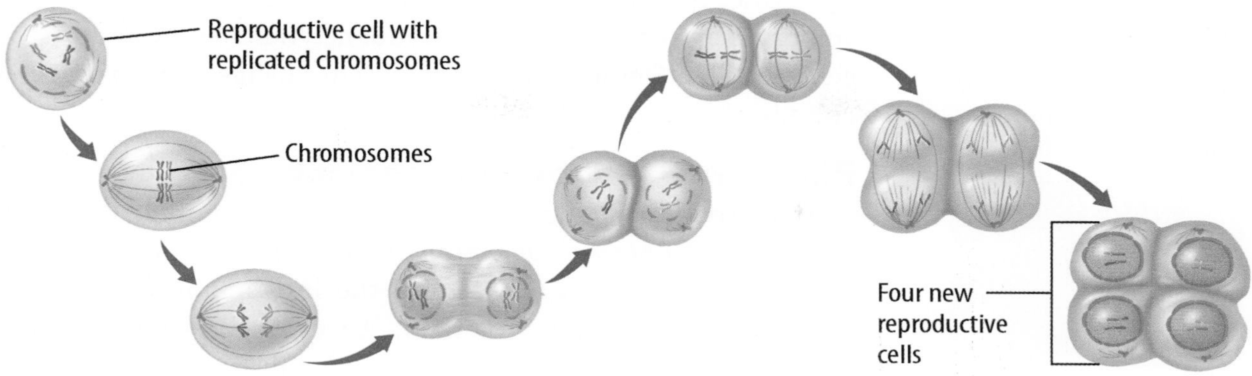

Figure 4 During meiosis, four new reproductive cells are formed, each with a single set of chromosomes.

Combining Genes

In sexual reproduction, an egg cell and a sperm cell each contribute one gene for a trait. Each gene for a single trait is called an allele (uh LEEL). How the alleles are sorted and combined into offspring during sexual reproduction is mostly a matter of chance.

Meiosis

Much of the randomness of sexual reproduction occurs during meiosis. Meiosis is the process during which sperm and egg cells form. During meiosis, the chromosomes in existing egg and sperm cells replicate and divide, as shown in **Figure 4.** Then they split into four separate cells, each with half the number of chromosomes–23 in human egg and sperm cells. Each sperm and egg cell contains a unique combination of genes on each chromosome.

Fertilization

During fertilization, a sperm and an egg unite. When this happens, the egg cell chromosomes and the sperm cell chromosomes combine to form an offspring with a full set of paired chromosomes. Because each sperm cell and egg cell is unique, the resulting offspring also is unique. In humans, there are many potential gene arrangements from the union of sperm and egg chromosomes. So many that if a mother and a father could have billions of offspring, each produced from a different fertilized egg, no two would be alike.

Reading Check Why is each sperm cell and egg cell produced by meiosis unique?

Math Skills

Use Probability

You can calculate the probability, or likelihood, of chromosome combinations using the formula 2^n where n = total number of chromosomes divided by 2. For example, fruit flies have 8 chromosomes. How many different combinations of chromosomes can be produced in the offspring?

1. Divide the number of chromosomes by 2.

   $$\frac{8 \text{ chromosomes}}{2} = 4$$

2. Replace n in the formula with the answer to step 1 and calculate.

   $$2^4 = 2 \times 2 \times 2 \times 2 = 16$$

Practice

The common housefly has 12 chromosomes. How many different chromosome combinations can form in the offspring?

Math Practice

Personal Tutor

FOLDABLES®

Make a horizontal shutterfold book. Label it as shown. Use it to organize your notes on factors that influence traits.

Physical Factors & Traits

Social Factors & Traits

Influencing Traits

An organism's complete set of genes is its **genotype** (JEE nuh tipe). Once inherited, this genotype remains unchanged. However, an organism's environment can influence traits expressed by the genotype. If a factor in the environment changes, the expression of a trait in an individual organism can be affected.

Phenotype and the Environment

Inherited traits are part of an organism's phenotype (FEE nuh tipe). *The* **phenotype** *of a trait is how the trait appears, or is expressed.* Phenotypes result from the interaction of an organism's genes and its environment. An organism's environment changes all the time. Light, temperature, moisture, nutrients, and social factors are not constant. These factors influence organisms in different ways. For example, because light is critical to plants, light levels have a strong effect on plant phenotype. Plants that grow tall in full sunlight might not grow as tall in low light.

Physical Factors There are many physical factors other than light that influence phenotype. For example, low levels of nutrients in soils, such as nitrogen or iron, might turn a plant's leaves yellow or cause them to fall off.

Nutrients can cause dramatic changes in the phenotype of some animals, too. The large honeybee shown in **Figure 5** has genes for the same traits as the smaller bees around it. But because it ate a special, nutrient-rich diet, it developed into the queen bee. Similarly, flamingos, also shown in **Figure 5,** are born white but turn pink because the food they eat, including algae and crustaceans, is rich in red pigment.

Figure 5 Phenotypes can change when the environment changes.

Visual Check How does diet affect the phenotype of flamingos?

The large bee in the middle—the queen—is larger because she ate a nutrient-rich diet.

Adult flamingos are pink because of their diet, but their offspring are born white.

(l)Steve Hopkin/Getty Images, (r)Steve Winter/National Geographic/Getty Images

Figure 6 When desert locusts are alone (solitary phase), they are green. When they are in a large group (social phase), their color changes to yellowish-brown, and they swarm.

Social Factors An organism's social group also can affect color, body structure, or behavior. Desert locusts usually are solitary insects, which means they live alone. But when these locusts are in a large group, they apply pressure on each other's legs. This causes them to change color and to swarm, as shown in **Figure 6.** Flamingos are another example of animals that are influenced by social factors. Through studies conducted in zoos, scientists have learned that the large social group in which flamingos live is important because it triggers breeding among them. A flock consisting of at least 20 flamingos is needed for breeding to occur in zoos. Studies have shown that adding more birds to the flock leads to increased breeding success. In the wild, flamingos live in flocks of up to 10,000 birds.

Key Concept Check What are some environmental factors that can influence phenotype?

MiniLab

20 minutes

How can the environment affect phenotypes?

You may have noticed that the same plant species grows better in a shady part of a yard than in a sunny part. The data in the table to the right describe the growth of 50 seeds over a period of 10 days. The seeds were planted in two different areas and began to emerge from the soil on day 4.

1. On **graph paper,** graph the average plant height for each area. Use different **colored markers** for each area and include a key.
2. Make a similar graph of the average number of leaves for each area over the 10 days.

Plant Height and Leaf Number

	Average Plant Height (cm)		Average Number of Leaves	
Day	**Area 1**	**Area 2**	**Area 1**	**Area 2**
4	2.9	2.9	2	2
5	6.1	5.1	3.5	4
6	9.0	8.3	5.5	6
7	11.0	9.9	7	9
8	12.0	11.1	8	10.5
9	12.8	11.4	9.5	11
10	13.6	12	9.5	12.5

Analyze and Conclude

1. **Infer** Based on the data, what inference can you make about the two areas?
2. **Key Concept** What environmental factors might have caused the differences in the phenotypes?

(t) Gianni Tortoli/Science Source, (c, b) Burrows et al. Neural Systems & Circuits 2011

Figure 7 The penguin on the right has a genetic mutation that affects the color of its feathers.

Visual Check How was the phenotype of the penguin on the right affected by a mutation?

Phenotype and Mutations

When an organism's phenotype changes in response to its environment, the organism's genes are not affected and the change cannot be passed on to the next generation. The only way that a trait can change so that it can be passed to the next generation is by mutation, or changing an organism's genes.

Random Changes *A* **mutation** *is a permanent change in the sequence of DNA in a gene.* It is an error in the DNA's arrangement in a gene. Have you ever made an error when you were typing or texting? For example, you might use one letter instead of another in a word. This could change the meaning of the word. Similarly, a mutation can change the trait for which the gene holds information.

Although all genes can mutate, only mutated genes in egg or sperm cells are inherited. Some mutations in egg or sperm cells occur if an organism is exposed to harsh chemicals or severe radiation. But most mutations occur randomly. The color of the lobster shown in the photo at the beginning of this lesson occurred as a result of a random mutation in an egg or a sperm cell. The feather color of the penguin on the right in **Figure 7** also is the result of a mutation.

ACADEMIC VOCABULARY
random
(adjective) without a definite aim, rule, or method

 **Key Concept Check** How can a mutation influence a trait?

Effects of Mutations Many mutations have no effect on an organism. They neither help nor hurt it. But some mutations change an organism's genes–and its traits–so much that they can affect an organism's ability to survive in its environment. Some mutations are harmful to an organism, but other mutations might help it survive. In Lesson 2, you will read how mutations that benefit an organism can spread to an entire population.

SCIENCE USE V. COMMON USE
population
Science Use all the members of a species living in a given area
Common Use the number of people living in a country or other defined area

Lesson 1 Review

Visual Summary

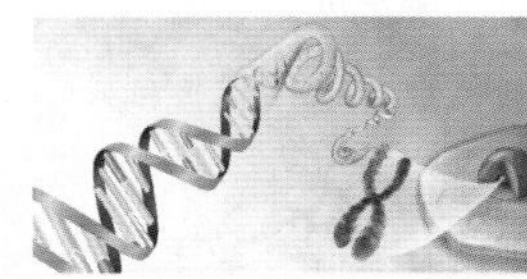

Traits are either inherited or acquired. Inherited traits are passed from one generation to the next.

An organism's phenotype can be influenced by factors in the environment, such as light, nutrients, or social interactions.

One result of a mutation could be a change in appearance, such as a change in feather color.

Use your lesson Foldable to review the lesson. Save your Foldable for the project at the end of the chapter.

What do you think NOW?

You first read the statements below at the beginning of the chapter.

1. Genes are made of chromosomes.
2. A mutation is a permanent change in a gene.
3. The environment cannot affect an inherited trait.

Did you change your mind about whether you agree or disagree with the statements? Rewrite any false statements to make them true.

Use Vocabulary

1. A distinguishing characteristic of an organism is a(n) ________.
2. A permanent change in the sequence of DNA in a gene is a(n) ________.
3. **Distinguish** between phenotype and genotype.

Understand Key Concepts

4. Which is an inherited trait?
   - **A.** learning to sing
   - **B.** losing a claw
   - **C.** having a hooked beak
   - **D.** learning a new trick
5. **Design an experiment** to determine the environmental factors that cause adult flamingos to turn pink.
6. **Compare and contrast** sexual reproduction and asexual reproduction.

Interpret Graphics

7. **Organize Information** Copy the graphic organizer below and use it to list *gene, chromosome, cell,* and *DNA* from smallest to largest.

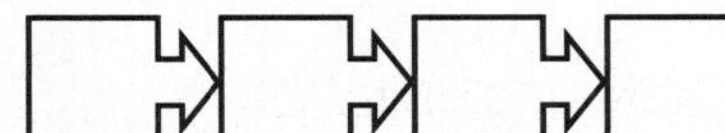

Critical Thinking

8. **Propose** an explanation for why this wallaby lacks normal coloration in its fur.

Math Skills

9. The common hamster has 20 chromosomes. How many different types of hamster offspring could form?

SCIENCE & SOCIETY

Down the Drain

Flushing Chemicals into Fish Habitats

Anytime you flush a toilet or run a faucet, you produce wastewater. When chemicals are added to waste water, they sometimes can end up in lakes and rivers, threatening wildlife. In recent years, scientists have discovered that when the chemical estrogen is in wastewater, it harms wild fish populations.

Estrogen is a chemical responsible for sexual development and reproduction in female vertebrates. It is produced naturally in the body, but it also is in medications that many women take. Like other chemicals in drugs, estrogen is released in urine and ends up in wastewater.

If even a small amount of estrogen gets into waterways, it can have a huge impact on local fish. The chemical disrupts the organs that enable them to reproduce. Most fish are not born male or female. Instead, physical factors in the environment, such as temperature and diet, determine which sex organs they develop. In a healthy habitat, some fish produce eggs and others produce the sperm to fertilize eggs. However, increased levels of estrogen in the water can affect the expression of traits in fish. Data collected in the field and in the laboratory show that estrogen affects how fish develop. When exposed to estrogen, males produce fewer or no sperm and some even produce eggs. Many females stop producing eggs. Without healthy males and females to reproduce, these fish populations drop.

Lake Experiment

To understand estrogen's effects on wild fish, scientists released small amounts of the chemical into a lake in Ontario, Canada. They observed drastic changes in the fathead minnows that had been thriving in the lake. The females produced fewer eggs. The males produced fewer sperm or began to develop eggs. After 3 years, the minnow population had nearly disappeared. Once scientists stopped adding estrogen, the fish population began to recover.

It's Your Turn

RESEARCH With a partner, research how pesticides and other chemicals are affecting the expression of traits in frogs. Include the name of the chemical, how it is used, and how it affects the frogs. Report your findings to the class.

(inset) blickwinkel/Alamy, (bkgd)Abrams/Lacagnina/Getty Images

Lesson 2

Adaptations in Species

Reading Guide

Key Concepts
ESSENTIAL QUESTIONS

- How do mutations cause variations?
- How does natural selection lead to adaptations in species?
- What are some ways adaptations help species survive in their environments?

Vocabulary

variation p. 226
adaptation p. 226
natural selection p. 227
selective breeding p. 228
camouflage p. 229
mimicry p. 229

Multilingual eGlossary

Science Video
What's Science Got to do With It?

Go to the resource tab in ConnectED to find the PBLs *Spot On* and *Population Probabilities.*

Inquiry Why Blend In?

This snake, called an eyelash viper, blends in well with its environment. How does this adaptation help the snake survive? What are some other adaptations that help organisms survive?

George Grall/Getty Images

Launch Lab

15 minutes

How alike are members of a population?

It is easy to see the differences among people, but what about plants or animals? Are all robins alike? What about sunflower seeds?

1. Read and complete a lab safety form.
2. Place 10 **sunflower seeds** on a **paper towel.** Number the seeds 1–10 by writing on the paper towel below each seed.
3. Use a **magnifying lens** to examine the seeds, focusing on how their coloration is alike and/or different. Record your observations in your Science Journal.
4. Copy the table on the right in your Science Journal. Perform the following steps and record your observations.
   - Use a **metric ruler** to measure the length of each seed.
   - Measure the thickness of each seed at its thickest point.
5. Compare the length and thickness of your 10 seeds with those of other teams.

Seed Variations

Seed	Length (mm)	Thickness (mm)
1		
2		
3		
4		
5		
6		
7		
8		
9		
10		

Think About This

1. Do all sunflower seeds have the same length and thickness? Why do you think the seeds differed in so many ways?
2. **Key Concept** If you were a bird, do you think you would be more or less attracted to any of the seeds? How might this affect the reproduction of the sunflowers?

What is adaptation?

In all species that reproduce sexually, offspring are different from their parents. The giraffes in **Figure 8** are members of the same species, yet each one has a slightly different pattern of spots on its coat. *Slight differences in inherited traits among individual members of a species are* **variations.**

Figure 8 Giraffes have variations in the patterns of their spots depending on the genes they inherit.

Variations occur through mutations. A mutation might harm an organism's chances of survival. However, many mutations, such as those that cause the unique pattern of spots on a giraffe, cause no harm. Still other mutations can benefit an organism. They produce traits that help an organism survive.

The giraffes in **Figure 8** have different spot patterns, but each has spots. The spots help the giraffes blend in with their environment–the grasslands of Africa. As a result, predators of giraffes, such as lions and hyenas, cannot see them as easily. The spotted coat of giraffes is an adaptation. *An* **adaptation** *is an inherited trait that helps a species survive in its environment.*

Key Concept Check How are mutations related to variations?

WLDavies/Getty Images

How Adaptations Occur

Giraffe spots were probably the result of a mutation that occurred in an individual giraffe many generations ago. The mutation produced a variation that helped the giraffe survive. Eventually, the mutated gene became part of the giraffe population genotype. How did this happen? How can a variation in a single individual become common to an entire population?

Natural Selection

Natural selection *is the process by which organisms with variations that help them survive in their environment live longer, compete better, and reproduce more than those that do not have the variation.* If a variation helps an organism survive or compete better in its environment, the organism with that variation lives longer. Because it lives longer, it has more offspring that also can have the variation. Over many generations, more and more offspring inherit the variation. Eventually, most of the population has the variation, and it becomes an adaptation, as shown in **Figure 9**.

Because mutations are random and occur continually, so do new variations. The variations that become adaptations depend on the environment. Over time, all environments change. Huge volcanic eruptions can change a climate rapidly. The movement of continents causes slow, gradual changes. When an environment changes, a population either adapts through natural selection or dies off. The repeated elimination of populations can lead to the extinction of a species.

 **Key Concept Check** How does a variation become an adaptation?

WORD ORIGIN

adaptation
from Latin *adaptare,* means "to adjust"

REVIEW VOCABULARY

species
a group of organisms that share similar characteristics and can reproduce among themselves, producing fertile offspring

 **Personal Tutor**

Figure 9 Through natural selection, a color variation in one or a few beetles can be inherited by many other beetles to become an adaptation.

❶ **Variation in Traits** In this population of beetles, some are yellow and some are brown. The color does not affect the ability of the beetles to survive in their environment.

❷ **Organisms Compete** A new predator eats yellow beetles more often because it sees the yellow beetles more easily than the brown beetles.

❸ **Traits are Inherited** The yellow beetles do not live as long as the brown beetles, and—since color is inherited—fewer yellow beetles hatch.

❹ **Adaptation over Time** Nearly all individuals in a population are brown. The color brown has become an adaptation that helps the beetles avoid predators in that environment.

Figure 10 The frizzle chicken is the result of breeding birds with a mutation—outward-curling feathers.

Selective Breeding

Watching natural selection in action is like watching mountains grow taller. It occurs over so many generations that it usually cannot be seen. It is easier to observe a type of selection practiced by humans. When humans breed organisms for food or for use as pets, they are selecting variations that occur naturally in populations. *The selection and breeding of organisms with desired traits is* **selective breeding.** Selective breeding is similar to natural selection except that humans, instead of nature, do the selecting. By breeding organisms with desired traits, humans change traits just as natural selection does. Cows with increased levels of milk production, dogs of different sizes, and roses of unique colors are products of selective breeding. So is the chicken shown in **Figure 10**.

Reading Check How is selective breeding different from natural selection?

Types of Adaptations

Through natural selection or selective breeding, all species on Earth are uniquely adapted to their environments. Chickens are adapted to life in a henhouse just as giraffes are adapted to life in the grasslands. Adaptations enable species to maintain homeostasis, avoid predators, find and eat food, and move. There are three main categories of adaptations: structural, behavioral, and functional. Examples of each are shown in **Table 1**.

Table 1 Types of adaptations include structural, behavioral, and functional.

Visual Check What are some examples of behavioral adaptations?

Table 1 Types of Adaptations

Type of Adaptation	Description	Example
Structural	a physical trait, such as color, shape, or internal structure, that increases survival	The color and shape of this insect's eyes are structural adaptations.
Behavioral	a behavior or action, such as migration, hibernation, hunting at night, or playing dead, that increases survival	This snake is playing dead, a behavioral adaptation to fool predators.
Functional	a biochemical change, such as hibernating, shedding, or spitting, that enables a species to increase survival or maintain homeostasis	Spraying venom, as this cobra is doing, is a functional adaptation.

Maintaining Homeostasis

The ability of an organism to keep its internal conditions within certain limits is homeostasis. Sweating on a hot day is an adaptation that helps you maintain your internal body temperature when external temperatures increase. All species have adaptations that help them survive temporary changes in their environments. Species also have adaptations specific to their environments. Plants living in deserts store water in their leaves. Fish in the ocean have gills that remove oxygen from water.

Figure 11 Can you see the fish? The stonefish is well-camouflaged in its environment.

Protection from Predators

Species also have adaptations that protect them from predators. For example, sharp quills protect porcupines. Sometimes, through natural selection, variations are selected that make an organism resemble something else. **Camouflage** (KAM uh flahj) *is an adaptation that enables a species to blend in with its environment.* The stonefish in Figure 11 resembles a rock. This makes it less visible to predators. **Mimicry** (MIH mih kree) *is an adaptation in which one species looks like another species.* The scarlet kingsnake is a nonpoisonous snake that looks like, or mimics, the poisonous coral snake. Predators often avoid the kingsnake because they cannot tell the two snakes apart.

Key Concept Check Give an example of how adaptations help species survive.

MiniLab

20 minutes

How do species' adaptations affect one another's survival?

In a predator-prey relationship, adaptations can greatly affect the survival of a population.

1. Read and complete a lab safety form.
2. Obtain a **bag with cards** according to your assigned group. Each card in your bag represents the speed of one individual in your population. Find and record the average speed of the ten individuals in your population.
3. Shuffle your cards and place them face down between your team and the opposing team (predator v. prey).
4. At the same time, each team turns over the top card. The team with the faster speed on its card places its card face-up in a separate pile, representing surviving individuals. The slower team places their card face-down in a separate pile. In case of a tie, flip a **coin.** Continue playing through ten rounds.
5. Count your surviving individuals, and calculate the new average speed of the survivors.

Analyze and Conclude

1. **Draw Conclusions** Based on your results, how did natural selection affect your population?
2. **Key Concept** How would the survival of either population be affected if one population developed a large adaptive advantage over the other? Explain your answer.

(t) johnandersonphoto/Getty Images, (b) Nicholas DeVore/Getty Images

Woodpeckers use their long, thin beaks to search for insects in tree bark.

Parrots have strong beaks that help them crack nuts and seeds.

The condor uses its long, powerful beak to tear the flesh from dead organisms.

Figure 12 Though all birds have wings, beaks, and feathers, each species is adapted to a different environment. Each uses its beak in a different way to gather food.

Visual Check How is a condor's beak adapted for the food it eats?

Food Gathering

As you have just read, camouflage and mimicry protect species from predators. These same adaptations also can help species find food. The camouflaged stonefish in **Figure 11** is hidden not only from predators, but also from its prey. Many other kinds of adaptations help species gather and eat food. An anteater has a long nose and a long tongue for gathering ants. Each of the birds shown in **Figure 12** has a beak that helps it gather a different type of food. Some plants also have adaptations that enable them to store food. Potatoes, onions, and tulips all have modified underground stems that store food for the plants.

As predators develop adaptations for hunting their prey, the species they hunt develop adaptations for avoiding them. A cheetah is a fast runner. But so are the gazelles it chases as prey. Over time, cheetahs might become even faster due to chance variations and natural selection. But faster gazelles also might arise from the same process. In this way, species adapt to each other.

FOLDABLES®

Make a vertical four-tab book, and label it as shown. Use it to organize your notes on benefits of adaptations.

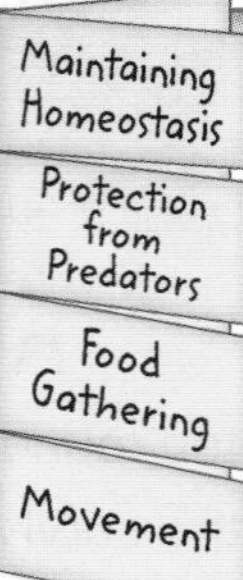

Movement

Cheetahs and gazelles have long, powerful legs adapted to running fast. Legs, wings, flippers, fins, and even tails are adaptations that help species move. Movement helps species search for food, avoid predators, and escape unpleasant stimuli. Even plants have adaptations for movement. Their leaves turn to face the Sun as it moves across the sky.

Lesson 2 Review

Visual Summary

Variations in populations occur because of mutations. Variations can lead to adaptations.

Through natural selection, a variation that helps organisms survive and reproduce eventually is inherited by most members of the population.

Adaptations may be structural, behavioral, or functional. Structural adaptations help organisms blend in with their environments.

Use your lesson Foldable to review the lesson. Save your Foldable for the project at the end of the chapter.

What do you think NOW?

You first read the statements below at the beginning of the chapter.

4. Mutations are a source of variation.

5. All species on Earth are uniquely adapted to their environments.

6. Plants have adaptations for movement.

Did you change your mind about whether you agree or disagree with the statements? Rewrite any false statements to make them true.

(t) W.Davies/Getty Images, (c) johnandersonphoto/Getty Images, (b) tips rf/age fotostock

Use Vocabulary

1. Slight differences in inherited traits are __________.
2. **Describe** natural selection in your own words.
3. **Distinguish** between mimicry and camouflage.

Understand Key Concepts 

4. A nonpoisonous butterfly has coloration and markings similar to a poisonous butterfly. This an example of
   - **A.** camouflage.
   - **B.** mimicry.
   - **C.** behavioral adaptation.
   - **D.** functional adaptation.
5. **Compare and contrast** natural selection and selective breeding.
6. **Explain** how two species might trigger adaptive changes in each other.

Interpret Graphics

7. **Identify** the type of adaptation the insect at right exhibits, and explain how the insect might benefit from the adaptation.

8. **Organize Information** Copy the graphic organizer below. Use it to list three ways that an organism you choose is adapted to its environment. Classify each adaptation as structural, behavioral, or functional.

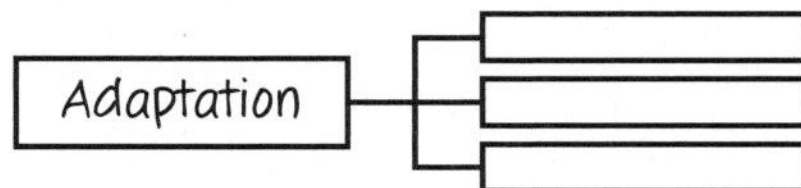

Critical Thinking

9. **Evaluate** the role of the environment in natural selection.
10. **Assess** the role of mutations in adaptations.

Lab

40 minutes

Materials

paper bag

marker

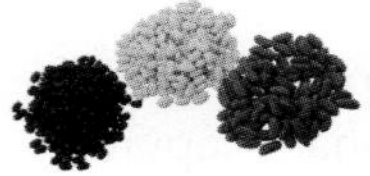

red beans, white beans, and black beans

Safety

⚠ Do not taste or eat any material used in the lab.

Model Natural Selection

The interaction between predators and their prey is a driving force for natural selection. There are many other forces in nature, such as changes in environment, that act on natural selection. However, predation provides a good model to explore how natural selection works.

Ask a Question

What happens to a population when a mutation occurs in an individual that helps it survive in its environment?

Make Observations

1. Read and complete a lab safety form.
2. Obtain a paper bag and write *Rabbit Gene Pool* on the front of the bag.
3. Place 10 red beans and 10 white beans in the bag.
4. Make a table like the one below in your Science Journal. Use the table to record the genotype and the phenotype of individuals in each generation of the population. Assume that a pair of beans is the genotype of an individual rabbit. The phenotype of a pair of red beans is brown fur. The phenotype of a red bean and a white bean is gray fur, and the phenotype of a pair of white beans is white fur.
5. Without looking in the bag, take out two beans to represent an offspring. Record the colors in your table. Continue taking beans out of the bag two at a time and recording the results for the first generation of rabbits. Determine and record the phenotype of each rabbit.
6. To model selection, predators eat 100 percent of the white rabbits, which do not blend in well with the environment, 50 percent of the gray rabbits, and 25 percent of the brown rabbits. In the case of an odd number of individuals, flip a coin to determine whether an individual will survive.
7. After you have eliminated the correct number of individuals, place two offspring per surviving individual in the bag, along with the surviving parent. In this activity, each offspring should have the same genotype as its parent.
8. Repeat steps 4–7. Then repeat steps 4–6 again.

Rabbit Population—Generation 1

Rabbit #	1st Bean Color	2nd Bean Color	Phenotype
1			
2			
3			
4			
5			

(t) Jacques Cornell/McGraw-Hill Education, (c) Ken Karp/McGraw-Hill Education, (b) McGraw-Hill Education

Form a Hypothesis

9 Review the data you have collected so far. Suppose one of the gray rabbits from the third generation had a mutation on one of its genes, and its fur is multicolored. A multicolored rabbit blends in well with its environment, and predators eat 0 percent of the multicolored rabbits in successive generations. Formulate a hypothesis that explains how the population will change over the next three generations of rabbits.

Test Your Hypothesis

10 Determine which rabbits in the third generation will survive predation. Choose a gray rabbit. Replace the red bean in its genotype with a black bean. This represents the multicolored rabbit.

11 Continue step 7 with the third generation by placing two offspring per surviving individual in the bag, along with the surviving parent. Each offspring should have the same genotype as its parent.

12 Repeat steps 4–7 three more times, using these new rules for eliminating rabbits: predators eat 100 percent of the white rabbits, 60 percent of the gray rabbits, 35 percent of the brown rabbits, and 0 percent of the multicolored rabbits. Assume that the multicolored gene is dominant over the other colors. End the model with six generations of rabbits.

Analyze and Conclude

13 **Analyze** How did the rabbit population change during the first three generations of rabbits?

14 **Analyze** How did the rabbit population change after the mutation occurred and multicolored fur became a phenotype? Was your hypothesis correct? Why or why not?

15 **Draw Conclusions** What explains the changes to the rabbit population over the six generations you tested?

16 **Describe** how this model is similar to and different from the way that natural selection occurs in nature.

17 **The Big Idea** Predict what the population of rabbits would be like after ten generations, assuming conditions remain the same as they are in step 12.

Communicate Your Results

Compare your results with those of other groups. Did any groups have results that were significantly different than yours? Why or why not?

Extension

Design your own model to mimic natural selection.

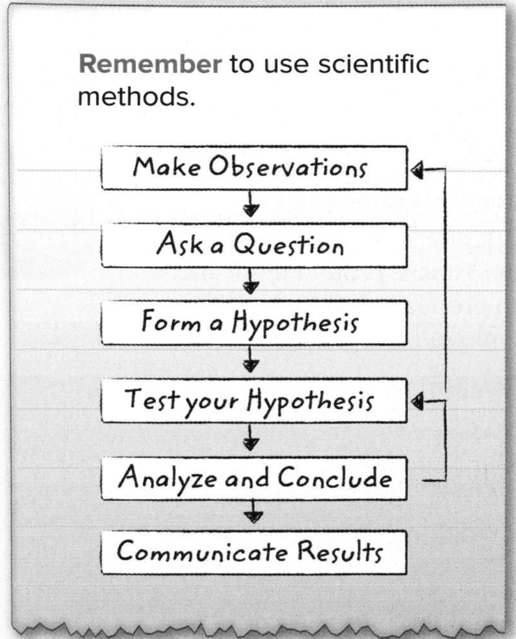

Hutchings Photography/Digital Light Source

Chapter 7 Study Guide

Inherited mutations can lead to variations, which can become adaptations through natural selection over many generations.

Key Concepts Summary

Lesson 1: Inheritance and Traits

- The passing of **traits** from generation to generation is **inheritance.**
- Information about traits is passed from parents to offspring on **genes.**
- An organism's **phenotype** can be influenced by environmental factors, such as temperature, nutrients, and social interaction.
- Only traits affected by **mutation** can be passed to offspring.

Vocabulary

trait p. 214
inheritance p. 217
gene p. 218
genotype p. 220
phenotype p. 220
mutation p. 222

Lesson 2: Adaptations in Species

- **Variations** arise when mutations cause changes in the sequence of an organism's DNA.
- **Natural selection** explains how variations that help organisms survive are passed to offspring and eventually become **adaptations.**
- Adaptations help species maintain homeostasis, protect themselves from predators, gather food, and move.

Vocabulary

variation p. 226
adaptation p. 226
natural selection p. 227
selective breeding p. 228
camouflage p. 229
mimicry p. 229

Vocabulary eFlashcards
Vocabulary eGames
Personal Tutor

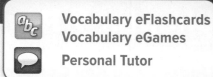

FOLDABLES® Chapter Project

Assemble your lesson Foldables as shown to make a Chapter Project. Use the project to review what you have learned in this chapter.

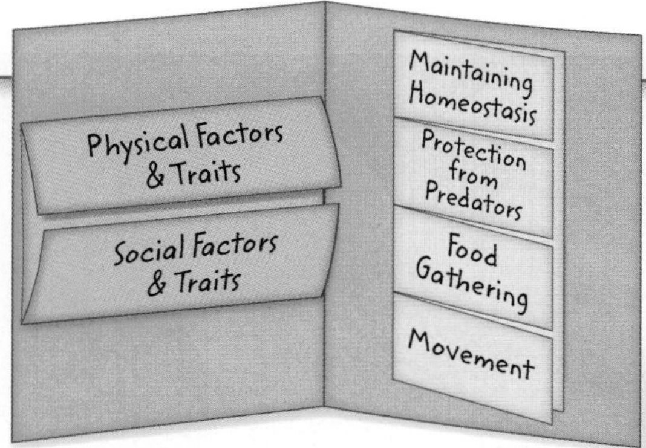

Use Vocabulary

Choose the vocabulary term that best matches the descriptions below.

1. slight differences in inherited traits
2. a distinguishing characteristic of an organism
3. all of an organism's genes
4. how a trait appears, or is expressed
5. adaptation that helps an organism blend in with its surroundings
6. the human practice of breeding organisms with desired characteristics
7. resembling another species

Link Vocabulary and Key Concepts

Interactive Concept Map

Copy this concept map, and then use vocabulary terms from the previous page to complete the concept map.

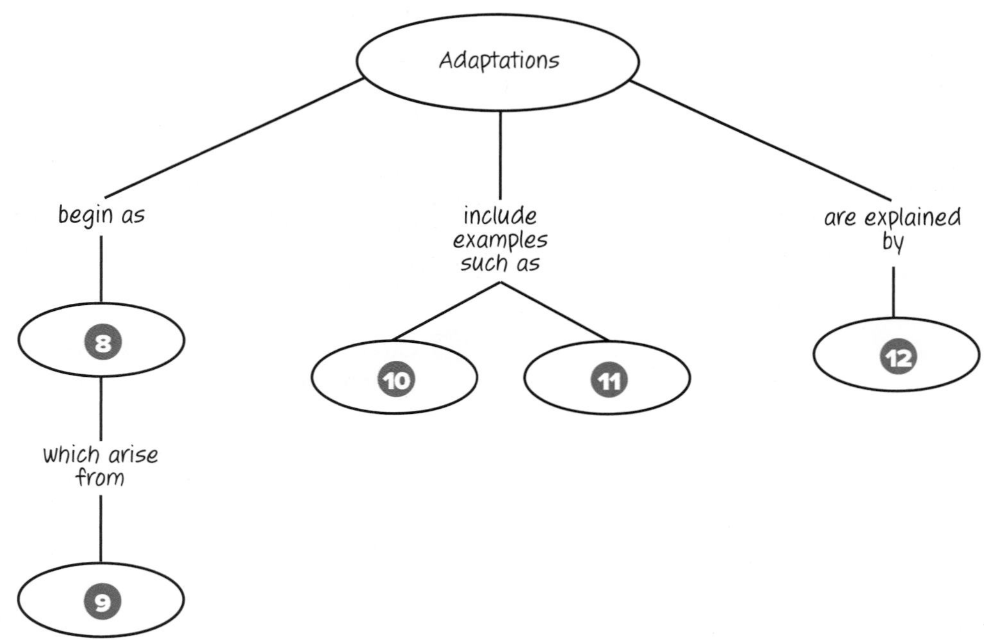

Chapter 7 Review

Understand Key Concepts 

1. In which way does asexual reproduction differ from sexual reproduction?
   - A. Genes are not involved in asexual reproduction.
   - B. No traits are passed to offspring in asexual reproduction.
   - C. Offspring are identical to the parent in asexual reproduction.
   - D. There are no mutations in asexual reproduction.

2. Which is a source of variations?
   - A. adaptations
   - B. mutations
   - C. phenotype
   - D. traits

3. Which is the sequence by which natural selection works?
   - A. selection ⟶ adaptation ⟶ variation
   - B. selection ⟶ variation ⟶ adaptation
   - C. variation ⟶ adaptation ⟶ selection
   - D. variation ⟶ selection ⟶ adaptation

4. Which adaptation is functional?
   - A. a lizard playing dead
   - B. a monkey swinging by its tail
   - C. a skunk spraying a predator
   - D. a wolf hunting in a pack

5. Which process is illustrated below?

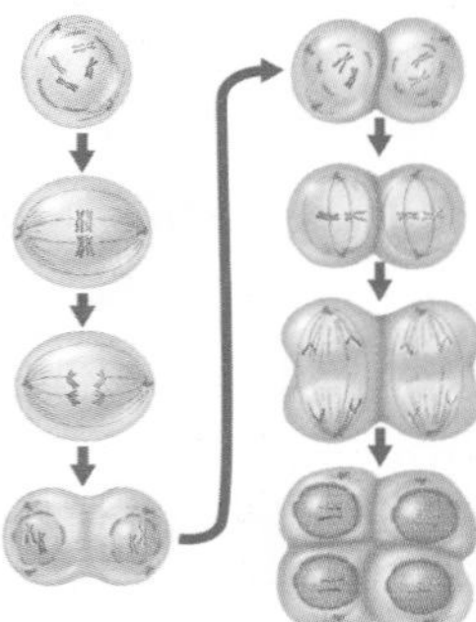

   - A. meiosis
   - B. mutation
   - C. asexual reproduction
   - D. natural selection

6. Which trait cannot be inherited?
   - A. scars
   - B. shyness
   - C. big feet
   - D. red hair

7. The photo below is a leaf butterfly. Which explains how the butterfly came to resemble a leaf?

   - A. The butterfly's shape is the result of an exchange of genes with plants over many generations.
   - B. The butterfly's shape is the result of the environment causing mutations over many generations.
   - C. The butterfly's shape is the result of the environment influencing its phenotype over many generations.
   - D. The butterfly's shape is the result of the environment selecting variations over many generations.

8. Giraffes range in color from orange to yellow. Which explains these color differences?
   - A. adaptations
   - B. variations
   - C. natural selection
   - D. selective breeding

Creatas Images/PictureQuest

Critical Thinking

9 **Design** an organism adapted to a murky lake with many plants. The organism's major predator is a large fish that swims slowly.

10 **Assess** how mutations can be beneficial.

11 **Differentiate** among mutation, variation, and adaptation, and explain how they are related to one another.

12 **Classify** the following adaptations as structural, behavioral, or functional: robins migrating, llamas spitting, bats hibernating, a beetle's color, wolves hunting in packs.

13 **Predict** what might happen to a species of ground plants over many generations when leaf-eating tortoises move into its range.

14 **Design an experiment** to test whether a trait in an animal is inherited or the result of an environmental factor.

15 **Interpret Graphics** The seal on the right has normal coloration. The seal on the left does not. What could explain why the seal on the left has abnormal coloration?

Writing in Science

16 **Write** Scientists have determined that all dogs were bred from wolves. Think about how wolves might have become tame enough to be pets. Then write a paragraph explaining how dogs became so different over time. Include a main idea, supporting details, and a concluding sentence.

REVIEW THE BIG IDEA

17 Adaptations help species survive in their environments. Choose two species that live near you, and list at least three ways—one structural, one behavioral, and one functional—that each is adapted to its environment. Explain how each adaptation helps the species survive.

18 In what ways does the juvenile mountain goat look like its mother? In what ways might the offspring be different from its mother? Explain how differences in individual mountain goats could help the species survive if its environment suddenly changed.

Math Skills

Math Practice

Use Probability

19 A dandelion has 24 chromosomes. How many possible combinations of chromosomes can form in the offspring?

20 A human has 46 chromosomes. How many different combinations of chromosomes can be produced during reproduction?

21 A radish has 18 chromosomes. How many possible combinations can the chromosomes make during reproduction?

(l) Michael Nolan/Getty Images, (r) FRANK HILDEBRAND/Getty Images

Standardized Test Practice

Record your answers on the answer sheet provided by your teacher or on a sheet of paper.

Multiple Choice

1 Two black dogs produce a litter of black puppies. This is an example of

- **A** camouflage.
- **B** chromosomes.
- **C** inheritance.
- **D** mimicry.

Use the image below to answer question 2.

2 The sunflower plants shown are the same species. The differences in height among the plants is an example of

- **A** adaptation.
- **B** fertilization.
- **C** population.
- **D** variation.

3 Which explains how variations arise within a population of organisms?

- **A** asexual reproduction
- **B** behavioral adaptation
- **C** natural selection
- **D** random mutation

4 Which carry information about traits from parent to offspring?

- **A** genes
- **B** meiosis
- **C** mutations
- **D** variations

5 Which results from the interaction of genes and environment?

- **A** genotype
- **B** phenotype
- **C** chromosome number
- **D** sequence of DNA

Use the image below to answer question 6.

6 Feather color is an inherited trait in penguins. What most likely caused the differences shown?

- **A** change in environment
- **B** DNA sequence error
- **C** physical factor
- **D** social factor

7 Which statement about mutations is NOT true?

- **A** Genes in any cell type can mutate.
- **B** Most mutations are harmful.
- **C** Most mutations occur randomly.
- **D** Some mutations help organisms survive.

8 The giraffe's long neck helps this species reach food that animals with short necks cannot reach. What type of adaptation is the long neck?

A behavioral adaptation
B biochemical adaptation
C functional adaptation
D structural adaptation

Use the diagram below to answer question 9.

9 The plant shown above is responding to light in its environment. This is an example of

A an adaptation.
B a population.
C selection.
D variation.

10 Which describes a mutation?

A a change in a gene's DNA sequence
B a trait that helps a species survive
C a change due to an environmental factor
D a distinguishing inherited characteristic

Constructed Response

Use the figure to answer questions 11 and 12.

11 Use the images to explain the process of natural selection. In your answer, briefly explain what happens in each step.

12 Classify the adaptation shown above as structural, behavioral, or functional. Briefly explain your reasoning.

13 Predators avoid the scarlet king snake because it looks like the coral snake. Is this similarity in coloring an example of camouflage or mimicry? Explain your reasoning.

14 Give an example of an adaptation that helps a species maintain homeostasis. In your response, briefly explain the environmental conditions that select for the adaptation.

NEED EXTRA HELP?														
If You Missed Question...	1	2	3	4	5	6	7	8	9	10	11	12	13	14
Go to Lesson...	1	2	2	1	1	1	1	2	2	1	2	2	2	2

Chapter 8

Introduction to Plants

What structures help ensure the survival of plants, and what is the function of each?

What Type of Tree?

This is a magnified cross section of a needle from a Scotch pine tree.

- How do the needles help the pine tree stay alive?
- What plant structures can you name that are common to plants? What function does each of these structures carry out to enable a plant's survival?

Get Ready to Read

What do you think?

Before you read, decide if you agree or disagree with each of these statements. As you read this chapter, see if you change your mind about any of the statements.

1. Humans could survive without plants.
2. Plant cells contain the same organelles as animal cells.
3. Plants can reproduce both sexually and asexually.
4. All plants have a two-stage life cycle.
5. Plants respond to their environments.
6. Because plants make their own food, they do not carry out cellular respiration.

Your one-stop online resource
connectED.mcgraw-hill.com

LearnSmart®

Chapter Resources Files, Reading Essentials, Get Ready to Read, Quick Vocabulary

Animations, Videos, Interactive Tables

Self-checks, Quizzes, Tests

Project-Based Learning Activities

Lab Manuals, Safety Videos, Virtual Labs & Other Tools

Vocabulary, Multilingual eGlossary, Vocab eGames, Vocab eFlashcards

Personal Tutors

Lesson 1

Reading Guide

Key Concepts
ESSENTIAL QUESTIONS

- How do a plant's structures ensure its survival?
- How are the different plant types alike and different?

Vocabulary

rhizoids p. 244
stomata p. 245
nonvascular plant p. 246
vascular plant p. 247
gymnosperm p. 248
angiosperm p. 249

Multilingual eGlossary

BrainPOP®

Plant Diversity

Inquiry Why Such Diversity?

There are a wide variety of plant species shown in this photo. Some of the plants are growing directly in the pond, while other plants are growing outside the pond. What similarities do all of these plants share? What differences do you observe?

joel-t/iStock/Getty Images

Launch Lab

15 minutes

What does a plant need to grow?

Plants grow in many different environments. What sorts of things do plants need to survive?

1. Read and complete a lab safety form.
2. Brainstorm things a plant needs to survive. List the items on a large sheet of **poster board** and **tape** it to a wall of your classroom.
3. Obtain several **radish seeds.** Select the materials you will need to grow the seeds from the materials provided.
4. Plant the radish seeds in a **petri dish.** Place the petri dish in an appropriate environment.
5. Write a brief plan in your Science Journal describing how to grow and care for your radish plants. Include information about what your seeds need and how you will meet these needs. Follow this plan for the next several days.

Think About This

1. What things do plants require to survive? Which of these requirements are similar to the things humans need to survive?
2. What did you use to grow your radish seeds? Explain your reasoning for each.
3. **Key Concept** What types of structures do you think plants have in order to obtain the things they need to survive?

What is a plant?

Humans depend on plants for food, oxygen, building materials, and many other things. Even the pages of this book are made from plant material! Plants are a vital part of the world.

Plant Cells

Did you know that your cells and all animal cells are similar to plant cells in many ways? They have many of the same organelles such as a nucleus, mitochondria, and ribosomes. However, a plant cell also contains chloroplasts, which are organelles that make food. Unlike an animal cell, a rigid cell wall surrounds a plant cell and helps protect and support it. As shown in **Figure 1**, a plant cell also contains a large central vacuole.

Reading Check What are some structures in plant cells that are not in animal cells?

Animation

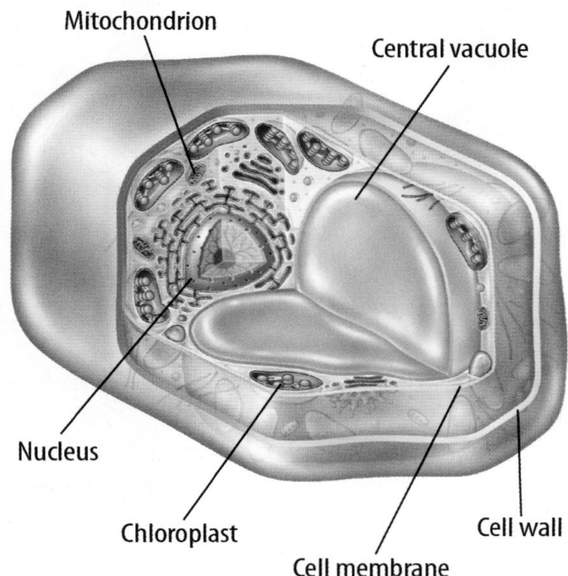

Figure 1 A plant cell has chloroplasts, a cell wall, and a large central vacuole.

Hutchings Photography/Digital Light Source

FOLDABLES®

Make a vertical three-tab book. Label and illustrate it as shown. Use it to organize your notes on the structures of a plant.

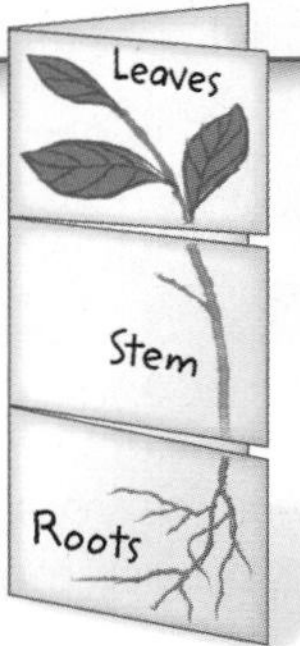

Plant Structures and Functions

Most plants have roots, stems, and leaves. These structures have functions that help plants survive. Root, stems, and leaves have specialized transport tissues. Some plants, such as the mosses you will read about later, don't have specialized transport tissues. However, they do have root-, stem-, and leaflike structures that perform similar functions.

Roots There are many different types of roots, as shown in **Figure 2**. Some plants have a large main root, called a taproot, with smaller roots growing from it. Some plants have additional small roots above ground, called prop roots, that help support the plant. Other plants have fibrous root systems that consist of many small branching roots. Roots anchor a plant in the soil and enable it to grow upright and not be blown away by wind or carried away by water.

Roots also absorb water and minerals from the soil, which plants require for cellular processes. Some roots store food such as sugar and starch. Plants that survive from one growing season to the next use this stored food for growing leaves at the beginning of the next season.

Some plants, such as mosses and hornworts, have rootlike structures called rhizoids. **Rhizoids** *are structures that anchor a plant without transport tissue to a surface.* Scientists do not consider rhizoids roots because they do not have the transport tissues that roots have.

Figure 2 Roots take many forms, each suited to the plant's needs.

Visual Check Why are fibrous roots better at absorbing water but prop roots are better at supporting a plant?

Plant Roots

The taproot can store food for the plant.

Prop roots can provide additional support for the corn plant.

Fibrous roots spread out and can absorb large amounts of water for the plant.

Stems Have you ever leaned against a tree? If so, you were leaning on a plant stem. Stems help support the leaves, and in some cases flowers, of a plant. There are two main types of stems, as shown in **Figure 3.** A woody stem is like the one you might have leaned against, and an herbaceous (hur BAY shus) stem is flexible and green, such as the stem of a vine.

Stems have tissues that help carry water and the minerals absorbed by the roots to a plant's leaves. These tissues also transport the sugar produced in chloroplasts during photosynthesis (the process by which cells convert light energy into food energy) to the roots. In some plants such as cacti, stems store water that the plants use during dry periods. Other plants such as potatoes have underground stems that store food.

▲ **Figure 3** The tree has a woody stem, while the vine has an herbaceous stem.

Leaves In most plants, leaves are the **major** sites for photosynthesis. Some cells in a leaf contain many chloroplasts where light energy is converted into the chemical energy stored in sugar during photosynthesis. There are many different sizes and shapes of leaves, as shown in **Figure 4.**

ACADEMIC VOCABULARY

major
(adjective) greater in number, quantity, or extent

In addition to making food, leaves also are involved in the exchange of gases with the environment. *The* **stomata** (singular, stoma) *are small openings in the surfaces of most plant leaves.* Water vapor, carbon dioxide, and oxygen can pass into and out of a leaf through stomata.

Key Concept Check How do plant structures such as roots, stems, and leaves ensure a plant's survival?

Figure 4 No matter the size or shape, all leaves have vascular tissue. ▼

Animation

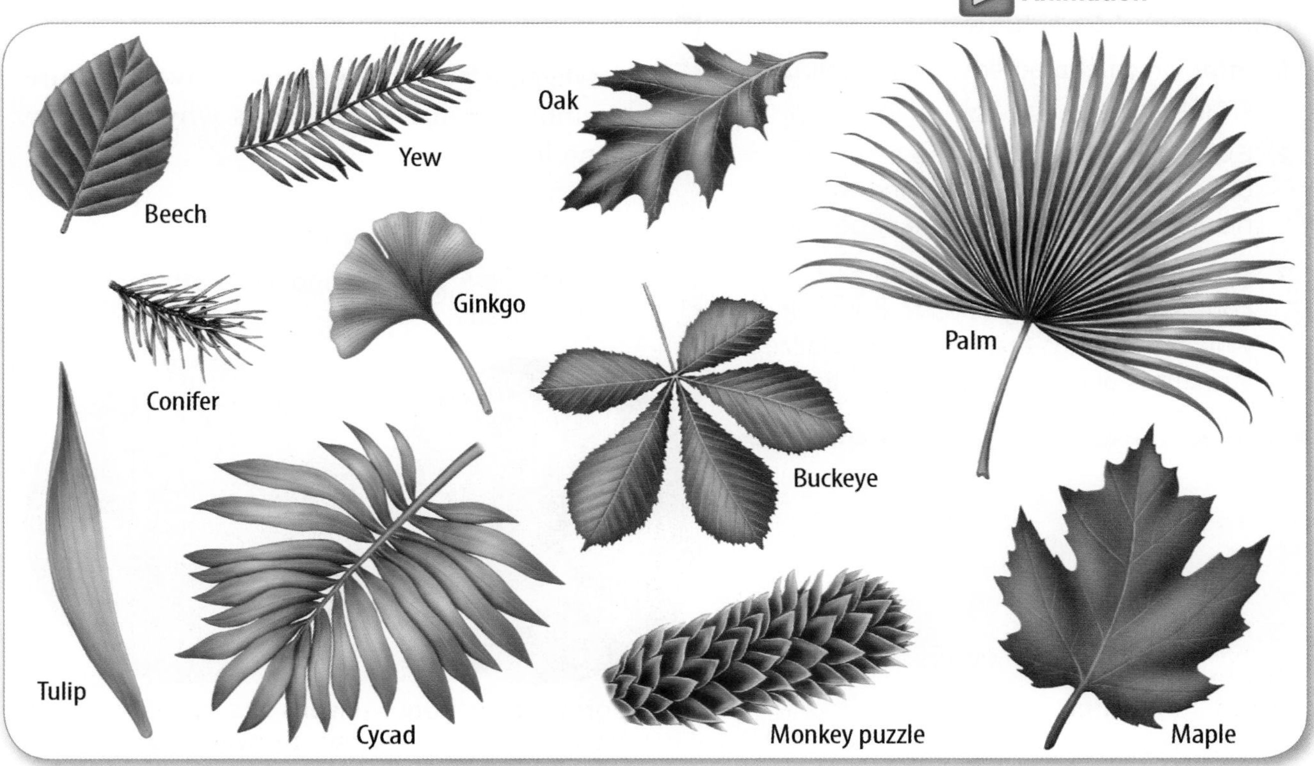

MiniLab

20 minutes

How can you compare and contrast plant groups?

Plants can be organized into different groups according to their characteristics using a dichotomous key.

1. List characteristics of nonvascular plants, vascular seedless plants, and vascular seed plants in your Science Journal.
2. Using these characteristics, create a dichotomous key to separate the three groups of plants. Write a yes/no question about the plants' characteristics that can separate the plants into two groups. For example, your first question might be about vascular tissue. Use the format below:

   1. Does the plant group have vascular tissue?
   Yes . . . . go to question ______
   No . . . . go to question ______
3. Repeat this process by splitting each group repeatedly until they can no longer be divided. Use more-specific questions with each split of the group.

Analyze and Conclude

1. **Infer** which characteristics are helpful in creating a key.
2. **Explain** how you could use a dichotomous key to determine which plant group an unknown plant belongs to.
3. **Key Concept** Compare and contrast the plant groups to determine which group has the fewest specialized structures and which group has the most specialized structures.

Nonvascular Plants

Have you ever noticed the tiny green plants that grow on the bark of a tree? These plants might be one of several types of nonvascular plants. *Plants that lack specialized tissues for transporting water and nutrients are* **nonvascular plants.**

You might recall that animals are grouped into phyla, and plants are grouped into divisions. The divisions of nonvascular plants include mosses, liverworts, and hornworts.

Mosses

One type of plant that lacks specialized transport tissues is mosses. Most mosses are less than 5 cm tall. As shown in Figure 5, mosses have tiny green leaflike structures. Scientists do not call these structures leaves because they do not have transport tissues. Recall that rootlike structures called rhizoids help anchor a moss and absorb water.

Liverworts and Hornworts

There are two types of liverworts. The thallose (THA lohs) form is flat and lobed, as shown in Figure 5. A leafy liverwort has small, leaflike structures attached to a central stalk.

Hornworts appear similar to liverworts. However, a hornwort's reproductive structure resembles a small horn. That's why it has the name *hornwort.*

Figure 5 Mosses and liverworts lack vascular tissue.

Mosses grow in a variety of habitats.

Liverworts tend to grow in damp environments.

Ferns grow in a variety of habitats, including forests and gardens.

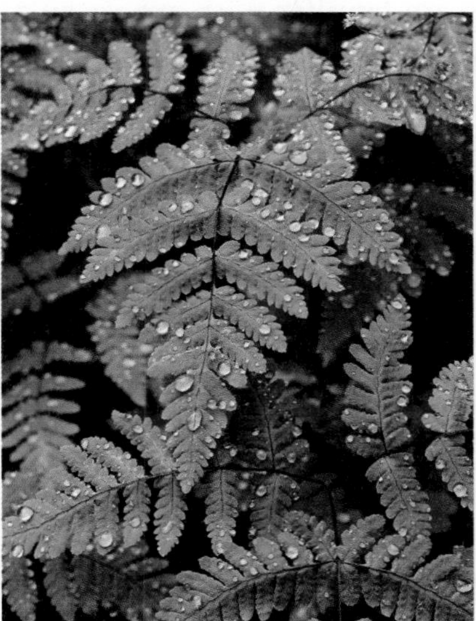

The name *horsetail* came from the bushy form of this plant.

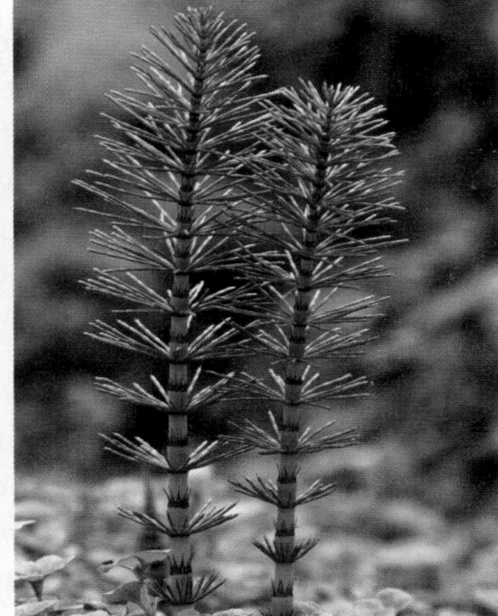

Some club mosses are called ground pines.

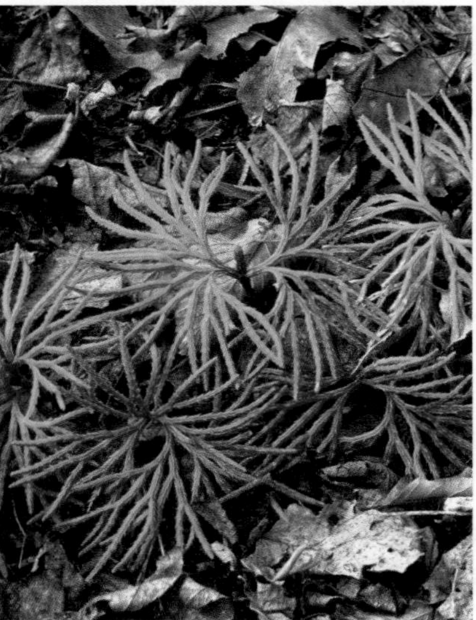

Figure 6 Ferns, horsetails, and club mosses have vascular tissue that transports water and nutrients.

Vascular Seedless Plants

Vascular plants *have specialized tissues, called vascular tissues, that transport water and nutrients throughout the plant.* Vascular plants are divided into two groups–those that produce seeds and those that do not. Ferns, horsetails, and club mosses do not produce seeds.

Ferns

Fossil evidence indicates that millions of years ago ferns grew as large as trees. Present-day ferns are much smaller, many growing to only 50 cm or less. A fern leaf is called a frond. It can have an intricate shape, as shown in **Figure 6.** A fern's fronds grow from an underground stem called a rhizome (RI zohm). Ferns usually grow in shady locations.

Horsetails

Horsetails get their name from a stage of their life cycle that looks like a horse's tail, as shown in **Figure 6.** Horsetails also are called scouring rushes due to an abrasive mineral called silica in the stems. This abrasiveness made the horsetail plant useful to early settlers and to campers for cleaning pots and pans.

 **Reading Check** How did horsetails get their name?

Club Mosses

This group of plants gets its name from its reproductive structure that resembles a club. They often look like small pine trees, as shown in **Figure 6.** These plants once were abundant in ancient forests. Much of the fossil fuels we use today come from the remains of these forests. Present-day club mosses grow in diverse locations that include tropical and arctic habitats.

(l)Jan Tove Johansson/Getty Images, (c)Simon Booth/Photoshot, (r)Michael P.Gadomski/Photo Researchers

Vascular Seed Plants

Probably most of the plants you are familiar with are vascular seed plants. They are the most common type of plants. Grasses, flowering shrubs, and trees are all examples of vascular seed plants. From tiny aquatic plants that are less than 1 mm across to towering redwood trees, all vascular plants share one important characteristic–they produce seeds. Scientists further organize vascular seed plants into two groups–those that produce flowers and those that do not.

WORD ORIGIN

gymnosperm
From French *gymnosperme*, means "naked seed"

SCIENCE USE V. COMMON USE

cone
Science Use a structure in most conifers or in cycads that contains reproductive structures

Common Use a crisp, usually cone-shaped wafer for holding ice cream

Nonflowering Seed Plants

For many plants, seeds are inside or on the surface of fruits. However, some plants produce seeds without fruits. **Gymnosperms** *are plants that produce seeds that are not part of a fruit.* As shown in **Figure 7**, gymnosperms are a diverse group. The most common gymnosperms are conifers. Conifers are usually evergreen, meaning they stay green all year. They have needlelike or scalelike leaves and most produce **cones.** The seeds are part of the cones. Conifers can grow in diverse habitats from near-arctic regions to tropical areas. Conifers also have many commercial uses. Lumber, paper products, and turpentine are products made from conifers.

Gymnosperms

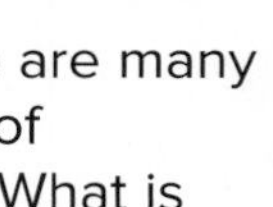

Figure 7 There are many different types of gymnosperms. What is one characteristic they all share?

Visual Check How do conifer leaves differ from ginkgo leaves?

Cycad

Ginkgo

Conifer

Gnetophyte

Flowering Seed Plants

How many flowering plants can you name? There are more than 260,000 species of flowering plants! **Angiosperms** *are plants that produce flowers and develop fruits.* Some of the different varieties of flowering seed plants are shown in **Figure 8.**

Flowering plants have many adaptations that enable them to survive in most habitats on Earth. Their specialized vascular tissues carry water and nutrients throughout the plants. Plants that live in dry areas have special adaptations that help prevent water loss. Perhaps the most amazing characteristic is the incredible diversity of their flowers. There are flowers that attract insects and birds of all kinds. Other flowers are specialized so wind or water can aid in reproduction. As you will read in the next lesson, flowers play a key role in plant reproduction.

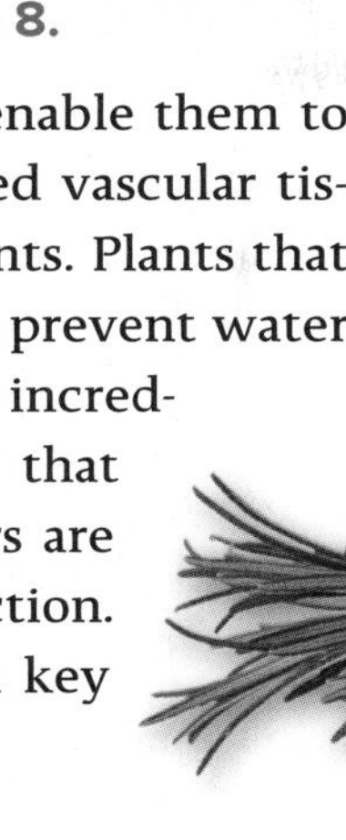

Squash

Rosemary

Key Concept Check How do the different plant types compare and contrast?

Figure 8 No group of plants is more diverse than flowering plants.

Wisteria vine

Poppies

Pear tree

Palm tree

Grass

Cactus

Lily pads

Lesson 1 Review

Visual Summary

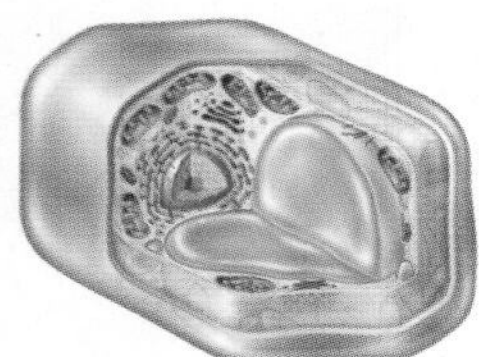

Unlike animal cells, a plant cell has a rigid cell wall, chloroplasts, and a large central vacuole.

Plants have structures that help ensure their survival.

Vascular plants have specialized tissues for transporting water and nutrients.

Use your lesson Foldable to review the lesson. Save your Foldable for the project at the end of the chapter.

What do you think NOW?

You first read the statements below at the beginning of the chapter.

1. Humans could survive without plants.
2. Plant cells contain the same organelles as animal cells.

Did you change your mind about whether you agree or disagree with the statements? Rewrite any false statements to make them true.

Use Vocabulary

1 **Distinguish** between vascular and nonvascular plants.

2 **Define** *stomata* in your own words.

3 **Write** a sentence using the terms *angiosperm* and *gymnosperm.*

Understand Key Concepts 

4 Which are NOT vascular plants?

A. angiosperms **C.** gymnosperms
B. ferns **D.** mosses

5 **Give an example** of a vascular seed plant.

6 **Compare** roots and rhizoids.

7 **Differentiate** between woody and herbaceous stems.

Interpret Graphics

8 **Describe** the function of the structure below.

9 **Summarize** Copy and fill in the table below to describe the function of roots, stems, and leaves.

Structure	Function

Critical Thinking

10 **Assess** the importance of vascular tissue in larger plants.

11 **Evaluate** the advantage to a plant of flower production.

A Life-Saving Plant

SCIENCE & SOCIETY

▲ Sweet wormwood tree leaves are the source of artemisinin.

How can a plant protect you from disease?

Nobody likes a mosquito bite—especially the itchy, red bump it leaves behind. But in some parts of the world, mosquito bites can be deadly. In many tropical countries, female *Anopheles* mosquitoes transmit a serious disease called malaria. Every year, hundreds of millions of people become sick with malaria, and more than 1 million die from the infection.

The real cause of malaria isn't the mosquito—it's a tiny parasite called *Plasmodium* that lives in the mosquito. Mosquitoes get infected with the parasite after biting a person who has malarial fever. When an infected mosquito then bites someone else, it transfers the parasite into that person's blood.

One of the most effective treatments for malaria comes from a plant—the sweet wormwood tree (*Artemisia annua).* The drug, called artemisinin (ar tah MIH sih nin), is a natural substance taken from the tree's leaves. Like many medicinal plants, the sweet wormwood tree has been used since ancient times. Around 2,000 years ago, the Chinese treated malaria with tea made from the tree's dried leaves. However, artemisinin was not scientifically studied until 1972 and has only become widely available since 2001.

Today, artemisinin is available in pill or shot form and is proving to be the critical weapon in the fight against malaria. Treatment with artemisinin has cured more than a million patients so far. Recently, however, scientists working in Cambodia discovered that artemisinin is becoming less effective. This could be a sign that the parasite is becoming resistant to the drug. Now more than ever, it's important that researchers keep investigating other plants for possible treatments.

AMERICAN MUSEUM of NATURAL HISTORY

It's Your Turn

RESEARCH The spices cinnamon, cayenne pepper, garlic, thyme, and ginger come from plants and can be used to cure ailments. Select one and research how it can be used medicinally. Report your findings to your class.

(t) Scott Bauer/USDA, (bkgd) Ira Block/National Geographic/Getty Images

Lesson 2

Plant Reproduction

Reading Guide

Key Concepts
ESSENTIAL QUESTIONS

- How do asexual and sexual reproduction in plants compare and contrast?
- What are the differences between the life cycles of seedless and seed plants?

Vocabulary
pollination p. 255
dormancy p. 255
pistil p. 256
stamen p. 256

Multilingual eGlossary

BrainPOP®

A Plant Sneeze?

It might look as if this plant has just sneezed, but what is being released from the plant might make you sneeze! These grass flowers are releasing pollen, which causes itchy noses in many people. However, pollen is important in the reproduction of this species.

Launch Lab

10 minutes

Do you need seeds to grow a plant?

You grew radish plants from seeds in Lesson 1. Can you grow a plant without using seeds?

1. Read and complete a lab safety form.
2. Pour water into a **glass** until it is half full. Place several **toothpicks** around the middle of a **potato.** Place the potato in the glass so that the bottom of it touches the water and the toothpicks hold the rest of the potato above the rim of the glass.
3. Place the glass and potato in a sunny area.
4. Using a **dissecting knife,** carefully cut a stem approximately 8 cm long from a **coleus plant.** Place the stem cutting in a glass of water so that only the cut portion of the stem is immersed in the water.
5. Place the coleus cutting in a sunny area.
6. Observe the potato and the coleus cutting after one week.

Think About This

1. How did the potato and the coleus plant change after one week?
2. How do you think the traits of the plantlets will compare with those of the parent plants?
3. **Key Concept** Compare and contrast the growth of the potato and the coleus plant with that of the radish plants from Lesson 1.

Asexual Reproduction

Some plants don't need seeds to make new plants. Some plants can be grown from a leaf, a stem, or another plant part, as shown in Figure 9. Asexual reproduction occurs when only one parent organism or part of that organism produces a new organism. The new organism is genetically identical to the parent. Farmers and florists often use asexual reproduction to produce multiple plants with desired traits.

REVIEW VOCABULARY
trait
a distinguishing characteristic of an organism

Sexual Reproduction

The process of sexual reproduction involves male sex cells and female sex cells. Each sex cell contributes genetic material to the offspring. Like animals, plants produce sperm, which are the male sex cells, and eggs, which are the female sex cells. Fertilization occurs when a sperm and an egg join, combining their genetic material. Sexual reproduction produces individuals that have a different genetic makeup than the parent organism or organisms. Both seedless plants and seed plants can reproduce sexually.

Key Concept Check How do asexual and sexual reproduction in plants compare and contrast?

Figure 9 These plants have grown at the edges of a leaf from a single parent plant.

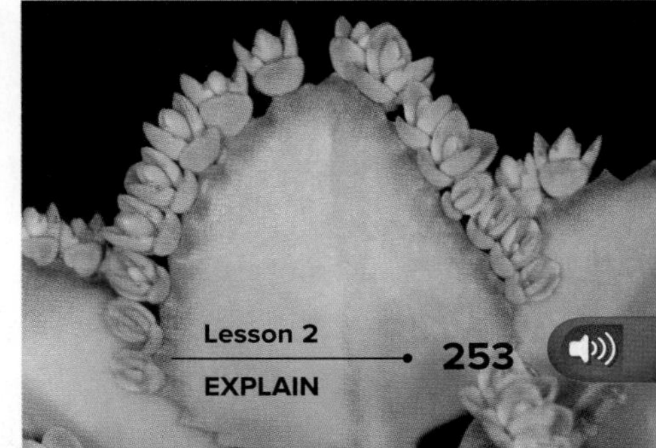

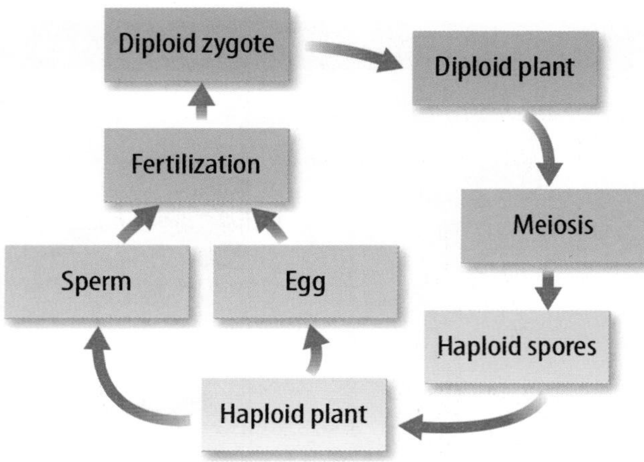

▲ **Figure 10** All plants have a life cycle that includes two stages.

Visual Check Identify the gametophyte and the sporophyte stages in this diagram.

Plant Life Cycles

There are two stages in the life cycle of every plant—the gametophyte (guh MEE tuh fite) stage and the sporophyte (SPO ruh fite) stage. The gametophyte stage begins with a spore, or haploid cell. Through mitosis and cell division, the spore produces a plant structure or an entire plant called a gametophyte. The gametophyte produces male and female sex cells through meiosis.

During sexual reproduction, a male sex cell and a female sex cell combine. If fertilization occurs, a diploid cell forms, as shown in **Figure 10.** That diploid cell is the beginning of the sporophyte stage. This cell divides through mitosis and cell division and forms the sporophyte. In some plants, the sporophyte is a small structure, but in others such as an apple tree, the sporophyte is the tree.

Seedless Plants

Plants that do not produce seeds are called seedless plants. They can reproduce either by asexual reproduction or by producing spores. Spores are produced by the sporophyte. Recall that the sporophyte results from sexual reproduction. Mosses, liverworts, and ferns are examples of seedless plants. The life cycle of a fern is shown in **Figure 11.**

Reading Check Name the two stages in the life cycle of a plant.

Animation

Figure 11 The life cycle of a fern alternates between a gametophyte and a sporophyte stage. ▼

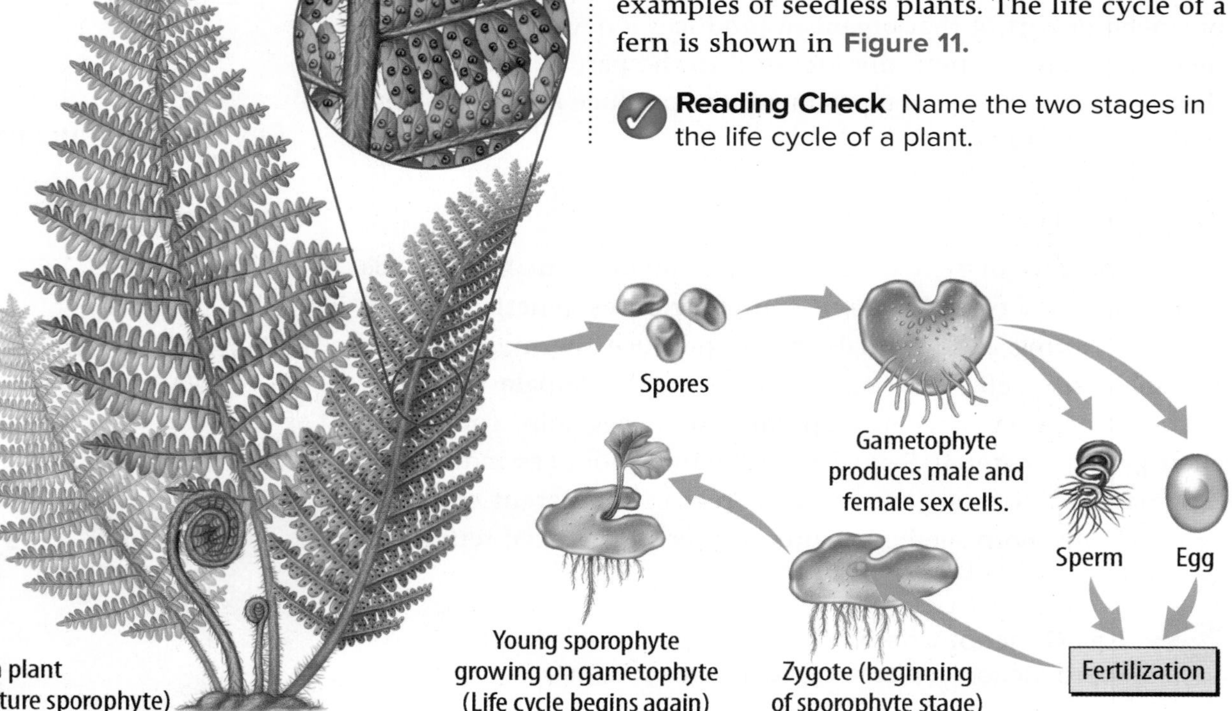

Seed Plants

Most plants produce seeds that result from sexual reproduction. The plants produce pollen grains, which contain sperm. They also produce female structures, which contain one or more eggs. *The process that occurs when pollen grains land on a female plant structure of a plant in the same species is* **pollination.** If a sperm from a pollen grain joins with an egg, this is called fertilization. Once fertilization occurs, the diploid cell undergoes many cell divisions, forming an embryo. The embryo is the beginning of the sporophyte stage of seed plants. The embryo and its food supply are enclosed within a protective coat. This is the seed, as shown in **Figure 12.**

In most seed plants, the seed will go through **dormancy,** *which is a period of no growth.* Dormancy might last days, weeks, or even years. Once environmental conditions are favorable, the seed will become active again. The process of a seed beginning to grow is called germination.

Gymnosperm Reproduction The life cycle shown in **Figure 12** is typical of a gymnosperm. Notice that pollen is produced by the male cone, while the eggs, and eventually the seeds, are contained within the female cone.

FOLDABLES®

Make a vertical three-tab Venn book. Label it as shown. Use it to compare and contrast the life cycles of seed plants and seedless plants.

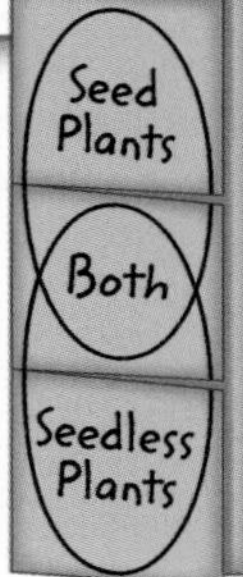

Gymnosperm Life Cycle

Animation

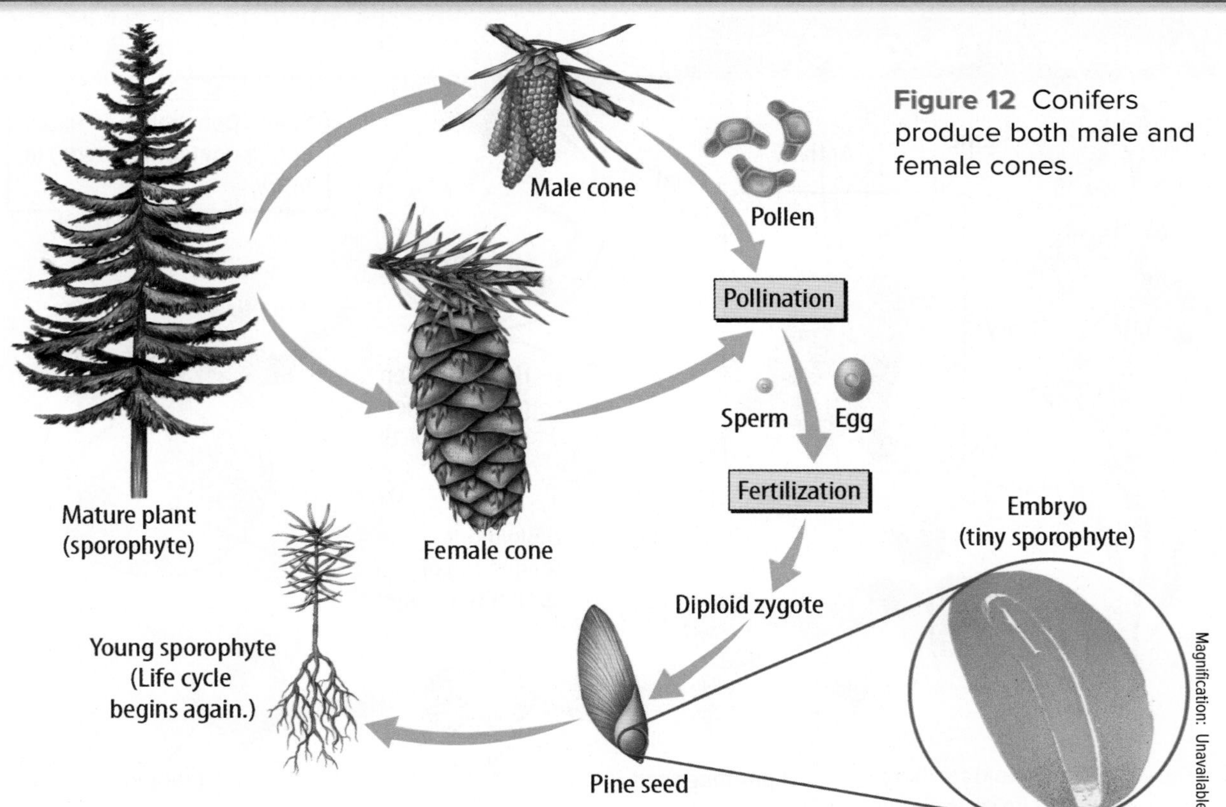

Figure 12 Conifers produce both male and female cones.

Angiosperm Reproduction Flowering plants often are seen in parks and around houses. You might not know that some plants, such as grasses and maple trees, produce flowers. Some aquatic flowering plants are less than 1 mm in length!

Most flowers have four main structures. The petals, which attract insect or animal pollinators, might be brightly colored. The sepals are usually located beneath the petals and help protect the flower when it is a bud. *The female reproductive organ of a flower is the* **pistil.** It contains the ovary, where the seed develops. *The* **stamen** *is the male reproductive organ of a flower.* The anthers of the stamen produce pollen. Examine **Figure 13** to see all the parts of a flower. Some plants have flowers that have only the male or only the female structures. These are called male flowers or female flowers.

WORD ORIGIN

pistil
from French *pistil,* means "female organ of a flower"

As shown in **Figure 13,** the life cycle of a flowering plant includes both gametophyte and sporophyte stages. The gametophyte stage lasts a short time and includes the production of eggs and sperm by a flower. When a sperm fertilizes an egg, the resulting diploid cell is the beginning of the sporophyte stage. In flowering plants, the sporophyte stage lasts much longer than the gametophyte stage.

Key Concept Check What are the differences between life cycles of seedless and seed plants?

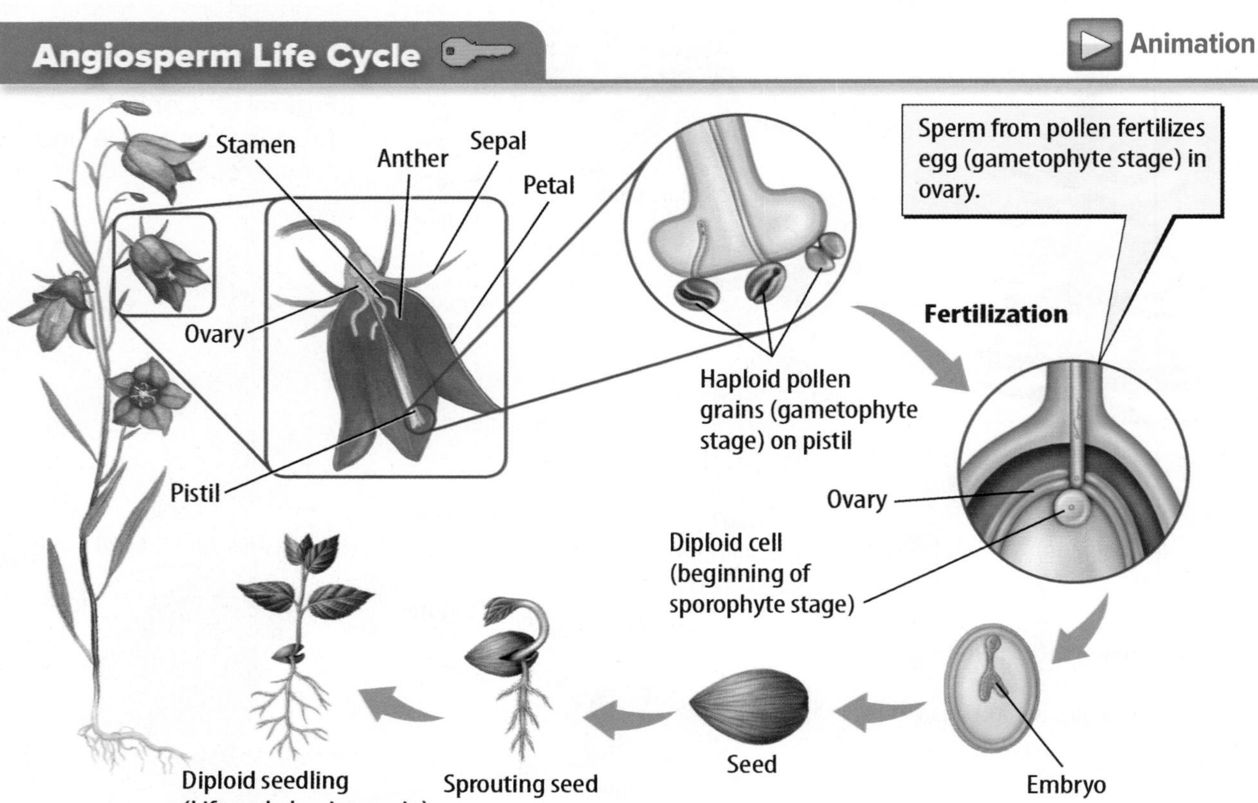

Figure 13 The life cycle of an angiosperm involves several steps.

Visual Check Describe what happens following fertilization.

Growth Cycles Plants that grow from a seed and produce flowers in one growing season are called annuals. They must be planted every year.

Some plants take two growing seasons to produce flowers, such as the mullein (MUH lun) in **Figure 14.** These are known as biennials and include carrots and beets. Biennials go through a period of dormancy between growing seasons. Many biennials have large roots that store food between growing seasons.

Perennials are plants that grow and bud for many years. Some perennials, such as trees, can grow for hundreds of years. In cold climates, some perennials lose their leaves and become dormant for several months. Once warmer temperatures return, the plant produces new leaves and begins capturing sunlight for photosynthesis.

First-year growth

Second-year growth

Figure 14 Plants that take two growing seasons to flower and produce seeds are biennials.

Visual Check Describe the differences between the plants in **Figure 14.** Why are there differences?

MiniLab

15 minutes

How do seeds differ?

Flowering plants and nonflowering plants both use seeds to reproduce. How are the seeds different?

1. Read and complete a lab safety form.
2. Examine a **pine cone** and find the seeds. Draw a sketch of a pine cone seed, and record the location of the seeds in your Science Journal.
3. Examine an **apple.** Using a **knife,** carefully cut the apple in half to locate the seeds. Draw a sketch of an apple seed that shows the location of the seeds.

Analyze and Conclude

1. **Analyze** the location of the seeds in the pine cone. Is a pine tree a flowering or a nonflowering plant?
2. **Describe** the location of the seeds in the apple. Is an apple tree a flowering or a nonflowering plant?
3. **Key Concept** Compare and contrast the seeds of flowering and nonflowering plants.

Lesson 2 Review

Visual Summary

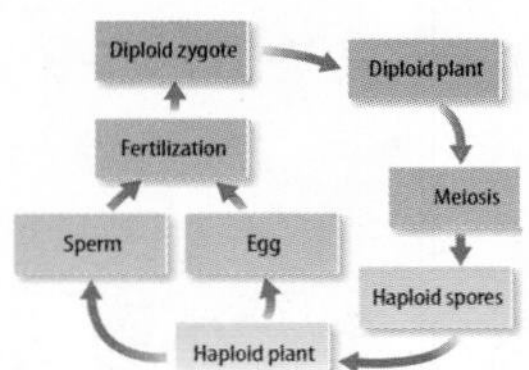

There are two stages in the life cycle of every plant—the gametophyte stage and the sporophyte stage.

Annuals, biennials, and perennials are the different growth cycles of plants.

Most seed plants produce flowers.

FOLDABLES®

Use your lesson Foldable to review the lesson. Save your Foldable for the project at the end of the chapter.

What do you think NOW?

You first read the statements below at the beginning of the chapter.

3. Plants can reproduce both sexually and asexually.

4. All plants have a two-stage life cycle.

Did you change your mind about whether you agree or disagree with the statements? Rewrite any false statements to make them true.

Use Vocabulary

1 A period of no growth is called ________.

2 **Define** *pollination* in your own words.

3 **Write** a sentence using the terms *pistil* and *stamen.*

Understand Key Concepts 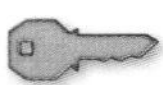

4 Which has a cone for its reproductive structure?

A. angiosperm
B. gymnosperm
C. hornwort
D. horsetail

5 **Compare** the life cycles of seedless plants and seed plants.

6 **Illustrate** and label the four parts of a flower.

7 **Contrast** sexual and asexual reproduction.

Interpret Graphics

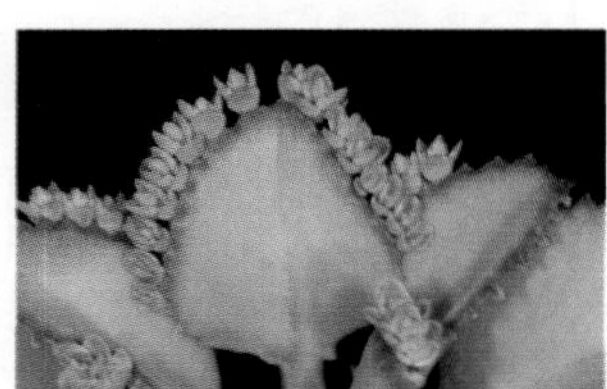

8 **Compare** the traits of the new plants produced above to those of the parent plant.

9 **Classify Information** Copy and fill in the table below to list differences between gymnosperms and angiosperms.

Division	Description

Critical Thinking

10 **Assess** the value of fruit production.

11 **Analyze** the difference between a fern's life cycle and that of a gymnosperm.

Skill Practice Compare and Contrast 40 minutes

How can you learn about plant structures?

Materials

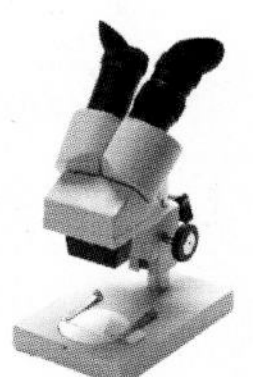

dissecting microscope

dissecting knife

flowering plant

Safety

Scientists learn more about plants by dissecting and observing their internal structures. A longitudinal section is made by cutting an object in half along its longest plane. A cross section is made by cutting an object in half at a right angle with its shortest plane.

Learn It

Comparing and contrasting structures and functions enables scientists to learn by analyzing similarities and differences.

Try It

1. Read and complete a lab safety form.
2. Cut off a 5-cm to 7-cm section of stem from a flowering plant.
3. Cut a cross section from the remaining stem.
4. Observe the cross section under a dissecting microscope. Draw what you observe in a table like the one below, listing *stem* under the Structure column.
5. Make a longitudinal cut of your stem by cutting the stem in half lengthwise. Observe the inside of one half of your longitudinal cut under the dissecting microscope. Draw what you observe in the Longitudinal View column of your table.
6. Repeat these procedures using the roots, leaves, and flowers of the plant. List each type of structure on your table and then make sketches of your observations in the table.

Apply It

7. **Create a diagram** that represents a three-dimensional view of each of the structures you observed using the information from both the cross-section and the longitudinal cuts.
8. **Analyze** the information from each view. Describe which types of measurements could be obtained from each one—for the entire section of the plant, as well as the details of any internal structures seen within the section.
9. **Key Concept** Evaluate how cross-sections and longitudinal sections provide information about a structure. Describe how sometimes one view might be more helpful than the other.

Structure	Cross-section View	Longitudinal View

Lesson 3

Reading Guide

Key Concepts
ESSENTIAL QUESTIONS

- What is the relationship between photosynthesis and cellular respiration?
- How do water and minerals move in vascular and nonvascular plants?
- How do plants respond to environmental changes?

Vocabulary

transpiration p. 264
stimulus p. 265
tropism p. 266

Multilingual eGlossary

Plant Processes

Inquiry Alien Life-Form?

This might look like a space creature with green skin and black eyes, but it is a color-enhanced magnification of openings on the surface of a leaf. These openings enable carbon dioxide, oxygen, and water vapor to pass into and out of a leaf. Why do you think this is important to a plant?

Biophoto/Associates/Science Source

Launch Lab

15 minutes

How important is light to the growth of plants?

All plants require light to grow, but just how important is it?

1. Read and complete a lab safety form.
2. Plant several **bean seeds** in two identical **cups** filled with **potting soil.** Add water to moisten the soil in both cups.
3. Place one cup in a sunny place, such as a windowsill. Place the other cup in a dark place, such as a cabinet.
4. Place a **two-week-old bean plant** in the dark location alongside the seeds you planted. Place another **two-week-old bean plant** in the sunny location alongside the other seeds you planted.
5. Check on all plants and seeds every 2 days for 10 days. Add water to keep the soil moist as needed. Record your observations in your Science Journal.

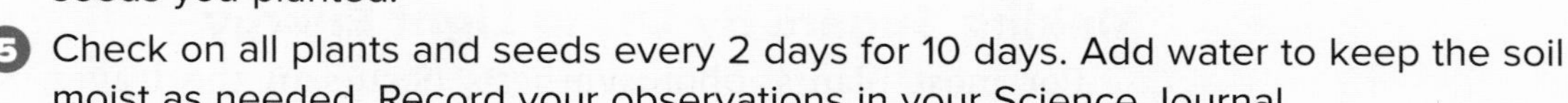

Think About This

1. How does the growth of the seeds exposed to light compare with those kept in the dark?
2. How does the appearance of the plant that was exposed to light compare with the plant that was kept in the dark?
3. **Key Concept** How do you think the presence or absence of light in the environment affects plant growth?

Photosynthesis and Cellular Respiration

Without plants, all animal life, including human beings, would not exist! Some of the many foods and products that are provided by plants are shown in **Figure 15.** Recall that plants absorb light energy from the Sun and convert it into chemical energy in a process called photosynthesis. During photosynthesis, a plant produces sugar that it uses as food. Even organisms that don't eat plants directly depend on plants because they eat other organisms that do eat plants.

Organisms need energy for growth, repair, movement, and other life processes. Where does this energy come from? Cellular respiration is the process of releasing energy by breaking down food.

Key Concept Check What is the relationship between photosynthesis and cellular respiration?

Figure 15 People consume many different foods from plants. People use many products made from plant materials such as paper, wood, and cotton cloth.

(t) Hutchings Photography/Digital Light Source, (b) Andersen Ross/Getty Images

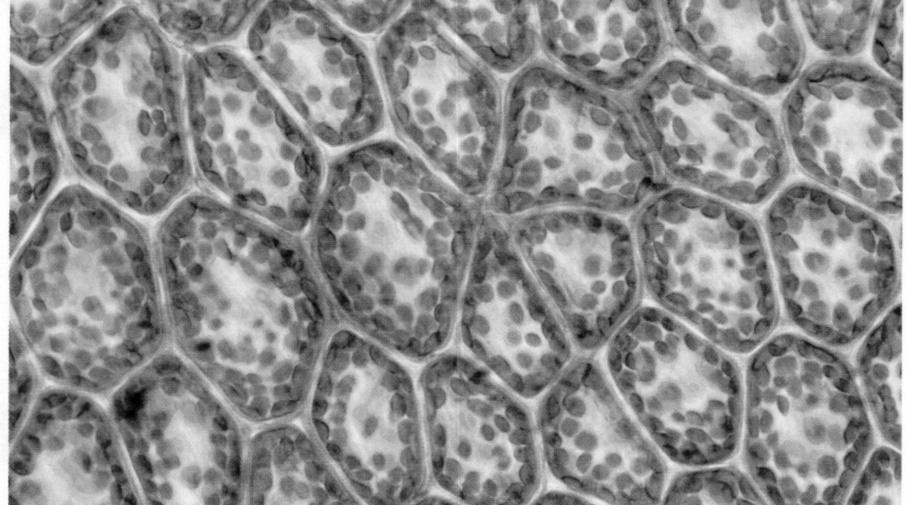

LM Magnification: 250×

Figure 16 The chlorophyll within chloroplasts is what gives most plants their green color.

Visual Check Where are the chloroplasts located in these plant cells?

John Durham/Science Source

Making Sugars By Using Light Energy

For most plants, photosynthesis occurs in the leaves. Some leaf cells contain chloroplasts. Photosynthesis occurs inside these organelles. As shown in **Figure 16**, chloroplasts contain chlorophyll, a green pigment that absorbs light energy. That energy splits apart water molecules into hydrogen atoms and oxygen atoms. Some of the oxygen leaves the plant through the stomata. Carbon dioxide, which entered the leaf through the stomata, combines with the hydrogen atoms and forms glucose, a type of sugar. Photosynthesis can be shown by the following equation:

$$6CO_2 + 6H_2O \xrightarrow[\text{Chlorophyll}]{\text{Light energy}} C_6H_{12}O_6 + 6O_2$$

REVIEW VOCABULARY

organelle
a structure in a eukaryotic cell that performs a specific function for the cell

A plant can store sugars, and it uses some of the oxygen in another process called cellular respiration.

Reading Check Write the equation for photosynthesis using words.

Breaking Down Sugars

The process of cellular respiration breaks down the glucose produced during photosynthesis and releases the sugar's energy. This process occurs in the cytoplasm and the mitochondria. As shown below, oxygen also is used during cellular respiration. The equation for cellular respiration is as follows:

$$C_6H_{12}O_6 + 6O_2 \longrightarrow 6CO_2 + 6H_2O + \text{ATP (Energy)}$$

During cellular respiration, glucose molecules release more energy than cells can use at one time. That energy is stored in a molecule called adenosine triphosphate (uh DEN uh seen • tri FAHS fayt), or ATP. It is used later for other cell processes.

The Importance of Photosynthesis and Cellular Respiration

Do you know why plants are so important to life on Earth? One answer can be found in the equations on the previous page. Organisms, such as humans, need oxygen. Each time you inhale, your lungs fill with air that contains oxygen. Your body uses that oxygen for cellular respiration. In your body's cells, cellular respiration breaks down food and stores the energy from food in ATP. Cellular processes such as growth, repair, and reproduction all use ATP. During cellular respiration, carbon dioxide and water are given off as waste products. Plants use these two compounds for photosynthesis.

Most organisms, including humans, use the products of photosynthesis–sugars and oxygen–during cellular respiration. Plants and some other organisms can use the waste products of cellular respiration–carbon dioxide and water–during photosynthesis, as shown in **Figure 17**. It is important to remember that plants also carry out cellular respiration, so they will use some of the oxygen released during photosynthesis.

Math Skills 

Use Proportions

A proportion is an equation with two ratios that are equivalent. The cross products of the ratios are equal. Proportions can be used to solve problems such as the following: In a cell, when one molecule of glucose breaks down completely to carbon dioxide and water, 36 ATP molecules are produced. How many ATP molecules are produced when 30 glucose molecules break down?

Set up the proportion.

$$\frac{\text{1 molecule glucose}}{\text{36 molecules ATP}} = \frac{\text{30 molecules glucose}}{x \text{ molecule ATP}}$$

Cross multiply.

$$x = 30 \times 36$$

$$x = 1{,}080 \text{ molecules ATP}$$

Practice

During photosynthesis, 18 ATP molecules are required to produce 1 glucose molecule. How many ATP molecules would be required to produce 2,500 glucose molecules?

 Math Practice

 Personal Tutor

Making and Using Energy

Photosynthesis

Cellular respiration

 Personal Tutor

Figure 17 The products of photosynthesis are used during cellular respiration. The products of cellular respiration are used during photosynthesis.

Visual Check Compare the equations for photosynthesis and cellular respiration.

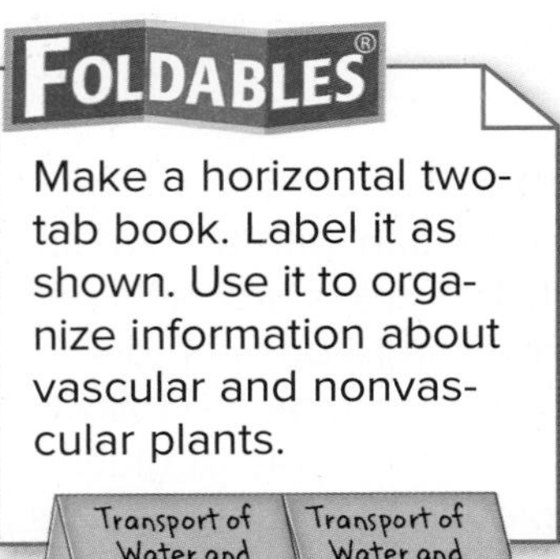

REVIEW VOCABULARY

diffusion
the movement of substances from an area of higher concentration to an area of lower concentration

osmosis
the diffusion of water molecules through a membrane

Movement of Nutrients and Water

In order for plants to carry out processes such as photosynthesis and cellular respiration, water and nutrients must move inside them. This movement or transport of materials occurs through **diffusion** and **osmosis** in nonvascular plants. In vascular plants, water and nutrients move inside specialized vascular tissues. Osmosis and diffusion also move materials once they are outside the vascular tissues.

Absorption

Roots and rhizoids of plants absorb water and nutrients from the soil. Once inside a plant, water and nutrients move to cells, where they are used in cellular processes. As you just read, water is used for photosynthesis. It also is part of many other chemical reactions inside cells. Nutrients from the soil, such as minerals, are used for making many of the compounds needed for cell growth and maintenance.

Transpiration

Recall that water is a waste product of cellular respiration. Plants release excess water as water vapor in a process called transpiration. **Transpiration** *is the release of water vapor from stomata in leaves.* This process helps move water from the roots, up through the vascular tissue, and to the leaves. This movement provides water for photosynthesis and helps cool a plant on hot days. Examine **Figure 18** and follow the path of water from the soil, up through the plant, and out of the leaves.

Key Concept Check How do water and nutrients move in a nonvascular plant? In a vascular plant?

Figure 18  Water is absorbed into the roots and travels to the leaves, where it is used for plant processes or released.

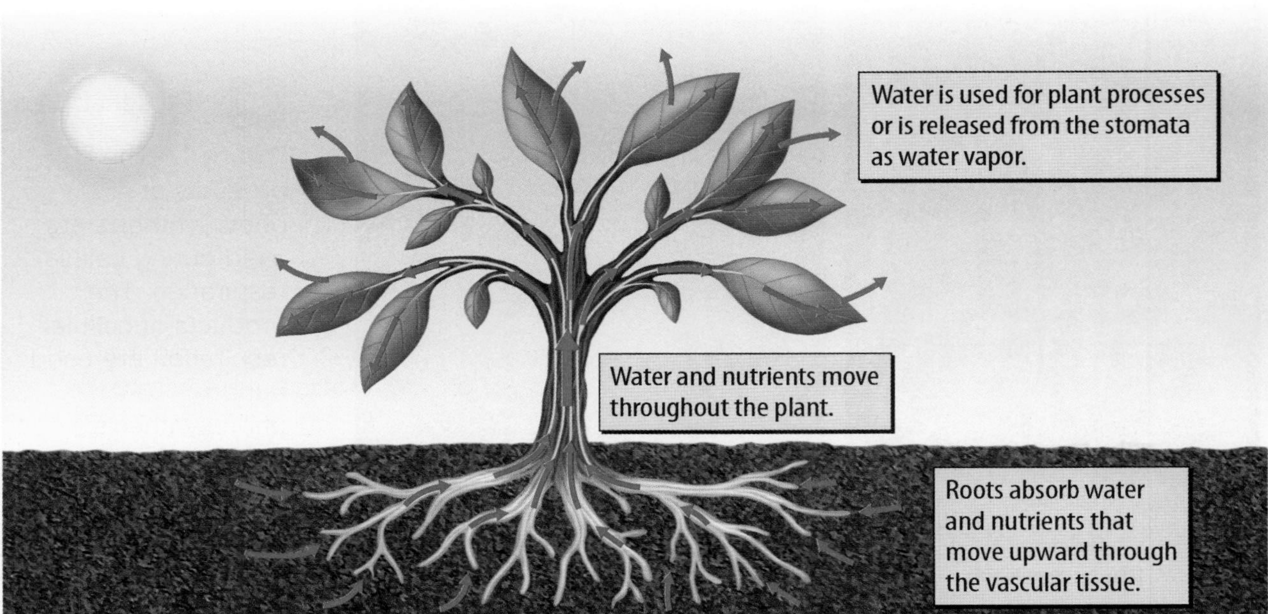

Plant Responses

Can you remember the last time a loud noise startled you? You might have jumped and turned around to see what made the noise. **Stimuli** (singular, stimulus) *are any changes in an organism's environment that cause a response.* Although a plant might not jump or turn around like a person would, a plant can respond to stimuli in a number of ways, one of which is shown in **Figure 19.**

Figure 19 The mimosa plant responds to the stimulus of being touched by collapsing its leaves.

Types of Stimuli

Plants respond to both external and internal stimuli. External stimuli include light, touch, and gravity. Internal stimuli occur inside a plant. These internal stimuli are chemicals, called hormones, that the plant produces. Plants produce many different hormones. These hormones can affect growth, seed germination, or fruit ripening. The hormones that promote growth increase the rate of mitosis and cell divisions. Some hormones slow growth and can be used to help control weeds. One type of hormone can cause seeds to germinate, or begin to grow, by starting the breakdown of the stored food in a seed. This releases energy needed for new growth. Another plant hormone often is used to speed up the ripening of fruit to be sold in grocery stores.

MiniLab

10 minutes

How does an external stimulus affect the growth of a plant?

A plant can grow toward or away from an external stimulus. You can observe the responses to a stimulus by changing the environment of a growing plant.

1. Read and complete a lab safety form.
2. Obtain the **petri dish** with the growing **radish plant** from Lesson 1. Turn the dish vertically so that the roots and the top of the plant are now both facing sideways in opposite directions. Make sure the dish is oriented so that the source of light comes from a different direction as well.
3. Place the dish on a shelf or windowsill in this new position.
4. Observe the radish plant over several days. Record your observations in your Science Journal.

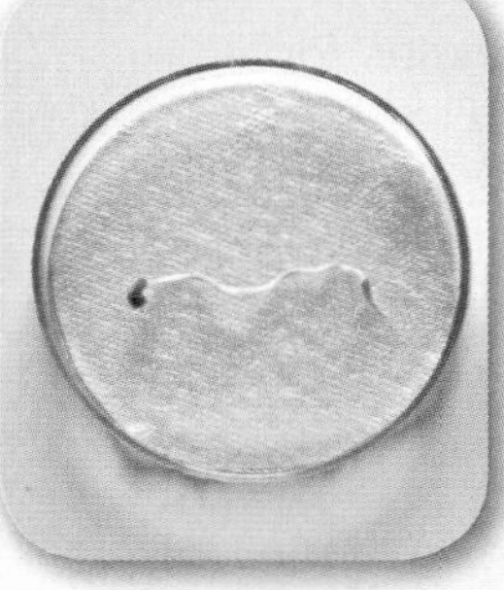

Analyze and Conclude

1. **Analyze** how the change in position affected the growth of the top of the radish plant.
2. **Infer** from your observations how the force of gravity affects the growth of plant roots.
3. **Key Concept** Infer how both responses to light and gravity are important to the survival of plants in a changing environment.

Tropisms

Any external environmental stimulus affects plants. This includes light, gravity, and touch. *Plant growth toward or away from an external stimulus is called* **tropism.**

Word Origin

tropism
from Greek *tropos,* means "a turn, change"

Phototropism When a plant grows toward a light source, it is called positive phototropism. As shown in **Figure 20,** growing toward a light source enables leaves and stems to receive the maximum amount of light for photosynthesis. The roots of a plant generally exhibit negative phototropism by growing into the soil away from light. By growing into the soil, the roots are able to anchor the plant.

Gravitropism A plant's response to gravity is called gravitropism. The first root produced by a germinating seed grows downward. This is positive gravitropism. It enables the new plant to be anchored in soil, where it can absorb water. A plant's stems and leaves grow upward and away from gravity, as shown in **Figure 20.** This is negative gravitropism. This response enables leaves to be exposed to light, making photosynthesis possible.

Thigmotropism Did you know plants have a sense of touch? A plant's response to touch is called thigmotropism. The coiling of a vine's tendrils around another plant, as shown in **Figure 20,** is an example of positive thigmotropism. A plant's roots exhibit negative thigmotropism when they grow around a rock in the soil.

Key Concept Check How do plants respond to environmental changes?

Figure 20 Plants respond to stimuli in a variety of ways. See if you can determine the stimulus for each of these pictures.

Plant Responses to External Stimuli

Positive phototropism

Negative gravitropism

Positive thigmotropism

Lesson 3 Review

Visual Summary

Plants make sugar through the process of photosynthesis. Plants break down sugar into usable energy through the process of cellular respiration.

All plants must be able to transport water and nutrients in order to survive.

Plants respond to internal and external stimuli.

Use your lesson Foldable to review the lesson. Save your Foldable for the project at the end of the chapter.

What do you think NOW?

You first read the statements below at the beginning of the chapter.

5. Plants respond to their environments.

6. Because plants make their own food, they do not carry out cellular respiration.

Did you change your mind about whether you agree or disagree with the statements? Rewrite any false statements to make them true.

Use Vocabulary

1. Plant growth toward or away from a stimulus is called ________.
2. **Define** *transpiration* in your own words.

Understand Key Concepts

3. A plant that is growing toward a window most likely is exhibiting ________.
   A. gravitropism
   B. hydrotropism
   C. phototropism
   D. thigmotropism
4. **Explain** how water and nutrients move in nonvascular plants.
5. **Compare** cellular respiration and photosynthesis.

Interpret Graphics

6. **Sequence** Draw a graphic organizer like the one below to illustrate important transpiration events, beginning with the absorption of water by roots.

7. **Identify** Where are the cells at right likely to be located in a plant? Justify your answer.

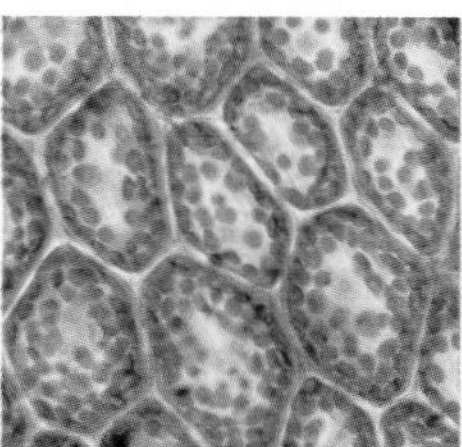

Critical Thinking

8. **Invent** a new type of tropism, and explain why it would be beneficial to plants.
9. **Reflect** on the relationship between photosynthesis and cellular respiration.

Math Skills

10. During one step in a cellular process, 9 molecules of ATP are produced from 2 starting molecules. How many molecules of ATP would be produced from 100 starting molecules?

 Lab

1–2 class periods

Safety

Model the Form and Function of Plant Structures

Plants use different structures to perform specialized functions and obtain materials for survival. Use your knowledge of plant structures to design a model of a plant using common materials to represent both form and function.

Question

What types of materials could be used to represent both the appearance and the function of the different parts of a plant?

Procedure

1. Read and complete a lab safety form.
2. Make a list of all the plant structures you will need to model in a vascular seed-producing plant and a vascular seedless plant.
3. Expand your list into a table by adding a column with the heading *Function*. In this column, write a brief, general description of the function of each plant structure. Use as many descriptive words as you can think of that fit the structure or function.

4. Add another column with the heading *Materials*.
5. Complete the Materials column of your table by writing down any materials that fit the descriptions in the previous columns.
6. Select the most appropriate materials from your table to create three-dimensional models of a vascular seedless plant and a vascular seed-producing plant. Use the materials to construct models that show form and function. For example, you might use drinking straws to represent vascular tissues in a stem.

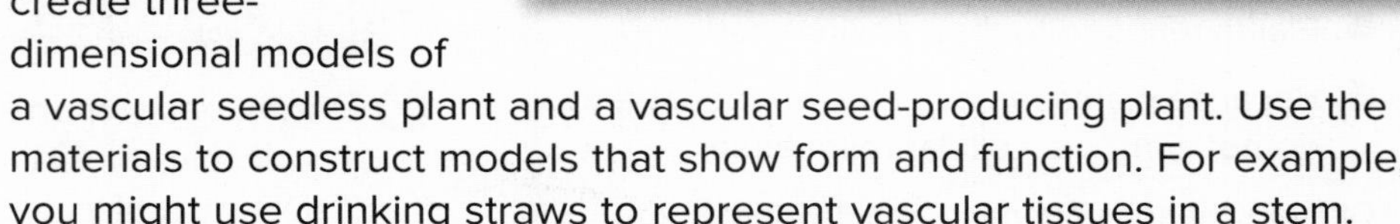

Hutchings Photography/Digital Light Source

Lab Tips

- Take your time filling out the first two columns of the table before beginning to brainstorm materials. Use as many descriptive terms as you can to describe the functions. Then think of as many different materials as you can that might model the functions you've listed.
- Discuss the possible materials with your lab partners. Remember that the materials should represent functions yet still be usable for construction.
- Don't limit yourself to obvious materials. You do not have to use objects that resemble plant structures or typical craft construction materials. Try to use some objects that will be unique, yet remind observers about different functions.

Analyze and Conclude

7. **Analyze** Was there a structure that was difficult to model using an appropriate material? Describe which structures were easy and which ones were a challenge.
8. **Contrast** How were some of the materials you chose for each of your models different from each other? How did these materials reflect differences in their functions?
9. **The Big Idea** How do the structures you modeled help ensure the survival of plants? Cite some examples in your answer.

Communicate Your Results

Share your results with the class. Discuss your choice of materials. Compare and contrast your choice of materials with those chosen by other students. Which materials were most common? Which choices were unique?

Extension

Think about how you would use appropriate materials to design a model of some nonvascular plants. As a class, design and build some models of nonvascular plants using materials that represent both function and structure.

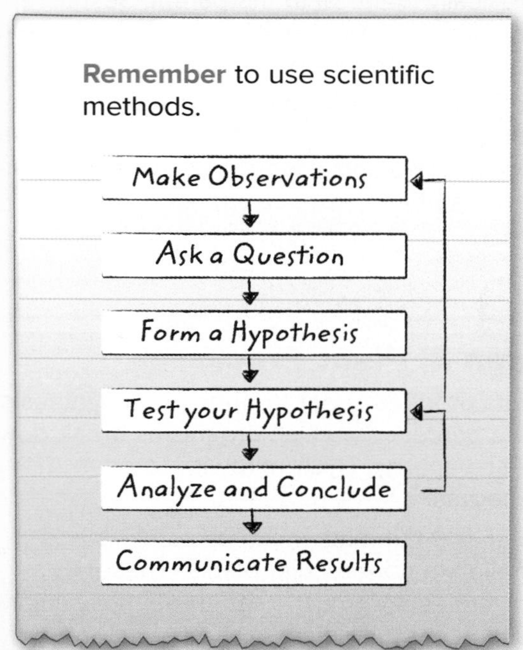

Hutchings Photography/Digital Light Source

Chapter 8 Study Guide

There are many different types of plants, but they all have structures and functions that help ensure survival.

Key Concepts Summary

Lesson 1: Plant Diversity

- Roots and **rhizoids** anchor a plant and absorb water and nutrients. Stems help support the leaves, and in some cases flowers, of a plant. Stems help carry water and nutrients throughout the plant. In most plants, leaves are the major sites for photosynthesis. In addition to making food, leaves also are involved in the exchange of gases with the environment through the **stomata.**
- Plants are classified into groups called divisions. The main divisions are **nonvascular plants** and **vascular plants.**

Vocabulary

rhizoid p. 244
stoma p. 245
nonvascular plant p. 246
vascular plant p. 247
gymnosperm p. 248
angiosperm p. 249

Lesson 2: Plant Reproduction

- Asexual reproduction does not involve sex cells. Offspring are genetically identical to the parent. Sexual reproduction involves sex cells and produces offspring that are not genetically identical to each other or the parent plant(s).
- The life cycles of seedless and seed plants both contain a gametophyte and a sporophyte stage. Seed plants produce seeds, and seedless plants produce spores.

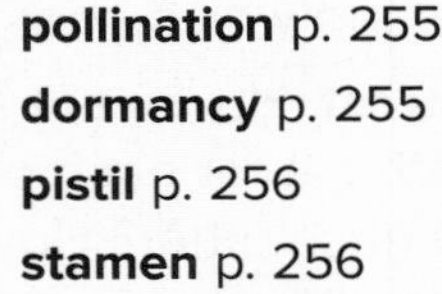

Vocabulary

pollination p. 255
dormancy p. 255
pistil p. 256
stamen p. 256

Lesson 3: Plant Processes

- Plants produce sugar through photosynthesis. Cellular respiration is the process by which organisms break down the sugar and release energy. This energy is stored in ATP. ATP is used for life processes.
- Water and nutrients move by osmosis and diffusion in nonvascular plants. These substances are transported through vascular tissue in vascular plants.
- Plants respond to **stimuli** in their environment. Growth toward or away from a stimulus is called a **tropism.**

Vocabulary

transpiration p. 264
stimulus p. 265
tropism p. 266

Vocabulary eFlashcards
Vocabulary eGames
Personal Tutor

FOLDABLES®

Chapter Project

Assemble your lesson Foldables as shown to make a Chapter Project. Use the project to review what you have learned in this chapter.

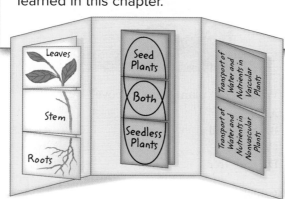

Leaves
Stem
Roots
Seed Plants
Both
Seedless Plants
Transport of Water and Nutrients in Vascular Plants
Transport of Water and Nutrients in Nonvascular Plants

Use Vocabulary

1. The release of water vapor from stomata in leaves is called ________.
2. Pollen is produced in the male reproductive structure of a plant, or the ________.
3. Distinguish between angiosperms and gymnosperms.
4. Changes in an organism's environment that cause a response are called ________.
5. Use the term *dormancy* in a sentence.

Link Vocabulary and Key Concepts

Interactive Concept Map

Copy this concept map, and then use vocabulary terms from the previous page to complete the concept map.

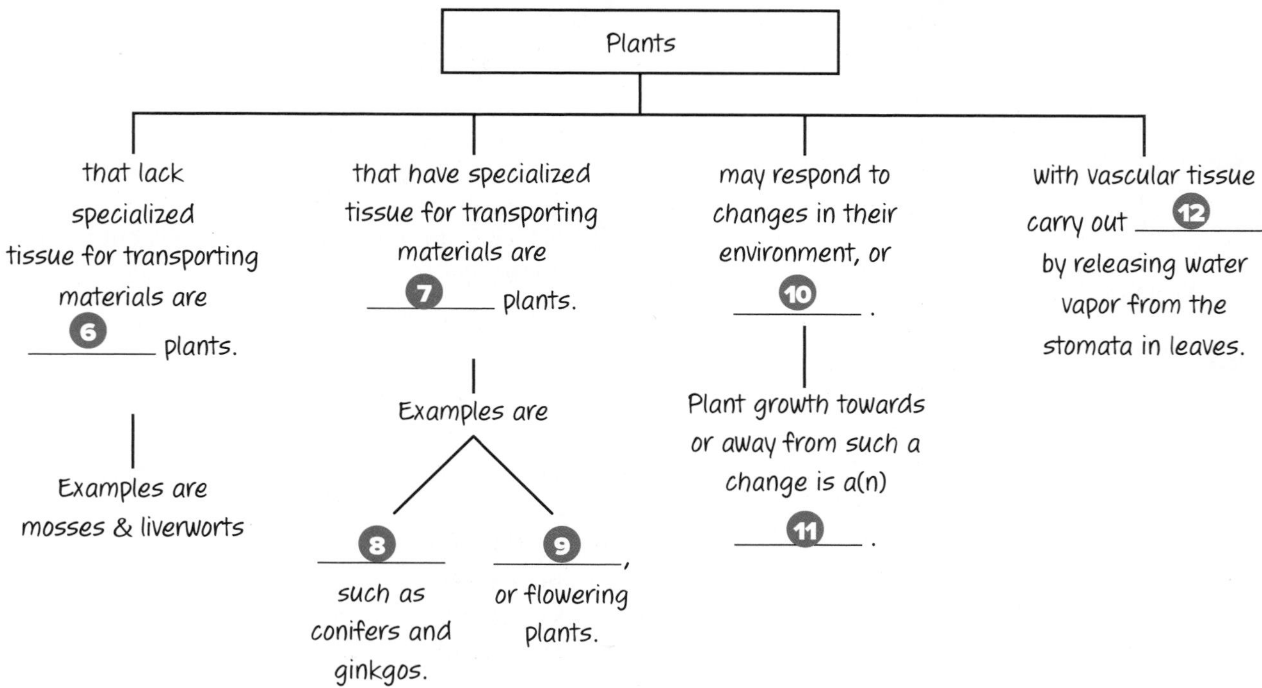

Plants
that lack specialized tissue for transporting materials are 6 ________ plants.
Examples are mosses & liverworts
that have specialized tissue for transporting materials are 7 ________ plants.
Examples are
8 ________ such as conifers and ginkgos.
9 ________, or flowering plants.
may respond to changes in their environment, or 10 ________.
Plant growth towards or away from such a change is a(n) 11 ________.
with vascular tissue carry out 12 ________ by releasing water vapor from the stomata in leaves.

Chapter 8 Review

Understand Key Concepts 

1 During which process are carbon dioxide, water, and ATP produced?
A. cellular respiration
B. photosynthesis
C. thigmotropism
D. transpiration

2 Which is the cause of the green color in plant leaves?
A. chlorophyll
B. flowers
C. glucose
D. oxygen

3 What do angiosperms produce?
A. cones
B. flowers
C. needles
D. rhizoids

Use the diagram below to answer questions 4 and 5.

4 In which flower part is an egg produced?
A. A
B. B
C. C
D. D

5 Which flower part is often brightly colored and helps attract insects?
A. A
B. B
C. C
D. D

6 The stomata on a leaf
A. allow gases to enter and leave the leaf.
B. allow water and energy into the leaf.
C. perform cellular respiration.
D. produce sugar and water vapor.

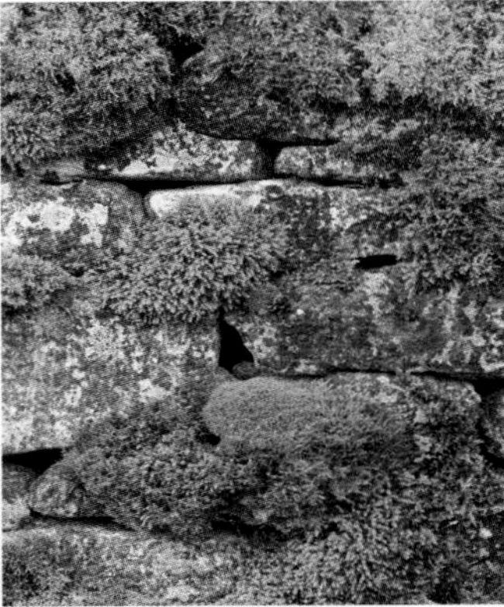

7 What is the plant shown above?
A. fern
B. horsetail
C. moss
D. pine tree

8 Which do ferns produce in order to reproduce?
A. cones
B. flowers
C. seeds
D. spores

9 All plants have a life cycle that includes a __________
A. cone and gametophyte.
B. cone and seed.
C. seed and sporophyte.
D. sporophyte and gametophyte.

10 Which is an organelle in plant cells but not in animal cells?
A. chlorophyll
B. chloroplast
C. mitochondria
D. nucleus

11 What is the major site of photosynthesis in plants?
A. flowers
B. leaves
C. stems
D. roots

Tim Graham/Getty Images

Critical Thinking

12 **Suggest** an environment where you might find succulents, or plants that store water in their leaves.

13 **Reflect** on the importance of flowers in plant reproduction.

14 **Assess** the advantages of sexual and asexual reproduction.

15 **Predict** the effect of germinating a seed without any light.

16 **Hypothesize** why natural selection has favored flowers with colorful petals.

17 **Analyze** the need for woody stems in some plants that live many years through many different seasons and weather conditions, such as some perennials.

18 **Hypothesize** why the type of plant shown below often grows in moist areas.

19 **Evaluate** the differences between the life cycles of a moss and a gymnosperm.

20 **Suggest** a reason for the great abundance and diversity of flowering plants.

Writing in Science

21 **Choose** a habitat near your home, and write a description of the plants in that habitat. Be sure to include a physical description of the plants, as well as how many of each kind of plant are present. See if you can identify the division and name of each plant.

REVIEW THE BIG IDEA

22 Make a list to summarize the different structures and functions of plants that you have learned about in this chapter. How does each structure and function from your list help plants survive?

23 The photo below shows a magnified cross section of a pine needle. How do pine needles help conifers live in their environment?

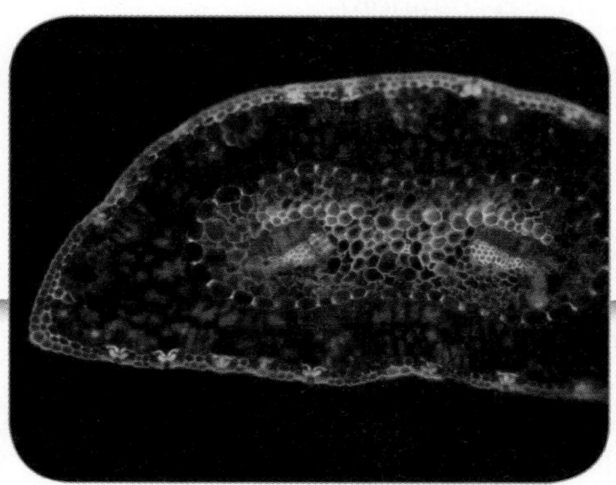

Math Skills

Math Practice

Use Proportions

24 If each ATP molecule in the body takes part in about three reactions every minute, how many reactions would this molecule take part in during one hour?

25 The human body contains about 0.05 kg of ATP. During each 24-hour period, the body's ATP is recycled about 3,600 times. How many kilograms of ATP would the body need in order to provide separate ATP molecules for each reaction?

Standardized Test Practice

Record your answers on the answer sheet provided by your teacher or on a sheet of paper.

Multiple Choice

1 Which structures enable a plant to exchange water vapor and gases such as carbon dioxide and oxygen with its environment?

A rhizoids

B roots

C seeds

D stomata

2 Which is true of photosynthesis and cellular respiration?

A They both occur in plants.

B They both occur in animals.

C They both produce sugars.

D They both require sunlight.

Use the image below to answer question 3.

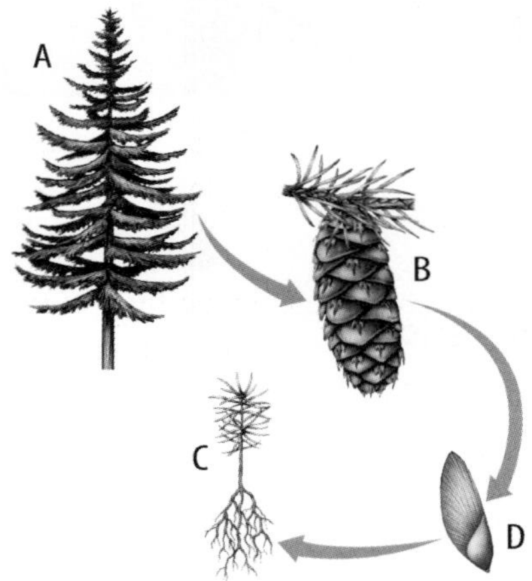

3 For the life cycle shown, which structures are part of the sporophyte stage?

A A and B

B A and C

C B and C

D B and D

4 Which is NOT a product of cellular respiration?

A energy

B glucose

C oxygen

D water

5 Which two divisions are used to classify vascular seed plants?

A conifers and nonconifers

B flowering and nonflowering

C mosses and liverworts

D sporophytes and gametophytes

Use the image below to answer question 6.

6 What term describes the plant response shown above?

A gravitropism

B hydrotropism

C phototropism

D thigmotropism

7 Which structures anchor nonvascular plants to surfaces?

A rhizoids

B roots

C stems

D xylem

Online Standardized Test Practice

8 Which processes do nonvascular plants use to transport water and nutrients through their tissues?

A absorption and photosynthesis

B cellular respiration and pollination

C diffusion and osmosis

D transpiration and reproduction

Use the diagram below to answer questions 9 and 10.

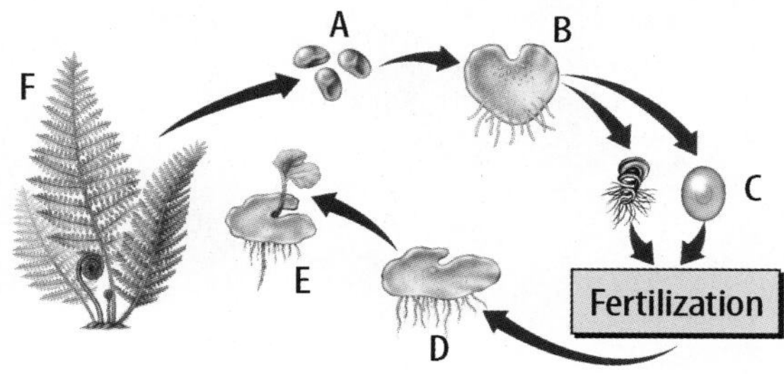

9 Which structures in the diagram are haploid?

A A, B, and F

B C, D, and E

C A, B, and C

D D, E, and F

10 Which structures in the diagram are diploid?

A A, B, and F

B C, D, and E

C A, B, and C

D D, E, and F

Constructed Response

Use the diagram below to answer question 11.

11 The diagram shows the path water takes in moving through a plant. Describe what happens to CO_2 in the plant. Use the terms *sunlight, sugar, leaves,* and *cellular respiration* in your answer.

12 Which structures in a plant contain vascular tissue? What is their function?

13 How are the life cycles of a fern and a pine similar? How are they different?

14 How do offspring produced by asexual reproduction differ from offspring produced by sexual reproduction?

NEED EXTRA HELP?														
If You Missed Question...	1	2	3	4	5	6	7	8	9	10	11	12	13	14
Go to Lesson...	1	3	2	3	1	3	1	1	2	2	3	1	2	2

Chapter 9

Introduction to Animals

THE BIG IDEA What are animals, and how are they classified?

What's the same?

This American robin is eating an earthworm. The earthworm provides energy for the robin's use.

- What animal characteristics do the robin and the worm have?
- What makes the robin and the worm different kinds of animals?
- What are animals, and how are they classified?

Abdolhamid Ebrahimi/E+/Getty Images

Get Ready to Read

What do you think?

Before you read, decide if you agree or disagree with each of these statements. As you read this chapter, see if you change your mind about any of the statements.

1. Animals must eat plants or other animals to live.
2. All animals have a left side and a right side that are similar.
3. A sponge is not an animal because it cannot move.
4. There are more arthropods on Earth than all other kinds of animals combined.
5. All young mammals take in milk from their mothers.
6. Birds are the only animals that lay shelled eggs.

Your one-stop online resource
connectED.mcgraw-hill.com

LearnSmart®

Chapter Resources Files, Reading Essentials, Get Ready to Read, Quick Vocabulary

Animations, Videos, Interactive Tables

Self-checks, Quizzes, Tests

Project-Based Learning Activities

Lab Manuals, Safety Videos, Virtual Labs & Other Tools

Vocabulary, Multilingual eGlossary, Vocab eGames, Vocab eFlashcards

Personal Tutors

Lesson 1

Reading Guide

Key Concepts
ESSENTIAL QUESTIONS

- What characteristics are common to all animals?
- How do scientists group animals?
- How are animal species adapted to their environments?

Vocabulary

bilateral symmetry p. 280
radial symmetry p. 280
asymmetry p. 280
adaptation p. 282
hydrostatic skeleton p. 282
exoskeleton p. 282
endoskeleton p. 282

Multilingual eGlossary

BrainPOP® Science Video

What are animals?

Inquiry Plant or Animal?

When ancient Greeks first classified living things, they thought sponges, such as the ones in this picture, were plants. What animal characteristics do you think sponges have? Why was the sponge thought to be a plant? How can we make similar mistakes in identifying living things today?

Launch Lab

10 minutes

What are animals?

The animal kingdom is diverse. Its members live everywhere from the depths of the ocean to the heights of the upper atmosphere. Though there are many differences among animals, they all have some of the same characteristics.

1. Examine the pictures of sea squirts, a bird, a crab, and a human.
2. In your Science Journal, make two lists. Title one *Similarities* and the other *Differences.*
3. Using what you know about the animals and what you can see in the pictures, list traits, behaviors, and processes the animals share or do not share under the appropriate heading.

Think About This

1. Which list was easier to create? Why?
2. **Key Concept** What are some characteristics that you think are common to all animals?

Animal Characteristics

Zoos are interesting to people of all ages. Why do people keep coming back to the zoo year after year? To see the animals of course! In fact, *zoo* comes from the Greek word *zoion,* which means "living being" or "animal."

All animals, like plants, are multicellular. Like plant cells, each animal cell has a nucleus at some point during its life. While cell walls support plant cells, a protein called collagen holds animal cells together. Animals are the only living things that have nerve cells. Nerve cells conduct nerve impulses. Most animals also have muscle cells that help them move.

Animals cannot transform light energy into food energy as most plants can. All animals get energy from the food they take into their bodies. In most animals, food passes through their stomachs, and then their intestines absorb nutrients from it.

All animals begin as a fertilized egg cell called a zygote. Recall that fertilization is the joining of an egg cell with a sperm cell. The zygote divides into more cells and forms an embryo. After many more cell divisions, the body of the animal is recognizable.

Key Concept Check What characteristics are common to all animals?

FOLDABLES®

Make a vertical two-column chart. Label it as shown. Use it to organize your notes on animal characteristics and identification.

How do scientists group animals?

When you first learned to talk, you probably grouped things by what they looked like. For example, you might have called any round object "ball." As you developed, however, you came to know the differences between things, such as trees and cats. You knew that dogs were not birds. This was your first experience with classification. Scientists classify animals in many different ways.

Symmetry

One way to group animals is by looking at their symmetry, or how body parts are arranged. The three types of symmetry are shown in **Figure 1**.

Some animals have **bilateral symmetry,** *a body plan in which an organism can be divided into two parts that are nearly mirror images of each other*. Humans, frogs, and the gecko in **Figure 1** are examples of organisms with bilateral symmetry.

An animal with **radial symmetry** *has a body plan which can be divided into two parts that are nearly mirror images of each other anywhere through its central axis.* The sand dollar in **Figure 1** has radial symmetry.

Some animals, such as the sponges in **Figure 1** and in the photo at the beginning of this lesson, do not have bilateral symmetry or radial symmetry. They have **asymmetry,** *meaning they have body plans which cannot be divided into any two parts that are nearly mirror images.*

WORD ORIGIN
asymmetry
from Greek *asymmetros,* means "not having a common measure"

Symmetry Personal Tutor

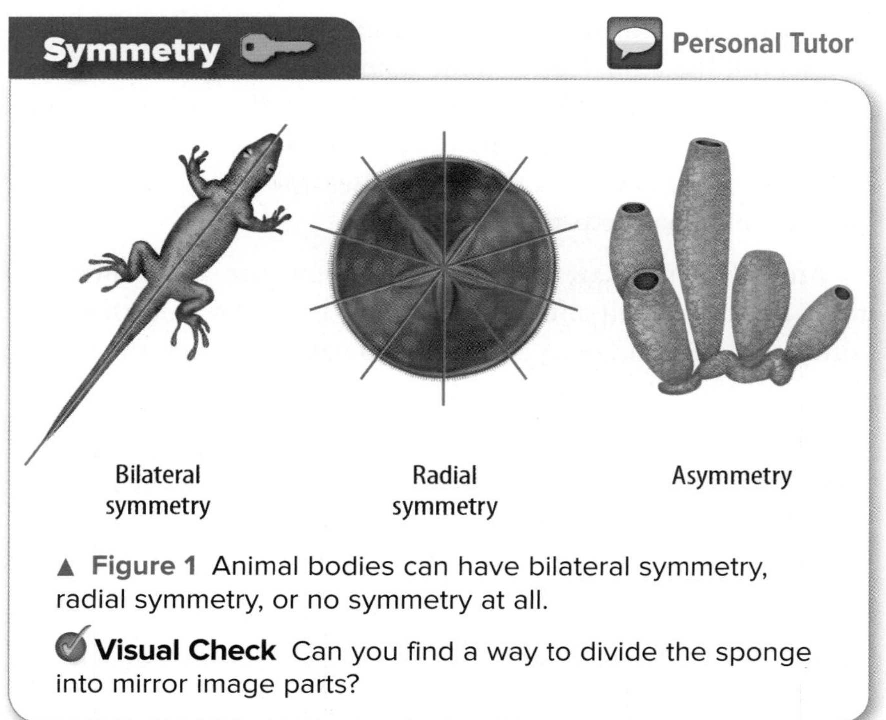

▲ **Figure 1** Animal bodies can have bilateral symmetry, radial symmetry, or no symmetry at all.

Visual Check Can you find a way to divide the sponge into mirror image parts?

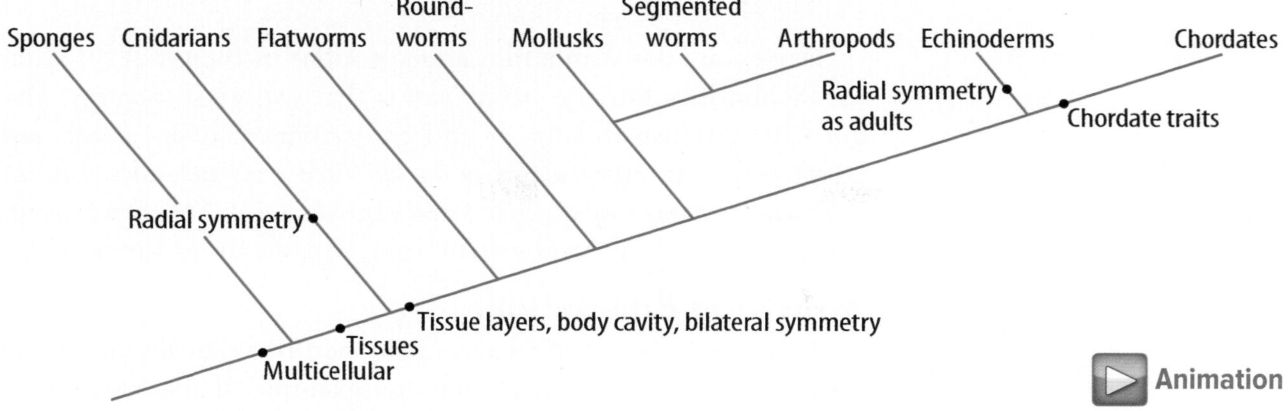

Animation

Figure 2 Scientists organize animals into groups based on characteristics they have in common. This tree diagram shows the relationships among animal groups based on common characteristics.

Visual Check Which animals are multicellular?

Groups of Animals

Scientists use a system called taxonomy to organize living things. This system groups living things into levels called taxons. Taxons are groups of living things that have certain traits in common.

Taxonomy The biggest groups in taxonomy are called domains. Because each animal cell has a nucleus at some point in its life, animals are in the Domain Eukarya. The next level consists of the kingdom taxons. Scientists use the traits you read about earlier to determine whether an organism belongs in the Kingdom Animalia. Animals then are classified into phyla (singular, phylum), genera (singular, genus), and species. Lessons 2 and 3 cover the nine most common animal phyla shown in **Figure 2.**

Family Tree A family tree shows the relationships among and within generations of a family. Animal phyla also are organized by how they are related through time. A tree of phyla shows the relationships among animals. As each new feature evolved over time, it was placed on a new branch of the tree, as shown in **Figure 2.** For example, all animals above sponges have tissues.

Key Concept Check How do scientists group animals?

MiniLab

10 minutes

What type of symmetry do animals have?

One way to group animals is by symmetry. First you must determine if an animal has symmetry. If so, determine the kind of symmetry—radial or bilateral—it has.

1. Copy the outlines of the animals shown below into your Science Journal.

Anemone | Ant | Penguin | Sponge

2. For each sketch, draw lines of symmetry on them if possible.
3. Identify the type of symmetry for each.

Analyze and Conclude

1. **Analyze Data** Think about the animals shown above, as well as other animals you know. Which body plan seems to be the most common among animals?
2. **Key Concept** Identify the relationship between the kind of symmetry an animal has and how the animal is classified.

Animal Adaptations

Have you ever seen a film about gorillas in the wild? A gorilla eats when it is hungry. It has hands that can grasp objects. The gorilla's response to hunger and the structure of its hands are adaptations. *An* **adaptation** *is an inherited trait that increases an organism's chances of surviving and reproducing in its environment.* Animal adaptations can be structural, behavioral, or functional.

REVIEW VOCABULARY

species
a group of organisms that have similar traits and are able to produce fertile offspring

Structural Adaptations

Animal species have structural adaptations that include their senses, skeletons, and circulation. For example, snakes can detect infrared light, and some insects can sense ultraviolet light. These adaptations can help them detect the presence of food or an enemy. Other animals have complex eyes that work like a camera. These eyes help the animals form accurate images of their environments.

Animals' skeletons have evolved into different types to support their bodies. An earthworm has a **hydrostatic skeleton,** *a fluid-filled internal cavity surrounded by muscle tissue.* A crab has soft internal structures. They are protected by *a thick, hard outer covering called an* **exoskeleton.** You probably are most familiar with *the internal rigid framework that supports you and other animals called an* **endoskeleton.** Your endoskeleton is made of bone. Your muscles attach to your bones and help you move.

Animal species also have structural adaptations for circulating blood. For example, ants have an open circulatory system. An ant's heart pumps blood into open spaces around its organs. However, an earthworm has a closed circulatory system. Many hearts pump blood through a system of vessels. Other animals with closed circulation have only one heart. More animal structural adaptations are shown in **Figure 3.**

Figure 3 Many animal species have evolved special tongue and teeth adaptations to help them obtain food and break it down.

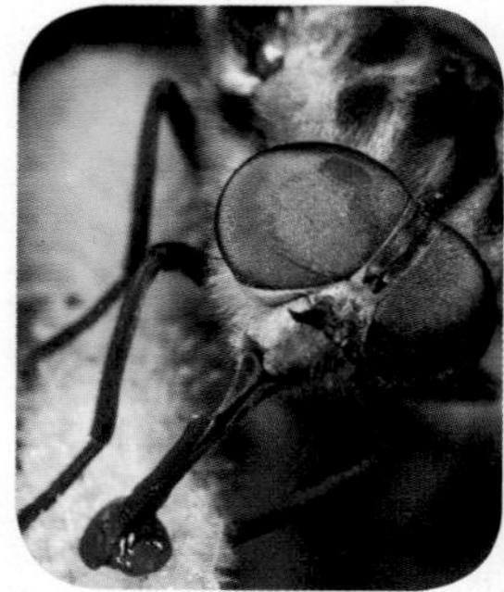

A fly uses its tongue to lap liquid.

A chameleon's tongue is long and sticky for catching insects.

A piranha has razor-sharp teeth for catching prey.

The sharp edge of a beaver's teeth can cut through tree trunks.

Behavioral Adaptations

Animals are born with behaviors called instincts. These behaviors have evolved over time and help species survive in their environments. A male fly instinctively waves its wings at a female to attract its attention. This action makes it more likely that the flies will breed and have offspring. Many tropical birds instinctively migrate when the number of daylight hours changes. These bird species have adapted to fly thousands of miles for food and habitats for breeding.

The ability to learn behaviors also is an important animal adaptation. For example, young songbirds learn how to sing their songs by listening to their parents. Baby geese also learn to follow their mothers soon after birth. This form of learned behavior is called imprinting. These behavioral adaptations increase an animal species' ability to survive and produce offspring.

Functional Adaptations

Animal species also have functional adaptations, which enable them to increase survival or maintain homeostasis. Some of these adaptations enable animals to reproduce successfully either in water or on land.

Most animals that live in water release large numbers of eggs or sperm, as shown in **Figure 4**. If fertilization occurs in the water, the process is called external fertilization. If fertilization occurs inside a female, it is called internal fertilization. Many eggs and sperm are produced because a water environment doesn't provide much protection for developing young. Many do not survive. Fertilizing many eggs ensures that a few will survive.

Most animal species that live on land use internal fertilization. Because the eggs are inside a female, only a few eggs need to be produced to ensure survival of the young.

Key Concept Check How are species adapted to their environments?

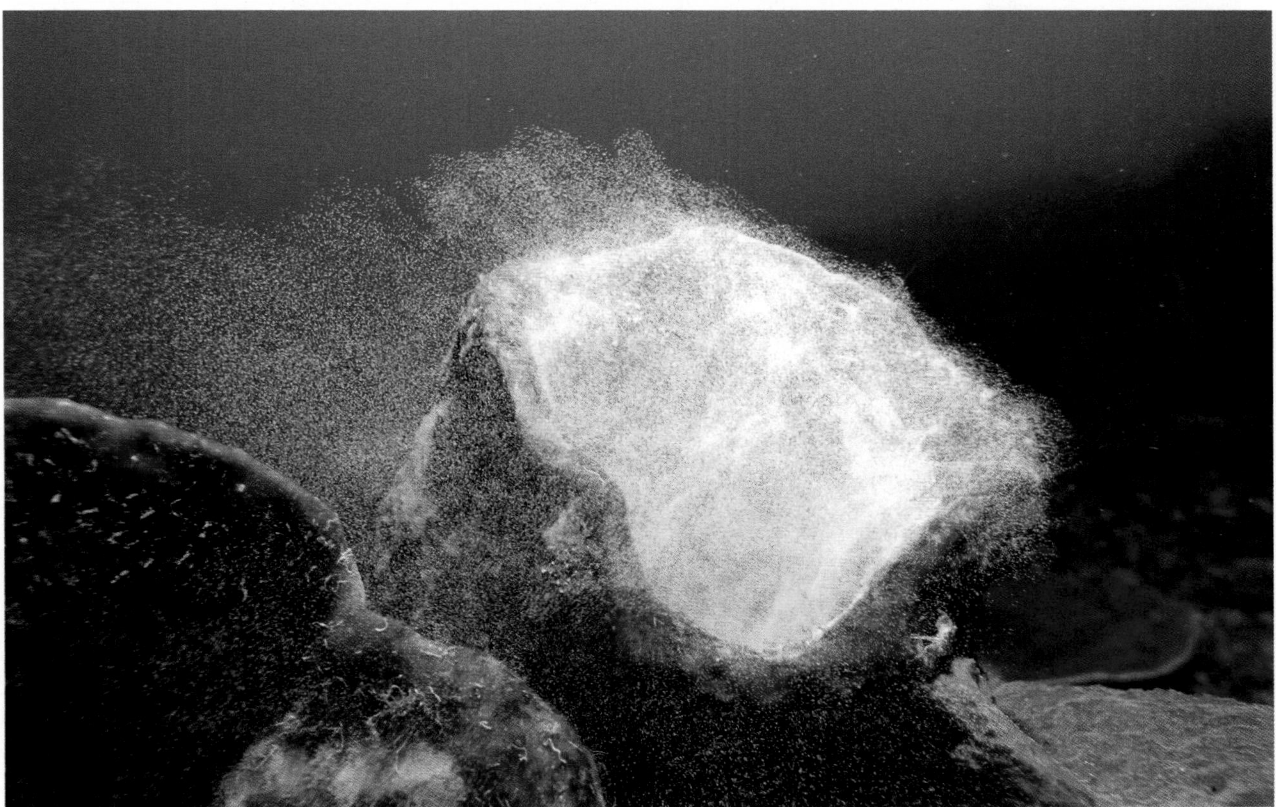

Figure 4 Sponges release large numbers of sperm into the water. Fertilization is usually internal.

Lesson 1 Review

Visual Summary

Animals are multicellular, have nerve cells, and cannot change light energy into food energy.

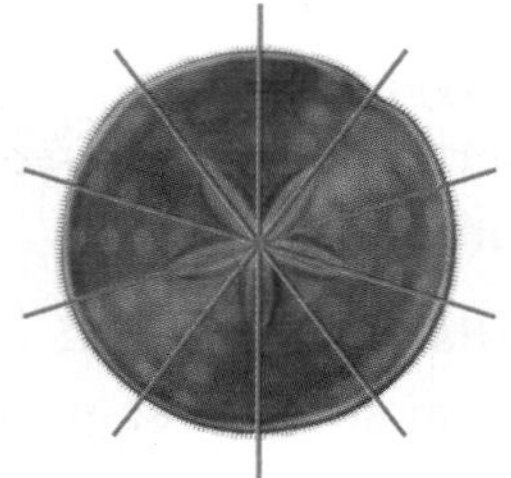

One way to group animals is by their symmetry.

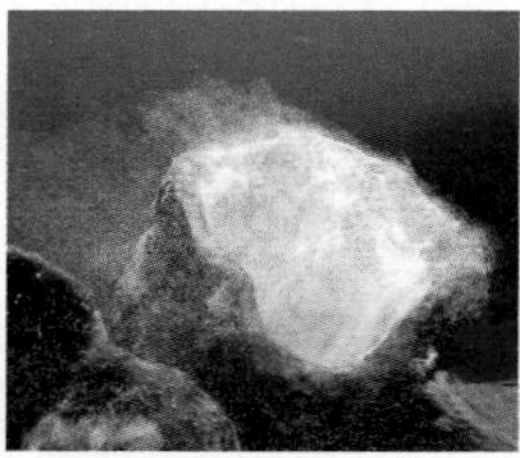

Structural, behavioral, and functional adaptations help animals survive.

FOLDABLES®

Use your lesson Foldable to review the lesson. Save your Foldable for the project at the end of the chapter.

What do you think NOW?

You first read the statements below at the beginning of the chapter.

1. Animals must eat plants or other animals to live.
2. All animals have a left side and a right side that are similar.

Did you change your mind about whether you agree or disagree with the statements? Rewrite any false statements to make them true.

Use Vocabulary

1. **Distinguish** between radial symmetry and bilateral symmetry.
2. **Define** *endoskeleton* and *exoskeleton* in your own words.
3. **Use the term** *adaptation* in a complete sentence.

Understand Key Concepts 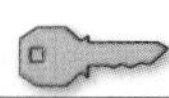

4. Which is NOT a characteristic common to all animals?
   - **A.** All animals are multicellular.
   - **B.** All animals have collagen in their bodies.
   - **C.** All animals have an endoskeleton.
   - **D.** All animals take in food.
5. **Summarize** how animals are grouped.
6. **Describe** some ways animal species are adapted to their environments.

Interpret Graphics

7. **Summarize** Copy and fill in the graphic organizer below to describe the types of symmetry.

8. **Explain** why the behavior shown below is a good adaptation.

Critical Thinking

9. **Design an experiment** to test whether a behavioral adaptation is an instinct or a learned behavior.
10. **Infer** Why are there no flying animals with asymmetry or radial symmetry?

(t) belizar73/Getty Images, (b) ©R. Dirscherl/age fotostock

45 minutes

How are adaptations useful?

Materials

rice

plastic cup (2)

forceps

tape

stopwatch

Safety

A physical adaptation is a feature an organism has that makes it better able to live in its environment. Some adaptations help an organism gather food. Others help an animal build a shelter or hide from a predator.

Learn It

A **prediction** makes a statement in advance, based on prior observation, experience, or scientific reasoning. In science, predictions are tested. Sometimes the predictions are supported after testing, but other times they are not.

Try It

1. Read and complete a lab safety form.
2. Count out 100 grains of rice and place them in a plastic cup. The rice is your food. The cup represents where you found your food.
3. Place an empty cup 15 cm away from the cup containing rice. The empty cup represents your shelter.
4. You will transfer rice from the first cup to the second cup two times—once using your fingers, and once using forceps. The forceps represent longer, pointier fingers or beaks that some animals have. You only can move one grain of rice at a time and only use one hand. The cups must remain standing up.
5. Based on your experiences and knowledge of forceps and rice, predict if using forceps or not using forceps will be more efficient in moving the rice. Write your prediction in your Science Journal and explain your reasoning.
6. Have your partner use the stopwatch to time you as you transfer the rice with your fingers. Record the time in your Science Journal.
7. Tape the forceps inside your thumb and index finger, as shown below.

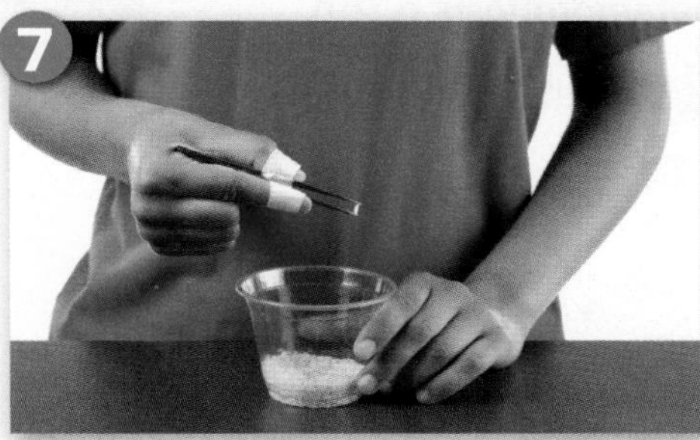

8. Have your partner time you again as you transfer the rice with forceps. Record the time.

Apply It

9. What were the independent and dependent variables and the controls in this activity?
10. Was your prediction accurate? What data supports your answer?
11. **Key Concept** Would the adaptation of longer, pointier fingers or beaks be an advantage or disadvantage for obtaining food if the food source were peanuts? What if the food were alive, such as a wiggling worm? Explain your reasoning.

Lesson 2

Invertebrates

Reading Guide

Key Concepts
ESSENTIAL QUESTIONS

- What characteristics do invertebrates have in common?
- How do the groups of invertebrates differ?

Vocabulary

parasite p. 287
mantle p. 290
molting p. 290
metamorphosis p. 291

Multilingual eGlossary

BrainPOP®

Inquiry Pretty Flowers?

No, these are animals called sea anemones. Would you believe that these sea anemones can paralyze and eat many other animals that swim or float by them? How do you think the animals in this picture support their bodies? How do they move?

joebelanger/iStock/Getty Images

Launch Lab

30 minutes

Who lives here?

The number of animals living in a small patch of soil can outnumber the human population of a large city. Some of these animals do not have a backbone; others might have a backbone. What types of animals can you find on a patch of ground?

1. Read and complete a lab safety form.
2. Throw a large **plastic hoop** on a **patch of ground.**
3. Search your patch of ground for animals and evidence of animal life. You may need to use a **magnifying lens.**
4. In your Science Journal, record your observations. Make a list of any animals or evidence of animals that you find.

Think About This

1. Which animals did you observe? What evidence did you find that suggests other animals had been in your sample area?
2. Did most of the animals you found have backbones or not have backbones?
3. **Key Concept** Describe three differences among the animals without backbones that you identified.

What is an invertebrate?

What was your first thought when you saw the picture on the facing page? Maybe you wondered why there was a picture of flowers in a chapter about animals. What you saw are anemones, which are animals. They trap food in fingerlike tentacles. Anemones lack a backbone.

Recall how animals support their bodies. Most animals with an endoskeleton have a backbone for support. These animals are called vertebrates. Animals without backbones are called invertebrates. Most invertebrates support their bodies with either a hydrostatic skeleton–a fluid-filled internal cavity–or an exoskeleton–a hard outer covering. Some invertebrates have endoskeletons.

Invertebrates are about 95 percent of all known animal species. In this lesson, you will read about eight of the most common invertebrate phyla. Recall phyla are one of the taxons.

Invertebrates have many adaptations for survival. Some invertebrates are **parasites,** *animals that survive by living inside or on another organism, get food from the organism, and do not help in the organism's survival.* Other invertebrates hunt their food. Some invertebrates can even change the color of their skin to match their environments.

Key Concept Check What characteristics do invertebrates have in common?

FOLDABLES®

Make a horizontal two-tab book with an extended tab. Label it as shown. Use it to identify similarities and differences in invertebrates.

Invertebrates

Common Characteristics | Differences

Hutchings Photography/Digital Light Source

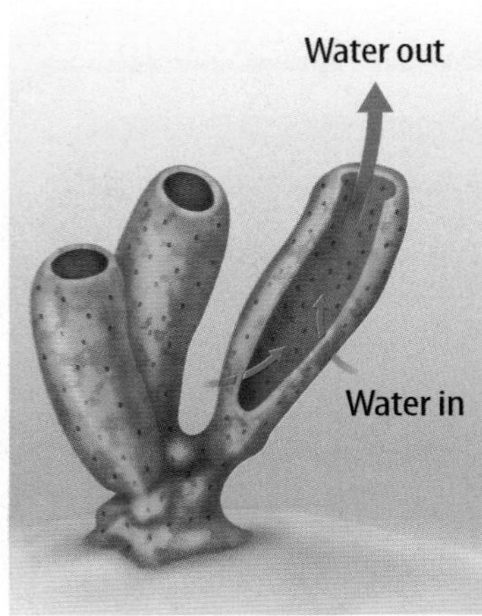

▲ **Figure 5** Water passes through sponges, and food particles are filtered out.

Sponges

The oldest branch of the animal family tree, phylum Porifera (puh RIH fuh ruh), includes the sponges. Sponges are often called simple animals because they have only a few types of cells and no true tissues. Sponges live in water and cannot move as adults. Instead, they **attach** to rocks and other underwater structures. Sponges take in food when water passes through their bodies, as shown in **Figure 5.** Special cells inside the sponge filter out food particles in the water.

Sponges have tiny, stiff fibers that support their bodies. Scientists group sponges by the kinds of materials that make up these fibers. The fibers in the most common group of sponges are made of the protein spongin or silica, or both. These sponges are sold as natural sponges. They are used to scrub surfaces, but the fibers can scratch shiny surfaces.

ACADEMIC VOCABULARY

attach
(verb) to fasten

Cnidarians

Corals, anemones, jellyfish, hydras, and Portuguese man-of-wars are members of phylum Cnidaria (nih DAYR ee uh). The name *Cnidaria* comes from special cells these animals use to catch their prey. The cells–called nematocysts (NE mah toh sihsts) and shown in **Figure 6**–can inject poison into animals that come in contact with them. Cnidarians also have radial symmetry.

Cnidarians are different from sponges because they have true tissues. Some cnidarians, such as corals and anemones, spend their adult lives attached to underwater surfaces. Others, such as jellyfish, can swim.

Reading Check Identify how cnidarians are different from sponges.

Figure 6 Cnidarians can inject paralyzing poison into their prey. ▼

**Visual Check** Where are the nematocysts on this cnidarian?

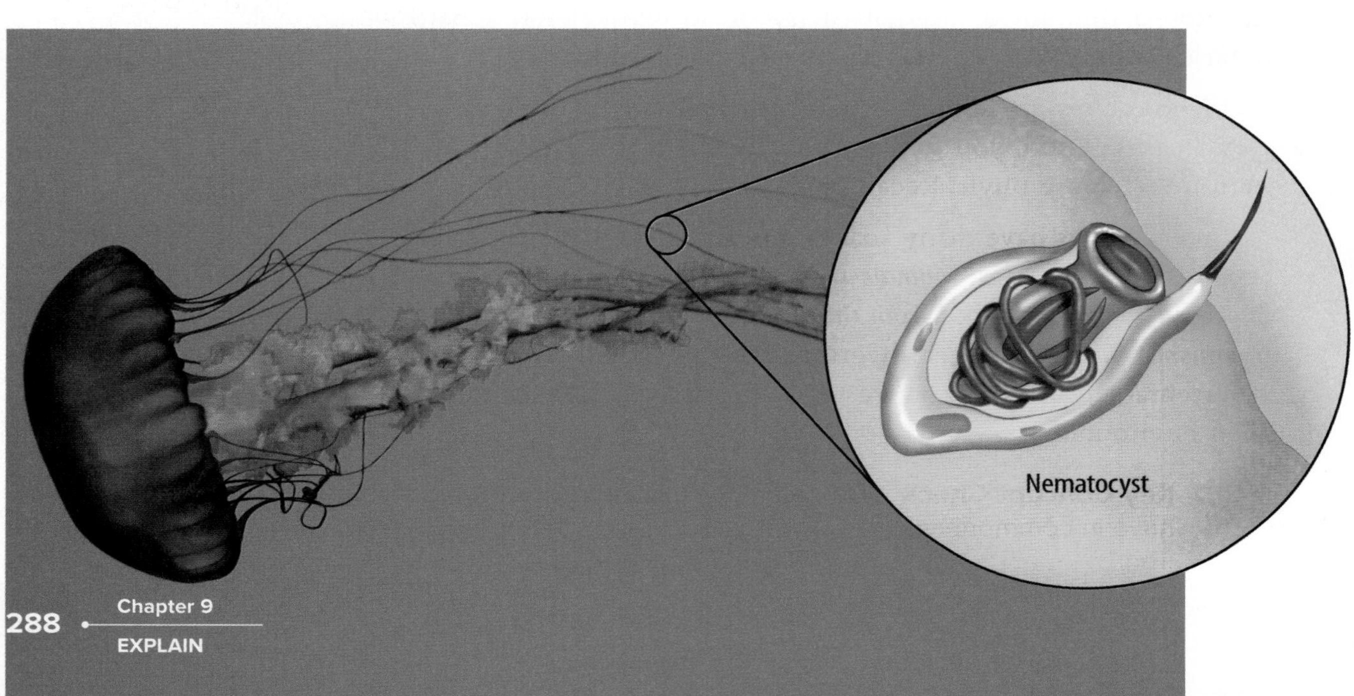

Flatworms

The common name for an animal in the phylum Platyhelminthes (pla tih hel MIHN theez), flatworm, describes it accurately. Its body shape is flat. Flatworms also have bilateral symmetry; each worm has a left side and a right side that are similar.

Most flatworms live in freshwater or salt water. Some flatworms are free-living. A planarian, for example, swims freely in water and ingests food through a tube on the underside of its body. Other flatworms, such as the liver fluke shown in **Figure 7**, are parasites and can infect humans.

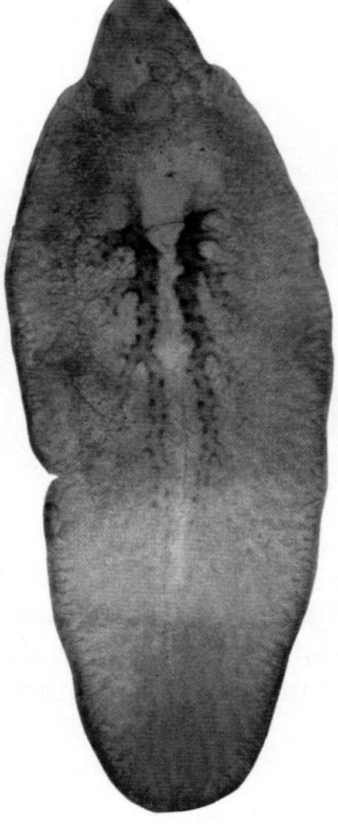

▲ **Figure 7** The liver fluke is a parasite that sometimes infects humans.

Segmented Worms

Have you ever handled an earthworm? Did you notice that its body was like a tube of tiny rings? The name for the phylum that includes earthworms, *Annelida* (ah NEL ud uh), means "little rings." These rings, shown in **Figure 8**, are called segments. Each segment is a fluid-filled compartment. Therefore, a segmented worm has a hydrostatic skeleton.

You also might have noticed that the sides of the earthworm's body felt prickly. The prickles are tiny, stiff hairs called setae (SEE tee). Setae help earthworms grip surfaces.

Earthworms tunnel through soil and take it into their bodies to absorb nutrients from the soil. Their tunnels help break up soil. Segmented worms also can be parasites. Leeches attach their mouths to other animals and suck blood.

Reading Check Why are Annelida called segmented worms?

Animation

Figure 8 Segmented worms have bilateral symmetry. Each segment is a compartment filled with fluid. The fluid helps give the worm internal support. ▼

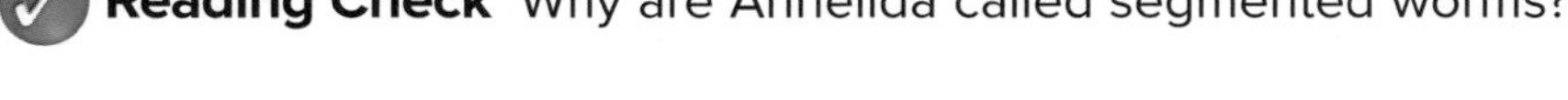

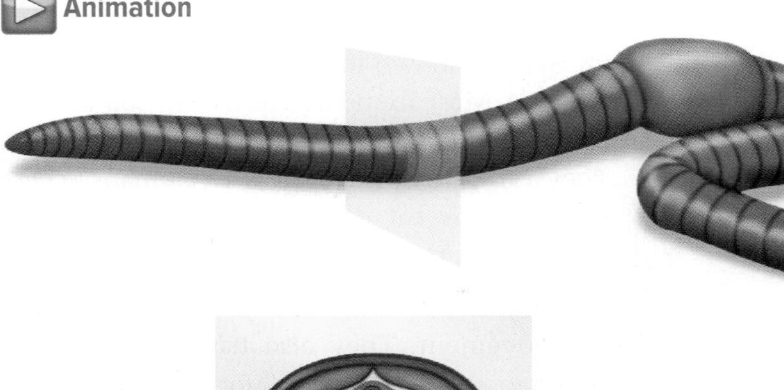

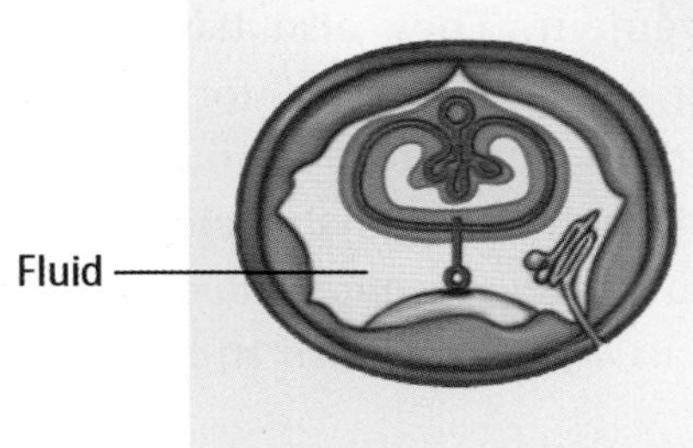

PERENNOU NURIDSA/Science Source

▲ **Figure 9** Mollusks have bilateral symmetry. Most mollusks, such as this snail, have external shells for protection.

Mollusks

On summer mornings, you might notice thin, slimy trails across a sidewalk. These trails were likely made by snails or slugs searching for food during the night. Snails, such as the one shown in **Figure 9,** and slugs are mollusks in phylum Mollusca (mah LUS kuh). Most mollusks have a footlike muscle that generally is used for movement. A mollusk also has a mass of tissue called a mantle. A **mantle** *is a thin layer of tissue that covers a mollusk's internal organs.* It also is involved in making the shell of most mollusks. A mollusk's shell supports and protects its soft body. Some mollusks, such as slugs, do not have shells. Other mollusks, such as squids, have internal shells.

Reading Check Name one type of mollusk that does not have a shell.

The eating methods of mollusks vary. Some mollusks, such as clams, oysters, and scallops, eat by filtering food particles from the water in which they live. Other mollusks, such as octopuses, are predators and catch their prey in long, strong tentacles.

Roundworms

Animals in phylum Nematoda (ne muh TOH duh) are called nematodes or roundworms. Some roundworms can infect humans, and others can infect plant roots. Others, such as the vinegar eel shown in **Figure 10,** are harmless to humans. Most roundworms live in soil and are too small to see without a magnifying lens. They typically are harmless to humans. These roundworms eat dead organisms and return nutrients to the soil.

Roundworms have a hydrostatic skeleton for movement. They also have a hard outer covering, called a cuticle, for protection. The cuticle does not grow. It must be shed and replaced with a larger cuticle for the roundworm to grow. *An outer covering is shed and replaced in a process called* **molting.**

Figure 10 The vinegar eel is a roundworm that feeds on organisms used in making vinegar. It is harmless but is removed from vinegar by the manufacturer. All roundworms have bilateral symmetry. ▼

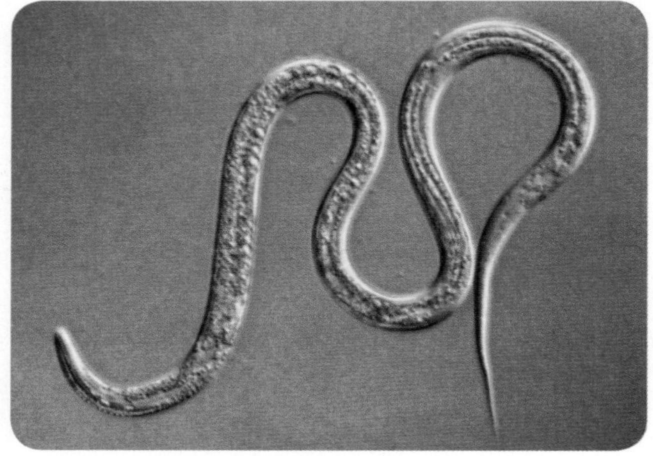

Arthropods

Can you imagine a billion billion of something? Scientists estimate that is how many individual arthropods there are on Earth. There are more animals in phylum Arthropoda (ar THRAH puh duh) than in all other animal phyla combined. Arthropods have bilateral symmetry.

Like a roundworm, an arthropod has a hard outer covering, so it must molt in order to grow. An arthropod has an exoskeleton for both movement and protection. Its muscles attach to the exoskeleton. An arthropod uses its muscles when moving its jointed appendages. An appendage is a structure, such as a leg or an arm, that extends from the central part of the body.

Arthropod bodies have three parts: a head, a thorax, and an abdomen. The head contains sense organs that can see, feel, and taste the environment. The thorax is where legs are attached. The abdomen contains intestines and reproductive organs. Arthropods have open circulation. This means their blood is not in vessels. Instead, it washes over internal organs.

Insects Most arthropods are insects, such as the one shown in **Figure 11.** Scientists call them hexapods because they have six legs. Insects are the only arthropods that have the ability to fly. Another trait of insects is metamorphosis. In **metamorphosis,** *the body form of an animal changes as it grows from an egg to an adult.* For example, a caterpillar eventually changes into a moth or a butterfly.

Figure 11 The Colorado potato beetle is a pest of potato crops.

WORD ORIGIN
metamorphosis
from Greek *metamorphoun,* means "to transform"

MiniLab

10 minutes

Bigger on Land or in Water?

Invertebrates come in many sizes. A fruit fly is about 3 mm long. The Chilean rose hair tarantula is a large spider. It can have a leg span of 12.5 cm as an adult. It gets its name from its rose-colored hair.

1. In your Science Journal, write the following list of invertebrates in order from smallest in size to largest in size: anemone, ant, bee, butterfly, earthworm, fly, jellyfish, lobster, octopus, spider, sponge, and sea star.
2. Identify each of the organisms as either an aquatic invertebrate (lives in water) or a terrestrial invertebrate (lives on land).

Analyze and Conclude

1. **Interpret Data** Based on your list of invertebrates, which tend to be larger—aquatic or terrestrial invertebrates?
2. **Infer** Why do you think the group you chose in question 1 is able to be larger?
3. **Key Concept** How does the relationship between invertebrate animal size and habitat compare to vertebrate animal size and habitat?

Other Arthropod Groups There are three other major groups of arthropods. Spiders and scorpions are one group. They have eight legs used for walking and grasping. Crabs and lobsters make up another group. Members of this group mostly live in water. They have chewing mouthparts and three or more pairs of legs. Some lobsters have as many as 19 pairs of appendages! Centipedes and millipedes are in another group. They have the most appendages of all. Generally, a centipede has one pair of legs per segment and a millipede has two pairs per segment. Millipedes eat dead plants, but centipedes are predators.

Echinoderms

Do the animals in **Figure 12** appear fuzzy and soft? Touch an echinoderm (ih KI nuh durm), from the phylum Echinodermata (ih kin uh DUR muh tuh), and you will find it is the opposite. *Echinoderm* means "spiny skin." An echinoderm feels spiny due to the hard endoskeleton just beneath its thin outer skin.

Sea star

Sea cucumber

Sea urchin

Figure 12 Echinoderms have hard, spiny skin. They move slowly using tube feet.

Visual Check For which echinoderm is it hard to identify radial symmetry?

All echinoderms live in salt water. They move slowly with tiny suction-cuplike feet, called tube feet. Their tube feet are connected to larger tubes called canals. These canals connect to a central ring that controls water movement within the animal. Water moves back and forth through the canals and tube feet. This movement enables echinoderms to grab onto or let go of any surface they are moving across. Echinoderms have bilateral symmetry when they are young, and radial symmetry as adults.

Echinoderms are more closely related to humans than all other invertebrates. Both echinoderm embryos and human embryos have similar early growth patterns.

Key Concept Check How do the invertebrate groups differ?

Lesson 2 Review

Visual Summary

Invertebrates do not have a backbone; they support their bodies with an exo-skeleton, an endo-skeleton, or a hydrostatic skeleton.

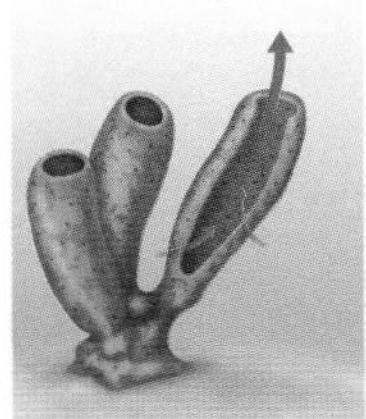

Sponges and cnidarians are the oldest branches on the animal family tree.

Insects have three body parts—the head, the thorax, and the abdomen.

FOLDABLES®

Use your lesson Foldable to review the lesson. Save your Foldable for the project at the end of the chapter.

What do you think NOW?

You first read the statements below at the beginning of the chapter.

3. A sponge is not an animal because it cannot move.

4. There are more arthropods on Earth than all other kinds of animals combined.

Did you change your mind about whether you agree or disagree with the statements? Rewrite any false statements to make them true.

Use Vocabulary

1. The tissue involved in making the shell of most mollusks is called the ________.
2. **Use the term** *metamorphosis* in a complete sentence.
3. **Define** *parasite* in your own words.

Understand Key Concepts

4. Which characteristic is common to all invertebrates?
   - **A.** backbone
   - **B.** mantle
   - **C.** cell walls
   - **D.** no backbone
5. **Explain** how a scientist might determine that a grasshopper is an arthropod.
6. **Describe** how an echinoderm is different from a cnidarian.

Interpret Graphics

7. **Summarize** Use the graphic organizer below to identify which invertebrate phyla have bilateral symmetry.

8. **Classify** the animal shown below as a mollusk, an arthropod, or a cnidarian. Explain your choice.

Critical Thinking

9. **Diagram** an ant, and label the three body parts that are common to all arthropods.
10. **Analyze** why some invertebrates that live in the sea grow to be larger than invertebrates that live on land.

CAREERS in SCIENCE

AMERICAN MUSEUM of NATURAL HISTORY

Investigating True Bugs

▲ Toby Schuh discusses a catch in Mexico with his daughter.

True Bugs Close-Up

Here's a look at a few remarkable true bugs:

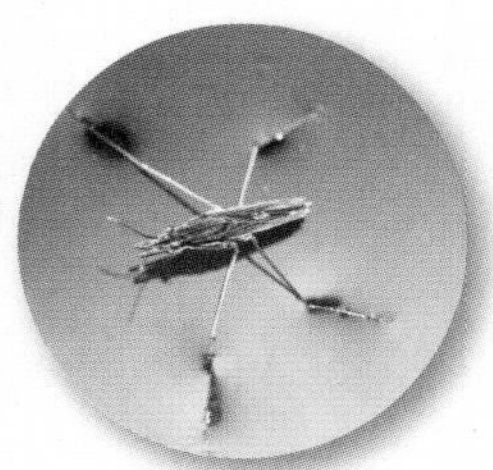

This sea skater (Gerris lacustris) spends its entire life on the ocean surface.

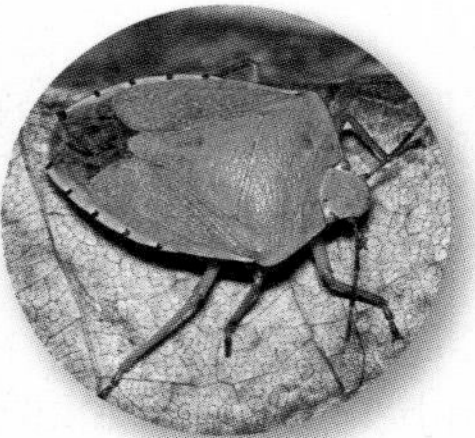

The southern green stink bug (Nezara viridula) gives off a strong odor when disturbed.

Meet Toby Schuh, a scientist who studies bugs.

Many of the amazing creatures people call bugs aren't really bugs at all. Bees, flies, butterflies, ground beetles, and ladybugs are examples of insects. However, scientists classify some insects into a group called "true bugs." This group includes assassin bugs that shoot poison into their prey, water striders that "skate" on water, and stink bugs that ooze smelly fluid.

How are true bugs different from other insects? Just ask Dr. Toby Schuh. He's an entomologist, a scientist who studies insects, at the American Museum of Natural History. He specializes in true bugs.

All insects have six legs, three body sections, and usually two pairs of wings. Bugs are a group of insects that share certain characteristics. They use slender, tubelike mouthparts to suck fluids from plants and sometimes from animals. Their saliva, or spit, has chemicals that break down food into liquid. They also have "stink glands" that release scents for finding mates, defending themselves, and communicating with each other. True bugs hatch from their eggs looking like small versions of adults. They don't go through a gradual metamorphosis as many insects, such as butterflies, do. During metamorphosis, an animal changes from one form to another as it develops from an egg to an adult.

Scientists have identified more than 40,000 species of true bugs, and they discover new ones all the time. Schuh is in charge of a worldwide database called the Plant Bug Planetary Biodiversity Inventory. It tracks information about true bugs. He also adds to this collection by searching for new species in places such as Africa, Australia, and South America. So far, Schuh and his colleagues have discovered over 1,200 new species of true bugs. Their search continues.

It's Your Turn

DIAGRAM Choose a true bug to research. Draw a picture of it. Label the features that make it an insect and the features that make it a true bug. Share your labeled diagram with your class.

Lesson 3

Reading Guide

Key Concepts

ESSENTIAL QUESTIONS

- What characteristics do chordates have in common?
- What is the difference between vertebrate and invertebrate chordates?
- How do the groups of vertebrate chordates differ?

Vocabulary

notochord p. 296
pharyngeal pouch p. 296
gill p. 298
amnion p. 300
ectotherm p. 300
endotherm p. 301
mammary gland p. 302

Multilingual eGlossary

Science Video

Chordates

Inquiry Are they related?

The two animals in this photo appear different. The fish has a backbone. The purple tunicate does not. However, both belong to the phylum Chordata. What similar traits do these animals display?

Launch Lab

10 minutes

What animals are around you?

Have you ever noticed all the animals around you in the park or in your neighborhood?

1. In your Science Journal, make a list of all the animals you have seen in your neighborhood or at a park.
2. Classify each of the animals you listed as either a vertebrate or an invertebrate.
3. Tally the number of vertebrates on your list. Do the same for the invertebrates.

Think About This

1. How do your totals compare with those of your classmates?
2. Did you think of more vertebrates or invertebrates? Why?
3. **Key Concept** What do you think are two main differences between land vertebrates and invertebrates?

FOLDABLES®

Make a horizontal three-tab Venn book. Label it as shown. Use it to compare and contrast the different types of chordates.

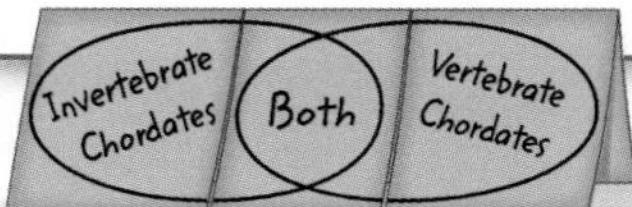

What is a chordate?

Think of the zoo you read about at the beginning of this chapter. It's likely that most of the animals at the zoo were chordates. It also is likely that most of the animals at the zoo were mammals, like you. Chordates are animals that are grouped in the phylum Chordata. Mammals are chordates.

There are two types of chordates–vertebrate chordates and invertebrate chordates. Recall that a vertebrate is an animal with a backbone. An invertebrate chordate shares many traits with vertebrates, but it has no backbone. All chordates have four traits in common: a notochord, a tail, a nerve cord, and pharyngeal pouches. These four traits exist at some time during the life of a chordate.

You are a chordate. When you were developing in the womb, you had a notochord. *A* **notochord** *is a flexible rod-shaped structure that supports the body of a developing chordate.* It was replaced by your backbone. You also had a tail. What is left of your tail is your tailbone. Before you had a brain and a spinal cord, you had a nerve cord. You also had pharyngeal (fuh run JEE uhl) pouches. **Pharyngeal pouches** *are grooves along the side of a developing chordate.* Your pharyngeal pouches developed into parts of your ears, head, and neck. Fish have pharyngeal slits that provide support for gills. That you had these characteristics at one time is evidence that you and other chordates have ancestors in common.

WORD ORIGIN

notochord
from Greek *notos*, means "back"; and Latin *chorda*, means "cord"

Key Concept Check Name characteristics all chordates share.

John Giustina/Getty Images

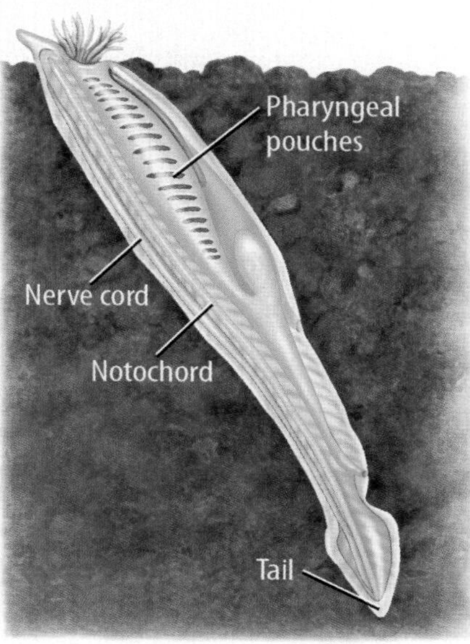

Figure 13 Lancelets live today, but they probably resemble the ancestor to all chordates.

Invertebrate Chordates

As you have read, some chordates never develop backbones. These animals are called invertebrate chordates. The tunicate in the picture at the beginning of this lesson is an invertebrate chordate. You will read about tunicates later. But first, what did the ancestor of chordates look like?

Lancelets

The earliest chordates probably looked similar to the lancelet in **Figure 13.** Lancelets are small animals found burrowed in the sand just off ocean shores. Lancelets grow only 5 cm in length. Lancelets can swim, but they often sit in the sand and catch food particles floating by. Lancelets have all four chordate traits, as shown in **Figure 13.**

Tunicates

Adult tunicates, shown in the photo at the beginning of this lesson, look like sponges. Like sponges, adult tunicates live in the sea attached to rocks or other stationary objects. However, adult tunicates have organized tissues and internal structures such as organs. They also have all the characteristics of chordates at some time in their lives. If you study a tunicate before it becomes an adult, you will see an animal that looks and acts like a tadpole. Young tunicates can swim and have all four chordate traits.

Lancelets look more like fish than tunicates. Therefore, scientists once thought lancelets were more closely related to vertebrates than tunicates were. But when scientists studied the DNA of all three groups, they discovered the opposite to be true. Vertebrate and tunicate DNA is more similar than vertebrate and lancelet DNA.

Vertebrate Chordates

Most of the animals you are familiar with are probably vertebrate chordates. This group includes cats, dogs, fish, snakes, frogs, and birds. Recall that vertebrates are animals with backbones. Most vertebrates also have jaws. As vertebrate bodies and skeletons continued to adapt, vertebrates became better at catching food and avoiding being eaten.

Key Concept Check What is the difference between invertebrate and vertebrate chordates?

Fish

When you think of a fish, you might think of a goldfish. You even might think of a shark. But would you think of a sea horse? All are fish and have traits that make them fish. All fish live in water and use gills for breathing. **Gills** *are organs that exchange carbon dioxide for oxygen in the water.* All fish have powerful tails, and most fish have paired fins. There are three major groups of fish, as shown in **Table 1**.

Table 1 Fish are classified into three different groups.

Visual Check Which group of fish does not have paired fins?

Table 1 Groups of Fish

Jawless Fish Lampreys are jawless fish. The skeleton of jawless fish is made of cartilage. The tip of your nose and the flaps of your ears are made of cartilage. Some jawless fish get their nutrition from other fish. They have a circle of teeth that attach to the sides of other fish and make a wound. They then slowly suck out blood and other body fluid from the fish.

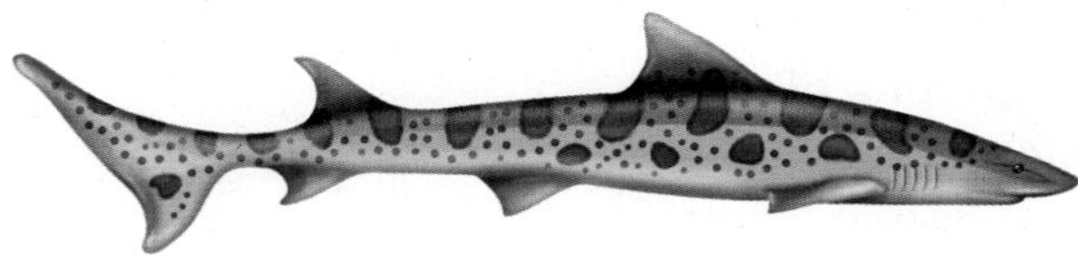

Sharks and Rays Most of the skeleton of sharks and rays is made of cartilage. However, shark skulls are made of bone. Sharks have paired fins. They are fast swimmers and also have powerful jaws. Their jaws make them dangerous predators of other animals, especially other fish.

Bony Fish All other fish have a bony skeleton, as well as paired fins and jaws. Bony fish, such as goldfish, also have a special sac called a swim bladder that the fish can fill with gas. This helps the fish move up and down in the water. Sea horses are unique bony fish because the males carry the young in their bodies as they develop.

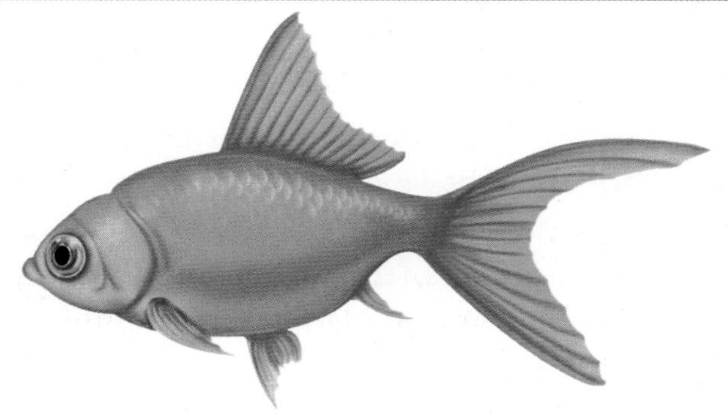

Amphibians

In Canada in 2004, scientists discovered the fossil of an animal that lived long ago in shallow water. The animal had both gills and lungs, a flexible neck, and fins with arm and hand bones. The fins could help the animal move in the water and on land. The scientists think they discovered one of the first tetrapods. A tetrapod is a vertebrate animal with four limbs.

Amphibians are a group of tetrapods that live on land. But like the fossil described above, amphibians still depend on water to survive and reproduce. The word *amphibian* means "both ways of life." Most amphibians must lay their eggs in water. In addition, young amphibians, such as tadpoles, have gills and must spend most of their time in water. Most adult amphibians have lungs for breathing on land. However, amphibian skin is thin and moist. On land, amphibians must live in moist habitats to keep their bodies from drying out.

Reading Check Summarize the characteristics of amphibians.

There are three types of amphibians, as shown in **Figure 14.** Salamanders and newts have tails and move by bending their bodies side-to-side. Frogs and toads do not have tails as adults. They have long legs that enable them to jump. Some frogs can jump several meters in one bound! Caecilians (sih SIHL yuhnz) are a group of amphibians that do not have legs. They look similar to earthworms and move by twisting their bodies back and forth like a snake.

Scientists are concerned about the survival of amphibians. Many amphibian populations have become smaller since 1980. Some types of amphibians have not been seen for years. Scientists think the amphibian population is decreasing because of disease, climate change, herbicides, and the destruction of amphibian habitat.

Some adult salamanders live on land.

Frog legs are longer and more powerful than toad legs.

Caecilians burrow in the soil.

Figure 14 The three types of amphibians look different, but all have similar characteristics.

Figure 15 Reptiles, birds, and mammals do not require water for reproduction. The amniotic egg protects the developing embryo.

Visual Check What surrounds the developing embryo?

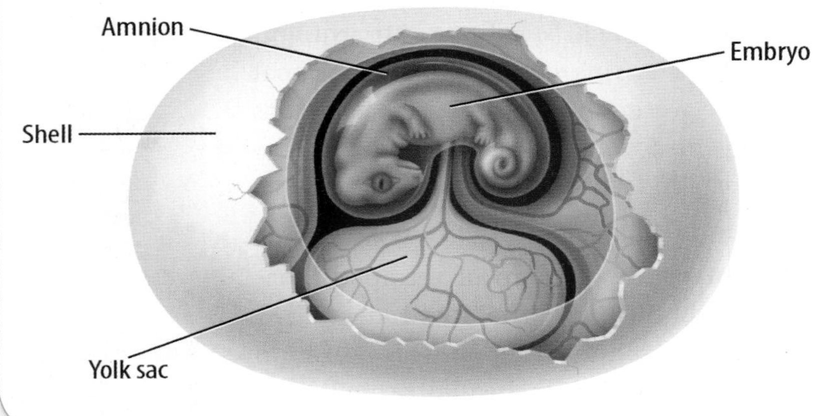

SCIENCE USE V. COMMON USE

scale

Science Use small, flat, rigid external body covering

Common Use an instrument for weighing

Reptiles

Lizards and snakes, turtles, and alligators and crocodiles are the three most common groups of reptiles. Most reptiles live on land, and all have lungs for breathing. Scales on their skin prevent reptiles from drying out. Reptiles also do not need water to lay their eggs. Most reptiles lay shelled eggs that don't dry out. Inside the egg is **amnion,** *a protective membrane that surrounds the embryo.* An egg with an amnion is called an amniotic egg, as shown in **Figure 15.** Reptiles, birds, and mammals all have amniotic eggs.

One group of reptiles includes both lizards and snakes. Most lizards are small and can fit in your hand. However, one lizard, the Komodo dragon, can grow up to 3 m. Snakes are legless reptiles. Many snakes also are small, but some can be several meters long. All snakes eat other animals. When snakes catch their prey, they can crush them or bite and poison them. Either way, most snakes swallow their prey whole!

Turtles are best known for their protective shells. Some turtles can live in the desert. Others, such as snapping turtles and sea turtles, live mostly in water.

Alligators and crocodiles are found in warm parts of the world. They live in or near water but lay their eggs in nests on the shore. They are fierce hunters and can move quickly for short distances.

Reptiles are **ectotherms,** *animals that heat their bodies from heat in their environments.* Warming, or basking, in sunlight is a behavioral adaptation of ectotherms. A reptile with a warm body can move faster and catch prey more easily. Reptiles move to cool, dark places to conserve energy when food is scarce.

Math Skills 

Use a Formula

The size of two organisms might be the same, but one floats while the other sinks. This is because the organisms have different densities. The formula for density is

$$\text{density} = \frac{\text{mass}}{\text{volume}}$$

For example, what is the density of a chicken's leg bone that has a mass of 5.5 g and a volume of 5.0 cm^3?

Replace the terms in the formula with the given values.

$$\text{density} = \frac{5.5\text{ g}}{5.0\text{ cm}^3}$$

Solve the problem.

$$\frac{5.5\text{ g}}{5.0\text{ cm}^3} = \frac{1.1\text{ g}}{\text{cm}^3}$$

Practice

A piece of a cow's leg bone with a volume of 10 cm^3 has a mass of 18 g. What is the density of the bone?

 Math Practice

 Personal Tutor

Birds

Can you name a unique trait of birds? Did you think of flight? Many insects fly, and so do certain mammals. Some birds, such as penguins and emus, don't fly. Maybe you thought of wings. But most insects and some mammals also have wings. The one trait that makes birds different from all other animals is their feathers.

Birds have many adaptations that enable them to fly. A bird does not have a urinary bladder to weigh the bird down when the bladder is full. Instead, birds concentrate their urine into crystals. The crystals are the white part of bird droppings. Birds have bones that are nearly hollow and filled with air. This makes the bones lighter than the bones of other vertebrates.

Wings and feathers are birds' major adaptations for flight. A bird's wings are connected to powerful chest muscles. Bird wings come in different shapes. Long, narrow wings enable a seagull to soar on long flights. The short, broad wings of a sparrow, shown in **Figure 16,** enable quick changes of direction to catch food or escape an enemy.

Reading Check Describe the different shapes of bird wings, and explain how the shapes help the birds survive.

Feathers also keep birds warm. Unlike reptiles, birds are **endotherms,** *animals that generate their body heat from the inside.* This enables birds to live in cold habitats. However, keeping their body temperatures high requires much energy. Like you, birds shiver when they get cold. Shivering muscles help produce more body heat.

Figure 16 Short, broad wings help a sparrow survive.

(l) Kimberly Hosey/Moment Open/Getty Images, (r) McGraw-Hill Education

MiniLab 30 minutes

Which bone is less dense?

The number of bones in a vertebrate varies from one species to another. Many birds have about 120 bones in their bodies. Humans have just over 200. Dogs have over 300! How do the bones of animals compare?

1. Read and complete a lab safety form.
2. Copy the table below into your Science Journal.

Bone Type	Mass (g)	Volume (mL)	Density (g/mL)
Chicken bone			
Mammal bone			

3. Use a **triple-beam balance** to find the mass of a **dry chicken bone.** Record the mass in your table.
4. Find the volume of the chicken bone using water displacement. Place 50 mL of water in a 100-mL **graduated cylinder.** Place the chicken bone in the water, and record the combined volume of water and chicken bone. Calculate the volume of the chicken bone by subtracting the volume of water from the combined volume of the water and the chicken bone. Record the volume in your table.
5. Repeat steps 3 and 4 for the **mammal bone.**

Analyze and Conclude

1. **Make and Use Tables** Complete the table by calculating the density for each of the bones. Recall that density = mass/volume. Be sure to include units. How do the densities of the chicken and mammal bones compare?
2. **Key Concept** Evaluate how the difference in densities might be considered a physical adaptation for birds.

An echidna is a monotreme.

This opossum raises its young in a pouch.

Like most mammals, this coyote is a placental mammal.

Figure 17 There are three different groups of mammals, but all have hair and nourish their young with milk.

Mammals

Maybe the main reason to go to the zoo is to see the mammals. Lemurs, lions, alpacas, and apes all are mammals. You are a mammal, too. All mammals have hair and **mammary glands,** *special tissues that produce milk for young mammals.* Like birds, mammals are endotherms. The hair of mammals is an adaptation that helps keep them warm. Milk production also is an adaptation. The milk helps the young grow and survive when they are too young to find their own food. There are three groups of mammals: monotremes, marsupials, and placental mammals, as shown in **Figure 17.**

Monotremes A few types of mammals lay eggs. When their young hatch, they are nourished by their mother's milk. These mammals include the platypus and the echidna.

Marsupials Mammals that raise their young in pouches are called marsupials. The young are not fully developed when born. After birth, they crawl through their mother's hair into a pouch. Here they can drink their mother's milk and continue to grow. Most marsupials are native to, or live in, Australia. Many marsupials resemble mammals that live in North America. There are marsupial squirrels, marsupial mice, and marsupial moles! The only marsupial native to North America is the opossum, shown in **Figure 17.**

Placental Mammals The last group of mammals is called placental mammals. They have a structure called a placenta that the young are attached to as they grow inside the mother. You are probably most familiar with different kinds of placental mammals, such as dogs, cats, horses, cows, and humans.

Key Concept Check How do the groups of vertebrate chordates differ?

Lesson 3 Review

Visual Summary

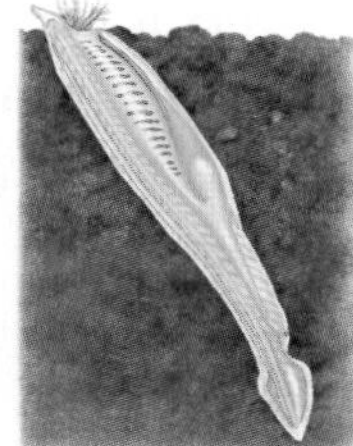

There are two types of chordates—vertebrate chordates and invertebrate chordates.

Invertebrate chordates do not have a backbone and include lancelets and tunicates.

Vertebrate chordates have a backbone and include fish, amphibians, reptiles, birds, and mammals.

FOLDABLES®

Use your lesson Foldable to review the lesson. Save your Foldable for the project at the end of the chapter.

What do you think NOW?

You first read the statements below at the beginning of the chapter.

5. All young mammals take in milk from their mothers.

6. Birds are the only animals that lay shelled eggs.

Did you change your mind about whether you agree or disagree with the statements? Rewrite any false statements to make them true.

Use Vocabulary

1. **Contrast** ectotherms and endotherms.
2. **Define** *pharyngeal pouches* and *gills.*
3. **Use the term** *amnion* in a sentence.

Understand Key Concepts 

4. Which is a characteristic all chordates have in common?

   A. amnion
   B. notochord
   C. mammary glands
   D. paired fins

5. **Summarize** the difference between invertebrate and vertebrate chordates.
6. **Explain** how amphibians are different from reptiles.

Interpret Graphics

7. **Summarize** Copy and fill in the graphic organizer below to summarize the four characteristics of all chordates.

Critical Thinking

8. **Assess** the benefits of the structure shown in the figure below.

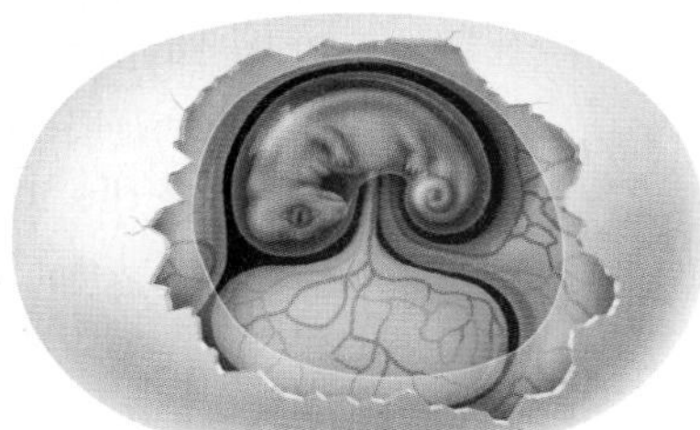

Math Skills 

 Math Practice

9. A wing bone from a flying bird has a mass of 1.8 g. The volume of the bone is 3.0 cm^3. What is the density of the bone?

Lab

2 class periods

Animal Adaptations

Did you know that some breeds of chickens are well-suited for living in warm environments, while others are better suited for living in cool environments? Chickens accustomed to a warmer climate tend to have few feathers near their legs and feet to help them stay cool. They also have large combs. These chickens are able to lose heat through their combs, much like we lose heat through our heads. Chickens accustomed to cooler climates have the opposite characteristics, feathery legs and feet and small combs. Most animals are well-suited for their environments. If you knew about a species' habitat and needs, what adaptations would help ensure the survival of the species?

Materials

number die

coin

drawing supplies

Safety

Question

What adaptations would help your animal survive in its environment?

Procedure

1. Read and complete a lab safety form.
2. Flip a coin to determine whether your animal is an invertebrate or a vertebrate. If you get heads, your animal will be an invertebrate. If you get tails, it will be a vertebrate. Record the result in your Science Journal.
3. Roll the die to determine your animal's habitat. If you roll a 1, your animal is aquatic. If you roll a 2, your animal is terrestrial. If you roll a 3, your animal lives in trees. If you roll a 4, your animal is primarily aerial and is able to fly. If you roll a 5 or a 6, roll again. Record the result.
4. Roll the die to determine what your animal eats. If you roll a 1 or a 2, your animal is an herbivore and eats only plants. If you roll a 3 or a 4, your animal is a carnivore and eats only other animals. If you roll a 5 or a 6, your animal is an omnivore and can eat both plants and other animals. Record the result.
5. Roll the die to determine your animal's body plan. If you roll a 1 or a 2, your animal has no appendages. If you roll a 3 or a 4, your animal can have 1–4 appendages. If you roll a 5 or a 6, your animal can have 5 or more appendages. Record the result.

6 Flip a coin to determine your animal's sleep pattern. If you get heads, your animal will be diurnal and is awake during the day. If you get tails, your animal will be nocturnal and is awake during the night. Record the result.

7 Review the list of restrictions/parameters that you have for your animal. Identify and write down physical features that you can give your animal to help them better survive where and how they live.

8 Using the materials provided, sketch a model of your animal and its adaptations. Write a paragraph explaining the adaptations your animal has and their usefulness.

Analyze and Conclude

9 **Predict** Choose a different habitat than the one you were assigned. Identify two things you would change about your animal's adaptations. Explain how the changes would make your animal better suited for its new habitat.

10 **Compare and Contrast** Choose an animal with adaptations designed by another classmate. Does it have any similar adaptations to your animal? Why do you think this is so?

11 **The Big Idea** How can adaptations be used to classify animals?

Communicate Your Results

Share your animal sketch with your classmates. Identify and explain the special adaptations that your animal has and why it has them.

Extension

Perform some additional research into periods of mass extinctions. Explore the adaptations, physical and behavioral, of the animals that survived during those periods, or create a three-dimensional model of your animal.

Lab Tips

- As you generate ideas for designing your animal's adaptations, think about how the animal is going to move around. How is it going to get its food? How is it going to consume its food? How is it going to obtain or build a shelter? How is it going to protect itself from predators? How big can it get where it lives?
- As you draw your animal, be sure to show the adaptations in the picture.

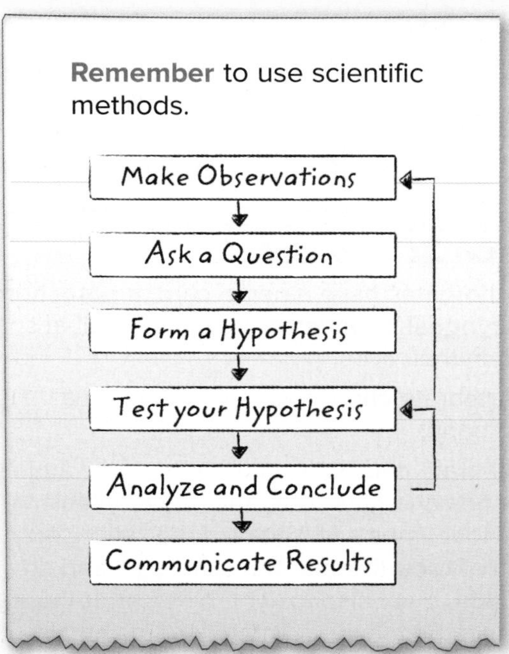

Chapter 9 Study Guide

Animals are multicellular, have muscle cells and nerve cells, cannot make their own food, and ingest their food. Animals are grouped by how they develop and by what type of symmetry they display.

Key Concepts Summary

Lesson 1: What are animals?

- All animals are multicellular, have collagen protein to hold their cells together, have muscle and nerve cells, and cannot make their own food.
- Animals are grouped by what kinds of symmetry they have and by how they develop.
- Animal species are adapted to their environments by having unique structures, exhibiting specialized behaviors, and reproducing by certain methods.

Vocabulary

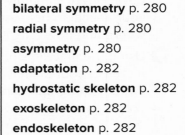

bilateral symmetry p. 280
radial symmetry p. 280
asymmetry p. 280
adaptation p. 282
hydrostatic skeleton p. 282
exoskeleton p. 282
endoskeleton p. 282

Lesson 2: Invertebrates

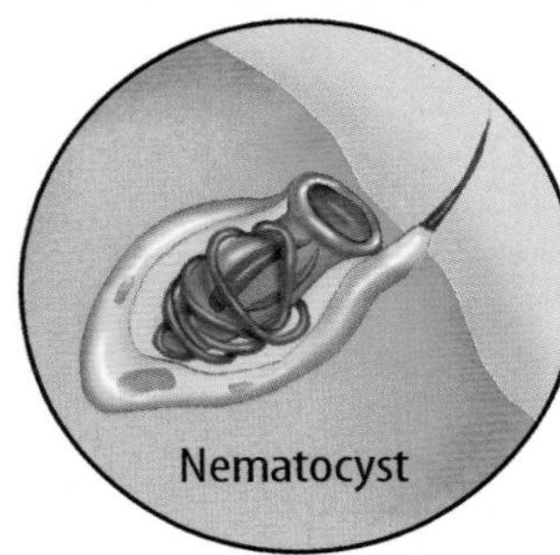

- Invertebrates do not have backbones, and instead they have hydrostatic skeletons, endoskeletons, or exoskeletons.
- Invertebrates differ in how they develop as embryos, how they support their soft bodies, how they live and obtain food, and in their symmetry.

Vocabulary

parasite p. 287
mantle p. 290
molting p. 290
metamorphosis p. 291

Lesson 3: Chordates

- All chordates have a nerve cord, a **notochord, pharyngeal pouches** or slits, and a tail at some time in their lives.
- Invertebrate chordates never develop a backbone.
- Vertebrate chordates can be fish, fully aquatic with fins and **gills**. They may be amphibians with four limbs and aquatic fertilization; or reptiles, **ectotherms** with scaly skin and an **amnion**. They also may be birds, **endotherms** with feathers and an amnion; or mammals, endotherms with an amnion, hair, and mammary glands.

Vocabulary

notochord p. 296
pharyngeal pouch p. 296
gill p. 298
amnion p. 300
ectotherm p. 300
endotherm p. 301
mammary gland p. 302

Study Guide

Vocabulary eFlashcards
Vocabulary eGames

Personal Tutor

Chapter Project

Assemble your lesson Foldables as shown to make a Chapter Project. Use the project to review what you have learned in this chapter.

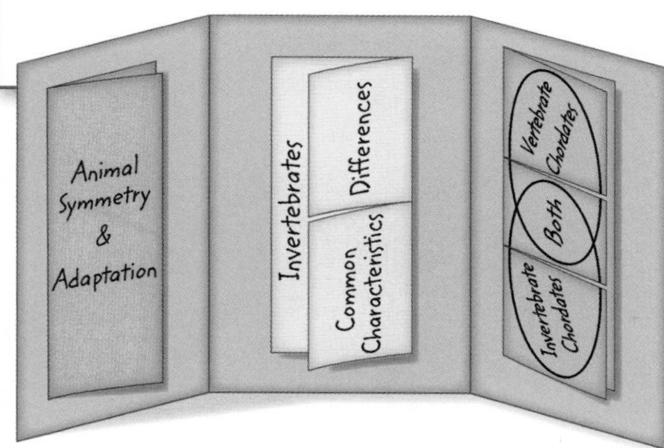

Use Vocabulary

1. Use *molting* in a sentence.
2. Define the term *adaptation* in your own words.
3. Use *hydrostatic skeleton* in a sentence.
4. Animals that heat their bodies from heat in their environments are called ________.
5. An animal that lives inside another animal, gets food from the other animal, and does not help it survive is called a(n) ________.
6. Special tissue that produces milk for young animals is called ________.

Link Vocabulary and Key Concepts

Interactive Concept Map

Copy this concept map, and then use vocabulary terms from the previous page to complete the concept map.

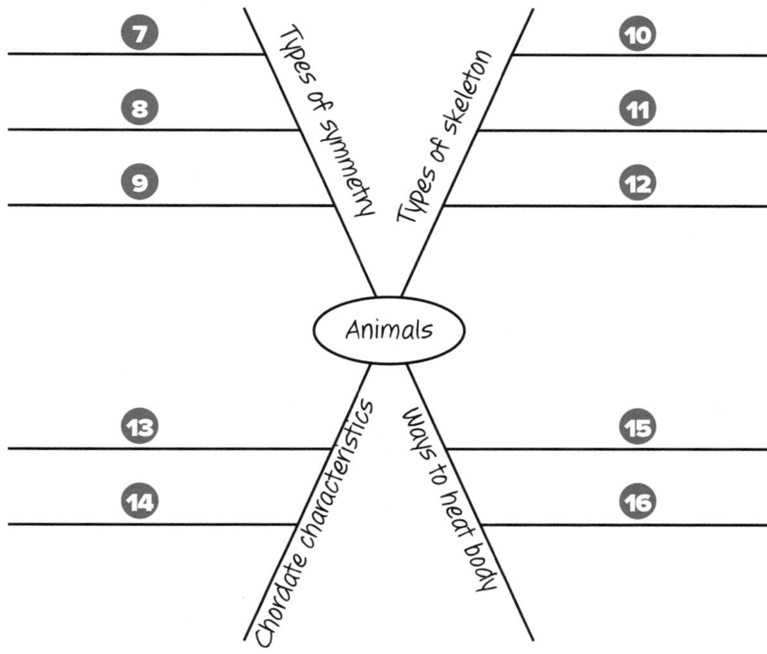

Chapter 9 Review

Understand Key Concepts 

1. Which is a characteristic common to all animals?
   A. asymmetry
   B. collagen
   C. endothermy
   D. exoskeleton

2. Which is NOT a type of body plan in animals?
   A. asymmetry
   B. ectothermy
   C. bilateral symmetry
   D. radial symmetry

3. Which best describes an adaptation?
   A. A trait that has no effect on the survival of an individual.
   B. A trait that makes it difficult for an individual to survive.
   C. A trait that makes a population a better match for its environment and helps it survive.
   D. A trait that shows up in an individual but is not passed on to its offspring.

4. Examine the branching tree diagram below.

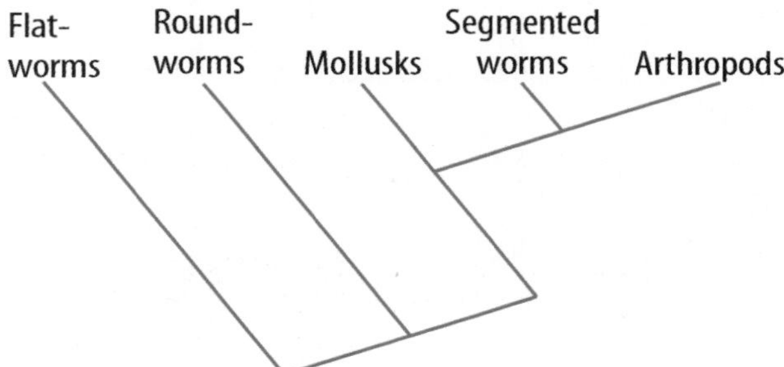

   Which group of animals is most closely related to segmented worms?
   A. arthropods
   B. flatworms
   C. mollusks
   D. roundworms

5. Which is true of invertebrates?
   A. They have no backbones.
   B. They have no muscle tissue.
   C. They have no nervous tissue.
   D. They have no skeletons.

6. In which group would you place the animal pictured below?

   A. arthropods
   B. echinoderms
   C. flatworms
   D. sponges

7. Which is NOT a typical chordate characteristic?
   A. nerve cord
   B. notochord
   C. scales
   D. tail

8. Which chordate might be confused with an invertebrate?
   A. marsupial
   B. sea horse
   C. snake
   D. tunicate

9. Which structural adaptation made it possible for vertebrates to reproduce on land?
   A. exoskeleton
   B. spiracles
   C. amniotic egg
   D. pharyngeal pouches

Don Farrall/Getty Images

Critical Thinking

10 **Evaluate** the importance of the amnion to life on land.

11 **Compare** ectotherms and endotherms.

12 **Summarize** the characteristics that are common to all chordates.

13 **Evaluate** why so many vertebrate animals have jaws.

14 **Infer** how you, a human, can be classified as a chordate when you have no tail.

15 **Justify** why the organism shown below is considered an animal.

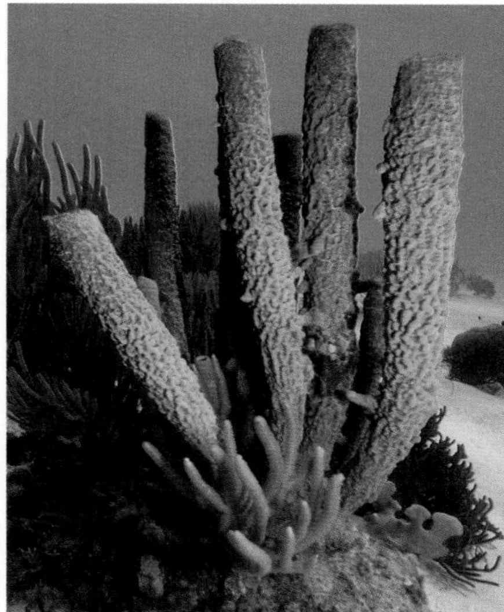

16 **Assess** the role of endothermy in an animal's ability to survive and reproduce.

17 **Formulate** a theory of why mammals are able to live in so many different kinds of environments.

Writing in Science

18 **Write** a five-sentence paragraph comparing the adaptation of feathers in birds to the adaptation of hair in mammals. Be sure to include a topic sentence and a concluding sentence in your paragraph.

REVIEW THE BIG IDEA

19 **Assess** how you would determine whether a new organism you found was an animal, and how you would decide what kind of animal it was.

20 What are animals and how are they classified? Use the animals in the photo below as examples.

Math Skills

Math Practice

Use a Formula

The table below shows the mass and the volume of several sample leg bones. Use the table to answer questions 21 and 22.

Animal	Mass (g)	Volume (cm^3)
Giraffe	76.8	40
Human	46.25	25

21 What is the density of the giraffe's leg bone?

22 What is the density of the human's leg bone?

23 A chicken's leg bone has a density of about 1.11 g/cm^3.

a. Would you expect the density of a buffalo's leg bone to be greater or less than a chicken's leg bone? Explain.

b. A buffalo's leg bone with a volume of 35 cm^3 has a mass of 68.6 g. What is the density of the buffalo's leg bone?

Standardized Test Practice

Record your answers on the answer sheet provided by your teacher or on a sheet of paper.

Multiple Choice

1 Which is a similarity between plants and animals?

A Both have cells with cell walls.

B Both have cells with nuclei.

C Both have nerve cells.

D Both use light to make energy.

2 Which is the main difference between vertebrates and invertebrates?

A Invertebrates have backbones, and vertebrates do not.

B Invertebrates only live in water, and vertebrates live in many environments.

C There are many more vertebrate species than invertebrates.

D Vertebrates have backbones, and invertebrates do not.

3 The human brain and spinal cord develop from which structure?

A nerve cord

B notochord

C pharyngeal pouch

D tail

Use the diagram to answer question 4.

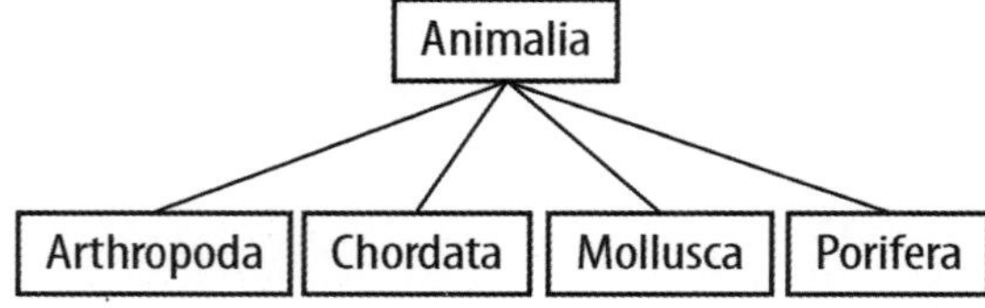

4 What taxon level is Porifera?

A genus

B kingdom

C phylum

D species

5 Which characteristic is found only in mammals?

A Mammals are endotherms and generate body heat.

B Mammals have lungs for respiration on land.

C Mammals produce milk for their young.

D Mammals use amniotic eggs for reproduction.

Use the diagram to answer question 6.

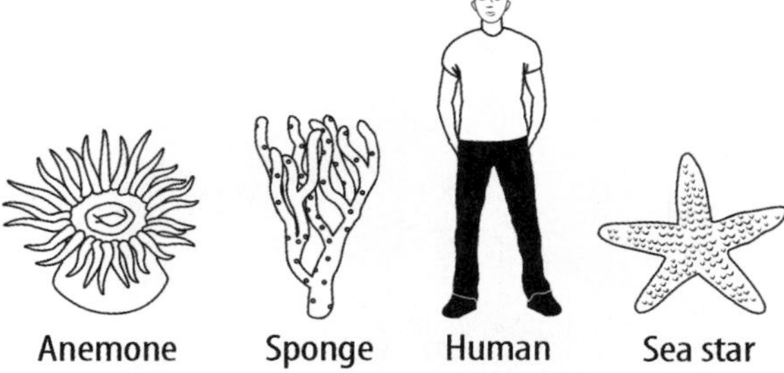

6 Which animal is an example of bilateral symmetry?

A anemone

B human

C sea star

D sponge

7 Which is NOT a characteristic of arthropods?

A Arthropods have hard exoskeletons.

B Arthropods have blood vessels.

C Arthropods are a very large group of species.

D Arthropods have a head, a thorax, and an abdomen.

8 Which is an example of a behavioral adaptation?

A a hydrostatic skeleton

B an open circulatory system

C external fertilization in water

D waving wings to attract a mate

Use the diagram to answer question 9.

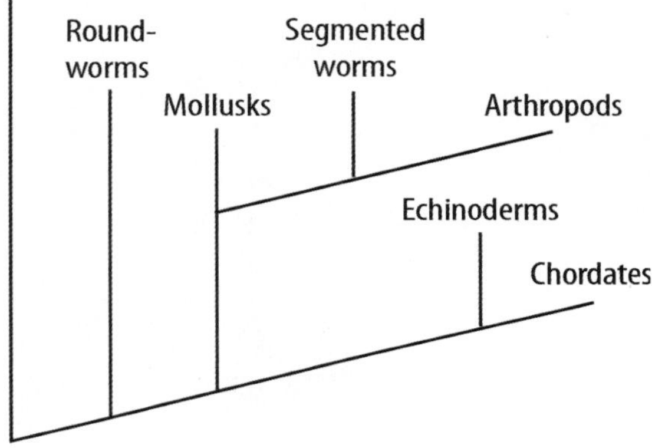

9 Which invertebrate phylum is most closely related to chordates?

A arthropods

B echinoderms

C mollusks

D sponges

10 Why is a lancelet classified as an invertebrate chordate?

A It can swim.

B It does not have a backbone.

C It lives near the ocean.

D It lives under the ground.

Constructed Response

11 What type of habitat do amphibians need to survive? What change in that habitat would cause amphibian populations to decline?

12 An explorer discovers a new species on the sea floor. The explorer isn't sure if this discovery is an animal. What would he or she look for to classify this discovery as an animal?

Use the diagram to answer question 13.

13 Based on the physical characteristics of this organism, what phylum could it belong to? Explain why you made this choice.

14 Name an adaptation of an animal you are familiar with. How does this adaptation help the animal survive or reproduce? Is the adaptation structural, behavioral, or functional? Explain why.

NEED EXTRA HELP?														
If You Missed Question...	1	2	3	4	5	6	7	8	9	10	11	12	13	14
Go to Lesson...	1	2	3	1	3	1	2	1	1	3	3	1	2	1

Chapter 10

Interactions of Life

THE BIG IDEA

How do living things interact with each other and the environment?

What is it doing?

This raccoon has found its way into a campsite. While there, it might interact with humans in several ways, including taking food from the campsite or moving items, such as the cookware shown here.

- In what other ways might living things interact with humans?
- How do living things, such as this raccoon, interact with other living things and the environment?

Get Ready to Read

What do you think?

Before you read, decide if you agree or disagree with each of these statements. As you read this chapter, see if you change your mind about any of the statements.

1. An ecosystem is all the animals that live together in a given area.
2. A layer of decayed leaves that covers the soil in a forest is an example of a living factor.
3. A niche is the place where an animal lives.
4. Symbiosis is a close relationship between two species.
5. Energy from sunlight is the basis for almost every food chain on Earth.
6. A plant creates matter when it grows.

Your one-stop online resource
connectED.mcgraw-hill.com

LearnSmart®

Chapter Resources Files, Reading Essentials, Get Ready to Read, Quick Vocabulary

Animations, Videos, Interactive Tables

Self-checks, Quizzes, Tests

Project-Based Learning Activities

Lab Manuals, Safety Videos, Virtual Labs & Other Tools

Vocabulary, Multilingual eGlossary, Vocab eGames, Vocab eFlashcards

Personal Tutors

Lesson 1

Ecosystems

Reading Guide

Key Concepts

ESSENTIAL QUESTIONS

- How can you describe an ecosystem?
- What are the similarities and differences between the abiotic and biotic parts of an ecosystem?
- In what ways can populations change?

Vocabulary

ecosystem p. 315
abiotic factor p. 316
biotic factor p. 317
habitat p. 318
population p. 319
community p. 319
population density p. 320

Multilingual eGlossary

BrainPOP®

Inquiry What lives here?

Look at all the organisms in this picture. This coastal reef provides a place for many organisms to live. How do you think each of these organisms survives? How do you think they interact with each other and the environment?

Launch Lab

10 minutes

What is an environment?

Earth contains many different environments. What makes these environments different?

1. Select a **postcard** from the ones provided. Look at the picture of the location on it.
2. Plan a vacation to the location on your postcard. In your Science Journal, describe the type of transportation you will need to use to get there, the clothing and other accessories you will need to pack, and the types of activities you will participate in at the location.
3. Write a note to a friend describing the environment that you might visit on your vacation. Describe the living and nonliving things you might find.

Think About This

1. How is your vacation environment different from the one where you live?
2. What types of living organisms will you see in your vacation environment? How are these organisms suited to this environment? What organisms are not suited to your vacation environment? Explain.
3. **Key Concept** What are the different ways you can describe the environment of your vacation spot?

What is an ecosystem?

Imagine that you are visiting a park. You sit on the grass in the warm sunshine. You watch a squirrel run down a tree trunk and chew an acorn. A robin pulls an earthworm from the soil. Traveling in a line, ants carry bits of dead insects to their underground nest. A breeze blows dandelion seeds through the air. These interactions are just a few of the many that can happen in an ecosystem, such as the park shown in **Figure 1**. *An* **ecosystem** *is all the living things and nonliving things in a given area.*

There are many kinds of ecosystems on Earth, including forests, deserts, grasslands, rivers, beaches, and coral reefs. Ecosystems that have similar climates and contain similar types of plants are grouped together into biomes. For example, the tropical rain forest biome includes ecosystems full of lush plant growth located near the equator in places where rainfall averages 200 cm per year and the temperature averages 25°C.

Key Concept Check How can you describe an ecosystem?

Figure 1 Ecosystems include the interactions among organisms and their environment.

Make a vertical shutterfold book. Label it as shown. Use it to organize your notes about abiotic and biotic factors.

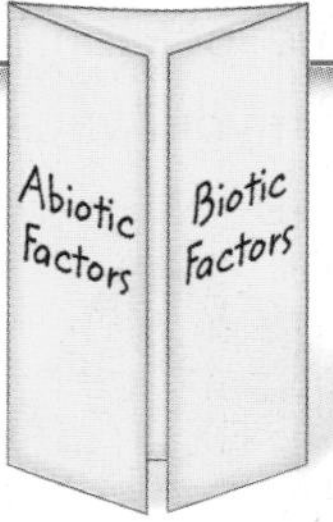

Abiotic Factors

The nonliving parts of an ecosystem are called **abiotic factors.** They include sunlight, temperature, air, water, and soil. Abiotic factors provide many of the resources organisms need for survival and reproduction.

Sunlight and Temperature Sunlight is essential for almost all life on Earth. It supplies the energy for photosynthesis–the chemical reactions that produce sugars and occur in most plants and some bacteria and protists.

Sunlight also provides warmth. An ecosystem's temperature depends in part on the amount of sunlight it receives. In some ecosystems, such as the hot, dry desert shown in **Figure 2,** temperatures can be around 49°C during the day and below freezing at night.

REVIEW VOCABULARY

atmosphere
the whole mass of air surrounding Earth

Atmosphere The gases in Earth's atmosphere include nitrogen, oxygen, and carbon dioxide. Nitrogen is needed for plant growth. Some bacteria in the soil take nitrogen from the air and convert it to a form that plants can use. Oxygen is needed by most organisms for cellular respiration–the process that releases energy in cells. Air also contains carbon dioxide that is needed for photosynthesis.

Figure 2 Abiotic factors in an ecosystem determine what kinds of organisms can live there.

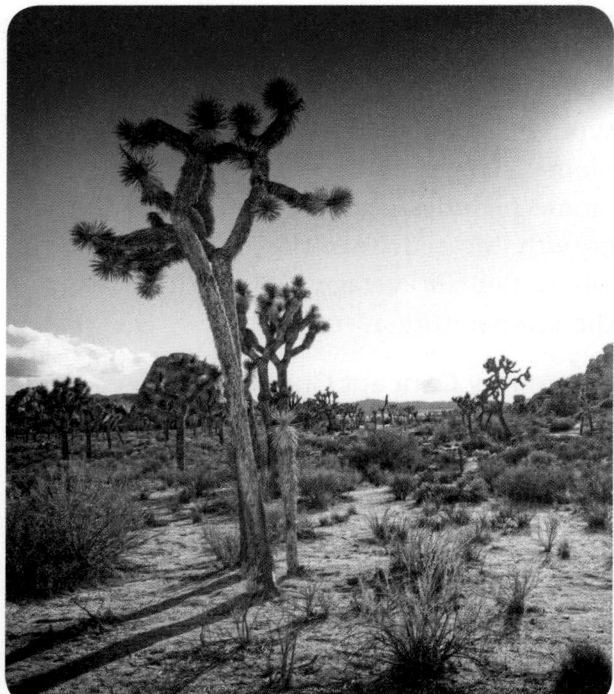

Desert life is limited to organisms that can survive with little water. Shade from plant life provides shelter from the heat of the Sun.

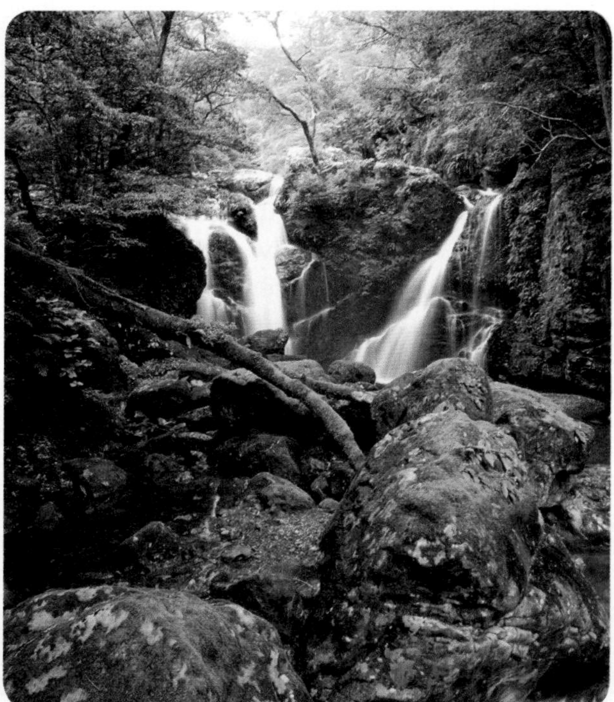

Stream life depends on a constant supply of water. Plants along stream banks provide shelter and food for hundreds of species.

(l)Silvia Otte/Getty Images, (r)Image Source/PunchStock

Water Without water, life would not be possible. Water is required for all the life processes that take place inside cells, including cellular respiration, digestion, and photosynthesis. The stream ecosystem shown in **Figure 2** can support many forms of life because water is plentiful. Areas with very little water support fewer organisms.

Soil If you ever have planted a garden, you might know about the importance of soil for healthy plants. Soil contains a mixture of living and nonliving things. The biotic part of soil is humus (HEW mus)–the decayed remains of plants, animals, bacteria, and other organisms. Deserts have thin soil with little humus. Forest soils usually are thick and fertile, with a higher humus content. Abiotic factors include minerals and particles of rock, sand, and clay. Many animals, including gophers, insects, and earthworms, such as those shown in **Figure 3**, make their homes in soil. Their tunnels help move water and air through the soil.

Figure 3 Plant roots and soil-dwelling organisms need oxygen and water. Gophers, earthworms, and insects loosen the soil. This helps move air and water through soil.

Biotic Factors

Living or once-living things in an ecosystem are called **biotic factors.** They include all living organisms–from the smallest bacterium (plural, bacteria) to the largest redwood tree. Biotic factors also include the remains of dead organisms, such as fallen leaves or decayed plant matter in soil.

Species are adapted to the abiotic and biotic factors of the ecosystems in which they live. Algae, fungi, and mosses live in moist ecosystems such as forests, ponds, and oceans. Many cactus species can survive in a desert because they have thick stems that can hold stored water. Gophers live in burrows underground. They have large front claws for digging and strong teeth for loosening soil and chewing plant roots.

Key Concept Check What are the similarities and differences between abiotic and biotic factors?

WORD ORIGIN

biotic
from Greek *bios,* means "life"

MiniLab 20 minutes

How do you describe an ecosystem? 

An ecosystem can be a small area or a very large environment. What makes up the ecosystem around your school yard?

1. Read and complete a lab safety form.
2. Walk around the outside of your school. Note the general appearance of your ecosystem.

3. In your Science Journal, list the abiotic factors you find around your ecosystem. Describe as many of the abiotic factors, in as much detail, as you can.
4. List and describe the biotic factors in your ecosystem. Examine the soil and the ground area closely for smaller organisms.
5. In your Science Journal, illustrate your ecosystem.

Analyze and Conclude

1. **Summarize** the interactions that occur between biotic and abiotic factors within your ecosystem.
2. **Define** the populations within your ecosystem.
3. **Describe** the community that inhabits your ecosystem.
4. **Key Concept** Diagram your ecosystem in terms of populations, community, and abiotic factors.

Habitats

Every organism in an ecosystem has its own place to live. *A* **habitat** *is the place within an ecosystem that provides food, water, shelter, and other biotic and abiotic factors an organism needs to survive and reproduce.*

Organisms have a variety of habitats. For example, house martins such as the one shown in **Figure 4** sometimes live in meadows or grasslands, but this bird has found a habitat under the eaves of a building. Crickets live in damp, dark places with plenty of plant material and fungi to eat. Skunks live in areas where they can find food such as mice, insects, eggs, and fruit. During the day, skunks take shelter near their food supply–in hollow logs, under brush piles, and underneath buildings.

Plants have their own habitats, too. You have read that cacti live in desert habitats. The wood sorrel is a plant species that grows in deep shade beneath redwood trees.

When biotic or abiotic factors in an ecosystem change, habitats can change or disappear. A wildfire quickly can destroy the habitats of thousands of animals that live in forests or grasslands. Erosion or flooding can wash away soil, destroying plant habitats.

Reading Check What is a habitat?

Figure 4 An organism's habitat provides shelter, food, and all the other resources it needs for survival.

Populations

Every ecosystem includes many individuals of many species. *A* **population** *is all the organisms of the same species that live in the same area at the same time.* For example, all the dandelions growing in a vacant lot form a population. All the ants in the vacant lot make up another population. *All the populations living in the same area at the same time form a* **community.** As shown in Figure 5, a vacant-lot community might include populations of grasses, dandelions, spiders, ants, and pigeons. A community combined with all the abiotic factors in the same area forms an ecosystem. The populations that make up the community interact in the ecosystem.

Figure 5 The community living in this vacant lot includes populations of dandelions, grasses, ants, spiders, and pigeons.

Visual Check What abiotic factors are included in this ecosystem?

Figure 6 Guppies are easy to care for, but overcrowding can lead to disease.

Math Skills 

Use a Formula

A formula shows the relationship among several factors. The formula for population density determines the number of individuals in a unit area or a volume of space. Population density = number of individuals / **unit area or volume of space.** For example, what is the population density of insects if 140 insects are found in a patch of ground measuring **3.0 m²**?

1. Replace the terms in the equation with the given values.

   Population density = $\frac{\text{140 insects}}{\text{3.0 m}^2}$

2. Solve the problem.

   $\frac{\text{140 insects}}{\text{3.0 m}^2} = \frac{\text{46.6 insects}}{\text{m}^2}$

3. Round the answer to significant figures.

   $\frac{\text{46.6 insects}}{\text{m}^2} = \frac{\text{47 insects}}{\text{m}^2}$

Practice

There are 20 small tropical fish swimming in a 55-gallon aquarium. What is the population density?

 Math Practice

 Personal Tutor

Population Density

Suppose your classroom has an aquarium like the one shown in **Figure 6.** It contains guppies, water ferns, and a few algae-eating snails. Keeping your aquarium community healthy includes cleaning the tank and feeding the fish. However, it also means making sure the fish don't get overcrowded. Overcrowding can lead to stress and disease.

How can you determine if the aquarium contains too many fish? You could calculate the population density. **Population density** *is the size of a population compared to the amount of space available.* It can be calculated using the following formula:

$$\text{Population density} = \frac{\text{number of individuals}}{\text{unit area or volume of space}}$$

An aquarium expert has recommended that you keep no more than 10 guppies in your 20-gallon aquarium. Using the formula, you can calculate the population density:

$$\text{0.5 fish per gallon} = \frac{\text{10 fish}}{\text{20 gallons}}$$

When population density is high, organisms live closer together and might not be able to obtain all the resources needed for life. Diseases also spread more easily when organisms are forced to live too close together.

 **Reading Check** How does population density affect organisms?

Population Change

On a hike one summer, you notice a few wild sunflowers growing among grasses in an abandoned field. Two years later you return and find the field completely covered with sunflowers. What caused the population to increase? Each sunflower plant produces hundreds of seeds. Even if only a few of the seeds from each plant sprout and grow, the number of sunflowers will increase. If a drought prevents seeds from sprouting or if a farmer plants the field with corn, soybeans, or another crop, the sunflower population will decrease.

Most populations change over time. Production of offspring increases the size of a population. The death of individuals reduces population size. If births outnumber deaths, the population grows.

Changes in the abiotic or biotic factors in an ecosystem can cause organisms to move away or die out. For example, if there is a forest fire, birds, deer, and other fast-moving animals can escape to another area. Others, such as the mountain beaver in **Figure 7**, could die out.

Key Concept Check In what ways can populations change?

Figure 7 Before fire swept through this forest ecosystem, it provided a habitat for about 5,000 mountain beavers. Fewer than 100 beavers survived the fire, but their habitat slowly grew back and the mountain beaver population increased again.

Mountain beaver

(t)Tom McHugh/Photo Researchers, (b)Gary Fellers/USGS

Lesson 1 Review

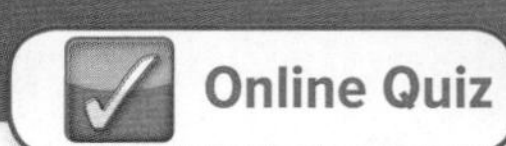

Visual Summary

Abiotic factors include sunlight, temperature, air, water, and soil.

Many organisms often share the same habitat.

Population density can affect the health of a population.

FOLDABLES®

Use your lesson Foldable to review the lesson. Save your Foldable for the project at the end of the chapter.

What do you think NOW?

You first read the statements below at the beginning of the chapter.

1. An ecosystem is all the animals that live together in a given area.
2. A layer of decayed leaves that covers the soil in a forest is an example of a living factor.

Did you change your mind about whether you agree or disagree with the statements? Rewrite any false statements to make them true.

Use Vocabulary

1. **Define** *ecosystem* in your own words.
2. **Distinguish** between a population and a community.
3. An organism lives in a(n) ________.

Understand Key Concepts

4. Air is an example of a(n)
   - **A.** community.
   - **B.** habitat.
   - **C.** abiotic factor.
   - **D.** biotic factor.
5. **Explain** why soil is considered both an abiotic and a biotic factor.

Interpret Graphics

6. **Classify** factors in the desert ecosystem shown at right as abiotic or biotic.

7. **Summarize Information** Copy the graphic organizer below and fill in the ways in which populations can change.

Critical Thinking

8. **Analyze** All organisms need living space to survive. Would you consider living space to be a biotic factor, an abiotic factor, or both? Explain your answer.

Math Skills

9. There are four foxes in 10 km^2 of a forest.
   - **a.** What is the population density?
   - **b.** How many foxes would you expect to find in 50 km^2?

20 minutes

What can analyzing data reveal about predator-prey populations?

Predators are organisms that hunt and kill other organisms, their prey, for food. You might think this means that predators control population size in an ecosystem. In reality, predators and their prey affect each other. You can observe this relationship by analyzing population data.

Learn It

Scientists make observations to learn about the world. However, scientists rely on data as a means to present explanations and prove hypotheses. They can **analyze data** from the results and form conclusions. Unlike general observations, collecting data is effective because data are less likely to be distorted by chance or misinterpretation.

Try It

1. Observe the data table showing the populations of hares and lynxes in an ecosystem. Note that the data show the changes that occur in the numbers of both populations over time.
2. Examine the blue line that represents the population levels of hares in the ecosystem. Note any patterns in the graph.
3. Examine the orange line that represents the population levels of lynxes in the ecosystem. Note any patterns in the graph.
4. Describe the patterns of the two population lines in relation to each other.

Apply It

5. **Explain** which population is the predator and which is the prey.
6. **Describe** the patterns you note in the individual population levels of the hares and the lynxes.
7. **Describe** the patterns you notice in the relationship between both populations. What might have caused these patterns?
8. **Infer** why there is a time lag between similar patterns in both populations.
9. **Key Concept** Use the information from this graph and infer what causes the changes in both populations. Use this information to describe the relationship that each population has with the other.

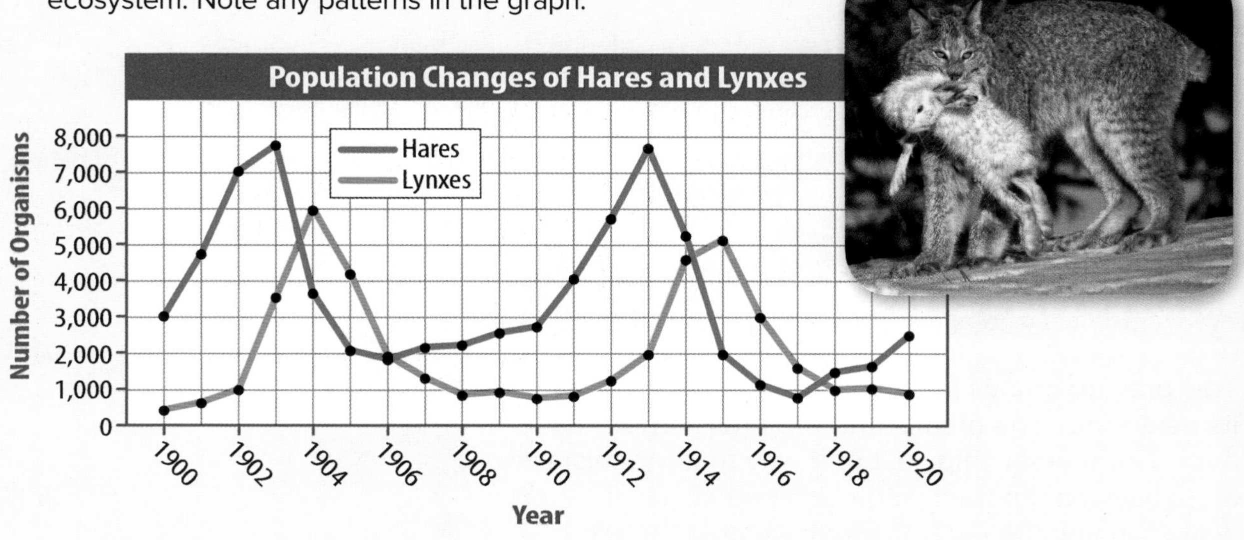

Lesson 2

Reading Guide

Key Concepts

ESSENTIAL QUESTIONS

- How does a niche differ from a habitat?
- In what ways can organisms interact in an ecosystem?

Vocabulary

niche p. 325
competition p. 326
overpopulation p. 327
predation p. 328
symbiosis p. 329
mutualism p. 329
commensalism p. 329
parasitism p. 329

Multilingual eGlossary

What's Science Got to do With It?

Go to the resource tab in ConnectED to find the PBLs *The Fox and the Hare* and *The Hungry Games: Eat or Be Eaten.*

Relationships Within Ecosystems

Inquiry What's it doing?

This praying mantis has captured a grasshopper for its next meal. The mantis and the grasshopper have a feeding relationship, just one way that organisms in an ecosystem interact with one another. What other ways can you think of for organisms to interact?

Launch Lab

10 minutes

How do organisms help plants grow?

Different organisms often rely on other organisms to survive. One such relationship exists between types of fungi and plants. The fungi live on the roots of plants and help them obtain water and minerals. In return, the plants supply the fungi with nutrients.

1. Examine the photograph of plant roots. The roots on the right have had a beneficial fungus added to them. The roots on the left were grown in sterilized soil. Record your observations in your Science Journal.

Think About This

1. What difference do you note in the plant roots grown with a fungus added to the soil compared to the plant roots grown in sterile soil?
2. Which plants do you think might grow larger or faster?
3. **Key Concept** How do you think the relationship between these organisms helps them use the resources in their environment?

Niches

Recall that a habitat is the area within an ecosystem that provides an organism with the resources it needs for life. Most organisms don't have a habitat all to themselves. A habitat usually is shared by many species.

Hundreds of species share the coral reef habitat shown in **Figure 8.** Spiny lobsters hide under the coral. They come out at night and feed on worms, shrimp, clams, and dead fish. Angelfish have rough teeth for scraping sponges and sea squirts from the surface of the coral. Filefish scrape algae from the coral to eat.

Each species that shares a habitat has a separate niche. *A* **niche** (NICH) *is the way a species interacts with abiotic and biotic factors to obtain food, find shelter, and fulfill other needs.* Species share habitats, but no two species share the same niche. For example, two species of crabs on a reef might share a habitat, but one might eat algae and the other might eat snails.

Key Concept Check How does a niche differ from a habitat?

Figure 8 Many organisms, such as the ones that live in this coral reef, share the same habitat.

adokon/Getty Images

Figure 9 Organisms in the same area sometimes compete for the same resources.

Competition

In springtime, robins find mates, build nests, and raise their young. A male robin chooses a safe nesting site with plenty of food and water nearby. It sings to attract a female and to keep other males away. If another male comes too close to its territory, it chases its competitor away.

Competition *describes the demand for resources, such as food, water, and shelter, in short supply in a community.* Competition can take place among the members of a population or between populations of different species. The plants shown in **Figure 9** are competing for nutrients and living space.

Competition helps limit population size. If a community has too many robins and too few nesting sites, competition for these sites increases, and some robins will leave the area. The availability of nesting sites limits the size of the robin population.

 **Reading Check** What is competition?

MiniLab

10 minutes

How does competition affect the growth of radish plants?

All living things must obtain resources in order to survive. In any environment, there is a limited amount of resources available. How does the number of organisms in an environment affect the ability of a specific organism, such as a plant, to obtain the resources it needs?

1. Read and complete a lab safety form.
2. Fill one **small planting pot** with **potting soil.** Place five **radish seeds** on the surface, and cover the seeds with a thin layer of potting soil. Water the soil so that it is damp.
3. Fill a second small planting pot with potting soil. Add a small patch of **sod** to the surface of the soil. Place five radish seeds along the surface of the sod and cover the seeds with a thin layer of potting soil. Water the soil so that it is damp.
4. Continue to water and observe the two pots for several days. Note any observations in your Science Journal.

Analyze and Conclude

1. **Compare and contrast** the number and size of radish seedlings in both pots after several days.
2. **Key Concept** Infer how competition between plant species affects the growth of all plants in an environment.

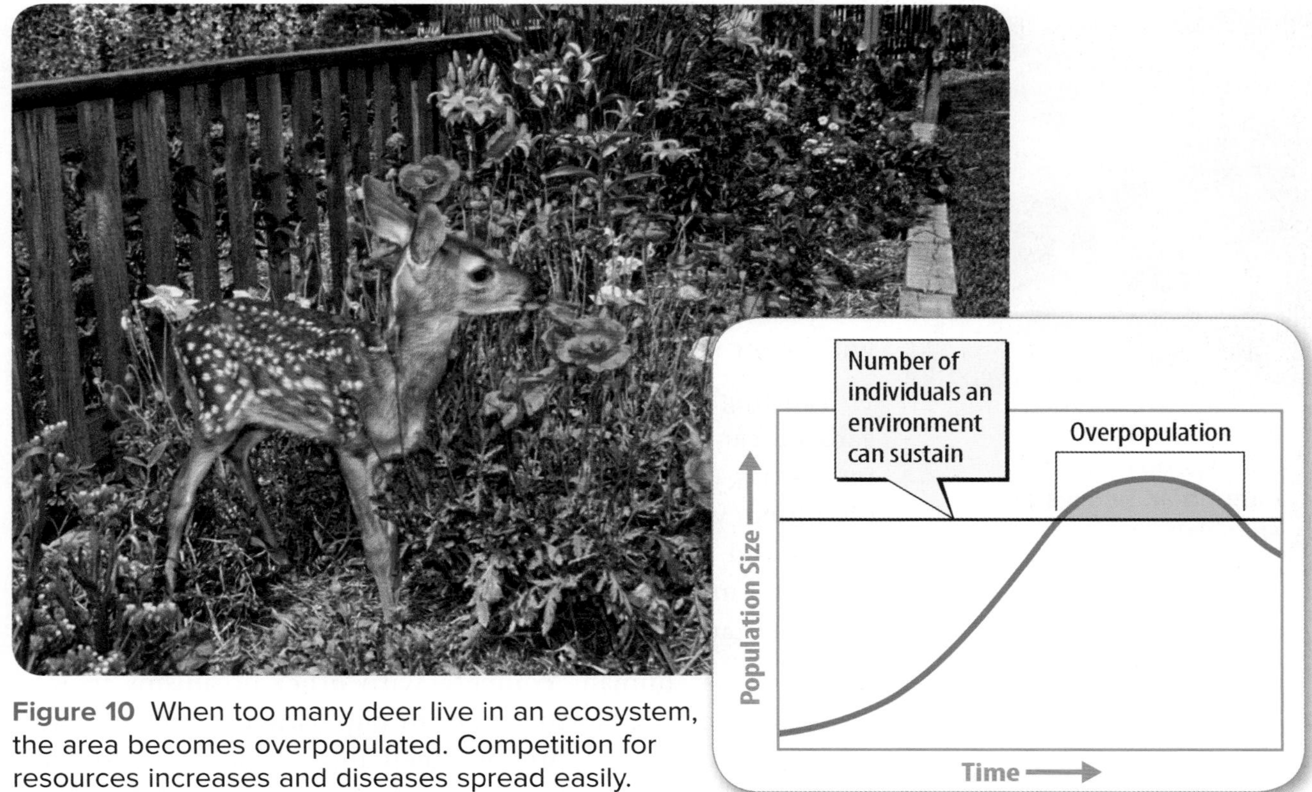

Figure 10 When too many deer live in an ecosystem, the area becomes overpopulated. Competition for resources increases and diseases spread easily.

Visual Check How might this deer affect this garden?

Overpopulation

White-tailed deer live near the edges of forests and meadows. They eat leaves, twigs, acorns, and fruit. Deer populations in some areas have become so large that they harm forest habitats, destroy crops, and even invade home gardens. **Overpopulation** *occurs when a population becomes so large that it causes damage to the environment.*

When too many deer live in an ecosystem, they eat plants at a faster rate than the plants can grow back. This reduces the available habitat for the deer and other species. The deer, as well as other organisms in the area, must compete for a limited amount of resources. Sometimes the deer move into areas where they are not normally found, such as the deer pictured in a home garden in Figure 10. If there is nowhere for deer to move, they are forced to live too close together. Disease can spread easily within populations when this happens.

Overpopulation is temporary. When food and other resources eventually run out, some animals will move elsewhere, starve, or die from disease. Then the population quickly shrinks, as shown in the graph in Figure 10. This allows the resources in the environment to slowly return to normal.

Reading Check Why is overpopulation temporary?

▲ **Figure 11** When humans build roads through an ecosystem, animals face new dangers.

Competing with Humans

Humans need some of the same biotic and abiotic factors as other organisms, including food, living space, and water. To meet these needs, people take certain actions. They plow grasslands to plant food crops. People clear forests and fill in wetlands to make room for roads and buildings. They divert water from lakes and streams to supply irrigation for crops and drinking water for cities and towns. Actions such as these put humans in competition with other species for the same resources.

You might have heard news reports about raccoons raiding garbage cans, snakes living under houses, or squirrels moving into attics. Natural habitats for these and other organisms are disrupted when humans replace natural environments with homes and other structures. As shown in **Figure 11**, roads can make it dangerous for animals to move safely from one part of their habitat to another.

Sometimes humans compete with other organisms in less-obvious ways. The North American population of monarch butterflies spends the winter in small forested areas in Mexico. Logging by humans endangers the monarch population. Without enough trees to live in, many monarchs do not survive for the return trip north in spring.

Predation

A predator is an organism that hunts and kills other organisms for food. Prey are the organisms hunted or eaten by a predator. **Predation** *is the act of one organism, a predator, feeding on another organism, its prey,* as shown in **Figure 12**. Predator and prey populations influence each other, as you learned in the Skill Practice lab. Predators help control the size of prey populations. When prey populations decrease, the number of predators usually decreases because less food is available.

Figure 12 In the relationship pictured here, the beetle is the predator and the caterpillar is the prey. ▶

Symbiosis

Competition and predation are two types of interactions that take place between organisms in an ecosystem. Another type of interaction that occurs is called symbiosis (sim bee OH sus). **Symbiosis** *is a close, long-term relationship between two species that usually involves an exchange of food or energy.* Examples of the three types of symbiosis are shown in **Figure 13.**

A symbiotic relationship in which both organisms benefit is **mutualism.** For example, fish benefit by having tiny organisms removed from their bodies by cleaner shrimp, and cleaner shrimp benefit by getting food. *A symbiotic relationship in which one organism benefits but the other neither benefits nor is harmed is* **commensalism.** Clumps of moss growing on the bark of a tree is an example of a commensal relationship. The moss benefits by having somewhere to grow, and the tree is neither benefited nor harmed. *A symbiotic relationship in which one organism benefits while the other is harmed is* **parasitism.** The organism that benefits is a parasite. For example, a parasitic wasp lays its eggs in a caterpillar's body. When the eggs hatch, the larvae develop and eventually chew their way out of the caterpillar and kill it. The organism that is harmed is the host, in this case the caterpillar that was attacked by the wasp.

Key Concept Check In what ways can organisms interact in an ecosystem?

WORD ORIGIN

symbiosis
from Greek *symbios,* means "living together"

FOLDABLES®

Make a horizontal three-tab concept map book. Label it as shown. Use it to organize your notes on symbiotic relationships.

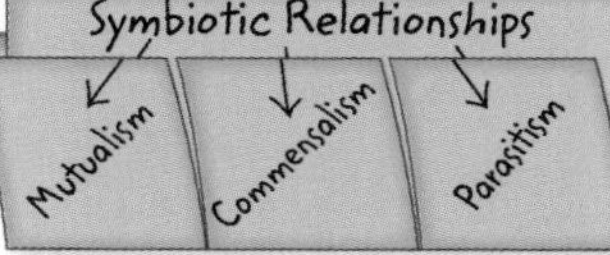

Figure 13 The three types of symbiosis are mutualism, commensalism, and parasitism.

Mutualism

Commensalism

Parasitism

Lesson 2 Review

Visual Summary

Each species that shares a habitat has a separate niche.

Overpopulation occurs when a population becomes so large that it causes damage to the environment.

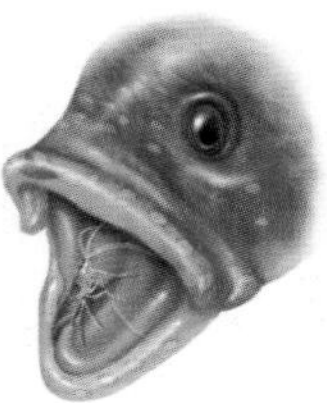

Symbiosis usually involves obtaining energy.

FOLDABLES®

Use your lesson Foldable to review the lesson. Save your Foldable for the project at the end of the chapter.

What do you think NOW?

You first read the statements below at the beginning of the chapter.

3. A niche is the place where an animal lives.

4. Symbiosis is a close relationship between two species.

Did you change your mind about whether you agree or disagree with the statements? Rewrite any false statements to make them true.

Use Vocabulary

1. **Define** *competition* in your own words.
2. **Distinguish** between predation and symbiosis.
3. **Distinguish** between commensalism and parasitism.

Understand Key Concepts 

4. Which is a symbiotic relationship in which both organisms benefit?
   - **A.** commensalism
   - **B.** mutualism
   - **C.** parasitism
   - **D.** predation
5. **Compare and contrast** a habitat and a niche.
6. **List** two ways in which human populations compete with populations of other species.

Interpret Graphics

7. **Organize** Copy and fill in the graphic organizer below. In each oval, list the types of interactions that can take place among organisms in an ecosystem.

8. **Describe** the relationship that the organisms shown at right have with each other.

Critical Thinking

9. **Analyze** Some biologists consider predation to be a kind of symbiosis. Explain why you agree or disagree.
10. **Apply** A mite species lives on the bodies of bees. The mites help keep beehives clear of fungus. The bees provide the mites with a place to live. What kind of relationship is this? Explain your reasoning.

(t) adokon/Getty Images, (c) Gay Bumgarner/Alamy, (b) Photo by Scott Bauer, USDA-ARS

GREEN SCIENCE

Purple Loosestrife: An Invasive Plant Species

Stamping Out the Purple Plague

Wetlands, such as swamps and marshes, are important ecosystems. These soggy areas control flooding, affect the flow of rivers, and filter pollution from water. They also are home to a diversity of wildlife, such as birds, fish, mammals, and plants. But not every species in a wetland is native to the habitat. In North America, one invasive species in particular has caused trouble for many wetland ecosystems.

In the early 1800s, European ships brought a hardy plant to America's shores—purple loosestrife. Settlers used it as a medicinal herb to treat digestive problems, such as diarrhea and ulcers. Before long, the tall plant with reddish-purple flowers was growing in wetlands across the United States.

The fast-growing plant is devastating for wetlands. Its thick roots crowd out native plants that provide food, shelter, and nesting sites for many animal species. Loosestrife also can disrupt the flow of water to rivers and canals and clog irrigation systems. The effect of loosestrife on biodiversity and local communities is so harmful that the plants have become known as the purple plague.

Scientists have tried many ways of controlling purple loosestrife, including plant-eating animals, bacteria, and herbicides. Cutting down the plants doesn't work because new plants sprout from even tiny pieces of root left in the soil. The best solution to date has been the introduction of organisms that eat purple loosestrife. Scientists have identified five species of beetles that eat purple loosestrife in its native range in Europe. These beetles do not harm other North American plants, so they have been released into the wetlands. Since 1996, the insects have successfully controlled the spread of purple loosestrife in many regions.

Scientists release leaf-eating beetles such as this one into a wetland invaded by purple loosestrife. ▼

▲ A sea of purple loosestrife overruns a wetland. It spreads quickly because one plant can produce up to three million seeds a year. The hardy seeds are scattered long distances by wind, water, animals, and even people.

It's Your Turn

RESEARCH AND REPORT Choose another invasive species. Describe how it was introduced into an ecosystem, its impact on the environment, and the steps taken to control it. Present your findings to the class.

Lesson 3

Reading Guide

Key Concepts
ESSENTIAL QUESTIONS

- How do matter and energy move through ecosystems?
- How do organisms obtain energy?
- What are the differences between a food chain and a food web?

Vocabulary

producer p. 334
consumer p. 335
food chain p. 336
food web p. 336

Multilingual eGlossary

Matter and Energy in Ecosystems

Inquiry Where's the energy?

This elephant gets its energy by eating plants. It uses this energy for life processes. Where do you think plants get their energy? How do you think energy moves through an ecosystem?

Launch Lab

15 minutes

Where does matter go?

Matter cannot be created or destroyed but is recycled. What happens to matter that seems to vanish?

1. Read and complete a lab safety form.
2. Half fill a small **paper cup** with water. Find the mass of the cup and water using a **balance,** and record it in your Science Journal.
3. Use the balance to find the mass of two **effervescent antacid tablets.** Add this mass to the mass from step 2 to find the total.

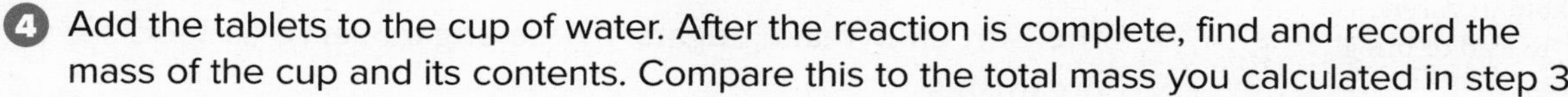

4. Add the tablets to the cup of water. After the reaction is complete, find and record the mass of the cup and its contents. Compare this to the total mass you calculated in step 3.
5. Find and record the mass of a **large self-sealing bag.** Repeat steps 2 and 3, but also add the mass of the bag to find total mass. Place the cup and tablets into the bag and seal it.
6. Holding the cup with one hand, pick up each tablet and drop it in the water.
7. After the reaction is complete, find the total mass.

Think About This

1. How did the mass compare between steps 3 and 4? Between steps 5 and 7?
2. **Key Concept** Where do you think the mass of the tablets went? What observation indicates that energy was involved?

Matter and Energy

A leaf drops to the ground. Over time, bacteria and fungi break apart the chemical bonds that hold together the atoms and the molecules of the leaf. This releases energy, water vapor, and other compounds. Carbon compounds and water molecules become part of the soil. When new seedlings grow in spring, these materials enter the seedlings.

Almost all of the matter on Earth today has been here since it formed. Matter can change form, but it cannot be created or destroyed. As shown in Figure 14, some matter cycles through ecosystems as organisms grow, die, and decompose.

Unlike matter, energy cannot be recycled. However, energy can be converted. The chemical energy in a log converts to thermal energy and light energy when it burns.

Key Concept Check How do matter and energy move through ecosystems?

Figure 14 Matter cycles through an ecosystem as organisms grow, die, and decompose.

Obtaining Energy

When you eat a sandwich, your body gets atoms and molecules that it needs to make new cells and tissues. Your cells also get the energy they need to make proteins and carry out other life processes. All organisms need a constant supply of energy to maintain life. Where does that energy come from?

Producers

Most of the energy used by all organisms on Earth comes from the Sun. Photosynthesis is the process during which some organisms use carbon dioxide, water, and light energy, usually from the Sun, and make sugars. These sugars serve as food for living organisms.

Word Origin
producer
from Latin *producere,* means "to lead or bring forth"

Producers *are organisms that use an outside energy source, such as the Sun, and produce their own food.* The energy in food molecules is in the chemical bonds that hold the molecules together. During cellular respiration, these bonds break. This releases energy that fuels the producer's life processes. As shown in **Figure 15,** photosynthesis and cellular respiration occur throughout ecosystems.

Figure 15 Producers use energy from the environment and make food molecules. They release waste products during cellular respiration.

Visual Check What abiotic factors does this producer require before storing energy?

Consumers

The energy-rich molecules formed by producers provide food for other organisms. **Consumers** *are organisms that cannot make their own food. Consumers obtain food by eating producers or other consumers.* Ecosystems include several different kinds of consumers, as shown in **Table 1.**

Table 1 Types of Consumers

	Herbivores eat only plants and other producers.
	Carnivores eat herbivores and other consumers.
	Omnivores eat producers and consumers.
	Decomposers break down dead organisms.

Herbivores eat plants and other producers. Examples of herbivores include snails, rabbits, deer, and bees. Carnivores eat herbivores and other consumers. Cats, snakes, hawks, frogs, and spiders are carnivores. Omnivores eat producers and consumers. Omnivores include bears, robins, pigs, rats, and humans. Decomposers break down the bodies of dead organisms into compounds that can be used by living organisms. Without decomposers, matter could not be recycled. Decomposers include fungi, bacteria, wood lice, termites, and earthworms.

Key Concept Check How do organisms obtain energy?

MiniLab 10 minutes

How do decomposers recycle nutrients in an ecosystem?

Nutrients and other materials do not simply disappear when they are used in an ecosystem. Organisms called decomposers break down waste, and it can be used again. You can observe decomposition in action.

1. Read and complete a lab safety form.
2. Half fill a **glass jar** with **whole, steel-cut oats.**
3. Add two small wedges of **apple** to the contents of the jar.
4. Place 3–4 **mealworms** in the jar.
5. Cover the contents of the jar with **strips of paper.** ⚠ Do not put a lid on the jar or cover it with anything.

6. Observe the activity in the jar over several days.

Analyze and Conclude

1. **Describe** what you observed in the jar over several days.
2. **Infer** what happened to the oats and the apple you placed in the jar.
3. **Key Concept** Analyze the importance of this activity to plants and other organisms present in an ecosystem. Predict what might result if this activity did not take place.

ACADEMIC VOCABULARY

transfer
(verb) to pass from one to another

Transferring Energy

Not only can energy be converted from one form to another, it also can be **transferred** from one organism to another. The transfer of energy takes place in an ecosystem when one organism eats another. Food chains and food webs are models used to describe these energy transfers.

FOLDABLES®

Make a vertical trifold Venn book. Label the front *Energy Transfer* and label the inside as shown. Use it to compare and contrast the transfer of energy in a food chain and a food web.

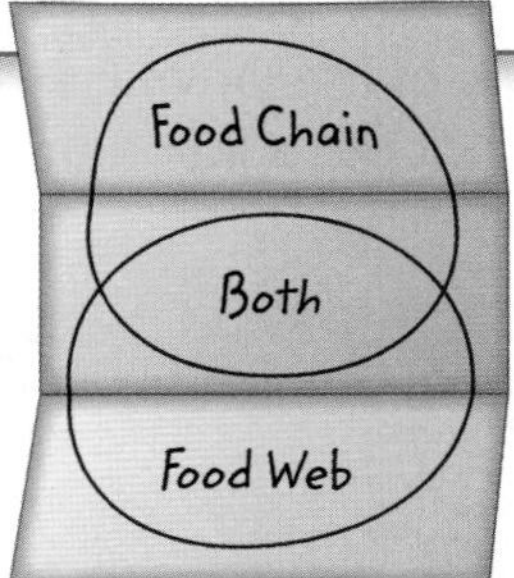

Food Chains

A model that shows how energy flows in an ecosystem through feeding relationships is called a **food chain.** A food chain always begins with a producer because producers are the source of energy for the rest of the organisms in a community. Energy moves from a producer to consumers such as herbivores or omnivores, and then on to other omnivores, carnivores, or decomposers.

A simple food chain from a community of organisms living in a vacant lot might look like this:

Grass → Mouse → Cat

The arrows show the directions of the energy transfer.

Food Webs

Most ecosystems contain many food chains. *A* **food web** *is a model of energy transfer that can show how the food chains in a community are interconnected.* For example, in the food web shown in **Figure 16**, pigeons eat berries and insects. They are prey for hawks and cats.

Figure 16 A vacant lot food web contains many food chains.

Personal Tutor

Key Concept Check What are the differences between a food chain and a food web?

Lesson 3 Review

Visual Summary

Matter cycles throughout an ecosystem.

Organisms obtain energy from the environment or by eating other organisms.

Many exchanges of energy occur among organisms in an ecosystem.

FOLDABLES®

Use your lesson Foldable to review the lesson. Save your Foldable for the project at the end of the chapter.

What do you think NOW?

You first read the statements below at the beginning of the chapter.

5. Energy from sunlight is the basis for almost every food chain on Earth.

6. Matter is created when a plant grows.

Did you change your mind about whether you agree or disagree with the statements? Rewrite any false statements to make them true.

Use Vocabulary

1 **Define** *producer* in your own words.

2 An organism that cannot make its own food is called a(n) ________.

3 **Define** *food chain* in your own words.

Understand Key Concepts

4 Which term describes a bacterium that uses light energy and makes energy-rich molecules?

A. consumer
B. decomposer
C. herbivore
D. producer

5 **Predict** why many species of orb-weaver spiders eat their old webs before spinning new ones.

6 **Distinguish** between a food chain and a food web.

Interpret Graphics

7 **List** the producers and the consumers in the food web shown below.

Critical Thinking

8 **Construct** a food chain that models energy transfer among the following organisms: an oak tree, a squirrel, and a hawk.

9 **Construct** a food web that includes bacteria, fungi, oak trees, deer, quail, crows, raccoons, foxes, hawks, and bobcats.

3 class periods

Design an Ecosystem

Materials

5–10-gallon aquarium

assorted ecosystem materials

Safety

You have read about the connections between biotic and abiotic factors of an ecosystem and how they depend on each other. An ecosystem requires abiotic factors to support the organisms that inhabit it. In addition, the organisms of the ecosystem serve different roles. Some organisms are producers, some are consumers, and some are decomposers. In this lab, you will create an ecosystem with abiotic and biotic factors that function together. Then you will use your observations to analyze the role of each part of the ecosystem you assembled.

Ask a Question

What biotic and abiotic factors can you assemble to create a functioning ecosystem?

Make Observations

1. Read and complete a lab safety form.
2. In your Science Journal, make three columns with the following headings: *Organism*, *Abiotic Needs*, and *Biotic Needs*.
3. Visit a pet store or go online to research the types of organisms that can live in your glass aquarium. As you research, list the organisms in your Science Journal. Along with the types of organisms, list the specific biotic and abiotic needs of each organism.
4. Using your research, design an ecosystem that you can build in your aquarium. The ecosystem can be aquatic or terrestrial. Your plan must provide for the needs of all organisms in your ecosystem.
5. Have your teacher approve your design before you create your ecosystem.

7

Form a Hypothesis

6. After receiving approval for your ecosystem design, formulate a hypothesis about what your ecosystem requires to function successfully.

Test Your Hypothesis

7. With your teacher's help, obtain the resources and organisms you need to construct your ecosystem. Work with your teacher to put your ecosystem together and add the living organisms. Take care to research the needs of each organism you use. You might consult a biologist or a specialist at a pet store to make sure the needs of all your organisms will be met.
8. Observe your ecosystem over several days. Record your observations. Provide details on the ways that the parts of your ecosystem interact and connect with each other.

Analyze and Conclude

9. **Analyze** Describe the parts and interactions of your ecosystem. How well does your ecosystem sustain itself? What things, if any, must you add to the ecosystem to maintain it? How do the populations in your ecosystem change?
10. **Classify** Write a list of the organisms in your ecosystem. Classify each organism as a producer, a consumer, or a decomposer.
11. **The Big Idea** Create a food-web diagram of your ecosystem. Diagram the connections that occur between the organisms. Describe how nutrients and resources cycle through the ecosystem. Explain how matter and energy are transformed.

Communicate Your Results

Prepare a scientific report on your ecosystem. Include descriptions of the niches occupied by the organisms, explanations of relationships between organisms, and data on population changes.

Extension

A successful ecosystem recycles all the things needed for survival. What additions or changes could you make to your ecosystem so that nothing would need to be added or removed for it to sustain itself? Write a brief plan.

Lab Tips

- ☑ Ask for help from several different sources for obtaining and working with the organisms in your ecosystem.
- ☑ Remember that your organisms will need an appropriate environment in which to live. Find a suitable location, with proper light and temperature, in which to keep your aquarium.

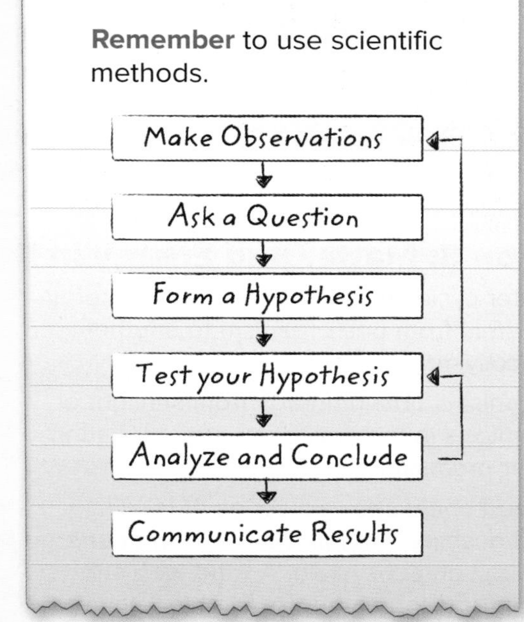

Chapter 10 Study Guide

Organisms depend on one another and on their environment for food, shelter, living space, and other needs. The nonliving parts of the environment–including sunlight, water, air, and soil nutrients–determine what kinds of organisms can live in a given area and how many organisms can live there.

Key Concepts Summary	Vocabulary
Lesson 1: Ecosystems • An **ecosystem** is all the interactions among the living and nonliving parts of the environment in a given area. • Both **abiotic factors** and **biotic factors** are parts of an ecosystem. Abiotic factors include nonliving things such as sunlight, water, and air. Biotic factors include all the living and once-living organisms in an ecosystem. • **Populations** can increase, decrease, move, and die out. 	**ecosystem** p. 315 **abiotic factor** p. 316 **biotic factor** p. 317 **habitat** p. 318 **population** p. 319 **community** p. 319 **population density** p. 320
Lesson 2: Relationships Within Ecosystems  • A habitat is the place within an ecosystem that provides the resources an organism needs. A **niche** is the way an organism interacts with the biotic and abiotic factors in its environment to meet its needs. • Organisms rely on each other for food and other resources, compete with each other for resources, and cooperate with each other to obtain resources.	**niche** p. 325 **competition** p. 326 **overpopulation** p. 327 **predation** p. 328 **symbiosis** p. 329 **mutualism** p. 329 **commensalism** p. 329 **parasitism** p. 329
Lesson 3: Matter and Energy in Ecosystems • Matter cycles through ecosystems. Energy transfers from one organism to another in ecosystems. • Organisms obtain energy from sunlight or chemicals in the environment or by eating other organisms. • A **food chain** shows a series of feeding relationships that follows the path of energy through an ecosystem. A **food web** shows the interaction among several food chains within an ecosystem. 	**producer** p. 334 **consumer** p. 335 **food chain** p. 336 **food web** p. 336

Jeremy Walker/Photo Researchers

Study Guide

Vocabulary eFlashcards
Vocabulary eGames

Personal Tutor

FOLDABLES® Chapter Project

Assemble your lesson Foldables as shown to make a Chapter Project. Use the project to review what you have learned in this chapter.

Use Vocabulary

Write the vocabulary term that best matches each phrase.

1. provides an organism with the abiotic and biotic factors needed for life
2. a number of individuals of a species living in the same area at the same time
3. a relationship in which one organism hunts another for food
4. a close relationship between two organisms in which one organism benefits and the other is harmed
5. an organism that can convert energy from sunlight into chemical energy
6. an organism that cannot make its own food

Link Vocabulary and Key Concepts

Interactive Concept Map

Copy this concept map, and then use vocabulary terms from the previous page to complete the concept map.

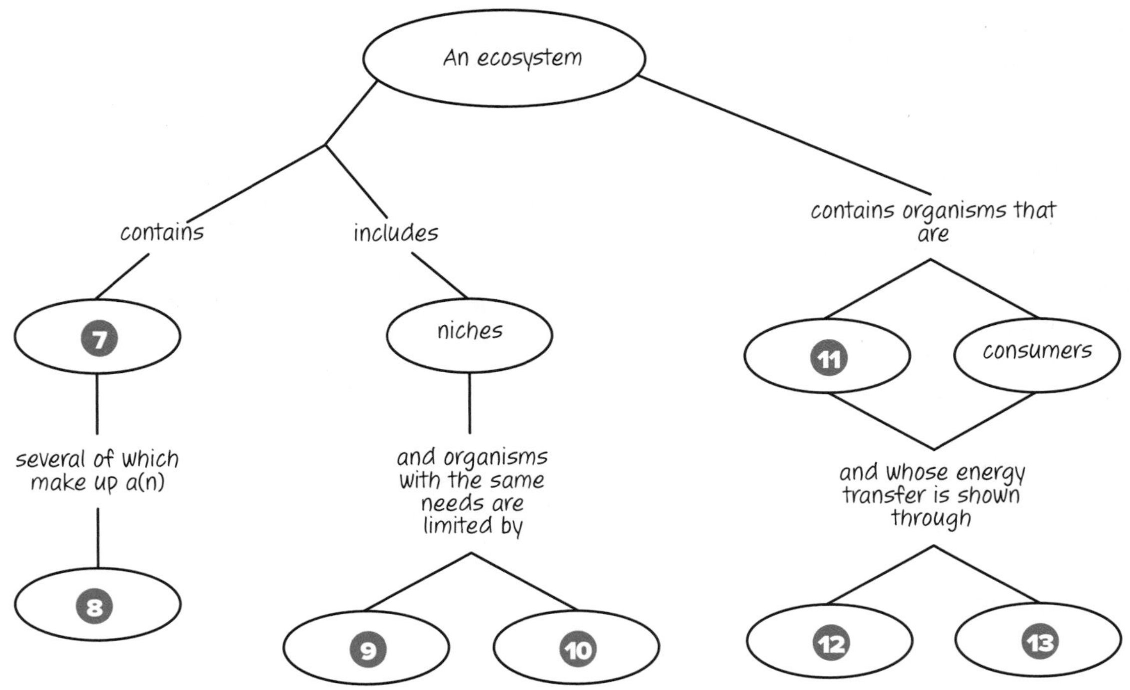

Chapter 10 Review

Understand Key Concepts 

1. Which two abiotic factors probably have the greatest effect on the organism living in the ecosystem shown below?

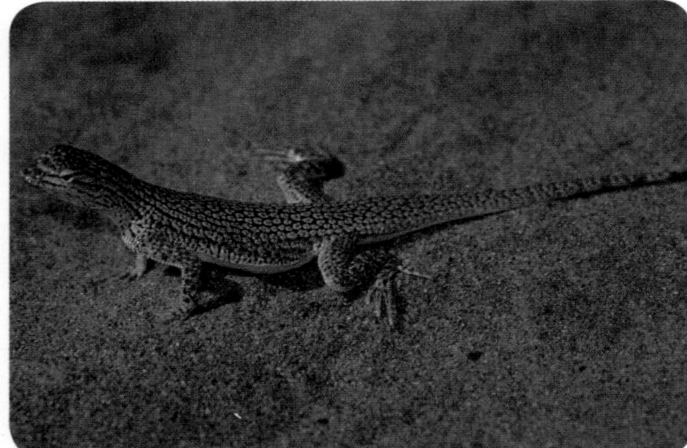

   A. carbon dioxide and water
   B. nitrogen and soil
   C. sunlight and oxygen
   D. water and temperature

2. Sunlight plus what other two abiotic factors are required for photosynthesis?
   A. soil and air
   B. soil and water
   C. carbon dioxide and oxygen
   D. carbon dioxide and water

3. Which is a biotic factor in the habitat of an insect?
   A. bark
   B. oxygen
   C. soil
   D. water

4. What do all individuals of all species living in an area form?
   A. a community
   B. an ecosystem
   C. a habitat
   D. a population

5. The way a robin builds a nest and finds food describes its
   A. habitat.
   B. niche.
   C. population.
   D. species.

6. What does X on the graph below show?

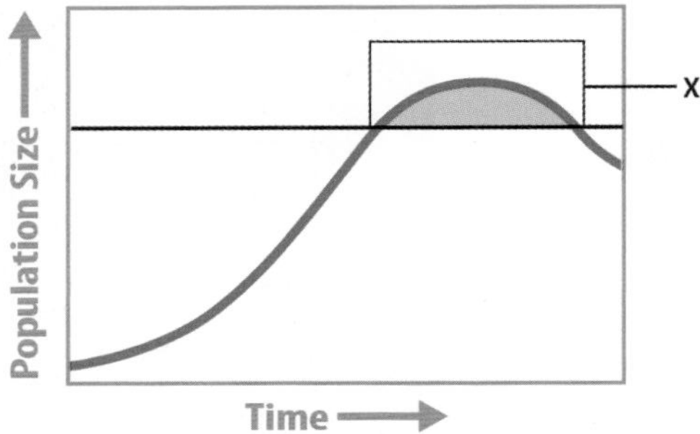

   A. overpopulation
   B. predation
   C. ecosystem size
   D. population density

7. Which term describes two organisms needing the same resources in the same place at the same time?
   A. competition
   B. niches
   C. predation
   D. symbiosis

8. Termites eat dead wood. What are they examples of?
   A. carnivores
   B. decomposers
   C. omnivores
   D. producers

9. Which is a fern an example of?
   A. carnivore
   B. decomposer
   C. omnivore
   D. producer

William Radke/U.S. Fish & Wildlife Service

Critical Thinking

10. **Describe** two ways in which loss of habitat could affect the size of a population.

11. **Predict** how a squirrel population would be affected if a pine forest that had been cut down grew again.

12. **Hypothesize** In what ways might a bird and a squirrel compete for resources?

13. **Summarize** Describe the ways in which the size of the fish population shown at right could be reduced by changes in the abiotic factors in its habitat.

14. **Compare** In what ways is predation similar to parasitism?

15. **Explain** A plant uses a carbon atom from a carbon dioxide molecule and makes a sugar molecule. How could that carbon molecule be in the body of a carnivore?

16. **Construct** a food chain that describes the feeding relationships between a bird, a wildflower, and a butterfly.

17. **Draw** a food web that describes the following relationships: A parasite sucks the blood of fish and eels. The fish feed on algae. Eels feed on the fish. Cleaner shrimp remove parasites from the fish and eat the parasites.

Writing in Science

18. **Write** a four- or five-sentence paragraph that explains the difference between the flow of matter through an ecosystem and the flow of energy through an ecosystem. Be sure to include a topic sentence and a concluding sentence in your paragraph.

REVIEW THE BIG IDEA

19. Describe five different ways in which the biotic factors in an ecosystem can interact, and give an example of each.

20. The photo below shows a raccoon that has gotten into a human's campsite. In what ways might the raccoon interact with other living things, including humans, and the environment?

Math Skills

Math Practice

Use a Formula

21. Between 1991 and 2000, the moose population in Poland decreased from 5,400 animals to 1,718 animals. The area of Poland is 3,115 km^2.

    a. What was the population density in 1991?

    b. What was the population density in 2000?

22. A total of 1,650 earthworms were counted in a 50-m^2 area of moist forest. What is the population density?

23. The recommended density for a freshwater aquarium is 2.5 cm of fish per gallon of water. How many fish should you put in a 30-gallon aquarium if each fish measures 5 cm?

Standardized Test Practice

Record your answers on the answer sheet provided by your teacher or on a sheet of paper.

Multiple Choice

1 What do abiotic and biotic factors have in common?

- A They both contain living parts.
- B They both contain nonliving parts.
- C They both include water for living things.
- D They both provide resources for living things.

Use the figure below to answer questions 2 and 3.

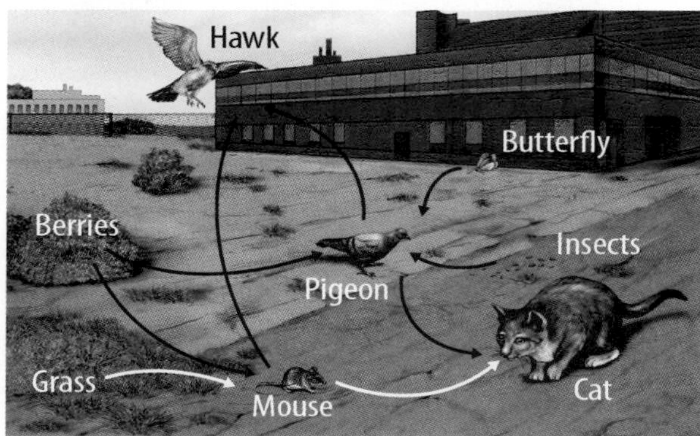

2 Which statement describes the food chain shown in the figure above?

- A Cats and hawks both eat pigeons.
- B Hawks eat pigeons, which eat cats.
- C Mice eat grass and are eaten by hawks.
- D Pigeons eat beetles and butterflies.

3 Which result is likely if the cat population moves away from the vacant lot in the figure?

- A The grass population will increase.
- B The hawk population will decrease.
- C The mouse population will increase.
- D The pigeon population will decrease.

4 Which is a biotic factor in an ecosystem?

- A atmosphere
- B plants
- C temperature
- D water

5 What do herbivores eat?

- A consumers and decomposers
- B only consumers
- C only producers
- D producers and consumers

6 Which is an example of parasitism?

- A A bat pollinates a cactus.
- B A bird builds a nest in a tree.
- C A flea ingests the blood of a dog.
- D A hawk eats a rabbit.

Use the food chain below to answer question 7.

Parsley → Rabbit → Fox → Bear

7 In the food chain shown, which organism eats a producer?

- A bear
- B fox
- C parsley
- D rabbit

8 Which relationship between organisms allows both species to benefit?

- A commensalism
- B mutualism
- C parasitism
- D predation

Use the figure below to answer question 9.

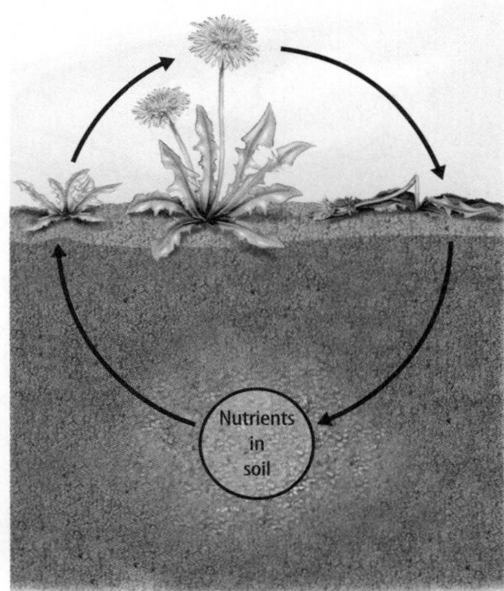

9 What is shown in the diagram above?

A Energy cycles through ecosystems as one organism eats another.

B Energy moves through ecosystems as it is transferred from one organism to another.

C The matter that makes up a plant is recycled as the plant grows, dies, and decomposes.

D The matter that makes up a plant moves through the ecosystem from producer to consumer.

10 How do decomposers benefit an ecosystem?

A They carry out photosynthesis.

B They control population growth.

C They produce energy.

D They recycle nutrients.

Constructed Response

Use the figure below to answer question 11.

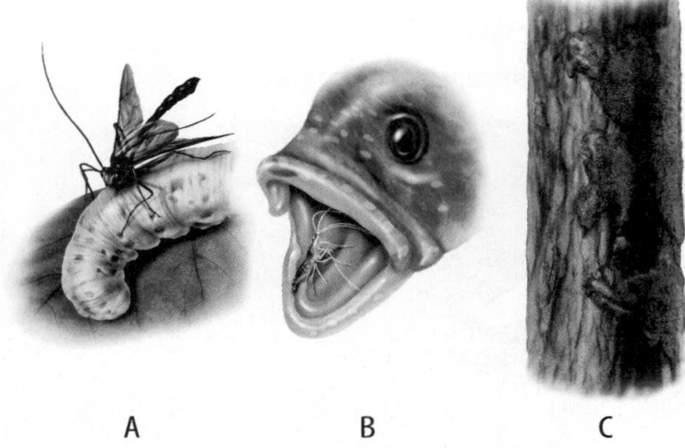

11 The figure above illustrates three examples of interactions between organisms in an ecosystem. Pick one diagram, and describe the type of interaction shown.

12 Explain how symbiosis differs from predation.

13 Competition between members of the same species is usually more intense than between members of different species. Explain why this is usually true based on your understanding of habitats and niches.

14 Explain how a food chain and a food web are related. Could they contain the same living things? Why or why not?

NEED EXTRA HELP?														
If You Missed Question...	1	2	3	4	5	6	7	8	9	10	11	12	13	14
Go to Lesson...	1	3	1	1	3	2	3	2	3	3	2	2	2	3

Unit 3

Understanding Matter

1000 B.C. — 1700 — 1800

350 B.C.
Greek philosopher Aristotle defines an element as "one of those bodies into which other bodies can decompose, and that itself is not capable of being divided into another."

1704
Isaac Newton proposes that atoms attach to each other by some type of force.

1869
Dmitri Mendeleev publishes the first version of the periodic table.

1874
G. Johnstone Stoney proposes the existence of the electron, a subatomic particle that carries a negative electric charge, after experiments in electrochemistry.

1897
J.J. Thompson demonstrates the existence of the electron, proving Stoney's claim.

Look at all the houses!

Dinner tiiiime!

By changing state once more to become plasma, she could be back in time for dinner...

...And no one would ever know!

1900 — 1950

1907
Physicists Hans Geiger and Ernest Marsden, under the direction of Ernest Rutherford, conduct the famous gold foil experiment. Rutherford concludes that the atom is mostly empty space and that most of the mass is concentrated in the atomic nucleus.

1918
Ernest Rutherford reports that the hydrogen nucleus has a positive charge, and he names it the proton.

1932
James Chadwick discovers the neutron, a subatomic particle with no electric charge and a mass slightly larger than a proton.

Technology

Figure 1 Nature cannot easily recycle many human-made materials. Much of our trash remains in landfills for years. Scientists are developing materials that degrade quickly. This will help decrease the amount of pollution.

Scientists use technology to develop materials with desirable properties. **Technology** is the practical use of scientific knowledge, especially for industrial or commercial use. In the late 1800s, scientists developed the first plastic material, called celluloid, from cotton. Celluloid quickly gained popularity for use as photographic film. In the 20th century, scientists developed other plastic materials, such as polystyrene, rayon, and nylon. These new materials were inexpensive, durable, lightweight, and could be molded into any shape.

New technologies can come with problems. For example, many plastics are made from petroleum and contain harmful chemicals. The high pressures and temperatures needed to produce plastics require large amounts of energy. Bacteria and fungi that easily break down natural materials do not easily decompose plastics. Often, plastics accumulate in landfills where they can remain for hundreds, or even thousands, of years, as shown in **Figure 1.**

Types of Materials

Figure 2 Some organisms produce materials with properties that are useful to people. Scientists are trying to replicate these materials for new technologies.

◄ Most human-made adhesives attach to some surfaces, but not others. Mussels, which are similar to clams, produce a "superglue" that is stronger than anything people can make. It also works on any surface, wet or dry. Chemists are trying to develop a technology that will replicate the mussel glue. This glue would provide solutions to difficult problems: Ships could be repaired under water. The glue also would work on teeth and could be used to set broken bones.

Abalone and other mollusks construct a protective shell from proteins and seawater. The material is four times stronger than human-made metal alloys and ceramics. Using technology, scientists are working to duplicate this material. They hope to use the new product in many ways, including hip and elbow replacements. Automakers could use these strong, lightweight materials for automobile body panels. ►

Consider the Possibilities!

Chemists are looking to nature for ideas for new materials. For example, some sea sponges have skeletons that beam light deep inside the animal, similar to the way fiber-optic cables work. A bacterium from a snail-like nudibranch contains compounds that stop other sea creatures from growing on the nudibranch's back. These compounds could be used in paints to stop creatures from forming a harmful crust on submerged parts of boats and docks. Chrysanthemum flowers produce a product that keeps ticks and mosquitoes away. **Figure 2** includes other organisms that produce materials with remarkable properties.

Chemists and biologists are teaming up to understand, and hopefully replicate, the processes that organisms use to survive. Hopefully, these processes can lead to technologies and materials with unique properties that are helpful to people.

MiniLab

15 minutes

How would you use it?

How would you use an adhesive that could stick to any surface? Invent a new purpose for mussel "superglue"!

1. Work with a partner to develop three tasks that could be accomplished using products produced by an organism.
2. Select one of your ideas, and develop it into an invention. Draw pictures of your new invention and explain how it works.
3. Write an advertisement for your invention including a description of the role of the material from nature used in your product.

Analyze and Conclude

1. **Explain** What task or problem did your invention solve?
2. **Infer** How did a material from an organism help you develop your invention?

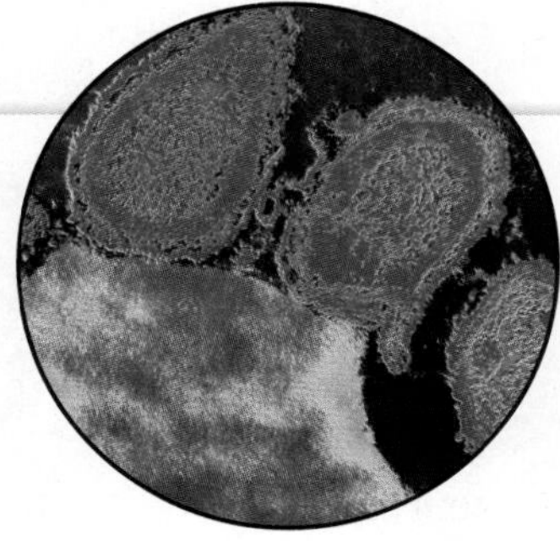

◄ A British company has developed bacteria that produce large amounts of hydrogen gas when fed a diet of sugar. Chemists are working to produce tanks of these microorganisms that produce enough hydrogen to replace other fuels used to heat homes. Bacteria may become the power plants of the future.

Under a microscope, the horn of a rhinoceros looks much like the material used to make the wings of a Stealth aircraft. However, the rhino horn is self-healing. Picture a car with technologically advanced fenders similar to the horn of a rhinoceros; such a car could repair itself if it were in a fender-bender! ►

◄ Spider silk begins as a liquid inside the spider's body. When ejected through openings, called spinnerets, it becomes similar to a plastic thread. However, its properties include strength five times greater than steel, stretchability greater than nylon, and toughness better than the material in bulletproof vests! Chemists are using technology to make a synthetic spider silk. They hope to someday use the material for cables strong enough to support a bridge or as reinforcing fibers in aircraft bodies.

(t) Dr. Kari Lounatmaa/Science Source, (c)PhotoAlto/PunchStock, (b)Amanda Hall/Robert Harding World Imagery/Getty Images

Chapter 11

Matter and Atoms

THE BIG IDEA How does the classification of matter depend on atoms?

Inquiry Tiny Parts?

From a distance, you might think this looks like a normal picture, but what happens when you look closely? Tiny photographs arranged in a specific way make a new image that looks very different from the individual pictures. The new image depends on the parts and the way they are arranged. Similarly, all the matter around you depends on its parts and the way they are arranged.

- How would the image be different if the individual pictures were arranged in another way?
- How does the image depend on the individual parts?

Scott Camazine/AndreaMosaic/Science Source

Get Ready to Read

What do you think?

Before you read, decide if you agree or disagree with each of these statements. As you read this chapter, see if you change your mind about any of the statements.

1. Things that have no mass are not matter.
2. The arrangement of particles is the same throughout a mixture.
3. An atom that makes up gold is exactly the same as an atom that makes up aluminum.
4. An atom is mostly empty space.
5. If an atom gains electrons, the atom will have a positive charge.
6. Each electron is a cloud of charge that surrounds the center of an atom.

Mc Graw Hill Education **connectED**

Your one-stop online resource

connectED.mcgraw-hill.com

LearnSmart®

Chapter Resources Files, Reading Essentials, Get Ready to Read, Quick Vocabulary

Animations, Videos, Interactive Tables

Self-checks, Quizzes, Tests

Project-Based Learning Activities

Lab Manuals, Safety Videos, Virtual Labs & Other Tools

Vocabulary, Multilingual eGlossary, Vocab eGames, Vocab eFlashcards

Personal Tutors

Lesson 1

Reading Guide

Key Concepts
ESSENTIAL QUESTIONS

- What is the relationship among atoms, elements, and compounds?
- How are some mixtures different from solutions?
- How do mixtures and compounds differ?

Vocabulary

matter p. 353
atom p. 353
substance p. 354
element p. 355
molecule p. 355
compound p. 356
mixture p. 358
heterogeneous mixture p. 359
homogeneous mixture p. 360

Multilingual eGlossary

BrainPOP®
What's Science Got to do With It?

Substances and Mixtures

Inquiry Is it pure?

This worker is making a trophy by pouring hot, liquid metal into a mold. The molten metal is bronze, which is a mixture of several metals blended to make the trophy stronger. Why do you think a bronze trophy would be stronger than a pure metal trophy?

David McNew/Getty Images

Launch Lab

10 minutes

Can you always see the parts of materials? 

If you eat a pizza, you can see the cheese, the pepperoni, and the other parts it is made from. Can you always see the individual parts when you mix materials?

1. Read and complete a lab safety form.
2. Observe the **materials** at the eight stations your teacher has set up.
3. Record in your Science Journal the name and a short description of each material.

Think About This

1. **Classify** Which materials have easily identifiable parts?
2. **Key Concept** Is it always easy to see the parts of materials that are mixed? Explain.

What is matter?

Imagine how much fun it would be to go windsurfing! As the force of the wind pushes the sail, you lean back to balance the board. You feel the heat of the Sun and the spray of water against your face. Whether you are windsurfing on a lake or sitting at your desk in a classroom, everything around you is made of matter. **Matter** *is anything that has mass and takes up space.* Matter is everything you can see, such as water and trees. It is also some things you cannot see, such as air. You know that air is matter because you can feel its mass when it blows against your skin. You can see that it takes up space when it inflates a sail or a balloon.

Anything that does not have mass or volume is not matter. Types of energy, such as heat, sound, and electricity, are not matter. Forces, such as magnetism and gravity, also are not forms of matter.

What is matter made of?

The matter around you, including all solids, liquids, and gases, is made of atoms. *An* **atom** *is a small particle that is the building block of matter.* In this chapter, you will read that an atom is made of even smaller particles. There are many types of atoms. Each type of atom has a different number of smaller particles. You also will read that atoms can combine with each other in many ways. It is the many kinds of atoms and the ways they combine that form the different types of matter.

Reading Check Why are there so many types of matter?

WORD ORIGIN

atom
from Greek *atomos,* means "uncut"

Figure 1 You can classify matter as a substance or a mixture. ▼

Matter
- Anything that has mass and takes up space
- Matter is made up of atoms.

Substances
- Matter with a composition that is always the same

Mixtures
- Matter that can vary in composition

Classifying Matter

Because all the different types of matter around you are made of atoms, they must have characteristics in common. But why do all types of matter look and feel different? How is the matter that makes up a pure gold ring similar to the matter that makes up your favorite soda or even the matter that makes up your body? How are these types of matter different?

As the chart in **Figure 1** shows, scientists place matter into one of two groups–substances or mixtures. Pure gold is in one group. Soda and your body are in the other. What determines whether a type of matter is a substance or a mixture? The difference is in the composition.

What is a substance?

What is the difference between a gold ring and a can of soda? What is the difference between table salt and trail mix? Pure gold is always made up of the same type of atom, but soda is not. Similarly, table salt, or sodium chloride, is always made up of the same types of atoms, but trail mix is not. This is because sodium chloride and gold are substances. *A* **substance** *is matter with a composition that is always the same.* A certain substance always contains the same kinds of atoms in the same combination. Soda and trail mix are another type of matter that you will read about later in this lesson.

Because gold is a substance, anything that is pure gold will have the same composition. Bars of gold are made of the same atoms as those in a pure gold ring, as shown in **Figure 2.** And, since sodium chloride is a substance, if you are salting your food in Alaska or in Ohio, the atoms that make up the salt will be the same. If the composition of a given substance changes, you will have a new substance.

Reading Check Why is gold classified as a substance?

Figure 2 A substance always contains the same kinds of atoms bonded in the same way. ▼

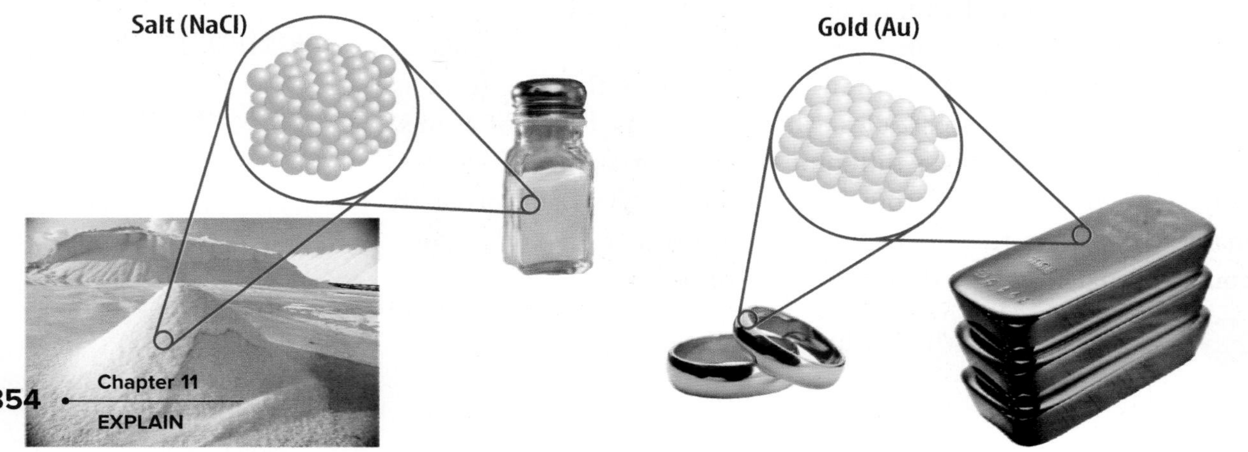

Elements

Some substances, such as gold, are made of only one kind of atom. Others, such as sodium chloride, are made of more than one kind. *An* **element** *is a substance made of only one kind of atom.* All atoms of an element are alike, but atoms of one element are different from atoms of other elements. For example, the element gold is made of only gold atoms, and all gold atoms are alike. But gold atoms are different from silver atoms, oxygen atoms, and atoms of every other element.

Key Concept Check How are atoms and elements related?

What is the smallest part of an element? If you could break down an element into its smallest part, that part would be one atom. Most elements, such as carbon and silver, consist of a large group of individual atoms. Some elements, such as hydrogen and bromine, consist of molecules. *A* **molecule** (MAH lih kyewl) *is two or more atoms that are held together by chemical bonds and act as a unit.* Examples of elements made of individual atoms and molecules are shown in **Figure 3.**

Elements on the Periodic Table You probably can name many elements, such as carbon, gold, and oxygen. Did you know that there are about 118 known elements? As shown in **Figure 4,** each element has a symbol, such as C for carbon, Au for gold, and O for oxygen. The periodic table printed in the back of this book gives other information about each element. You will learn more about elements in the next lesson.

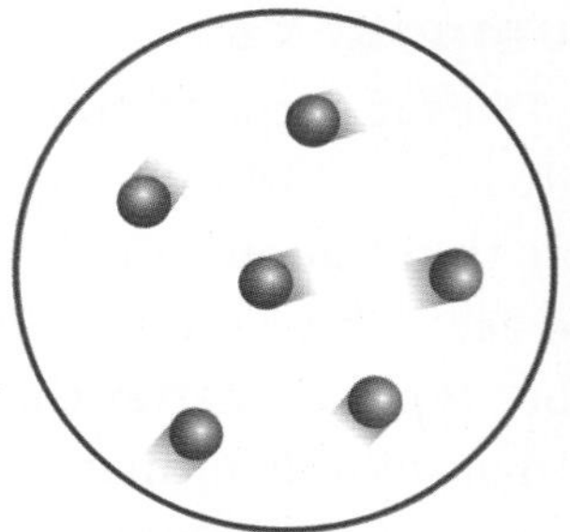

Individual atoms

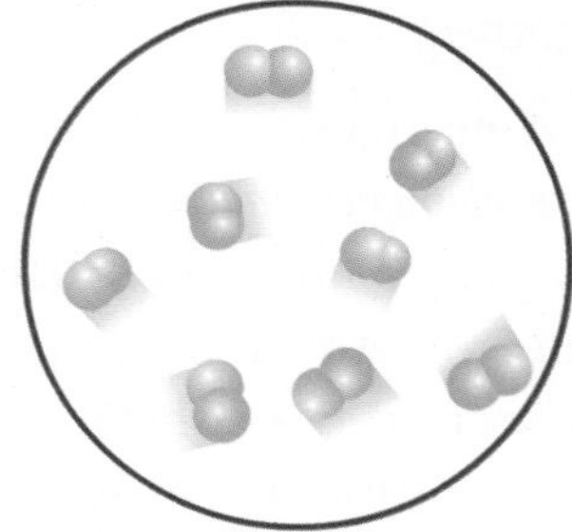

Molecules

Figure 3  The smallest part of all elements is an atom. In some elements, the atoms are grouped into molecules.

Elements

Animation

Figure 4 Element symbols have either one or two letters. Temporary symbols have three letters.

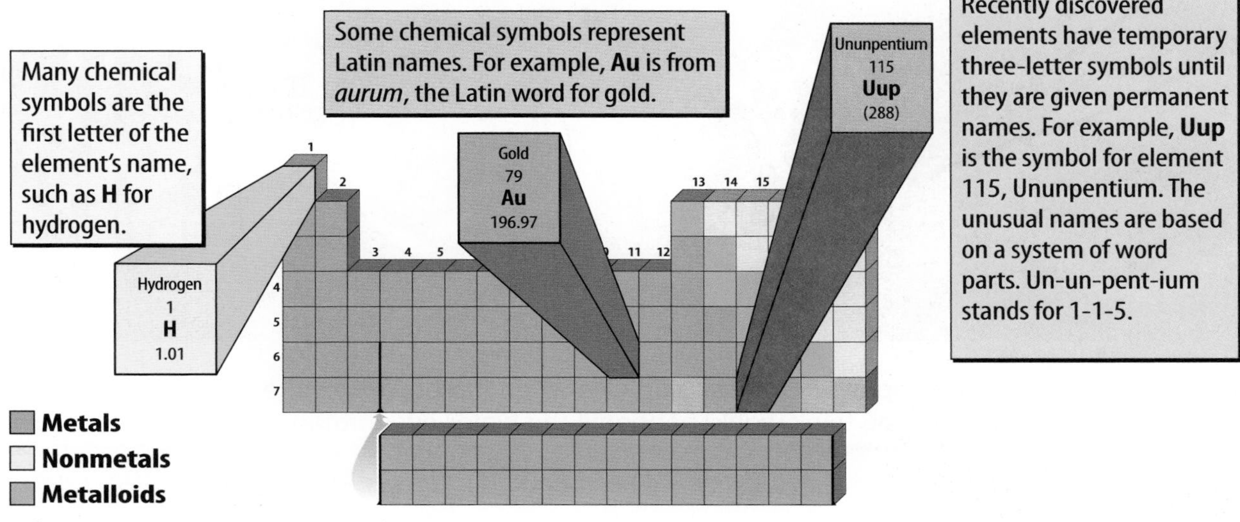

Visual Check What color are the blocks used for elements that have not yet been verified?

Compounds

Does it surprise you to learn that there are only about 118 different elements? After all, if you think about all the different things you see each day, you could probably name many more types of matter than this. Why are there so many kinds of matter when there are only about 118 elements? Most matter is made of atoms of different types of elements bonded together.

A **compound** *is a substance made of two or more elements that are chemically joined in a specific combination.* Because each compound is made of atoms in a specific combination, a compound is a substance. Pure water (H_2O) is a compound because every sample of pure water contains atoms of hydrogen and oxygen in the same combination–two hydrogen atoms to every oxygen atom. There are many types of matter because elements can join to form compounds.

Molecules Recall that a molecule is two or more atoms that are held together by chemical bonds and that act as a unit. Is a molecule the smallest part of a compound? For many compounds, this is true. Many compounds exist as molecules. An example is water. In water, two hydrogen atoms and one oxygen atom always exist together and act as a unit. Carbon dioxide (CO_2) and table sugar ($C_6H_{12}O_6$) are also examples of compounds that are made of molecules.

However, as shown in **Figure 5,** some compounds are not made of molecules. In some compounds, such as table salt, or sodium chloride, no specific atoms travel together as a unit. However, table salt (NaCl) is still a substance because it always contains only sodium (Na) and chlorine (Cl) atoms.

Key Concept Check How do elements and compounds differ?

Figure 5 Sugar particles are molecules because they always travel together as a unit. Salt particles do not travel together as a unit.

Sugar

Salt

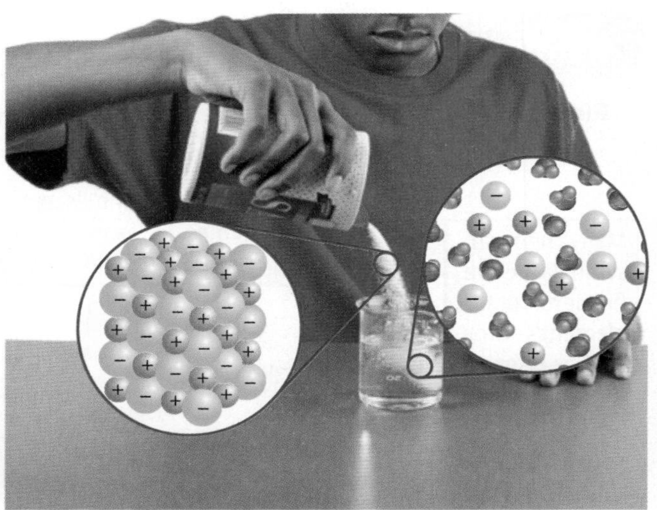

Visual Check What happens to the salt particles when the boy mixes the salt in the water? What do you think would happen if the water evaporated?

Hutchings Photography/Digital Light Source

Properties of Compounds How would you describe sodium chloride, or table salt? The properties of a compound, such as table salt, are usually different from the properties of the elements from which it is made. Table salt, for example, is made of the elements sodium and chlorine. Sodium is a soft metal, and chlorine is a poisonous, green gas. These properties are much different from the table salt you sprinkle on food!

Chemical Formulas Just as elements have chemical symbols, compounds have chemical formulas. A formula includes the symbols of each element in the compound. It also includes numbers, called subscripts, that show the ratio of the elements in the compound. You can see the formulas for some compounds in **Table 1.**

Different Combinations of Atoms Sometimes the same elements combine to form different compounds. For example, nitrogen and oxygen can form six different compounds. The chemical formulas are N_2O, NO, N_2O_3, NO_2, N_2O_4, and N_2O_5. They contain the same elements, but because the combinations of atoms are different, each compound has different properties, as shown in **Table 1.**

FOLDABLES®

Make a vertical two-tab book, and label it as shown. Use it to review properties of elements and compounds.

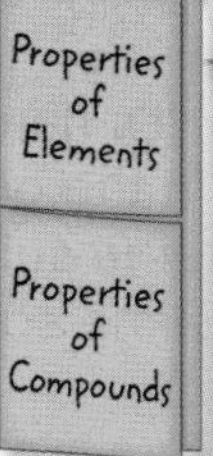

Table 1 Atoms can combine in different ways and form different compounds.

Personal Tutor

Table 1

Formula and Molecular Structure	Properties/Functions
N_2O Nitrous oxide	colorless gas used as an anesthetic
NO_2 Nitrogen dioxide	brown gas, toxic, air pollutant
N_2O_3 Dinitrogen trioxide	blue liquid

Figure 6 It's hard to tell which is in the glass—pure water (a substance) or lemon-lime soda (a mixture).

What is a mixture?

By looking at the glass of clear liquid in **Figure 6**, can you tell whether it is lemon-lime soda or water? Lemon-lime soda is almost clear, and someone might confuse it with water, which is a substance. Recall that a substance is matter with a composition that is always the same. However, sodas are a combination of substances such as water, carbon dioxide, sugar, and other compounds. In fact, most solids, liquids, and gases you experience each day are mixtures. *A* **mixture** *is matter that can vary in composition.* It is made of two or more substances that are blended but are not chemically bonded.

What would happen if you added more sugar to a glass of soda? You would still have soda, but it would be sweeter. Changing the amount of one substance in a mixture does not change the identity of the mixture or its individual substances.

Air and tap water are also mixtures. Air is a mixture of nitrogen, oxygen, and other substances. However, the composition of air can vary. Air in a scuba tank usually contains more oxygen and less of the other substances. Tap water might look like pure water, but it is a mixture of pure water (H_2O) and small amounts of other substances. Since the substances that make up tap water are not bonded together, the composition of tap water can vary. This is true for all mixtures.

MiniLab

20 minutes

How do elements, compounds, and mixtures differ?

The elements in a compound cannot be separated easily. However, you often can use the properties of the substances in a mixture to separate them.

1. Read and complete a lab safety form.
2. Observe samples of **sand** and **iron filings** with a **magnifying lens.** Record your observations in your Science Journal.
3. Combine the sand and iron filings in a **clear cup.** Stir with a **toothpick.** Observe the mixture with the magnifying lens. Record your observations.
4. Cover one end of a **magnet** with **plastic wrap.** Stir the mixture with the covered magnet. Record your observations.

Analyze and Conclude

1. **Classify** The formula for sand is SiO_2. The symbol for iron is Fe. Use this to classify each as an element, a compound, or a mixture.
2. **Key Concept** What are two ways you could tell from your observations that the combination of sand and iron filings is a mixture and not a substance?

Types of Mixtures

How do trail mix, soda, and air differ? One difference is that trail mix is a solid, soda is a liquid, and air is a gas. This tells you that a mixture can be any state of matter. Another difference is that you can see the individual parts that make up trail mix, but you cannot see the parts that make up soda or air. This is because trail mix is a different type of mixture than soda and air. There are two types of mixtures–heterogeneous (he tuh roh JEE nee us) and homogeneous (hoh muh JEE nee us). The prefix *hetero-* means "different," and the prefix *homo-* means "the same." Heterogeneous and homogeneous mixtures differ in how evenly the substances that compose them are mixed.

Academic Vocabulary

individual
(adjective) single; separate

Heterogeneous Mixtures

Suppose you take a bag of trail mix and pour it into two identical bowls. What might you notice? At first glance, each bowl appears the same. However, if you look closely, you might notice that one bowl has more nuts and another bowl has more raisins. The contents of the bowls differ because trail mix is a heterogeneous mixture. *A* **heterogeneous mixture** *is a mixture in which the substances are not evenly mixed.* Therefore, if you take two samples from the same mixture, such as trail mix, the samples might have different amounts of the individual substances. The mixtures shown in **Figure 7** are examples of heterogeneous mixtures.

Reading Check Explain why vegetable soup is classified as a heterogeneous mixture.

Figure 7 The different parts of a heterogeneous mixture are not evenly mixed.

Heterogeneous Mixtures

The numbers of peanuts, pretzels, raisins, and other types of food in trail mix could change, and it still would be trail mix.

You know that granite is a heterogeneous mixture because you can see the different minerals from which it is made.

With a microscope, you would be able to see that smoke is a heterogeneous mixture of gas and solid particles.

▲ **Figure 8** Salt is soluble in water. Pepper is insoluble in water. The pepper and water is a mixture, but not a solution.

Homogeneous Mixtures

If you pour soda into two glasses, the amounts of water, carbon dioxide, sugar, and other substances in the mixture would be the same in both glasses. Soda is an example of a **homogeneous mixture**–*a mixture in which two or more substances are evenly mixed, but not bonded together.*

Evenly Mixed Parts In a homogeneous mixture, the substances are so small and evenly mixed that you cannot see the boundaries between substances in the mixture. Brass, a mixture of copper and zinc, is a homogeneous mixture because the copper atoms and the zinc atoms are evenly mixed. You cannot see the boundaries between the different types of substances, even under most microscopes. Lemonade and air are also examples of homogeneous mixtures for the same reason.

Solution Another name for a homogeneous mixture is a solution. A solution is made of two parts–a solvent and one or more solutes. The solvent is the substance that is present in the largest amount. The solutes dissolve, or break apart, and mix evenly in the solvent. In **Figure 8**, water is the solvent, and salt is the solute. Salt is soluble in water. Notice also in the figure that pepper does not dissolve in water. No solution forms between pepper and water. Pepper is insoluble in water.

Other examples of solutions are described in **Figure 9**. Note that all three states of matter–solid, liquid, and gas–can be a solvent or a solute in a solution.

Key Concept Check How are some mixtures different from solutions?

Figure 9  Solids, liquids, and gases can combine to make solutions. ▼

Homogeneous Mixtures

A trumpet is made of brass, a solution of solid copper and solid zinc.

The natural gas used in a gas stove is a solution of methane, ethane, and other gases.

This ammonia cleaner is a solution of water and ammonia gas.

(t) Hutchings Photography/Digital Light Source, (bl) C Squared Studios/Getty Images, (bc) Steve Allen/Brand X Pictures/Alamy, (br) McGraw-Hill Education

Compounds v. Mixtures

Think again about putting trail mix into two bowls. If you put more peanuts in one of the bowls, you still have trail mix in both bowls. Since the substances that make up a mixture are not bonded, adding more of one substance does not change the identity or the properties of the mixture. It also does not change the identity or the properties of each individual substance. In a heterogeneous mixture of peanuts, raisins, and pretzels, the properties of the individual parts don't change if you add more peanuts. The peanuts and the raisins don't bond together and become something new.

Similarly, in a solution such as soda or air, the substances do not bond together and form something new. Carbon dioxide, water, sugar, and other substances in soda are mixed together. Nitrogen, oxygen, and other substances in air also keep their separate properties because air is a mixture. If it were a compound, the parts would be bonded and would not keep their separate properties.

Key Concept Check How do mixtures and compounds differ?

Compounds and Solutions Differ

Compounds and solutions are alike in that they both look like pure substances. Look back at the lemon-lime soda and the water in **Figure 6.** The soda is a solution. A solution might look like a substance because the elements and the compounds that make up a solution are evenly mixed. However, compounds and solutions differ in one important way. The atoms that make up a given compound are bonded together. Therefore, the composition of a given compound is always the same. Changing the composition results in a new compound.

However, the substances that make up a solution, or any other mixture, are not bonded together. Therefore, adding more of one substance will not change the composition of the solution. It will just change the ratio of the substances in the solution. These differences are described in **Table 2.**

Table 2 Differences Between Solutions and Compounds

	Solutions	Compound
Composition	Made up of substances (elements and compounds) evenly mixed together; the composition can vary in a given mixture.	Made up of atoms bonded together; the combination of atoms is always the same in a given compound.
Changing the composition	The solution is still the same with similar properties. However, the relative amounts of substances might be different.	Changing the composition of a compound changes it into a new compound with new properties.
Properties of parts	The substances keep their own properties when they are mixed.	The properties of the compound are different from the properties of the atoms that make it up.

Separating Mixtures

Have you ever picked something you did not like off a slice of pizza? If you have, you have separated a mixture. Because the parts of a mixture are not combined chemically, you can use a physical process, such as removing them by hand, to separate the mixture. The identity of the parts does not change. Separating the parts of a compound is more difficult. The elements that make up a compound are combined chemically. Only a **chemical change** can separate them.

REVIEW VOCABULARY

chemical change
a change in matter in which the substances that make up the matter change into other substances with different chemical and physical properties

Separating Heterogeneous Mixtures Separating the parts of a pizza is easy because the pizza has large, solid parts. Two other ways to separate heterogeneous mixtures are shown in **Figure 10.** The strainer in the figure filters larger rocks from the mixture of rocks and dirt. The oil and vinegar is also a heterogeneous mixture because the oil floats on the vinegar. You can separate this mixture by carefully removing the floating oil.

Other properties also might be useful for separating the parts. For example, if one of the parts is magnetic, you could use a magnet to remove it. In a mixture of solid powders, you might dissolve one part in water and then pour it out, leaving the other part behind. In each case, to separate a heterogeneous mixture, you use differences in the physical properties of the parts.

Reading Check Name three methods of separating heterogeneous mixtures.

Figure 10 You can separate heterogeneous and homogeneous mixtures.

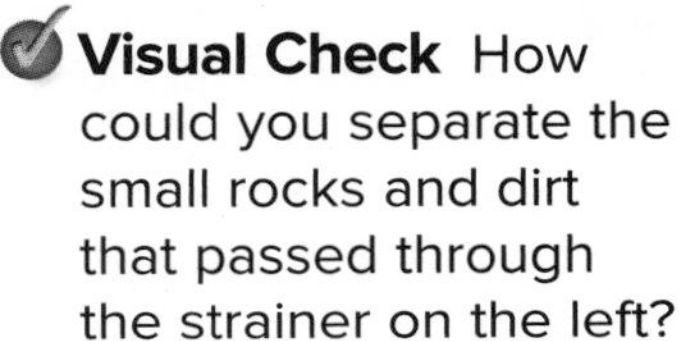

Visual Check How could you separate the small rocks and dirt that passed through the strainer on the left?

Separating Mixtures

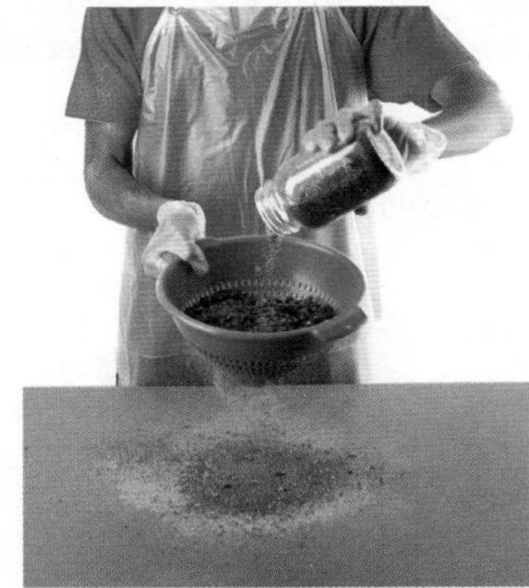

A strainer removes large parts of the heterogeneous mixture of rocks and sediment. Only small rocks and dirt fall through.

In this heterogeneous mixture of oil and vinegar, the oil floats on the vinegar. You can separate them by lifting off the oil.

Making rock candy is a way of separating a solution. Solid sugar crystals form as a mixture of hot water and sugar cools.

(l)(c)Hutchings Photography/Digital Light Source, (r)Andy Crawford and Tim Ridley/Getty Images

Separating Homogeneous Mixtures Imagine trying to separate soda into water, carbon dioxide, sugar, and other substances it is made from. Because the parts are so small and evenly mixed, separating a homogeneous mixture such as soda can be difficult. However, you can separate some homogeneous mixtures by boiling or evaporation. For example, if you leave a bowl of sugar water outside on a hot day, the water will evaporate, leaving the sugar behind. An example of separating a homogeneous mixture by making rock candy is shown in **Figure 10.**

Visualizing Classification of Matter

Think about all the types of matter you have read about in this lesson. As shown in **Figure 11,** matter can be classified as either a substance or a mixture. Substances are either elements or compounds. The two kinds of mixtures are homogeneous mixtures and heterogeneous mixtures. Notice that all substances and mixtures are made of atoms. Matter is classified according to the types of atoms and the arrangement of atoms in matter. In the next lesson, you will study the structure of atoms.

Figure 11 You can classify matter based on its characteristics.

Classifying Matter

Matter
- Anything that has mass and takes up space
- Matter on Earth is made up of atoms.
- Two classifications of matter: substances and mixtures

Substances
- Matter with a composition that is always the same
- Two types of substances: elements and compounds

Element
- Consists of just one type of atom
- Organized on the periodic table
- Each element has a chemical symbol.

Compound
- Two or more types of atoms bonded together
- Properties are different from the properties of the elements that make it up
- Each compound has a chemical formula.

Substances physically combine to form mixtures.

Mixtures can be separated into substances by physical methods.

Mixtures
- Matter that can vary in composition
- Substances are not bonded together.
- Two types of mixtures: heterogeneous and homogeneous

Heterogeneous Mixture
- Two or more substances unevenly mixed
- Different substances are visible by an unaided eye or a microscope.

Homogeneous Mixture—Solution
- Two or more substances evenly mixed
- Different substances cannot be seen even by a microscope.

Lesson 1 Review

Visual Summary

An element is a substance made of only one kind of atom.

The substances that make up a mixture are blended but not chemically bonded.

Homogeneous mixtures have the same makeup of substances throughout a given sample.

FOLDABLES®

Use your lesson Foldable to review the lesson. Save your Foldable for the project at the end of the chapter.

What do you think NOW?

You first read the statements below at the beginning of the chapter.

1. Things that have no mass are not matter.

2. The arrangement of particles is the same throughout a mixture.

3. An atom that makes up gold is exactly the same as an atom that makes up aluminum.

Did you change your mind about whether you agree or disagree with the statements? Rewrite any false statements to make them true.

Use Vocabulary

1. A small particle that is the building block of matter is a(n) ________.
2. **Use the term** *substance* in a sentence.
3. **Define** *molecule* in your own words.

Understand Key Concepts 

4. **Describe** the relationship among atoms, elements, and compounds.
5. **Explain** how some mixtures are different from solutions.
6. How does changing the amount of one substance affect a mixture's identity and a compound's identity.

Interpret Graphics

7. **Observe** Does the model at the right represent a mixture or a substance? How do you know?

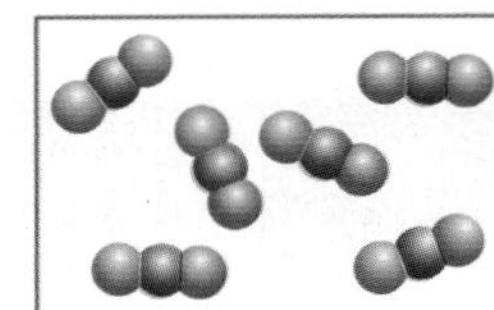

8. **Organize Information** Copy and fill in the graphic organizer below with details about substances and mixtures.

Substances	Mixtures

Critical Thinking

9. **Design** a method to separate a mixture of sugar, sand, and bits of iron.
10. **Decide** During a science investigation, a sample of matter breaks down into two kinds of atoms. Was the original sample an element or a compound? Explain.

Crude Oil

HOW IT WORKS

Separating Out Gasoline

Have you ever wondered where the gasoline used in automobiles comes from? Gasoline is part of a mixture of fuels called crude oil. How can workers separate gasoline from this mixture?

One way to separate a mixture is by boiling it. Crude oil is separated by a process called fractional distillation. First, the oil is boiled and allowed to cool. As the crude oil cools, each part changes from a gas to a liquid at a different temperature. Workers catch each fuel just as it changes back to a liquid. Eventually the crude oil is refined into all its useful parts.

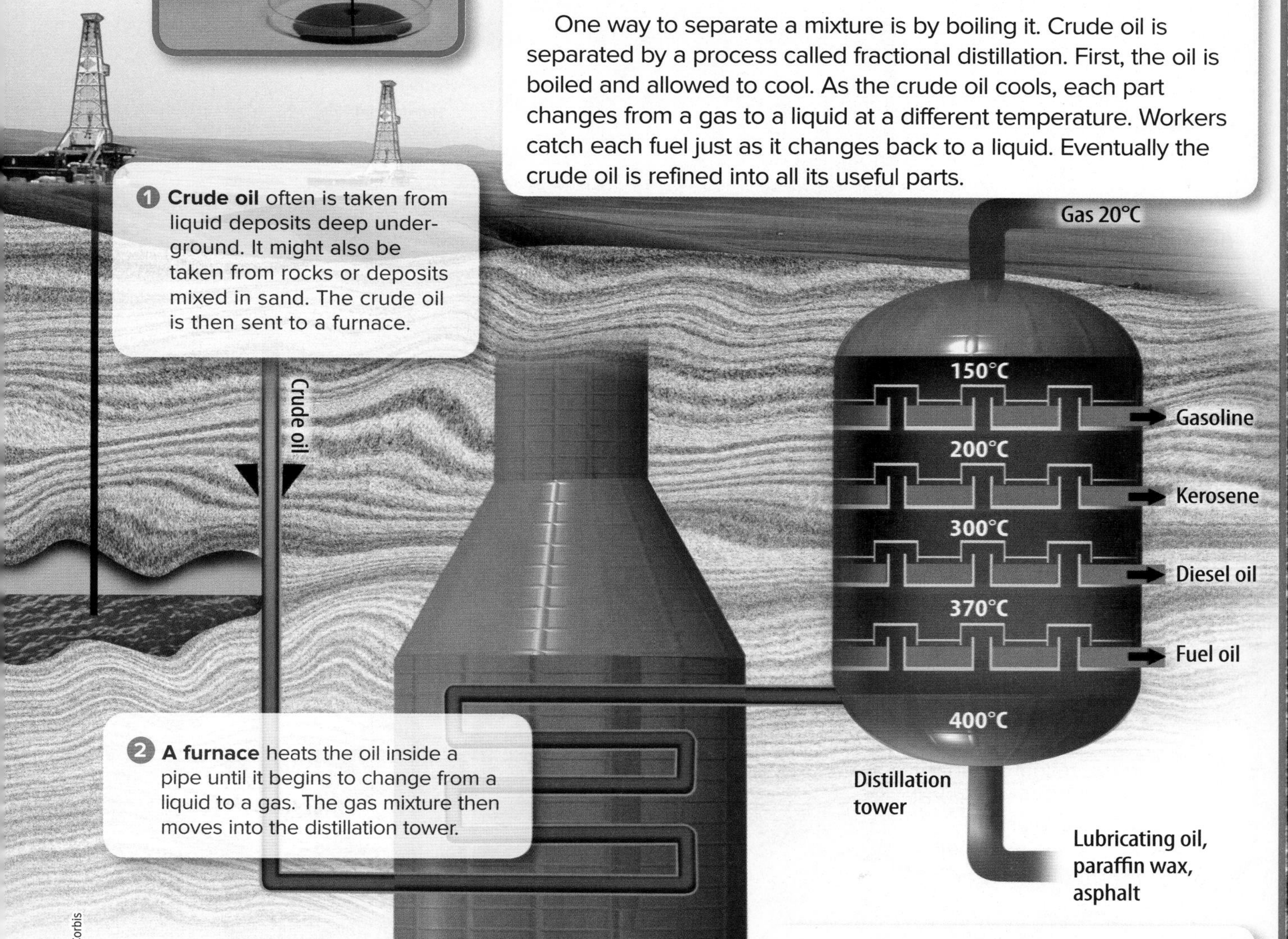

1 **Crude oil** often is taken from liquid deposits deep underground. It might also be taken from rocks or deposits mixed in sand. The crude oil is then sent to a furnace.

2 **A furnace** heats the oil inside a pipe until it begins to change from a liquid to a gas. The gas mixture then moves into the distillation tower.

3 **The distillation tower** is hot at the bottom and cooler higher up. As the gas mixture rises to fill the tower, it cools. It also passes over trays at different levels. Each fuel in the mixture changes to a liquid when it cools to a temperature that matches its boiling point. Gasoline changes to a liquid at the level in the tower at 150°C. A tray then catches the gasoline and moves it away.

It's Your Turn

CREATE A POSTER Blood is a mixture, too. Donated blood often is refined in laboratories to separate it into parts. What are those parts? What are they used for? How are they separated? Find the answers, and create a poster based on your findings.

Lesson 2

Reading Guide

Key Concepts
ESSENTIAL QUESTIONS

- Where are protons, neutrons, and electrons located in an atom?
- How is the atomic number related to the number of protons in an atom?
- What effect does changing the number of particles in an atom have on the atom's identity?

Vocabulary

nucleus p. 368
proton p. 368
neutron p. 368
electron p. 368
electron cloud p. 369
atomic number p. 370
isotope p. 371
ion p. 371

Multilingual eGlossary

BrainPOP®

Go to the resource tab in ConnectED to find the PBL *Model Molecules.*

The Structure of Atoms

Inquiry What makes them different?

This ring is made of two of the most beautiful materials in the world—diamond and gold. Diamond is a clear, sparkling crystal made of only carbon atoms. Gold is a shiny, yellow metal made of only gold atoms. How can they be so different if each is made of just one type of atom? The structure of atoms makes significant differences in materials.

Launch Lab

20 minutes

How can you make different things from the same parts?

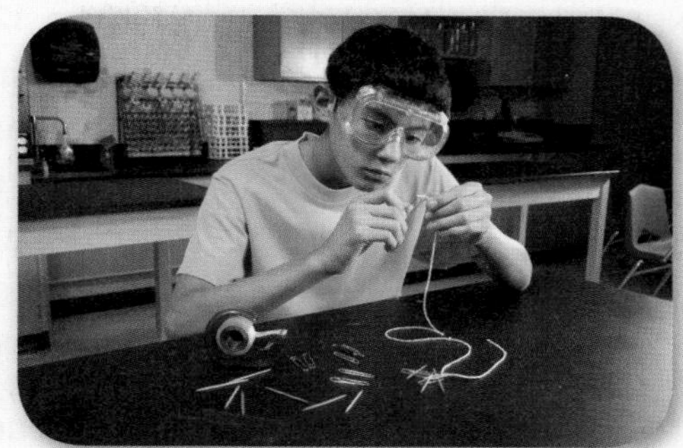

Atoms are all made of the same parts. Atoms can be different from each other because they have different numbers of these parts. In this lab, you will investigate how you can make things that are different from each other even though you use the same parts to make them.

1. Read and complete a lab safety form.
2. Think about how you can join **paper clips, toothpicks,** and **string** to make different types of objects. You must use at least one of each item, but not more than five of any kind.
3. Make the object. Use **tape** to connect the items.
4. Plan and make two more objects using the same three items, varying the numbers of each item.
5. In your Science Journal, describe how each of the objects you made are alike and different.

Think About This

1. **Observe** What do the objects you made have in common? In what ways are they different?
2. **Key Concept** What effect do you think increasing or decreasing the number of items you used would have on the objects you made?

The Parts of an Atom

Now that you have read about ways to classify matter, you can probably recognize the different types you see each day. You might see pure elements, such as copper and iron, and you probably see many compounds, such as table salt. Table salt is a compound because it contains the atoms of two different elements–sodium and chlorine–in a specific combination. You also probably see many mixtures. The silver often used in jewelry is a homogeneous mixture of metals that are evenly mixed, but not bonded together.

As you read in Lesson 1, the many types of matter are possible because there are about 118 different elements. Each element is made up of a different type of atom. Atoms can combine in many different ways. They are the basic parts of matter.

What makes the atoms of each element different? Atoms are made of several types of particles. The number of each of these particles in an atom is what makes atoms different from each other. It is what makes so many types of matter possible.

Reading Check What makes the atoms of different elements different from each other?

FOLDABLES®

Make a vertical two-column chart book. Label it as shown. Use it to organize information about the particles in an atom.

Hutchings Photography/Digital Light Source

The Nucleus—Protons and Neutrons

The basic structure of all atoms is the same. As shown in **Figure 12,** an atom has a center region with a positive charge. One or more negatively charged particles move around this center region. *The* **nucleus** *is the region at the center of an atom that contains most of the mass of the atom.* Two kinds of particles make up the nucleus. *A* **proton** *is a positively charged particle in the nucleus of an atom. A* **neutron** *is an uncharged particle in the nucleus of an atom.*

SCIENCE USE V. COMMON USE

charge

Science Use an electrical property of some objects that determines whether the object is positive, negative, or neutral

Common Use buying something with a credit card

Reading Check Why does a nucleus always have a positive charge?

WORD ORIGIN

proton

from Greek *protos,* means "first"

Electrons

Atoms have no electric charge unless they change in some way. Therefore, there must be a negative charge that balances the positive charge of the nucleus. *An* **electron** *is a negatively charged particle that occupies the space in an atom outside the nucleus.* Electrons are so small and move so quickly that scientists are unable to tell exactly where a given electron is located at any specific time. Therefore, scientists describe their positions around the nucleus as a cloud rather than specific points. A model of an atom and its parts is shown in **Figure 12.**

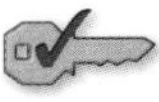

Key Concept Check Where are protons, neutrons, and electrons located in an atom?

Figure 12 All atoms have a positively charged nucleus surrounded by one or more electrons.

Parts of an Atom

Animation

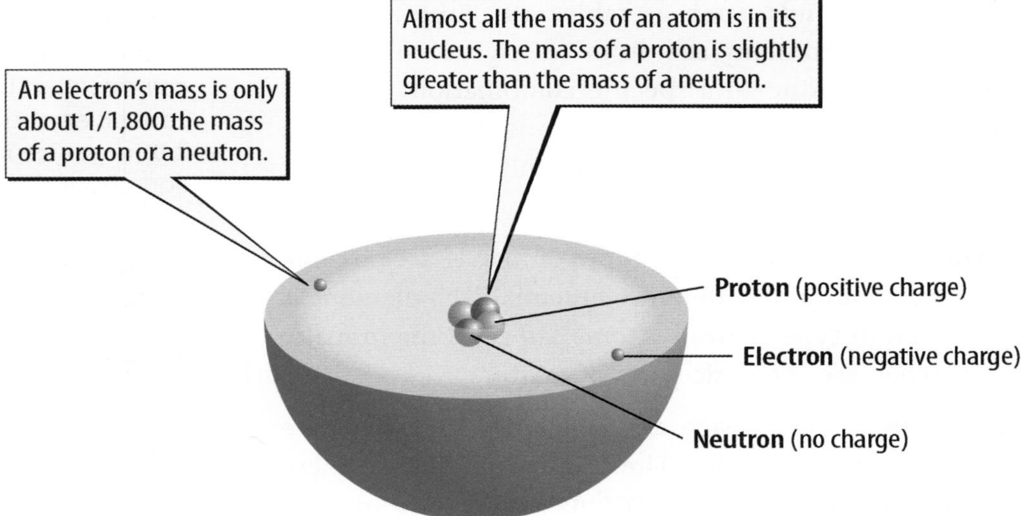

An electron's mass is only about 1/1,800 the mass of a proton or a neutron.
Almost all the mass of an atom is in its nucleus. The mass of a proton is slightly greater than the mass of a neutron.
Proton (positive charge)
Electron (negative charge)
Neutron (no charge)

Visual Check How many protons and how many electrons does this atom have?

An Electron Cloud Drawings of an atom, such as the one in Figure 13, often show electrons circling the nucleus like planets orbiting the Sun. Scientists have conducted experiments that show the movement of electrons is more complex than this. The modern idea of an atom is called the electron-cloud model. *An* **electron cloud** *is the region surrounding an atom's nucleus where one or more electrons are most likely to be found.* It is important to understand that an electron is not a cloud of charge. An electron is one tiny particle. An electron cloud is mostly empty space. At any moment in time, electrons are located at specific points within that area.

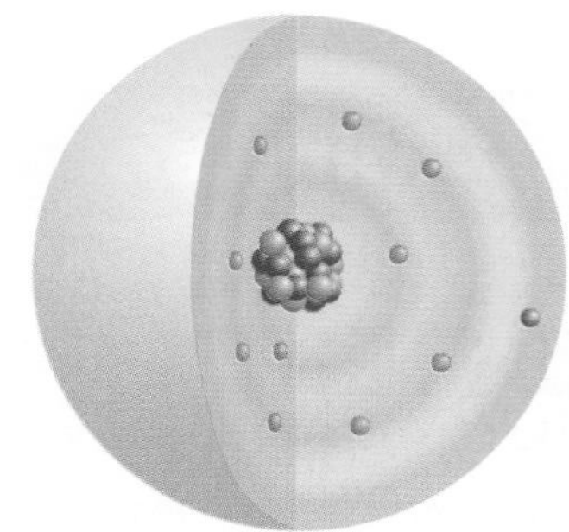

▲ **Figure 13** Electrons farther from the nucleus have more energy.

Electron Energy You have read that electrons are constantly moving around the nucleus in a region called the electron cloud. However, some electrons are closer to the nucleus than others. Electrons occupy certain areas around the nucleus according to their energy, as shown in Figure 13. Electrons close to the nucleus are strongly attracted to it and have less energy. Electrons farther from the nucleus are less attracted to it and have more energy.

The Size of Atoms

It might be difficult to visualize an atom, but every solid, liquid, and gas is made of millions and millions of atoms. Your body, your desk, and the air you breathe are all made of tiny atoms. To understand how small an atom is, look at Figure 14. Suppose you could increase the size of everything around you. If you could multiply the width of an atom by 100 million, or 1×10^8, it would be the size of an orange. An orange would then increase to the size of Earth!

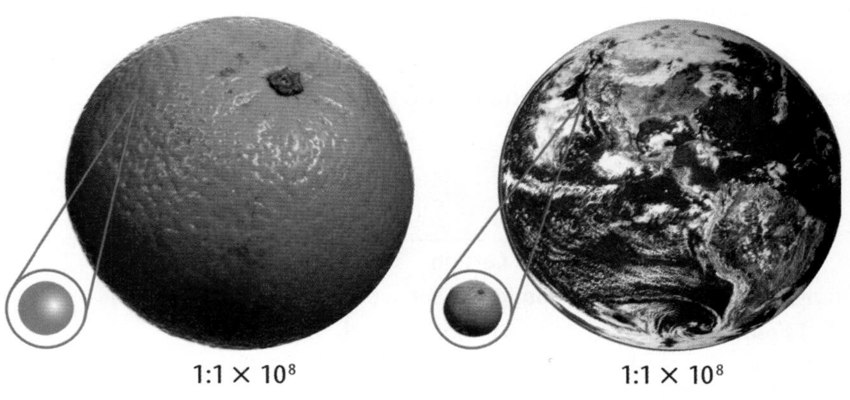

▲ **Figure 14** If an orange were the size of Earth, then an atom would be the size of an orange.

Math Skills

Use Scientific Notation

Scientists write very large and very small values using scientific notation. A gram of carbon has about 50,000,000,000,000,000,000 atoms. Express this in scientific notation.

1. Move the decimal until one nonzero digit remains on the left:

   5.0000000000000000000

2. Count the places you moved. Here it is 19 left.
3. Show that number as a power of 10. The exponent is negative if the decimal moves right and positive if it moves left. Answer: 5×10^{19}
4. Reverse the process to change scientific notation back to a whole number.

Practice

The diameter of a carbon atom is 2.2×10^{-8} cm. Write this as a whole number.

 Math Practice

 Personal Tutor

MiniLab

20 minutes

How can you model atoms?

You can use models to study parts of atoms.

Element	Protons	Neutrons	Electrons
Boron	5	6	
		5	4
Carbon		6	6
	2	2	
Nitrogen	7	6	

1. Read and complete a lab safety form.
2. Copy the table above into your Science Journal. Fill in the blanks in the table.
3. Use pieces of **toothpicks** and **colored marshmallows** to model the nucleus of an atom of each element. Use pink for protons and green for neutrons.
   ⚠ *Do not eat any food you use for a lab.*
4. On a desk, use yellow marshmallows to surround each nucleus with electrons.

Analyze and Conclude

1. **Decide** Which model element's atomic number is greatest? How do you know?
2. **Key Concept** What would change if the last model element had eight protons?

Differences in Atoms

In some ways atoms are alike. Each has a positively charged nucleus surrounded by a negatively charged electron cloud. But atoms can differ from each other in several ways. Atoms can have different numbers of protons, neutrons, or electrons.

Protons and Atomic Number

Look at the periodic table in the back of this book. In each block, the number under the element name shows how many protons each atom of the element has. For example, each oxygen atom has eight protons. *The* **atomic number** *is the number of protons in the nucleus of an atom of an element.* If there are 12 protons in the nucleus of an atom, that element's atomic number is 12. Examine **Figure 15.** Notice that the atomic number of magnesium is the whole number above its symbol. The atomic number of carbon is 6. This means that each carbon atom has 6 protons.

Every element in the periodic table has a different atomic number. You can identify an element if you know either its atomic number or the number of protons its atoms have. If an atom has a different number of protons, it is a different element.

Key Concept Check How is the atomic number related to the number of protons in an atom?

Figure 15 An atomic number is the number of protons in each atom of the element.

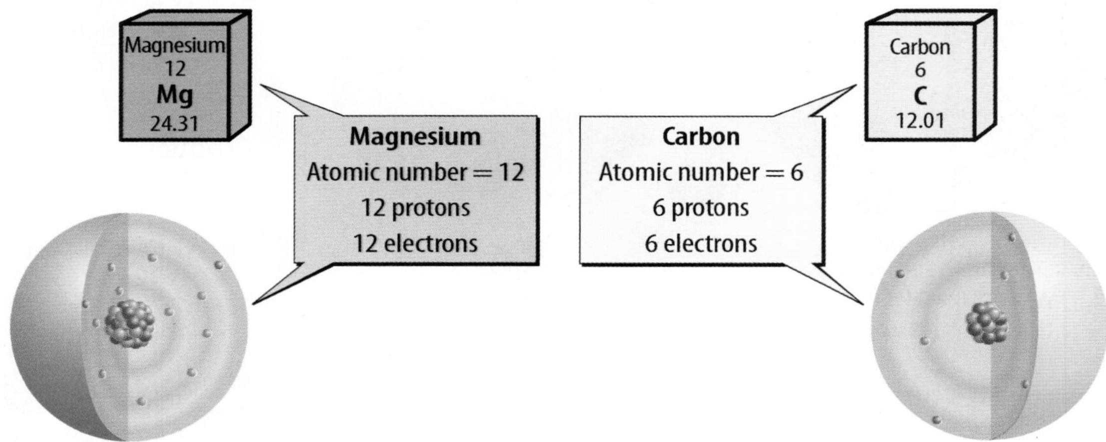

Neutrons and Isotopes

Each atom of an element contains the same number of protons, but the number of neutrons can vary. *An* **isotope** (I suhtohp) ***is one of two or more atoms of an element having the same number of protons, but a different number of neutrons.*** Boron-10 and boron-11 are isotopes of boron, as shown in **Figure 16.** Notice that boron-10 has ten particles in its nucleus. Boron-11 has 11 particles in its nucleus.

 **Reading Check** How do fluorine-19 and fluorine-20 differ?

Electrons and Ions

You read that atoms can differ by the number of protons or neutrons they have. **Figure 17** illustrates a third way atoms can differ–by the number of electrons. A neutral, or uncharged, atom has the same number of positively charged protons and negatively charged electrons. As atoms bond, their numbers of electrons can change. Because electrons are negatively charged, a neutral atom that has lost an electron has a positive charge. A neutral atom that has gained an electron has a negative charge. *An* **ion** (I ahn) ***is an atom that has a charge because it has gained or lost electrons.*** Because the number of protons is unchanged, an ion is the same element it was before.

In the previous lesson, you read that each particle of a compound is two or more atoms of different elements bonded together. One of the ways compounds form is when one or more electrons move from an atom of an element to an atom of a different element. This results in a positive ion for one element and a negative ion for the other element.

Isotopes

Boron-10

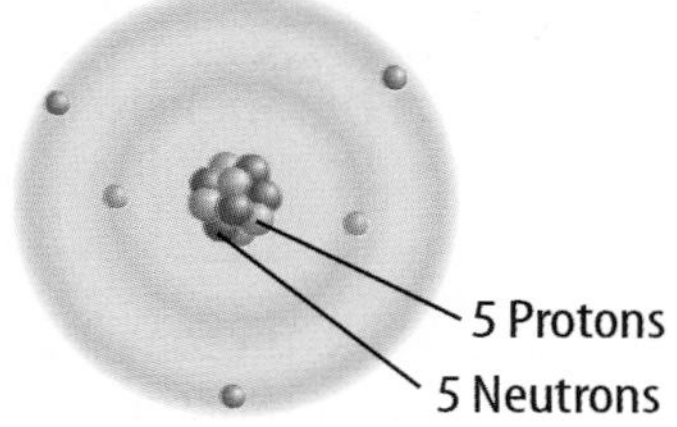

Boron-11

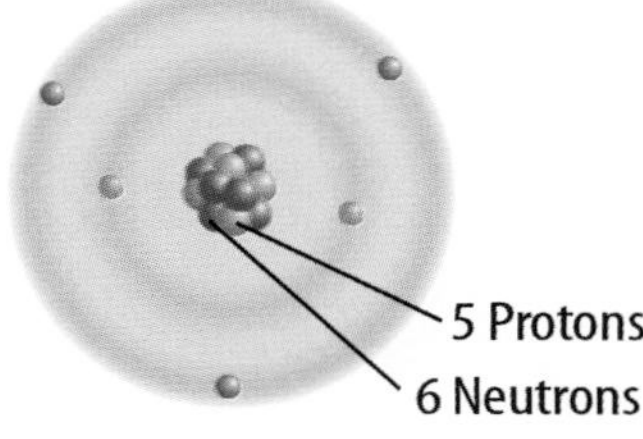

▲ **Figure 16** Boron-10 and boron-11 are isotopes. The number of protons is the same, but the number of neutrons is different.

Figure 17 A positive ion has fewer electrons than protons. A negative ion has more electrons than protons. ▼

Ions

 Animation

Beryllium

4 Protons
4 Electrons

Neutral atom

A neutral atom has the same number of electrons and protons. The atom has no charge.

Sodium

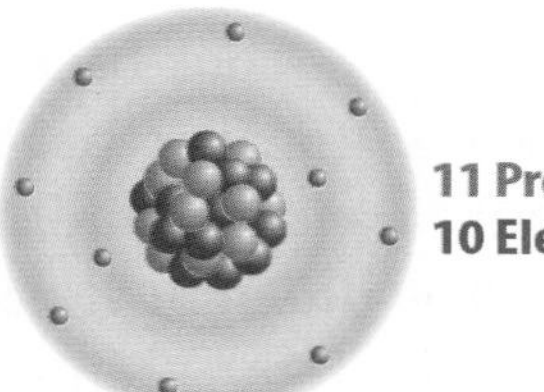

Positive ion (Na^+)

If an atom loses an electron, it has more protons than electrons. It is positively charged.

Chlorine

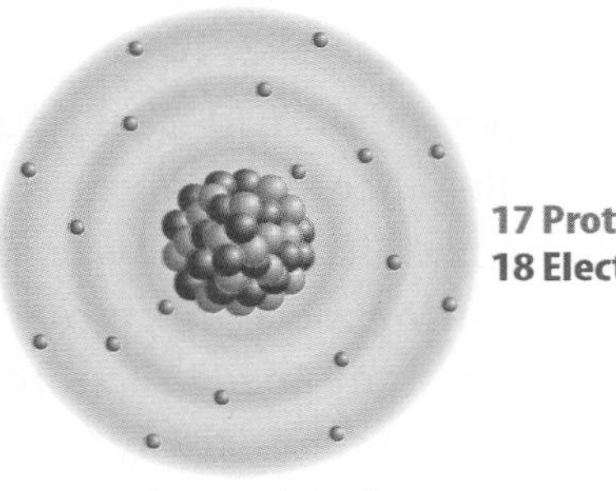

Negative ion (Cl^-)

If an atom gains an electron, it has more electrons than protons. It is negatively charged.

Visual Check Would a nitrogen atom be a positive or a negative ion if it had 10 electrons? Why?

Table 3 Possible Changes in Atoms

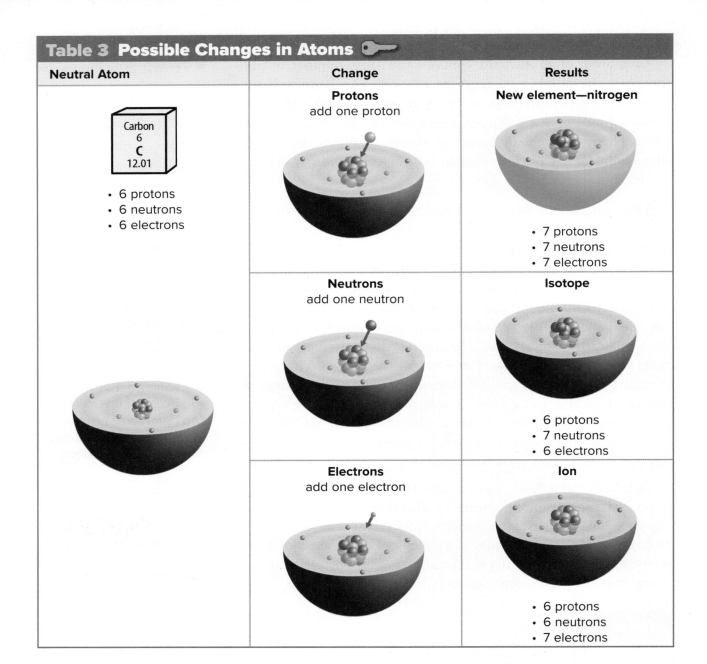

Neutral Atom	Change	Results
Carbon 6 C 12.01 • 6 protons • 6 neutrons • 6 electrons	**Protons** add one proton	**New element—nitrogen** • 7 protons • 7 neutrons • 7 electrons
	Neutrons add one neutron	**Isotope** • 6 protons • 7 neutrons • 6 electrons
	Electrons add one electron	**Ion** • 6 protons • 6 neutrons • 7 electrons

Atoms and Matter

You have now read that matter can be either a substance or a mixture. A substance has a composition that is always the same, but the composition of a mixture can vary. All types of matter are made of atoms. The atoms of a certain element always have the same number of protons, but the number of neutrons can vary. When elements combine to form compounds, the number of electrons in the atoms can change. The different ways in which atoms can change are summarized in **Table 3.**

Look back at the diamond and gold ring on the first page of this lesson. Now can you answer the question of how they can be so different if each is made of just one type of atom? Each carbon atom in diamond has six protons. Each gold atom has 79 protons. The parts of an atom give an element its identity. The ways in which the atoms combine result in the many different kinds of matter.

Key Concept Check What effect does changing the number of particles in an atom have on the atom's identity?

Lesson 2 Review

Visual Summary

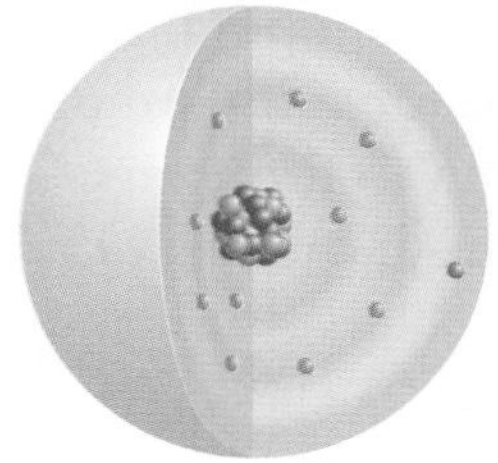

All matter is made of atoms. Atoms are made of protons, electrons, and neutrons.

An orange is about 100 million times wider than an atom.

Atoms of the same element can have different numbers of neutrons.

FOLDABLES®

Use your lesson Foldable to review the lesson. Save your Foldable for the project at the end of the chapter.

What do you think 

You first read the statements below at the beginning of the chapter.

4. An atom is mostly empty space.

5. If an atom gains electrons, the atom will have a positive charge.

6. Each electron is a cloud of charge that surrounds the center of an atom.

Did you change your mind about whether you agree or disagree with the statements? Rewrite any false statements to make them true.

Use Vocabulary

1. **Distinguish** between a proton and a neutron.
2. An atom that has lost one or more electrons is a(n) ________.
3. **Use the term** *isotope* in a complete sentence.

Understand Key Concepts 

4. Which is located outside the nucleus of an atom?
   A. electron
   B. ion
   C. neutron
   D. proton
5. **Identify** the element that has nine protons.
6. **Explain** how atomic number relates to the number of particles in an atom's nucleus.

Interpret Graphics

7. **Organize** Copy and fill in the graphic organizer below to summarize what you have learned about the parts, the sizes, and the differences of atoms.

Properties of Atoms	
Parts	
Sizes	
Differences	

Critical Thinking

8. **Decide** Can you tell which element an atom is if you know its charge and the number of electrons it has? Explain.

Math Skills

9. The diameter of an atomic nucleus is about 0.0000000000000016 cm. Express this number in scientific notation.
10. The mass of a hydrogen atom is about 1.67×10^{-27} kg. Express this as a whole number.

 Lab

50 minutes

Balloon Molecules

Materials

balloons

tape

black marker

index cards

Safety

Knowing how atoms join to form the smallest parts of a compound can be useful. It can sometimes help you predict properties of compounds. It also can help you understand how compounds combine to form mixtures. In this lab, you will connect small balloons to make models of molecules.

Question

How do atoms combine to make molecules?

Procedure

1. Read and complete a lab safety form.
2. Look at the molecule models in the table below. Each molecule is made of two or more atoms. Each type of atom is drawn in a different color.
3. Notice that a water molecule—H_2O—consists of two hydrogen atoms and one oxygen atom.

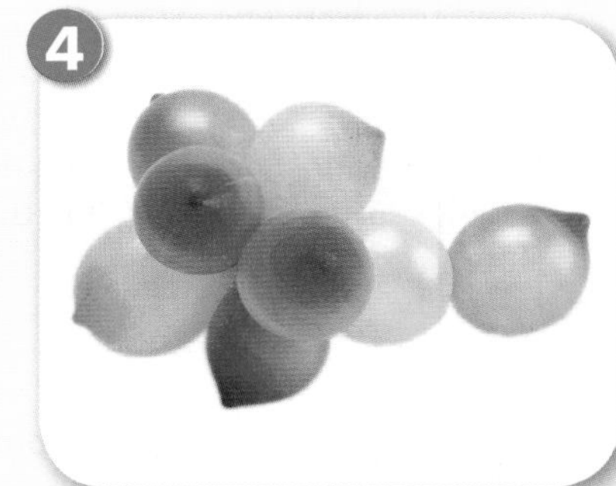

4. Inflate three balloons as models of the three atoms that make up a water molecule. Choose one color for the two hydrogen atoms and a different color for the oxygen atom. Inflate each balloon until it is about 4 cm wide.
5. Look at the shape of the water molecule in the table. Use tape to connect your model atoms in that shape.
6. Use a black marker to write *H* on each hydrogen balloon and *O* on the oxygen balloon.
7. Write *Water H_2O* on an index card, and place the card next to your model.

Water H_2O	Carbon dioxide CO_2	Bromine Br_2	Dinitrogen tetraoxide N_2O_4
Chloroform $CHCl_3$	Ammonia NH_3	Hydrogen peroxide H_2O_2	Ozone O_3

(t to b) (1, 2, 5) Hutchings Photography/Digital Light Source, (3) Ken Karp/McGraw-Hill Education, (4) Joe Polillio/McGraw-Hill Education

8 Look at the molecules in the table. Choose three molecules that you would like to model. Notice the types of atoms that make up the molecules you have chosen to model.

9 Choose a different color balloon for each type of atom. If possible, use the same colors for hydrogen and oxygen that you used for your water molecule.

10 Use tape to connect the atoms in the same arrangements shown in the table. Then use a marker to write the chemical symbol of each element on the balloon for that type of atom.

11 Label an index card for each molecule, just as you did for the water molecule. Display each of your models together.

Analyze and Conclude

12 **Analyze** Which, if any, of the molecules you modeled represent the smallest particles of a substance? Which, if any, represent the smallest particles of an element? Explain.

13 **The Big Idea** How do the molecules you modeled depend on atoms?

Communicate Your Results

Use a digital camera to take photographs of each model you made. Then, use the photos to make a computer presentation explaining the atoms that join to make each molecule you modeled.

Inquiry Extension

Make models for the other compounds shown in this chapter, including any that you did not previously make in the table on the previous page. Remember that the smallest parts of some compounds, such as NaCl, are not molecules because the same atoms do not always travel together. You can still model these particles as long as you keep in mind that they are not called molecules.

10

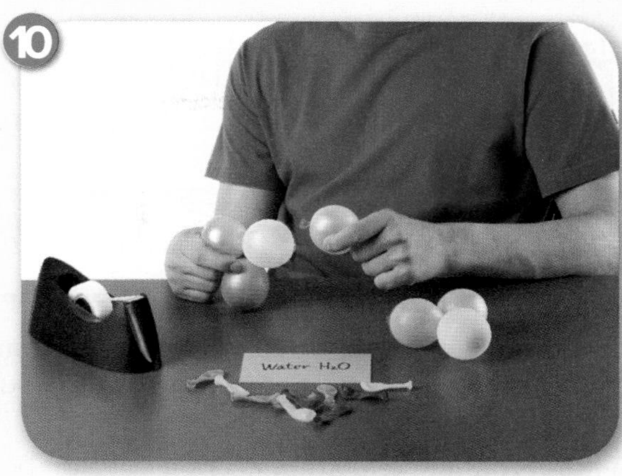

Lab Tips

- When making your models, it is best to have all the balloons inflated to the same size, but keep in mind that real atoms have different diameters.
- Press down lightly when writing the chemical symbols on the model atoms to avoid popping the balloons.

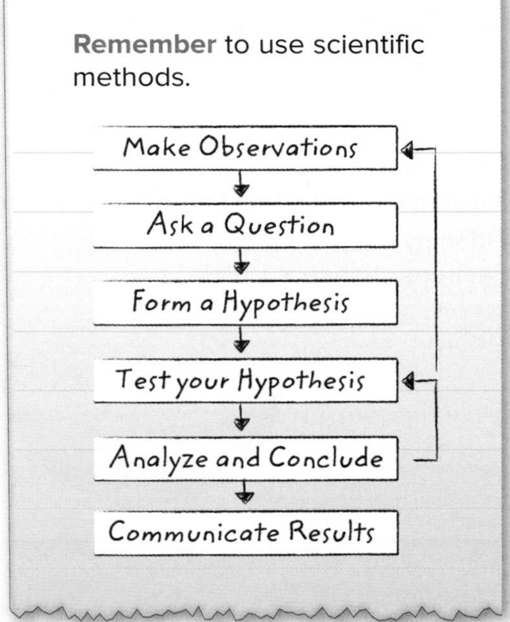

Chapter 11 Study Guide

Matter is classified according to the type and arrangement of atoms from which it is made.

Key Concepts Summary

Lesson 1: Substances and Mixtures

- An **atom** is a building block of **matter.** An **element** is matter made of only one type of atom. A **compound** is a **substance** that contains two or more elements.
- A **heterogeneous mixture** is not a solution because the substances that make up a heterogeneous mixture are not evenly mixed. The substances that make up a solution, or a **homogeneous mixture,** are evenly mixed.
- **Mixtures** differ from compounds in their composition, whether their parts join, and the properties of their parts.

Vocabulary

matter p. 353
atom p. 353
substance p. 354
element p. 355
molecule p. 355
compound p. 356
mixture p. 358
heterogeneous mixture p. 359
homogeneous mixture p. 360

Lesson 2: The Structure of Atoms

- The center of an atom is the **nucleus.** The nucleus contains **protons** and **neutrons. Electrons** occupy the space in an atom outside the nucleus.
- The identity of an atom is determined by its **atomic number.** The atomic number is the number of protons in the atom.
- The identity of an atom stays the same if the number of neutrons or electrons changes.

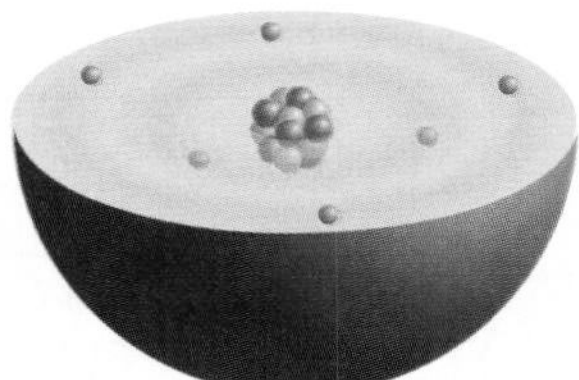

Vocabulary

nucleus p. 368
proton p. 368
neutron p. 368
electron p. 368
electron cloud p. 369
atomic number p. 370
isotope p. 371
ion p. 371

Steve Allen/Brand X Pictures/Alamy

Study Guide

Vocabulary eFlashcards
Vocabulary eGames
Personal Tutor

FOLDABLES® Chapter Project

Assemble your lesson Foldables as shown to make a Chapter Project. Use the project to review what you have learned in this chapter.

Properties of Elements
Properties of Compounds
Particles INSIDE the Nucleus
Particles OUTSIDE the Nucleus

Use Vocabulary

1. A particle that consists of two or more atoms bonded together is a(n) ________.
2. A salad is an example of a(n) ________ because it is a mixture in which you can easily remove the individual parts.
3. Matter is classified as a(n) ________ if it is made of two or more substances that are physically blended but are not chemically bonded.
4. A positively charged particle in the nucleus of an atom is a(n) ________.
5. Almost all of the mass of an atom is found in the ________ of an atom.
6. If a chlorine atom gains an electron, it becomes a(n) ________ of chlorine.

Link Vocabulary and Key Concepts

Interactive Concept Map

Copy this concept map, and then use vocabulary terms from the previous page to complete the concept map.

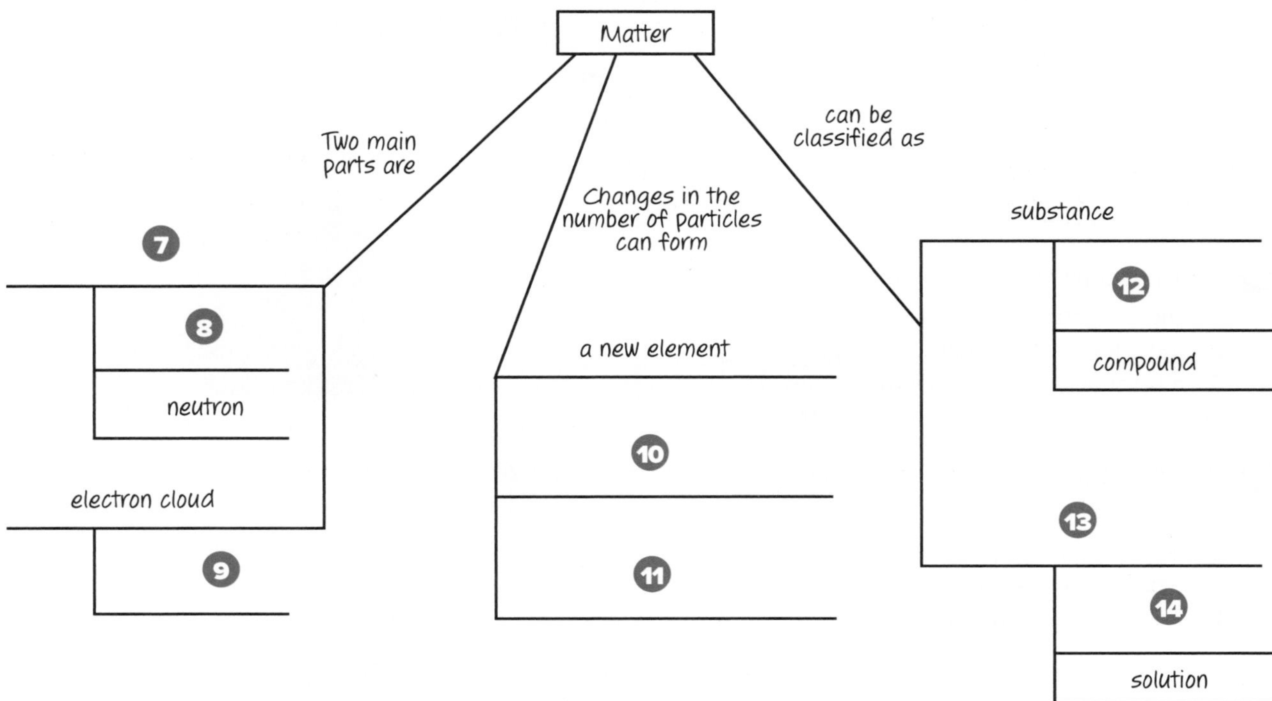

Chapter 11 Review

Understand Key Concepts 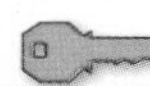

1. Which is a substance?
   - A. fruit salad
   - B. granola cereal
   - C. spaghetti
   - D. table salt

2. Which is the best model for a homogeneous mixture?
   - A. 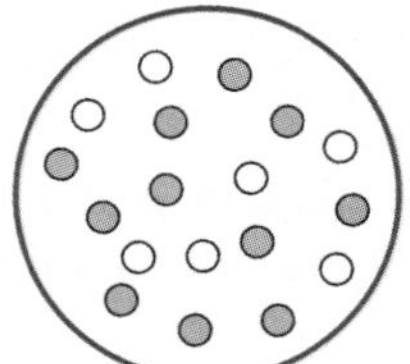
   - B. 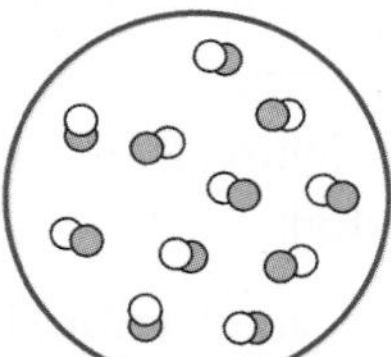
   - C. 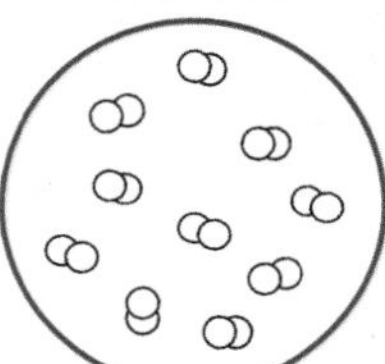
   - D. 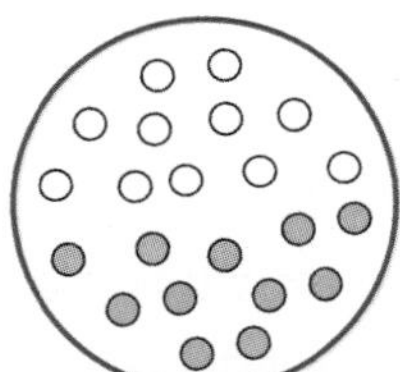

3. Which is a property of all atoms?
   - A. more electrons than protons
   - B. a nucleus with a positive charge
   - C. a positively charged electron cloud
   - D. same number of protons as neutrons

4. Which is another name for a solution?
   - A. element
   - B. compound
   - C. heterogeneous mixture
   - D. homogeneous mixture

5. Which would you most likely be able to separate into its parts by filtering?
   - A. heterogeneous mixture of two liquids
   - B. heterogeneous mixture of two solids
   - C. homogeneous mixture of two liquids
   - D. homogeneous mixture of two solids

6. Where is almost all the mass of an atom located?
   - A. in the electrons
   - B. in the neutrons
   - C. in the nucleus
   - D. in the protons

7. Which best describes an electron cloud?
   - A. an area of charged particles with a fixed boundary
   - B. electrons on a fixed path around the nucleus
   - C. mostly empty space with tiny charged particles in it
   - D. a solid mass of charge around the nucleus

8. Which is true about carbon-12 compared with carbon-13?
   - A. Carbon-12 has more neutrons.
   - B. Carbon-12 has more protons.
   - C. Carbon-13 has more neutrons.
   - D. Carbon-13 has more protons.

9. Look at the periodic table block below for potassium. How many electrons does an uncharged atom of potassium have?

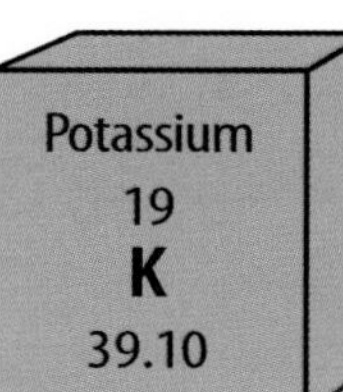

   - A. 19
   - B. 20
   - C. 39
   - D. 40

Online Test Practice

Critical Thinking

10 **Classify** Look at the illustration below. Is this a model of a substance or a mixture? How do you know?

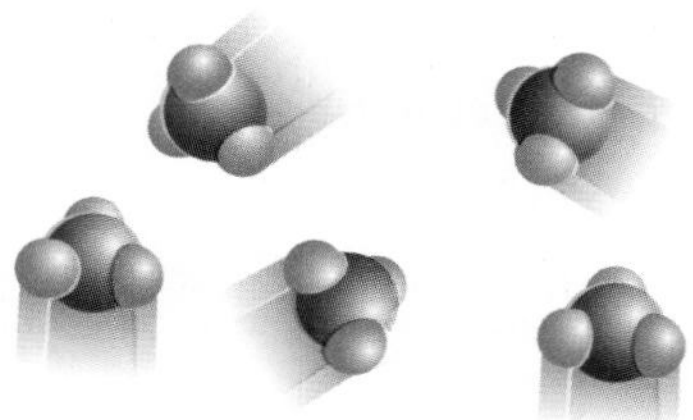

11 **Deduce** Each atom of protium has one proton, no neutrons, and one electron. Each atom of deuterium has one proton, two neutrons, and one electron. Are these the same or different elements? Why?

12 **Decide** Suppose you mix several liquids in a jar. After a few minutes, the liquids form layers. Is this a homogeneous mixture or a heterogeneous mixture? Why?

13 **Describe** a method for separating a mixture of salt water.

14 **Generalize** Consider the substances N_2O_5, H_2, CH_4, H_2O, KCl, and O_2. Is it possible to tell just from the symbols and the numbers which are elements and which are compounds? Explain.

15 **Suggest** how you can define an electron cloud differently from the chapter.

16 **Analyze** A substance has an atomic number of 80. How many protons and electrons do atoms of the substance have? What is the substance?

Writing in Science

17 **Write** a paragraph in which you explain the modern atomic model to an adult who has never heard of it before. Include two questions he or she might ask, and write answers to the questions.

REVIEW THE BIG IDEA

18 Explain how compounds, elements, heterogeneous mixtures, homogeneous mixtures, matter, and substances are related.

19 The photograph below depends on its parts. This is similar to the relationship of matter and atoms. How does the classification of matter depend on atoms?

Scott Camazine/AndreaMosaic/Science Source

Math Skills

Math Practice

Use Scientific Notation

20 The mass of one carbon atom is 0.00000000000000000000001994 g. Express this number in scientific notation.

21 The mass of an electron is about 9.11×10^{-31} kg. Write this as a whole number.

22 In 1 L of hydrogen gas, there are about 54,000,000,000,000,000,000,000 hydrogen atoms. Express the number of atoms using scientific notation.

23 Particles in chemistry are often described by the unit mole. One mole is defined as about 6.022×10^{23} particles. Write this as a whole number.

24 The mass of hydrogen-3, tritium, is about 5.01×10^{-27} kg. Write this as a whole number.

Standardized Test Practice

Record your answers on the answer sheet provided by your teacher or on a sheet of paper.

Multiple Choice

Use the figure below to answer questions 1 and 2.

1 How many atoms are in the particle?

A 1

B 2

C 3

D 5

2 Which kind of matter might contain only this type of particle?

A a compound

B an element

C a heterogeneous mixture

D a homogeneous mixture

3 Which class of matter is the least evenly mixed?

A compounds

B heterogeneous mixtures

C homogeneous mixtures

D solutions

4 Which correctly describes a compound but not a mixture?

A All the atoms are of the same element.

B All the molecules have at least two atoms.

C The combination of substances never changes.

D The substances can be separated without breaking bonds.

5 A girl pours a spoonful of sugar into a glass of warm water. She stirs the water until the sugar disappears. When she tastes the water, she notices that it is now sweet. Which describes the kind of matter in the glass?

A a compound

B an element

C a solution

D a substance

6 How could you separate a mixture of stone and wooden beads that are all the same size?

A Add water to the mixture and skim off the wooden beads, which float.

B Heat the mixture until the stone beads melt.

C Strain the mixture to separate out the stone beads.

D Use a magnet to pull out the wooden beads.

Use the figure below to answer question 7.

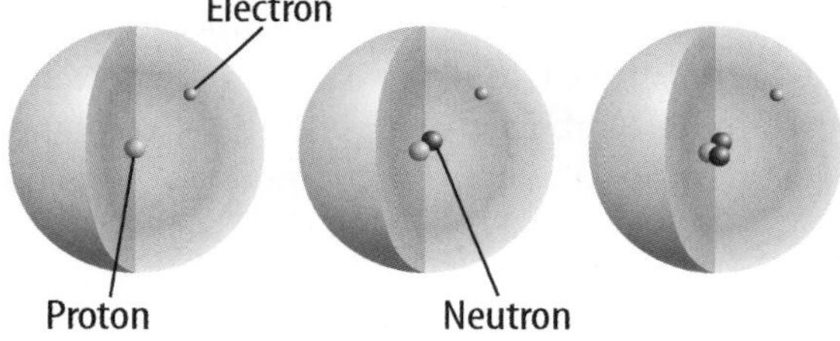

7 The figure shows models of three different atoms. What can you conclude about the three models shown in the figure?

A They all show positive ions.

B They all show negative ions.

C They all show the same element.

D They all show the same isotope.

8 What is the atomic number of an atom that has 2 electrons, 3 protons, and 4 neutrons?

A 2

B 3

C 4

D 7

Use the table below to answer questions 9 and 10.

	Number of Protons	Number of Neutrons	Number of Electrons
A	8	8	8
B	8	8	10
C	8	9	8
D	9	10	9

9 The table shows the numbers of protons, neutrons, and electrons for four atoms. Which atom has a negative charge?

A A

B B

C C

D D

10 Which of the atoms is a different element than the others?

A A

B B

C C

D D

Constructed Response

11 How do protons, electrons, and neutrons differ in charge and location in the atom?

Use the figures below to answer questions 12 and 13.

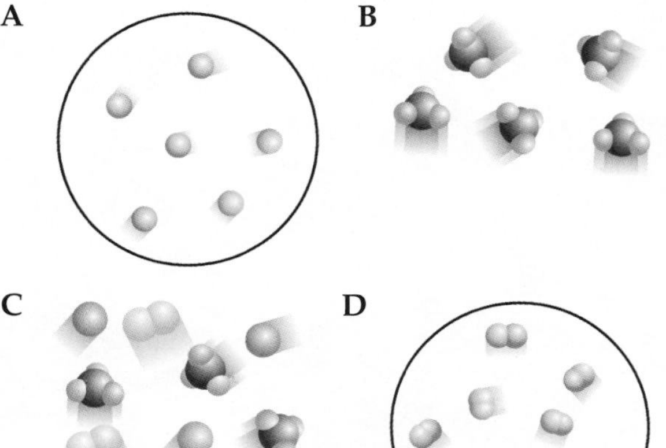

12 Classify each model A–D as either an element, a compound, or a mixture. Explain your reasoning for each answer.

13 Imagine that samples A and D were reacted and formed a compound. Then imagine that the same samples were combined to form a mixture. How would the two combinations differ?

14 Suppose a neutral atom has 5 protons, 5 neutrons, and 5 electrons. List the number of protons, electrons, and neutrons for the following:

a. a positive ion of the same element

b. a negative ion of the same element

c. a neutral isotope of the same element

NEED EXTRA HELP?														
If You Missed Question...	1	2	3	4	5	6	7	8	9	10	11	12	13	14
Go to Lesson...	1	1	1	1	1	1	2	2	2	2	1	1	2	2

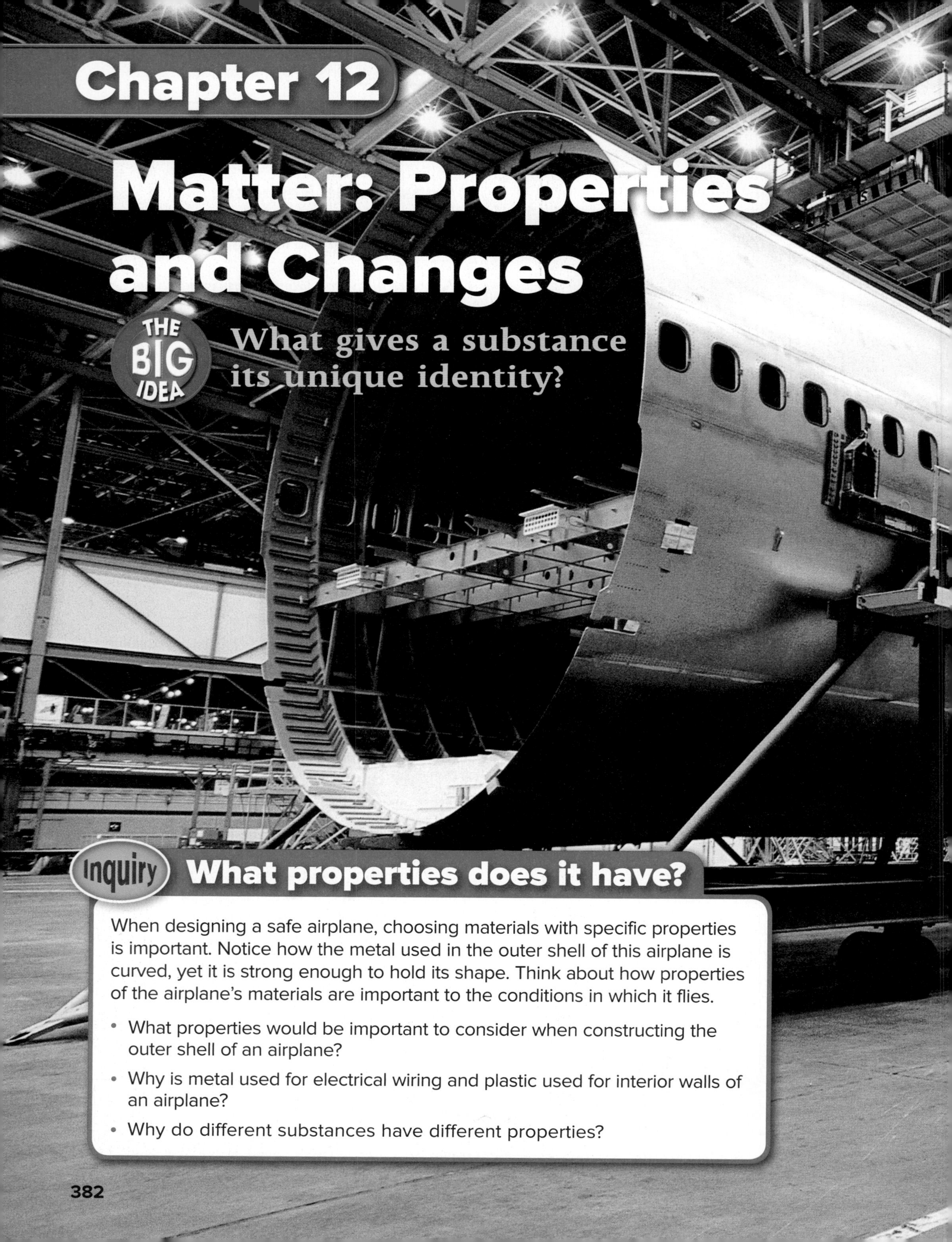

Chapter 12

Matter: Properties and Changes

THE BIG IDEA What gives a substance its unique identity?

Inquiry What properties does it have?

When designing a safe airplane, choosing materials with specific properties is important. Notice how the metal used in the outer shell of this airplane is curved, yet it is strong enough to hold its shape. Think about how properties of the airplane's materials are important to the conditions in which it flies.

- What properties would be important to consider when constructing the outer shell of an airplane?
- Why is metal used for electrical wiring and plastic used for interior walls of an airplane?
- Why do different substances have different properties?

Get Ready to Read

What do you think?

Before you read, decide if you agree or disagree with each of these statements. As you read this chapter, see if you change your mind about any of the statements.

1. The particles in a solid object do not move.
2. Your weight depends on your location.
3. The particles in ice are the same as the particles in liquid water.
4. Mixing powdered drink mix with water causes a new substance to form.
5. If you combine two substances, bubbling is a sign that a new type of substance might be forming.
6. If you stir salt into water, the total amount of matter decreases.

Mc Graw Hill Education connectED

Your one-stop online resource

connectED.mcgraw-hill.com

- LearnSmart®
- Chapter Resources Files, Reading Essentials, Get Ready to Read, Quick Vocabulary
- Animations, Videos, Interactive Tables
- Self-checks, Quizzes, Tests
- Project-Based Learning Activities
- Lab Manuals, Safety Videos, Virtual Labs & Other Tools
- Vocabulary, Multilingual eGlossary, Vocab eGames, Vocab eFlashcards
- Personal Tutors

age fotostock/SuperStock

Lesson 1

Matter and Its Properties

Reading Guide

Key Concepts
ESSENTIAL QUESTIONS

- How do particles move in solids, liquids, and gases?
- How are physical properties different from chemical properties?
- How are properties used to identify a substance?

Vocabulary
volume p. 386
solid p. 386
liquid p. 386
gas p. 386
physical property p. 388
mass p. 388
density p. 389
solubility p. 390
chemical property p. 391

Multilingual eGlossary

BrainPOP®

What makes this possible?

White-water rafting is a lot of fun, but you have to be prepared. The ride down the rapids can be dangerous, and you need good equipment. What properties must the helmets, the raft, the oars, and the life vests have to make a safe white-water ride possible?

Pixtal/age fotostock

Launch Lab

15 minutes

How can you describe a substance?

Think about the different ways you can describe a type of matter. Is it hard? Can you pour it? What color is it? Answering questions like these can help you describe the properties of a substance. In this lab, you will observe how the properties of a mixture can be very different from the properties of the substances it is made from.

1. Read and complete a lab safety form.
2. Using a **small plastic spoon,** measure two spoonfuls of **cornstarch** into a **clear plastic cup.** What does the cornstarch look like? What does it feel like?
3. Slowly stir one spoonful of **water** into the cup containing the cornstarch. Gently roll the new substance around in the cup with your finger.

Think About This

1. What were some properties of the cornstarch and water before they were mixed?
2. **Key Concept** How were the properties of the mixture different from the original properties of the cornstarch and water?

What is matter?

Imagine the excitement of white-water rafting through a mountain pass. As your raft plunges up and down through the rushing water, you grip your oar. You hope that the powerful current will lead you safely past the massive boulders. Only after you reach a quiet pool of water can you finally take a breath and enjoy the beautiful surroundings.

Imagine looking around and asking yourself, "What is matter?" Trees, rocks, water, and all the things you might see on a rafting trip are matter because they have mass and take up space. Air, even though you can't see it, is also matter because it has mass and takes up space. Light from the Sun is not matter because it does not have mass and does not take up space. Sounds, forces, and energy also are not matter.

REVIEW VOCABULARY

matter
anything that has mass and takes up space

Think about the properties of matter you would see on your white-water rafting trip. The helmet you wear is hard and shiny. The rubber raft is soft and flexible. The water is cool and clear. Matter has many different properties. You will learn about some physical properties and chemical properties of matter in this chapter. You will also read about how these properties help to identify many types of matter.

States of Matter

One property that is useful when you are describing different materials is the state of matter. Three familiar states of matter are solids, liquids, and gases. You can determine a material's state of matter by answering the following questions:

- Does it have a definite shape?
- Does it have a definite volume?

Volume *is the amount of space a sample of matter occupies.* As shown in **Table 1,** a material's state of matter determines whether its shape and its volume change when it is moved from one container to another.

Solids, Liquids, and Gases

Notice in **Table 1** that *a* **solid** *is a state of matter with a definite shape and volume.* The shape and volume of a solid do not change regardless of whether it is inside or outside a container. *A* **liquid** *is a state of matter with a definite volume but not a definite shape.* A liquid changes shape if it is moved to another container, but its volume does not change. *A state of matter without a definite shape or a definite volume is a* **gas.** A gas changes both shape and volume depending on the size and shape of its container.

Reading Check Which state of matter has a definite shape and a definite volume?

Table 1 Solids, Liquids, and Gases

Solid Solids, such as rocks, do not change shape or volume regardless of whether they are inside or outside a container.	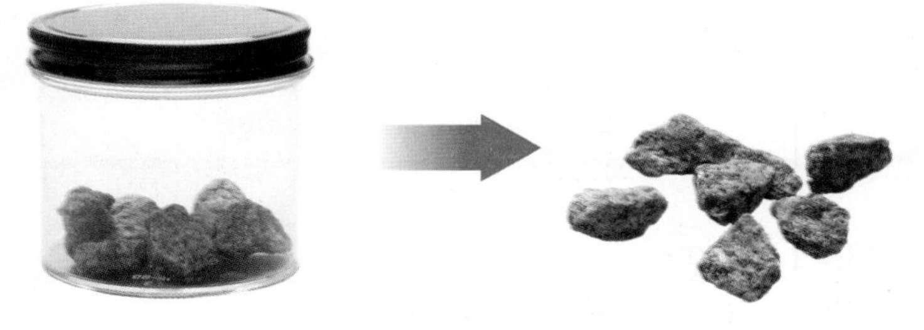
Liquid A liquid, such as fruit juice, changes shape if it is moved from one container to another. Its volume does not change.	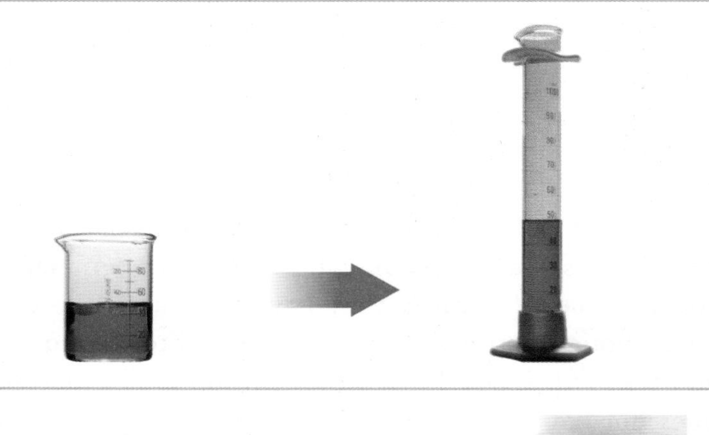
Gas A gas, such as nitrogen dioxide, changes both shape and volume if it is moved from one container to another. If the container is not closed, the gas spreads out of the container.	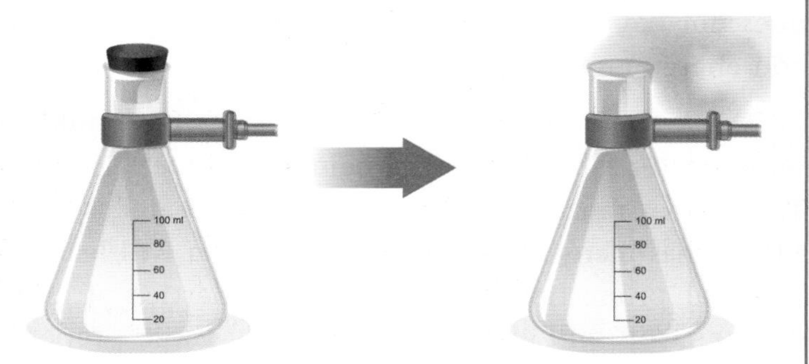

Hutchings Photography/Digital Light Source

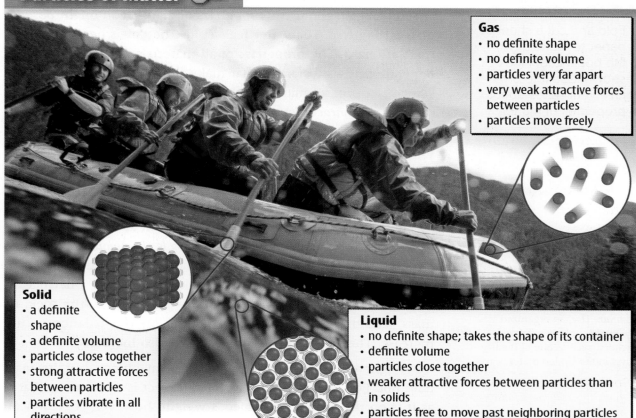

Particles of Matter

Gas
- no definite shape
- no definite volume
- particles very far apart
- very weak attractive forces between particles
- particles move freely

Solid
- a definite shape
- a definite volume
- particles close together
- strong attractive forces between particles
- particles vibrate in all directions

Liquid
- no definite shape; takes the shape of its container
- definite volume
- particles close together
- weaker attractive forces between particles than in solids
- particles free to move past neighboring particles

Figure 1 The movement and attraction between particles are different in solids, liquids, and gases.

Visual Check How does the force between particles differ in a solid, a liquid, and a gas?

Animation

Moving Particles

All matter is made of tiny particles that are constantly moving. Notice in **Figure 1** how the movement of particles is different in each state of matter. In solids, particles vibrate back and forth in all directions. However, particles in a solid cannot move from place to place. In liquids, the distance between particles is greater. Particles in liquids can slide past one another, similar to the way marbles in a box slide around. In a gas, particles move freely rather than staying close together.

Key Concept Check How do particles move in solids, liquids, and gases?

Attraction Between Particles

Particles of matter that are close together exert an attractive force, or pull, on each other. The strength of the attraction depends on the distance between particles. Think about how this attraction affects the properties of the objects in **Figure 1.** A strong attraction holds particles of a solid close together in the same position. Liquids can flow because forces between the particles are weaker. Particles of a gas are so spread apart that they are not held together by attractive forces.

Fold and cut a sheet of paper to make a two-tab book. Label it as shown. Use it to organize your notes about properties of matter.

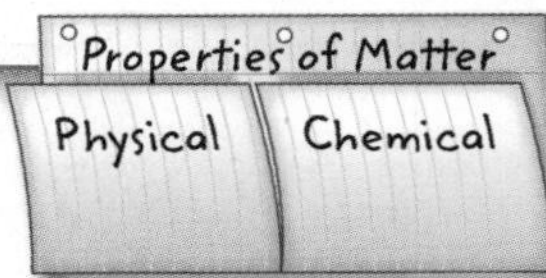

What are physical properties?

Think again about the properties of matter you might observe on a rafting trip. The water feels cold. The raft is heavy. The helmets are hard. The properties of all materials, or types of matter, depend on the substances that make them up. Recall that a substance is a type of matter with a composition that is always the same. *Any characteristic of matter that you can observe without changing the identity of the substances that make it up is a* **physical property.** State of matter, temperature, and the size of an object are all examples of physical properties.

SCIENCE USE V. COMMON USE

state

Science Use a condition or physical property of matter

Common Use an organized group of people in a defined territory, such as one of the fifty states in the United States

Mass and Weight

Some physical properties of matter, such as mass and weight, depend on the size of the sample. **Mass** *is the amount of matter in an object.* Weight is the gravitational pull on the mass of an object. To measure the mass of a rock, you can use a balance, as shown in **Figure 2.** If more particles were added to the rock, its mass would increase, and the reading on the balance would increase. The weight of the rock would also increase.

Weight depends on the location of an object, but its mass does not. For example, the mass of an object is the same on Earth as it is on the Moon. The object's weight, however, is greater on Earth because the gravitational pull on the object is greater on Earth than on the Moon.

Figure 2 You can measure a material's mass and volume and then calculate its density.

Reading Check How do mass and weight differ?

Mass, Volume, and Density

Personal Tutor

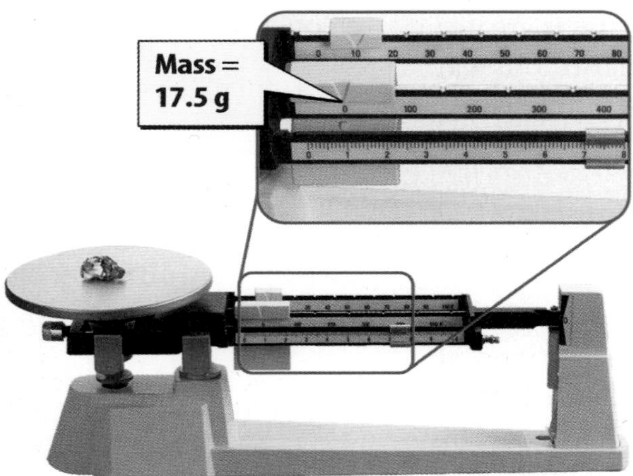

Mass
A balance measures an object's mass by comparing it to the known mass of the slides on the balance. Common units for measuring mass are the kilogram (kg) and the gram (g).

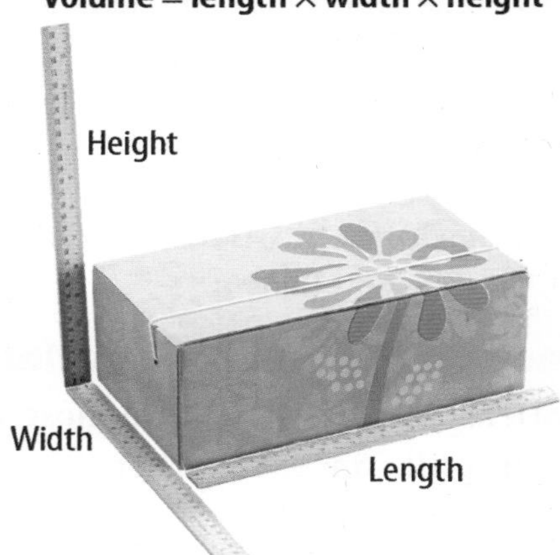

Volume of a Rectangular-Shaped Solid
If a solid has a rectangular shape, you can find its volume by multiplying its length, its width, and its height together. A common unit of volume for a solid is the cubic centimeter (cm^3).

Hutchings Photography/Digital Light Source

Volume

Another physical property of matter that depends on the amount or size of the sample is volume. You can measure the volume of a liquid by pouring it into a graduated cylinder or a measuring cup and reading the volume mark. Two ways to measure the volume of a solid are shown in **Figure 2.** If a solid has a regular geometric shape, you can calculate its volume by using the correct formula. If a solid has an irregular shape, you can use the displacement method to measure its volume.

Density

Density is a physical property of matter that does not depend on the size or amount of the sample. **Density** *is the mass per unit volume of a substance.* Density is useful when identifying unknown substances because it is constant for a given substance, regardless of the size of the sample. For example, imagine hiking in the mountains and finding a shiny yellow rock. Is it gold? Suppose you calculate that the density of the rock is 5.0 g/cm^3. This rock cannot be gold because the density of gold is 19.3 g/cm^3. A sample of pure gold, regardless of the size, will always have a density of 19.3 g/cm^3.

MiniLab — 10 minutes

How can you find an object's mass and volume?

1. Read and complete a lab safety form.
2. Obtain a small sample of **modeling clay.**
3. Using a **balance,** find the mass of the sample. Record it in your Science Journal.

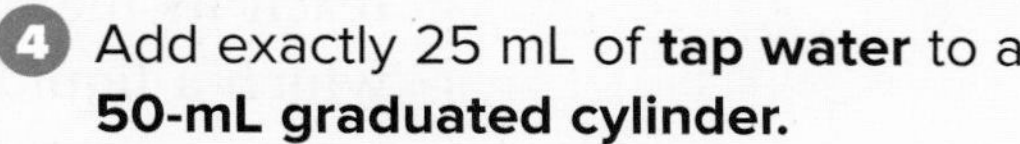

4. Add exactly 25 mL of **tap water** to a **50-mL graduated cylinder.**
5. Shape the clay so that it can be placed into the graduated cylinder.
6. Slide the clay into the graduated cylinder. Record the new volume of the water.

Analyze and Conclude

1. **Compare** the volume of the water with the total volume of the water and the clay. What is the volume of the clay?
2. **Key Concept** Why are mass and volume considered physical properties?

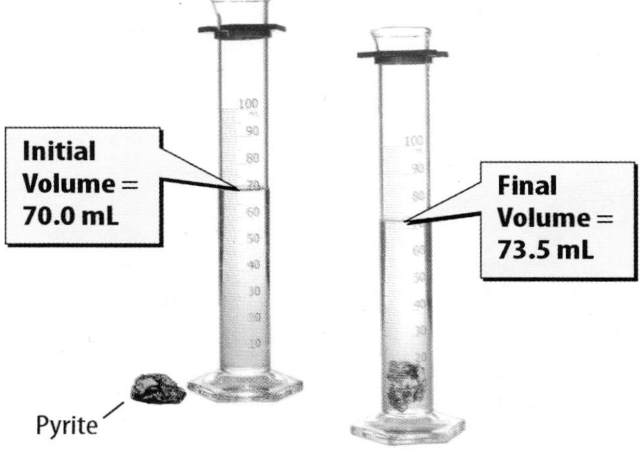

Volume of an Irregular-Shaped Solid
The volume of an irregular-shaped object can be measured by displacement. The volume of the object is the difference between the water level before and after placing the object in the water. The common unit for liquid volume is the milliliter (mL).

Density Equation

$$\text{Density (in g/mL)} = \frac{\text{mass (in g)}}{\text{volume (in mL)}}$$

$$D = \frac{m}{V}$$

To find the density of the rock, first determine the mass and the volume of the rock:

mass: $m = 17.5$ g

volume: $V = 73.5 \text{ mL} - 70.0 \text{ mL} = 3.5 \text{ mL}$

Then, divide the mass by the volume:

$$D = \frac{17.5 \text{ g}}{3.5 \text{ mL}} = 5.0 \text{ g/mL}$$

Density Calculation
Density can be calculated using the density equation. The common units of density are grams per milliliter (g/mL) or grams per cubic centimeter (g/cm^3). 1 mL = 1 cm^3.

Figure 3 The drink mix is soluble in water. The sand is not soluble in water.

Solubility

You can observe another physical property of matter if you stir a powdered drink mix into water. The powder dissolves, or mixes evenly, in the water. **Solubility** *is the ability of one material to dissolve in another.* You cannot see the drink mix powder in the left glass in **Figure 3** because the powder is soluble in water. The liquid is red because of the food coloring in the powder. The sand settles in the glass because it is not soluble in water.

WORD ORIGIN

solubility
from Latin *solubilis,* means "capable of being dissolved"

Melting and Boiling Point

Melting point and boiling point also are physical properties. The melting point is the temperature at which a solid changes to a liquid. Ice cream, for example, melts when it warms enough to reach its melting point. The boiling point is the temperature at which a liquid changes to a gas. If you heat a pan of water, the water will boil, or change to a gas, at its boiling point. Different materials have different melting and boiling points. These temperatures do not depend on the size or amount of the material.

Reading Check How does a substance change at its melting point and at its boiling point?

Additional Physical Properties

Several other physical properties–magnetism, malleability, and electrical conductivity–are shown in **Figure 4.** Notice how the physical properties of each material make it useful. Can you think of other examples of materials chosen for certain uses because of their physical properties?

Physical Properties

Figure 4 Physical properties include magnetism, malleability, and electrical conductivity.

Magnetism is a physical property that allows some materials to attract certain metals.

A malleable material, such as aluminum foil used in cooking, is useful because it can be hammered or rolled into thin sheets.

Some metals, such as copper, are used in electrical wire because of their high electrical conductivity.

(t) Hutchings Photography/Digital Light Source, (bl)Dorling Kindersley/Getty Images, (bc)John A. Rizzo/Getty Images, (br) Dave King/Dorling Kindersley/Getty Images

Chemical Properties

Figure 5 Flammability and the ability to rust are examples of chemical properties.

Flammability
In 1937 the airship *Hindenburg* caught fire and crashed. It was filled with hydrogen, a highly flammable gas.

Ability to rust
The metal parts of an old car soon rust because the metal contains iron. The ability to rust is a chemical property of iron.

What are chemical properties?

Have you ever seen an apple turn brown? When you bite into or cut open apples or other fruits, substances that make up the fruit react with oxygen in the air. When substances react with each other, their particles combine to form a new, different substance. The ability of substances in fruit to react with oxygen is a chemical property of the substances. *A* **chemical property** *is the ability or inability of a substance to combine with or change into one or more new substances.* A chemical property is a characteristic of matter that you observe as it reacts with or changes into a different substance. For example, copper on the roof of a building turns green as it reacts with oxygen in the air. The ability to react with oxygen is a chemical property of copper. Two other chemical properties–flammability and the ability to rust–are shown in **Figure 5.**

Key Concept Check How do chemical properties and physical properties differ?

Flammability

Flammability is the ability of a type of matter to burn easily. Suppose you are on a camping trip and want to light a campfire. You see rocks, sand, and wood. Which would you choose for your fire? Wood is a good choice because it is flammable. Rocks and sand are not flammable.

Materials are often chosen for certain uses based on flammability. For example, gasoline is used in cars because it burns easily in engines. Materials that are used for cooking pans must not be flammable. The tragedy shown in **Figure 5** resulted when hydrogen, a highly flammable gas, was used in the airship *Hindenburg.* Today, airships are filled with helium, a nonflammable gas.

Ability to Rust

You have probably seen old cars that have begun to rust like the one in **Figure 5.** You might also have seen rust on bicycles or tools left outside. Rust is a substance that forms when iron reacts with water and oxygen in the air. The ability to rust is a chemical property of iron or metals that contain iron.

Table 2 Identifying an Unknown Material by its Physical Properties

Substance	Color	Mass (g)	Melting Point (°C)	Density (g/cm^3)
Table salt	white	14.5	801	2.17
Sugar	white	11.5	148	1.53
Baking soda	white	16.0	50	2.16
Unknown	white	16.0	801	2.17

Math Skills

Solve a One-Step Equation

A statement that two expressions are equal is an equation. For example, examine the density equation:

$$D = \frac{m}{V}$$

This equation shows that density, D, is equal to mass, m, divided by volume, V. To solve a one-step equation, place the variables you know into the equation. Then solve for the unknown variable. For example, if an object has a mass of **52 g** and a volume of **4 cm³**, calculate the density as follows:

$$D = \frac{52 \text{ g}}{4 \text{ cm}^3} = 13 \text{ g/cm}^3$$

Practice

A cube of metal measures 3 cm on each side. It has a mass of 216 g. What is the density of the metal?

 Math Practice

 Personal Tutor

Identifying Matter Using Physical Properties

Physical properties are useful for describing types of matter, but they are also useful for identifying unknown substances. For example, look at the substances in **Table 2.** Notice how their physical properties are alike and how they are different. How can you use these properties to identify the unknown substance?

You cannot identify the unknown substance by its color. All of the substances are white. You also cannot identify the unknown substance by its mass or volume. Mass and volume are properties of matter that change with the amount of the sample present. However, recall that melting point and density are properties of matter that do not depend on the size or the amount of the sample. They are more reliable for identifying an unknown substance. Notice that both the melting point and the density of the unknown substance match those of table salt. The unknown substance must be table salt.

When you identify matter using physical properties, consider how the properties are alike and how they are different from known types of matter. It is important that the physical properties you use to identify an unknown type of matter are properties that do not change for any sample size. A cup of salt and a spoonful of salt will have the same melting point and density even though the mass and volume for each will be different. Therefore, melting point and density are physical properties that are reliable when identifying an unknown substance.

 **Key Concept Check** How are properties used to identify a substance?

Hutchings Photography/Digital Light Source

Sorting Materials Using Properties

Both physical properties and chemical properties are useful for sorting materials. The beads in **Figure 6** are sorted by color and shape–two physical properties. When you bring groceries home from the store, you might put crackers in a cupboard, but you probably put milk and yogurt in the refrigerator to keep them from spoiling. The tendency to spoil is a chemical property of the milk and yogurt. You probably often sort other types of matter by physical or chemical properties without realizing it.

▲ **Figure 6** These beads are sorted by color and shape.

Separating Mixtures Using Physical Properties

Physical properties are useful for separating different types of matter that are mixed. For example, suppose you have a frozen juice pop on a stick. How could you separate the frozen juice from the stick? If you set the freezer pop on a counter, the frozen juice will melt and separate from the stick. The melting point of the juice is much lower than the melting point of the stick. Melting point is a physical property you can use to separate mixtures. Other ways that you can use physical properties to separate mixtures are shown in **Figure 7**.

Reading Check How could you separate a mixture of sand and small pebbles?

Figure 7 Physical properties, such as state of matter, boiling point, and magnetism, can be used to separate mixtures. ▼

Separating Mixtures

Separation by State of Matter

▲ Water can flow through the holes in the strainer because it is a liquid. The pasta cannot flow through because the pieces are solid and too large.

Separation by Boiling Point

▲ If you boil a mixture of salt and water, the liquid water changes to a gas when it reaches its boiling point. The salt is left behind.

Separation by Magnetism

▲ Iron filings, which have the property of magnetism, can be separated from the sand using a magnet. The magnet attracts the iron filings but not the sand.

Visual Check How could you separate a mixture of salt, sand, and iron filings?

Lesson 1 Review

Visual Summary

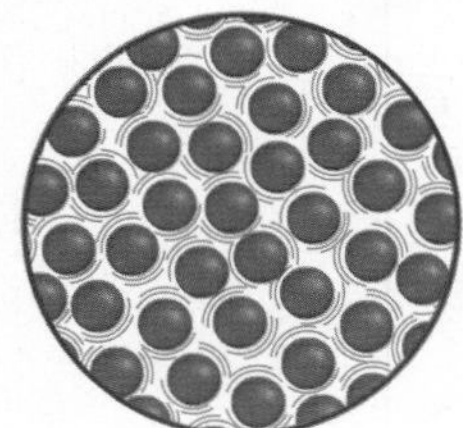

The movement of particles is different in a solid, a liquid, and a gas.

Physical properties and chemical properties are used to describe types of matter.

Physical properties such as magnetism can be used to separate mixtures.

FOLDABLES®

Use your lesson Foldable to review the lesson. Save your Foldable for the project at the end of the chapter.

What do you think 

You first read the statements below at the beginning of the chapter.

1. The particles in a solid object do not move.
2. Your weight depends on your location.
3. The particles in ice are the same as the particles in liquid water.

Did you change your mind about whether you agree or disagree with the statements? Rewrite any false statements to make them true.

Use Vocabulary

1. A state of matter that has a definite volume but not a definite shape is a ___________.
2. **Distinguish** between a physical property and a chemical property.

Understand Key Concepts 

3. **Analyze** Which can be used to identify an unknown substance: mass, melting point, density, volume, state of matter?
4. **Contrast** the movement of particles in a solid, a liquid, and a gas.
5. Which of these is a chemical property?
   A. boiling point
   B. density
   C. flammability
   D. solubility

Interpret Graphics

6. **Explain** Use the drawing to explain why a gas has no definite shape or volume.

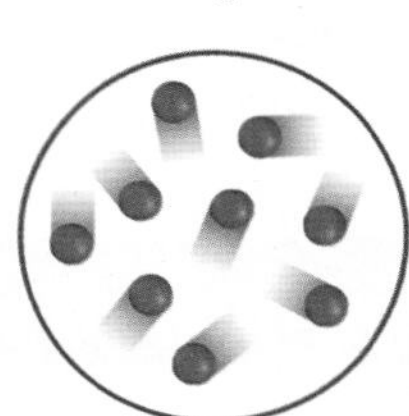

7. **Calculate** Copy the table below and calculate the density of each object.

Object	Mass	Volume	Density
1	6.50 g	1.25 cm^3	
2	8.65 g	2.50 mL	

Critical Thinking

8. **Design** an investigation you could use to find the density of a penny.

Math Skills   Math Practice

9. The mass of a mineral is 9.6 g. The mineral is placed in a graduated cylinder containing 8.0 mL of water. The water level rises to 16.0 mL. What is the mineral's density?

Skill Practice **Measure** 15 minutes

How can you calculate density?

Materials

metal block

100-mL graduated cylinder

metric ruler

triple-beam balance

Safety

Density is the mass per unit volume of a substance. In this lab, you will measure the mass of a solid block. Next you will measure the volume in two different ways. Then you will calculate the density of the block for each volume measurement.

Learn It

Scientists take measurements when collecting data. In this lab, you will **measure** mass and volume, then use these data to calculate density.

Try It

1. Read and complete a lab safety form.
2. Copy the data table in your Science Journal. Use a triple beam balance to measure the mass of the metal block. Record your measurements.
3. Use a ruler to measure the length, width, and height of the block. Record your measurements.
4. Pour 30 mL of water into a 100-mL graduated cylinder. Record the volume of the water.
5. Carefully slide the metal block into the graduated cylinder. Record the total volume.
6. Using the measurements from step 3, determine the volume of the block using this equation: *volume = length × width × height*
7. Calculate the volume of the block using displacement. Subtract the volume of the water in step 4 from the volume of the water and block in step 5.

Apply It

8. **Calculate** Using the mass and each volume measurement of the block, calculate the density of the block.
9. **Compare** the density of the block calculated by the two different volumes. *Hint:* 1 mL = 1 cm^3. Are they the same? Why or why not?
10. **Key Concept** Why is density a physical property of the block?

Measurements	
Mass (g)	
Length (cm)	
Width (cm)	
Height (cm)	
Volume of water (mL)	
Volume of water and block (mL)	

Hutchings Photography/Digital Light Source

Lesson 2

Reading Guide

Key Concepts
ESSENTIAL QUESTIONS

- How are physical changes different from chemical changes?
- How do physical and chemical changes affect mass?

Vocabulary
physical change p. 398
chemical change p. 400
law of conservation of mass p. 403

Multilingual eGlossary

BrainPOP®

Go to the resource tab in ConnectED to find the PBL *A Tale of Two Changes.*

Matter and Its Changes

Inquiry Why is it orange?

Streams are usually filled with clear freshwater. What happened to this water? Chemicals from a nearby mine seeped through rocks before flowing into the stream. These chemicals combined with metals in the rocks, causing orange rust to form in the water.

BRUCE DALE/National Geographic Image Collection

Launch Lab

10 minutes

What does a change in the color of matter show?

Matter has many different properties. Chemical properties can only be observed if the matter changes from one type to another. How can you tell if a chemical property has changed? Sometimes a change in the color of matter shows that its chemical properties have changed.

1. Read and complete a lab safety form.
2. Obtain the **red indicator sponge** and the **red acid solution** from your teacher. Predict what will happen if the red acid solution touches the red sponge.
3. Use a **dropper** to remove a few drops of acid solution from the **beaker.** Place the drops on the sponge. ⚠ *Be careful not to splash the liquid onto yourself or your clothing.*
4. Record your observations in your Science Journal.

Think About This

1. Compare the properties of the sponge before and after you placed the acid solution onto the sponge. Was your prediction correct?
2. **Key Concept** How do you know that physical properties and chemical properties changed?

Changes of Matter

Imagine going to a park in the spring and then going back to the same spot in the fall. What changes do you think you might see? The changes would depend on where you live. An example of what a park in the fall might look like in many places is shown in **Figure 8.** Leaves that are green in the spring might turn red, yellow, or brown in the fall. The air that was warm in the spring might be cooler in the fall. If you visit the park early on a fall morning, you might notice a thin layer of frost on the leaves. Matter, such as the things you see at a park, can change in many ways. These changes can be either physical or chemical.

Reading Check What are some examples of matter changing in winter?

Figure 8 The physical and chemical properties of matter change in a park throughout the year.

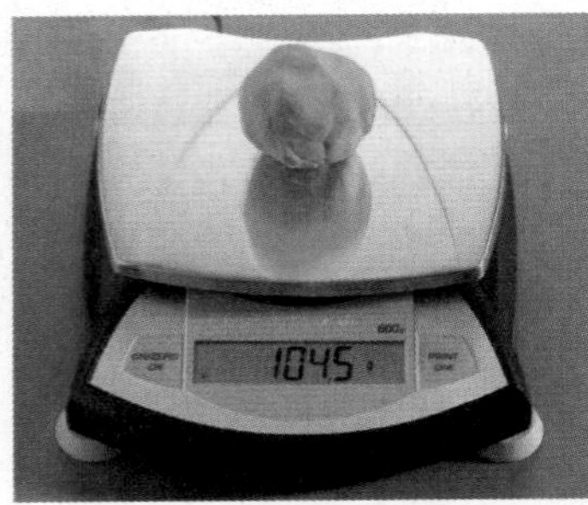

104.5

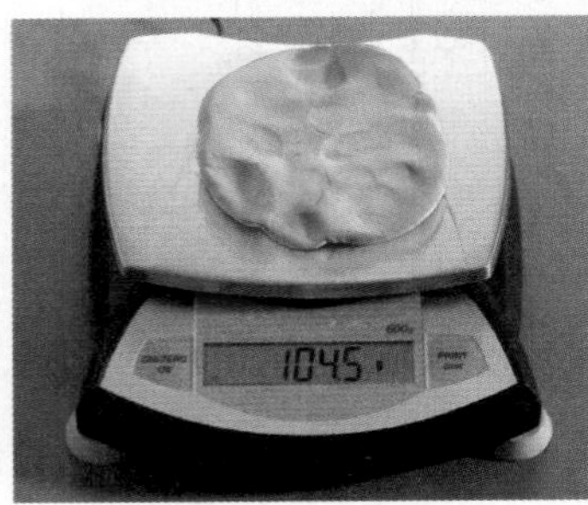

104.5

Figure 9 Changing the shape of the modeling clay does not change its mass.

What are physical changes?

A change in the size, shape, form, or state of matter that does not change the matter's identity is a **physical change**. You can see an example of a physical change in **Figure 9.** Recall that mass is an example of a physical property. Notice that the mass of the modeling clay is the same before and after its shape was changed. When a physical change occurs, the chemical properties of the matter stay the same. The substances that make up matter are exactly the same both before and after a physical change.

Dissolving

One of the physical properties you read about in Lesson 1 was solubility–the ability of one material to dissolve, or mix evenly, in another. Dissolving is a physical change because the identities of the substances do not change when they are mixed. As shown in **Figure 10,** the identities of the water molecules and the sugar molecules do not change when sugar crystals dissolve in water.

Reading Check Explain why dissolving is classified as a physical change.

Dissolving—A Physical Change

Key
Sugar crystal
1 Sugar molecule
$C_{12}H_{22}O_{11}$
1 Water molecule
H_2O
Crystals of sugar are made up of many sugar molecules. The crystals are surrounded by molecules of water.
As the sugar begins to dissolve, the crystals break apart.
Individual sugar and water molecules remain unchanged even after all sugar crystals have dissolved.

Figure 10 The sugar crystals dissolve because they are soluble in water.

Hutchings Photography/Digital Light Source

Changing State

In Lesson 1 you read about three states of matter–solid, liquid, and gas. Can you think of examples of matter changing from one state to another? A layer of ice might form on a lake in the winter. A glassblower melts glass into a liquid so that it can be formed into shapes. Changes in the state of matter are physical changes.

Melting and Boiling If you heat ice cubes in a pot on the stove, the ice will melt, forming water that soon begins to boil. When a material melts, it changes from a solid to a liquid. When it boils, it changes from a liquid to a gas. The substances that make up the material do not change during a change in the state of matter, as shown in **Figure 11.** The particles that make up ice (solid water) are the same as the particles that make up water as a liquid or as a gas.

Energy and Change in State The energy of the particles and the distances between the particles are different for a solid, a liquid, and a gas. Changes in energy cause changes in the state of matter. For example, energy must be added to a substance to change it from a solid to a liquid or from a liquid to a gas. Adding energy to a substance can increase its temperature. When the temperature reaches the substance's melting point, the solid changes to a liquid. At the boiling point, the liquid changes to a gas.

What would happen if you changed the rate at which you add energy to a substance? For example, what would happen if you heated an ice cube in your hand instead of in a pot on the stove? The ice would reach its melting point more slowly in your hand. The rate at which one state of matter changes to another depends on the rate at which energy is added to or taken away from the substance.

Changing State

Figure 11 The particles that make up ice (solid water), liquid water, and water vapor (water in the gaseous state) are the same. Changing from one state to another changes only the amount of energy of the particles and the distances between the particles.

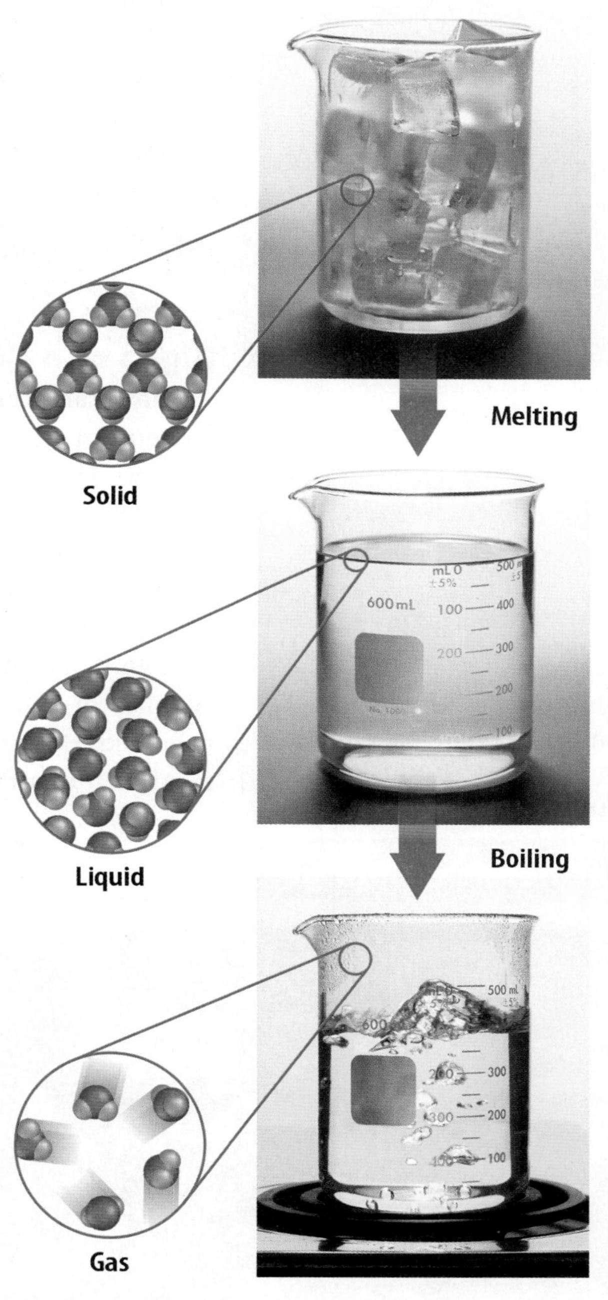

Visual Check Describe the change in the energy and motion of particles of a substance if the substance changes from a gas to a liquid.

Make a half book from a sheet of paper. Use it to record and compare information about physical and chemical changes.

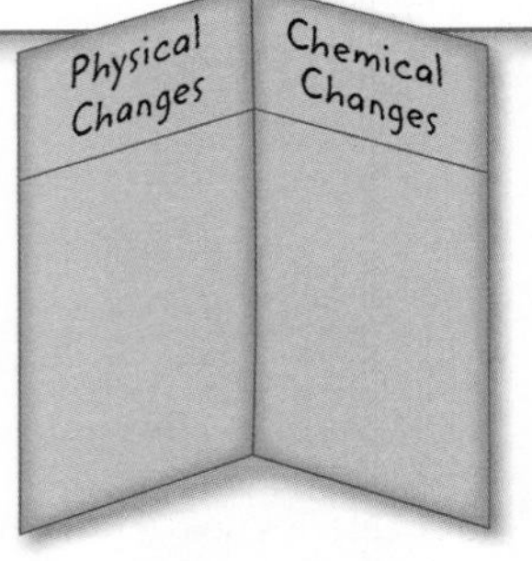

What are chemical changes?

Some changes in matter involve more than just changing physical properties. A **chemical change** *is a change in matter in which the substances that make up the matter change into other substances with different chemical and physical properties.* Recall that a chemical property is the ability or inability of a substance to combine with or change into one or more new substances. During a physical change, only the physical properties of matter change. However, the new substance produced during a chemical change has different chemical and physical properties. Another name for a chemical change is a chemical reaction. The particles that make up two or more substances react, or combine, with each other and form a new substance.

 **Key Concept Check** How are chemical changes different from physical changes?

Signs of a Chemical Change

How can you tell that the burning of the trees in **Figure 12** is a chemical change? The reaction produces two gases–carbon dioxide and water vapor–even though you cannot see them. After the fire, you can see that any part of the trees that remains is black, and you can see ash–another new substance. But with some changes, the only new substance formed is a gas you cannot see. As trees burn in a forest fire, light and heat are signs of a chemical change. For many reactions, changes in physical properties, such as color or state of matter, are signs that a chemical change has occurred. However, the only sure sign of a chemical change is the formation of a new substance.

Figure 12 A forest fire causes a chemical change in the trees, producing new substances.

Chemical Change

Light and heat during a forest fire are signs that a chemical change is occurring.

After the fire, the formation of new substances shows that a chemical change has taken place.

Visual Check Why is the smoke produced during a forest fire a sign of a chemical change?

Formation of Gas Bubbles of gas can form during both a physical change and a chemical change. When you heat a substance to its boiling point, the bubbles show that a liquid is changing to a gas—a physical change. When you combine substances, such as the medicine tablet and the water in **Figure 13**, gas bubbles show that a chemical change is occurring. Sometimes you cannot see the gas produced, but you might be able to smell it. The aroma of freshly baked bread, for example, is a sign that baking bread causes a chemical reaction that produces a gas.

Reading Check How can you determine whether the formation of bubbles is the result of a physical change or a chemical change?

Formation of a Precipitate Some chemical reactions result in the formation of a precipitate (prih SIH puh tut). As shown in the middle photo in **Figure 13**, a precipitate is a solid that sometimes forms when two liquids combine. When a liquid freezes, the solid formed is not a precipitate. A precipitate is not a state change from a liquid to a solid. Instead, the particles that make up two liquids react and form the particles that make up the solid precipitate, a new substance.

Color Change Suppose you want your room to be a different color. You would simply apply paint to the walls. The change in color is a physical change because you have only covered the wall. A new substance does not form. But notice the color of the precipitate in the middle photo of **Figure 13**. In this case, the change in color is a sign of a chemical change. The photo in the bottom of the figure shows that marshmallows change from white to brown when they are toasted. The change in the color of the marshmallows is also a sign of a chemical change.

Reading Check What are some signs that a chemical change has occurred?

Signs of Chemical Change

Figure 13 Formation of a gas, formation of a precipitate, and color change are all signs of a chemical change.

Formation of gas bubbles

Formation of a precipitate

Color change

Visual Check What is a sign besides color change that indicates that the marshmallow is undergoing a chemical change?

(t) Charles D. Winters/Photo Researchers, Inc., (c) Stephen Frisch/McGraw-Hill Education, (b) Dennis Lane Photography/Getty Images

▲ **Figure 14** The flames, the light, and the sound of a fireworks display are signs of a chemical change.

Energy and Chemical Change

Think about a fireworks show. Again and again, you hear loud bangs as the fireworks burst into a display of colors, as in **Figure 14.** The release of thermal energy, light, and sound are signs that the fireworks result from chemical changes. All chemical reactions involve energy changes.

Thermal energy is often needed for a chemical reaction to take place. Suppose you want to bake pretzels, as shown in **Figure 15.** What would happen if you placed one pan of unbaked pretzel dough in the oven and another pan of unbaked pretzel dough on the kitchen counter? Only the dough in the hot oven would become pretzels. Thermal energy is needed for the chemical reactions to occur that bake the pretzels.

ACADEMIC VOCABULARY

expose
(verb) to uncover; to make visible

Energy in the form of light is needed for other chemical reactions. Photosynthesis is a chemical reaction by which plants and some unicellular organisms produce sugar and oxygen. This process only occurs if the organisms are **exposed** to light. Many medicines also undergo chemical reactions when exposed to light. You might have seen some medicines stored in orange bottles. If the medicines are not stored in these light-resistant bottles, the ingredients can change into other substances.

Figure 15 Thermal energy is needed for the chemical reactions that take place when baking pretzels. ►

Can changes be reversed?

Think again about the way matter changes form during a fireworks display. Once the chemicals combine and cause the explosions, you cannot get back the original chemicals. Like most chemical changes, the fireworks display cannot be reversed.

Grating a carrot and cutting an apple are physical changes, but you cannot reverse these changes either. Making a mixture by dissolving salt in a pan of water is also a physical change. You can reverse this change by boiling the mixture. The water will change to a gas, leaving the salt behind in the pan. Some physical changes can be easily reversed, but others cannot.

Reading Check Identify one physical change that can be reversed and one that cannot be reversed.

Conservation of Mass

Physical changes do not affect the mass of substances. When ice melts, for example, the mass of the ice equals the mass of the resulting liquid water. If you cut a piece of paper into strips, the total mass of the paper remains the same. Mass is conserved, or unchanged, during a physical change.

Mass is also conserved during a chemical change. Antoine Lavoisier (AN twon · luh VWAH zee ay) (1743–1794), a French chemist, made this discovery. Lavoisier carefully measured the masses of materials before and after chemical reactions. His discovery is now a scientific law. *The* **law of conservation of mass** *states that the total mass before a chemical reaction is the same as the total mass after the chemical reaction.* Weight also is the same because it depends on mass. For example, the mass of an unburned match plus the mass of the oxygen it reacts with equals the mass of the ashes plus the masses of all the gases given off when the match burns.

Key Concept Check How do physical and chemical changes affect mass?

MiniLab 10 minutes

Is mass conserved during a chemical reaction?

If you have ever seen the glow of a light stick, you have observed a chemical change. How does the chemical reaction affect the mass of the light stick?

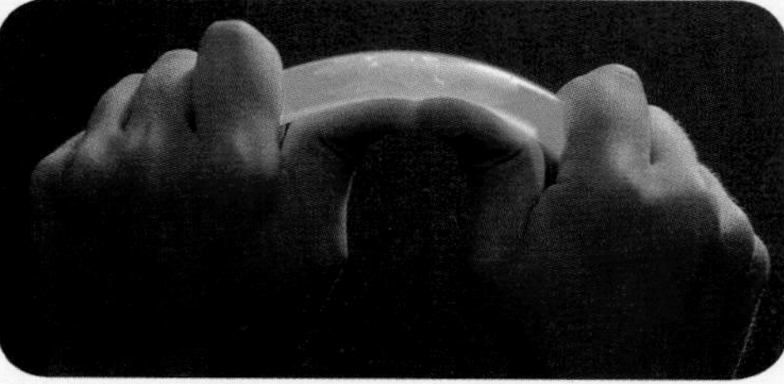

1. Read and complete a lab safety form.
2. Obtain a **light stick** from your teacher. Carefully remove it from the packaging.
3. Observe the structure of the light stick. Record your observations in your Science Journal.
4. Measure and record the mass of the light stick using a **balance.**
5. Grasp the ends of the light stick. Gently bend it to break the inner vial. Shake the stick gently to start the reaction.
6. Use a **stopwatch** to time the reaction for 3 minutes. Record your observations.
7. Repeat step 4.

Analyze and Conclude

1. **Explain** the purpose of the inner vial in the light stick.
2. **Describe** what occurred when the inner vial was broken.
3. **Key Concept** What effect did the chemical reaction have on the mass? Why?

WORD ORIGIN

conservation
from Latin *conservare,* means "to keep, preserve"

Hutchings Photography/Digital Light Source

Comparing Physical and Chemical Changes

Suppose you want to explain to a friend the difference between a physical change and a chemical change. What would you say? You could explain that the identity of matter does not change during a physical change, but the identity of matter does change during a chemical change. However, you might not be able to tell just by looking at a substance whether its identity changed. You cannot tell whether the particles that make up the matter are the same or different.

Sometimes deciding if a change is physical or chemical is easy. Often, however, identifying the type of change is like being a detective. You have to look for clues that will help you figure out whether the identity of the substance has changed. For example, look at the summary of physical changes and chemical changes in **Table 3.** A change in color can occur during a chemical change or when substances are mixed (a physical change). Bubbles might indicate the formation of gas (a chemical change) or boiling (a physical change). You must consider many factors when comparing physical and chemical changes.

Reading Check What are some clues you can use to decide if a change is a physical change or a chemical change?

Table 3 Chemical changes produce a new substance, but physical changes do not.

Interactive Table

Table 3 Comparing Physical and Chemical Changes

Type of Change	Examples	Characteristics
Physical change	• melting • boiling • changing shape • mixing • dissolving • increasing or decreasing in temperature	• Substance is the same before and after the change. • Only physical properties change.
Chemical change	• changing color • burning • rusting • formation of gas • formation of a precipitate • spoiling food • tarnishing silver • digesting food	• Substance is different after the change. • Both physical and chemical properties change.

Lesson 2 Review

Visual Summary

The identity of a substance does not change during a physical change such as a change in the state of matter.

A new substance is produced during a chemical change.

The law of conservation of mass states that the total mass of the materials does not change during a chemical change.

FOLDABLES®

Use your lesson Foldable to review the lesson. Save your Foldable for the project at the end of the chapter.

What do you think

You first read the statements below at the beginning of the chapter.

4. Mixing powdered drink mix with water causes a new substance to form.

5. If you combine two substances, bubbling is a sign that a new type of substance might be forming.

6. If you stir salt into water, the total amount of mass decreases.

Did you change your mind about whether you agree or disagree with the statements? Rewrite any false statements to make them true.

Use Vocabulary

1. The particles that make up matter do not change during a(n) ___________.

Understand Key Concepts

2. **Explain** how physical and chemical changes affect the mass of a material.

3. Which is a physical change?
   A. burning wood
   B. melting ice
   C. rusting iron
   D. spoiling food

Interpret Graphics

4. **Analyze** Suppose you mix 12.8 g of one substance with 11.4 g of another. The picture shows the mass you measure for the mixture. Is this reasonable? Explain.

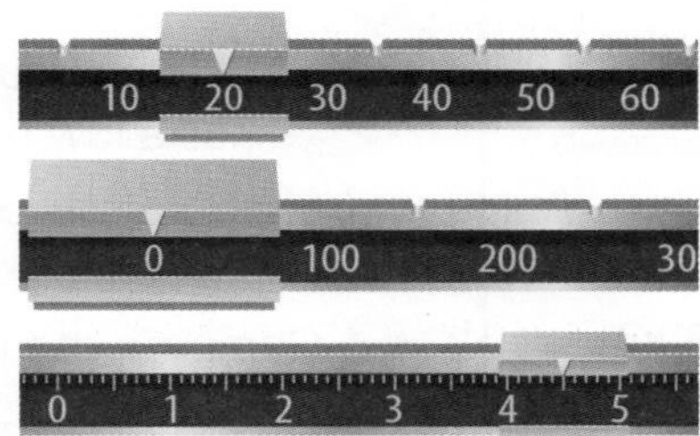

5. **Organize Information** Copy the graphic organizer below, and list an example of each type of change.

Type of Change	Examples
Physical change with formation of bubbles	
Chemical change with formation of bubbles	

Critical Thinking

6. **Consider** Suppose you mix baking soda and white vinegar. What signs might indicate that a chemical change occurs?

7. **Evaluate** You read that a physical change is a change in physical properties, and a chemical change is a change in chemical properties. Do you agree? Explain your answer.

(t) Hutchings Photography/Digital Light Source, (c) Stephen Frisch/McGraw-Hill Education, (b) Dennis Lane Photography/Getty Images

Materials

mineral samples

nail

100-mL graduated cylinder

triple-beam balance

Safety

Identifying Unknown Minerals

Imagine you are a geologist digging for minerals. You find one that you would like to identify. What properties of the mineral would help you? Geologists consider many physical properties of a mineral when determining its identification.

Question

How can you use physical properties to identify unknown minerals?

Procedure

1. Read and complete a lab safety form.
2. Select a mineral sample to observe. Record its color in your Science Journal.
3. Observe the hardness of your mineral.
   - **a.** Scratch your mineral with your fingernail. If it scratches, then your mineral has a low hardness. Go to step 4. If it does not scratch, go to step 3b.
   - **b.** Scratch your mineral with a nail. If it scratches, it has a moderate hardness. If it does not scratch, it has a high hardness.
4. Compare the properties of your mineral with the properties in the chart.

Physical Properties of Minerals

Mineral	Color	Typical Density (g/cm^3)	Hardness
Fluorite	white or light green	3.1	moderate
Gypsum	white or brown	2.3	fairly soft
Hornblende	black or grayish brown	3.2	moderate
Magnetite (iron ore)	dark gray	5.2	moderate
Quartz	white or colorless	2.6	fairly hard
Sphalerite (zinc ore)	black or reddish brown	4.1	fairly soft

(2) Jacques Cornell/McGraw-Hill Education, (others) Hutchings Photography/Digital Light Source

5 Think about the properties you observed so far. Are you able to determine which mineral you have based on your initial observations? Explain why or why not in your Science Journal.

6 Look back through the chapter to review the physical property *density*.

7 Design an experiment using mass and volume to determine the density of your mineral.

8 Share your procedure with your teacher for approval before conducting your experiment.

9 Compare your results with information in the Physical Properties of Minerals table.

Analyze and Conclude

10 **Infer** the identity of your mineral sample.

11 **The Big Idea** Which physical property was most useful in identifying the mineral? Why?

12 **Predict** Suppose you have another sample of the same mineral. What properties would you expect to be the same? What properties would be different?

Lab TIPS

☑ To measure the water in a cylinder accurately, first put your eye at the level of the liquid. Then observe the level at the meniscus (the center or bottom of the curve in the surface of the liquid).

☑ 1 mL = 1 cm^3

Communicate Your Results

In a small group, share your experiences and your results. How did you collect and record data? What was successful? Did others use different techniques or get different results? Did anything surprise you?

Inquiry Extension

Choose a different unknown sample to test that looks similar to the one you tested. Which properties might be different? Test your sample in the same way you tested the first one. Were the results the same or different? What can you conclude from this?

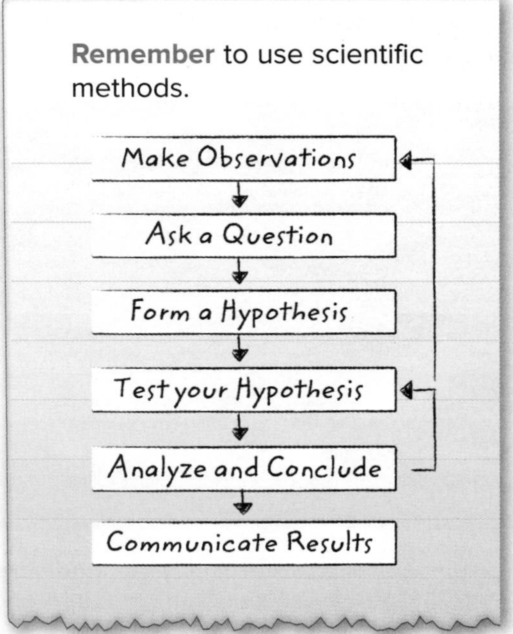

Hutchings Photography/Digital Light Source

Chapter 12 Study Guide

Physical and chemical properties give a substance its unique identity.

Key Concepts Summary 

Lesson 1: Matter and Its Properties

- Particles of a **solid** vibrate about a definite position. Particles of a **liquid** can slide past one another. Particles of a **gas** move freely within their container.
- A **physical property** is a characteristic of matter that you can observe without changing the identity of the substances that make it up. A **chemical property** is the ability or inability of a substance to combine with or change into one or more new substances.
- Some properties of matter do not depend on size or amount of the sample. You can identify a substance by comparing these properties to those of other known substances.

Vocabulary

volume p. 386
solid p. 386
liquid p. 386
gas p. 386
physical property p. 388
mass p. 388
density p. 389
solubility p. 390
chemical property p. 391

Lesson 2: Matter and Its Changes

- A change in the size, shape, form, or state of matter in which the identity of the matter stays the same is a **physical change.** A change in matter in which the substances that make it up change into other substances with different chemical and physical properties is a **chemical change.**
- The **law of conservation of mass** states that the total mass before a chemical reaction is the same as the total mass after the reaction.

Vocabulary

physical change p. 398
chemical change p. 400
law of conservation of mass p. 403

Study Guide

Vocabulary eFlashcards
Vocabulary eGames

Personal Tutor

FOLDABLES® Chapter Project

Assemble your lesson Foldables as shown to make a Chapter Project. Use the project to review what you have learned in this chapter.

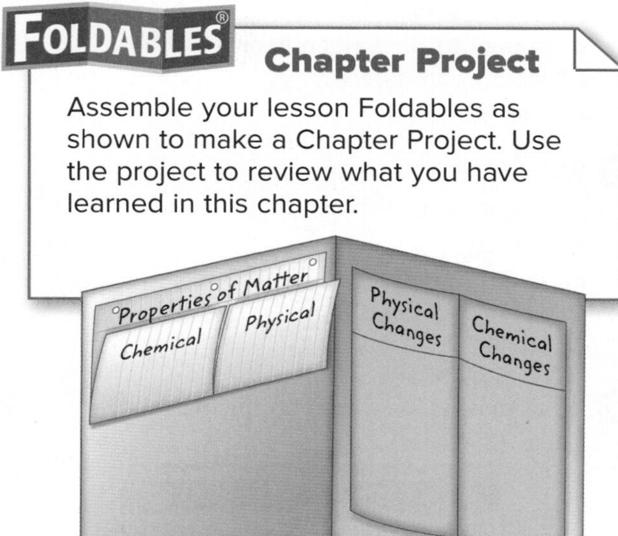

Use Vocabulary

1. A state of matter with a definite volume and a definite shape is a ______.
2. Flammability is an example of a ______ of wood because when wood burns, it changes to different materials.
3. A drink mix dissolves in water because of its ______ in water.
4. The rusting of a metal tool left in the rain is an example of a ______.
5. According to the ______, the mass of an untoasted marshmallow equals its mass after it is toasted plus the mass of any gases produced as it was toasting.
6. Slicing an apple into sections is an example of a ______ that cannot be reversed.

Link Vocabulary and Key Concepts

Interactive Concept Map

Copy this concept map, and then use vocabulary terms from the previous page to complete the concept map.

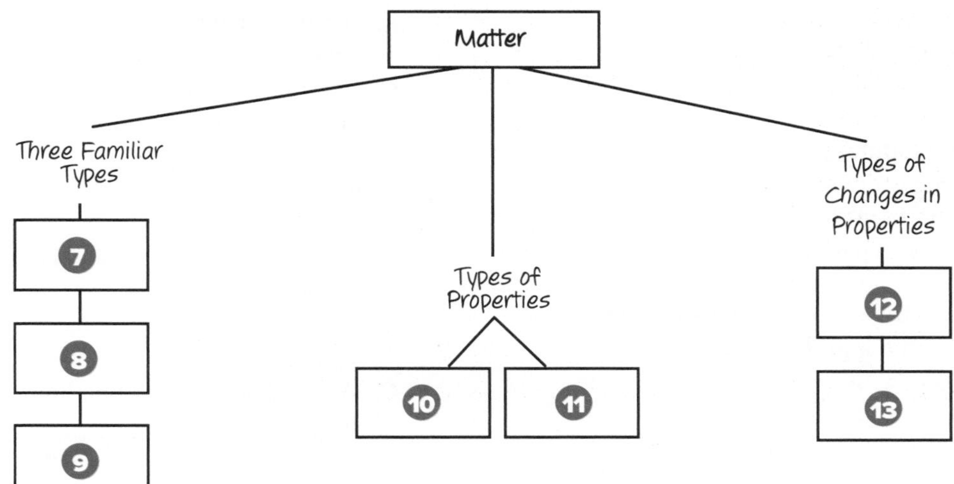

Chapter 12 Review

Understand Key Concepts

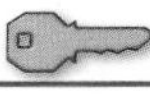

1. Which is a property of all solids?
   - A. Particles are far apart.
   - B. Particles vibrate in all directions.
   - C. Volume and shape can easily change.
   - D. Weak forces exist between particles.

2. Which characteristic is a chemical property?
   - A. highly flammable
   - B. mass of 15 kg
   - C. woolly texture
   - D. golden color

3. Which property of an object depends on its location?
   - A. density
   - B. mass
   - C. volume
   - D. weight

4. How are the particles of a gas different from the particles of a liquid shown here?

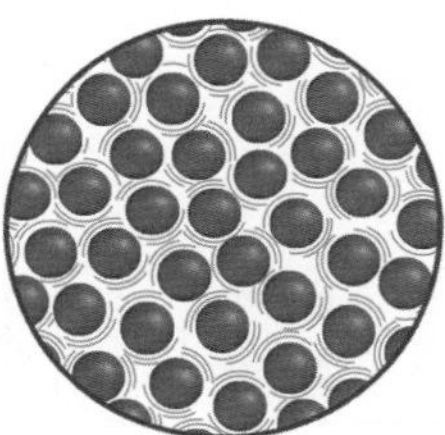

   - A. They move more slowly.
   - B. They are farther apart.
   - C. They have less energy.
   - D. They have stronger attractions.

5. Which is a physical change?
   - A. burning natural gas
   - B. chopping onions
   - C. digesting food
   - D. exploding dynamite

6. Which stays the same when a substance changes from a liquid to a gas?
   - A. density
   - B. mass
   - C. forces between particles
   - D. distance between particles

7. Which is a chemical change?
   - A. boiling water
   - B. copper turning green in air
   - C. freezing fruit juice
   - D. slicing a potato

8. Which would be most useful for identifying an unknown liquid?
   - A. density
   - B. mass
   - C. volume
   - D. weight

9. What mass is measured on this balance?

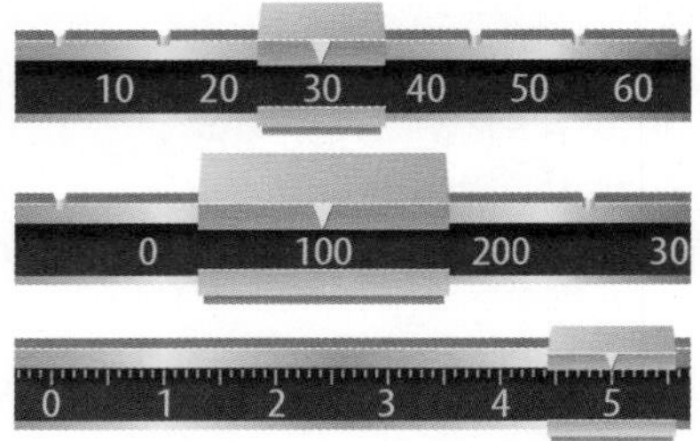

   - A. 35 g
   - B. 45 g
   - C. 135 g
   - D. 145 g

10. What causes a chemical reaction when you prepare scrambled eggs?
    - A. removing the eggs from the shells
    - B. mixing the egg yolks and the egg whites together
    - C. heating the eggs in a pan
    - D. sprinkling pepper onto the cooked eggs

11. Which describes the formation of a precipitate?
    - A. A gas forms when a solid is placed in a liquid.
    - B. A liquid forms when a block of metal is heated.
    - C. A solid forms when one liquid is poured into another.
    - D. Bubbles form when an acid is poured onto a rock.

Online Test Practice

Critical Thinking

12 **Apply** Suppose you find a gold-colored ring. Explain why you could use some physical properties but not others to determine whether the ring is actually made of gold.

13 **Reason** You make lemonade by mixing lemon juice, sugar, and water. Is this a physical change or a chemical change? Explain.

14 **Give an example** of a physical change you might observe at your school that is reversible and a physical change that is not reversible.

15 **Defend** A classmate defines a liquid as any substance that can be poured. Use the picture below to explain why this is not an acceptable definition.

16 **Suggest** a way that you could use displacement to determine the volume of a rock that is too large to fit into a graduated cylinder.

17 **Hypothesize** A scientist measures the mass of two liquids before and after combining them. The mass after combining the liquids is less than the sum of the masses before. Where is the missing mass?

Writing in Science

18 **Write** a four-sentence description of an object in your home or classroom. Be sure to identify both physical properties and chemical properties of the object.

REVIEW THE BIG IDEA

19 What gives a substance its unique identity?

20 What are some physical and chemical properties that an airplane manufacturer must consider when choosing materials to be used in constructing the shell of the aircraft shown below?

Math Skills

Math Practice

21 Use what you have learned about density to complete the table below. Then, determine the identities of the two unknown metals.

Metal	Mass (g)	Volume (cm^3)	Density (g/cm^3)
Iron	42.5	5.40	
Lead	28.8	2.55	
Tungsten	69.5	3.60	
Zinc	46.4	6.50	
	61.0	5.40	
	46.4	2.40	

Standardized Test Practice

Record your answers on the answer sheet provided by your teacher or on a sheet of paper.

Multiple Choice

1 Which describes the particles in a substance with no definite volume or shape?

A Particles are close but can move freely.

B Particles are close but can vibrate in all directions.

C Particles are far apart and cannot move.

D Particles are far apart and move freely.

2 Which diagram shows a chemical change?

A

B

C

D

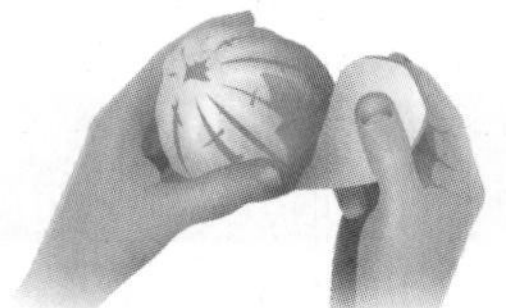

3 Which is NOT true about firewood that burns completely?

A Ashes and gases form from the substances in the wood.

B Oxygen from the air combines with substances in the wood.

C The total mass of substances in this process decreases.

D The wood gives off thermal energy and light.

Use the diagram below to answer question 4.

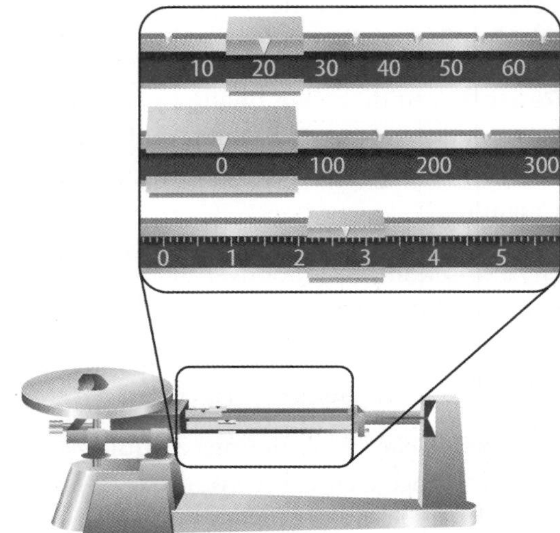

4 What is the mass of the object on the balance scale?

A 22 g

B 22.5 g

C 22.7 g

D 30 g

5 Which is true when an ice cube melts?

A Volume and mass increase.

B Volume and mass do not change.

C Volume decreases, but mass does not change.

D Volume increases, but mass decreases.

6 What is the BEST way to separate and save the parts of a sand-and-water mixture?

A Boil the mixture and collect the steam.

B Pour the mixture through a filter that only the water can pass through.

C Lift the sand out of the mix with a spoon.

D Pour a strong acid into the mixture to dissolve the sand.

Use the table below to answer questions 7 and 8.

Action	Time	Result
Heated	30 minutes	solid
Heated	60 minutes	liquid
Not heated	30 minutes	solid
Not heated	60 minutes	solid

7 Based on the results of this experiment, what can you conclude about heating this unknown substance?

A Heating melted it in 30 minutes.

B Heating melted it in 60 minutes.

C Heating made it solid in 60 minutes.

D Heating caused no changes.

8 What can you conclude about the original state of the substance?

A It is part solid and part liquid.

B It is a liquid.

C It is a solid.

D It is part liquid and part gas.

9 Which is a sign of a physical change?

A Bread gets moldy with age.

B Ice forms on a puddle in winter.

C The metal on a car starts to rust.

D Yeast causes bread dough to rise.

Constructed Response

Use the table below to answer questions 10–13.

Properties	Substance 1	Substance 2	Substance 3
Color	yellow	yellow	yellow
State	solid	solid	solid
Mass	217 g	217 g	75 g
Melting point	505°C	230°C	505°C
Density	3.78 g/cm^3	2.76 g/cm^3	3.78 g/cm^3
Flammable	yes	yes	yes

10 Identify each property of the unknown substances as either chemical or physical. Explain your reasoning.

11 Of the three unknown substances tested, two are the same substance and one is different. Which two substances do you think are the same? Explain your reasoning.

12 Which properties in the table helped you determine your answer in number 11? Which properties were not helpful? Explain your reasoning.

13 What additional physical and chemical properties of substances might the table have included?

NEED EXTRA HELP?													
If You Missed Question...	1	2	3	4	5	6	7	8	9	10	11	12	13
Go to Lesson...	1	2	2	1	1, 2	1	2	2	2	1	1	1	1

Unit 4
UNDERSTANDING ENERGY
IT'S 9:30, DID YOU CALL YOUR FATHER?!
WATER?! Nah, I don't need any water.

1600 | 1700 | 1800

1660
Robert Hooke publishes the wave theory of light, comparing light's movement to that of waves in water.

1705
Francis Hauksbee experiments with a clock in a vacuum and proves that sound cannot travel without air.

1820
Danish physicist Hans Christian Ørsted publishes his discovery that an electric current passing through a wire produces a magnetic field.

1878
Thomas Edison develops a system to provide electricity to homes and businesses using locally generated and distributed direct current (DC) electricity.

1882
Thomas Edison develops and builds the first electricity-generating plant in New York City, which provides 110 V of direct current to 59 customers in lower Manhattan.

1900 2000

1883
The first standardized incandescent electric lighting system using overhead wires begins service in Roselle, New Jersey.

1890s
Physicist Nikola Tesla introduces alternating current (AC) by inventing the alternating current generator, allowing electricity to be transmitted at higher voltages over longer distances.

1947
Chuck Yeager becomes the first pilot to travel faster than the speed of sound.

Visit ConnectED for this unit's STEM activity.

Unit 4

Nature of SCIENCE

Graphs

Have you ever felt a shock from static electricity? The electric energy that you feel is similar to the electric energy you see as a flash of lightning, such as in **Figure 1**, only millions of times smaller. Scientists are still investigating what causes lightning and where it will occur. They use graphs to learn about the risk of lightning in different places and at different times. A **graph** is a type of chart that shows relationships between variables. Graphs organize and summarize data in a visual way. Three of the most common graphs are circle graphs, bar graphs, and line graphs.

Types of Graphs

Line Graphs

A line graph is used when you want to analyze how a change in one variable affects another variable. This line graph shows how the average number of lightning flashes changes over time in Illinois. Time is plotted on the *x*-axis. The average numbers of lightning flashes are plotted on the *y*-axis. Each dot, called a data point, indicates the average number of flashes recorded during that hour. A line connects the data points so a trend can be analyzed.

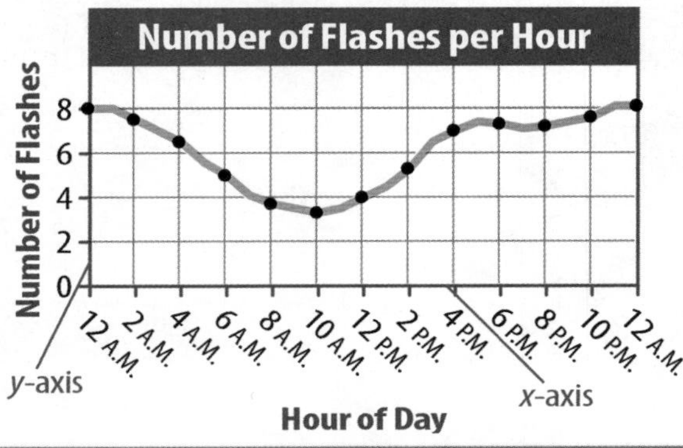

Number of Flashes per State

Number of Flashes per Square Kilometer per Year

0 2 4 6 8 10

AL AR FL GA LA MS NC OK SC TX

State

Bar Graphs

When you want to compare amounts in different categories, you use a bar graph. The horizontal axis often contains categories instead of numbers. This bar graph shows the average number of lightning flashes that occur in different states. On average, about 9.8 lightning flashes strike each square kilometer of land in Florida every year. Florida has more lightning flashes per square kilometer than all other states shown on the graph.

Circle Graphs

If you want to show how the parts of something relate to the whole, use a circle graph. This circle graph shows the average percentage of lightning flashes each U.S. region receives in a year. The graph shows that the mountain region of the United States receives about 14 percent of all lightning flashes that strike the country each year. From the graph, you can also determine that the southeast receives the most lightning in a given year.

Lightning Flashes by US Region

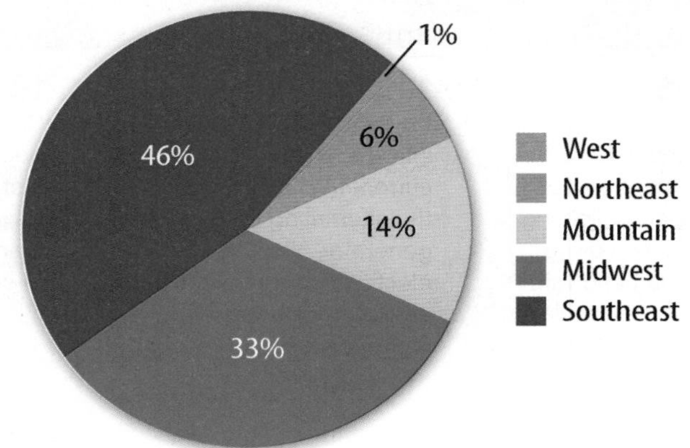

Line Graphs and Trends

Suppose you are planning a picnic in an area that experiences quite a bit of lightning. When would be the safest time to go? First, you gather data about the average number of lightning flashes per hour. Next, you plot the data on a line graph and analyze trends. Trends are patterns in data that help you find relationships among the data and make predictions.

Follow the orange line on the line graph from 12 A.M. to 10 A.M. in **Figure 2.** Notice that the line slopes downward, indicated by the green arrow. A downward slope means that as measurements on the *x*-axis increase, measurements on the *y*-axis decrease. So, as time passes from 12 A.M. to 10 A.M., the number of lightning flashes decreases.

If you follow the orange line on the line graph from 12 P.M. to 5 P.M., you will notice that the line slopes upward. This is indicated by the blue arrow. An upward slope means that as the measurements on the *x*-axis increase, the measurements on the *y*-axis also increase. So, as time passes from 12 P.M. to 5 P.M., the number of lightning flashes increases.

The line graph shows you that between about 8 A.M. and 12 P.M., you would have the least risk of lightning during your picnic.

Figure 2 The slope of a line in a line graph shows the relationship between the variables on the *x*-axis and the variables on the *y*-axis. ▼

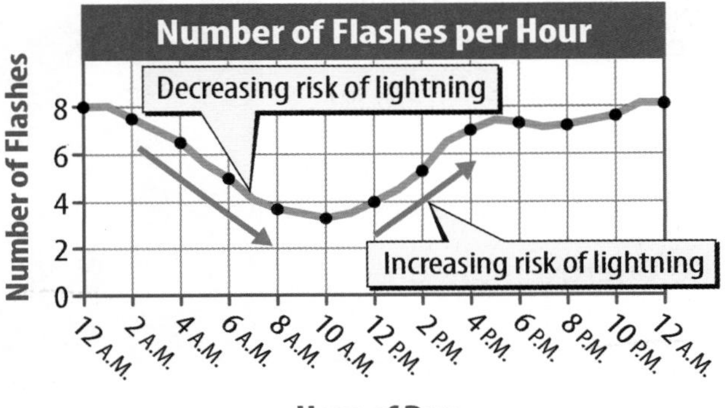

▲ **Figure 1** Scientists study lightning to get a better understanding of what causes it and to predict when it will occur.

MiniLab 25 minutes

When does lightning strike?

Meteorologists in New Mexico collected data on the number of lightning flashes throughout the day. How can you use a line graph to plan the safest day trip?

1. Make a line graph of the data in the table.
2. Find the trends that show when the risk for lightning is increasing and when it is decreasing.

Hour	# of Flashes
12:00 A.M.	2
3:00 A.M.	2
6:00 A.M.	2
9:00 A.M.	2
12:00 P.M.	5
3:00 P.M.	21
6:00 P.M.	36
9:00 P.M.	22
12:00 A.M.	2

Analyze and Conclude

Decide How could you use your graph to plan a day trip in New Mexico with the least risk of lightning?

Chapter 13

Energy and Energy Transformations

What is energy, and what are energy transformations?

Inquiry Which objects have energy?

If your answer is everything in the photo, you are right. All objects contain energy. Some objects contain more energy than other objects. The Sun contains so much energy that it is considered an energy resource.

- From where do you think the energy that powers the cars comes?
- Do you think the energy in the Sun and the energy in the green plants are related?
- What do the terms *energy* and *energy transformations* mean to you?

Get Ready to Read

What do you think?

Before you read, decide if you agree or disagree with each of these statements. As you read this chapter, see if you change your mind about any of the statements.

1. A fast-moving baseball has more kinetic energy than a slow-moving baseball.
2. A large truck and a small car moving at the same speed have the same kinetic energy.
3. A book sitting on a shelf has no energy.
4. Energy can change from one form to another.
5. Energy is destroyed when you apply the brakes on a moving bicycle or a moving car.
6. The Sun releases radiant energy.

Your one-stop online resource
connectED.mcgraw-hill.com

LearnSmart®

Chapter Resources Files, Reading Essentials, Get Ready to Read, Quick Vocabulary

Animations, Videos, Interactive Tables

Self-checks, Quizzes, Tests

Project-Based Learning Activities

Lab Manuals, Safety Videos, Virtual Labs & Other Tools

Vocabulary, Multilingual eGlossary, Vocab eGames, Vocab eFlashcards

Personal Tutors

Lesson 1

Forms of Energy

Reading Guide

Key Concepts

ESSENTIAL QUESTIONS

- What is energy?
- What are potential and kinetic energy?
- How is energy related to work?
- What are the different forms of energy?

Vocabulary

energy p. 421
kinetic energy p. 422
potential energy p. 422
work p. 424
mechanical energy p. 425
sound energy p. 425
thermal energy p. 425
electric energy p. 425
radiant energy p. 425
nuclear energy p. 425

Multilingual eGlossary

BrainPOP®

Go to the resource tab in ConnectED to find the PBLs *Energy in Motion* and *Physics Day at the Amusement Park.*

Inquiry Why is this cat glowing?

A camera that detects temperature made this image. Dark colors represent cooler temperatures, and light colors represent warmer temperatures. Temperatures are cooler where the cat's body emits less radiant energy and warmer where the cat's body emits more radiant energy.

Launch Lab

20 minutes

Can you change matter?

You observe many things changing. Birds change their positions when they fly. Bubbles form in boiling water. The filament in a lightbulb glows when you turn on a light. How can you cause a change in matter?

1. Read and complete the lab safety form.
2. Half-fill a **foam cup** with **sand.** Place the bulb of a **thermometer** about halfway into the sand. *Do not stir.* Record the temperature in your Science Journal.
3. Remove the thermometer and place a **lid** on the cup. Hold down the lid and shake the cup vigorously for 10 min.
4. Remove the lid. Measure and record the temperature of the sand.

Think About This

1. What change did you observe in the sand?
2. How could you change your results?
3. **Key Concept** What do you think caused the change you observed in the sand?

What is energy?

It might be exciting to watch a fireworks display, such as the one shown in **Figure 1.** Over and over, you hear the crack of explosions and see bursts of colors in the night sky. Fireworks release energy when they explode. **Energy** is the ability to cause change. The energy in the fireworks causes the changes you see as bursting flashes of light and hear as loud booms.

Energy also causes other changes. The plant in **Figure 1** uses the energy from the Sun and makes food that it uses for growth and other processes. Energy can cause changes in the motions and positions of objects, such as the nail in **Figure 1.** Can you think of other ways energy might cause changes?

Word Origin
energy
from Greek *energeia,* means "activity"

 **Key Concept Check** What is energy?

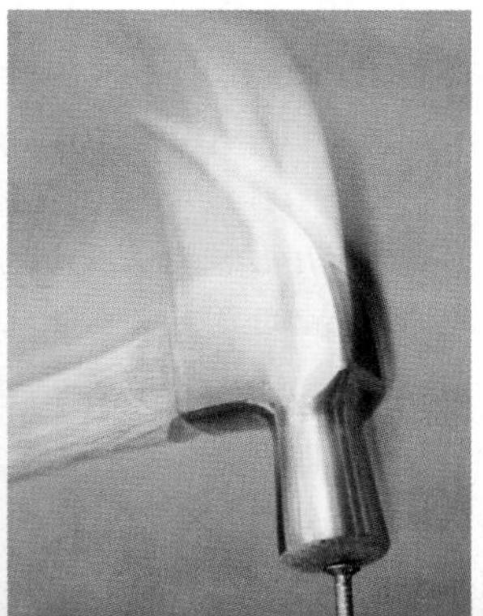

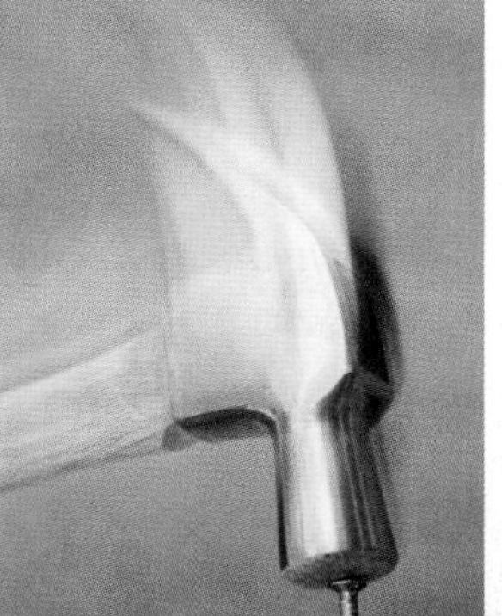

Figure 1 The explosion of fireworks, the growth of a plant, and the motion of a hammer all involve energy.

(t)Hutchings Photography/Digital Light Source, (bl)Thinkstock/Corbis, (bc)Sharon Dominick/Getty Images, (br)Dimitri Vervitsiotis/Getty Images

Figure 2  The kinetic energy (KE) of an object depends on its speed and its mass. The vertical bars show the kinetic energy of each vehicle.

Kinetic Energy—Energy of Motion

Have you ever been to a bowling alley? When you rolled the ball and it hit the pins, a change occurred–the pins fell over. This change occurred because the ball had a form of energy called kinetic (kuh NEH tik) energy. **Kinetic energy** *is energy due to motion.* All moving objects have kinetic energy.

Kinetic Energy and Speed

An object's kinetic energy depends on its speed. The faster an object moves, the more kinetic energy it has. For example, the blue car has more kinetic energy than the green car in Figure 2 because the blue car is moving faster.

Kinetic Energy and Mass

A moving object's kinetic energy also depends on its mass. If two objects move at the same speed, the object with more mass has more kinetic energy. For example, the truck and the green car in Figure 2 are moving at the same speed, but the truck has more kinetic energy because it has more mass.

Key Concept Check What is kinetic energy?

FOLDABLES®

Make a 18-cm fold along the long edge of a sheet of paper to make a two-pocket book. Label it as shown. Organize information about the forms of energy on quarter sheets of paper, and put them in the pockets.

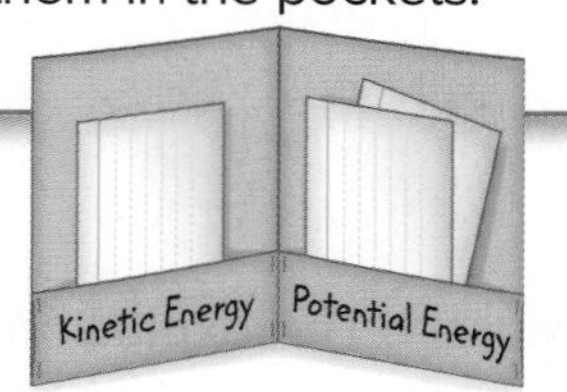

Potential Energy—Stored Energy

Energy can be present even if objects are not moving. If you hold a ball in your hand and then let it go, the gravitational interaction between the ball and Earth causes a change to occur. Before you dropped the ball, it had a form of energy called potential (puh TEN chul) energy. **Potential energy** *is stored energy due to the interactions between objects or particles.* Gravitational potential energy, elastic potential energy, and chemical potential energy are all forms of potential energy.

Gravitational Potential Energy

Even when you are just holding a book, gravitational potential energy is stored between the book and Earth. The girl shown in **Figure 3** increases the gravitational potential energy between her backpack and Earth by lifting the backpack higher from the ground.

The gravitational potential energy stored between an object and Earth depends on the object's weight and height. Dropping a bowling ball from a height of 1 m causes a greater change than dropping a tennis ball from 1 m. Similarly, dropping a bowling ball from 3 m causes a greater change than dropping the same bowling ball from only 1 m.

Reading Check What factors determine the gravitational potential energy stored between an object and Earth?

Elastic Potential Energy

When you stretch a rubber band, as in **Figure 3**, another form of potential energy, called elastic (ih LAS tik) potential energy, is being stored in the rubber band. Elastic potential energy is energy stored in objects that are compressed or stretched, such as springs and rubber bands. When you release the end of a stretched rubber band, the stored elastic potential energy is transformed into kinetic energy. This transformation is obvious when it flies across the room.

Chemical Potential Energy

Food, gasoline, and other substances are made of atoms joined together by chemical bonds. Chemical potential energy is energy stored in the chemical bonds between atoms, as shown in **Figure 3**. Chemical potential energy is released when chemical reactions occur. Your body uses the chemical potential energy in foods for all its activities. People also use the chemical potential energy in gasoline to power cars and buses.

Key Concept Check In what way are all forms of potential energy the same?

Potential Energy

Figure 3 There are different forms of potential energy.

Gravitational Potential Energy
Gravitational potential energy increases when the girl lifts her backpack.

Elastic Potential Energy
The rubber band's elastic potential energy increases when it is stretched.

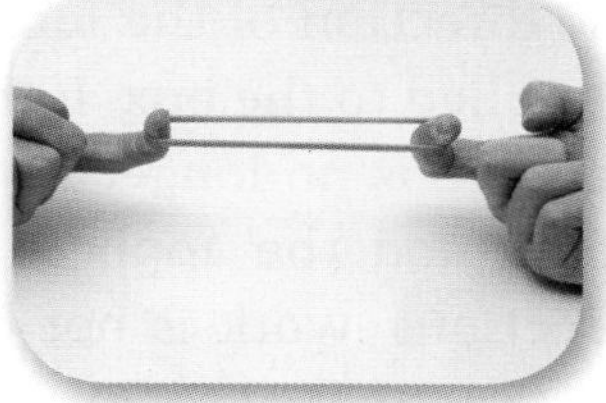

Chemical Potential Energy
Foods and other substances, including glucose, have chemical potential energy stored in the bonds between atoms.

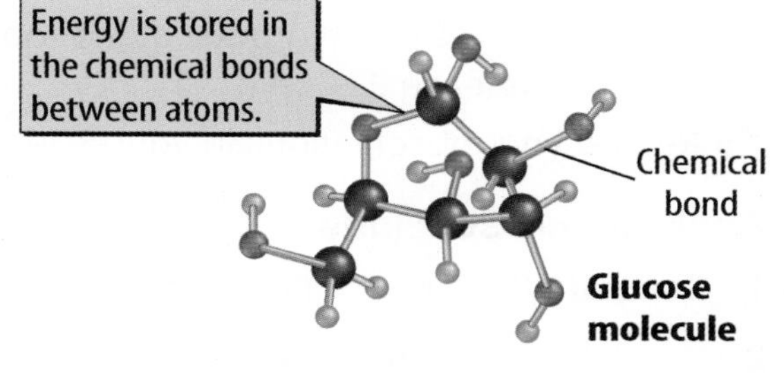
Energy is stored in the chemical bonds between atoms.
Chemical bond
Glucose molecule

Figure 4 The girl does work on the box as she lifts it and increases its gravitational potential energy. The colored bars show the work that the girl does (W) and the box's potential energy (PE).

Energy and Work

You can transfer energy by doing work. **Work** *is the transfer of energy that occurs when a force makes an object move in the direction of the force while the force is acting on the object.* For example, as the girl lifts the box onto the shelf in **Figure 4**, she transfers energy from herself to the box. She does work only while the box moves in the direction of the force and while the force is applied to the box. If the box stops moving, the force is no longer applied, or the box movement and the applied force are in different directions, work is not done on the box.

Key Concept Check How is energy related to work?

An object that has energy also can do work. For example, when a bowling ball collides with a bowling pin, the bowling ball does work on the pin. Some of the ball's kinetic energy is transferred to the bowling pin. Because of this connection between energy and work, energy is sometimes described as the ability to do work.

Other Forms of Energy

Some other forms of energy are shown in **Table 1**. All energy can be measured in joules (J). A softball dropped from a height of about 0.5 m has about 1 J of kinetic energy just before it hits the floor.

MiniLab

20 minutes

Can a moving object do work?

Is work done when a moving object hits another object?

1. Read and complete a lab safety form.
2. **Tape** one end of a **30-cm grooved ruler** to the edge of a stack of **books** about 8 cm high. Put the lower end of the ruler in a **paper cup** lying on its side.
3. Place a **marble** in the groove at the top end of the ruler and release it.
4. Record your observations in your Science Journal.

Analyze and Conclude

1. **Compare** the kinetic energy of the marble just before and after it hit the cup.
2. 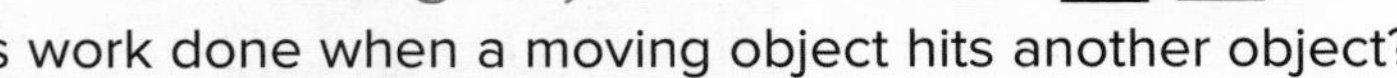 **Key Concept** Is work being done on the cup? Explain your answer.

Hutchings Photography/Digital Light Source

Table 1 Forms of Energy

Mechanical Energy *The sum of potential energy and kinetic energy in a system of objects is* **mechanical energy.** For example, the mechanical energy of a basketball increases when a player shoots the basketball. Both the kinetic energy and gravitational potential energy of the ball increases in the player-ball-ground system.	
Sound Energy When you pluck a guitar string, the string vibrates and produces sound. *The energy that sound carries is* **sound energy.** Vibrating objects emit sound energy. However, sound energy cannot travel through a vacuum, such as the space between Earth and the Sun.	
Thermal Energy All objects and materials are made of particles that have energy. **Thermal energy** *is the sum of kinetic energy and potential energy of the particles that make up an object.* Mechanical energy is due to large-scale motions and interactions in a system and thermal energy is due to atomic-scale motions and interactions of particles. Thermal energy moves from warmer objects, such as burning logs, to cooler objects, such as air.	
Electric Energy An electrical fan uses another form of energy–electric energy. When you turn on a fan, there is an electric current through the fan's motor. **Electric energy** *is the energy an electric current carries.* Electrical appliances, such as fans and dishwashers, change electric energy into other forms of energy.	
Radiant Energy—Light Energy The Sun gives off energy that travels to Earth as electromagnetic waves. Unlike sound waves, electromagnetic waves can travel through a vacuum. Light waves, microwaves, and radio waves are all electromagnetic waves. *The energy that electromagnetic waves carry is* **radiant energy.** Radiant energy sometimes is called light energy.	
Nuclear Energy At the center of every atom is a nucleus. **Nuclear energy** *is energy that is stored and released in the nucleus of an atom.* In the Sun, nuclear energy is released when nuclei join together. In a nuclear power plant, nuclear energy is released when the nuclei of uranium atoms are split apart.	

Key Concept Check Describe three forms of energy.

Lesson 1 Review

Visual Summary

Energy is the ability to cause change.

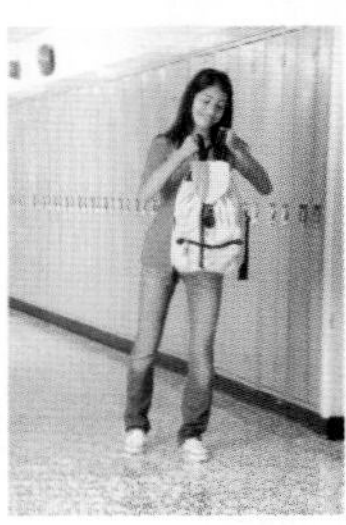

The gravitational potential energy between an object and Earth increases when you lift the object.

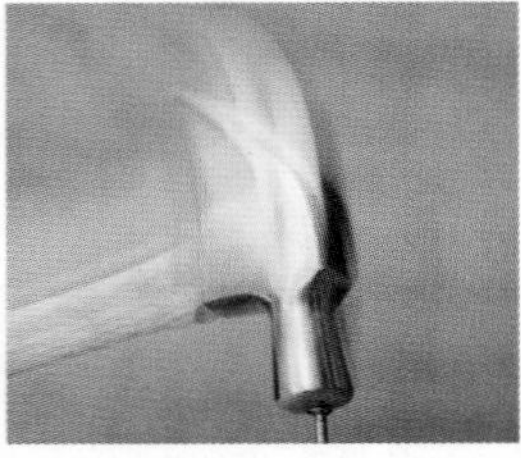

You do work on an object when you apply a force to that object over a distance.

FOLDABLES®

Use your lesson Foldable to review the lesson. Save your Foldable for the project at the end of the chapter.

What do you think NOW?

You first read the statements below at the beginning of the chapter.

1. A fast-moving baseball has more kinetic energy than a slow-moving baseball.
2. A large truck and a small car moving at the same speed have the same kinetic energy.
3. A book sitting on a shelf has no energy.

Did you change your mind about whether you agree or disagree with the statements? Rewrite any false statements to make them true.

Use Vocabulary

1. **Distinguish** between kinetic energy and potential energy.

Understand Key Concepts 

2. **Write** a definition of work.
3. Which type of energy increases when you compress a spring?
   - **A.** elastic potential energy
   - **B.** kinetic energy
   - **C.** radiant energy
   - **D.** sound energy
4. **Infer** How could you increase the gravitational potential energy between yourself and Earth?
5. **Infer** how a bicycle's kinetic energy changes when that bicycle slows down.
6. **Compare and contrast** radiant energy and sound energy.

Interpret Graphics

7. **Identify** Copy and fill in the graphic organizer below to identify three types of potential energy.

8. **Describe** where chemical potential energy is stored in the molecule shown below.

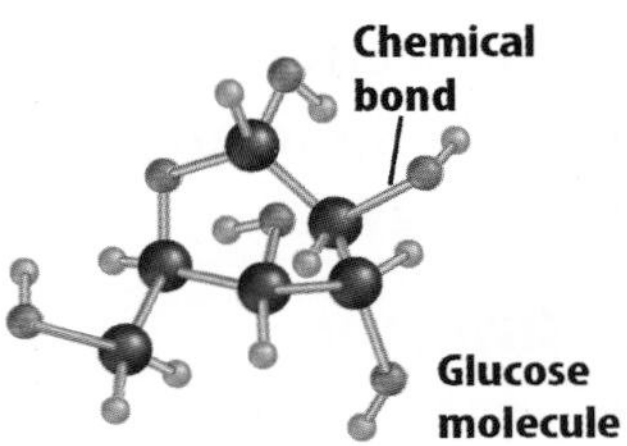

Critical Thinking

9. **Analyze** Will pushing on a car always change the car's mechanical energy? What must happen for the car's kinetic energy to increase?

45 minutes

Can you identify potential and kinetic energy?

Materials

string

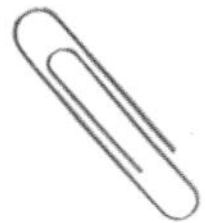
paper clip

three large washers

meterstick

tape

small box

ruler

Safety

Have you ever watched the pendulum move in a grandfather clock? The pendulum has energy because it causes change as it moves back and forth. What kind of energy does a moving pendulum have? Can it do work on an object? In this lab, you will analyze the movement and energy of a pendulum.

Learn It

Before you can draw valid conclusions from any scientific experiment, you must **analyze the results** of that experiment. This means you must look for patterns in the results.

Try It

1. Read and complete a lab safety form.
2. Use the photo below as a guide to make a pendulum. Hang one washer on a paper clip. Place a box so it will block the swinging pendulum. Mark the position of the box with tape.
3. Pull the pendulum back until the bottom of the washer is 15 cm from the floor. Release the pendulum. Measure and record the distance the box moves in your Science Journal. Repeat two more times.
4. Repeat step 3 using pendulum heights of 30 cm and 45 cm.
5. Repeat steps 3 and 4 with two washers, then with three washers.

Apply It

6. Does the pendulum have potential energy? Explain.
7. Does it have kinetic energy? How do you know?
8. How does the gravitational potential energy depend on the pendulum's weight and height?
9. How does the distance the box travels depend on the initial gravitational potential energy?
10. Does the pendulum do work on the box? Explain your answer.
11. **Key Concept** Determine when the pendulum had maximum potential energy and maximum kinetic energy. Explain your reasoning.

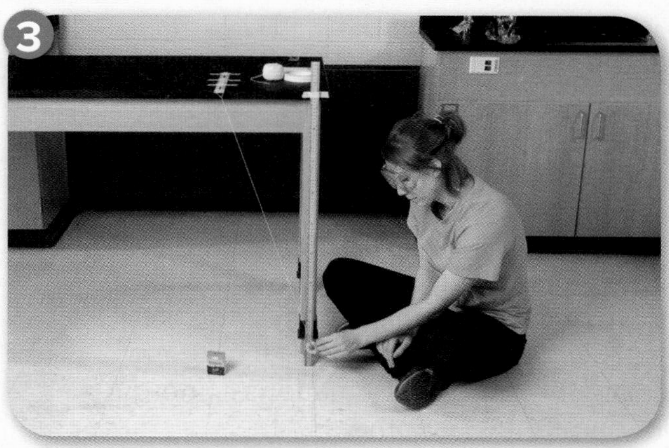

Lesson 2

Reading Guide

Key Concepts
ESSENTIAL QUESTIONS

- What is the law of conservation of energy?
- How does friction affect energy transformations?
- How are different types of energy used?

Vocabulary

law of conservation of energy p. 430

friction p. 431

Multilingual eGlossary

Go to the resource tab in ConnectED to find the PBL *Tearin' It Up!*

Energy Transformations

Inquiry What's that sound?

Blocks of ice breaking off the front of this glacier can be bigger than a car. Imagine the loud rumble they make as they crash into the sea. But after the ice falls into the sea, it will melt gradually. All of these processes involve energy transformations—energy changing from one form to another.

sorincolac/Getty Images

Launch Lab

15 minutes

Is energy lost when it changes form?

Energy can have different forms. What happens when energy changes from one form to another?

1. Read and complete a lab safety form.
2. Three students should sit in a circle. One student has 30 **buttons,** one has 30 **pennies,** and one has 30 **paper clips.**
3. Each student should exchange 10 items with the student to the right and 10 items with the student to the left.
4. Repeat step 3.

Think About This

1. If the buttons, the pennies, and the paper clips represent different forms of energy, what represents changes from one form of energy to another?
2. **Key Concept** If each button, penny, and paper clip represents one unit of energy, does the total amount of energy increase, decrease, or stay the same? Explain your answer.

Changes Between Forms of Energy

It is the weekend and you are ready to make some popcorn in the microwave and watch a movie. Energy changes form when you make popcorn and watch TV. As shown in **Figure 5,** a microwave changes electric energy into radiant energy. Radiant energy changes into thermal energy in the popcorn kernels.

The changes from electric energy to radiant energy to thermal energy are called energy transformations. As you watch the movie, energy transformations also occur in the television. A television transforms electric energy into sound energy and radiant energy.

Science Use v. Common Use

radiant

Science Use energy transmitted by electromagnetic waves

Common Use bright and shining; glowing

Figure 5 Energy changes from one form to another when you use a microwave oven to make popcorn.

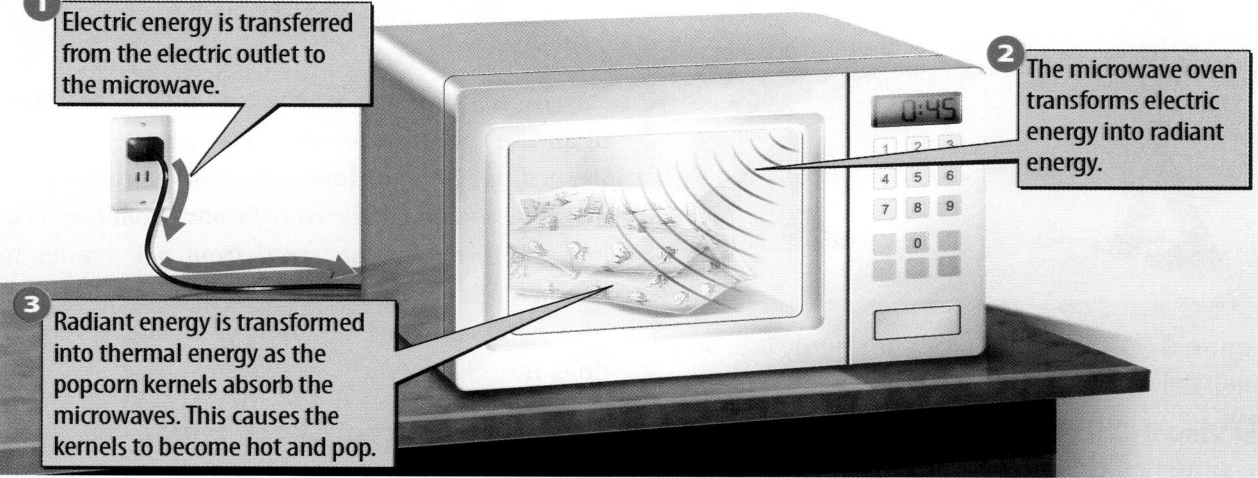

1 Electric energy is transferred from the electric outlet to the microwave.
2 The microwave oven transforms electric energy into radiant energy.
3 Radiant energy is transformed into thermal energy as the popcorn kernels absorb the microwaves. This causes the kernels to become hot and pop.
0:45
1 2 3
4 5 6
7 8 9
0

Hutchings Photography/ Digital Light Source

Conservation of Energy

Animation

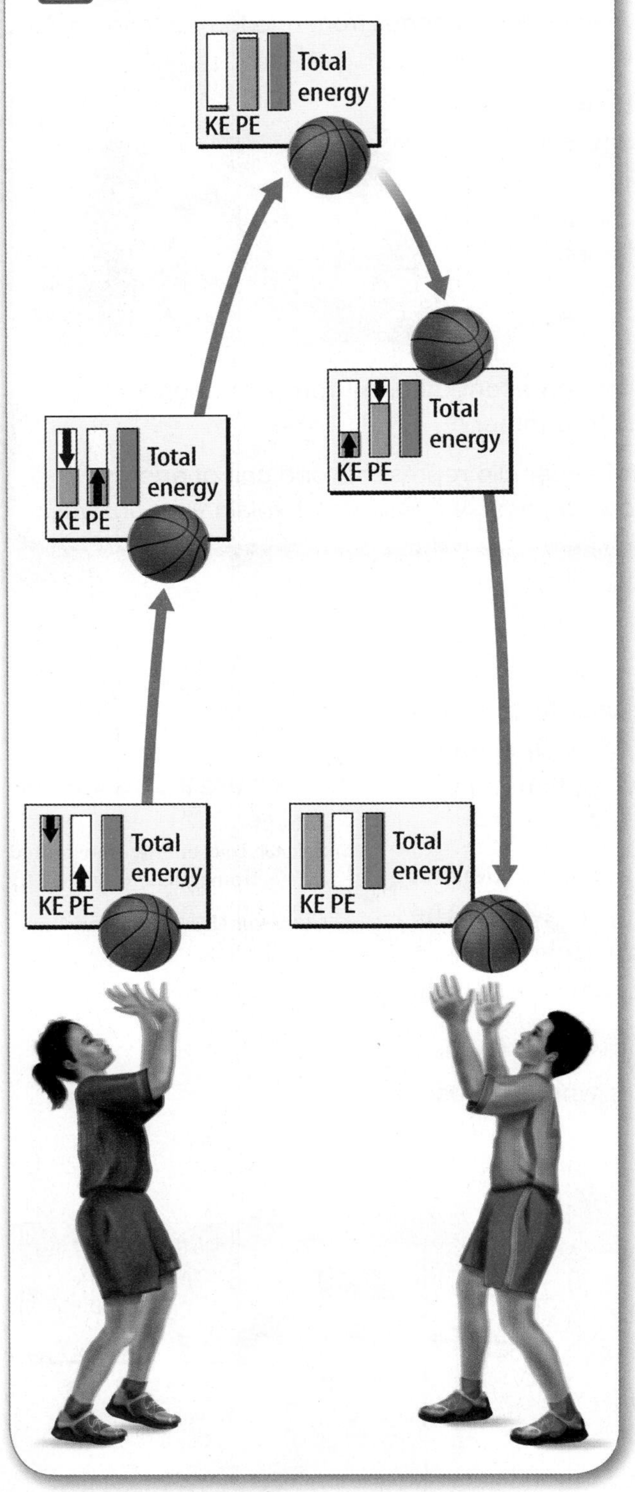

Figure 6 The ball's kinetic energy (KE) and potential energy (PE) change as it moves.

Visual Check When is the gravitational potential energy the greatest?

Changes Between Kinetic and Potential Energy

Energy transformations also occur when you toss a ball upward, as shown in **Figure 6.** The ball slows down as it moves upward and then speeds up as it moves downward. The ball's speed and height change as energy changes from one form to another.

Kinetic Energy to Potential Energy

The ball is moving fastest and has the most kinetic energy as it leaves your hand, as shown in **Figure 6.** As the ball moves upward, its speed and kinetic energy decrease. However, the potential energy is increasing because the ball's height is increasing. Kinetic energy is changing into potential energy. At the ball's highest point, the gravitational potential energy is at its greatest, and the ball's kinetic energy is at its lowest.

Potential Energy to Kinetic Energy

As the ball moves downward, its potential energy decreases. At the same time, the ball's speed increases. Therefore, the ball's kinetic energy increases. Potential energy is transformed into kinetic energy. When the ball reaches the other player's hand, its kinetic energy is at the maximum value again.

Reading Check Why does the potential energy decrease as the ball falls?

The Law of Conservation of Energy

The total energy in the universe is the sum of all the different forms of energy everywhere. *According to the* **law of conservation of energy,** *energy can be transformed from one form into another or transferred from one region to another, but energy cannot be created or destroyed.* The total amount of energy in the universe does not change.

Key Concept Check What is the law of conservation of energy?

Friction and the Law of Conservation of Energy

Sometimes it may seem as if the law of conservation of energy is not accurate. Imagine riding a bicycle, as in **Figure 7**. The moving bicycle has mechanical energy. What happens to this mechanical energy when you apply the brakes and the bicycle stops?

When you apply the brakes, the bicycle's mechanical energy is not destroyed. Instead the bicycle's mechanical energy is transformed to thermal energy, as shown in **Figure 7**. The total amount of energy never changes. The additional thermal energy causes the brakes, the wheels, and the air around the bicycle to become slightly warmer.

Friction between the bicycle's brake pads and the moving wheels transforms mechanical energy into thermal energy. **Friction** *is a force that resists the sliding of two surfaces that are touching.*

Key Concept Check How does friction affect energy transformations?

There is always some friction between any two surfaces that are rubbing against each other. As a result, some mechanical energy always is transformed into thermal energy when two surfaces rub against each other.

It is easier to pedal a bicycle if there is less friction between the bicycle's parts. With less friction, less of the bicycle's mechanical energy is transformed into thermal energy. One way to reduce friction is to apply a lubricant, such as oil, grease, or graphite, to surfaces that rub against each other.

FOLDABLES®

Cut three sheets of paper in half. Use the six half sheets to make a side-tab book with five tabs and a cover. Use your book to organize your notes on energy transformations.

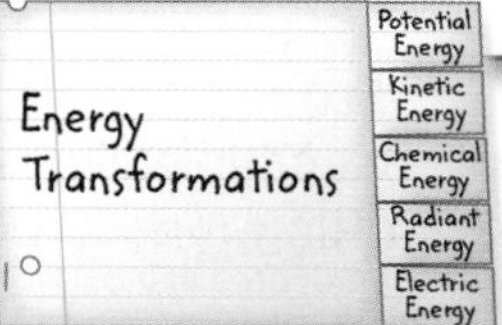

WORD ORIGIN

friction
from Latin *fricare,* means "to rub"

Friction and Thermal Energy

Personal Tutor

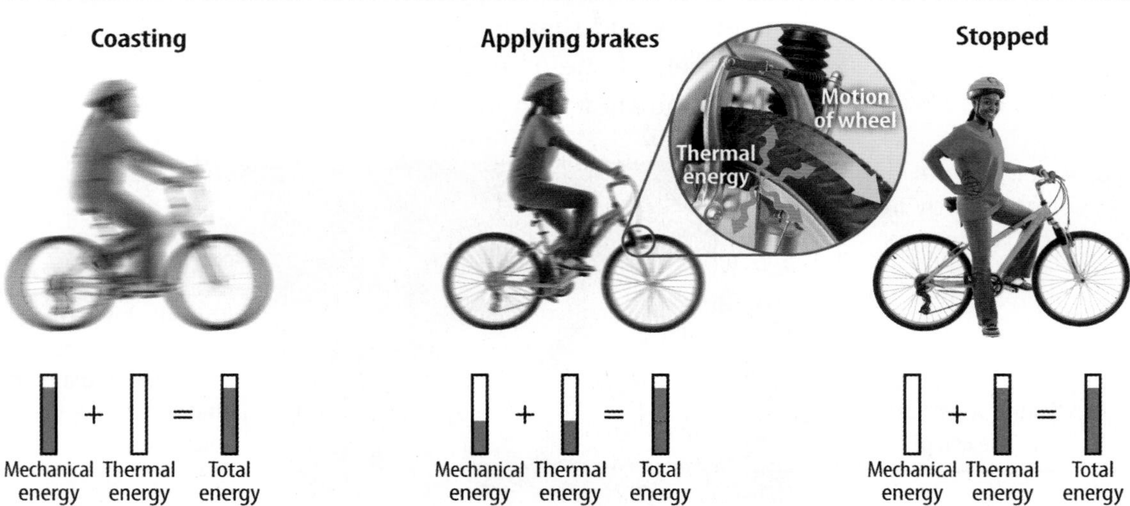

Figure 7 When the girl applies the brakes, friction between the bicycle's brake pads and its wheels transforms mechanical energy into thermal energy. As mechanical energy changes into thermal energy, the bicycle slows down. The total amount of energy does not change.

Math Skills

Solve a One-Step Equation

Electric energy often is measured in units called kilowatt-hours (kWh). To calculate the electric energy used by an appliance in kWh, use this equation:

$$\text{kWh} = \left(\frac{\text{watts}}{1{,}000}\right) \times \text{hours}$$

Appliances typically have a power rating measured in watts (W).

Practice

A hair dryer is rated at 1,200 W. If you use the dryer for 0.25 h, how much electric energy do you use?

Math Practice

Personal Tutor

Using Energy

Every day you use different forms of energy to do different things. You might use the radiant energy from a lamp to light a room, or you might use the chemical energy stored in your body to run a race. When you use energy, you usually change it from one form into another. For example, the lamp changes electric energy into radiant energy and thermal energy.

Using Thermal Energy

All forms of energy can be transformed into thermal energy. People often use thermal energy to cook food or provide warmth. A gas stove transforms the chemical energy stored in natural gas into the thermal energy that cooks food. An electric space heater transforms the electric energy from a power plant into the thermal energy that warms a room. In a jet engine, burning fuel releases thermal energy that the engine transforms into mechanical energy.

Using Chemical Energy

During photosynthesis, a plant transforms the Sun's radiant energy into chemical energy that it stores in chemical compounds. Some of these compounds become food for other living things. Your body transforms the chemical energy from your food into the kinetic energy necessary for movement. Your body also transforms chemical energy into the thermal energy necessary to keep you warm.

Using Radiant Energy

The cell phone in **Figure 8** sends and receives radiant energy using microwaves. When you are listening to someone on a cell phone, that cell phone is transforming radiant energy into electric energy and then into sound energy. When you are speaking into a cell phone, it is transforming sound energy into electric energy and then into radiant energy.

Sound waves carry energy into the cell phone.
The cell phone converts the energy carried by sound waves into radiant energy that is carried away by microwaves.

Figure 8 A cell phone changes sound energy into radiant energy when you speak.

Using Electric Energy

Many of the devices you might use every day, such as handheld video games, MP3 players, and hair dryers, use electric energy. Some devices, such as hair dryers, use electric energy from electric power plants. Other appliances, such as handheld video games, transform the chemical energy stored in batteries into electric energy.

 **Key Concept Check** How are different types of energy used?

Waste Energy

When energy changes form, some thermal energy is always released. For example, a lightbulb converts some electric energy into radiant energy. However, the lightbulb also transforms some electric energy into thermal energy. This is what makes the lightbulb hot. Some of this thermal energy moves into the air and cannot be used.

Scientists often refer to thermal energy that cannot be used as waste energy. Whenever energy is used, some energy is transformed into useful energy and some is transformed into waste energy. For example, we use the chemical energy in gasoline to make cars, such as those in **Figure 9**, move. However, most of that chemical energy ends up as waste energy—thermal energy that moves into the air.

 **Reading Check** What is waste energy?

MiniLab 20 minutes

How does energy change form?

When an object falls, energy changes form. How can you compare energies for falling objects?

1. Read and complete a lab safety form.
2. Place a piece of **clay** about 10 cm wide and 3 cm thick on a **small paper plate.**
3. Drop a **marble** onto the clay from a height of about 20 cm, and measure the depth of the depression caused by the marble. Record the measurement in your Science Journal.
4. Repeat step 3 with a heavier marble.

Analyze and Conclude

1. **Infer** Which marble had more kinetic energy just before it hit the clay? Explain your answer.
2. **Key Concept** For which marble was the potential energy greater just before the marble fell? Explain your answer using the law of conservation of energy.

Figure 9 Cars transform most of the chemical energy in gasoline into waste energy.

Lesson 2 Review

Visual Summary

Energy can change form, but according to the law of conservation of energy, energy can never be created or destroyed.

Friction transforms mechanical energy into thermal energy.

Different forms of energy, such as sound and radiant energy, are used when someone talks on a cell phone.

Use your lesson Foldable to review the lesson. Save your Foldable for the project at the end of the chapter.

What do you think NOW?

You first read the statements below at the beginning of the chapter.

4. Energy can change from one form to another.

5. Energy is destroyed when you apply the brakes on a moving bicycle or a moving car.

6. The Sun releases radiant energy.

Did you change your mind about whether you agree or disagree with the statements? Rewrite any false statements to make them true.

Use Vocabulary

1 **Use the term** *friction* in a complete sentence.

Understand Key Concepts 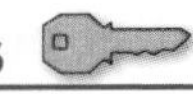

2 **Explain** the law of conservation of energy in your own words.

3 **Describe** the energy transformations that occur when a piece of wood burns.

4 **Identify** the energy transformation that takes place when you apply the brakes on a bicycle.

5 Which energy transformation occurs in a toaster?

A. chemical to electric
B. electric to thermal
C. kinetic to chemical
D. thermal to potential

Interpret Graphics

6 **Organize Information** Copy and fill in the graphic organizer below to show how kinetic and potential energy change when a ball is thrown straight up and then falls down.

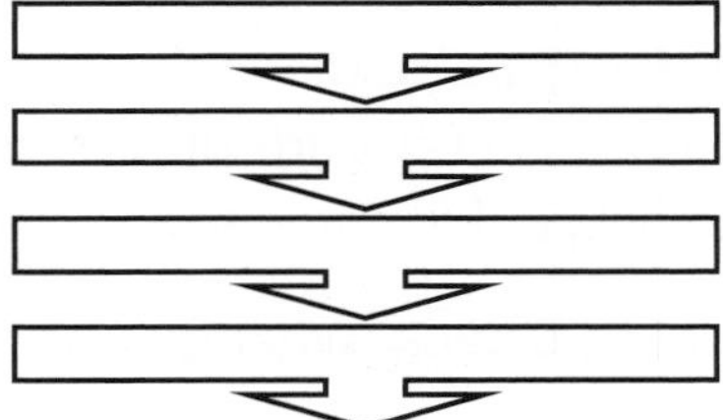

Critical Thinking

7 **Judge** An advertisement states that a machine with moving parts will continue moving forever without having to add any energy. Can this be correct? Explain.

Math Skills   Math Practice

8 **Calculate** If you use a 1,000-W microwave for 0.15 h, how much electric energy do you use?

(t)Hutchings Photography/Digital Light Source, (b)Peter Cade/Getty Images

GREEN SCIENCE

Fossil Fuels and Rising CO_2

Investigate the link between energy use and carbon dioxide in the atmosphere.

You use energy every day—when you ride in a car or on a bus, turn on a television or a radio, and even when you send an e-mail.

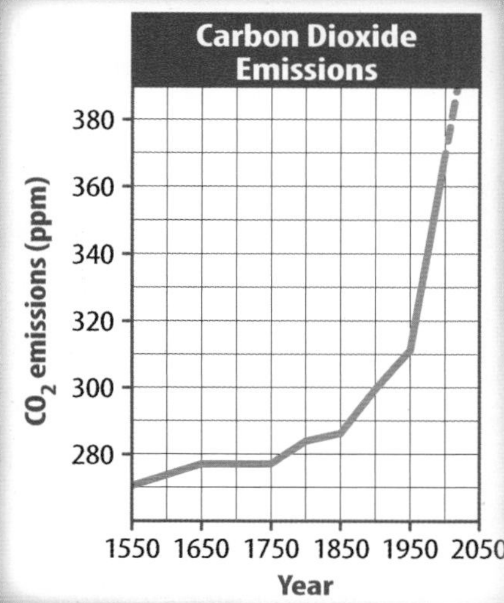

Much of the energy that produces electricity, heats and cools buildings, and powers engines, comes from burning fossil fuels—coal, oil, and natural gas. When fossil fuels burn, the carbon in them combines with oxygen in the atmosphere and forms carbon dioxide gas (CO_2). Carbon dioxide is a greenhouse gas. Greenhouse gases absorb energy. This causes the atmosphere and Earth's surface to become warmer. Greenhouse gases make Earth warm enough to support life. Without greenhouse gases, Earth's surface would be frozen.

However, over the past 150 years, the amount of CO_2 in the atmosphere has increased faster than at any time in the past 800,000 years. Most of this increase is the result of burning fossil fuels. More carbon dioxide in the atmosphere might cause average global temperatures to increase. As temperatures increase, weather patterns worldwide could change. More storms and heavier rainfall could occur in some areas, while other regions could become drier. Increased temperatures could also cause more of the polar ice sheets to melt and raise sea levels. Higher sea levels would cause more flooding in coastal areas.

Developing other energy sources such as geothermal, solar, nuclear, wind, and hydroelectric power would reduce the use of fossil fuels and slow the increase in atmospheric CO_2.

It's Your Turn

MAKE A LIST How can CO_2 emissions be reduced? Work with a partner. List five ways people in your home, school, or community could reduce their energy consumption. Combine your list with your classmates' lists to make a master list.

300 Years of Carbon Dioxide

- **1712**
  A new invention, the steam engine, is powered by burning coal that heats water to produce steam.

- **Early 1800s**
  Coal-fired steam engines, able to pull heavy trains and power steamboats, transform transportation.

- **1882**
  Companies make and sell electricity from coal for everyday use. Electricity is used to power the first lightbulbs, which give off 20 times the light of a candle.

- **1908**
  The first mass-produced automobiles are made available. By 1915, Ford is selling 500,000 cars a year. Gasoline becomes the fuel of choice for car engines.

- **Late 1900s**
  Electrical appliances transform the way we live, work, and communicate. Most electricity is generated by coal-burning power plants.

- **2007**
  There are more than 800 million cars and light trucks on the world's roads.

 Lab

45 minutes

Pinwheel Power

Moving air, or wind, is an energy source. In some places, wind turbines transform the kinetic energy of wind into electric energy. This electric energy can be used to do work by making an object move. In this lab, you will construct a pinwheel turbine and observe how changes in wind speed affect the rate at which your wind turbine does work.

Ask a Question

How does wind speed affect the rate at which a wind turbine does work?

Make Observations

1. Read and complete a lab safety form.
2. Construct a pinwheel from a manila folder using the diagram below.
3. Use a plastic pushpin to carefully attach the pinwheel to the eraser of a pencil.
4. Use a hole punch to make holes on opposite sides of the top of a container. Use your ruler to make sure the holes are exactly opposite each other. Weigh down the container with sand or small rocks.
5. Put the pencil through the holes, and make sure the pinwheel spins freely. Blow against the blades of the pinwheel with varying amounts of force to observe how the pinwheel moves. Record your observations in your Science Journal.
6. Measure and cut 100 cm of thread. Tie two washers to one end of the thread. Tape the other end of the thread to the pencil.

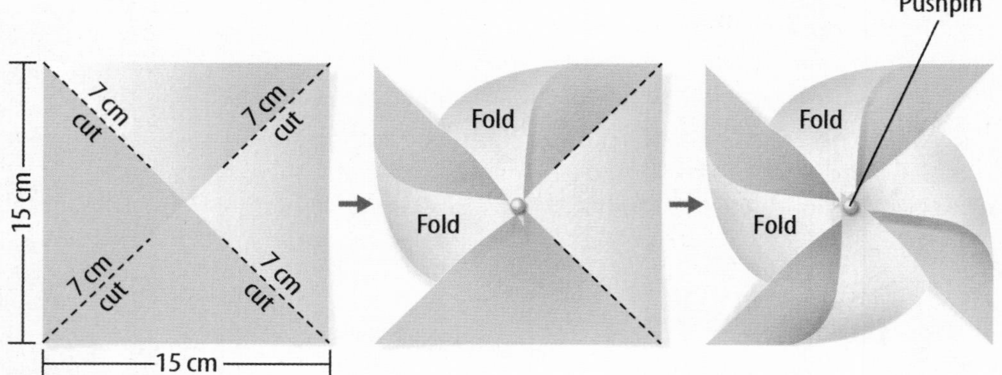

Materials

round pencil with unused eraser

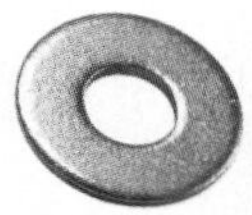

metal washers

cardboard container

sand or small rocks

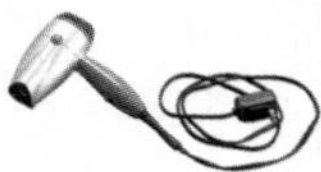

three-speed hair dryer

stopwatch

Also needed: manila folder, metric ruler, scissors, hole punch, thread, pushpin

Safety

Form a Hypothesis

7. Use your observations from step 5 to form a hypothesis about how wind speed will affect the rate at which the wind turbine does work.

Test Your Hypothesis

8. Work with two other students to test your hypothesis. One person will use the hair dryer to model a slow wind speed. Another person will stop the pencil's movement after 5 seconds on the stopwatch. The third person will measure the length of thread remaining between the pencil and the top of the washers. Then someone will unwind the thread and the group will repeat this procedure four more times with the dryer on low. Record all data in your Science Journal.
9. Repeat step 8 with the dryer on medium.
10. Repeat step 8 with the dryer on high.

Analyze and Conclude

11. **Interpret Data** Did your hypothesis agree with your data and observations? Explain.
12. **Sequence** Describe how energy was transformed from one form into another in this lab.
13. **Draw Conclusions** What factors might have affected the rate at which your pinwheel turbine did work?
14. **The Big Idea** Explain how wind is used as an energy resource.

Communicate Your Results

Use your data and observations to write a paragraph explaining how wind speed affects the rate at which a wind turbine can do work.

Inquiry Extension

Research the designs of real wind turbines. Create a model of an actual wind turbine. Write a short explanation of its advantages and disadvantages compared to other wind turbines.

Lab Tips

☑ You measure the rate at which the wind turbine does work by measuring how fast the turbine lifts the metal washers.

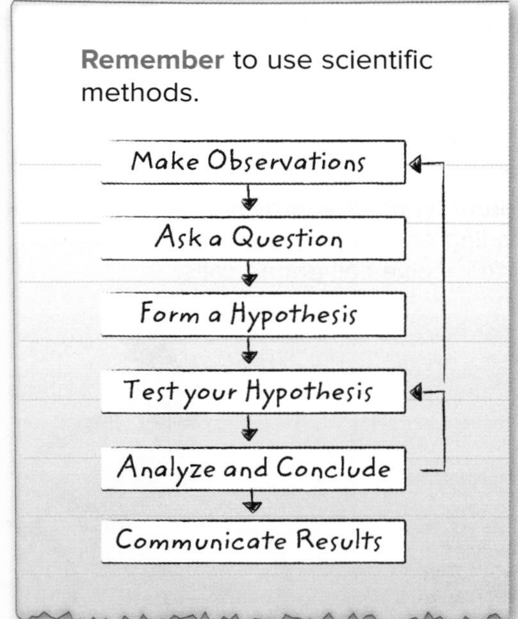

Chapter 13 Study Guide

Energy is the ability to cause change. Energy transformations occur when one form of energy changes into another form of energy.

Key Concepts Summary

Lesson 1: Forms of Energy

- **Energy** is the ability to cause change.
- **Kinetic energy** is the energy an object has because of its motion. **Potential energy** is stored energy.
- **Work** is the transfer of energy that occurs when a force makes an object move in the direction of the force while the force is acting on the object.
- Different forms of energy include **thermal energy** and **radiant energy.**

Vocabulary

energy p. 421
kinetic energy p. 422
potential energy p. 422
work p. 424
mechanical energy p. 425
sound energy p. 425
thermal energy p. 425
electric energy p. 425
radiant energy p. 425
nuclear energy p. 425

Lesson 2: Energy Transformations

- According to the **law of conservation of energy,** energy can be transformed from one form into another or transferred from one region to another, but energy cannot be created or destroyed.
- **Friction** transforms mechanical energy into thermal energy.
- Different types of energy are used in many ways including providing energy to move your body, to light a room, and to make and to receive cell phone calls.

Vocabulary

law of conservation of energy p. 430
friction p. 431

(t)David Madison/Getty Images, (b)Hutchings Photography/Digital Light Source

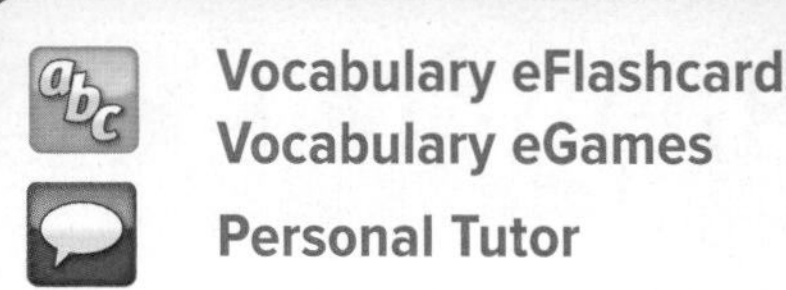

FOLDABLES® Chapter Project

Assemble your Lesson Foldables as shown to make a Chapter Project. Use the project to review what you have learned in this chapter.

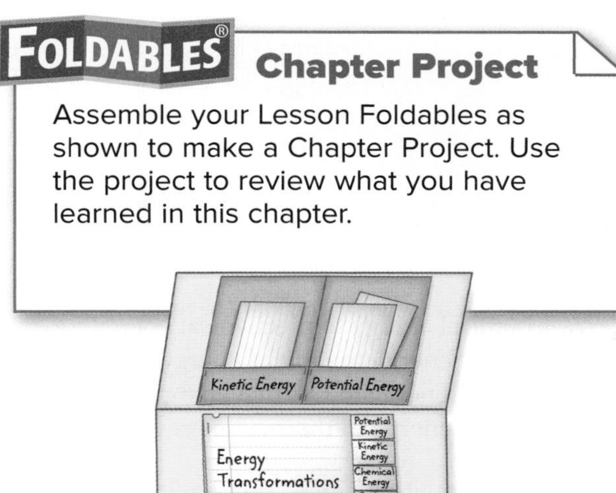

Use Vocabulary

Each of the following sentences is false. Make the sentence true by replacing the italicized word with a vocabulary term.

1. *Thermal energy* is the form of energy carried by an electric current.
2. The *chemical potential energy* of an object depends on its mass and its speed.
3. *Friction* is the transfer of energy that occurs when a force is applied over a distance.
4. A lubricant, such as oil, grease, or graphite, reduces *radiant energy* between rubbing objects.
5. *Radiant energy* is energy that is stored in the nucleus of an atom.

Interactive Concept Map

Link Vocabulary and Key Concepts

Copy this concept map, and then use vocabulary terms from the previous page to complete the concept map.

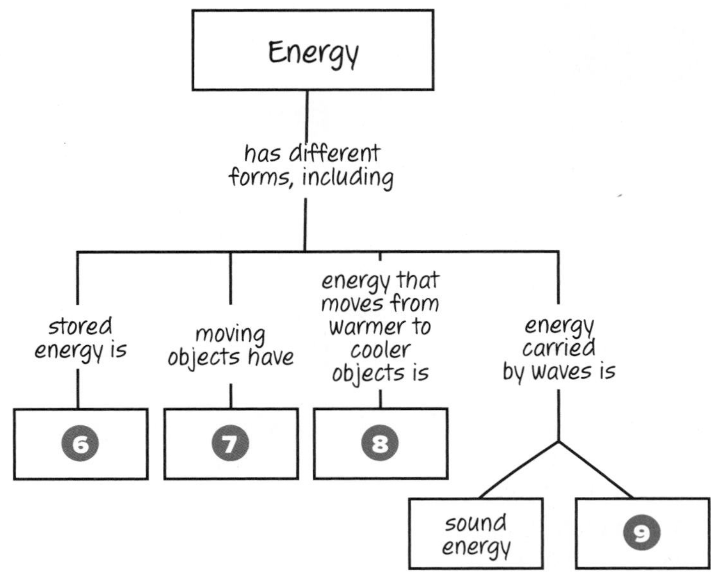

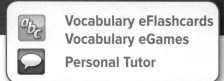

Chapter 13 Review

Understand Key Concepts 

1. What factors determine an object's kinetic energy?
   - A. its height and its mass
   - B. its mass and its speed
   - C. its size and its weight
   - D. its speed and its height

2. The gravitational potential energy stored between an object and Earth depends on
   - A. the object's height and weight.
   - B. the object's mass and speed.
   - C. the object's size and weight.
   - D. the object's speed and height.

3. When a ball is thrown upward, where does it have the least kinetic energy?
   - A. at its highest point
   - B. at its lowest point when it is moving downward
   - C. at its lowest point when it is moving upward
   - D. midway between its highest point and its lowest point

4. Which type of energy is released when the string in the photo below is plucked?

   - A. electric energy
   - B. nuclear energy
   - C. radiant energy
   - D. sound energy

5. According to the law of conservation of energy, which is always true?
   - A. Energy can never be created or destroyed.
   - B. Energy is always converted to friction in moving objects.
   - C. The universe is always gaining energy in many different forms.
   - D. Work is done when a force is exerted on an object.

6. Which energy transformation is occurring in the food below?

   - A. chemical energy to mechanical energy
   - B. electric energy to radiant energy
   - C. nuclear energy to thermal energy
   - D. radiant energy to thermal energy

7. In which situation would the gravitational potential energy between you and Earth be greatest?
   - A. You are running down a hill.
   - B. You are running up a hill.
   - C. You stand at the bottom of a hill.
   - D. You stand at the top of a hill.

8. When you speak into a cell phone which energy conversion occurs?
   - A. chemical energy to radiant energy
   - B. mechanical energy to chemical energy
   - C. sound energy to radiant energy
   - D. thermal energy to sound energy

9. Which type of energy is released when a firecracker explodes?
   - A. chemical potential energy
   - B. elastic potential energy
   - C. electric energy
   - D. nuclear energy

10. Inside the engine of a gasoline-powered car, chemical energy is converted primarily to which kind of energy?
    - A. kinetic
    - B. potential
    - C. sound
    - D. waste

Andrew Lambert Photography/Photo Researchers, Inc.

Critical Thinking

11 **Determine** if work is done on the nail shown below if a person pulls the handle to the left and the handle moves. Explain.

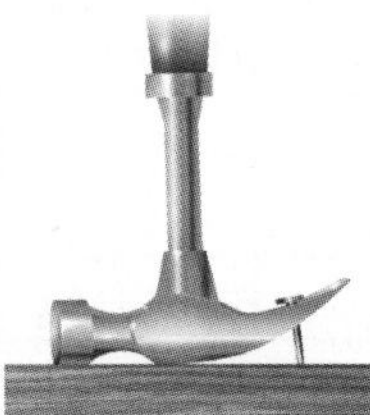

12 **Contrast** the energy transformations that occur in a electrical toaster oven and in an electrical fan.

13 **Infer** A red box and a blue box are on the same shelf. There is more gravitational potential energy between the red box and Earth than between the blue box and Earth. Which box weighs more? Explain your answer.

14 **Infer** Juanita moves a round box and a square box from a lower shelf to a higher shelf. The gravitational potential energy for the round box increases by 50 J. The gravitational potential energy for the square box increases by 100 J. On which box did Juanita do more work? Explain your reasoning.

15 **Explain** why a skateboard coasting on a flat surface slows down and comes to a stop.

16 **Describe** how energy is conserved when a basketball is thrown straight up into the air and falls back into your hands.

17 **Decide** Harold stretches a rubber band and lets it go. The rubber band flies across the room. Harold says this demonstrates the transformation of kinetic energy to elastic potential energy. Is Harold correct? Explain.

Writing in Science

18 **Write** a short essay explaining the energy transformations that occur in an incandescent lightbulb.

REVIEW THE BIG IDEA

19 Write an explanation of energy and energy transformations for a fourth grader who has never heard of these terms.

20 Identify five energy transformations occurring in the photo below.

Math Skills

Math Practice

Solve One-Step Equations

21 An electrical water heater is rated at 5,500 W and operates for 106 h per month. How much electric energy in kWh does the water heater use each month?

22 A family uses 1,303 kWh of electric energy in a month. If the power company charges $0.08 cents per kilowatt hour, what is the total electric energy bill for the month?

Standardized Test Practice

Record your answers on the answer sheet provided by your teacher or on a sheet of paper.

Multiple Choice

1 Which is true when a player throws a basketball toward a hoop?

A Kinetic energy is constant.

B Potential energy is constant.

C Work is done on the player.

D Work is done on the ball.

Use the diagram below to answer questions 2 and 3.

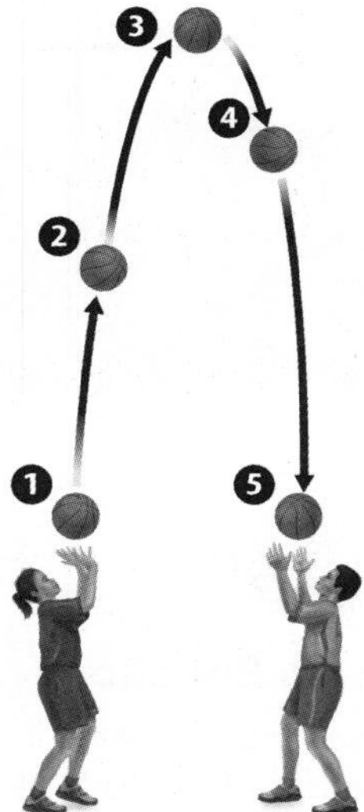

2 At which points is the kinetic energy of the basketball greatest?

A 1 and 5

B 2 and 3

C 2 and 4

D 3 and 4

3 At which point is the gravitational potential energy at its maximum?

A 1

B 2

C 3

D 4

Use the table below to answer question 4.

Vehicle	Mass	Speed
Car 1	1,200 kg	20 m/s
Car 2	1,500 kg	20 m/s
Truck 1	4,800 kg	20 m/s
Truck 2	6,000 kg	20 m/s

4 Which vehicle has the most kinetic energy?

A car 1

B car 2

C truck 1

D truck 2

5 When you compress a spring, which type of energy increases?

A kinetic

B nuclear

C potential

D radiant

6 Sound energy cannot travel through

A a vacuum.

B a wooden table.

C polluted air.

D pond water.

7 A bicyclist uses brakes to slow from 3 m/s to a stop. What stops the bike?

A friction

B gravity

C kinetic energy

D thermal energy

Online Standardized Test Practice

Use the diagram below to answer question 8.

8 The work being done in the diagram above transfers energy to
- **A** the box.
- **B** the floor.
- **C** the girl.
- **D** the shelf.

9 Which is true of energy?
- **A** It cannot be created or destroyed.
- **B** It cannot change form.
- **C** Most forms cannot be conserved.
- **D** Most forms cannot be traced to a source.

10 Which energy transformation occurs when you light a gas burner?
- **A** chemical to thermal
- **B** electric to chemical
- **C** nuclear to chemical
- **D** radiant to thermal

Constructed Response

Use the table below to answer questions 11 and 12.

Form of Energy	Definition

11 Copy the table above, and list six forms of energy. Briefly define each form.

12 Provide real-life examples of each of the listed forms of energy.

Use the diagram below to answer question 13.

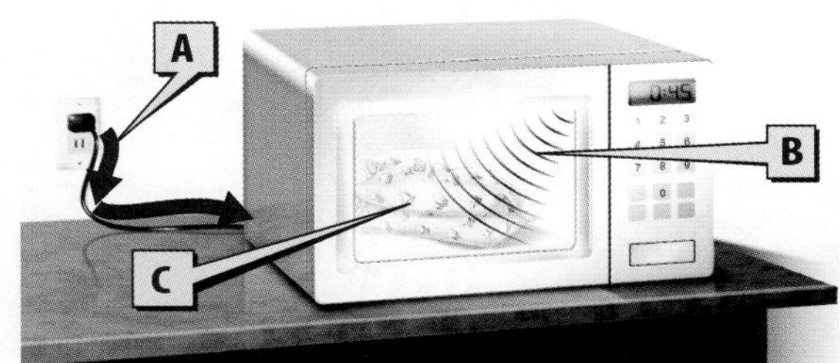

13 Describe the energy transformations that occur at locations A, B, and C.

NEED EXTRA HELP?													
If You Missed Question...	1	2	3	4	5	6	7	8	9	10	11	12	13
Go to Lesson...	1	2	2	1	1	1	2	1	2	2	1	2	2

Chapter 14

Waves, Light, and Sound

How do waves transfer energy through matter and through empty space?

Inquiry What do the colors mean?

Have you ever seen weather reports that show a map with colorful images? Clear skies produce a clear weather map, but watch out if you see lots of blue, green, yellow, and red on the map!

- What do the different colors on the map mean?
- How do meteorologists get the information they display on a weather map?
- How do waves transfer energy through matter and through empty space?

imac/Alamy

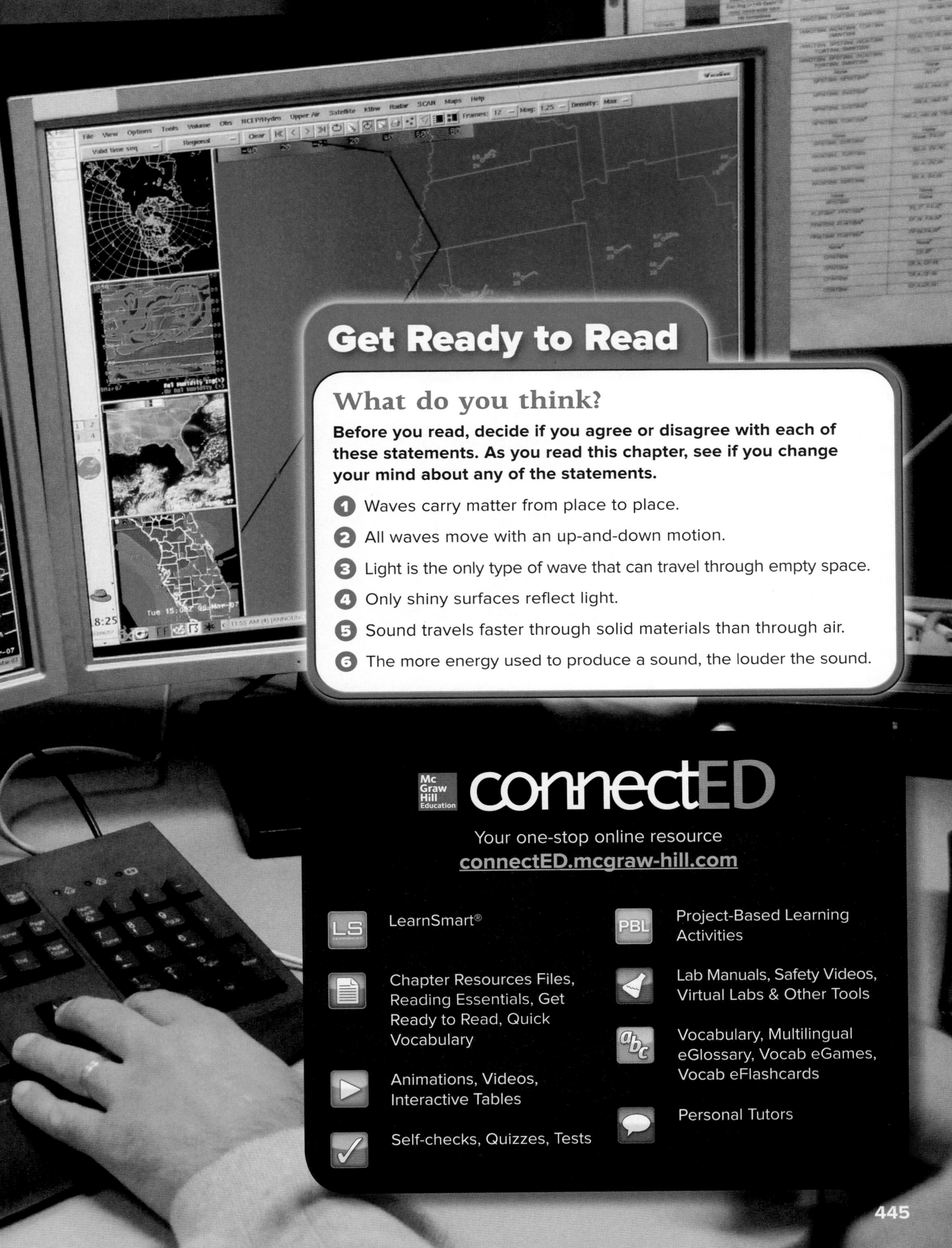

Get Ready to Read

What do you think?

Before you read, decide if you agree or disagree with each of these statements. As you read this chapter, see if you change your mind about any of the statements.

1. Waves carry matter from place to place.
2. All waves move with an up-and-down motion.
3. Light is the only type of wave that can travel through empty space.
4. Only shiny surfaces reflect light.
5. Sound travels faster through solid materials than through air.
6. The more energy used to produce a sound, the louder the sound.

Mc Graw Hill Education connectED

Your one-stop online resource

connectED.mcgraw-hill.com

- LS LearnSmart®
- Chapter Resources Files, Reading Essentials, Get Ready to Read, Quick Vocabulary
- Animations, Videos, Interactive Tables
- Self-checks, Quizzes, Tests
- PBL Project-Based Learning Activities
- Lab Manuals, Safety Videos, Virtual Labs & Other Tools
- Vocabulary, Multilingual eGlossary, Vocab eGames, Vocab eFlashcards
- Personal Tutors

Lesson 1

Waves

Reading Guide

Key Concepts
ESSENTIAL QUESTIONS

- What are waves, and how are waves produced?
- How can you describe waves by their properties?
- What are some ways in which waves interact with matter?

Vocabulary

mechanical wave p. 448
electromagnetic wave p. 448
transverse wave p. 449
longitudinal wave p. 449
frequency p. 451
amplitude p. 452
refraction p. 454

Multilingual eGlossary

Science Video

Go to the resource tab in ConnectED to find the PBL *Don't Make Waves!*

Inquiry What causes the waves?

Have you ever watched a surfer ride the waves? Ocean waves are produced by winds far out at sea. By the time they reach shore, some waves have so much energy that they are taller than a person or even a house. Why do waves get taller as they approach the shore? What properties do water waves have in common with other types of waves?

Launch Lab

15 minutes

How do waves form?

You probably have seen water waves on the surface of a lake or a swimming pool. How are the waves produced?

1. Read and complete a lab safety form.
2. Place **books** under opposite edges of a **glass pan**. Add about 5 mm of water to the pan. Place a **sheet of white paper** under the pan. Wait until the water is still.
3. Place a **cork** in the water about halfway between the center and the edge of the pan. Dip your **pencil** tip into the center of the water one time. What happens to the cork? Record your observations of the water and the cork in a data table in your Science Journal.
4. Repeatedly tap your pencil tip on the surface of the water slowly. Record your observations.
5. Repeat step 4, tapping your pencil tip faster this time. Record your observations.

Think About This

1. How are the waves you produced in steps 3 and 4 alike? How are they different?
2. How does the behavior of the cork change in steps 4 and 5?
3. **Key Concept** What do you think is the source of the waves that you made?

What are waves?

A flag waves in the breeze. Ocean waves break onto a beach. You wave your hand at a friend. All of these actions have something in common. Waves always begin with a source of energy that causes a back-and-forth or up-and-down disturbance, or movement. In **Figure 1**, energy of the wind causes a disturbance in the flag. This disturbance moves along the length of the flag as a wave. A wave is a disturbance that transfers energy from one place to another without transferring matter.

 **Key Concept Check** What are waves?

Energy Transfer

Wind transfers energy to the fabric in the flag. The flag ripples back and forth as the energy travels along the fabric. Notice that each point on the flag moves back and forth, but the fabric does not move along with the wave. Recall, waves only transfer energy, not matter, from place to place.

When you lift a pebble, you transfer energy to it. Suppose you drop the pebble into a pond. The pebble's energy transfers to the water. Waves carry the energy away from the point where the pebble hit the water. The water itself moves up and down as the wave passes, but the water does not move along with the wave.

Figure 1 The wave is a disturbance that transfers energy along the flag.

▲ **Figure 2**  The energy of the falling pebble produces a mechanical wave.

Two Main Types of Waves

Some waves carry energy only through matter. Other types of waves can carry energy through matter or empty space.

Mechanical Waves *A wave that travels only through matter is a* **mechanical wave.** A medium is the matter through which a mechanical wave travels. A mechanical wave forms when a source of energy causes particles that make up a medium to vibrate. For example, a pebble falling into water transfers its kinetic energy to particles of the water, as shown in **Figure 2.** The water particles vibrate and push against nearby particles, transferring the energy outward. After each particle pushes the next particle, it returns to its original rest position. Energy is transferred, but the water particles are not.

Word Origin

mechanical
from Greek *mekhanikos,* means "like a machine"

Electromagnetic Waves *A wave that can travel through empty space or through matter is an* **electromagnetic wave.** This type of wave forms when a charged particle, such as an electron, vibrates. For example, electromagnetic waves transfer the Sun's energy to Earth through empty space. Once the waves reach Earth, they travel through matter, such as the atmosphere or a glass window of your house.

Key Concept Check How are waves produced?

Review Vocabulary

perpendicular
at right angles

Describing Wave Motion

Some waves move particles of a medium up and down or side to side, perpendicular to the direction the wave travels. For example, the waves in a flag move side to side, perpendicular to the direction the wind. Other wave disturbances move particles of the medium forward then backward in same direction, or parallel, to the motion of the wave. And last, some waves are a combination of both of these two types of motion. **Table 1** summarizes these three types of wave motion–transverse, longitudinal, or a combination of both.

Table 1 Electromagnetic waves are always transverse. Mechanical waves can be either transverse, longitudinal, or a combination of both. ▼

Table 1 Types of Wave Motion

Type of Wave Motion	Mechanical Waves	Electromagnetic Waves
Transverse—perpendicular to the direction the wave travels	✓ example: flag waving in a breeze	✓ example: light waves
Longitudinal—parallel to the direction the wave travels	✓ example: sound waves	
Combination—both transverse and longitudinal	✓ example: water waves	

Don Farrall/Getty Images

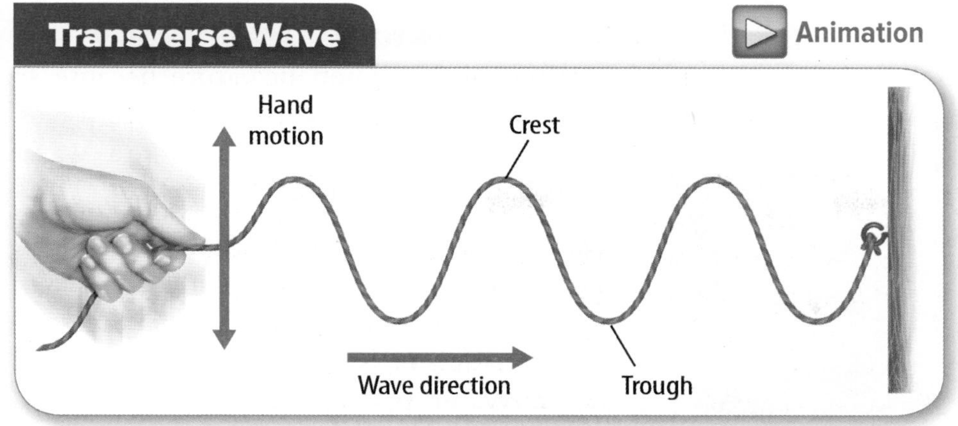

◀ **Figure 3** A transverse wave moves perpendicular to the hand's motion.

Transverse Waves *A wave in which the disturbance is perpendicular to the direction the wave travels is a* **transverse wave.** A breeze produces transverse waves in a flag. You can make transverse waves by attaching one end of a rope to a hook and holding the other end, as in **Figure 3.** When you move your hand up and down, transverse waves travel along the rope. High points on a wave are called crests. Low points are called troughs.

Recall that a vibrating charge, such as an electron, produces an electromagnetic wave. Electromagnetic waves are transverse waves. The electric and magnetic wave disturbances are perpendicular to the motion of the vibrating charge. You read that light is a form of energy transferred by transverse electromagnetic waves. X-rays and radio waves are two other examples.

Longitudinal Waves *A wave that makes the particles of a medium move back and forth parallel to the direction the wave travels is a* **longitudinal wave.** Longitudinal waves are mechanical waves. Like a transverse wave, a longitudinal wave disturbance passes energy from particle to particle of a medium. For example, when you knock on a door, energy of your hand transfers to the particles that make up the door. The energy of the vibrating particles of the door is then transferred to the air in the next room. Also, you can make a longitudinal wave by pushing or pulling on a coiled spring toy, as in **Figure 4.** Pushing moves the coils closer together. Pulling spreads the coils apart.

Make a vertical three-tab Venn book. Label it as shown. Use it to compare and contrast transverse and longitudinal waves.

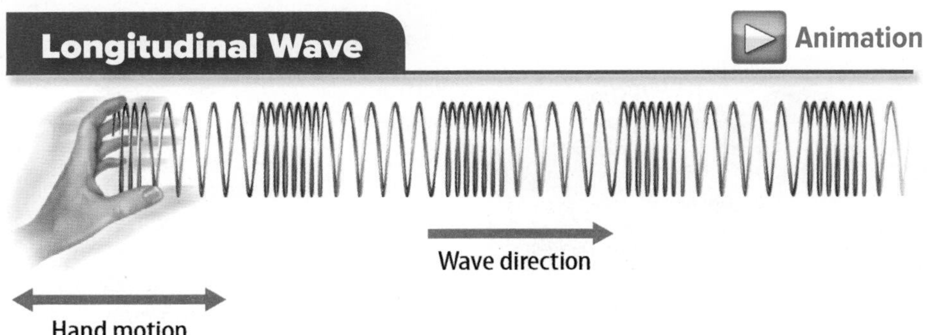

◀ **Figure 4** The back-and-forth motion of the hand causes a back-and-forth motion in the spring. The longitudinal waves move parallel to the hand's motion.

Water Waves

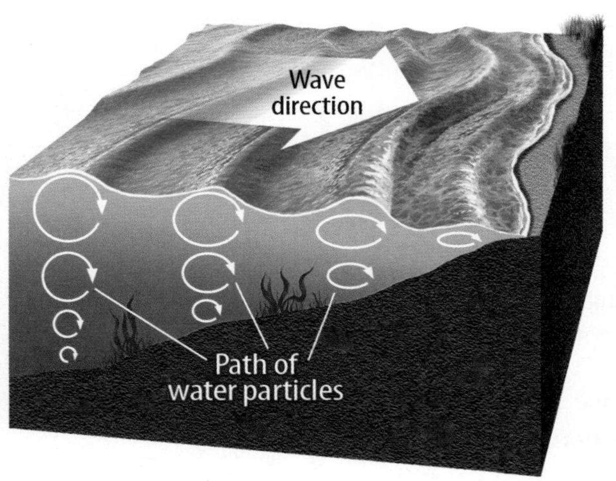

Figure 5 Waves cause water particles to move in small circles.

Visual Check How is the path of the water particles near the water's surface different from the path near the ocean floor?

Waves in Nature

Waves are common in nature because so many different energy sources produce waves. Two common waves in nature are water waves and seismic waves.

Water Waves Although water waves look like transverse waves, water particles move in circles, as shown in **Figure 5.** Water waves are a combination of transverse and longitudinal waves. Water particles move forward and backward. They also move up and down. The result is a circular path that gets smaller as the wave approaches land.

Water waves form because there is friction between the wind at sea and the water. Energy from the wind transfers to the water as the water moves toward land. Like all waves, water waves only transport energy. Because the waves move only through matter, water waves are mechanical waves.

Seismic Waves When layers of rock of Earth's crust suddenly shift, an earthquake occurs. The movement of rock sends out waves that travel to Earth's surface. An earthquake wave is called a seismic wave. As shown in **Figure 6,** there are different types of seismic waves. Seismic waves are mechanical waves because they move through matter.

Figure 6 Seismic waves can be longitudinal, transverse, or a combination of the two. ▼

Visual Check Which seismic wave is similar to a water wave?

Animation

Seismic Waves

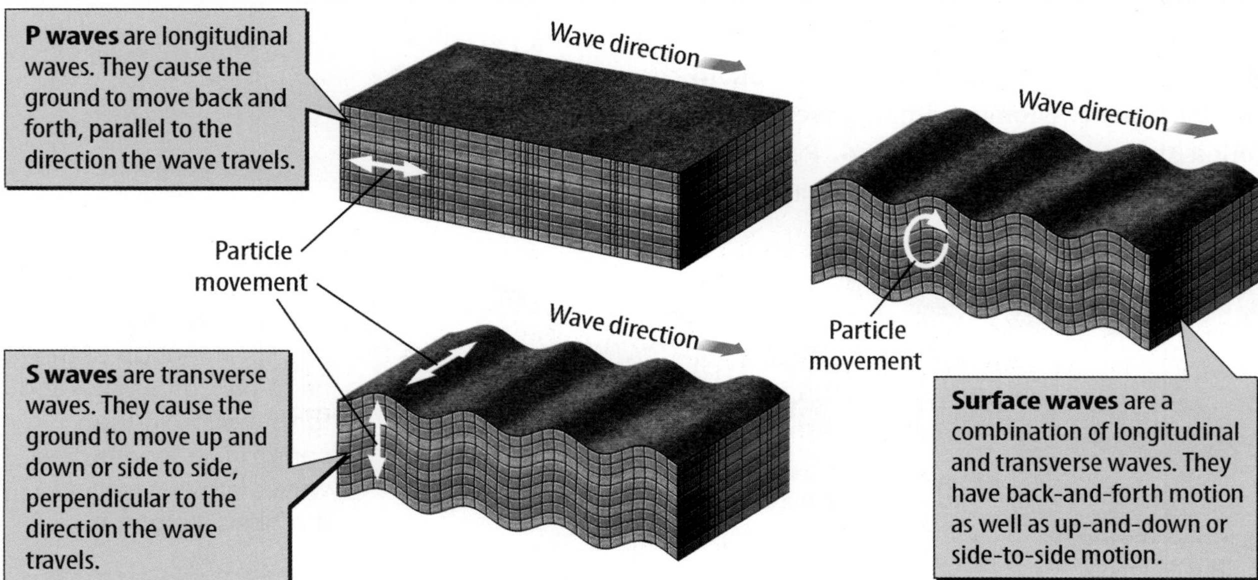

P waves are longitudinal waves. They cause the ground to move back and forth, parallel to the direction the wave travels.
Wave direction
Wave direction
Particle movement
Wave direction
Particle movement
S waves are transverse waves. They cause the ground to move up and down or side to side, perpendicular to the direction the wave travels.
Surface waves are a combination of longitudinal and transverse waves. They have back-and-forth motion as well as up-and-down or side-to-side motion.

MiniLab

15 minutes

How can you make waves with different properties? 

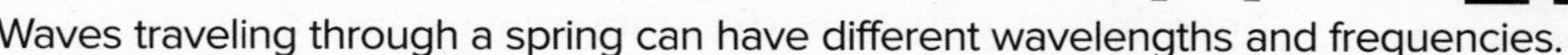

Waves traveling through a spring can have different wavelengths and frequencies.

1. Read and complete a lab safety form.
2. Use **tape** to secure one end of a **spring toy** to your desk and the other end to the floor. Tie pieces of **string** to the spring 1/4, 1/2, and 3/4 of the way between the floor and the desk.
3. Pull a few of the lowest coils on the spring toy to the right. Release. Record your observations in your Science Journal.
4. Slowly tap the bottom of the spring toward the right. Repeat, this time doubling your rate of tapping. Record your observations.
5. Push down the bottom 5 cm of the spring toy. Release. Repeat, this time pushing down the bottom 10 cm. Record your observations.

Analyze and Conclude

1. **Compare** the movement of the pieces of string in step 4 and in step 5.
2. **Classify** the types of waves you made in steps 3–5 as transverse or longitudinal.
3. **Key Concept** What do the waves transfer up and down the spring toy?

Properties of Waves

How could you describe water waves at a beach? You might describe properties such as the height or the speed of the wave. When scientists describe waves, they describe the properties of wavelength and frequency.

Wavelength

The distance between a point on one wave, such as the crest, and the same point on the next wave is called the wavelength. Different types of waves can have wavelengths that range from thousands of kilometers to less than the size of an atom!

Frequency

The number of wavelengths that pass a point each second is a wave's **frequency.** Frequency is measured in hertz (Hz). One hertz equals one wave per second. As shown in **Figure 7**, the longer the wavelength, the lower the frequency. As the distance between the crests gets shorter, the number of waves passing a point each second increases.

Reading Check What is frequency?

Figure 7 You can describe the wavelength and the frequency of both transverse and longitudinal waves.

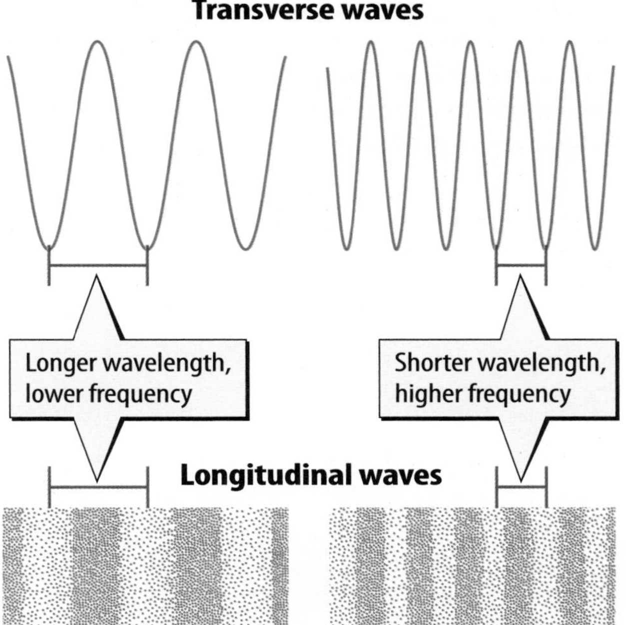

Table 2 Wave Speeds	
Type of Wave	**Typical wave speed (m/s)**
Ocean wave	25
Sound wave in air	340
Transverse seismic wave (S wave)	1,000 to 8,000
Longitudinal seismic wave (P wave)	1,000 to 14,000
Electromagnetic wave through empty space	300,000,000

▲ **Table 2** The speed of a wave depends on the type of wave and the medium through which the wave travels.

Wave Speed

A wave's speed depends on the medium, or type of material, through which it travels. Electromagnetic waves always travel through empty space at the same speed, 3×10^8 m/s. That's 300 million meters each second! They travel slower through a medium, or matter, because they must interact with particles. Mechanical waves also travel slower through matter because the waves transfer energy from one particle to another. For example, sound waves travel about one-millionth the speed of light waves. The speed of water waves depends on the strength of the wind that produces them. **Table 2** compares the speeds of different types of waves.

Amplitude and Energy

Different waves carry different amounts of energy. Some earthquakes, for example, are catastrophic because they carry so much energy. A shift in Earth's crust can cause particles in the crust to vibrate back and forth very far from their rest positions, producing seismic waves. In April 2015, seismic waves in Nepal transferred enough energy to destroy entire cities.

A wave's **amplitude** *is the maximum distance a wave varies from its rest position.* For mechanical waves, amplitude is the maximum distance the particles of the medium move from their rest positions as a wave passes. The more energy a mechanical wave has, the larger its amplitude. The amplitude of a transverse mechanical wave is shown in **Figure 8.**

 **Key Concept Check** How can you describe waves?

Figure 8 As more energy is used to produce a mechanical wave, particles of a medium vibrate farther from their rest positions. ►

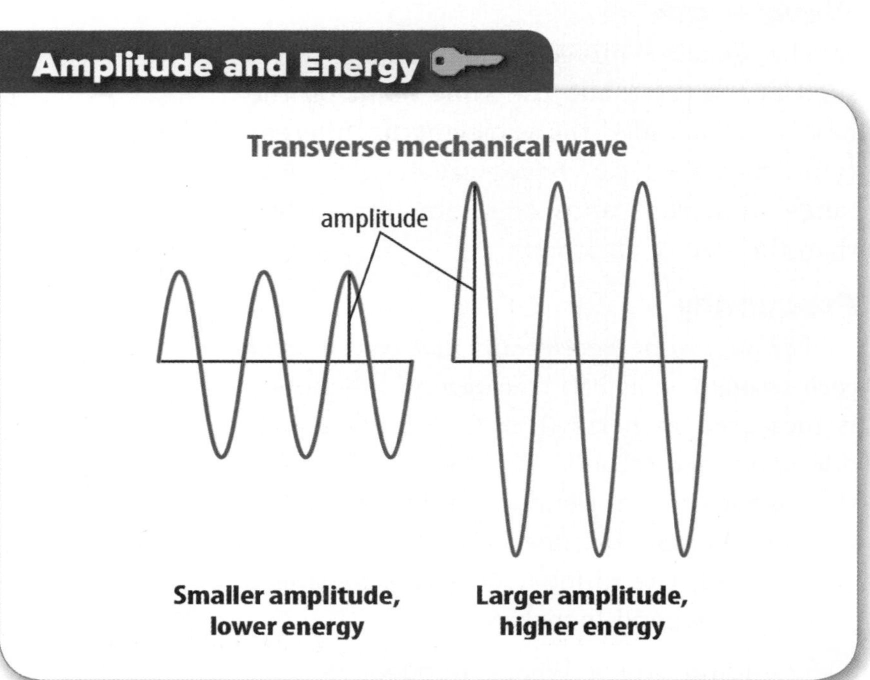

Wave Interaction with Matter

You have read that when you knock on a door, longitudinal sound waves transfer the energy of the knock through the door. However, when a person in the next room hears the knock, it is not as loud as the sound on your side of the door. The sound is weaker after it passes through the door because the waves interact with the matter that makes up the door.

Transmission

Some of the sound from your knock passes through the door. The waves transmit, or carry, the energy all the way through the door. The energy then passes into air particles, and the person on the other side hears the knock.

Absorption

Some of the sound is absorbed by the particles that make up the door. Instead of passing through the door, the energy increases the motion of the particles of the wood. The sound energy changes to thermal energy within the door. Therefore, less sound energy passes into the air in the next room.

Reflection

Some of the energy you used to knock on the door reflects, or bounces back, into the room you are in. Sound waves in the air transfer sound back to your ears. **Figure 9** shows how the energy of electromagnetic waves can also be transmitted, absorbed, or reflected.

Reading Check What are transmission, absorption, and reflection?

Transmission, Absorption, and Reflection

Figure 9 As waves travel, some of the energy they carry is transmitted, some is absorbed, and some is reflected by the particles in matter.

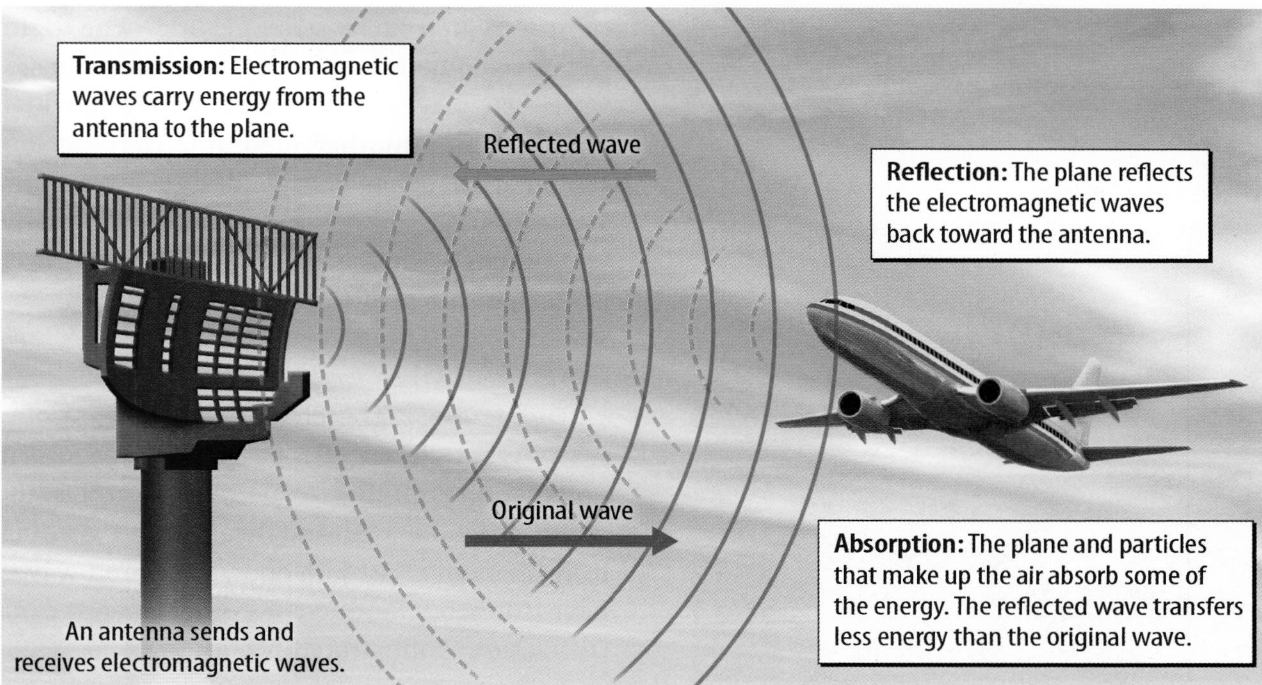

Transmission: Electromagnetic waves carry energy from the antenna to the plane.
Reflected wave
Reflection: The plane reflects the electromagnetic waves back toward the antenna.
Original wave
Absorption: The plane and particles that make up the air absorb some of the energy. The reflected wave transfers less energy than the original wave.
An antenna sends and receives electromagnetic waves.

Visual Check Does all of the energy reflected from the plane return to the antenna? Why or why not?

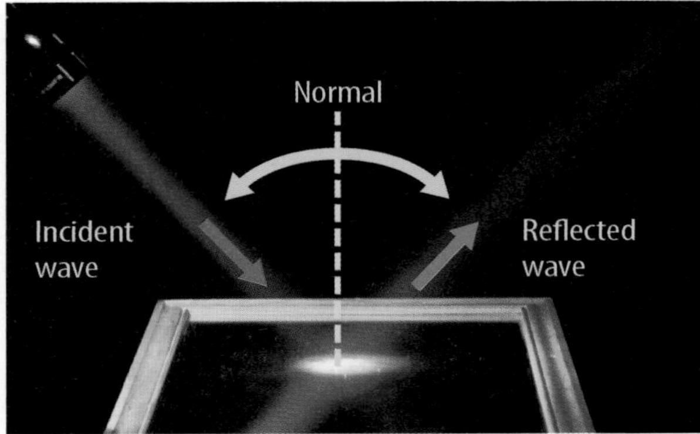

▲ **Figure 10** The law of reflection describes the direction of a reflected wave.

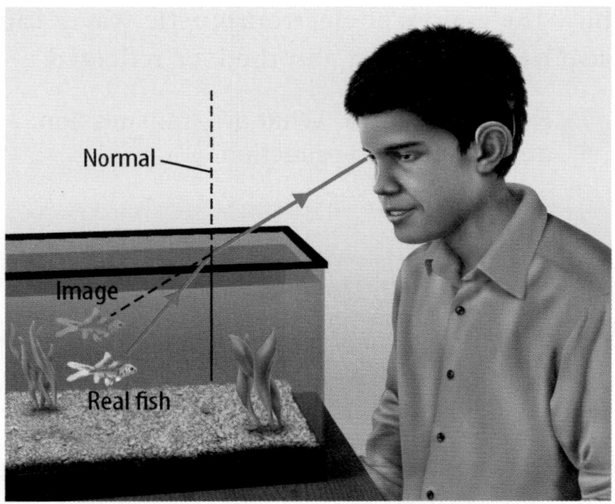

▲ **Figure 11** Refraction causes the fish to appear in a place different from its real location.

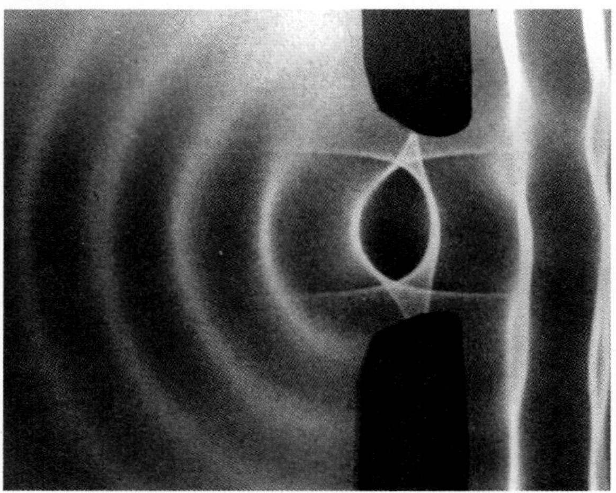

▲ **Figure 12** Diffraction causes waves to spread around barriers and through openings.

Law of Reflection

You can predict how waves will reflect from a smooth surface. The red arrow in **Figure 10** represents a light wave approaching a surface at an angle. This is called the incident wave. The blue arrow represents the reflected wave. The dotted line perpendicular to the surface at the point where the wave hits the surface is the normal. The law of reflection states that the angle between the incident wave and the normal always equals the angle between the reflected wave and the normal. If the incident angle in **Figure 10** increases, the reflected angle also increases.

Refraction

The change in direction of a wave as it changes speed, moving from one medium into another, is called **refraction.** The image of the fish in **Figure 11** is an example. Light reflects off the fish in all directions. The light speeds up as it moves from water into air. Notice that the light refracts away from the normal, or the line perpendicular to the surface at which the wave moves from one medium to other. This is the light the boy sees. His brain assumes the light traveled in a straight line. The light seems to come from the position of the image. Note that waves only refract if they move at an angle into another medium. They do not refract if they move straight into a medium. Waves refract toward the normal if they slow down when entering a medium and away from the normal if they speed up.

Diffraction

Diffraction is the change in direction of a wave when it travels past the edge of an object or through an opening. If you walk down a school hall and hear sound coming from an open classroom door, the sound waves have diffracted around the corner to your ears. Diffraction is illustrated in **Figure 12.**

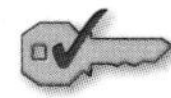

Key Concept Check What are some ways in which waves interact with matter?

Lesson 1 Review

Online Quiz | Virtual Lab

Visual Summary

A wave is a disturbance that transfers energy from one place to another without transferring matter.

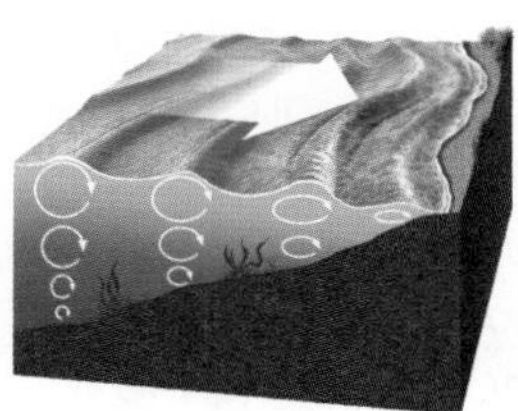

A wave can have a disturbance parallel or perpendicular to the direction the wave travels. Some waves are a combination of the two directions.

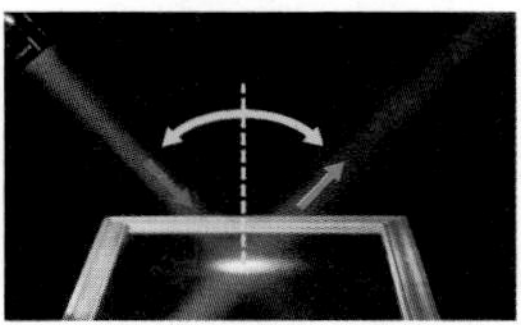

Waves can interact with matter by reflection, refraction, and diffraction.

FOLDABLES®

Use your lesson Foldable to review the lesson. Save your Foldable for the project at the end of the chapter.

What do you think NOW?

You first read the statements below at the beginning of the chapter.

1. Waves carry matter from place to place.

2. All waves move with an up-and-down motion.

Did you change your mind about whether you agree or disagree with the statements? Rewrite any false statements to make them true.

Use Vocabulary

1. **Define** *longitudinal wave* in your own words.
2. A wave that can travel through both matter and empty space is a(n) ________.

Understand Key Concepts 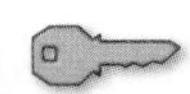

3. In which type of wave does the medium travel in a circular motion?
   A. electromagnetic **C.** transverse
   B. longitudinal **D.** water
4. **Identify** what produces a mechanical wave. An electromagnetic wave?
5. **Compare and contrast** how transmission, reflection, and absorption affect a wave.

Interpret Graphics

6. **Identify** The picture below shows a light ray bouncing off a flat surface. What is the correct scientific term for this interaction?

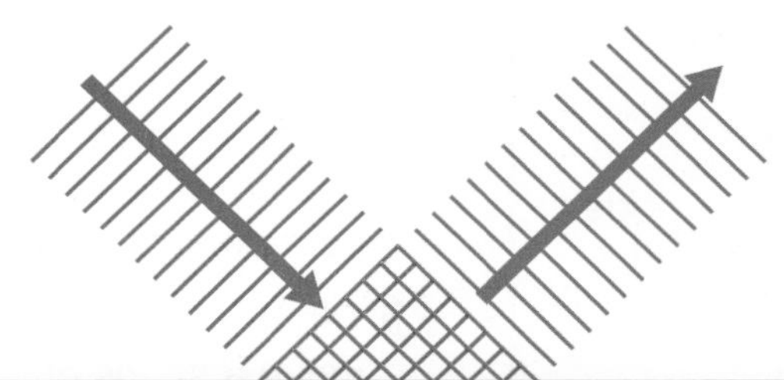

7. **Organize** Copy and fill in the graphic organizer below. In each oval, list a way in which waves can interact with matter.

Critical Thinking

8. **Decide** A forest fire makes a loud roaring sound. The explosive processes that release energy from the Sun occur at a much higher temperature. Why don't you hear a roaring sound from the Sun?

Skill Practice Observe

30 minutes

How do water waves interact with matter?

Materials

9-in. × 13-in. glass pan

sponges cut into thin strips

plastic snap-together blocks

adhesive putty

2-cm wooden dowel

Also needed: 2 books, white paper

Safety

You can use a ripple tank to observe waves with different properties. Think about how waves interact with matter and with each other. How can you change the properties of the waves?

Learn It

Scientists **observe** items and events and then record what they see. When you make observations, you should carefully look for details and then record your observations accurately and completely.

Try It

1. Read and complete a lab safety form.
2. Make a ripple tank by placing books under opposite edges of a glass pan. Secure the edges with putty. Pour water into the pan until it is about 5 mm deep. Place a sheet of white paper under the pan. Lay strips of sponge inside the short ends of the pan to absorb wave energy. Lay a dowel in the opposite end of the pan.
3. Tap the dowel with your finger to make a series of waves. Observe properties of the waves. Increase and decrease the wavelength of your waves. Explain in your Science Journal how you changed the wavelength.
4. Make barriers from snap-together blocks. Place the barriers end-to-end in your ripple tank at an angle to the dowel, as shown in the photo. What will happen to waves that hit against the barrier? Try it, and then change the angle of the barrier and repeat. Record your observations.
5. Place the barrier in the middle of the pan, parallel to the dowel, with a small space between the two parts. Demonstrate diffraction by making waves with different frequencies move through the space between the barriers. Observe how the waves change when they move through the space. Repeat, increasing the distance between the barriers. Record your observations.

Apply It

6. **Describe** How does changing the barrier's angle change the waves?
7. **Draw a diagram** of the waves passing between the two barriers.
8. **Predict** Place the barriers in a new formation in your tank. Predict the behavior of the waves. Draw a diagram of your setup and your prediction. How well were you able to foresee what the waves were going to do?
9. **Key Concept** Summarize your observations of wave reflection and diffraction in your investigation.

(t to b, 5, 6) Hutchings Photography/Digital Light Source, (2, 4) McGraw-Hill Education, (3) Ken Karp/McGraw-Hill Education

Lesson 2

Light

Reading Guide

Key Concepts
ESSENTIAL QUESTIONS

- How does light differ from other forms of electromagnetic waves?
- What are some ways in which light interacts with matter?
- How do eyes change light waves into the images you see?

Vocabulary

radio wave p. 459
infrared wave p. 460
ultraviolet wave p. 460
transparent p. 462
translucent p. 462
opaque p. 462
intensity p. 464

Multilingual eGlossary

BrainPOP®
Science Video
What's Science Got to do With It?

Inquiry Spreading Light?

Thick trees in a forest can block much of the sunlight, but some light still shines through. Why do you see bands of dim and bright light? Like all electromagnetic waves, light travels in straight lines. But light that moves past the trees can scatter and spread out.

Design Pics/Don Hammond

Launch Lab

15 minutes

Can you see the light?

When light travels through a medium, it interacts with the particles of the medium. Each material affects light differently.

1. Read and complete a lab safety form.
2. Obtain a **collection of materials** from your teacher. Make a two-column data table in your Science Journal. Write the headings *Material* above the left column and *Estimated Percentage of Light That Passes Through* above the right column. List each of your materials in the left column.
3. Shine a **flashlight** through one of the materials. Observe how much of the light passes through.
4. Estimate the percentage of light that passes through the material. Record your estimate in the data table.
5. Repeat steps 3 and 4 for each of the remaining materials.
6. Rank each material in order from the one that allows the most light to pass through to the one that allows the least amount of light to pass through.

Think About This

1. Which material allows the most light to pass through? Why?
2. What happens to the light when you shine your flashlight on the material you ranked number 3?
3. **Key Concept** Summarize ways in which you think the materials affect the light.

What are light waves?

You have read that there are two main types of waves—mechanical and electromagnetic. Mechanical waves move only through matter, but electromagnetic waves can move through matter and through empty space. Now you will read about different types of electromagnetic waves. The most familiar type of electromagnetic wave is light.

Recall that vibrating charged particles produce electromagnetic waves with many different wavelengths. Only a narrow range of these wavelengths are detected by most people's eyes. This small range of electromagnetic waves is what is known as light. Light waves and other forms of electromagnetic waves differ in wavelength and frequency.

ACADEMIC VOCABULARY

range
(noun) a set of values from least to greatest

An object that produces light is a luminous object. The Sun is Earth's major source of visible light. Almost half the Sun's energy that reaches Earth is visible light. Other luminous objects include lightbulbs and objects that produce light as they burn, such as a campfire.

Key Concept Check How does light differ from other forms of electromagnetic waves?

Hutchings Photography/Digital Light Source

The Electromagnetic Spectrum

Light is just a one type of electromagnetic wave. There is a wide range of electromagnetic waves that make up the electromagnetic spectrum, shown in **Figure 13.** Besides light, you encounter several other types of electromagnetic waves every day, and they probably play an important role in your life.

Types of Electromagnetic Waves

The electromagnetic spectrum consists of seven main types of waves. These waves range from low-energy, long-wavelength radio waves to very high-energy, short-wavelength gamma rays. Notice the relationship between wavelength, frequency, and energy indicated by the arrows in **Figure 13.** As the wavelength of electromagnetic waves decreases, the wave frequency increases. Low-frequency electromagnetic waves carry low amounts of energy, and high-frequency waves carry high amounts of energy.

Radio Waves *A low-frequency, low-energy electromagnetic wave that has a wavelength longer than about 30 cm is called a* **radio wave.** Radio waves have the least amount of energy of any electromagnetic wave. On Earth, radio and television transmitters produce radio waves that carry radio and television signals.

Microwaves You might use microwaves to cook your food. Microwaves also carry cell phone signals. Wavelengths of microwaves range from about 1 mm to 30 cm. Microwaves easily pass through smoke, light rain, and clouds, which makes them useful for transmitting information by satellites. Weather radar systems reflect microwaves off rain or storm clouds to detect and calculate the storm's distance and motion. Then, these calculations are used to make weather maps like the one shown on the first page of this chapter.

The Electromagnetic Spectrum

Figure 13 Electromagnetic waves have different wavelengths, frequencies, and energy.

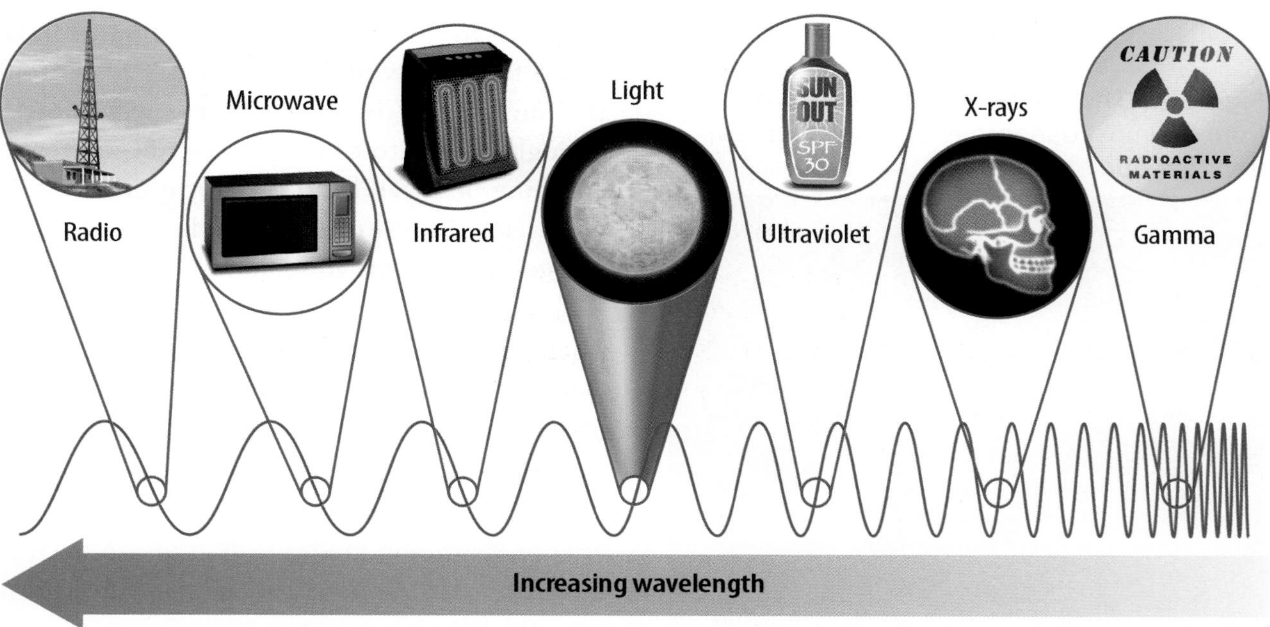

▲ **Figure 14** Infrared waves travel outward in all directions from the campfire.

WORD ORIGIN

infrared
from Latin *infra*, means "below"; and *ruber*, means "red"

ultraviolet
from Latin *ultra*, means "beyond"; and *viola*, means "violet"

Figure 15 The ozone layer protects Earth from the most dangerous ultraviolet waves from the Sun. ▼

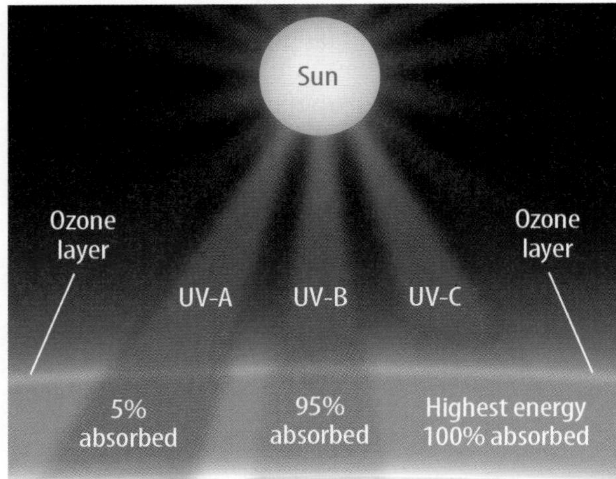

Light When you turn on a lamp or stand in sunshine, you probably don't think about waves entering your eyes. However, as you have read, light is a type of electromagnetic wave that the eyes detect. Light includes a range of wavelengths. You will read later in this lesson how this range of wavelengths relates to various properties of light.

Infrared Waves *An electromagnetic wave with a wavelength shorter than a microwave but longer than light is an* **infrared wave.** When you sit near a heater or a campfire, as in **Figure 14,** infrared waves transfer energy to your skin, and you feel warm. The Sun is Earth's major source of infrared waves. However, vibrating molecules in any type of matter, including your body, emit infrared waves.

Reading Check How do infrared waves and microwaves differ?

Ultraviolet Waves *An electromagnetic wave with a slightly shorter wavelength and higher frequency than light is an* **ultraviolet wave.** Electromagnetic waves with shorter wavelengths carry more energy than those with longer wavelengths and, therefore, can be harmful to living things. You might have heard that ultraviolet waves, or UV rays, from the Sun can be dangerous. These waves carry enough energy to cause particles of matter to combine or break apart and form other types of matter. Exposure to high levels of these waves can damage your skin.

Ultraviolet waves from the Sun are sometimes labeled UV-A, UV-B, or UV-C based on their wavelengths. UV-A have the longest wavelengths and the least energy. UV-C are the most dangerous because they have the shortest wavelengths and carry the most energy. As shown in **Figure 15,** the ozone layer in Earth's atmosphere blocks the Sun's most harmful UV rays from reaching Earth.

Reading Check Why can ultraviolet waves be dangerous?

imagebroker/Alamy

▲ **Figure 16** X-rays are useful for security scans because they have enough energy to pass through soft parts of luggage.

Visual Check How do the views of hard parts and soft parts of luggage differ in this X-ray image?

X-rays High-energy electromagnetic waves that have slightly shorter wavelengths and higher frequencies than ultraviolet waves are X-rays. These waves can be very powerful. They have enough energy to pass through skin and muscle, but denser bone can stop them. This makes them useful for taking pictures of the inside of the body. Airport scanners, as in **Figure 16,** sometimes use X-rays to take pictures of the contents of luggage.

Gamma Rays Electromagnetic waves produced by vibrations within the nucleus of an atom are called gamma rays. They have shorter wavelengths and higher frequencies than any other form of electromagnetic wave. Gamma rays carry so much energy that they can penetrate about 10 cm of lead, one of the densest elements. On Earth, gamma rays are produced by radioactive elements and nuclear reactions.

Reading Check Why do you think gamma rays cannot be used for communication in the same way radio waves are used?

Electromagnetic Waves from the Sun

The Sun produces an enormous amount of energy that is carried outward in all directions as electromagnetic waves. Because Earth is so far from the Sun, Earth receives less than one-billionth of the Sun's energy. However, if all the Sun's energy that reaches Earth in a 20-minute period could be transformed to useful energy, it could power the entire Earth for a year!

As shown in **Figure 17,** about 44 percent of the Sun's energy that reaches Earth is carried by light waves, and about 49 percent is carried by infrared waves. About 7 percent is carried by ultraviolet waves. Radio waves, microwaves, X-rays, and gamma rays carry less than 1 percent of the Sun's energy.

Figure 17 Infrared waves, light, and ultraviolet waves carry almost all of the Sun's energy. ▼

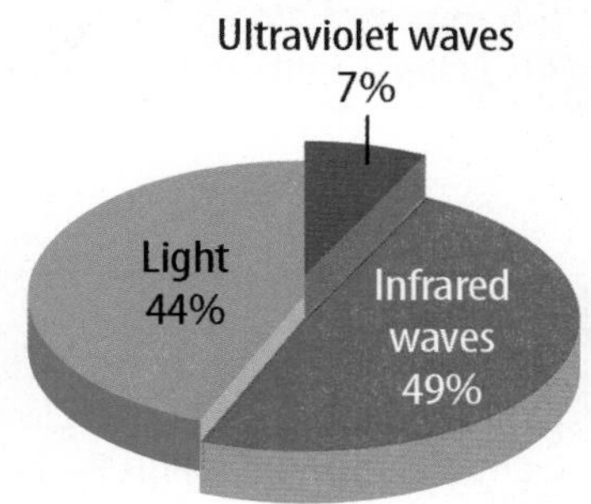

Photo montage by Nick Veasey/Getty Images

Speed, Wavelength, and Frequency

How could you describe the light from stars or the lights in a city at night? You might use words like *bright* or *dim*, or you might describe the color of the lights. You also could say how easily the light moves through a material. People use properties to describe light and to distinguish one color of light from another.

Like all types of electromagnetic waves, light travels at a speed of 3×10^8 m/s in empty space. When light enters a medium, or matter, it slows down. This is because of the interaction between the waves and the particles that make up the matter.

The wavelength and the frequency of a light wave determine the color of the light. The average human eye can distinguish among millions of wavelengths, or colors. Reds have the longest wavelengths and the lowest frequencies of light. Colors at the violet end of the visible light spectrum have the shortest wavelengths and the highest frequencies.

Light and Matter Interact

In Lesson 1, you read that matter can transmit, absorb, or reflect waves. How do these interactions affect light that travels from a source to your eyes?

Transmission

Air and clear glass, as shown in **Figure 18**, transmit light with little or no distortion. *A material that allows almost all of the light striking it to pass through, and through which objects can be seen clearly is* **transparent.**

Materials such as waxed paper or frosted glass also transmit light, but you cannot see through them clearly. *A material that allows most of the light that strikes it to pass through, but through which objects appear blurry is* **translucent.**

Absorption

Some materials absorb most of the light that strikes them. They transmit no light. Therefore, you cannot see objects through them. *A material through which light does not pass is* **opaque.**

Interactions of Light and Matter

Figure 18 Materials transmit, absorb, and reflect different amounts of light. This determines whether the material is transparent, translucent, or opaque.

Transparent
You can see clearly through a material such as this window glass because it is transparent. Light moves through the material without being scattered.

Opaque
You cannot see through the window frame because the material is opaque. All of the light that strikes the material is either absorbed or reflected.

Translucent
The lower part of this window contains panes of translucent frosted glass. Light that moves through the glass is scattered. Sometimes you can see colors and vague images through translucent materials, but it is difficult to determine what the shapes are.

Reflection

Why can you see your reflection clearly in a mirror, but not in the wall of your room? Recall that waves reflect off surfaces according to the law of reflection. Parallel rays that reflect from a smooth surface remain parallel and form a clear image. Light that reflects from a bumpy surface scatters in many directions. A wall seems smooth, but up close it is too bumpy to form a clear image.

Different types of matter interact with light in different ways. For example, the window in **Figure 18** both transmits and reflects light. Some of the light that strikes an opaque object, such as a book, is absorbed and reflected at the same time. Reflected light allows an object to be seen.

 **Key Concept Check** How does light interact with matter?

Color

The colors of an object depends on the wavelengths of light that enters the eye. A luminous object, such as a campfire, is the color of light that it emits. If an object is not luminous, its perceived color depends on other factors.

Opaque Objects Suppose white light strikes an American flag. The blue background absorbs all wavelengths of light except blue. The blue wavelengths reflect back to your eye. The red stripes absorb all colors but red, and red reflects to your eye. The white stars and stripes reflect all colors. You see white. An opaque object is the color it reflects, as shown in **Figure 19.**

Transparent and Translucent Objects If you look at a white lightbulb through a filter of red plastic wrap, only red wavelengths are transmitted through the plastic. The red plastic absorbs other wavelengths. Therefore, the lightbulb appears to be red.

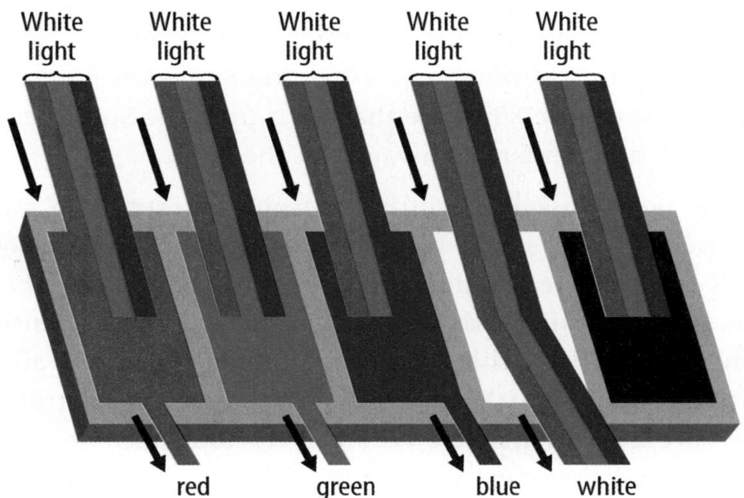

Figure 19 The color of an opaque object is the color of the light that reflects off the object. White objects reflect all colors of light. Black objects absorb all colors. Common black objects are visible because they actually reflect a small amount of light.

MiniLab

20 minutes

What color is the puppet?

An object's color depends on the materials it is made of. Does the color of light shining on an object have any effect?

1. Read and complete a lab safety form.
2. Copy the data table into your Science Journal.
3. Make a red filter by adding three drops of **red food coloring** to 100 mL of water in a **beaker.** Make blue and green filters using **blue and green food coloring.** With **scissors,** cut puppets out of **white and green paper.**
4. Turn off the lights. Shine a **flashlight** through the red water filter onto the white puppet. Record the puppet's color in the table.
5. Repeat step 4 using the colored filters listed in the table. When using more than one color of light, align the beakers so light moves through both and onto the puppet. Record your observations in the data table.

Light	Original Puppet Color	Observed Puppet Color
Red	White	
Green	White	
Blue	White	
Red	Green	
Green	Green	
Blue	Green	
Red and blue	White	
Red and blue	Green	
Red and green	White	
Red and green	Green	

Analyze and Conclude

1. **Model** Draw a picture showing whether each color of light was reflected or absorbed by the green puppet.
2. **Key Concept** How does the interaction of light and matter affect the puppet colors?

Figure 20 Light from nearby buildings and other sources can prevent you from seeing stars in the sky.

Intensity of Light

Another property you can use to describe light is intensity. **Intensity** *is the amount of energy that passes through a square meter of space in one second.* Intensity depends on the amount of energy a source emits. Light from a flashlight, for example, has a much lower intensity than light from the Sun. Intensity also depends on the light's distance from the source. When near a lamp, you probably notice that the intensity of the light is greater closer to the lamp than it is farther away. Many of the stars in **Figure 20** emit as much energy as the Sun. However, the light from the stars is less intense than light from the Sun, because the stars are so much farther away than the Sun.

The brightness of a light is a person's perception of intensity. One person's eyes might be more sensitive to light than someone else's eyes. As a result, different people might describe the intensity of a light differently. In addition, eyes are more sensitive to some colors than others. The environment also can affect the brightness of a light. Many stars are visible in the bottom photo of **Figure 20.** Few stars are visible in the top photo because there is so much light near the ground.

Interaction of Sunlight and Matter

Have you ever wondered why the sky is blue or the Sun is yellow? The interaction of light and matter causes interesting effects such as these when sunlight travels through air.

Scattering of Sunlight

As sunlight moves through Earth's atmosphere, most of the light reaches the ground. However, blue wavelengths are shorter than red wavelengths. The particles that make up the air scatter the shorter blue wavelengths more than they scatter longer wavelengths. The sky appears blue because the blue wavelengths spread out in every direction. They eventually reach the eye from all parts of the sky.

A light source, such as the Sun, that emits all colors of light should appear white. Why does the Sun often appear yellow instead of white? As shown in **Figure 21**, after the blue wavelengths of light scatter, the remaining colors appear yellow.

Reading Check Why is the sky blue? Why is the Sun yellow?

Refraction of Sunlight

Another interesting effect of sunlight occurs because of refraction. Recall that light changes speed as it travels from one medium into another. If light enters a new medium at an angle, the light wave refracts, or changes direction.

As shown in **Figure 22**, the refraction of light can affect the appearance of the setting Sun. The Sun's rays slow down when they enter Earth's atmosphere. The light rays refract toward Earth's surface. The brain assumes the rays that reach your eyes have traveled in a straight line, and the Sun seems to be higher in the sky than it actually is. This refraction causes you to see the Sun even after it has set below Earth's horizon.

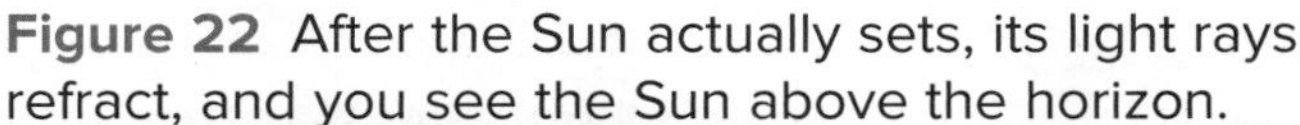

Figure 22 After the Sun actually sets, its light rays refract, and you see the Sun above the horizon.

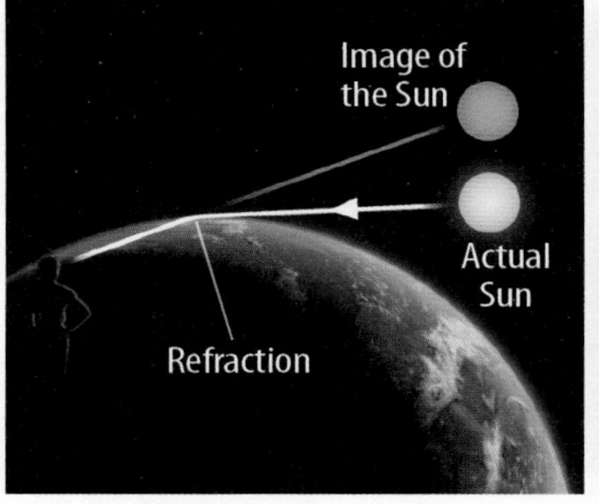

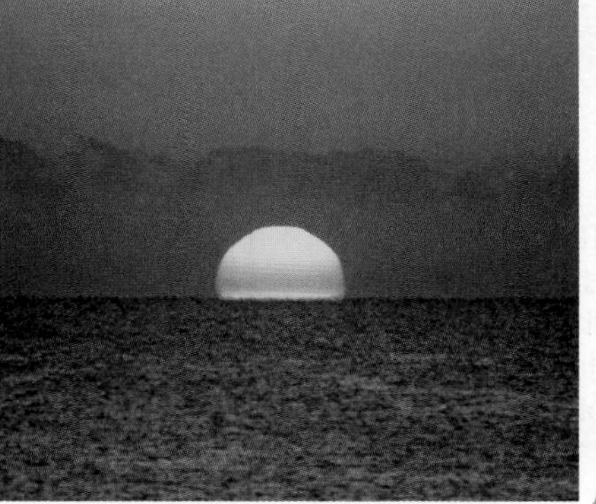

Alan Bolesta/age fotostock

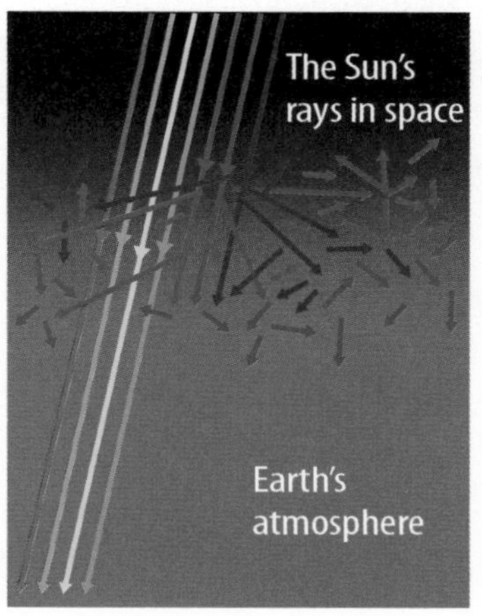

▲ **Figure 21** The Sun appears yellow because only longer wavelengths of light travel through the air in a straight line.

FOLDABLES®

Make a vertical two-tab book using the labels shown. Use it to organize your notes on scattering and refraction.

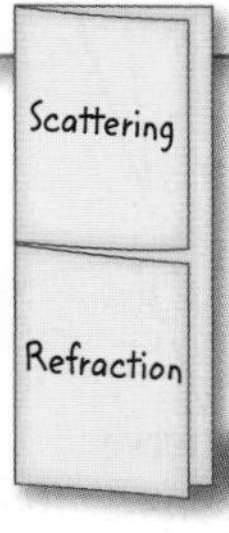

Vision and the Eye

Light enables objects to be seen. Light from luminous objects travels directly from the object to the viewer. Objects also are seen when they reflect light to the eyes. What happens to light after it enters the eyes? How do eyes and the brain transform light waves into information about people, places, and things?

As shown in **Figure 23,** light enters the eye through the cornea. The cornea and the lens focus light onto the retina. Cells in the retina absorb the light and send signals about the light to the brain. Follow the steps in **Figure 23** to learn more about how the eye works.

Key Concept Check How do eyes change light waves into the images you see?

The Eye

Animation

Figure 23 The parts of the eye work together to change light waves into signals your brain interprets as images.

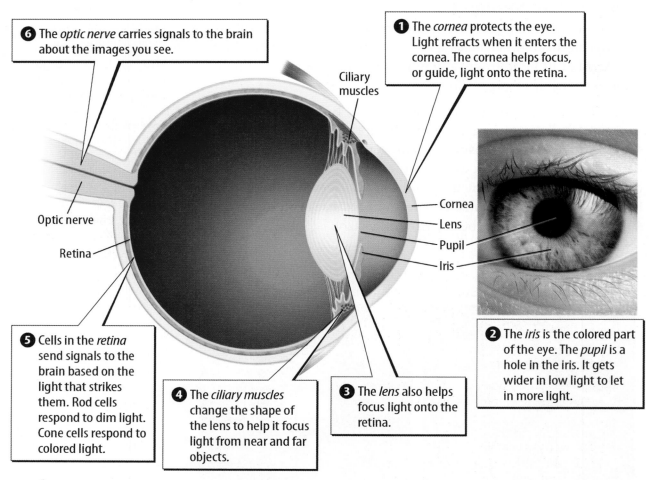

1. The *cornea* protects the eye. Light refracts when it enters the cornea. The cornea helps focus, or guide, light onto the retina.
2. The *iris* is the colored part of the eye. The *pupil* is a hole in the iris. It gets wider in low light to let in more light.
3. The *lens* also helps focus light onto the retina.
4. The *ciliary muscles* change the shape of the lens to help it focus light from near and far objects.
5. Cells in the *retina* send signals to the brain based on the light that strikes them. Rod cells respond to dim light. Cone cells respond to colored light.
6. The *optic nerve* carries signals to the brain about the images you see.

Ciliary muscles
Optic nerve
Retina
Cornea
Lens
Pupil
Iris

Momentum Creative Group/Alamy Images

Visual Check What part of the eye responds to color?

Lesson 2 Review

Visual Summary

The different types of electromagnetic waves play important roles in your life.

Materials transmit, absorb, and reflect different amounts of light.

Interaction with matter produces interesting effects in sunlight. You can see the Sun even after it sets below the horizon.

FOLDABLES®

Use your lesson Foldable to review the lesson. Save your Foldable for the project at the end of the chapter.

What do you think NOW?

You first read the statements below at the beginning of the chapter.

3. Light is the only type of wave that can travel through empty space.

4. Only shiny surfaces reflect light.

Did you change your mind about whether you agree or disagree with the statements? Rewrite any false statements to make them true.

Use Vocabulary

1. **Contrast** radio waves, infrared waves, and ultraviolet waves.
2. **Explain** the difference between a transparent and a translucent material.

Understand Key Concepts 

3. Which eye part responds to colored light?
   A. cones
   B. cornea
   C. iris
   D. lens
4. **Compare** the ways light interacts with a red book and a red stained-glass window.
5. **Describe** how light waves and ultraviolet waves differ.

Interpret Graphics

6. **Explain** the diagram below in terms of the interaction of light waves with matter.

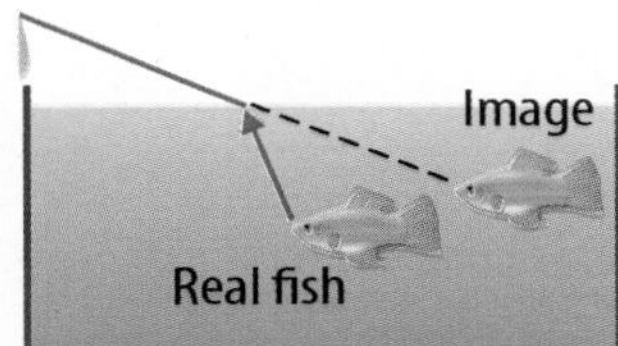

7. **Sequence** Copy and fill in a graphic organizer like the one below that shows the sequence of wave types in the electromagnetic spectrum. Add boxes as necessary.

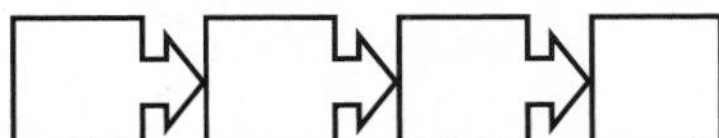

Critical Thinking

8. **Decide** If you turn on an electric stove and stand to the side of it, what type of electromagnetic wave causes you to feel heat from the burner?
9. **Construct** a drawing of the major parts of the eye and describe how each part helps turn light waves into visual information.

GREEN SCIENCE

Light

Is it keeping you from sleeping at night?

This image was created using data gathered by satellites. It shows light pollution generated by human populations around the world.

◀ The lights used to keep this road safe contribute to light pollution.

Imagine trying to sleep in this house! Light shining in bedroom windows at night is a form of light pollution. ▶

Trash on the sidewalk, automobile exhaust in the air, and fertilizer in a river's water are all types of pollution. But did you know that light also can be considered pollution? Light pollution is a serious problem in many urban areas worldwide.

Artificial lighting can be very useful. It can help keep areas free from crime and allow people to work and drive safely after dark. However, the lights people use often shine out into surrounding areas or up into the night sky. This is called light pollution.

Light pollution is a term that refers to the negative effects of artificial lighting. For example, light pollution can disrupt the daily cycles of nocturnal animals. Also, light that escapes into the atmosphere is wasted energy. In some areas, observing the night sky is very difficult because of light pollution.

Awareness of light pollution is increasing. Groups such as the American Medical Association (AMA) have recognized the negative impact of light pollution. The AMA has passed resolutions advocating energy-efficient, fully shielded streetlight design. Individuals can take steps to decrease light pollution by carefully choosing outdoor lights with light-pollution reduction in mind.

It's Your Turn

OBSERVE AND DRAW Observe the night sky near your home, and make a drawing of what you observe. Then, discuss how light pollution in your area might compare with light pollution in other parts of the country.

Lesson 3

Reading Guide

Key Concepts
ESSENTIAL QUESTIONS

- What are some properties of sound waves?
- How do ears enable people to hear sounds?

Vocabulary

compression p. 471
rarefaction p. 471
pitch p. 471
decibel p. 473

Multilingual eGlossary

Science Video

Go to the resource tab in ConnectED to find the PBL *Build a Better Room.*

Sound

Inquiry How does it make sounds?

Have you ever stood nearby as a marching band plays or carefully watched musicians during a concert? The notes they play can be high or low, loud or soft, or anything in between. Why are the sounds so different? How are sounds perceived?

Launch Lab

10 minutes

How can you change the sound of a straw?

Sounds are longitudinal waves that travel through matter. If you blow across a straw, you can make different wavelengths of sound. How do different wavelengths change the sounds you hear?

1. Read and complete a lab safety form.
2. Using **scissors,** cut a **straw** in half. Cut one of the halves into two equal parts. Cut one of those parts into two equal parts.
3. Blow across the top of each straw. How do the sounds differ? Make a data table in your Science Journal, and then record your observations in your data table.
4. Repeat step 3, this time covering the bottom of each straw with your finger.

Think About This

1. What is the source of energy that creates the sound waves?
2. How does covering the bottom of the straw change the sound?
3. **Key Concept** How do the sounds made by a long straw and a short straw differ? Why do you think this is?

What are sound waves?

Just as light is a type of wave that can be seen, sounds are a type of wave that can be heard. Sound waves are longitudinal mechanical waves. Unlike light waves, sound waves must travel through a medium.

Figure 24 People and animals hear different ranges of sound frequencies.

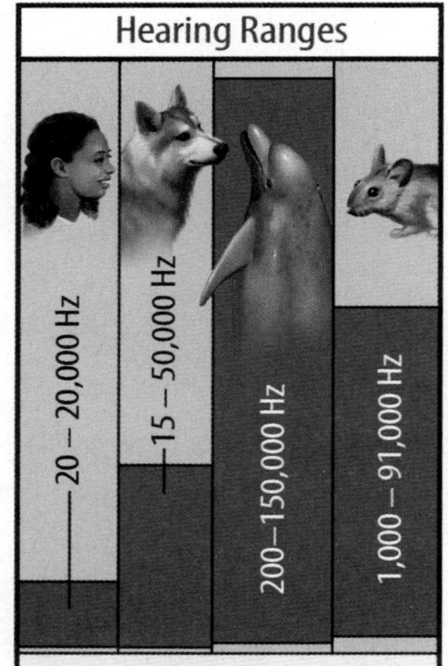

Audible Vibrations

Suppose you strike two metal pans together. Now, suppose you strike two pillows together. How would the two sounds differ? Sound waves are vibrations the ear can detect. You hear a loud sound when you hit the pans together because they vibrate so much. You barely hear the pillows because they vibrate so little. Healthy, young humans can hear sound waves produced by vibrations with frequencies between about 20 Hz and 20,000 Hz. As people age, their ability to hear the higher and lower frequencies of sound decreases. The human ear is most sensitive to frequencies between 1,000 Hz and 4,000 Hz.

Animals have ranges of hearing that help them catch prey or avoid predators. For example, elephants hear sounds as low as 15 Hz. Chickens hear sounds between 125 Hz and 2,000 Hz. Porpoises can hear sounds between 75 Hz and 150,000 Hz! Ranges for other animals are listed in **Figure 24.**

Hutchings Photography/Digital Light Source

Figure 25 A sound wave produces compressions and rarefactions as it passes through matter.

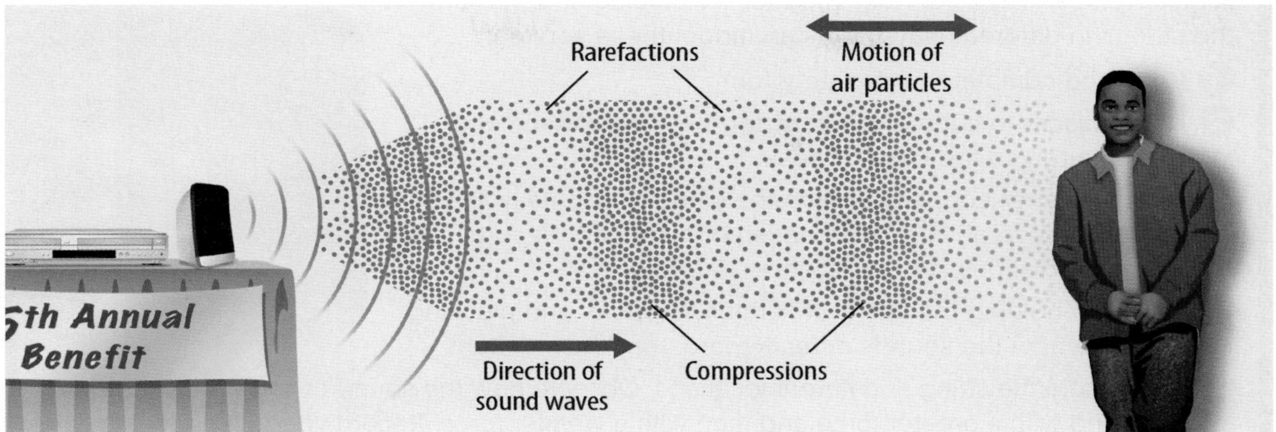

Compressions and Rarefactions

Sound waves usually travel to your ears through air. Air particles are in constant motion. As the particles bounce off objects, they exert a force, or pressure. Figure 25 shows how sound waves moving through air change the air pressure by causing air particles to move toward and then away from each other.

Suppose you pluck a guitar string. As the string springs back, it pushes air particles forward, forcing them closer together. This increases the air pressure near the string. *A* **compression** *is the region of a longitudinal wave where the particles of the medium are closest together.* As the string vibrates, it moves in the other direction. This leaves behind a region with lower pressure. *A* **rarefaction** *is the region of a longitudinal wave where the particles are farthest apart.*

Reading Check How do compressions and rarefactions differ?

Properties of Sound Waves

A sound wave is described by its wavelength, frequency, amplitude, and speed. These properties of sound waves depend on the compressions and rarefactions of the sound waves.

Wavelength, Frequency, and Pitch

Recall that the wavelength of a wave becomes shorter as the wave's frequency increases. How does the frequency of a sound wave affect what is heard?

The perception of how high or low a sound seems is called **pitch.** The higher the frequency, the higher the pitch of the sound. For example, a female voice generally produces higher-pitched sounds than a male voice. This is because the female voice has a higher range of frequencies. Figure 26 shows the range of frequencies produced by several instruments and voices.

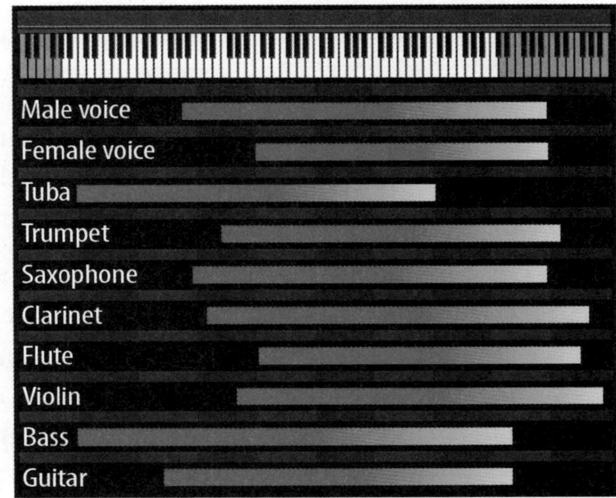

▲ **Figure 26** People and instruments have different ranges of sound frequencies.

MiniLab

10 minutes

Can you make different sounds with string?

A guitar player makes different sounds by holding and plucking the strings in different ways. You can model these sounds.

1. Read and complete a lab safety form.
2. Use **scissors** to cut a piece of **string** 1 m long. Attach one end securely to the leg of a desk.
3. Hold the other end, and stretch the string horizontally. Pluck the string several times, and observe the sound. Record your observations in your Science Journal.
4. Continue holding the string at various locations and plucking it. Notice how the sounds differ. Record your observations.
5. Again, hold the string at different locations. Observe how the sound changes as you pull the string with a greater force and then with a weaker force. Record your observations.

Analyze and Conclude

1. **Interpret** How does pulling the string tighter or changing its length affect the string's sound?
2. **Key Concept** Explain how you changed the frequency, wavelength, pitch, amplitude, and energy of the sound you made with the string.

SCIENCE USE V. COMMON USE

rest position
Science Use position of an undisturbed particle; particles are still in motion here

Common Use the state of something not moving

Amplitude and Energy

You use more energy to shout than to whisper. The more energy you put into your voice, the farther the particles of air move as they vibrate. The distance a vibrating particle moves from its **rest position** is the amplitude. The more energy used to produce the sound wave, the greater the amplitude.

Speed

Sound waves travel much slower than electromagnetic waves. With sound, the transmitted energy must pass from particle to particle. The type of medium and the temperature affect the speed of sound.

Type of Medium Gas particles are far apart and collide less often than particles in a liquid or a solid. As shown in **Table 3**, a gas takes longer to transfer sound energy between particles.

Table 3 The Speed of Sound

Material	Speed (m/s)
Air (0°C)	331
Air (20°C)	343
Water (20°C)	1,481
Water (0°C)	1,500
Seawater (25°C)	1,533
Ice (0°C)	3,500
Iron	5,130
Glass	5,640

Temperature Particles move faster and collide more often as the temperature of a gas increases. This increase in the number of collisions transfers more energy in less time. Temperature has the opposite effect on liquids and sounds. As liquids and solids cool, the molecules move closer together. They collide more often and transfer energy faster.

Key Concept Check What are some properties of sound waves?

Intensity and Loudness

Generally, the greater the amplitude of a sound wave, the louder the sound seems. But what happens if you move away from a sound source? As you move away, the wave's amplitude decreases and the sound seems quieter. This is because as a sound wave moves farther from the source, more and more particles collide, and the energy from the wave spreads out among more particles. Therefore, the farther you move from the source, the less energy present in the same area of space. Recall that the amount of energy that passes through a square meter of space in one second is the intensity of a wave. Loudness is your ear's perception of intensity.

The Decibel Scale

The unit used to measure sound intensity, or loudness, is the **decibel (dB).** The decibel levels of common sounds are shown in Figure 27. Each increase of 10 dB causes a sound about twice as loud. As the decibel level goes up, the amount of time you can listen to the sound without risking hearing loss gets shorter and shorter. People who work around loud sounds wear protective hearing devices to prevent hearing loss.

Decibel Levels

Figure 27 The decibel scale helps you understand safe limits of different types of sounds.

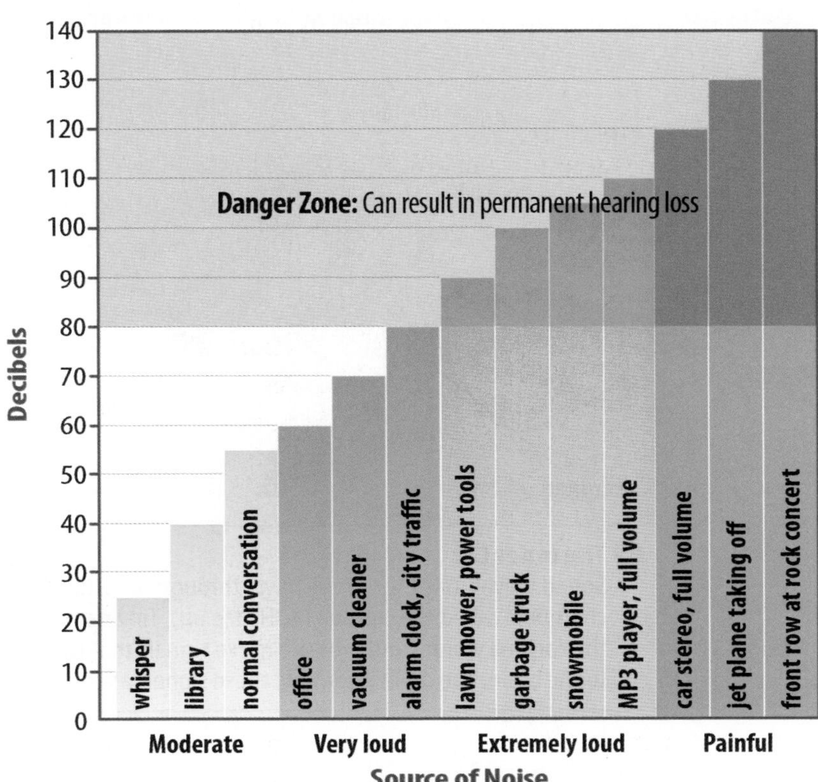

Math Skills 

Use a Fraction

Because sound energy travels out in all directions from the source, the intensity of the sound decreases as you move away. You can calculate the fraction by which the sound intensity changes.

The fraction $= \left(\frac{r_1}{r_2}\right)^2$, where r_1 is the starting distance and r_2 is the ending distance from the source. For example, by what fraction does sound intensity decrease if you move from **3 m** to **6 m** from a source?

1. Replace the variables with given values.
   fraction $= \left(\frac{3}{6}\right)^2$
2. Solve the problem.
   $\left(\frac{3}{6}\right)^2 = \left(\frac{1}{2}\right)^2 = \frac{1}{4}$, so the intensity decreases to $\frac{1}{4}$ of its original value.

Practice

You are standing at a distance of 2 m from a sound source. How does the sound intensity change if you move to a distance of 6 m?

 Math Practice

 Personal Tutor

WORD ORIGIN

decibel
from Latin *decibus,* means "tenth"

FOLDABLES®

Make a horizontal four-tab book, and label it as shown. Use it to review properties of sound waves.

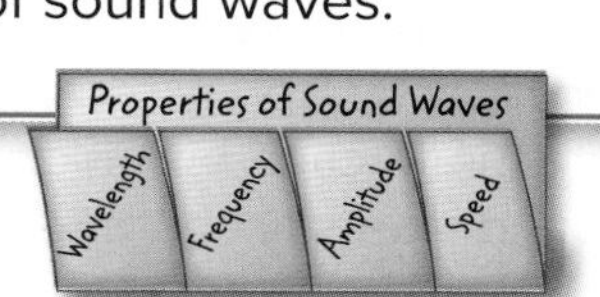

Hearing and the Ear

Typically, objects are seen when light enters the eyes. Similarly, sound waves enter the ears with information about the environment. The human ear has three main parts, as shown in **Figure 28.** First, the external outer ear collects sound waves. Next, the middle ear amplifies, or intensifies, the sound waves. The middle ear includes the ear drum and three small bones–the hammer, the anvil, and the stirrup, Finally, the inner ear contains the cochlea (KOH klee uh). The cochlea converts sound waves to nerve signals. These nerve signals are typically processed by the brain, creating the perception of sound.

Key Concept Check How do your ears enable you to hear sounds?

Parts of the Human Ear

Figure 28 The different parts of the ear work together to gather and interpret sound waves.

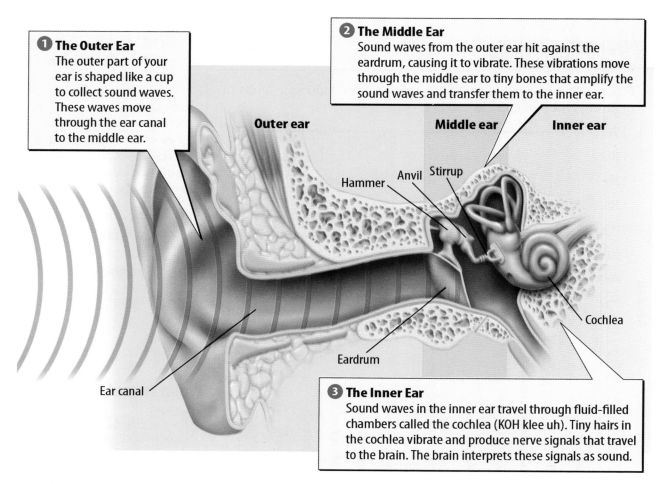

1 The Outer Ear
The outer part of your ear is shaped like a cup to collect sound waves. These waves move through the ear canal to the middle ear.

2 The Middle Ear
Sound waves from the outer ear hit against the eardrum, causing it to vibrate. These vibrations move through the middle ear to tiny bones that amplify the sound waves and transfer them to the inner ear.

Outer ear

Middle ear

Inner ear

Hammer

Anvil

Stirrup

Cochlea

Eardrum

Ear canal

3 The Inner Ear
Sound waves in the inner ear travel through fluid-filled chambers called the cochlea (KOH klee uh). Tiny hairs in the cochlea vibrate and produce nerve signals that travel to the brain. The brain interprets these signals as sound.

Visual Check Which part of the ear has a spiral shape?

Lesson 3 Review

Visual Summary

Sound waves are produced when an energy source causes matter to vibrate.

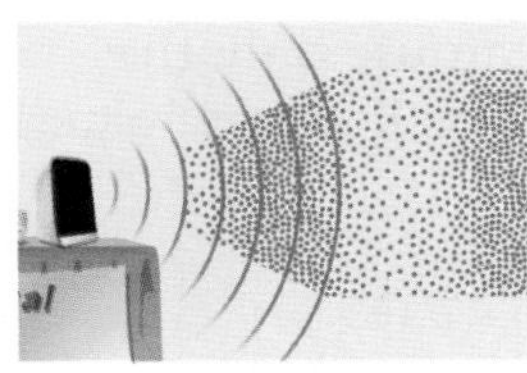

Sound waves are compressions and rarefactions that move away from a sound source.

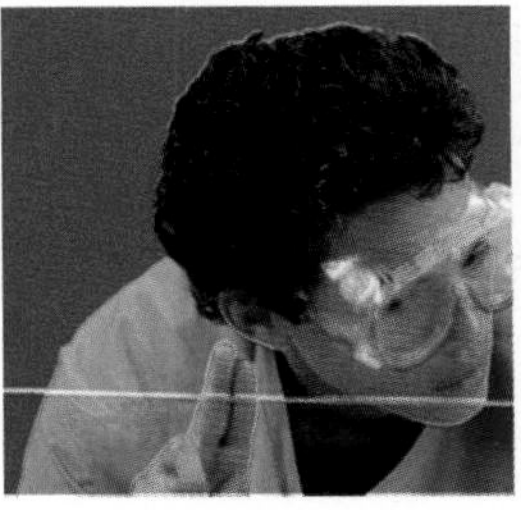

You hear sounds when your ears capture sound waves and produce signals that travel to your brain.

Use your lesson Foldable to review the lesson. Save your Foldable for the project at the end of the chapter.

What do you think NOW?

You first read the statements below at the beginning of the chapter.

5. Sound travels faster through solid materials than through air.

6. The more energy used to produce a sound, the louder the sound.

Did you change your mind about whether you agree or disagree with the statements? Rewrite any false statements to make them true.

Use Vocabulary

1. The property of a sound wave that relates to a high or low musical note is the sound's __________.

2. **Explain** the difference between a compression and a rarefaction in a sound wave.

Understand Key Concepts

3. Which property of a sound wave describes the amount of energy that passes through a square meter of space each second?
   - **A.** amplitude
   - **B.** frequency
   - **C.** intensity
   - **D.** wavelength

4. **Describe** how the three main parts of the ear enable people to hear.

Interpret Graphics

5. **Sequence** Copy and fill in a graphic organizer like the one below to describe the path of a sound wave from when it is produced by a source until is interpreted by the brain. Describe the function of each part of the path.

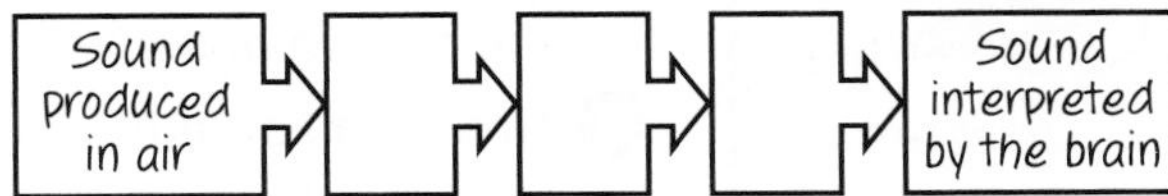

Critical Thinking

6. **Construct** a diagram of four sound waves. Two of the waves should have the same amplitude but different frequencies. The other two waves should have the same wavelength but different amplitudes. Label the properties of the waves.

Math Skills

7. A student is standing a distance of 4 m from the school bell. If the student moves to a distance 20 m away, what fraction of the original intensity of the bell's sound will the student hear?

Lab

2–3 class periods

Check the sound! Cue the lights!

Materials

coiled spring toy

snap-together plastic blocks

food coloring

flashlights

string

scissors

Also needed: variety of beakers, dowel, large glass pan, colored paper, sponge, water, white paper, office supplies

Safety

You are part of an entertainment firm that creates sound and light shows like the kind you see during a music concert or at halftime of a sports event. You have been asked to produce an exciting and entertaining wave show that is between 2 and 4 minutes long. The show must use at least three types of waves, and you must be able to identify at least three properties or behaviors of waves. Unfortunately, you do not have a big budget or high-tech equipment like professional show designers. You can use only the materials you have used in other labs in this chapter, as well as any materials your teacher approves.

Question

How can you use different waves to build an exciting and entertaining show?

Procedure

1. Read and complete a lab safety form.
2. As a group, decide on a concept or idea around which you will focus your show.
3. Develop a script for the show. Assign the different roles to the members of your group.
4. Make a table of the different waves in your show, as shown below. Consider the source of the wave and the medium through which the wave travels.
5. Create a list of the different physical concepts related to waves that you will include in your show.
6. Build all the instruments and lighting equipment required for your show. Practice your show so that everyone in your group is able to smoothly perform the different parts together for the class.

Title of the show: ____________________

Wave	Source	Medium
1.		
2.		
3.		
4.		
5.		

7 Show the script of your show to another group. Ask the group to evaluate how well you use different waves and wave properties. Get feedback on ways to make your show even more entertaining.

8 As a group, think about how you can use the suggestions you received from the other group. Consider what supplies you will need and how you will make the changes.

9 Record these suggestions and your ideas about how to make modifications to your setup in your Science Journal.

10 Make modifications to your show based on your ideas. You might need to test different parts to see how ideas work. When you have made all the changes, perform your entire show.

Analyze and Conclude

11 **Explain** How did you incorporate three types of waves into your show?

12 **Assess** What are some benefits and some challenges of using waves to create a show?

13 **The Big Idea** How did you use the physical properties of waves in your show?

Communicate Your Results

Perform your show for your class. After your show, have your class give you feedback on how well you used different types of waves and different properties of waves to create an exciting and entertaining show.

Inquiry Extension

What were some of the problems you encountered because of the limited materials that were available for making your show? What kind of equipment would you have liked to use? Write a proposal that you might submit to a client to explain why you would like to have the funds to purchase improved sound and light equipment. Explain how a larger budget might improve the show and make people pay to see your show.

John Bentley/Alamy

Lab Tips

- Be sure to decide on a theme around which to focus your show. For example, will it be a music concert? A sideshow at a sports event? A display at a theme park?
- Look back at the labs for this chapter for ideas on how to use waves.
- Try to make your show as creative as possible. How can you make your show exciting?

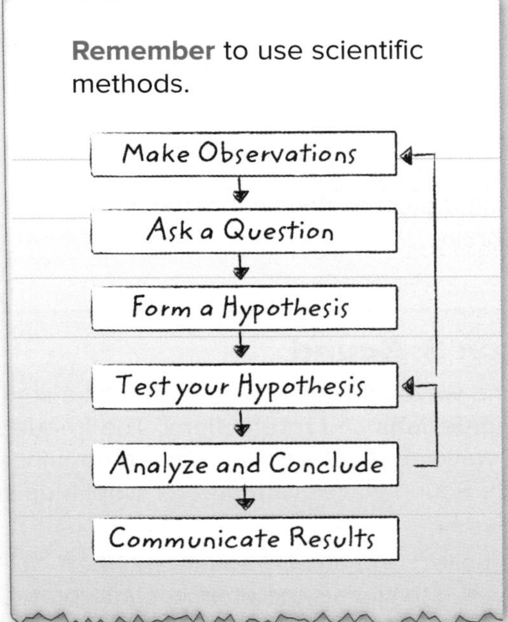

Chapter 14 Study Guide

Mechanical waves transfer energy from particle to particle in matter. Electromagnetic waves transfer energy through either matter or empty space.

Key Concepts Summary 

Lesson 1: Waves

- Waves are disturbances that transfer energy from place to place. A **mechanical wave** forms when a source of energy causes particles of a medium to vibrate. A vibrating electric charge produces an **electromagnetic wave.**
- You can describe wavelength, **frequency,** speed, **amplitude,** and energy of waves.
- Matter can transmit, absorb, or reflect a wave. It also can change a wave's direction by **refraction** or diffraction.

Vocabulary

mechanical wave **p. 448**
electromagnetic wave **p. 448**
transverse wave **p. 449**
longitudinal wave **p. 449**
frequency **p. 451**
amplitude **p. 452**
refraction **p. 454**

Lesson 2: Light

- Light differs from other forms of electromagnetic waves by its frequency, wavelength, and energy. Light is the type of electromagnetic wave that is visible with the human eye.
- Matter can transmit, absorb, and reflect light. These interactions differ in how much light the matter transmits and how it changes the direction of light.
- Cells in the retina of the eyes change light into electric signals that travel to the brain.

Vocabulary

radio wave **p. 459**
infrared wave **p. 460**
ultraviolet wave **p. 460**
transparent **p. 462**
translucent **p. 462**
opaque **p. 462**
intensity **p. 464**

Lesson 3: Sound

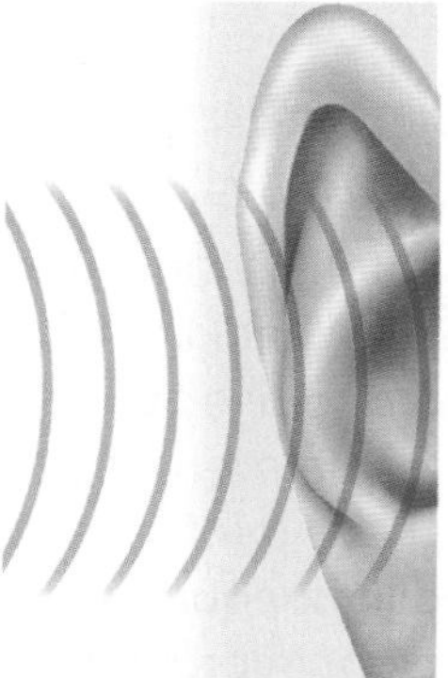

- Sound waves travel through matter as a series of **compressions** and **rarefactions.** The frequency and wavelength of a sound wave determines the **pitch.** Sound waves with greater amplitude sound louder.
- Ears collect and amplify sound and then convert it to signals the brain can interpret.

Vocabulary

compression **p. 471**
rarefaction **p. 471**
pitch **p. 471**
decibel **p. 473**

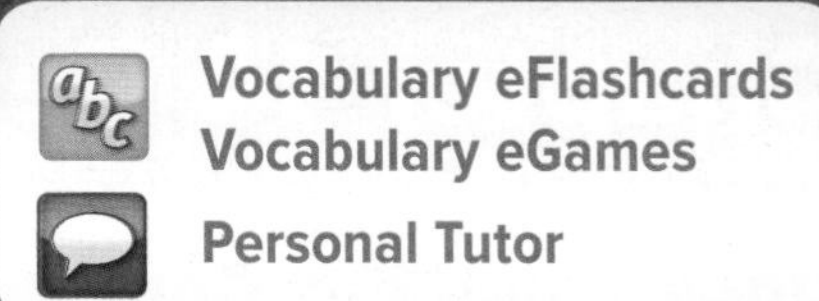

FOLDABLES® Chapter Project

Assemble your lesson Foldables as shown to make a Chapter Project. Use the project to review what you have learned in this chapter.

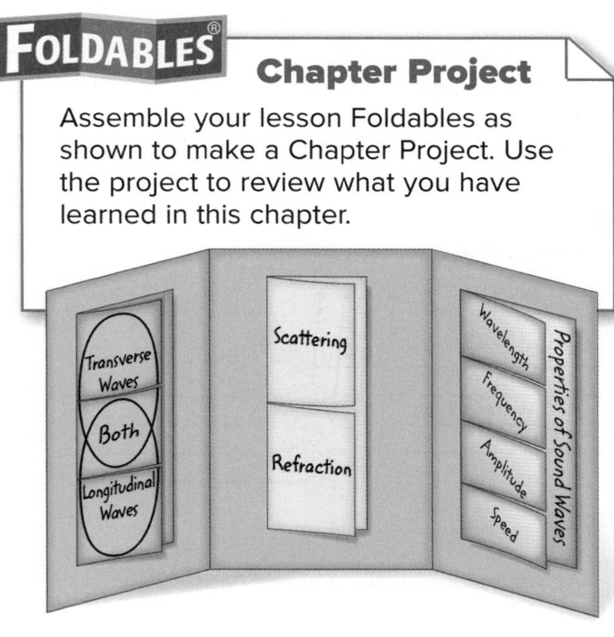

Use Vocabulary

1. The property of waves that is measured in hertz (Hz) is ______.
2. A change in direction, or ______, can occur as a wave moves into a medium.
3. A material that transmits light but through which objects appear blurry is ______.
4. An object that does not allow light to pass through it is ______.
5. The portion of a sound wave with higher-than-normal pressure is called a(n) ______.
6. A unit that describes the intensity or loudness of sound is the ______.

Link Vocabulary and Key Concepts

Interactive Concept Map

Copy this concept map, and then use vocabulary terms from the previous page to complete the concept map.

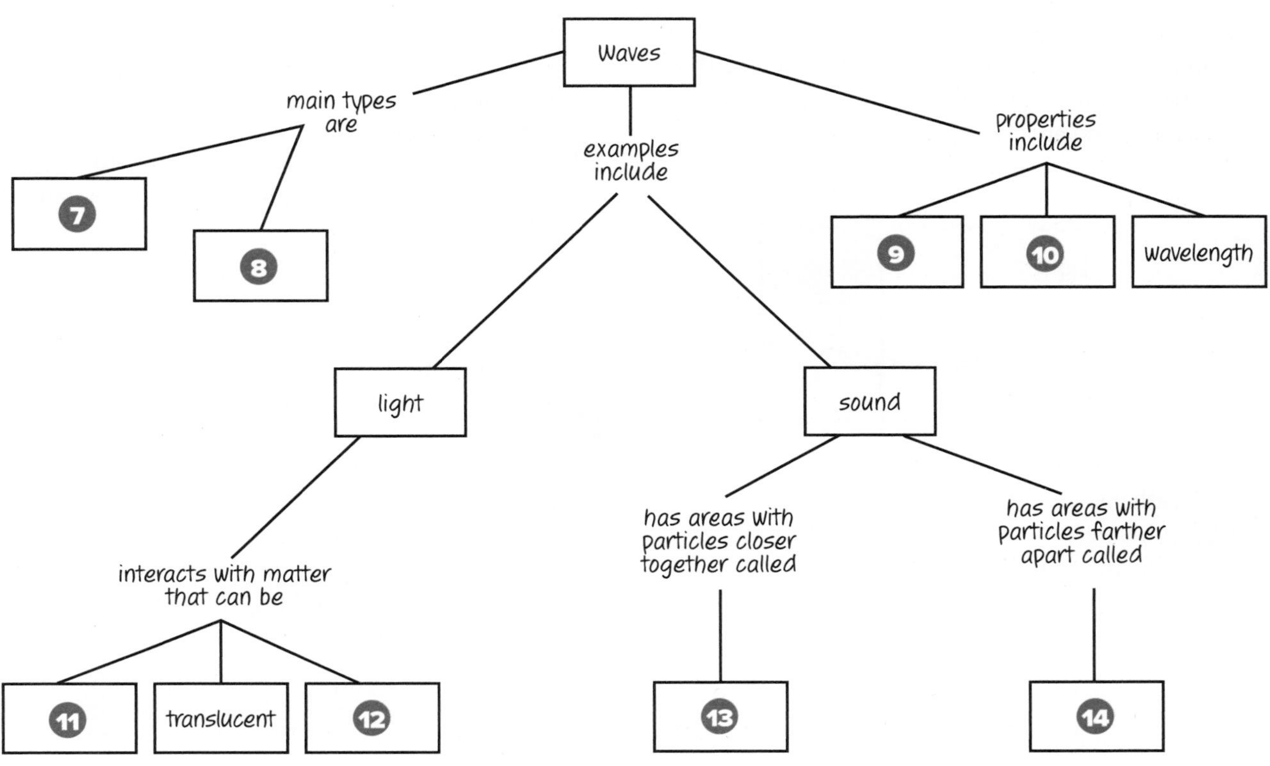

Chapter 14 Review

Understand Key Concepts 

1 As a water wave passes, the particles that make up the water move

A. back and forth, parallel to the wave.
B. in circles around the same point.
C. up and down at right angles to the wave.
D. whichever direction the wave moves.

2 The refraction of a wave is caused by a change in

A. amplitude.
B. frequency.
C. speed.
D. wavelength.

3 Which is always a transverse wave?

A. microwave
B. seismic wave
C. sound wave
D. water wave

4 Wave frequency is measured in

A. decibels.
B. hertz.
C. meters.
D. seconds.

5 The arrow in the diagram below shows a point on a light wave that stops as it interacts with matter. Which type of interaction does the arrow represent?

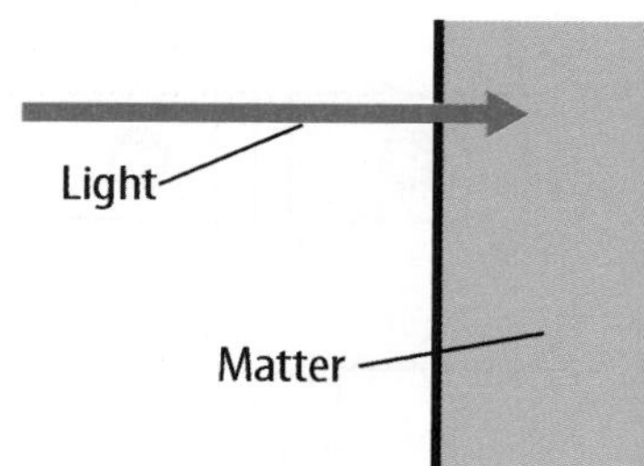

A. absorption
B. reflection
C. refraction
D. transmission

6 The distance between one point on a wave and the nearest point just like it is the

A. amplitude.
B. frequency.
C. pitch.
D. wavelength.

7 Which interactions of light with matter are taking place in the picture below?

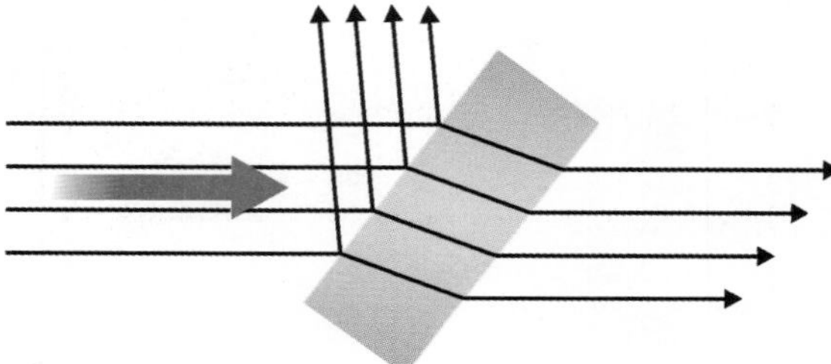

A. diffraction, reflection, and absorption
B. reflection, refraction, and transmission
C. reflection, scattering, and diffraction
D. translucent, transparent, and opaque

8 Which of the following colors of light has the longest wavelength?

A. red
B. green
C. violet
D. yellow

9 You turn up the volume on the car radio. Which of the following properties of the sound changes?

A. amplitude
B. frequency
C. speed
D. wavelength

10 If a sound is loud and low-pitched, the sound wave also has which of the following properties?

A. low frequency and high amplitude
B. low frequency and low amplitude
C. high frequency and high amplitude
D. high frequency and low amplitude

Critical Thinking

11 **Construct** Make a diagram that shows how interactions of light waves with matter cause a flower to appear orange.

12 **Synthesize** An MP3 player at maximum volume produces sound at 110 dB. The table shows the time exposure before a risk of hearing damage. How many hours a day could you listen to your MP3 player at full volume before a hearing loss risk? Explain.

Recommended Noise Exposure Limits	
Sound Level (dB)	Time Permitted (h)
90	8
95	4
100	2
105	1

13 **Summarize** What is the process by which an object can be seen and recognized? Be sure to include the interactions of light waves and matter in your summary.

14 **Hypothesize** Why does a 200-W lightbulb appear brighter than a 100-W lightbulb? Mention properties of light in your explanation.

15 **Apply** The passage of lightning through air produces thunder. Why is lightning seen before thunder is heard?

16 **Compare and Contrast** How does the motion of the medium in transverse mechanical waves, longitudinal waves, water waves, and seismic waves differ?

Writing in Science

17 **Write** a paragraph describing an example of sound waves and an example of light most people use each day. Identify a way you could change the properties of each wave.

REVIEW THE BIG IDEA

18 Explain various ways in which waves transfer energy through matter and empty space. Include correct terms to describe the various interactions of waves with matter.

19 Using the picture below, describe how the transfer of energy through matter and empty space helps a meteorologist predict the weather.

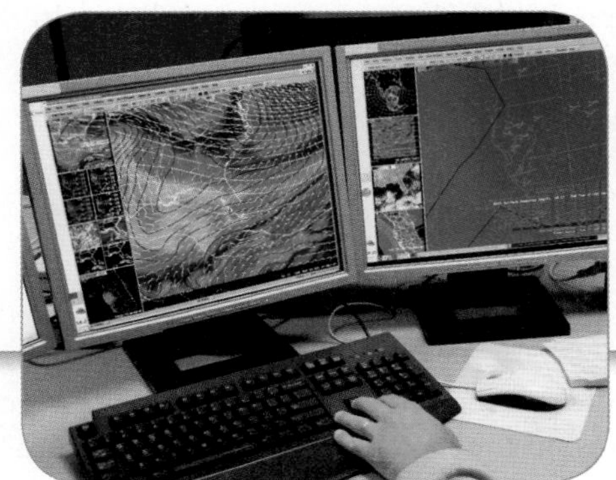

Math Skills

Math Practice

Use a Fraction

20 By what fraction does the sound intensity change if you move from 2 m away from a source to 10 m away from the source?

21 You are standing 3 m from someone who is using a lawn mower. How will the sound intensity change if the person moves the mower to a distance 12 m from you?

22 A car 5 m away from you beeps its horn. How would the intensity of the beep change if you moved to a distance 40 m from the car?

imac/Alamy

Standardized Test Practice

Record your answers on the answer sheet provided by your teacher or on a sheet of paper.

Multiple Choice

Use the figure to answer questions 1–3.

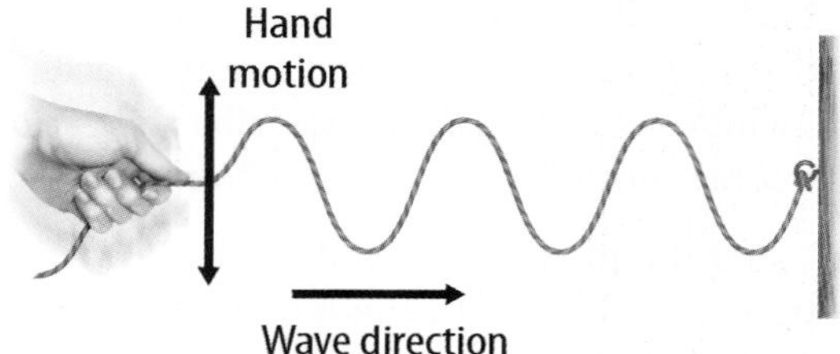

1 The figure above shows waves generated on a rope. Which type of waves are shown in the figure?

A combination

B electromagnetic

C longitudinal

D mechanical

2 Which statement best describes the correct relationship for the wave shown in the figure?

A The disturbance is parallel to the direction the wave travels.

B The disturbance is perpendicular to the direction the wave travels.

C The disturbance carries matter and energy in the same direction along the wave.

D The disturbance has both back-and-forth and up-and-down motion.

3 Which describes how the wave would change if the person's hand moved at a faster rate?

A The amplitude would decrease.

B The amplitude would increase.

C The frequency would decrease.

D The frequency would increase.

4 If two waves are traveling at the same speed, which description is most accurate?

A The wave with the longer wavelength has the higher frequency.

B The wave with the shorter wavelength has the higher frequency.

C Both waves must have equal wavelengths.

D Both waves must have equal frequencies.

5 Wood is opaque. Which describes how light waves can interact with wood?

A absorption and reflection

B diffraction and transmission

C reflection and refraction

D transmission and refraction

6 Which property is unique to electromagnetic waves?

A the ability to interact with matter

B the ability to travel through matter

C the ability to have different intensities

D the ability to travel through empty space

Use the table to answer question 7.

Incoming Light	Color of Filter	Outgoing Light
white	red	red
red	blue	none
white	blue	blue
green	green	?

7 The table above shows the interactions of different colors of light with different colors of filters. Which is the correct color to complete the table?

A green

B none

C red

D white

8 Which must be true of the cornea for the eye to work properly in sending a message to the brain?

A It must absorb light.

B It must block out light.

C It must reflect light.

D It must transmit light.

Use the table to answer question 9.

Material	Speed of Sound (m/s)
Air (0°C)	331
Air (20°C)	343
Water (0°C)	1,500
Water (20°C)	1,481
Ice (0°C)	3,500
Iron	5,130

9 Based on the data in the table, which of the following statements is most likely true?

A Sound travels fastest through gases because they are less dense.

B Sound travels fastest through liquids because they are most fluid.

C Sound travels fastest through solids because they are most dense.

D Sound travels fastest through materials that have higher temperatures.

10 Which color of light could you shine on a green object to make it appear black?

A green

B red

C white

D yellow

Constructed Response

11 You are standing outside and hear a jet flying overhead. You look up toward the direction of the sound, but you notice that the jet is far ahead of where the sound seems to come from. Explain why you can hear a jet only after it passes overhead.

Use the figure to answer question 12.

12 The figure above shows light rays striking a flat surface. Describe how the figure would change if the surface the light rays hit against were bumpy instead of flat.

13 People sometimes confuse the pitch of a sound with the sound's intensity. How would you explain the difference between these two properties to a classmate?

14 What roles do the outer ear, the middle ear, and the inner ear play in hearing?

NEED EXTRA HELP?														
If You Missed Question...	1	2	3	4	5	6	7	8	9	10	11	12	13	14
Go to Lesson...	1	1	1	1	2	2	2	2	3	2	3	2	3	3

Chapter 15

Electricity and Magnetism

How are an electric current and a magnet related?

Inquiry Fill it up?

For thousands of years, people have known about natural, invisible forces. Scientists continue to develop technologies that harness two of these forces—electricity and magnetism. For example, before this electric-powered sports car could be mass-produced, batteries needed to be developed to be long-lasting, lightweight, and able to be charged quickly.

- Why should people rely on electricity as an energy source?
- What roles do magnetic forces play in your life?
- How are electric currents and magnets used in this sports car?

Get Ready to Read

What do you think?

Before you read, decide if you agree or disagree with each of these statements. As you read this chapter, see if you change your mind about any of the statements.

1. Electrically charged objects always attract each other.
2. Electric fields apply magnetic forces on other electric fields.
3. A battery in an electric circuit produces an electric current.
4. Every magnet has one magnetic pole.
5. Earth is magnetic but is not a magnet.
6. A magnet moving within a wire loop produces an electric current.

Mc Graw Hill Education connectED

Your one-stop online resource

connectED.mcgraw-hill.com

LearnSmart®

Project-Based Learning Activities

Chapter Resources Files, Reading Essentials, Get Ready to Read, Quick Vocabulary

Lab Manuals, Safety Videos, Virtual Labs & Other Tools

Vocabulary, Multilingual eGlossary, Vocab eGames, Vocab eFlashcards

Animations, Videos, Interactive Tables

Personal Tutors

Self-checks, Quizzes, Tests

Lesson 1

Electric Charges and Electric Forces

Reading Guide

Key Concepts

ESSENTIAL QUESTIONS

- How do electrically charged objects differ?
- How do objects become electrically charged?
- How do electrically charged objects interact?

Vocabulary

electrically neutral p. 488
electrically charged p. 488
electric discharge p. 490
electric insulator p. 490
electric conductor p. 490
electric force p. 491
electric field p. 491

Multilingual eGlossary

What's Science Got to do With It?

Go to the resource tab in ConnectED to find the PBL *Hands Off!*

Inquiry Lightning in a Bottle?

Whether you call them nebula spheres, plasma lamps, or lightning balls, these devices put on a fascinating display. Are you really looking at miniature bolts of lightning? This lesson will help you understand the electric charges that produce these mysterious streams of light. But, what are electric charges, and how can they be useful to you?

David Wall/Alamy

Launch Lab

10 minutes

Why do they move?

Have you ever pulled a sweater out of a clothes dryer and found other items of clothing clinging to it? Maybe you heard a crackling sound or even saw sparks when you pulled the items apart. When different materials come into contact with each other, such as the clothes in the dryer, something happens to the materials. How do the materials interact? What causes their strange behaviors?

1. Read and complete a lab safety form.
2. Break a handful of **polystyrene packing pellets** into 2/3-cm pieces. Place the pieces in a **2-liter soda bottle,** and place the **cap** on the bottle.
3. Touch the bottle with a piece of **wool cloth.** Record your observations in your Science Journal.
4. Now, rub the sides of the bottle vigorously with the wool for 3 minutes. Record your observations.
5. Open and partially squeeze the bottle. Exhale into the bottle to return it to its original shape. Repeat several times to add moisture to the air in the bottle. Repeat step 4.

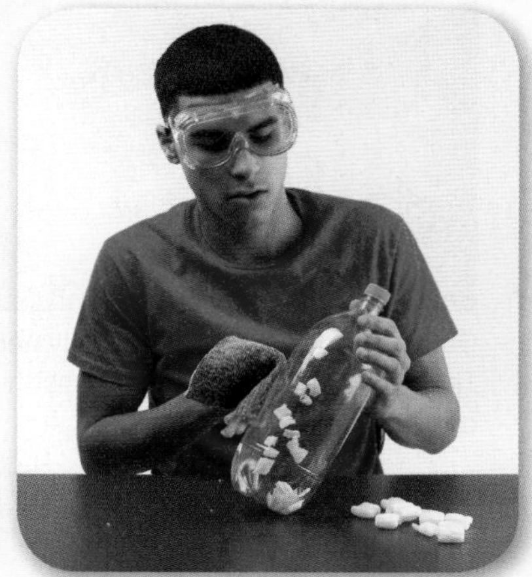

Think About This

1. How do the packing pellets act when you rub the bottle with the wool compared to when you simply touch the bottle with the wool?
2. How does adding moisture to the air in the bottle affect the behavior of the pellets?
3. **Key Concept** Hypothesize how the wool affects the packing pellets even though the wool does not touch the packing pellets.

Electric Charges

Have you ever walked across a carpeted floor, reached for a metal doorknob, and received a small shock? The shock comes from electric charges jumping between your fingers and the doorknob. What are electric charges? Where do they come from? Why do they jump from one object to another? In this lesson, you will learn the answers to these questions.

Recall that atoms are the tiny particles that make up all the matter around you. An atom has a **nucleus** made up of two kinds of smaller particles. These particles are protons and neutrons. An atom also is made up of electrons. Electrons move around the atom's nucleus, as shown in **Figure 1.** Protons and electrons have a property called electric charge. Neutrons do not have electric charge.

Reading Check What particles found in an atom have the property of electric charge?

ACADEMIC VOCABULARY

nucleus
(noun) basic or essential part; core

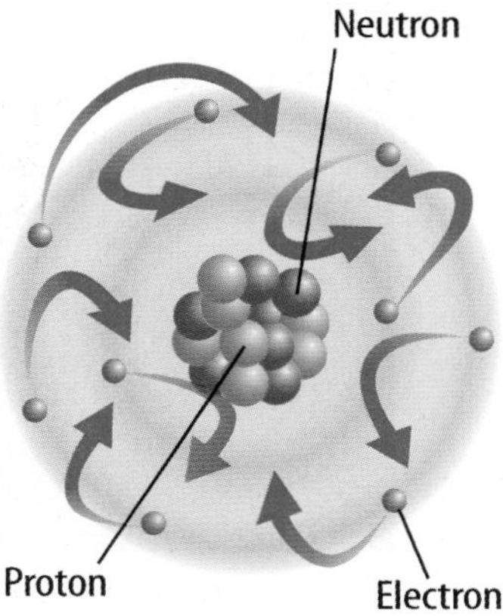

Figure 1 Atoms are made of protons, electrons, and neutrons.

Personal Tutor

Figure 2 An electrically neutral object becomes charged when it touches a different material.

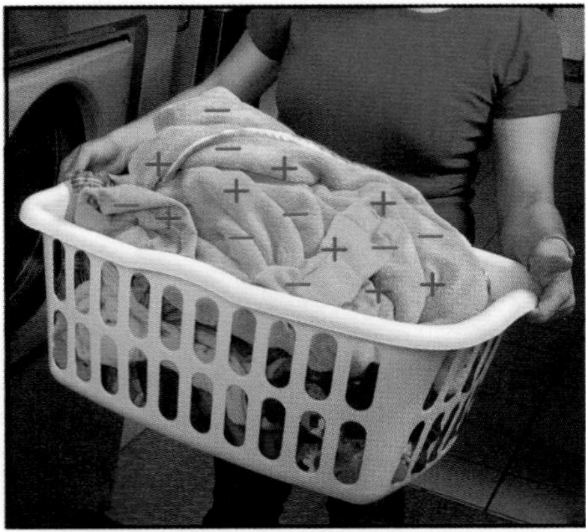

Objects normally are electrically neutral. They have equal amounts of positive charge and negative charge.

When objects made of different materials come in contact with each other, negatively charged electrons move from one object to another.

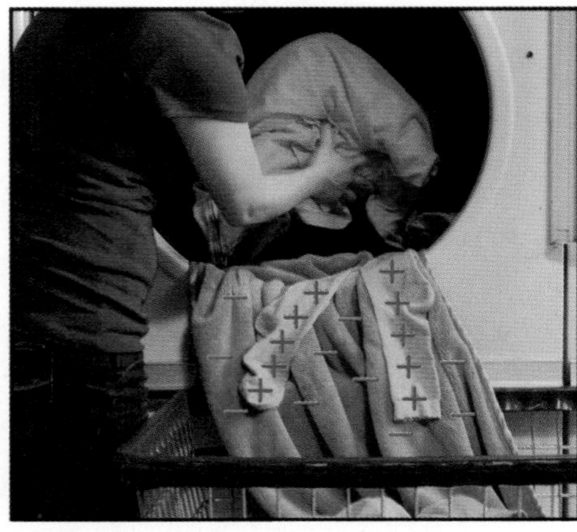

Objects that lose electrons are positively charged. Objects that gain electrons are negatively charged. Oppositely charged objects attract each other.

Positive and Negative Charge

There are two types of electric charge–positive and negative. Here, *positive* and *negative* do not mean more or less. The terms are simply names scientists use to talk about the two types of electric charge.

Protons have positive charge. Electrons have negative charge. The amount of positive charge of a proton equals the amount of negative charge of an electron.

Atoms have equal numbers of positive protons and negative electrons. *A particle with equal amounts of positive charge and negative charge is* **electrically neutral.** Electrically neutral atoms make up all objects. Therefore, objects are normally electrically neutral, too. However, electrons sometimes transfer between objects. How does transferring electrons affect objects?

Reading Check Why are atoms electrically neutral?

When electrons transfer from one electrically neutral object to another, both objects become electrically charged. An **electrically charged** object has an unbalanced amount of positive charge or negative charge. **Figure 2** shows that objects can be either positively charged or negatively charged.

Positively Charged An object that has lost one or more electrons has more protons than electrons. Thus, the object has more positive charge than negative charge. The object is positively charged.

Negatively Charged An object that has gained one or more electrons has more electrons than protons. Thus, the object has more negative charge than positive charge. The object is negatively charged.

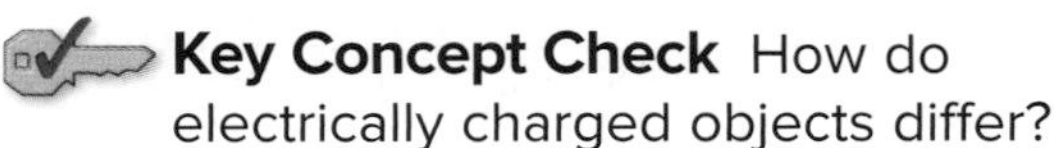

Key Concept Check How do electrically charged objects differ?

Materials and Electric Charge

How do electrically neutral objects become electrically charged? The table in the MiniLab below lists some common materials in the order of how tightly they hold electrons. Wool is above rubber on the list. This means that wool does not hold electrons as tightly as rubber. Look at **Figure 3.** As the rubber balloon touches the wool toy, electrons transfer from the toy to the balloon. The balloon becomes negatively charged and the toy becomes positively charged.

On the other hand, glass is above wool on the table in the MiniLab. **Figure 3** shows that as a glass cup touches the toy, electrons transfer from the glass to the wool. In this case, the glass becomes positively charged, and the wool becomes negatively charged.

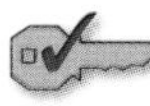 **Key Concept Check** How do the balloon and the stuffed toy become charged?

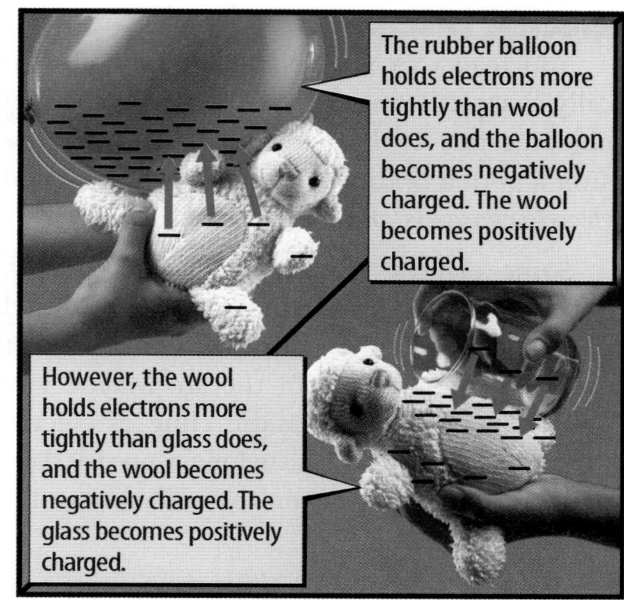

The rubber balloon holds electrons more tightly than wool does, and the balloon becomes negatively charged. The wool becomes positively charged.
However, the wool holds electrons more tightly than glass does, and the wool becomes negatively charged. The glass becomes positively charged.

Figure 3 Whether an object becomes positively charged or negatively charged depends on the material it contacts.

MiniLab

20 minutes

Can you create an electric charge?

In the table below, rubber is listed below nylon. This means that when a rubber object contacts a nylon object, electrons transfer from the nylon to the rubber. The rubber object becomes negatively charged, and the nylon becomes positively charged.

1. Read and complete a lab safety form.
2. Inflate and tie-off **two balloons.** Mark an *X* on both balloons with a **permanent marker.**
3. Select one **material** from the list above rubber and one **material** from the list below rubber.
4. Rub the marked area of each balloon on one of your materials.
5. Hold the balloons by their knots, and bring the two marked areas together. Record your observations in your Science Journal.
6. Now, rub the marked area of one of the balloons on one chosen material. Rub the marked area of the other balloon on the other chosen material. Repeat step 5.

Becomes Positive
Glass
Human hair
Nylon
Wool
Silk
Aluminum
Paper
Cotton
Wood
Rubber
Copper
Polyester
Polystyrene
Polyvinyl chloride
Becomes Negative

Analyze and Conclude

1. **Compare** the behavior of the balloons after they rub against the same material and when they rub against different materials.
2. **Key Concept** Explain how you can tell when the balloons receive the same charge and when they receive different charges.

Hutchings Photography/Digital Light Source

Electric Discharge

You read that objects can become electrically charged. However, an electrically charged object tends to lose its unbalanced charge after a period of time. *The loss of an unbalanced electric charge is an* **electric discharge.**

Some electric discharges happen slowly. For example, electrons on negatively charged objects discharge, or move, from the object onto water molecules in the air. Maybe you have noticed that the static cling of electrically charged clothing lasts longer on dry days than on humid days when there is more water vapor in the air.

Some electric discharges happen quickly. For example, lightning is the sudden loss of unbalanced electric charges that build up in thunderstorm clouds. **Figure 4** describes other examples of electric discharges.

Electric Insulators and Conductors

Different materials become electrically charged as they come in contact. With some materials, charges remain where the materials touched. With other materials, the charges evenly spread out over the object.

For example, after a balloon touches a sweater, charges from the sweater stay in the area of the balloon that touched the sweater. However, after you walk across a carpet, charges from the carpet spread over your entire body. Your hand receives an electric shock as you reach for a metal doorknob.

Electric charges do not spread over the balloon because electric charges cannot easily move in rubber. *A material in which electric charges cannot easily move is an* **electric insulator.** Plastic, wood, and glass are examples of electric insulators.

A material in which electric charges easily move is an **electric conductor.** Some of the best conductors are metals, such as copper.

Personal Tutor

Figure 4 Electric discharges occur all around you.

A sudden electric discharge through the gas-filled tube of the camera flash causes the gas to produce a burst of light. ▶

◀ A steady electric discharge between the metal rod and the steel plates produces enough thermal energy to melt the metals.

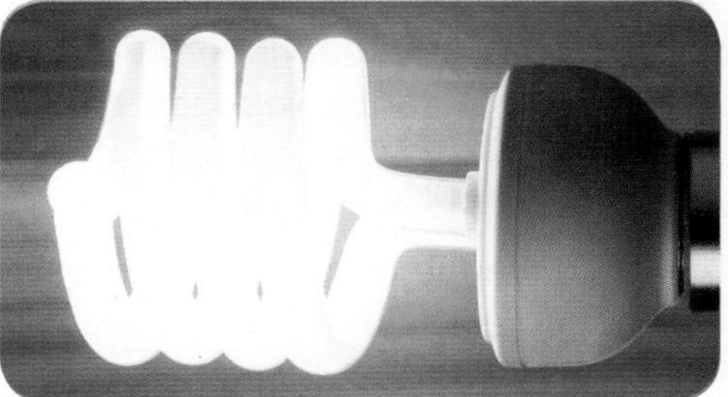

A continual electric discharge through the fluorescent light causes a powder inside the tube to glow brightly. ▶

Visual Check What are some careers that use an electric discharge?

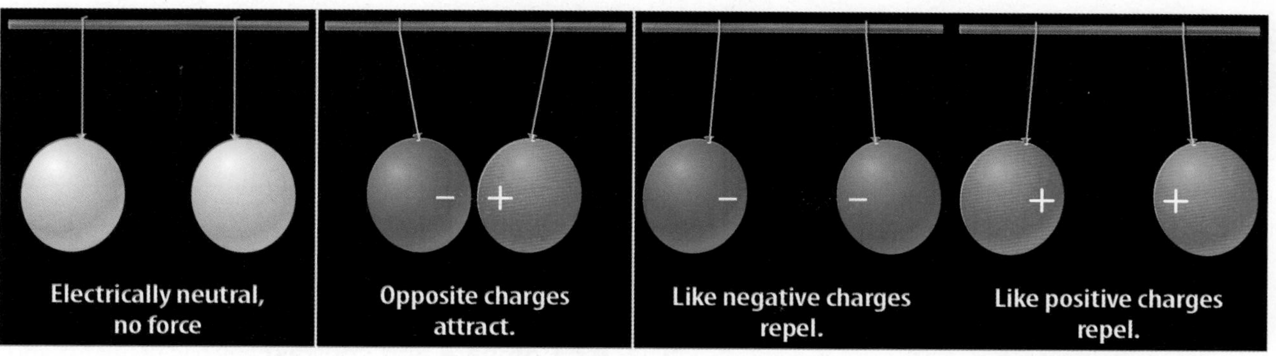
Electrically neutral, no force
Opposite charges attract.
Like negative charges repel.
Like positive charges repel.

Figure 5 Charged objects push or pull on other charged objects.

Visual Check In the figure above, why do the green and red balloons attract each other?

Electric Fields and Electric Forces

Suppose you rub two balloons on a wool sweater. Electrons transfer from the sweater to the balloons. Both balloons become negatively charged. The sweater becomes positively charged. You notice that the sweater attracts, or applies a pulling force on, the balloons. However, the balloons repel, or apply a pushing force to each other. *The force that two electrically charged objects apply to each other is an* **electric force.**

Electric fields surround charged objects.

To open a door, your hand must touch the door to apply a force to it. However, an electrically charged object does not have to touch another charged object to apply an electric force to it. For example, the two charged balloons in the example above repel each other even though they do not touch.

How do charged objects apply electric forces to each other without touching? The answer is a bit of a mystery. However, scientists know there is a region around a charged object that applies an electric force to other charged objects. *This invisible region around any charged object where an electric force is applied is an* **electric field.**

WORD ORIGIN

force
from Latin *fortis,* means "strong"

Electric force depends on the types of charge.

Whether an electric force is a push or a pull depends on the types of charge on the objects, as shown in **Figure 5.** If both objects are positively charged or if both objects are negatively charged, the two objects push each other away. In other words, objects with the same type of electric charge repel each other.

If one object is positively charged and the other object is negatively charged, the two objects pull each other together. In other words, objects with opposite types of electric charge attract each other.

Key Concept Check How do electrically charged objects interact?

FOLDABLES®

Make a horizontal three-tab book with an extended tab for the title. Label it as shown. Use it to organize your notes on the relationships between electric forces.

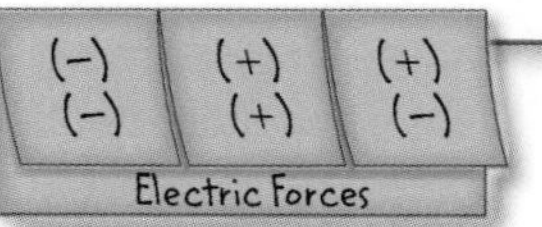
(–) (–)
(+) (+)
(+) (–)
Electric Forces

Lesson 1 Review

Visual Summary

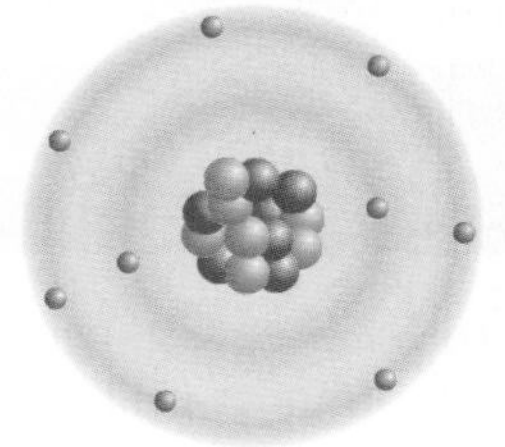

Atoms are the source of all electric charge.

Electric discharges occur all around you.

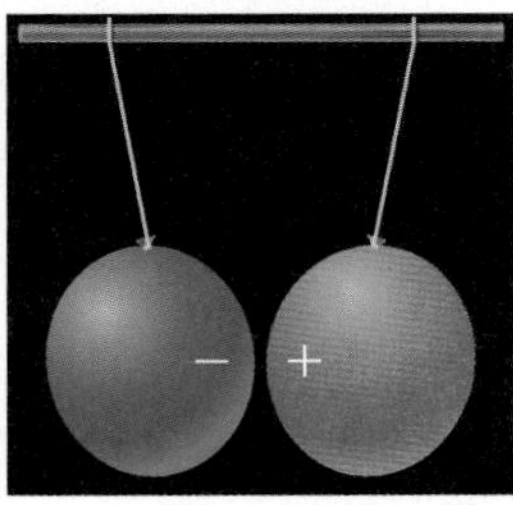

The electric force between two charged objects depends on the type of charge on each object.

Use your lesson Foldable to review the lesson. Save your Foldable for the project at the end of the chapter.

What do you think NOW?

You first read the statements below at the beginning of the chapter.

1. Electrically charged objects always attract each other.
2. Electric fields apply magnetic forces on other electric fields.

Did you change your mind about whether you agree or disagree with the statements? Rewrite any false statements to make them true.

Use Vocabulary

1. **Define** *electric discharge.*
2. **Utilize** the terms *electric force* and *electric field* in a single sentence.
3. **Compare and contrast** electric insulator and electric conductor.

Understand Key Concepts 

4. **Name** What are the two types of electric charge?
5. **Explain** two ways an object can become electrically charged.
6. Cotton holds electrons more tightly than wool. If a wool sweater touches a cotton shirt, the electric charge of the shirt will be
   - **A.** negative.
   - **B.** neutral.
   - **C.** polarized.
   - **D.** positive.

Interpret Graphics

7. **Organize** Copy and fill in the table below describing whether the particles in the left column will attract, repel, or exert no force on each other.

Particles	Type of Electric Force
Proton and proton	
Electron and electron	
Electron and proton	

Critical Thinking

8. **Infer** A rubber balloon and a glass cup are each rubbed with a polyester shirt. Will the cup and the balloon attract or repel each other? Explain your thinking.
9. **Assess** Is the following statement true or false? Write a short paragraph supporting your opinion. "An object with an excess of neutrons has no electric charge."

Dimitri Vervitsiotis/Getty Images

Skill Practice Follow a Procedure

40 minutes

How can you build your own electroscope?

Materials

paper clip

aluminum foil

nail

hammer

fabric samples

glass jar with metal lid

Also needed: assorted items to be electrically charged

Safety

Have you noticed how a brush and your hair become less electrically charged on humid days when there is a lot of moisture in the air? That is because water molecules in the air carry away the electric charge instead of letting it build up on surfaces, such as your hair. You can use an electroscope to detect the presence of an electric charge. In this activity, you will follow a procedure to construct your own electroscope, electrically charge it, and observe the ways it discharges.

Learn It

In order to construct and use a scientific device, you need to understand the properties of the materials to be used. Also, you need to know how and in what order to use the materials. To build a reliable device, you need to **follow a procedure.**

Try It

1. Read and complete a lab safety form.
2. Use a hammer and nail to poke a small hole in a jar lid.
3. Unfold a paper clip, and bend it into an L shape.
4. Push the long end of the paper clip L up through the hole in the lid. Bend the paper clip above the lid so it hangs through the lid without falling out.
5. Smooth the aluminum foil with your finger. Cut off a strip 4 cm by 1 cm. Fold the strip in half, and hang it from the paper clip below the lid.
6. Screw the lid on the jar.
7. Touch uncharged and charged objects to the paper clip. Record your observations of the hanging foil strip in your Science Journal.

5

8. Bring charged and uncharged objects near the paper clip, but do not touch it. Record your observations.

Apply It

9. **Explain** the importance of following a procedure.
10. **Classify** the components of your electroscope as conductors or insulators.
11. **Key Concept** Contrast the behavior of the foil to the ways the foil is charged.

Lesson 2

Reading Guide

Key Concepts
ESSENTIAL QUESTIONS

- How are electric current and electric charge related?
- What are the parts of a simple electric circuit?
- How do the two types of electric circuits differ?

Vocabulary

electric current p. 495
electric circuit p. 497
generator p. 498
electric resistance p. 499
voltage p. 501

Multilingual eGlossary

BrainPOP®

Electric Current and Electric Circuits

Inquiry Do you need it?

People used to plan their days, and lives, around the rising and setting of the Sun. Then, Thomas Edison invented the lightbulb. Suddenly, convenient, cheap electric power made life easier and better. How has electricity improved your life?

Launch Lab

20 minutes

What are two ways to light two lightbulbs? 

There are two ways to connect a battery and two lightbulbs so both bulbs light. Each way uses an electric conductor to create a path from the battery to each lightbulb and back to the battery.

1. Read and complete a lab safety form.
2. Examine the first diagram. Using **two lightbulbs in bases, one battery in a holder,** and **several wires,** construct the setup so both bulbs light.
3. Remove one bulb from its base. Observe the behavior of the other bulb. Record your observations in your Science Journal.
4. Examine the second diagram, and construct the setup. Both bulbs should be lit.
5. Unscrew one bulb from its base. Again, observe the behavior of the other bulb. Record your observations.

1.

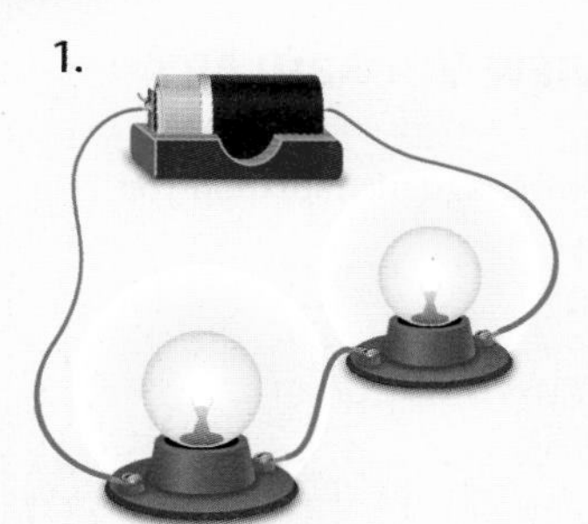

2.

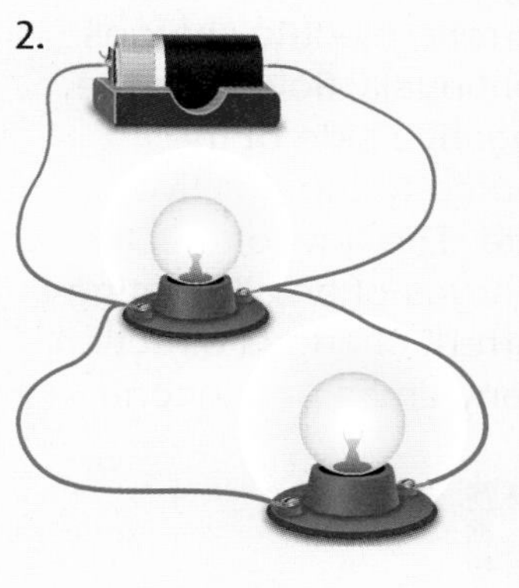

Think About This

1. Sketch each setup, and draw the path that connects the lightbulbs and the battery.
2. Describe how the brightness of the lightbulbs in each setup differs. Why do you think there is this difference?
3.  **Key Concept** How do the two setups differ?

Electric Current—Moving Electrons

You read in Lesson 1 that electrons have the property of negative electric charge. Recall that negatively charged electrons are the tiny particles that move around the nuclei of atoms. Also, recall that many of the electrons of an electric conductor, such as a metal wire, easily move from atom to atom. When free electrons move in the same direction, an electric current is produced. *An* **electric current** *is the movement of electrically charged particles, such as electrons.*

Like all moving objects, moving electrons have kinetic energy. As electrons move from atom to atom, their kinetic energy transforms to other useful energy forms, such as light and thermal energy. Moving electrons, or an electric current, is one of the most common forms of energy. In this lesson, you will read about what causes electrons to move. You also will read about how their movement is controlled to make an electric current useful.

 **Key Concept Check** How are electric current and electric charge related?

SCIENCE USE V COMMON USE

charge
Science Use a definite quantity of electricity

Common Use the price demanded for something

Two Types of Electric Current

An electric current, or the movement of electrons, carries energy near the speed of **light**. However, the negatively charged electrons themselves move more slowly.

REVIEW VOCABULARY
light
electromagnetic radiation you can see

Imagine a tube filled with marbles. Extra marbles pushed into one end of the tube cause other marbles to pop out the other end. The first marbles do not instantly move the length of the tube. Similarly, as electrons move into one end of a wire, other electrons leave the other end of the wire almost instantly. Each electron does not suddenly move the length of the wire.

Direct Current In the example above, marbles added continually to one end of the tube produce a steady stream of marbles flowing out the other end of the tube. **Figure 6** shows that, in a similar way, electrons continually added to one end of a wire create a constant one-way flow of electrons. This is known as direct current. Some energy sources, including batteries, produce only direct current. Many portable devices, such as flashlights, operate using direct current.

Alternating Current If marbles are repeatedly added to one end of the tube and then to the other end, the marbles in the tube would move back and forth, never moving far from their original positions. An electric current that also continually reverses direction is known as an alternating current. Large generators in power plants supply homes and businesses with alternating current.

Figure 6 With direct current, electric charges continually flow from the negative side of the source to the positive side. The flow of electric charges of an alternating current changes direction many times per second.

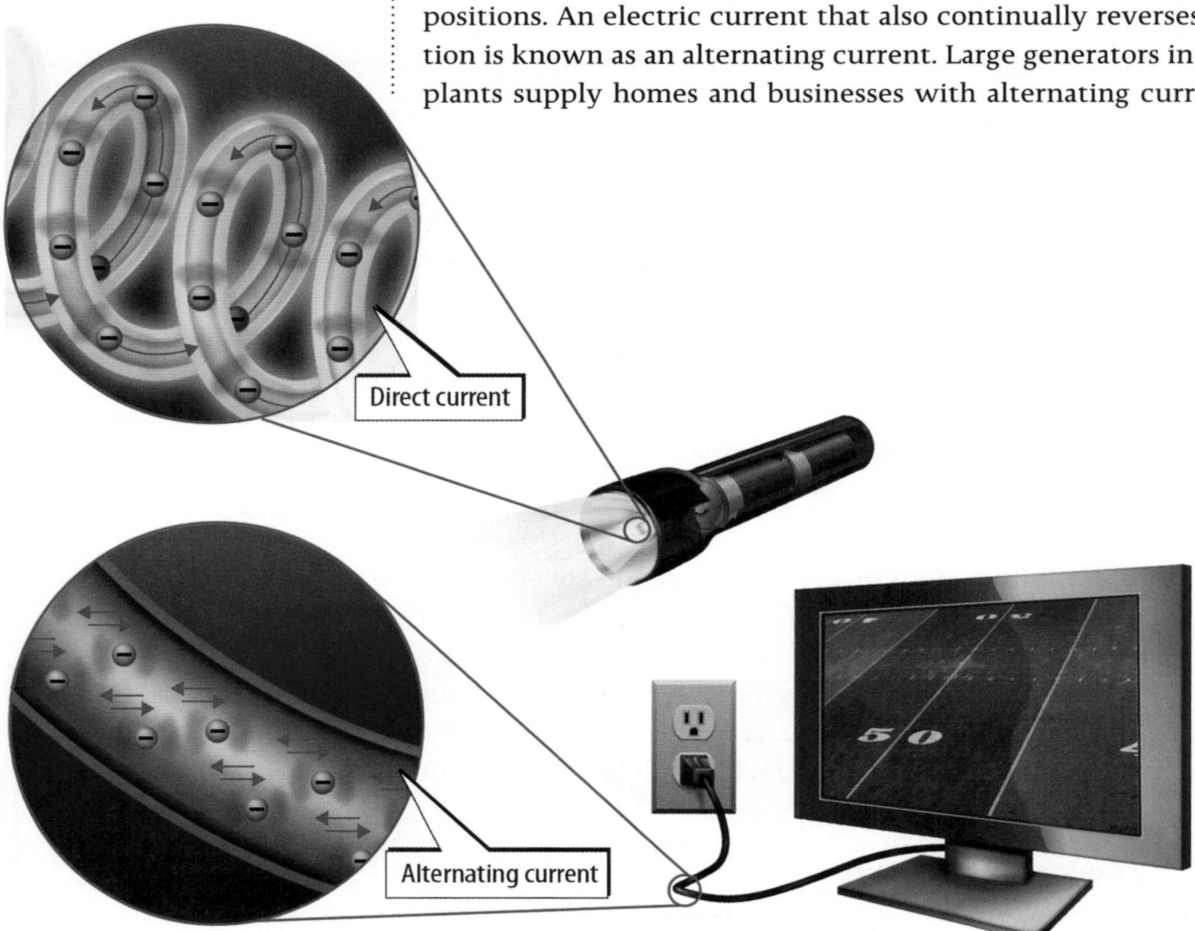

The Circuit—A Path for Electric Current

Electric circuits transform the energy of an electric current to useful forms of energy. An **electric circuit** *is a closed, or complete, path in which an electric current flows.* Electric circuits are all around you.

A Useful Circuit

Electric circuits are designed to transform electric energy to specific forms. For example, the electric circuits in a microwave oven transform electric energy to the radiant energy that cooks your food. Figure 7 illustrates an electric circuit designed to transform the electric energy of a battery to the light energy emitted by a lightbulb. As shown, the circuit is complete, or closed, and the lightbulb is lit. If the circuit is broken, or open, at any point, the electric current stops and the lightbulb does not light.

A Simple Circuit

Some electric circuits, such as those found in computers, are very complicated and have hundreds of parts. However, many common and useful circuits include only a few components. Simple circuits are found in flashlights, doorbells, and many kitchen appliances. All simple circuits contain: 1) a source of electric energy, such as a battery, 2) an electric device, such as a lightbulb, and 3) an electric conductor, such as wire. In addition to these basic components of all circuits, a switch is often included in a circuit. How do these basic components interact to make a useful electric current?

Key Concept Check What are the parts of a simple electric circuit?

FOLDABLES®

Create a horizontal three-tab book. Illustrate and label a simple electric circuit as shown. Use it to explain the components of a circuit.

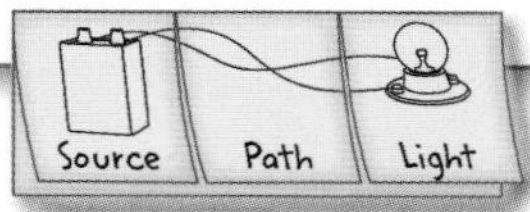

WORD ORIGIN

circuit
from Latin *circuire,* means "to go around"

Figure 7 A practical electric circuit may have only a few components.

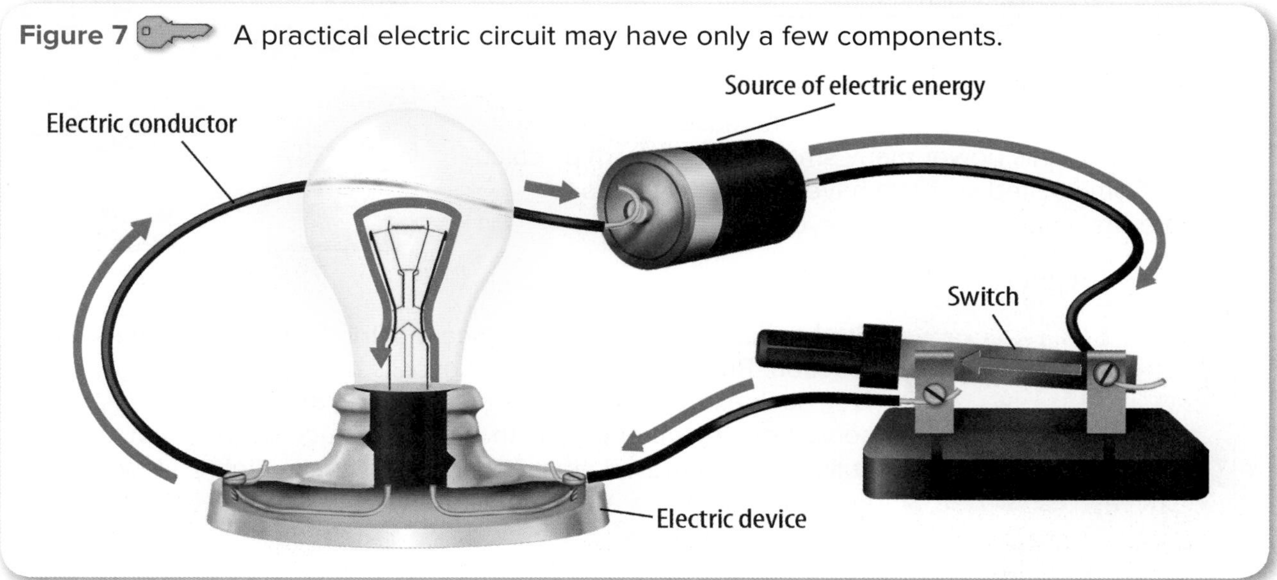

Figure 8 Many electric energy sources are being developed and improved.

Visual Check Why are fuel cells possibly a good source of electric energy?

Sources of Electric Energy There are many uses of electric energy. Most uses require specific types of sources of electric energy. For example, a flashlight requires a small, portable source. Cities need sources that produce large amounts of electric energy that are nonpolluting. Figure 8 includes some of the technologies now being developed and improved to help meet the world's growing demand for electric energy.

Batteries often are used when an electric energy source needs to be small and portable. A battery is simply a can of chemicals. Chemical reactions within a battery move electrons from one end of the battery (the positive terminal) to the other end (the negative terminal). Outside the battery, the electrons flow through a closed circuit from the negative terminal back to the positive terminal. As the chemical reactions continue, electrons keep moving through the battery and circuit.

Generators *are machines that transform mechanical energy to electric energy.* Many power plants use fossil fuels or nuclear energy to power large generators. These fuels provide thermal energy to boil water into steam. The steam flows through and rotates a turbine that, in turn, rotates a generator. These types of turbine-powered generators provide most of the electric energy used in the United States. Other generators use wind or moving water for power. You will read more about generators in the next lesson.

Solar cells change sunlight into electric energy. Often cells are connected into solar panels to increase energy output. Simple solar cells power many small devices, such as calculators. Complicated systems have enabled humans to explore the solar system and beyond.

Fuel cells, like batteries, produce electric energy by a chemical reaction. But, unlike batteries, fuel cells need a constant flow of fuel, such as hydrogen gas. An advantage of using fuel cells as a source of electric energy is that they produce no pollution. Fuel cells have generated electric energy on space flights. Now, scientists and engineers are developing ways to use fuel cells in people's everyday lives.

Electric devices transform energy. An electric device is a part of a circuit designed to transform electric energy to a useful form of energy. For example, a lightbulb is designed to transform electric energy to light. Transformation of electric energy occurs wherever there is electric resistance in a circuit. **Electric resistance** *is a measure of how difficult it is for an electric current to flow in a material.* Electric devices with greater electric resistance transform greater amounts of electric energy. What causes a transformation of electric energy?

Think of an electric lightbulb. As electrons move in the high-resistance wire filament of the lightbulb, they collide with atoms of the filament. The atoms absorb some of the electrons' kinetic energy, then release the energy as light.

Electric Conductors and Electric Circuits
An electric conductor, such as a wire, is used to complete the circuit by connecting the energy source to the electric device. Copper and aluminum make good wires for electric circuits because they are excellent conductors. A good conductor has little electric resistance.

Recall that an electric current easily flows through an electric conductor. However, even the best conductors, such as copper wire, resist an electric current a little. All conductors, including a device's power cord, have some electric resistance. Small amounts of electric energy in a circuit's conductors always transform to wasted thermal energy.

Reading Check Why are wires in an electric circuit often made of copper?

20 minutes

How can you determine whether a material is a conductor?

Conductors allow electrons to travel freely through them, while insulators prevent electrons from moving through them. Can you design a device that will determine whether a material is a conductor?

1. Read and complete a lab safety form.
2. Using a **battery, two wires,** and a **bulb in a screw base,** construct a circuit that lights the bulb.
3. Develop a plan to alter your circuit to test whether an item is a conductor. Record your ideas in your Science Journal.
4. Use your device to test **wood** and **aluminum foil.** Record your observations.
5. Select two other **items.** Predict and then test them. Record your observations.

Hutchings Photography/Digital Light Source

Analyze and Conclude

1. **Compare** the materials which you found to be conductors.
2. **Classify** all the materials found in your device.
3. **Key Concept** Explain how your device works.

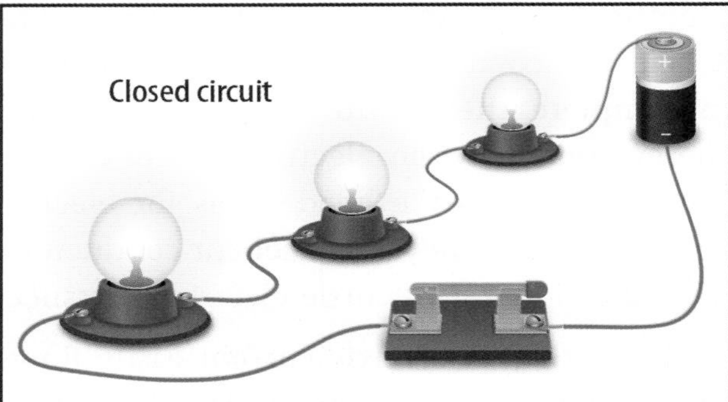

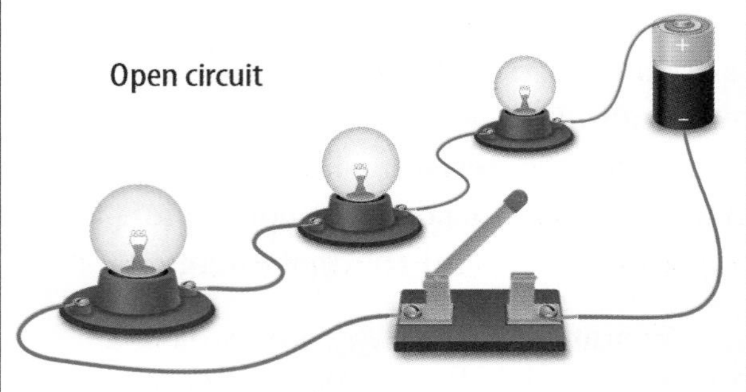

▲ **Figure 9** In a series circuit, all components are connected in a single loop.

Figure 10 In a parallel circuit, opening one branch does not affect devices in the other branches. ▼

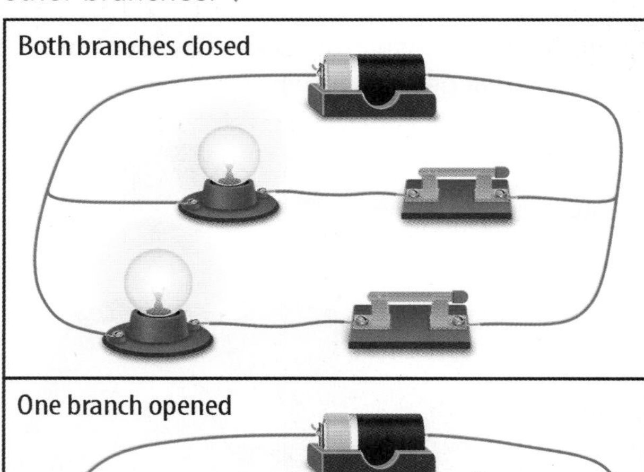

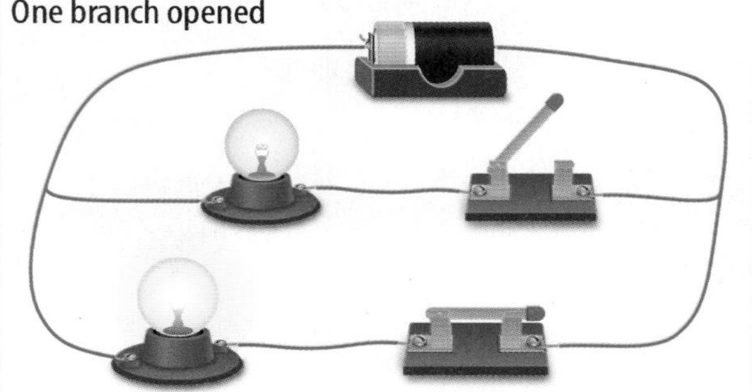

Series and Parallel Circuits

An electric circuit can have more than one device. For example, a string of holiday lights is a circuit that has many lightbulbs, or devices. Recall the circuits you built in the Launch Lab at the beginning of this lesson. With some holiday lights, if you remove one of the bulbs from its socket, all of the lightbulbs go out!

Now, think of the electric lights in the rooms of your home. These lights are devices connected in an electric circuit, too. However, if you remove the lightbulb from the lamp in your room, what happens to the light in the kitchen? Nothing. It remains lit.

How can you explain this difference in the two circuits? The answer is that there are two types of electric circuits.

Series Circuit In the example at the top of this page, the string of holiday lights is a series circuit. A series circuit is an electric circuit that has only one path through which an electric current can flow. In other words, all of the devices in a series circuit are connected end to end. As shown at the top of **Figure 9**, the same electric current flows through all the lightbulbs in the string. Breaking, or opening, a series circuit causes the electric current to stop flowing through the entire circuit.

Parallel Circuit A different type of circuit connects the devices in your home. Houses do not use series circuits. Instead, they use parallel circuits. A parallel circuit is an electric circuit where each device connects to the electric source with a separate path, or branch. The bottom of **Figure 10** shows two lightbulbs connected to a battery as a parallel circuit. If any one of the branches is opened, the other lightbulbs still have a complete path in which current flows.

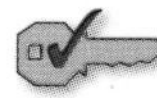
Key Concept Check How do the two types of electric circuits differ?

Voltage and Electric Energy

You may know the term *voltage*. For example, your home has 120-V outlets. To understand what this means, you must first know how to count electrons. But, there are so many electrons in a circuit! It is impossible to count them individually. Therefore, just as we quickly count eggs by the dozen, we count electrons by the coulomb (KEW lahm). One coulomb of electrons is a huge quantity–approximately 6,000,000,000,000,000,000 electrons!

Voltage of an Entire Circuit

Recall that all parts of an electric circuit have electric resistance. Therefore, energy is required to move electrons through a circuit. The **voltage** *of an electric circuit is the amount of energy used to move one coulomb of electrons through the circuit.*

Think of two identical lightbulbs. One is powered by a 3-V battery. The other is powered by a 6-V battery. The 6-V lightbulb is lit brighter than 3-V lightbulb. But why?

The definition of voltage tells you that the 6-V battery uses twice as much energy than the 3-V battery as it moves electrons through a circuit. Thus, the 6-V circuit transforms twice the electric energy to light.

Voltage of Part of a Circuit

You also can measure the voltage of part of a circuit. The voltage measured across a part of a circuit tells you how much energy is used by moving electrons through that part of the circuit. **Figure 11** shows the voltages across a lightbulb and a wire in the same circuit. The higher voltage across the lightbulb tells you that the lightbulb transforms more electric energy than the wire.

The sum of the voltages across all the parts of an electric circuit equals the voltage of the energy source. This means that all the parts of an electric circuit transform all the energy produced by the energy source.

WORD ORIGIN

voltage
from *Alessandro Volta,* (1745-1827), Italian physicist

Figure 11 Portions of an electric circuit with higher voltage readings transform more of the battery's energy.

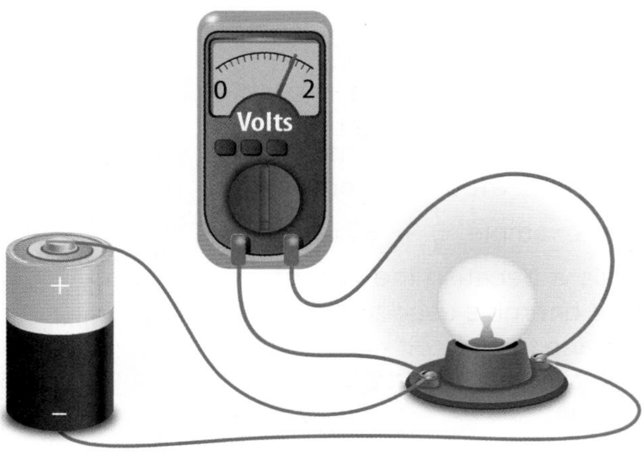

Higher voltage across lightbulb

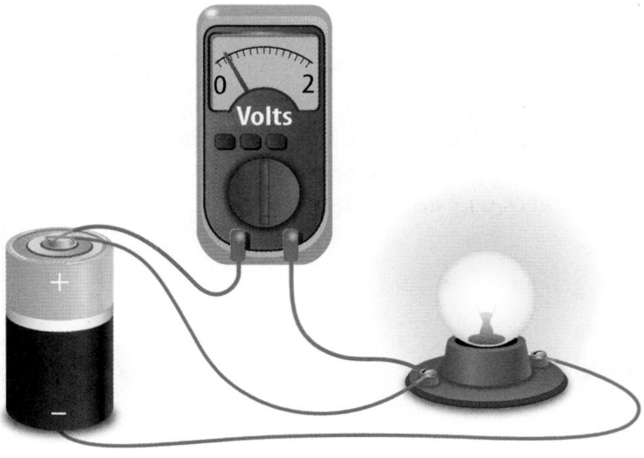

Lower voltage across wire

Visual Check Which part of the circuit is transforming most of the battery's energy into some other form?

Math Skills 

Using Fractions

Imagine a 9-V battery and two lightbulbs in a series circuit. The voltage across one lightbulb is 6 V. The second lightbulb reads 3 V. What part of the circuit's total energy is used by each lightbulb?

Divide the voltage reading across one of the lightbulbs by the voltage across the entire circuit (across the battery).

First bulb: $\frac{6\text{ V}}{9\text{ V}} = \frac{2}{3}$

Second bulb: $\frac{3\text{ V}}{9\text{ V}} = \frac{1}{3}$.

If you add the fractions together, they equal one. For example: $\frac{2}{3} + \frac{1}{3} = 1$ This is because the sum of the energies used by each device in a circuit equals the total energy in the circuit.

Practice

A 12-V battery powers a series circuit that contains two lightbulbs. The voltage across one of the lightbulbs is 8 V. What fractional part of the circuit's total energy is transformed in the second lightbulb?

 Math Practice

 Personal Tutor

A Practical Electric Circuit

Recall that a simple circuit can function with only a few basic parts–a lightbulb can be lit with just a battery and a couple of wires. However, most useful circuits include additional components to make them more useful and safer. **Figure 12** illustrates and describes some electric components of a hair dryer you might not be familiar with.

Figure 12 Useful electric circuits are simple circuits with a couple extra parts.

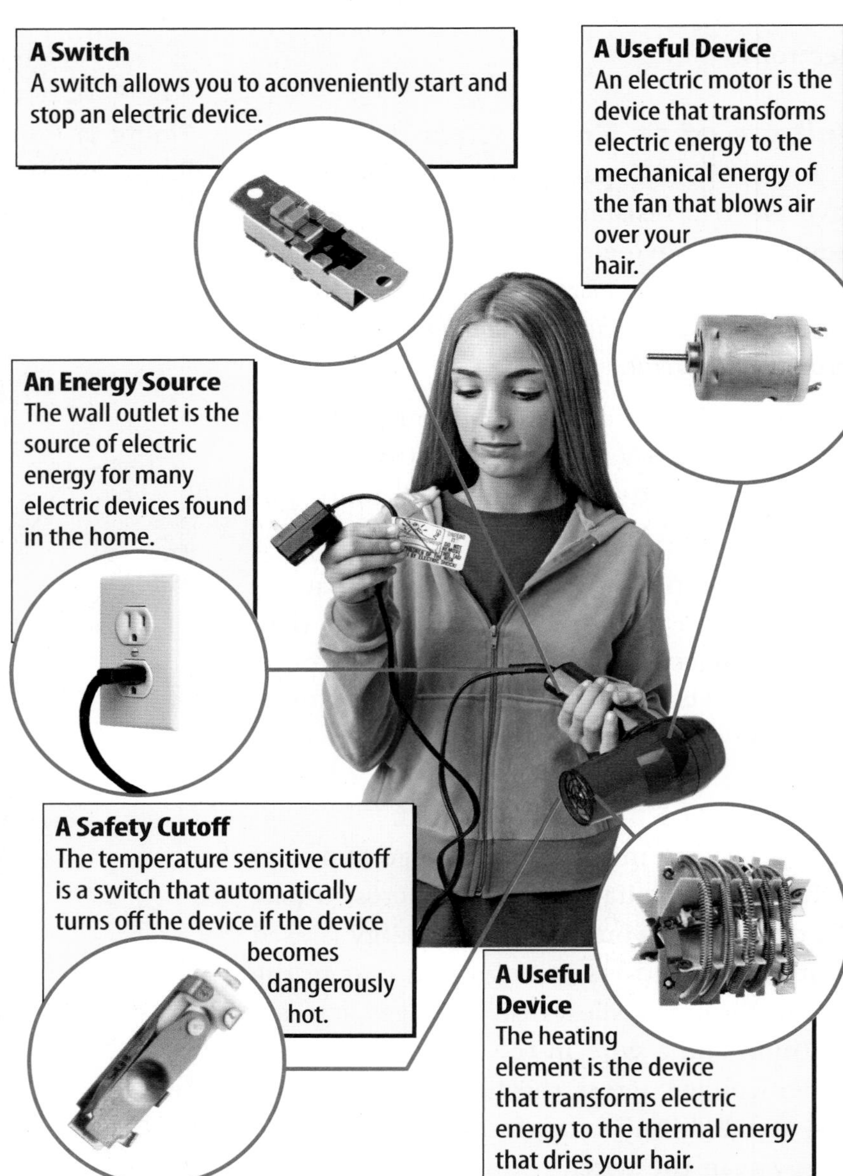

A Switch
A switch allows you to aconveniently start and stop an electric device.
A Useful Device
An electric motor is the device that transforms electric energy to the mechanical energy of the fan that blows air over your hair.
An Energy Source
The wall outlet is the source of electric energy for many electric devices found in the home.
A Safety Cutoff
The temperature sensitive cutoff is a switch that automatically turns off the device if the device becomes dangerously hot.
A Useful Device
The heating element is the device that transforms electric energy to the thermal energy that dries your hair.

Visual Check What is the function of a safety cutoff?

Lesson 2 Review

Visual Summary

A series circuit is one of two types of electric circuit.

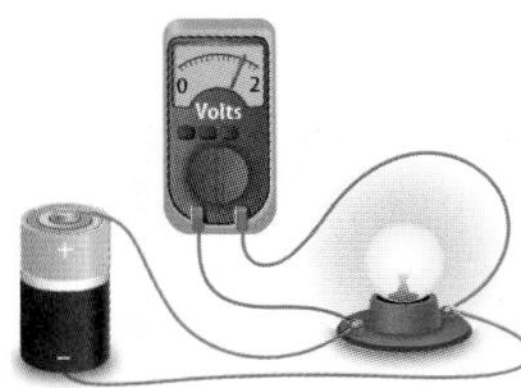

Voltage is related to the amount of electric energy transformed in a circuit.

A switch makes a simple circuit useful.

FOLDABLES®

Use your lesson Foldables to review the lesson. Save your Foldables for the project at the end of the chapter.

What do you think NOW?

You first read the statements below at the beginning of the chapter.

3. A battery in an electric circuit produces an electric current.

4. Every magnet has one magnetic pole.

Did you change your mind about whether you agree or disagree with the statements? Rewrite any false statements to make them true.

Use Vocabulary

1 **Distinguish** between electric resistance and voltage.

2 **Make up** a sentence using the terms *electric circuit* and *electric current.*

Understand Key Concepts 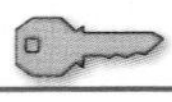

3 **Summarize** how the two types of electric circuits differ.

4 **List** the basic parts of a simple circuit.

5 An electric current is the movement of

A. atoms.
B. charged particles.
C. neutral particles.
D. neutrons.

Interpret Graphics

6 **Determine** In the circuit below, which switch will turn off only lights 2 and 3?

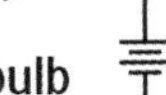

Battery
Lightbulb
Switch
Wire

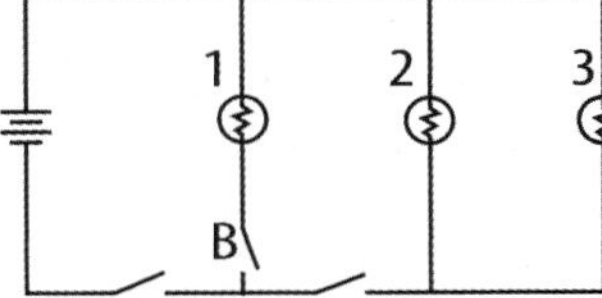

7 **Compare and Contrast** Copy and fill in the graphic organizer below. Compare and contrast the two types of electric current.

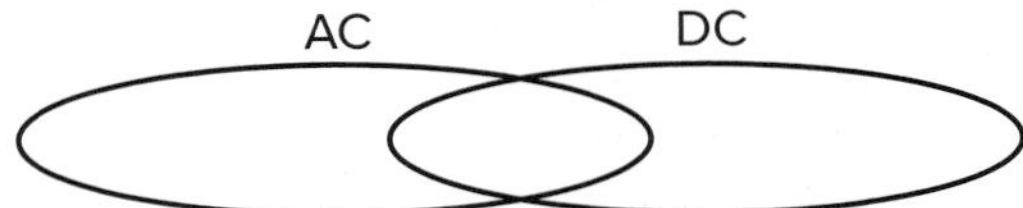

Critical Thinking

8 **Contrast** How might the circuits of a 6-V flashlight and a 1.5-V flashlight differ? Explain your reasoning.

Math Skills 

Math Practice

9 A string of ten holiday lights connected as a series circuit is plugged into to a 120-V outlet. All the lightbulbs are identical and are lit. What is the voltage across each bulb?

SCIENCE & SOCIETY

A Smart Grid?

Electric Energy for the Future

The North American power grid is a system of interconnected electric transmission wires that reaches across the continent. This network of transmission wires includes smaller, regional grids in the eastern United States, the western United States, and Texas. The grid is the electric super highway that delivers electric energy to all our communities.

The grid, shown as red lines on the map, is aging quickly. Many transmission wires are too small to carry all the electric energy people demand. Parts of the grid often are overloaded. As electric current becomes too great in one part of the grid, that part shuts down to prevent damage to generators and transmission wires. The electric current then shifts to other transmission wires that become overloaded, too. This type of cascading overload can cause power failures and blackouts over large areas of the country. One solution to our electric distribution problem is to build a smart grid, shown as green lines on the map.

Computers at distribution centers throughout the grid would constantly analyze electric energy needs across the country. The system could route electric power from where it is produced, anywhere in the country, to where it is needed.

Also, consumers would have smart meters at their homes. Each smart meter would be connected to a personal computer to allow homeowners to see how much energy each of their household's electrical devices uses. They quickly could see where they unwisely use electric energy. People could adjust their use of electric energy to save money and decrease their demands on the grid.

It's Your Turn

RESEARCH AND REPORT Many experts agree that we must soon build a smart grid for electric energy distribution. Research why a smart grid is necessary for the development of alternative energy sources.

Lesson 3

Reading Guide

Key Concepts
ESSENTIAL QUESTIONS

- What causes a magnetic force?
- How are magnets and magnetic domains related?
- How are electric currents and magnetic fields related?

Vocabulary

magnet p. 506
magnetic material p. 506
magnetic force p. 507
magnetic domain p. 509
electromagnet p. 512

Multilingual eGlossary

BrainPOP®

Magnetism

Go to the resource tab in ConnectED to find the PBL *The Great Metal Pick-Up Machine*.

Inquiry Is it science fiction?

Can a train travel at almost 600 km/h? Could it run on just magnetic forces, with no wheels and no engine, and create no pollution? Magnetic Levitation (Maglev) trains, such as the one shown above, are becoming a reality around the world. Airplanes revolutionized twentieth century transportation. Will these high-speed trains do the same in this century?

Launch Lab

10 minutes

What is magnetic?

For thousands of years, people have recognized that some rocks attract others. The word *magnet* comes from the area of ancient Greece called Magnesia where magnetic rocks could be found. Now it is your turn! With what types of objects do magnets interact?

1. Read and complete a lab safety form.
2. Count the number of **paper clips** your **magnet** will pick up. Try both ends. Record your observations in your Science Journal.
3. Cover the end of the magnet with a **penny,** a **nickel,** a **craft stick,** and **two items of your choosing.** Test the number of paper clips the magnet will pick up each time. Record your observations in a data table.

Think About This

1. How are the two ends of a magnet the same?
2. 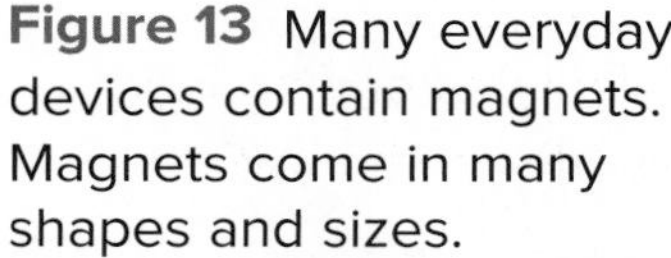 **Key Concept** What types of materials are magnetic?

Figure 13 Many everyday devices contain magnets. Magnets come in many shapes and sizes.

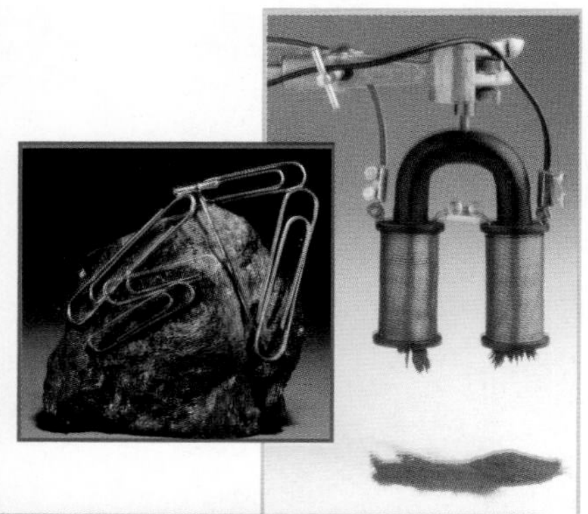

What is a magnet?

How many magnets could you find in your house? You might think of the magnets holding notes and papers to a refrigerator. However, many magnets are not so obvious. For example, almost every item in your home entertainment center, including the television, the DVD player, and the computer, use magnets. Refrigerators, vacuum cleaners, and telephones use magnets too. ATM cards and credit cards use magnetized strips to hold personal information. **Figure 13** shows that magnets often are found in factories, scientific laboratories, and even in nature. What is a magnet?

If you use magnets, you might know that magnets attract some objects, such as paper clips, but not other objects, such as pieces of paper. *A* **magnet** *is an object that attracts iron and other materials that have magnetic qualities similar to iron.* A magnet attracts paper clips and some nails because they contain iron. Magnets also attract other metals, such as nickel, cobalt, and alnico, which is an aluminum-nickel-cobalt alloy. *Any material that a magnet attracts is a* **magnetic material.**

Reading Check Why is the metal cobalt a magnetic material?

Magnetic Fields and Magnetic Forces

Recall from Lesson 1 that an invisible electric field surrounds an electrically charged object. In the same way, an invisible magnetic field surrounds a magnet and electric current. Even though magnetic fields are invisible, they can be detected by the forces they apply. *A* **magnetic force** *is a push or a pull a magnetic field applies to either a magnetic material or an electric current.* First, you will read about magnetic forces applied to magnetic materials. Later in this lesson, you will read about magnetic forces applied to electric currents.

WORD ORIGIN

magnetic
from Greek *magnēs,* means "stone from Magnesia," ancient city in Asia Minor

Seeing a Magnetic Field

A magnet's magnetic field applies a magnetic force to a magnetic material even if the magnet and magnetic material do not touch. A magnetic field and its force are stronger closer to the magnet and weaker farther away from the magnet.

Figure 14 helps you visualize a magnetic field. Because iron is a magnetic material, if you sprinkle iron filings around a magnet, they will line up with the magnet's magnetic field. These curved lines are called magnetic field lines.

FOLDABLES®

Create a horizontal two-tab book. Label it as shown. Use it to describe and to collect examples of magnetic and non-magnetic materials.

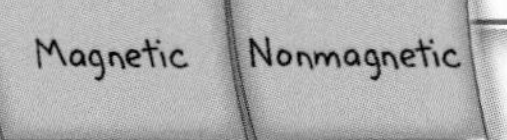

Magnetic Poles

Magnets are made in many sizes and shapes. However, all magnets have something in common–every magnet has two magnetic poles. One pole of a magnet is called the magnetic north pole. The other pole is called the magnetic south pole. The magnetic poles are the two places on a magnet where the magnetic field lines are closest together. This is also where the magnetic field applies the strongest force. Magnetic field lines point away from the magnet's magnetic north pole and toward the magnet's magnetic south pole. For a bar magnet, as shown in **Figure 14**, the ends of the magnet are the magnetic poles.

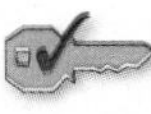

Key Concept Check What causes the forces applied by magnets?

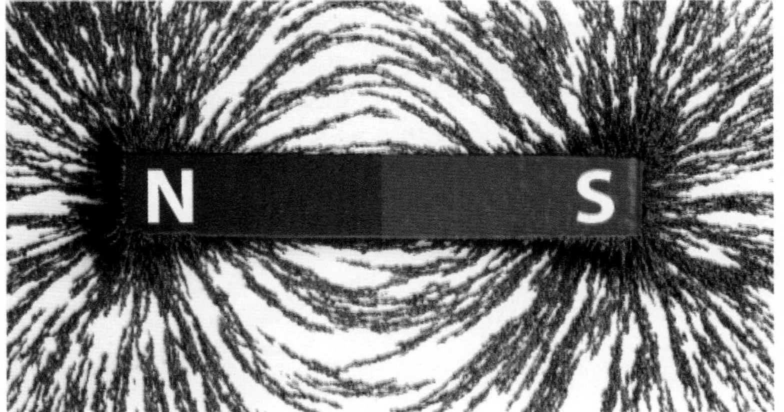

◀ **Figure 14** An invisible magnetic field can be shown with iron filings.

▲ **Figure 15** Similar, or like, magnetic poles repel; opposite magnetic poles attract.

Visual Check How do the disc magnets illustrate a magnetic attraction?

Animation

Magnetic Poles and Magnetic Forces

The forces that magnets apply to each other depend on which magnetic poles are near each other. **Figure 15** shows the interaction of the magnetic poles of several disc magnets that are near each other. If two magnetic south poles or two magnetic north poles are close to each other, the magnets repel, or push away from each other. This repulsion causes the one disc magnet to "float" on the invisible magnetic field. If a magnet's magnetic north pole is near another magnet's magnetic south pole, the magnets attract each other. In **Figure 15,** this attraction causes the magnets to come together. In other words, similar poles repel, opposite poles attract.

Earth as a Magnet

How does a magnetic compass help you find Earth's geographic North Pole? A compass needle is just a small bar magnet. As with all magnets, a magnetic field surrounds a compass needle.

Flowing molten iron and nickel in Earth's outer core create a magnetic field around Earth. Therefore, Earth has a magnetic north pole and a magnetic south pole, too. Recall that the opposite poles of two magnets attract each other. Thus, the compass needle's magnetic north pole points toward Earth's magnetic south pole, as shown in **Figure 16.** This means that Earth's magnetic south pole is actually near Earth's geographic North Pole.

Reading Check Which of Earth's geographic poles is about in the same location as Earth's magnetic south pole?

Figure 16 Earth is surrounded by a magnetic field. Earth's magnetic south pole is near Earth's geographic North Pole. ▼

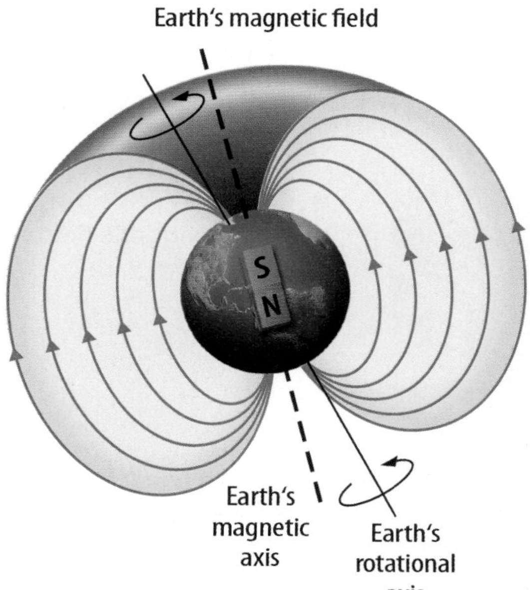

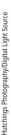

Hutchings Photography/Digital Light Source

Magnets

Why do magnets attract only some materials? Remember, all matter is made of atoms. A magnetic field surrounds each atom. In some materials, atoms are grouped in magnetic domains. *A* **magnetic domain** *is a region in a magnetic material in which the magnetic fields of the atoms all point in the same direction.* The magnetic fields of the atoms within a domain combine into a single field around the domain. Think of a magnetic domain as a tiny magnet within a material.

Nonmagnetic Materials

Most materials, including aluminum and plastic, do not have atoms grouped in magnetic domains. Part a) of **Figure 17** shows how the magnetic fields of the atoms of the plastic comb point in many different directions. The random magnetic fields cancel out the magnetic effects of each other. These nonmagnetic materials cannot be made into magnets.

Magnetic Materials

In some materials, such as iron and steel, atoms are grouped in magnetic domains. These materials are called magnetic materials. However, not all magnetic materials are magnets. As shown in part b) of **Figure 17**, the magnetic fields of the domains of the steel nail point in different directions. The magnetic fields of these domains cancel out the magnetic effects of each other. Here, the magnetic material is not a magnet.

A magnetic material becomes a magnet as the magnetic fields of the material's magnetic domains line up to point in the same direction. Part c) of **Figure 17** shows the aligned magnetic fields of the magnetic domains of a bar magnet. The magnetic fields of the domains combine to form a single magnetic field around the entire object. In this case, the magnetic material is a magnet.

 **Key Concept Check** How are magnets and magnetic domains related?

Magnetic Domains

Figure 17 Atoms of magnetic materials are grouped in magnetic domains.

Animation

a)

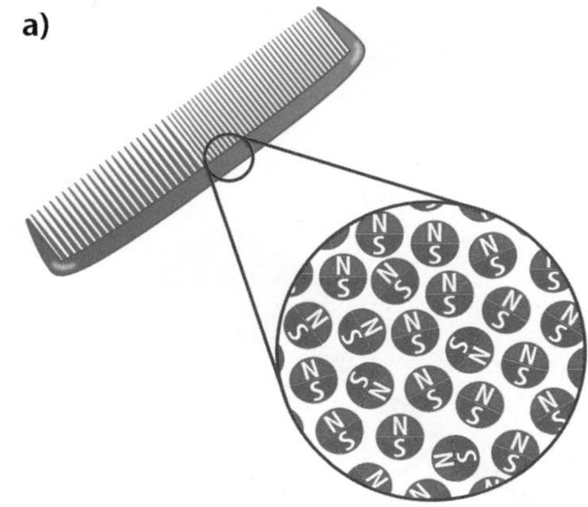

b)

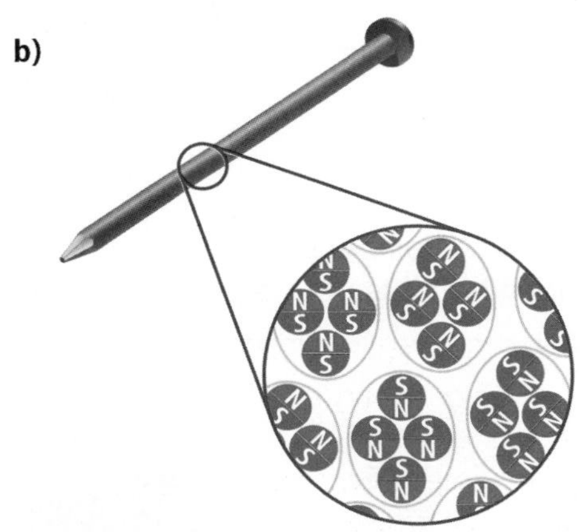

c)

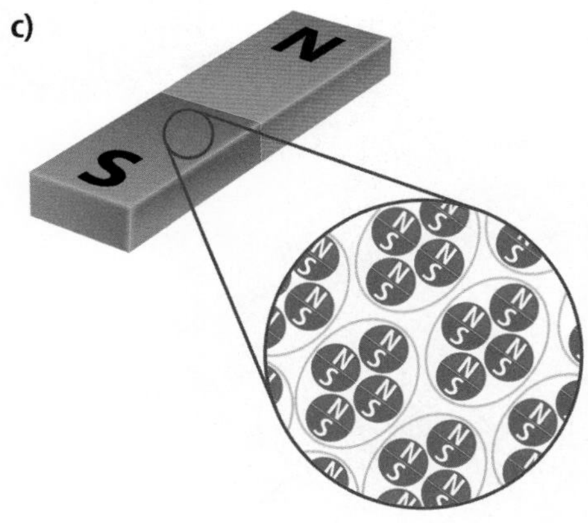

Visual Check How are magnetic domains and nonmagnetic materials related?

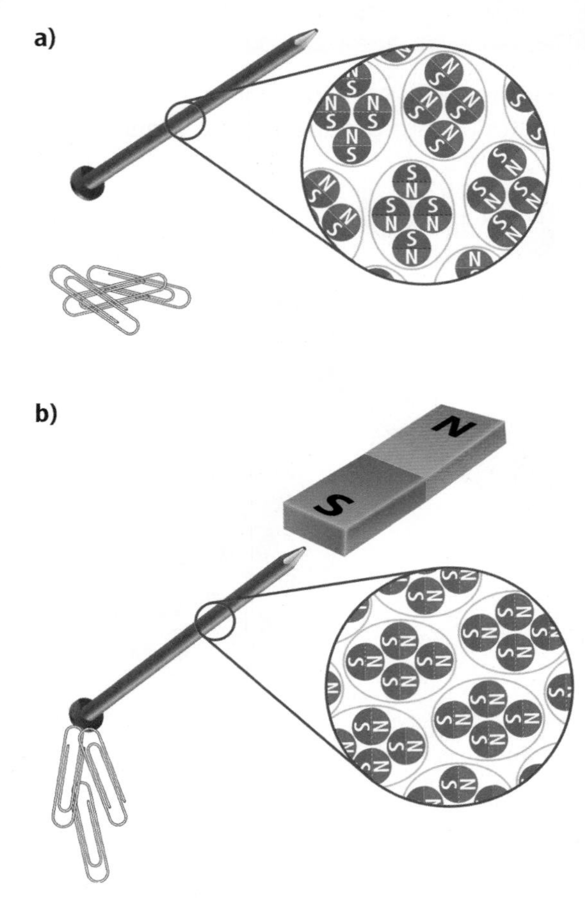

Figure 18 A nail becomes a temporary magnet when it is near a permanent magnet.

▲ Figure 19 Lodestone is a natural permanent magnet. A permanent magnet can be made in a laboratory with an electric magnetizer.

Temporary and Permanent Magnets

Some magnetic materials lose their magnetic fields quickly. Others keep their magnetic fields for a long time. How long a magnet remains a magnet depends partially on the material from which it is made. A soft magnetic material is not soft to the touch. It is called soft because it quickly loses its magnetic field. A material that keeps its magnetic field for long periods of time is called a hard magnetic material.

Temporary Magnets Placing a soft magnetic material, such as iron, in a strong magnetic field causes the material's magnetic domains to line up. This makes the material a magnet. When the object is moved away from the magnetic field, its domains return to their random positions, and the material is no longer a magnet. In part a) of **Figure 18,** the nail is not a magnet. However, in part b), the nail is a magnet. This is because the bar magnet's field causes the magnetic fields of the nail's domains to line up. Thus, the nail becomes a magnet. The nail is a temporary magnet because it attracts other magnetic materials only as long as it is within the magnetic field of another magnet.

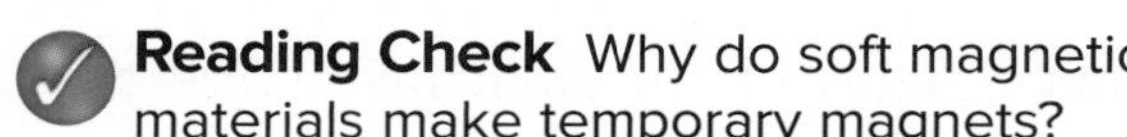

Reading Check Why do soft magnetic materials make temporary magnets?

Permanent Magnets Hard magnetic materials are mixtures of iron, nickel, and cobalt combined with other elements. When placed in an extremely strong magnetic field, the magnetic domains of a hard magnetic material align and lock into place. Unlike a temporary magnet, when a magnet made this way is removed from the strong magnetic field, the object remains a magnet permanently. A naturally occurring permanent magnet called lodestone is found in Earth's crust. Other permanent magnets can be made with electric devices called magnetizers, as shown in **Figure 19.**

Combining Electricity and Magnetism

In 1820, the Danish scientist Hans Christian Ørsted noticed that a compass needle moved when a nearby electric current was switched on. He was convinced there was a relationship between electricity and magnetism. Today, we call this relationship electromagnetism. Almost all the electrical devices in your home–anything that uses an electric motor–depend on electromagnetism.

Magnetic fields produce electric currents.

Recall that a generator is a machine that produces an electric current. Figure 20 shows how you can make a simple generator. All you need is a small wire coil connected in a circuit and a magnet. If you move the magnet through the center of the coil, the magnet's magnetic field moves over the loops of the coil. As the magnetic field moves over the coil, this forces an electric current to flow through the circuit. If the magnet stops moving, the current stops, too.

More complex generators use wire coils with more loops and stronger magnets that rotate in place. The students to the left in Figure 21 are using a hand-cranked generator. Turning the crank rotates a magnet within a small wire coil. This produces enough electric current for them to complete their experiment. However, huge generators, such as the one shown to the right of Figure 21, use coils with several kilometers of wire and giant magnets to produce the electric current that is supplied to homes, buildings, and cities.

 **Key Concept Check** How do electric currents and magnetic fields interact?

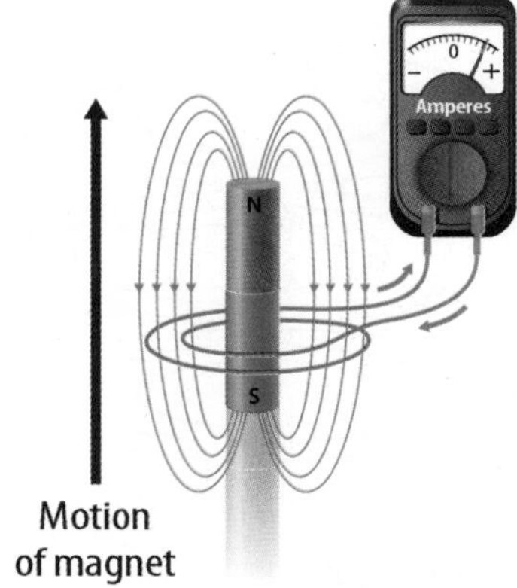

▲ **Figure 20**  A magnetic field moving over a wire coil produces a current in a circuit.

Animation

Figure 21 An electric generator uses a magnet and wire coil to produce an electric current.

 Personal Tutor

(l) Hutchings Photography/Digital Light Source, (r) Harris Shiffman/Getty Images

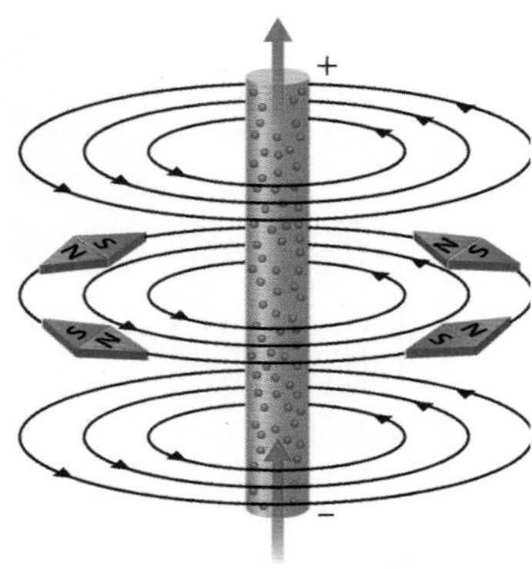

Figure 22 The magnetic field around the current-carrying wire is shown as a series of circles.

Electric currents produce magnetic fields.

You read that some magnetic materials become temporary magnets when placed in the magnetic field of another magnet. There is another type of temporary magnet that is very common and useful.

Hans Ørsted discovered that a magnetic field surrounds a current-carrying wire, as shown in **Figure 22.** If a current-carrying wire is wound into a coil, the magnetic field becomes stronger. If you place a soft magnetic material within the coil, the magnetic field becomes even stronger. *A temporary magnet made with a current-carrying wire coil wrapped around a magnetic core is an* **electromagnet.**

Electromagnets are useful because they can be controlled in ways other magnets cannot. First, an electromagnet's magnetic field can be turned off and on. Turning off the electric current in the coil turns off the magnetic field. Second, the north and south poles of the electromagnet reverse if the current reverses. And finally, the strength of an electromagnet can be controlled with the number of loops in the coil and the amount of electric current in the coil.

MiniLab

20 minutes

What determines the strength of an electromagnet?

Electromagnets do things that permanent magnets cannot. They can be turned off and on, and their strength can be changed. As a result, many modern electrical devices use electromagnets.

1. Read and complete a lab safety form.
2. Wrap half of 150 cm of **enamel-coated magnet wire** around a **straw.** Leave a short tail at each end of the wire.
3. How many **paper clips** can the coil pick up? How does it interact with another magnet? Record your observations in your Science Journal.
4. With **sandpaper,** scrape the insulation off the ends of the wire.
5. Connect the wire to a **D-cell battery.**
6. Now, how many paper clips can the coil pick up? How does it interact with another magnet? Disconnect the battery.
7. Write a plan to test the effect of the following on your electromagnet: amount of current in the coil, number of loops in the coil, and the direction in which the battery is connected. Record your plan.

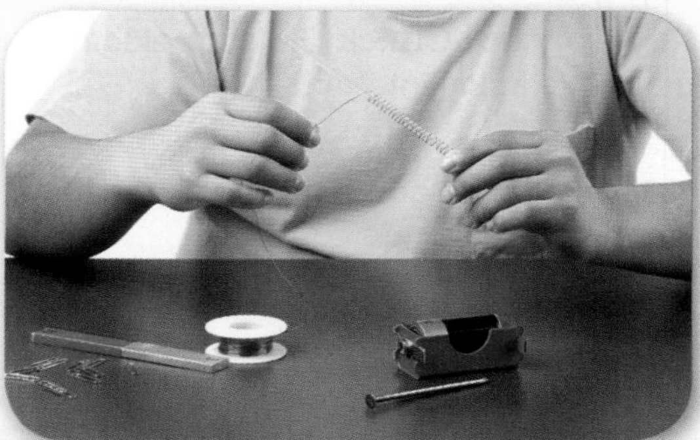

8. When your teacher has approved your plan, conduct your tests.

Analyze and Conclude

1. Critique your plans for testing your electromagnet.
2. **Key Concept** Explain the relationship between electric current and the strength of the magnetic field.

Hutchings Photography/Digital Light Source

Lesson 3 Review

Visual Summary

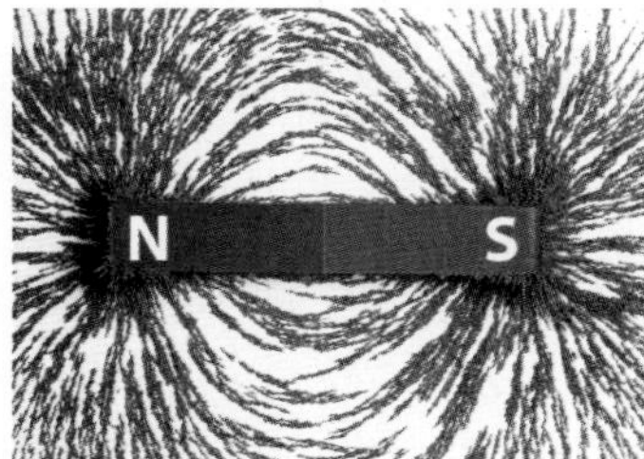

An invisible magnetic field can be shown with iron filings.

An electromagnet is a current-carrying wire coil wrapped around a magnetic core.

Magnets occur in nature. They also can be made from magnetic materials.

FOLDABLES®

Use your lesson Foldable to review the lesson. Save your Foldable for the project at the end of the chapter.

What do you think NOW?

You first read the statements below at the beginning of the chapter.

5. Earth is magnetic but is not a magnet.

6. A magnet moving within a wire loop produces an electric current.

Did you change your mind about whether you agree or disagree with the statements? Rewrite any false statements to make them true.

Use Vocabulary

1. An object that attracts iron is a(n) ________.
2. **Distinguish** between magnetic materials and nonmagnetic materials.
3. **State** in a sentence the relationship between an electric current and an electromagnet.

Understand Key Concepts 

4. **Explain** what causes a magnetic force.
5. **State** the relationship between magnetic domains and magnetic materials.
6. An electric current produces
   - **A.** a magnetic field.
   - **B.** electric charges.
   - **C.** magnetic domains.
   - **D.** magnetic materials.

Interpret Graphics

7. **Organize** Copy and fill in the graphic organizer below to compare and contrast temporary and permanent magnets.

8. **Describe** the strength of the magnetic field at points A, B, and C in the image below. Explain your answer in terms of the magnetic field lines.

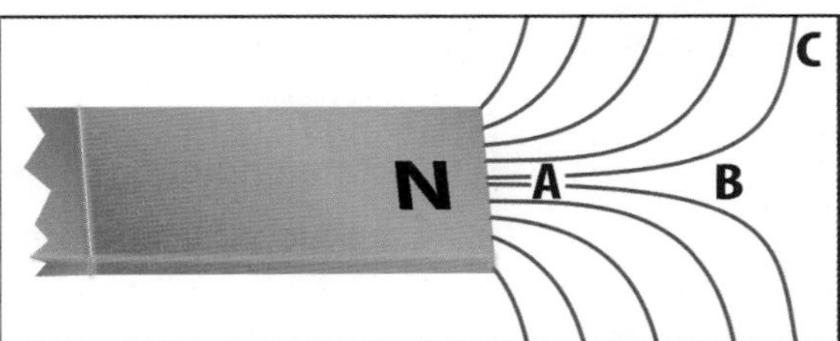

Critical Thinking

9. **Infer** why soft magnetic materials are used as cores in electromagnets.

Lab

2 class periods

How can you control the speed of an electric motor?

Materials

large paper clips

fine sand paper

alligator clip wires

D-cell battery in holder

foam board

pushpins

magnet wire

strong magnet

Safety

Electric motors are everywhere! They convert electric energy to motion. In most electric motors, magnetic forces between the fields of a permanent magnet and a rotating electromagnet make the electromagnet turn. Fans, hair dryers, and many battery-operated toys use electric motors. What other examples can you think of?

Question

How do the different parts of an electric motor affect its performance?

Procedure

1. Read and complete a lab safety form.
2. Wrap the magnet wire around a D-cell battery. Leave 5-cm tails at each end of the coil. Remove the coil from the battery.
3. Wrap the tails once around the coil so that the coil is held together and the wires stick straight out perpendicular to the coil.
4. Lay the coil flat on the table. Using sand paper, scrape the insulation off the visible side of the tails. Flip the coil over and scrape the insulation off one of the two tails.
5. Unfold each paper clip to form two S shapes. Using pushpins, attach the paper clips to the board, as shown in the photograph on the next page.
6. Place the magnet on the board, between the paper clips.
7. Suspend the coil by its tails in the hooks formed in the paper clips.
8. Using alligator clip wires, connect the paper clips to the terminals of the D-cell's holder. Give the coil a twist, and watch it spin rapidly.

2

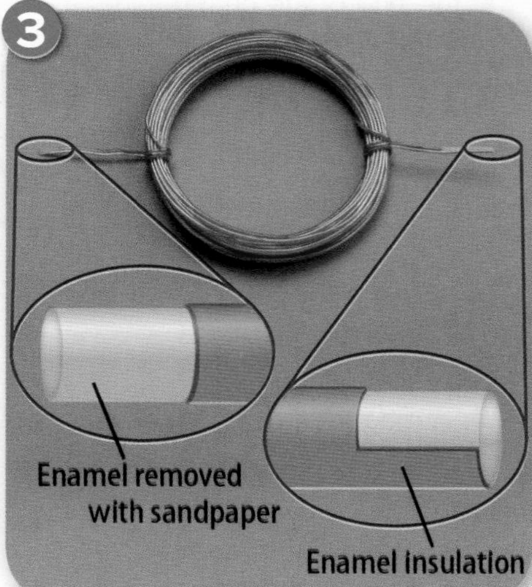

3

(tr, br)Hutchings Photography/ Digital Light Source, (t to b) Jacques Cornell/McGraw-Hill Education, (2, 5)Hutchings Photography/Digital Light Source, (3,7,8) McGraw-Hill Education, (4) Ken Cavanagh/McGraw-Hill Education, (6) Amos Morgan/Getty Images

Form a Hypothesis

9 After observing the behavior of your motor, formulate a hypothesis about how you could alter the speed of your motor.

Test Your Hypothesis

10 Develop a plan to alter the speed of your motor. Record your plans in your Science Journal.

11 Have your teacher approve your plans and procedure.

12 Test your hypothesis and record your results.

Analyze and Conclude

13 **Create** a flow chart showing the transfer of energy in this system.

14 **Explain** how you altered the speed of the motor.

15 **The Big Idea** How are moving electric charges and magnetic fields related in a motor?

Communicate Your Results

Imagine you are an entrepreneur living in the time shortly after electric motors were invented. You want people to give you money so you can invent and sell a device that contains an electric motor. Create a brochure that explains to your investors how motors work and what you plan to do with motors.

There are a number of factors that might be important in the design of an electric motor. For example, does it matter if the coil is round? How does the number of loops affect the motor? Will an electromagnet work in place of a permanent magnet? Write a short report that lists some possible variables that could affect the performance of an electric motor. Include your predictions of the effects of changing those variables.

8

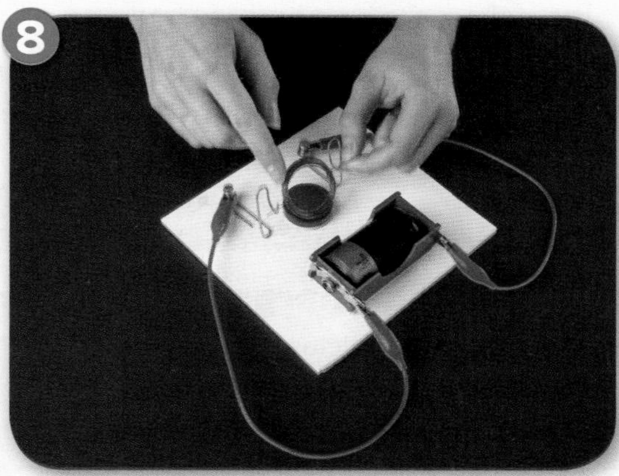

Lab Tips

- Hold the tails and spin the coil between your fingers. It should spin easily and not feel lopsided.
- Do not scrape the insulation off the coil!
- If your motor does not spin, be sure that the insulation has been scraped completely off one of the coil tails and half off the other tail.
- If your motor spins haphazardly, be sure that the tails are centered on the coil.

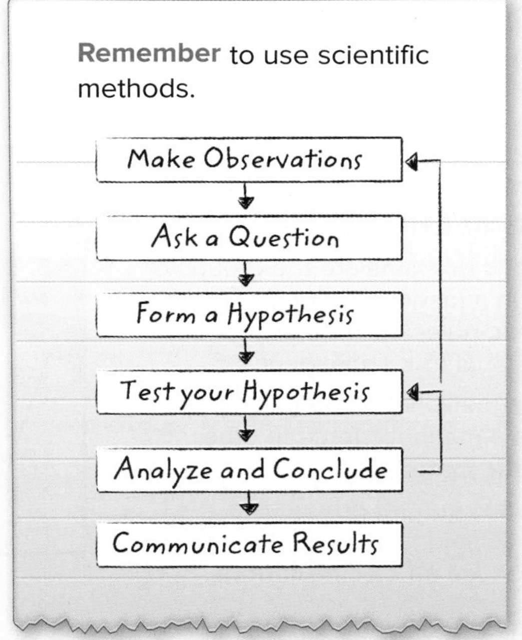

Hutchings Photography/Digital Light Source

Chapter 15 Study Guide

Electromagnetism is the term that describes the relationship between electricity and magnetism. An electric current produces a magnetic field, and a magnet's magnetic field can cause an electric current.

Key Concepts Summary

Lesson 1: Electric Charges and Electric Forces

- **Electrically charged** particles can be negatively charged or positively charged.
- When different materials touch, negatively charged electrons will move from one of the objects to the other.
- Electrically charged objects apply an **electric force** to each other. Objects with similar charges repel each other and objects with opposite charges attract each other.

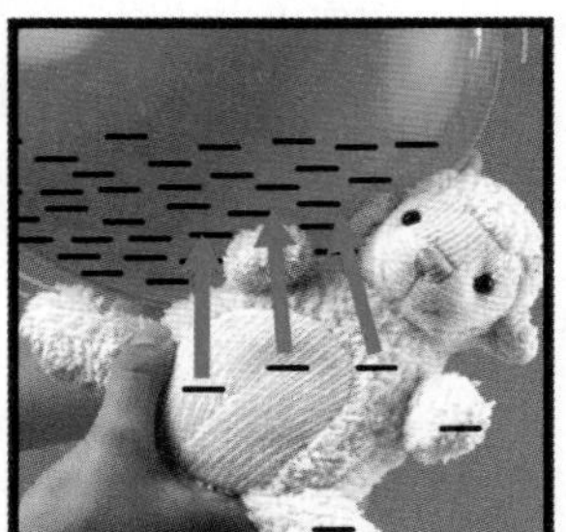

Vocabulary

electrically neutral p. 488
electrically charged p. 488
electric discharge p. 490
electric insulator p. 490
electric conductor p. 490
electric force p. 491
electric field p. 491

Lesson 2: Electric Current and Electric Circuits

- An **electric current** is the flow of electric charges.
- Most **electric circuits** contain the following basic components: a source of electric energy, a useful device, a closed path, a switch, and a safety cutoff.
- A series circuit has only one conducting path for all devices in the circuit. A parallel circuit has a separate path, or branch, for each device.

Vocabulary

electric current p. 495
electric circuit p. 497
generator p. 498
electric resistance p. 499
voltage p. 501

Lesson 3: Magnetism

- **Magnetic domains** are the groups of atoms in a magnetic material whose magnetic poles must be aligned for the material to be a magnet.
- The magnetic field around a magnet applies a magnetic force to other **magnetic materials.**
- An electric current produces a magnetic field, and a magnetic field can produce an electric current.

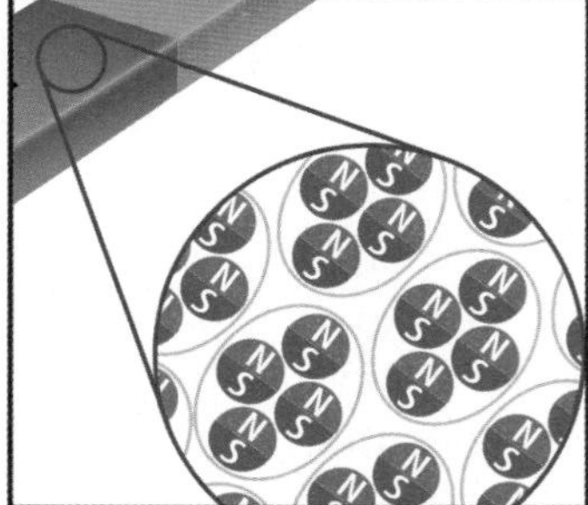

Vocabulary

magnet p. 506
magnetic material p. 506
magnetic force p. 507
magnetic domain p. 509
electromagnet p. 512

Study Guide

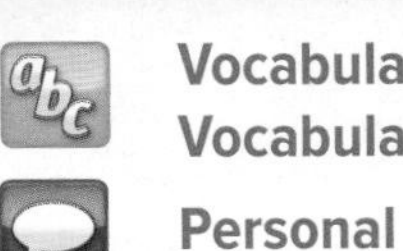

Vocabulary eFlashcards
Vocabulary eGames
Personal Tutor

FOLDABLES® Chapter Project

Assemble your lesson Foldables as shown to make a Chapter Project. Use the project to review what you have learned in this chapter.

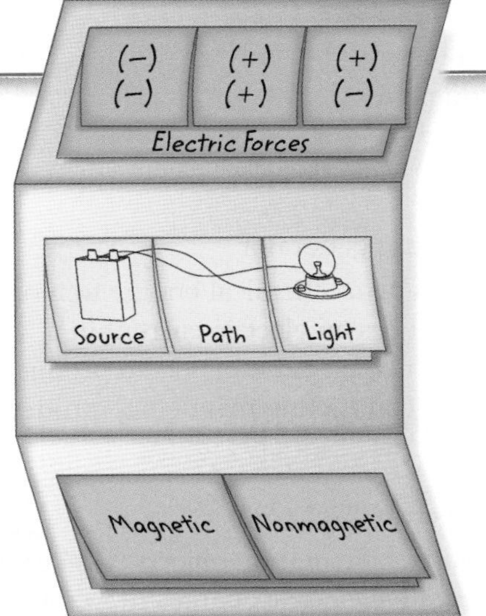

Use Vocabulary

1. A(n) ________ object has balanced positive and negative charges.
2. Define the term *electric insulator* in your own words.
3. Distinguish between electrically charged and electrically neutral.
4. A measure of the energy transferred by the flow of one coulomb of electrons in a circuit is ________.
5. Use the terms *generator* and *magnet* in one complete sentence.
6. Describe the effect of electric resistance on an electric current.
7. A closed path in which electric charges can flow is a(n) ________.
8. Explain two ways an electromagnet can be controlled.

Interactive Concept Map

Link Vocabulary and Key Concepts

Copy this concept map, and then use vocabulary terms from the previous page and other terms from the chapter to complete the concept map.

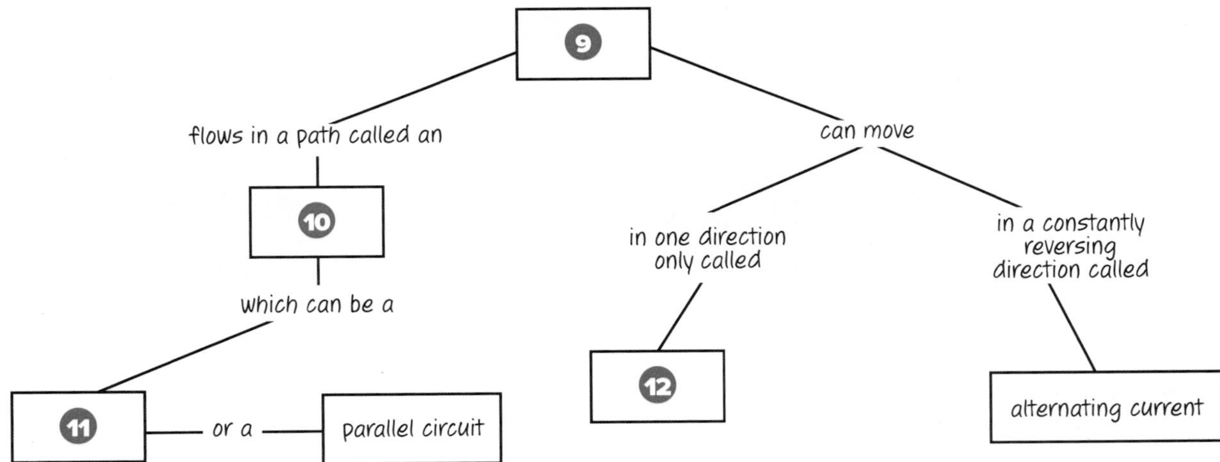

Chapter 15 Review

Understand Key Concepts 

1. Which is a measure of the electric energy a coulomb of electric charge transforms by flowing through a circuit?
   A. voltage
   B. resistance
   C. electric force
   D. electric current

2. When the switch in a circuit opens, which of the following stops?
   A. current
   B. resistance
   C. static charge
   D. total charge

3. A magnet sticks to a refrigerator door. Therefore, the door must be
   A. a magnet.
   B. electrically charged.
   C. made of a magnetic material.
   D. probably not electrically grounded.

4. An electric current
   A. flows easily in an insulator.
   B. flows through an open switch.
   C. is produced by a generator.
   D. is produced by an electric motor.

5. The picture below shows the magnetic domains of a(n)

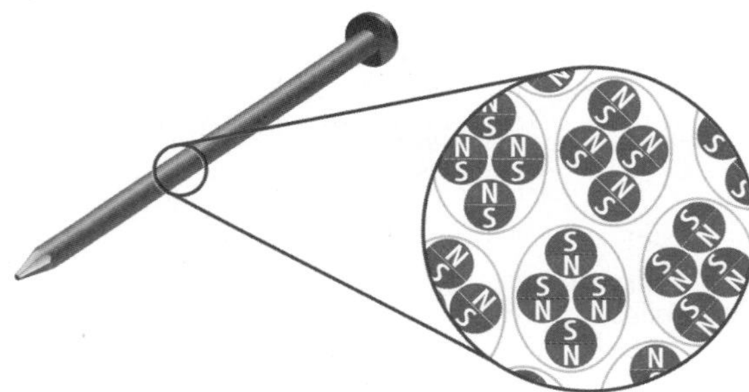

   A. insulator.
   B. magnet.
   C. magnetic material.
   D. nonmagnetic material.

6. In the diagram below, the arrow points to the electromagnet's

   A. coil.
   B. domain.
   C. hard magnetic core.
   D. soft magnetic core.

7. An electric generator
   A. transforms chemical energy to motion.
   B. produces an electric current in a wire coil.
   C. uses two electromagnets to produce motion.
   D. uses conducting magnets to produce a current.

8. An electric discharge occurs as
   A. electrically neutral objects repel each other.
   B. negative electric charges move onto a negatively charged object.
   C. positive electric charges move onto a positively charged object.
   D. unbalanced electric charges become balanced.

9. Which lightbulb(s) in the diagram below will remain lit if the wire breaks at point A?

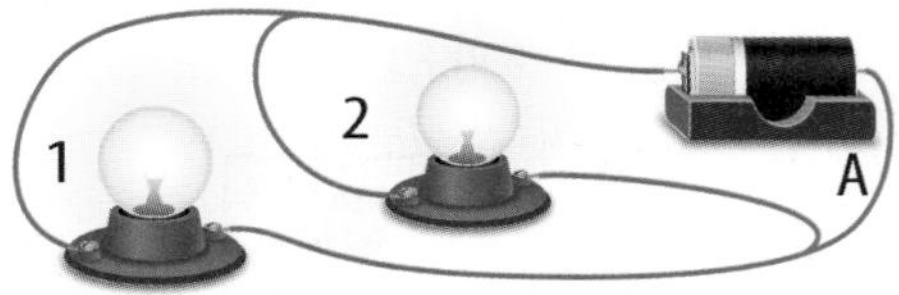

   A. both
   B. bulb 1 only
   C. bulb 2 only
   D. neither

Online Test Practice

Critical Thinking

10 **Create** Design a graphic organizer that compares and contrasts electric charges and magnetic poles.

11 **Evaluate** Dry air is a better electric insulator than humid air. Would the electric discharge from a charged balloon happen more slowly in dry air or humid air? Explain your answer.

12 **Modify** How could you change the electric circuit shown below to allow lightbulb A to stay lit even if lightbulb B is removed from its base?

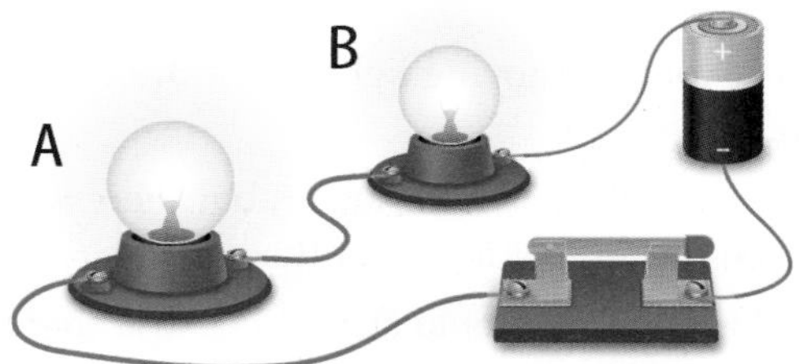

13 **Hypothesize** Both soft magnetic materials and hard magnetic materials are hard to the touch. What then is the difference between these two types of materials?

14 **Solve** Suppose all the lights in a room go out when you plug an electric heater into a wall outlet. What would you do to get the lights back on? Explain your thinking.

15 **Assess** Suppose lightbulb A and lightbulb B are connected in a series circuit. The voltage across lightbulb A is greater than the voltage across lightbulb B. Which lightbulb would you expect to be brighter? Explain your answer.

Writing in Science

16 **Write** Picture yourself as an electric charge flowing through an electric circuit. Write a three-paragraph story describing your trip through the entire circuit. Use as many Lesson 2 vocabulary words as possible.

REVIEW THE BIG IDEA

17 How are electricity and magnetism related?

18 How are electricity and magnetism used together in this sports car?

Math Skills

Using Fractions

19 Four identical lightbulbs are connected in series to a 30-V battery. What is the voltage across each lamp?

20 An electric motor and a lightbulb are connected in a series circuit that is plugged into a 120-V wall outlet. The voltage across the motor is 100 V.

- **a.** What is the voltage across the lightbulb?
- **b.** What fraction of the energy from the wall outlet transforms in the electric motor?
- **c.** What fraction of the energy coming from the wall outlet transforms in the lightbulb?

21 Three lights connected to a battery in a series circuit have voltage of 3 V, 4 V, and 5 V.

- **a.** What is the voltage of the battery?
- **b.** What portion of the battery's energy is transformed in each of the lights?

Standardized Test Practice

Record your answers on the answer sheet provided by your teacher or on a sheet of paper.

Multiple Choice

1 Which statement best describes how a balloon could become positively charged?

A Positive electrons are rubbed off the balloon and onto another object.

B Negative electrons are rubbed off the balloon and onto another object.

C Positive electrons are rubbed off another object and onto the balloon.

D Negative electrons are rubbed off another object and onto the balloon.

Use the figure below to answer question 2.

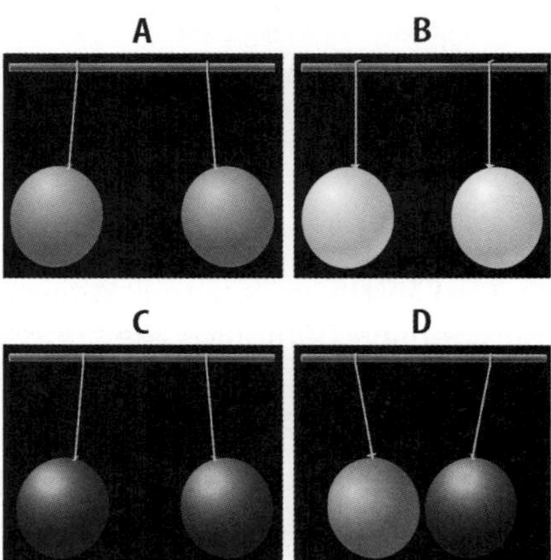

2 Which balloons have opposite charges?

A the pair in figure B

B the pair in figure D

C the pairs in figures A and C

D the pairs in figures B and C

3 Octavio pulls a sock from the clothes dryer. It is electrically charged. What must be true of the sock?

A It has lost all of its electrons.

B It can never again become electrically neutral.

C It would not interact with other charged objects.

D It has unequal amounts of positive and negative charges.

Use the figure below to answer questions 4 and 5.

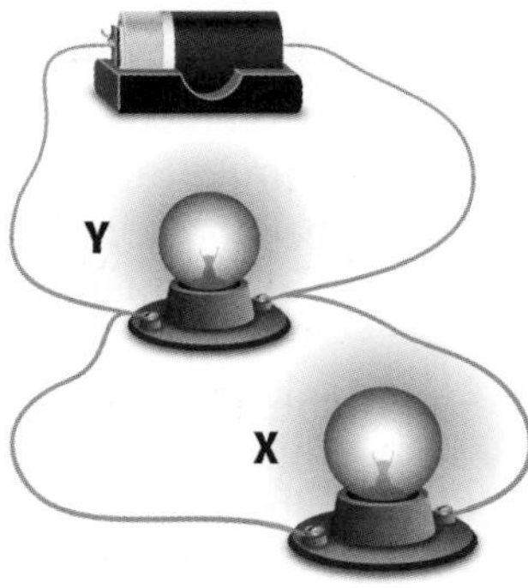

4 How would removing lightbulb X affect the circuit?

A Lightbulb Y would stay lit, but there would be no current in any of the wires.

B Lightbulb Y would stay lit because there still would be current through it.

C Lightbulb Y would go out because there would be current in the smaller loop.

D Lightbulb Y would go out because there would be no current in any of the wires.

5 Which best describes lightbulb Y?

A It is an electric insulator.

B It is a source of electric energy.

C It is a device that transforms light energy to electric energy.

D It is a device that transforms electric energy to light energy.

6 How does a battery generate electric current in a circuit?

A It moves the negative electric charges that are already in the circuit.

B It creates positive electric charges and pushes them into the circuit.

C It creates negative electric charges and pushes them into the circuit.

D It destroys positive electric charges, which it pulls from the circuit.

7 There are two lightbulbs in Emily's garage. When she turns off the light switch, only one lightbulb goes dark. The lightbulbs are connected

A as a series circuit.

B on an electrically insulated circuit.

C as a parallel circuit.

D on a circuit with a broken switch.

Use the figure below to answer question 8.

8 What kind of device is shown in the figure?

A a generator

B an electromagnet

C a parallel circuit

D alternating current

9 Zelda holds the north pole of a magnet near a compass needle. She notices that the end of the compass needle that usually points to geographic north is repelled by the magnet's north pole. What does this mean about the north-pointing end of a compass needle?

A It was never magnetic.

B It has lost its magnetic field.

C It is the north pole of a magnet.

D It is the south pole of a magnet.

Constructed Response

10 A particle called an anion forms when an electrically neutral atom gains one or more electrons. What kind of charge does an anion have? Explain your answer.

11 How do conductors and insulators differ? Which would be best for connecting the different parts of an electric circuit?

Use the figures below to answer questions 12 and 13.

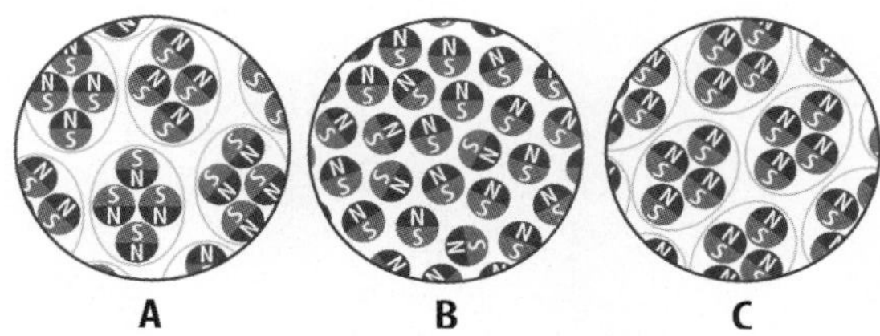

12 Identify whether each figure represents a nonmagnetic material, a magnetic material, or a magnet.

13 Under what circumstances would the material modeled by figure A temporarily become like that in figure C? Give an example of such a situation

14 Explain why a piece of iron is a magnetic material, even if it is not a magnet.

15 Most simple electric circuits contain a source of electric energy, a switch, and electric conductors. What is the function of each of these components?

NEED EXTRA HELP?															
If You Missed Question...	1	2	3	4	5	6	7	8	9	10	11	12	13	14	15
Go to Lesson...	1	1	1	2	2	2	2	3	3	1	1	3	3	3	2

Student Resources

For Students and Parents/Guardians

These resources are designed to help you achieve success in science. You will find useful information on laboratory safety, math skills, and science skills. In addition, science reference materials are found in the Reference Handbook. You'll find the information you need to learn and sharpen your skills in these resources.

Table of Contents

SCIENCE SKILL HANDBOOK
MATH SKILL HANDBOOK
FOLDABLES HANDBOOK
REFERENCE HANDBOOK
GLOSSARY/GLOSARIO
INDEX

Scientific Methods

Scientists use an orderly approach called the scientific method to solve problems. This includes organizing and recording data so others can understand them. Scientists use many variations in this method when they solve problems.

Identify a Question

The first step in a scientific investigation or experiment is to identify a question to be answered or a problem to be solved. For example, you might ask which gasoline is the most efficient.

Gather and Organize Information

After you have identified your question, begin gathering and organizing information. There are many ways to gather information, such as researching in a library, interviewing those knowledgeable about the subject, and testing and working in the laboratory and field. Fieldwork is investigations and observations done outside of a laboratory.

Researching Information Before moving in a new direction, it is important to gather the information that already is known about the subject. Start by asking yourself questions to determine exactly what you need to know. Then you will look for the information in various reference sources, like the student is doing in **Figure 1.** Some sources may include textbooks, encyclopedias, government documents, professional journals, science magazines, and the Internet. Always list the sources of your information.

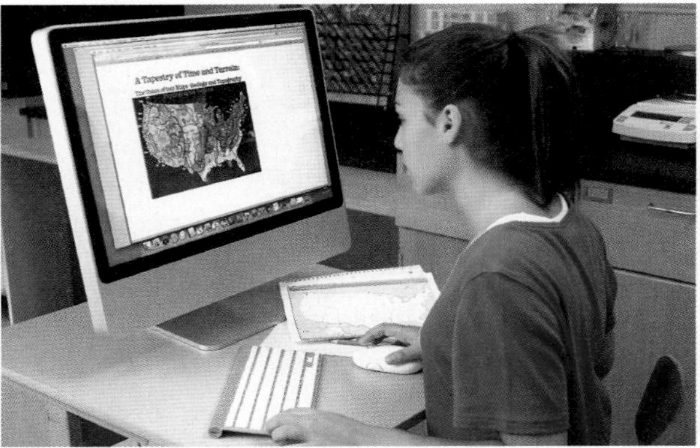

Figure 1 The Internet can be a valuable research tool.

Evaluate Sources of Information Not all sources of information are reliable. You should evaluate all of your sources of information, and use only those you know to be dependable. For example, if you are researching ways to make homes more energy efficient, a site written by the U.S. Department of Energy would be more reliable than a site written by a company that is trying to sell a new type of weatherproofing material. Also, remember that research always is changing. Consult the most current resources available to you. For example, a 1985 resource about saving energy would not reflect the most recent findings.

Sometimes scientists use data that they did not collect themselves, or conclusions drawn by other researchers. This data must be evaluated carefully. Ask questions about how the data were obtained, if the investigation was carried out properly, and if it has been duplicated exactly with the same results. Would you reach the same conclusion from the data? Only when you have confidence in the data can you believe it is true and feel comfortable using it.

Interpret Scientific Illustrations As you research a topic in science, you will see drawings, diagrams, and photographs to help you understand what you read. Some illustrations are included to help you understand an idea that you can't see easily by yourself, like the tiny particles in an atom in **Figure 2.** A drawing helps many people to remember details more easily and provides examples that clarify difficult concepts or give additional information about the topic you are studying. Most illustrations have labels or a caption to identify or to provide more information.

Nucleus
Neutron
Proton
Electrons

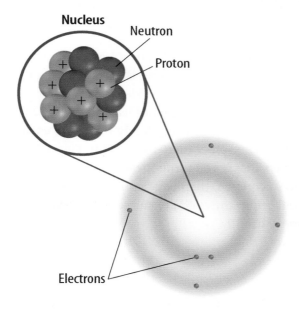

Figure 2 This drawing shows an atom of carbon with its six protons, six neutrons, and six electrons.

Concept Maps One way to organize data is to draw a diagram that shows relationships among ideas (or concepts). A concept map can help make the meanings of ideas and terms more clear, and help you understand and remember what you are studying. Concept maps are useful for breaking large concepts down into smaller parts, making learning easier.

Network Tree A type of concept map that not only shows a relationship, but how the concepts are related is a network tree, shown in **Figure 3.** In a network tree, the words are written in the ovals, while the description of the type of relationship is written across the connecting lines.

When constructing a network tree, write down the topic and all major topics on separate pieces of paper or notecards. Then arrange them in order from general to specific. Branch the related concepts from the major concept and describe the relationship on the connecting line. Continue to more specific concepts until finished.

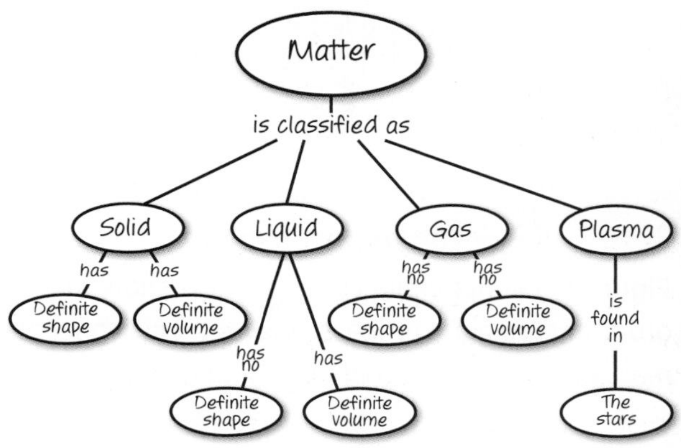

Figure 3 A network tree shows how concepts or objects are related.

Events Chain Another type of concept map is an events chain. Sometimes called a flow chart, it models the order or sequence of items. An events chain can be used to describe a sequence of events, the steps in a procedure, or the stages of a process.

When making an events chain, first find the one event that starts the chain. This event is called the initiating event. Then, find the next event and continue until the outcome is reached, as shown in **Figure 4** on the next page.

SCIENCE SKILL HANDBOOK
MATH SKILL HANDBOOK
FOLDABLES HANDBOOK
REFERENCE HANDBOOK
GLOSSARY/ GLOSARIO
INDEX

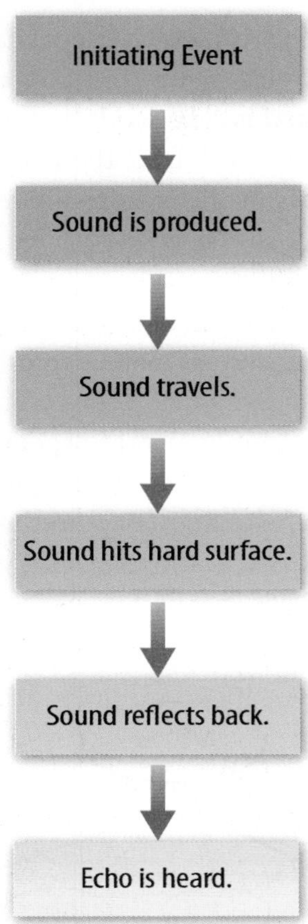

Figure 4 Events-chain concept maps show the order of steps in a process or event. This concept map shows how a sound makes an echo.

Cycle Map A specific type of events chain is a cycle map. It is used when the series of events do not produce a final outcome, but instead relate back to the beginning event, such as in **Figure 5.** Therefore, the cycle repeats itself.

To make a cycle map, first decide what event is the beginning event. This is also called the initiating event. Then list the next events in the order that they occur, with the last event relating back to the initiating event. Words can be written between the events that describe what happens from one event to the next. The number of events in a cycle map can vary, but usually contain three or more events.

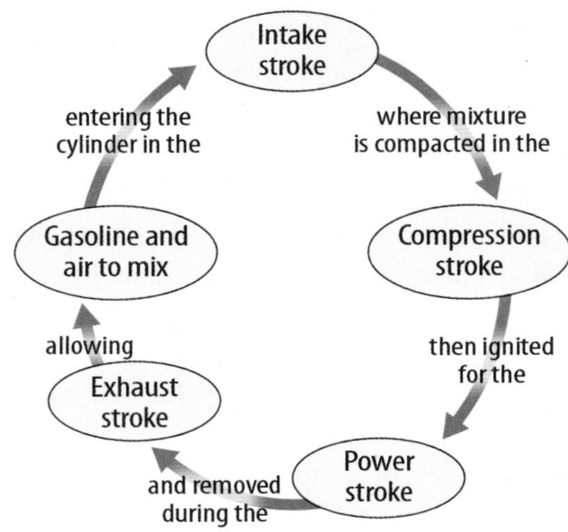

Figure 5 A cycle map shows events that occur in a cycle.

Spider Map A type of concept map that you can use for brainstorming is the spider map. When you have a central idea, you might find that you have a jumble of ideas that relate to it but are not necessarily clearly related to each other. The spider map on sound in **Figure 6** shows that if you write these ideas outside the main concept, then you can begin to separate and group unrelated terms so they become more useful.

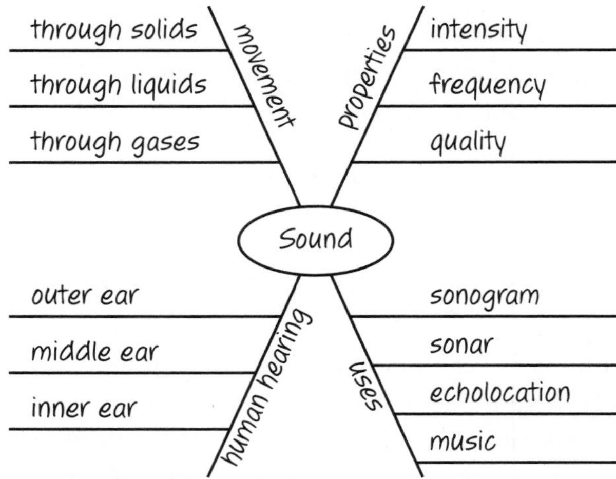

Figure 6 A spider map allows you to list ideas that relate to a central topic but not necessarily to one another.

SCIENCE SKILL HANDBOOK
MATH SKILL HANDBOOK
FOLDABLES HANDBOOK
REFERENCE HANDBOOK
GLOSSARY/GLOSARIO
INDEX

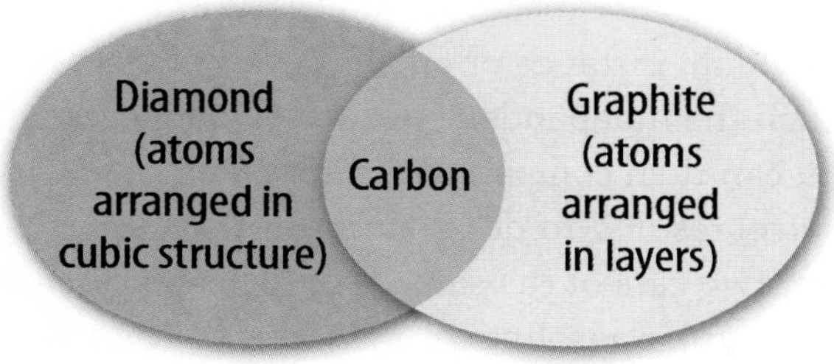

Diamond (atoms arranged in cubic structure) | Carbon | Graphite (atoms arranged in layers)

Figure 7 This Venn diagram compares and contrasts two substances made from carbon.

Venn Diagram To illustrate how two subjects compare and contrast you can use a Venn diagram. You can see the characteristics that the subjects have in common and those that they do not, shown in **Figure 7.**

To create a Venn diagram, draw two overlapping ovals that are big enough to write in. List the characteristics unique to one subject in one oval, and the characteristics of the other subject in the other oval. The characteristics in common are listed in the overlapping section.

Make and Use Tables One way to organize information so it is easier to understand is to use a table. Tables can contain numbers, words, or both.

To make a table, list the items to be compared in the first column and the characteristics to be compared in the first row. The title should clearly indicate the content of the table, and the column or row heads should be clear. Notice that in **Table 1** the units are included.

Table 1 Recyclables Collected During Week

Day of Week	Paper (kg)	Aluminum (kg)	Glass (kg)
Monday	5.0	4.0	12.0
Wednesday	4.0	1.0	10.0
Friday	2.5	2.0	10.0

Make a Model One way to help you better understand the parts of a structure, the way a process works, or to show things too large or small for viewing is to make a model. For example, an atomic model made of a plastic-ball nucleus and chenille stem electron shells can help you visualize how the parts of an atom relate to each other. Other types of models can be devised on a computer or represented by equations.

Form a Hypothesis

A possible explanation based on previous knowledge and observations is called a hypothesis. After researching gasoline types and recalling previous experiences in your family's car, you form a hypothesis—our car runs more efficiently because we use premium gasoline. To be valid, a hypothesis has to be something you can test by using an investigation.

Predict When you apply a hypothesis to a specific situation, you predict something about that situation. A prediction makes a statement in advance, based on prior observation, experience, or scientific reasoning. People use predictions to make everyday decisions. Scientists test predictions by performing investigations. Based on previous observations and experiences, you might form a prediction that cars are more efficient with premium gasoline. The prediction can be tested in an investigation.

Design an Experiment A scientist needs to make many decisions before beginning an investigation. Some of these include: how to carry out the investigation, what steps to follow, how to record the data, and how the investigation will answer the question. It also is important to address any safety concerns.

Test the Hypothesis

Now that you have formed your hypothesis, you need to test it. Using an investigation, you will make observations and collect data, or information. This data might either support or not support your hypothesis. Scientists collect and organize data as numbers and descriptions.

Follow a Procedure In order to know what materials to use, as well as how and in what order to use them, you must follow a procedure. **Figure 8** shows a procedure you might follow to test your hypothesis.

Procedure

Step 1	Use regular gasoline for two weeks.
Step 2	Record the number of kilometers between fill-ups and the amount of gasoline used.
Step 3	Switch to premium gasoline for two weeks.
Step 4	Record the number of kilometers between fill-ups and the amount of gasoline used.

Figure 8 A procedure tells you what to do step-by-step.

Identify and Manipulate Variables and Controls In any experiment, it is important to keep everything the same except for the item you are testing. The one factor you change is called the independent variable. The change that results is the dependent variable. Make sure you have only one independent variable, to assure yourself of the cause of the changes you observe in the dependent variable. For example, in your gasoline experiment the type of fuel is the independent variable. The dependent variable is the efficiency.

Many experiments also have a control—an individual instance or experimental subject for which the independent variable is not changed. You can then compare the test results to the control results. To design a control you must have two cars of the same type. The control car uses regular gasoline for four weeks. After you are done with the test, you can compare the experimental results to the control results.

Collect Data

Whether you are carrying out an investigation or a short observational experiment, you will collect data, as shown in **Figure 9**. Scientists collect data as numbers and descriptions and organize them in specific ways.

Observe Scientists observe items and events, then record what they see. When they use only words to describe an observation, it is called qualitative data. Scientists' observations also can describe how much there is of something. These observations use numbers, as well as words, in the description and are called quantitative data. For example, if a sample of the element gold is described as being "shiny and very dense" the data are qualitative. Quantitative data on this sample of gold might include "a mass of 30 g and a density of 19.3 g/cm^3."

Figure 9 Collecting data is one way to gather information directly.

Figure 10 Record data neatly and clearly so it is easy to understand.

When you make observations, you should examine the entire object or situation first, and then look carefully for details. It is important to record observations accurately and completely. Always record your notes immediately as you make them, so you do not miss details or make a mistake when recording results from memory. Never put unidentified observations on scraps of paper. Instead they should be recorded in a notebook, like the one in **Figure 10.** Write your data neatly so you can easily read it later. At each point in the experiment, record your observations and label them. That way, you will not have to determine what the figures mean when you look at your notes later. Set up any tables that you will need to use ahead of time, so you can record any observations right away. Remember to avoid bias when collecting data by not including personal thoughts when you record observations. Record only what you observe.

Estimate Scientific work also involves estimating. To estimate is to make a judgment about the size or the number of something without measuring or counting. This is important when the number or size of an object or population is too large or too difficult to accurately count or measure.

Sample Scientists may use a sample or a portion of the total number as a type of estimation. To sample is to take a small, representative portion of the objects or organisms of a population for research. By making careful observations or manipulating variables within that portion of the group, information is discovered and conclusions are drawn that might apply to the whole population. A poorly chosen sample can be unrepresentative of the whole. If you were trying to determine the rainfall in an area, it would not be best to take a rainfall sample from under a tree.

Measure You use measurements every day. Scientists also take measurements when collecting data. When taking measurements, it is important to know how to use measuring tools properly. Accuracy also is important.

Length The SI unit for length is the meter (m). Smaller measurements might be measured in centimeters or millimeters.

Length is measured using a metric ruler or meterstick. When using a metric ruler, line up the 0-cm mark with the end of the object being measured and read the number of the unit where the object ends. Look at the metric ruler shown in **Figure 11.** The centimeter lines are the long, numbered lines, and the shorter lines are millimeter lines. In this instance, the length would be 4.50 cm.

Figure 11 This metric ruler has centimeter and millimeter divisions.

Mass The SI unit for mass is the kilogram (kg). Scientists can measure mass using units formed by adding metric prefixes to the unit gram (g), such as milligram (mg). To measure mass, you might use a triple-beam balance similar to the one shown in **Figure 12.** The balance has a pan on one side and a set of beams on the other side. Each beam has a rider that slides on the beam.

When using a triple-beam balance, place an object on the pan. Slide the largest rider along its beam until the pointer drops below zero. Then move it back one notch. Repeat the process for each rider proceeding from the larger to smaller until the pointer swings an equal distance above and below the zero point. Sum the masses on each beam to find the mass of the object. Move all riders back to zero when finished.

Instead of putting materials directly on the balance, scientists often take a tare of a container. A tare is the mass of a container into which objects or substances are placed for measuring their masses. To find the mass of objects or substances, find the mass of a clean container. Remove the container from the pan, and place the object or substances in the container. Find the mass of the container with the materials in it. Subtract the mass of the empty container from the mass of the filled container to find the mass of the materials you are using.

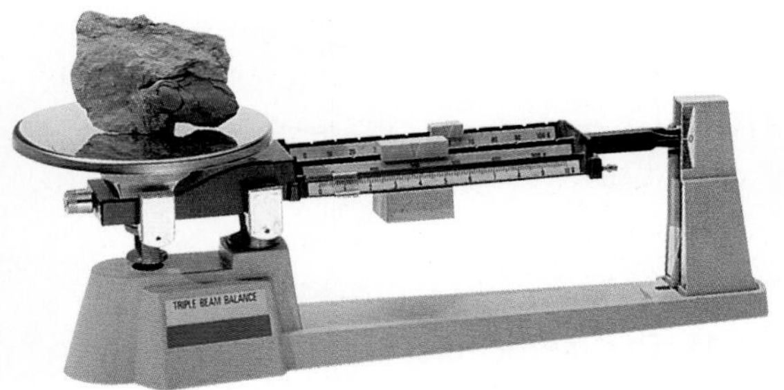

Figure 12 A triple-beam balance is used to determine the mass of an object.

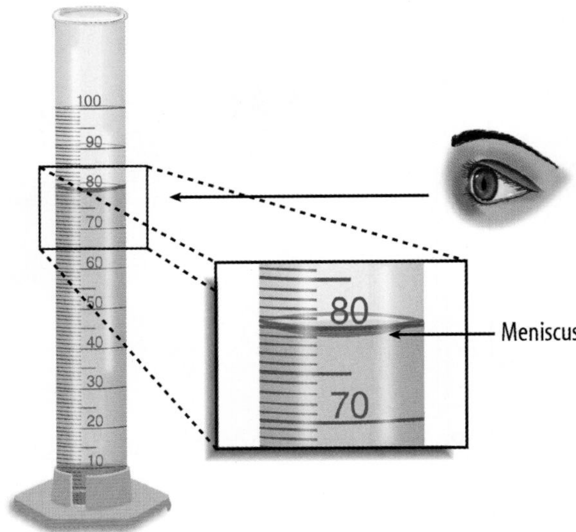

Figure 13 Graduated cylinders measure liquid volume.

Liquid Volume The SI unit for measuring liquids is the liter (l). When a smaller unit is needed, scientists might use a milliliter. Because a milliliter takes up the volume of a cube measuring 1 cm on each side it also can be called a cubic centimeter ($cm^3 = cm \times cm \times cm$).

You can use beakers and graduated cylinders to measure liquid volume. A graduated cylinder, shown in **Figure 13,** is marked from bottom to top in milliliters. In lab, you might use a 10-mL graduated cylinder or a 100-mL graduated cylinder. When measuring liquids, notice that the liquid has a curved surface. Look at the surface at eye level, and measure the bottom of the curve. This is called the meniscus. The graduated cylinder in **Figure 13** contains 79.0 mL, or 79.0 cm^3, of a liquid.

Temperature Scientists often measure temperature using the Celsius scale. Pure water has a freezing point of 0°C and boiling point of 100°C. The unit of measurement is degrees Celsius. Two other scales often used are the Fahrenheit and Kelvin scales.

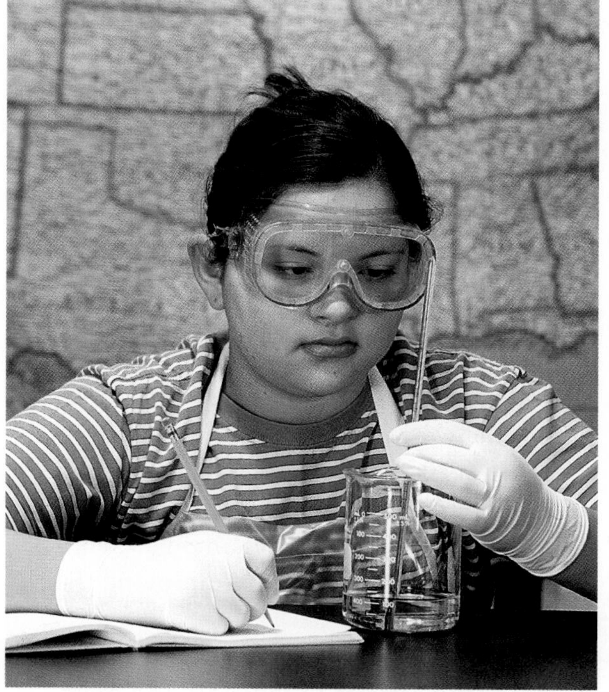

Figure 14 A thermometer measures the temperature of an object.

Scientists use a thermometer to measure temperature. Most thermometers in a laboratory are glass tubes with a bulb at the bottom end containing a liquid such as colored alcohol. The liquid rises or falls with a change in temperature. To read a glass thermometer like the thermometer in **Figure 14,** rotate it slowly until a red line appears. Read the temperature where the red line ends.

Form Operational Definitions An operational definition defines an object by how it functions, works, or behaves. For example, when you are playing hide and seek and a tree is home base, you have created an operational definition for a tree.

Objects can have more than one operational definition. For example, a ruler can be defined as a tool that measures the length of an object (how it is used). It can also be a tool with a series of marks used as a standard when measuring (how it works).

Analyze the Data

To determine the meaning of your observations and investigation results, you will need to look for patterns in the data. Then you must think critically to determine what the data mean. Scientists use several approaches when they analyze the data they have collected and recorded. Each approach is useful for identifying specific patterns.

Interpret Data The word *interpret* means "to explain the meaning of something." When analyzing data from an experiment, try to find out what the data show. Identify the control group and the test group to see whether changes in the independent variable have had an effect. Look for differences in the dependent variable between the control and test groups.

Classify Sorting objects or events into groups based on common features is called classifying. When classifying, first observe the objects or events to be classified. Then select one feature that is shared by some members in the group, but not by all. Place those members that share that feature in a subgroup. You can classify members into smaller and smaller subgroups based on characteristics. Remember that when you classify, you are grouping objects or events for a purpose. Keep your purpose in mind as you select the features to form groups and subgroups.

Compare and Contrast Observations can be analyzed by noting the similarities and differences between two or more objects or events that you observe. When you look at objects or events to see how they are similar, you are comparing them. Contrasting is looking for differences in objects or events.

Recognize Cause and Effect A cause is a reason for an action or condition. The effect is that action or condition. When two events happen together, it is not necessarily true that one event caused the other. Scientists must design a controlled investigation to recognize the exact cause and effect.

Draw Conclusions

When scientists have analyzed the data they collected, they proceed to draw conclusions about the data. These conclusions are sometimes stated in words similar to the hypothesis that you formed earlier. They may confirm a hypothesis, or lead you to a new hypothesis.

Infer Scientists often make inferences based on their observations. An inference is an attempt to explain observations or to indicate a cause. An inference is not a fact, but a logical conclusion that needs further investigation. For example, you may infer that a fire has caused smoke. Until you investigate, however, you do not know for sure.

Apply When you draw a conclusion, you must apply those conclusions to determine whether the data supports the hypothesis. If your data do not support your hypothesis, it does not mean that the hypothesis is wrong. It means only that the result of the investigation did not support the hypothesis. Maybe the experiment needs to be redesigned, or some of the initial observations on which the hypothesis was based were incomplete or biased. Perhaps more observation or research is needed to refine your hypothesis. A successful investigation does not always come out the way you originally predicted.

Avoid Bias Sometimes a scientific investigation involves making judgments. When you make a judgment, you form an opinion. It is important to be honest and not to allow any expectations of results to bias your judgments. This is important throughout the entire investigation, from researching to collecting data to drawing conclusions.

Communicate

The communication of ideas is an important part of the work of scientists. A discovery that is not reported will not advance the scientific community's understanding or knowledge. Communication among scientists also is important as a way of improving their investigations.

Scientists communicate in many ways, from writing articles in journals and magazines that explain their investigations and experiments, to announcing important discoveries on television and radio. Scientists also share ideas with colleagues on the Internet or present them as lectures, like the student is doing in **Figure 15**.

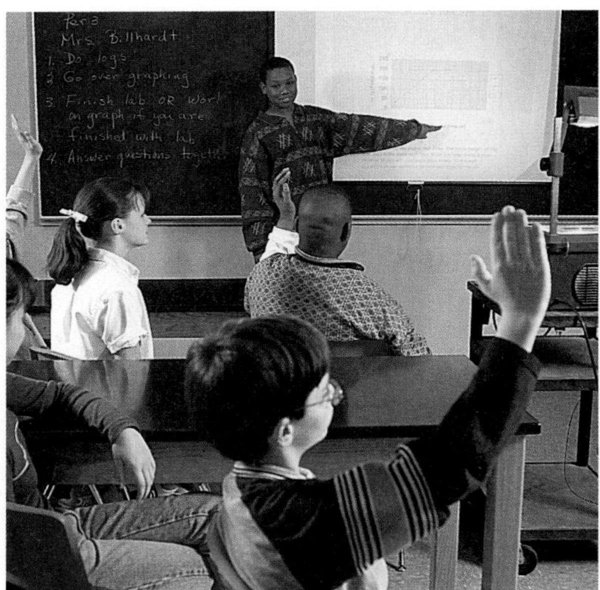

Figure 15 A student communicates to his peers about his investigation.

These safety symbols are used in laboratory and field investigations in this book to indicate possible hazards. Learn the meaning of each symbol and refer to this page often. *Remember to wash your hands thoroughly after completing lab procedures.*

PROTECTIVE EQUIPMENT

Do not begin any lab without the proper protection equipment.

GOGGLES 	Proper eye protection must be worn when performing or observing science activities that involve items or conditions as listed below.	**APRON** 	Wear an approved apron when using substances that could stain, wet, or destroy cloth.
SOAP 	Wash hands with soap and water before removing goggles and after all lab activities.	**GLOVES** 	Wear gloves when working with biological materials, chemicals, animals, or materials that can stain or irritate hands.

LABORATORY HAZARDS

Symbols	Potential Hazards	Precaution	Response
DISPOSAL	contamination of classroom or environment due to improper disposal of materials such as chemicals and live specimens	• **DO NOT** dispose of hazardous materials in the sink or trash can. • Dispose of wastes as directed by your teacher.	• If hazardous materials are disposed of improperly, notify your teacher immediately.
EXTREME TEMPERATURE	skin burns due to extremely hot or cold materials such as hot glass, liquids, or metals; liquid nitrogen; dry ice	• Use proper protective equipment, such as hot mitts and/or tongs, when handling objects with extreme temperatures.	• If injury occurs, notify your teacher immediately.
SHARP OBJECTS	punctures or cuts from sharp objects such as razor blades, pins, scalpels, and broken glass	• Handle glassware carefully to avoid breakage. • Walk with sharp objects pointed downward, away from you and others.	• If broken glass or injury occurs, notify your teacher immediately.
ELECTRICAL	electric shock or skin burn due to improper grounding, short circuits, liquid spills, or exposed wires	• Check condition of wires and apparatus for fraying or uninsulated wires, and broken or cracked equipment. • Use only GFCI-protected outlets	• **DO NOT** attempt to fix electrical problems. Notify your teacher immediately.
CHEMICAL	skin irritation or burns, breathing difficulty, and/or poisoning due to touching, swallowing, or inhalation of chemicals such as acids, bases, bleach, metal compounds, iodine, poinsettias, pollen, ammonia, acetone, nail polish remover, heated chemicals, mothballs, and any other chemicals labeled or known to be dangerous	• Wear proper protective equipment such as goggles, apron, and gloves when using chemicals. • Ensure proper room ventilation or use a fume hood when using materials that produce fumes. • **NEVER** smell fumes directly. • **NEVER** taste or eat any material in the laboratory.	• If contact occurs, immediately flush affected area with water and notify your teacher. • If a spill occurs, leave the area immediately and notify your teacher.
FLAMMABLE	unexpected fire due to liquids or gases that ignite easily such as rubbing alcohol	• Avoid open flames, sparks, or heat when flammable liquids are present.	• If a fire occurs, leave the area immediately and notify your teacher.
OPEN FLAME	burns or fire due to open flame from matches, Bunsen burners, or burning materials	• Tie back loose hair and clothing. • Keep flame away from all materials. • Follow teacher instructions when lighting and extinguishing flames. • Use proper protection, such as hot mitts or tongs, when handling hot objects.	• If a fire occurs, leave the area immediately and notify your teacher.
ANIMAL SAFETY	injury to or from laboratory animals	• Wear proper protective equipment such as gloves, apron, and goggles when working with animals. • Wash hands after handling animals.	• If injury occurs, notify your teacher immediately.
BIOLOGICAL	infection or adverse reaction due to contact with organisms such as bacteria, fungi, and biological materials such as blood, animal or plant materials	• Wear proper protective equipment such as gloves, goggles, and apron when working with biological materials. • Avoid skin contact with an organism or any part of the organism. • Wash hands after handling organisms.	• If contact occurs, wash the affected area and notify your teacher immediately.
FUME	breathing difficulties from inhalation of fumes from substances such as ammonia, acetone, nail polish remover, heated chemicals, and mothballs	• Wear goggles, apron, and gloves. • Ensure proper room ventilation or use a fume hood when using substances that produce fumes. • **NEVER** smell fumes directly.	• If a spill occurs, leave area and notify your teacher immediately.
IRRITANT	irritation of skin, mucous membranes, or respiratory tract due to materials such as acids, bases, bleach, pollen, mothballs, steel wool, and potassium permanganate	• Wear goggles, apron, and gloves. • Wear a dust mask to protect against fine particles.	• If skin contact occurs, immediately flush the affected area with water and notify your teacher.
RADIOACTIVE	excessive exposure from alpha, beta, and gamma particles	• Remove gloves and wash hands with soap and water before removing remainder of protective equipment.	• If cracks or holes are found in the container, notify your teacher immediately.

Safety in the Science Laboratory

Introduction to Science Safety

The science laboratory is a safe place to work if you follow standard safety procedures. Being responsible for your own safety helps to make the entire laboratory a safer place for everyone. When performing any lab, read and apply the caution statements and safety symbol listed at the beginning of the lab.

General Safety Rules

1. Complete the *Lab Safety Form* or other safety contract BEFORE starting any science lab.

2. Study the procedure. Ask your teacher any questions. Be sure you understand safety symbols shown on the page.

3. Notify your teacher about allergies or other health conditions that can affect your participation in a lab.

4. Learn and follow use and safety procedures for your equipment. If unsure, ask your teacher.

5. Never eat, drink, chew gum, apply cosmetics, or do any personal grooming in the lab. Never use lab glassware as food or drink containers. Keep your hands away from your face and mouth.

6. Know the location and proper use of the safety shower, eye wash, fire blanket, and fire alarm.

Prevent Accidents

1. Use the safety equipment provided to you. Goggles and a safety apron should be worn during investigations.

2. Do NOT use hair spray, mousse, or other flammable hair products. Tie back long hair and tie down loose clothing.

3. Do NOT wear sandals or other open-toed shoes in the lab.

4. Remove jewelry on hands and wrists. Loose jewelry, such as chains and long necklaces, should be removed to prevent them from getting caught in equipment.

5. Do not taste any substances or draw any material into a tube with your mouth.

7. Proper behavior is expected in the lab. Practical jokes and fooling around can lead to accidents and injury.

8. Keep your work area uncluttered.

Laboratory Work

1. Collect and carry all equipment and materials to your work area before beginning a lab.

2. Remain in your own work area unless given permission by your teacher to leave it.

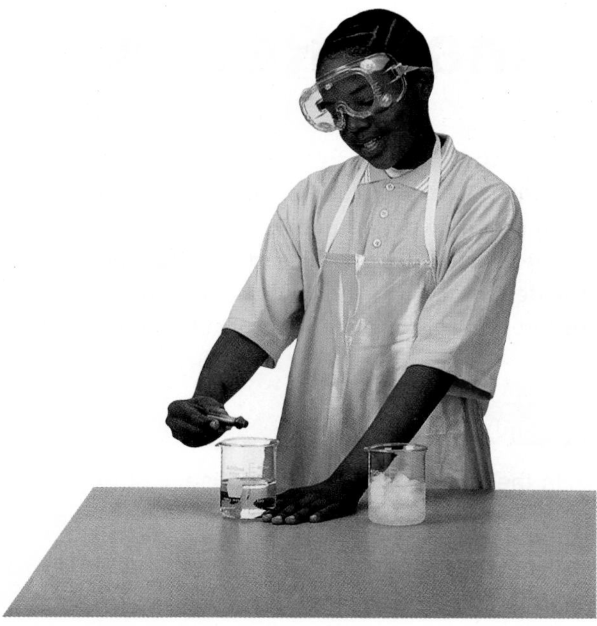

3. Always slant test tubes away from yourself and others when heating them, adding substances to them, or rinsing them.
4. If instructed to smell a substance in a container, hold the container a short distance away and fan vapors toward your nose.
5. Do NOT substitute other chemicals/substances for those in the materials list unless instructed to do so by your teacher.
6. Do NOT take any materials or chemicals outside of the laboratory.
7. Stay out of storage areas unless instructed to be there and supervised by your teacher.

Laboratory Cleanup

1. Turn off all burners, water, and gas, and disconnect all electrical devices.
2. Clean all pieces of equipment and return all materials to their proper places.
3. Dispose of chemicals and other materials as directed by your teacher. Place broken glass and solid substances in the proper containers. Never discard materials in the sink.
4. Clean your work area.
5. Wash your hands with soap and water thoroughly BEFORE removing your goggles.

Emergencies

1. Report any fire, electrical shock, glassware breakage, spill, or injury, no matter how small, to your teacher immediately. Follow his or her instructions.
2. If your clothing should catch fire, STOP, DROP, and ROLL. If possible, smother it with the fire blanket or get under a safety shower. NEVER RUN.
3. If a fire should occur, turn off all gas and leave the room according to established procedures.
4. In most instances, your teacher will clean up spills. Do NOT attempt to clean up spills unless you are given permission and instructions to do so.
5. If chemicals come into contact with your eyes or skin, notify your teacher immediately. Use the eyewash, or flush your skin or eyes with large quantities of water.
6. The fire extinguisher and first-aid kit should only be used by your teacher unless it is an extreme emergency and you have been given permission.
7. If someone is injured or becomes ill, only a professional medical provider or someone certified in first aid should perform first-aid procedures.

Math Review

Use Fractions

A fraction compares a part to a whole. In the fraction $\frac{2}{3}$, the 2 represents the part and is the numerator. The 3 represents the whole and is the denominator.

Reduce Fractions To reduce a fraction, you must find the largest factor that is common to both the numerator and the denominator, the greatest common factor (GCF). Divide both numbers by the GCF. The fraction has then been reduced, or it is in its simplest form.

Example

Twelve of the 20 chemicals in the science lab are in powder form. What fraction of the chemicals used in the lab are in powder form?

Step 1 Write the fraction.

$\frac{\text{part}}{\text{whole}} = \frac{12}{20}$

Step 2 To find the GCF of the numerator and denominator, list all of the factors of each number.

Factors of 12: 1, 2, 3, 4, 6, 12 (the numbers that divide evenly into 12)

Factors of 20: 1, 2, 4, 5, 10, 20 (the numbers that divide evenly into 20)

Step 3 List the common factors.

1, 2, 4

Step 4 Choose the greatest factor in the list. The GCF of 12 and 20 is 4.

Step 5 Divide the numerator and denominator by the GCF.

$\frac{12 \div 4}{20 \div 4} = \frac{3}{5}$

In the lab, $\frac{3}{5}$ of the chemicals are in powder form.

Practice Problem At an amusement park, 66 of 90 rides have a height restriction. What fraction of the rides, in its simplest form, has a height restriction?

Add and Subtract Fractions with Like Denominators To add or subtract fractions with the same denominator, add or subtract the numerators and write the sum or difference over the denominator. After finding the sum or difference, find the simplest form for your fraction.

Example 1

In the forest outside your house, $\frac{1}{8}$ of the animals are rabbits, $\frac{3}{8}$ are squirrels, and the remainder are birds and insects. How many are mammals?

Step 1 Add the numerators.

$\frac{1}{8} + \frac{3}{8} = \frac{(1 + 3)}{8} = \frac{4}{8}$

Step 2 Find the GCF.

$\frac{4}{8}$ (GCF, 4)

Step 3 Divide the numerator and denominator by the GCF.

$\frac{4 \div 4}{8 \div 4} = \frac{1}{2}$

$\frac{1}{2}$ of the animals are mammals.

Example 2

If $\frac{7}{16}$ of the Earth is covered by freshwater, and $\frac{1}{16}$ of that is in glaciers, how much freshwater is not frozen?

Step 1 Subtract the numerators.

$\frac{7}{16} - \frac{1}{16} = \frac{(7 - 1)}{16} = \frac{6}{16}$

Step 2 Find the GCF.

$\frac{6}{16}$ (GCF, 2)

Step 3 Divide the numerator and denominator by the GCF.

$\frac{6 \div 2}{16 \div 2} = \frac{3}{8}$

$\frac{3}{8}$ of the freshwater is not frozen.

Practice Problem A bicycle rider is riding at a rate of 15 km/h for $\frac{4}{9}$ of his ride, 10 km/h for $\frac{2}{9}$ of his ride, and 8 km/h for the remainder of the ride. How much of his ride is he riding at a rate greater than 8 km/h?

Add and Subtract Fractions with Unlike Denominators To add or subtract fractions with unlike denominators, first find the least common denominator (LCD). This is the smallest number that is a common multiple of both denominators. Rename each fraction with the LCD, and then add or subtract. Find the simplest form if necessary.

Example 1

A chemist makes a paste that is $\frac{1}{2}$ table salt (NaCl), $\frac{1}{3}$ sugar ($C_6H_{12}O_6$), and the remainder is water (H_2O). How much of the paste is a solid?

Step 1 Find the LCD of the fractions.

$\frac{1}{2} + \frac{1}{3}$ (LCD, 6)

Step 2 Rename each numerator and each denominator with the LCD.

Step 3 Add the numerators.

$\frac{3}{6} + \frac{2}{6} = \frac{(3 + 2)}{6} = \frac{5}{6}$

$\frac{5}{6}$ of the paste is a solid.

Example 2

The average precipitation in Grand Junction, CO, is $\frac{7}{10}$ inch in November, and $\frac{3}{5}$ inch in December. What is the total average precipitation?

Step 1 Find the LCD of the fractions.

$\frac{7}{10} + \frac{3}{5}$ (LCD, 10)

Step 2 Rename each numerator and each denominator with the LCD.

Step 3 Add the numerators.

$\frac{7}{10} + \frac{6}{10} = \frac{(7 + 6)}{10} = \frac{13}{10}$

$\frac{13}{10}$ inches total precipitation, or $1\frac{3}{10}$ inches.

Practice Problem On an electric bill, about $\frac{1}{8}$ of the energy is from solar energy and about $\frac{1}{10}$ is from wind power. How much of the total bill is from solar energy and wind power combined?

Example 3

In your body, $\frac{7}{10}$ of your muscle contractions are involuntary (cardiac and smooth muscle tissue). Smooth muscle makes $\frac{3}{15}$ of your muscle contractions. How many of your muscle contractions are made by cardiac muscle?

Step 1 Find the LCD of the fractions.

$\frac{7}{10} - \frac{3}{15}$ (LCD, 30)

Step 2 Rename each numerator and each denominator with the LCD.

$\frac{7 \times 3}{10 \times 3} = \frac{21}{30}$

$\frac{3 \times 2}{15 \times 2} = \frac{6}{30}$

Step 3 Subtract the numerators.

$\frac{21}{30} - \frac{6}{30} = \frac{(21 - 6)}{30} = \frac{15}{30}$

Step 4 Find the GCF.

$\frac{15}{30}$ (GCF, 15)

$\frac{1}{2}$

$\frac{1}{2}$ of all muscle contractions are cardiac muscle.

Example 4

Tony wants to make cookies that call for $\frac{3}{4}$ of a cup of flour, but he only has $\frac{1}{3}$ of a cup. How much more flour does he need?

Step 1 Find the LCD of the fractions.

$\frac{3}{4} - \frac{1}{3}$ (LCD, 12)

Step 2 Rename each numerator and each denominator with the LCD.

$\frac{3 \times 3}{4 \times 3} = \frac{9}{12}$

$\frac{1 \times 4}{3 \times 4} = \frac{4}{12}$

Step 3 Subtract the numerators.

$\frac{9}{12} - \frac{4}{12} = \frac{(9 - 4)}{12} = \frac{5}{12}$

$\frac{5}{12}$ of a cup of flour

Practice Problem Using the information provided to you in Example 3 above, determine how many muscle contractions are voluntary (skeletal muscle).

Multiply Fractions To multiply with fractions, multiply the numerators and multiply the denominators. Find the simplest form if necessary.

Example

Multiply $\frac{3}{5}$ by $\frac{1}{3}$.

Step 1 Multiply the numerators and denominators.

$$\frac{3}{5} \times \frac{1}{3} = \frac{(3 \times 1)}{(5 \times 3)} \frac{3}{15}$$

Step 2 Find the GCF.

$\frac{3}{15}$ (GCF, 3)

Step 3 Divide the numerator and denominator by the GCF.

$$\frac{3 \div 3}{15 \div 3} = \frac{1}{5}$$

$\frac{3}{5}$ multiplied by $\frac{1}{3}$ is $\frac{1}{5}$.

Practice Problem Multiply $\frac{3}{14}$ by $\frac{5}{16}$.

Find a Reciprocal Two numbers whose product is 1 are called multiplicative inverses, or reciprocals.

Example

Find the reciprocal of $\frac{3}{8}$.

Step 1 Inverse the fraction by putting the denominator on top and the numerator on the bottom.

$\frac{8}{3}$

The reciprocal of $\frac{3}{8}$ is $\frac{8}{3}$.

Practice Problem Find the reciprocal of $\frac{4}{9}$.

Divide Fractions To divide one fraction by another fraction, multiply the dividend by the reciprocal of the divisor. Find the simplest form if necessary.

Example 1

Divide $\frac{1}{9}$ by $\frac{1}{3}$.

Step 1 Find the reciprocal of the divisor.

The reciprocal of $\frac{1}{3}$ is $\frac{3}{1}$.

Step 2 Multiply the dividend by the reciprocal of the divisor.

$$\frac{\frac{1}{9}}{\frac{1}{3}} = \frac{1}{9} \times \frac{3}{1} = \frac{(1 \times 3)}{(9 \times 1)} = \frac{3}{9}$$

Step 3 Find the GCF.

$\frac{3}{9}$ (GCF, 3)

Step 4 Divide the numerator and denominator by the GCF.

$$\frac{3 \div 3}{9 \div 3} = \frac{1}{3}$$

$\frac{1}{9}$ divided by $\frac{1}{3}$ is $\frac{1}{3}$.

Example 2

Divide $\frac{3}{5}$ by $\frac{1}{4}$.

Step 1 Find the reciprocal of the divisor.

The reciprocal of $\frac{1}{4}$ is $\frac{4}{1}$.

Step 2 Multiply the dividend by the reciprocal of the divisor.

$$\frac{\frac{3}{5}}{\frac{1}{4}} = \frac{3}{5} \times \frac{4}{1} = \frac{(3 \times 4)}{(5 \times 1)} = \frac{12}{5}$$

$\frac{3}{5}$ divided by $\frac{1}{4}$ is $\frac{12}{5}$ or $2\frac{2}{5}$.

Practice Problem Divide $\frac{3}{11}$ by $\frac{7}{10}$.

Use Ratios

When you compare two numbers by division, you are using a ratio. Ratios can be written 3 to 5, 3:5, or $\frac{3}{5}$. Ratios, like fractions, also can be written in simplest form.

Ratios can represent one type of probability, called odds. This is a ratio that compares the number of ways a certain outcome occurs to the number of possible outcomes. For example, if you flip a coin 100 times, what are the odds that it will come up heads? There are two possible outcomes, heads or tails, so the odds of coming up heads are 50:100. Another way to say this is that 50 out of 100 times the coin will come up heads. In its simplest form, the ratio is 1:2.

Example 1

A chemical solution contains 40 g of salt and 64 g of baking soda. What is the ratio of salt to baking soda as a fraction in simplest form?

Step 1 Write the ratio as a fraction.

$$\frac{\text{salt}}{\text{baking soda}} = \frac{40}{64}$$

Step 2 Express the fraction in simplest form. The GCF of 40 and 64 is 8.

$$\frac{40}{64} = \frac{40 \div 8}{64 \div 8} = \frac{5}{8}$$

The ratio of salt to baking soda in the chemical solution is 5:8.

Example 2

Sean rolls a 6-sided die 6 times. What are the odds that the side with a 3 will show?

Step 1 Write the ratio as a fraction.

$$\frac{\text{number of sides with a 3}}{\text{number of possible sides}} = \frac{1}{6}$$

Step 2 Multiply by the number of attempts.

$\frac{1}{6} \times 6$ attempts $= \frac{6}{6}$ attempts $= 1$ attempt

1 attempt out of 6 will show a 3.

Practice Problem Two metal rods measure 100 cm and 144 cm in length. What is the ratio of their lengths in simplest form?

Use Decimals

A fraction with a denominator that is a power of ten can be written as a decimal. For example, 0.27 means $\frac{27}{100}$. The decimal point separates the ones place from the tenths place.

Any fraction can be written as a decimal using division. For example, the fraction $\frac{5}{8}$ can be written as a decimal by dividing 5 by 8. Written as a decimal, it is 0.625.

Add or Subtract Decimals When adding and subtracting decimals, line up the decimal points before carrying out the operation.

Example 1

Find the sum of 47.68 and 7.80.

Step 1 Line up the decimal places when you write the numbers.

$$\begin{array}{r} 47.68 \\ +\ 7.80 \\ \hline \end{array}$$

Step 2 Add the decimals.

$$\begin{array}{r} {}^{1\,1} \\ 47.68 \\ +\ 7.80 \\ \hline 55.48 \end{array}$$

The sum of 47.68 and 7.80 is 55.48.

Example 2

Find the difference of 42.17 and 15.85.

Step 1 Line up the decimal places when you write the number.

$$\begin{array}{r} 42.17 \\ -15.85 \\ \hline \end{array}$$

Step 2 Subtract the decimals.

$$\begin{array}{r} {}^{3\,11} \\ 42.17 \\ -15.85 \\ \hline 26.32 \end{array}$$

The difference of 42.17 and 15.85 is 26.32.

Practice Problem Find the sum of 1.245 and 3.842.

Multiply Decimals To multiply decimals, multiply the numbers like numbers without decimal points. Count the decimal places in each factor. The product will have the same number of decimal places as the sum of the decimal places in the factors.

Example

Multiply 2.4 by 5.9.

Step 1 Multiply the factors like two whole numbers.

$24 \times 59 = 1416$

Step 2 Find the sum of the number of decimal places in the factors. Each factor has one decimal place, for a sum of two decimal places.

Step 3 The product will have two decimal places.

14.16

The product of 2.4 and 5.9 is 14.16.

Practice Problem Multiply 4.6 by 2.2.

Divide Decimals When dividing decimals, change the divisor to a whole number. To do this, multiply both the divisor and the dividend by the same power of ten. Then place the decimal point in the quotient directly above the decimal point in the dividend. Then divide as you do with whole numbers.

Example

Divide 8.84 by 3.4.

Step 1 Multiply both factors by 10.

$3.4 \times 10 = 34$, $8.84 \times 10 = 88.4$

Step 2 Divide 88.4 by 34.

$$\begin{array}{r} 2.6 \\ 34\overline{)88.4} \\ -68 \\ \hline 204 \\ -204 \\ \hline 0 \end{array}$$

8.84 divided by 3.4 is 2.6.

Practice Problem Divide 75.6 by 3.6.

Use Proportions

An equation that shows that two ratios are equivalent is a proportion. The ratios $\frac{2}{4}$ and $\frac{5}{10}$ are equivalent, so they can be written as $\frac{2}{4} = \frac{5}{10}$. This equation is a proportion.

When two ratios form a proportion, the cross products are equal. To find the cross products in the proportion $\frac{2}{4} = \frac{5}{10}$, multiply the 2 and the 10, and the 4 and the 5. Therefore $2 \times 10 = 4 \times 5$, or $20 = 20$.

Because you know that both ratios are equal, you can use cross products to find a missing term in a proportion. This is known as solving the proportion.

Example

The heights of a tree and a pole are proportional to the lengths of their shadows. The tree casts a shadow of 24 m when a 6-m pole casts a shadow of 4 m. What is the height of the tree?

Step 1 Write a proportion.

$$\frac{\text{height of tree}}{\text{height of pole}} = \frac{\text{length of tree's shadow}}{\text{length of pole's shadow}}$$

Step 2 Substitute the known values into the proportion. Let h represent the unknown value, the height of the tree.

$$\frac{h}{6} \times \frac{24}{4}$$

Step 3 Find the cross products.

$h \times 4 = 6 \times 24$

Step 4 Simplify the equation.

$4h = 144$

Step 5 Divide each side by 4.

$$\frac{4h}{4} = \frac{144}{4}$$

$h = 36$

The height of the tree is 36 m.

Practice Problem The ratios of the weights of two objects on the Moon and on Earth are in proportion. A rock weighing 3 N on the Moon weighs 18 N on Earth. How much would a rock that weighs 5 N on the Moon weigh on Earth?

Use Percentages

The word *percent* means "out of one hundred." It is a ratio that compares a number to 100. Suppose you read that 77 percent of Earth's surface is covered by water. That is the same as reading that the fraction of Earth's surface covered by water is $\frac{77}{100}$. To express a fraction as a percent, first find the equivalent decimal for the fraction. Then, multiply the decimal by 100 and add the percent symbol.

Example 1

Express $\frac{13}{20}$ as a percent.

Step 1 Find the equivalent decimal for the fraction.

$$\begin{array}{r} 0.65 \\ 20\overline{)13.00} \\ \underline{12\,0} \\ 1\,00 \\ \underline{1\,00} \\ 0 \end{array}$$

Step 2 Rewrite the fraction $\frac{13}{20}$ as 0.65.

Step 3 Multiply 0.65 by 100 and add the % symbol.

$0.65 \times 100 = 65 = 65\%$

So, $\frac{13}{20} = 65\%$.

This also can be solved as a proportion.

Example 2

Express $\frac{13}{20}$ as a percent.

Step 1 Write a proportion.

$\frac{13}{20} = \frac{x}{100}$

Step 2 Find the cross products.

$1300 = 20x$

Step 3 Divide each side by 20.

$\frac{1300}{20} = \frac{20x}{20}$

$65 = x = 65\%$

So, $13/20 = 65\%$

Practice Problem In one year, 73 of 365 days were rainy in one city. What percent of the days in that city were rainy?

Solve One-Step Equations

A statement that two expressions are equal is an equation. For example, $A = B$ is an equation that states that A is equal to B.

An equation is solved when a variable is replaced with a value that makes both sides of the equation equal. To make both sides equal the inverse operation is used. Addition and subtraction are inverses, and multiplication and division are inverses.

Example 1

Solve the equation $x - 10 = 35$.

Step 1 Find the solution by adding 10 to each side of the equation.

$$x - 10 = 35$$
$$x - 10 + 10 = 35 + 10$$
$$x = 45$$

Step 2 Check the solution.

$$x - 10 = 35$$
$$45 - 10 = 35$$
$$35 = 35$$

Both sides of the equation are equal, so $x = 45$.

Example 2

In the formula $a = bc$, find the value of c if $a = 20$ and $b = 2$.

Step 1 Rearrange the formula so the unknown value is by itself on one side of the equation by dividing both sides by b.

$$a = bc$$
$$\frac{a}{b} = \frac{bc}{b}$$
$$\frac{a}{b} = c$$

Step 2 Replace the variables a and b with the values that are given.

$$\frac{a}{b} = c$$
$$\frac{20}{2} = c$$
$$10 = c$$

Step 3 Check the solution.

$$a = bc$$
$$20 = 2 \times 10$$
$$20 = 20$$

Both sides of the equation are equal, so $c = 10$ is the solution when $a = 20$ and $b = 2$.

Practice Problem In the formula $h = gd$, find the value of d if $g = 12.3$ and $h = 17.4$.

SCIENCE SKILL HANDBOOK | MATH SKILL HANDBOOK | FOLDABLES HANDBOOK | REFERENCE HANDBOOK | GLOSSARY/GLOSARIO | INDEX

Use Statistics

The branch of mathematics that deals with collecting, analyzing, and presenting data is statistics. In statistics, there are three common ways to summarize data with a single number—the mean, the median, and the mode.

The **mean** of a set of data is the arithmetic average. It is found by adding the numbers in the data set and dividing by the number of items in the set.

The **median** is the middle number in a set of data when the data are arranged in numerical order. If there were an even number of data points, the median would be the mean of the two middle numbers.

The **mode** of a set of data is the number or item that appears most often.

Another number that often is used to describe a set of data is the range. The **range** is the difference between the largest number and the smallest number in a set of data.

Example

The speeds (in m/s) for a race car during five different time trials are 39, 37, 44, 36, and 44.

To find the mean:

Step 1 Find the sum of the numbers.

39 + 37 + 44 + 36 + 44 = 200

Step 2 Divide the sum by the number of items, which is 5.

200 ÷ 5 = 40

The mean is 40 m/s.

To find the median:

Step 1 Arrange the measures from least to greatest.

36, 37, 39, 44, 44

Step 2 Determine the middle measure.

36, 37, 39, 44, 44

The median is 39 m/s.

To find the mode:

Step 1 Group the numbers that are the same together.

44, 44, 36, 37, 39

Step 2 Determine the number that occurs most in the set.

44, 44, 36, 37, 39

The mode is 44 m/s.

To find the range:

Step 1 Arrange the measures from greatest to least.

44, 44, 39, 37, 36

Step 2 Determine the greatest and least measures in the set.

44, 44, 39, 37, 36

Step 3 Find the difference between the greatest and least measures.

44 – 36 = 8

The range is 8 m/s.

Practice Problem Find the mean, median, mode, and range for the data set 8, 4, 12, 8, 11, 14, 16.

A **frequency table** shows how many times each piece of data occurs, usually in a survey. **Table 1** below shows the results of a student survey on favorite color.

Table 1 Student Color Choice

Color	Tally	Frequency
red	IIII	4
blue	~~IIII~~	5
black	II	2
green	III	3
purple	~~IIII~~ II	7
yellow	~~IIII~~ I	6

Based on the frequency table data, which color is the favorite?

Use Geometry

The branch of mathematics that deals with the measurement, properties, and relationships of points, lines, angles, surfaces, and solids is called geometry.

Perimeter The **perimeter** (P) is the distance around a geometric figure. To find the perimeter of a rectangle, add the length and width and multiply that sum by two, or $2(l + w)$. To find perimeters of irregular figures, add the length of all the sides.

Example 1

Find the perimeter of a rectangle that is 3 m long and 5 m wide.

Step 1 You know that the perimeter is 2 times the sum of the width and length.

$P = 2(3 \text{ m} + 5 \text{ m})$

Step 2 Find the sum of the width and length.

$P = 2(8 \text{ m})$

Step 3 Multiply by 2.

$P = 16 \text{ m}$

The perimeter is 16 m.

Example 2

Find the perimeter of a shape with sides measuring 2 cm, 5 cm, 6 cm, 3 cm.

Step 1 You know that the perimeter is the sum of all the sides.

$P = 2 + 5 + 6 + 3$

Step 2 Find the sum of the sides.

$P = 2 + 5 + 6 + 3$

$P = 16$

The perimeter is 16 cm.

Practice Problem Find the perimeter of a rectangle with a length of 18 m and a width of 7 m.

Practice Problem Find the perimeter of a triangle measuring 1.6 cm by 2.4 cm by 2.4 cm.

Area of a Rectangle The **area** (A) is the number of square units needed to cover a surface. To find the area of a rectangle, multiply the length times the width, or $l \times w$. When finding area, the units also are multiplied. Area is given in square units.

Example

Find the area of a rectangle with a length of 1 cm and a width of 10 cm.

Step 1 You know that the area is the length multiplied by the width.

$A = (1 \text{ cm} \times 10 \text{ cm})$

Step 2 Multiply the length by the width. Also multiply the units.

$A = 10 \text{ cm}^2$

The area is 10 cm^2.

Practice Problem Find the area of a square whose sides measure 4 m.

Area of a Triangle To find the area of a triangle, use the formula:

$A = \frac{1}{2}(\text{base} \times \text{height})$

The base of a triangle can be any of its sides. The height is the perpendicular distance from a base to the opposite endpoint, or vertex.

Example

Find the area of a triangle with a base of 18 m and a height of 7 m.

Step 1 You know that the area is $\frac{1}{2}$ the base times the height.

$A = \frac{1}{2}(18 \text{ m} \times 7 \text{ m})$

Step 2 Multiply $\frac{1}{2}$ by the product of 18×7.

Multiply the units.

$A = \frac{1}{2}(126 \text{ m}^2)$

$A = 63 \text{ m}^2$

The area is 63 m^2.

Practice Problem Find the area of a triangle with a base of 27 cm and a height of 17 cm.

Circumference of a Circle The **diameter** (d) of a circle is the distance across the circle through its center, and the **radius** (r) is the distance from the center to any point on the circle. The radius is half of the diameter. The distance around the circle is called the **circumference** (C). The formula for finding the circumference is:

$C = 2\pi r$ or $C = \pi d$

The circumference divided by the diameter is always equal to 3.1415926... This nonterminating and nonrepeating number is represented by the Greek letter π (pi). An approximation often used for π is 3.14.

Example 1

Find the circumference of a circle with a radius of 3 m.

Step 1 You know the formula for the circumference is 2 times the radius times π.

$C = 2\pi(3)$

Step 2 Multiply 2 times the radius.

$C = 6\pi$

Step 3 Multiply by π.

$C \approx 19$ m

The circumference is about 19 m.

Example 2

Find the circumference of a circle with a diameter of 24.0 cm.

Step 1 You know the formula for the circumference is the diameter times π.

$C = \pi(24.0)$

Step 2 Multiply the diameter by π.

$C \approx 75.4$ cm

The circumference is about 75.4 cm.

Practice Problem Find the circumference of a circle with a radius of 19 cm.

Area of a Circle The formula for the area of a circle is: $A = \pi r^2$

Example 1

Find the area of a circle with a radius of 4.0 cm.

Step 1 $A = \pi(4.0)^2$

Step 2 Find the square of the radius.

$A = 16\pi$

Step 3 Multiply the square of the radius by π.

$A \approx 50 \text{ cm}^2$

The area of the circle is about 50 cm^2.

Example 2

Find the area of a circle with a radius of 225 m.

Step 1 $A = \pi(225)^2$

Step 2 Find the square of the radius.

$A = 50625\pi$

Step 3 Multiply the square of the radius by π.

$A \approx 159043.1$

The area of the circle is about 159043.1 m^2.

Example 3

Find the area of a circle whose diameter is 20.0 mm.

Step 1 Remember that the radius is half of the diameter.

$A = \pi\left(\frac{20.0}{2}\right)^2$

Step 2 Find the radius.

$A = \pi(10.0)^2$

Step 3 Find the square of the radius.

$A = 100\pi$

Step 4 Multiply the square of the radius by π.

$A \approx 314 \text{ mm}^2$

The area of the circle is about 314 mm^2.

Practice Problem Find the area of a circle with a radius of 16 m.

SCIENCE SKILL HANDBOOK
MATH SKILL HANDBOOK
FOLDABLES HANDBOOK
REFERENCE HANDBOOK
GLOSSARY/GLOSARIO
INDEX

Volume The measure of space occupied by a solid is the **volume** (V). To find the volume of a rectangular solid, multiply the length times width times height, or $V = l \times w \times h$. It is measured in cubic units, such as cubic centimeters (cm^3).

Example

Find the volume of a rectangular solid with a length of 2.0 m, a width of 4.0 m, and a height of 3.0 m.

Step 1 You know the formula for volume is the length times the width times the height.

$V = 2.0\text{ m} \times 4.0\text{ m} \times 3.0\text{ m}$

Step 2 Multiply the length times the width times the height.

$V = 24\text{ m}^3$

The volume is 24 m^3.

Practice Problem Find the volume of a rectangular solid that is 8 m long, 4 m wide, and 4 m high.

To find the volume of other solids, multiply the area of the base times the height.

Example 1

Find the volume of a solid that has a triangular base with a length of 8.0 m and a height of 7.0 m. The height of the entire solid is 15.0 m.

Step 1 You know that the base is a triangle, and the area of a triangle is $\frac{1}{2}$ the base times the height, and the volume is the area of the base times the height.

$V = \left[\frac{1}{2}(b \times h)\right] \times 15$

Step 2 Find the area of the base.

$V = \left[\frac{1}{2}(8 \times 7)\right] \times 15$

$V = \left(\frac{1}{2} \times 56\right) \times 15$

Step 3 Multiply the area of the base by the height of the solid.

$V = 28 \times 15$

$V = 420\text{ m}^3$

The volume is 420 m^3.

Example 2

Find the volume of a cylinder that has a base with a radius of 12.0 cm, and a height of 21.0 cm.

Step 1 You know that the base is a circle, and the area of a circle is the square of the radius times π, and the volume is the area of the base times the height.

$V = (\pi r^2) \times 21$

$V = (\pi 12^2) \times 21$

Step 2 Find the area of the base.

$V = 144\pi \times 21$

$V = 452 \times 21$

Step 3 Multiply the area of the base by the height of the solid.

$V \approx 9{,}500\text{ cm}^3$

The volume is about 9,500 cm^3.

Example 3

Find the volume of a cylinder that has a diameter of 15 mm and a height of 4.8 mm.

Step 1 You know that the base is a circle with an area equal to the square of the radius times π. The radius is one-half the diameter. The volume is the area of the base times the height.

$V = (\pi r^2) \times 4.8$

$V = \left[\pi\left(\frac{1}{2} \times 15\right)^2\right] \times 4.8$

$V = (\pi 7.5^2) \times 4.8$

Step 2 Find the area of the base.

$V = 56.25\pi \times 4.8$

$V \approx 176.71 \times 4.8$

Step 3 Multiply the area of the base by the height of the solid.

$V \approx 848.2$

The volume is about 848.2 mm^3.

Practice Problem Find the volume of a cylinder with a diameter of 7 cm in the base and a height of 16 cm.

Science Applications

Measure in SI

The metric system of measurement was developed in 1795. A modern form of the metric system, called the International System (SI), was adopted in 1960 and provides the standard measurements that all scientists around the world can understand.

The SI system is convenient because unit sizes vary by powers of 10. Prefixes are used to name units. Look at **Table 2** for some common SI prefixes and their meanings.

Table 2 Common SI Prefixes

Prefix	Symbol	Meaning	
kilo–	k	1,000	thousandth
hecto–	h	100	hundred
deka–	da	10	ten
deci–	d	0.1	tenth
centi–	c	0.01	hundreth
milli–	m	0.001	thousandth

Example

How many grams equal one kilogram?

Step 1 Find the prefix *kilo–* in **Table 2.**

Step 2 Using **Table 2,** determine the meaning of *kilo–*. According to the table, it means 1,000. When the prefix *kilo–* is added to a unit, it means that there are 1,000 of the units in a "kilounit."

Step 3 Apply the prefix to the units in the question. The units in the question are grams. There are 1,000 grams in a kilogram.

Practice Problem Is a milligram larger or smaller than a gram? How many of the smaller units equal one larger unit? What fraction of the larger unit does one smaller unit represent?

Dimensional Analysis

Convert SI Units In science, quantities such as length, mass, and time sometimes are measured using different units. A process called dimensional analysis can be used to change one unit of measure to another. This process involves multiplying your starting quantity and units by one or more conversion factors. A conversion factor is a ratio equal to one and can be made from any two equal quantities with different units. If 1,000 mL equal 1 L then two ratios can be made.

$$\frac{1{,}000\ \text{mL}}{1\ \text{L}} = \frac{1\ \text{L}}{1{,}000\ \text{mL}} = 1$$

One can convert between units in the SI system by using the equivalents in **Table 2** to make conversion factors.

Example

How many cm are in 4 m?

Step 1 Write conversion factors for the units given. From **Table 2,** you know that 100 cm = 1 m. The conversion factors are

$$\frac{100\ \text{cm}}{1\ \text{m}} \text{ and } \frac{1\ \text{m}}{100\ \text{cm}}$$

Step 2 Decide which conversion factor to use. Select the factor that has the units you are converting from (m) in the denominator and the units you are converting to (cm) in the numerator.

$$\frac{100\ \text{cm}}{1\ \text{m}}$$

Step 3 Multiply the starting quantity and units by the conversion factor. Cancel the starting units with the units in the denominator. There are 400 cm in 4 m.

$$4\ \cancel{\text{m}} = \frac{100\ \text{cm}}{1\ \cancel{\text{m}}} = 400\ \text{cm}$$

Practice Problem How many milligrams are in one kilogram? (Hint: You will need to use two conversion factors from **Table 2.**)

SCIENCE SKILL HANDBOOK
MATH SKILL HANDBOOK
FOLDABLES HANDBOOK
REFERENCE HANDBOOK
GLOSSARY/GLOSARIO
INDEX

Table 3 Unit System Equivalents

Type of Measurement	Equivalent
Length	1 in = 2.54 cm 1 yd = 0.91 m 1 mi = 1.61 km
Mass and weight*	1 oz = 28.35 g 1 lb = 0.45 kg 1 ton (short) = 0.91 tonnes (metric tons) 1 lb = 4.45 N
Volume	1 in^3 = 16.39 cm^3 1 qt = 0.95 L 1 gal = 3.78 L
Area	1 in^2 = 6.45 cm^2 1 yd^2 = 0.83 m^2 1 mi^2 = 2.59 km^2 1 acre = 0.40 hectares
Temperature	$°C = \frac{(°F - 32)}{1.8}$ K = °C + 273

*Weight is measured in standard Earth gravity.

Convert Between Unit Systems **Table 3** gives a list of equivalents that can be used to convert between English and SI units.

Example

If a meterstick has a length of 100 cm, how long is the meterstick in inches?

Step 1 Write the conversion factors for the units given. From **Table 3,** 1 in = 2.54 cm.

$$\frac{1 \text{ in}}{2.54 \text{ cm}} \textit{ and } \frac{2.54 \text{ cm}}{1 \text{ in}}$$

Step 2 Determine which conversion factor to use. You are converting from cm to in. Use the conversion factor with cm on the bottom.

$$\frac{1 \text{ in}}{2.54 \text{ cm}}$$

Step 3 Multiply the starting quantity and units by the conversion factor. Cancel the starting units with the units in the denominator. Round your answer to the nearest tenth.

$$100 \cancel{\text{cm}} \times \frac{1 \text{ in}}{2.54 \cancel{\text{cm}}} = 39.37 \text{ in}$$

The meterstick is about 39.4 in long.

Practice Problem 1 A book has a mass of 5 lb. What is the mass of the book in kg?

Practice Problem 2 Use the equivalent for in and cm (1 in = 2.54 cm) to show how 1 $in^3 \approx$ 16.39 cm^3.

SCIENCE SKILL HANDBOOK
MATH SKILL HANDBOOK
FOLDABLES HANDBOOK
REFERENCE HANDBOOK
GLOSSARY/GLOSARIO
INDEX

SCIENCE SKILL HANDBOOK
MATH SKILL HANDBOOK
FOLDABLES HANDBOOK
REFERENCE HANDBOOK
GLOSSARY/GLOSARIO
INDEX

Precision and Significant Digits

When you make a measurement, the value you record depends on the precision of the measuring instrument. This precision is represented by the number of significant digits recorded in the measurement. When counting the number of significant digits, all digits are counted except zeros at the end of a number with no decimal point such as 2,050, and zeros at the beginning of a decimal such as 0.03020. When adding or subtracting numbers with different precision, round the answer to the smallest number of decimal places of any number in the sum or difference. When multiplying or dividing, the answer is rounded to the smallest number of significant digits of any number being multiplied or divided.

Example

The lengths 5.28 and 5.2 are measured in meters. Find the sum of these lengths and record your answer using the correct number of significant digits.

Step 1 Find the sum.

5.28 m	2 digits after the decimal
+ 5.2 m	1 digit after the decimal
10.48 m	

Step 2 Round to one digit after the decimal because the least number of digits after the decimal of the numbers being added is 1.

The sum is 10.5 m.

Practice Problem 1 How many significant digits are in the measurement 7,071,301 m? How many significant digits are in the measurement 0.003010 g?

Practice Problem 2 Multiply 5.28 and 5.2 using the rule for multiplying and dividing. Record the answer using the correct number of significant digits.

Scientific Notation

Many times numbers used in science are very small or very large. Because these numbers are difficult to work with, scientists use scientific notation. To write numbers in scientific notation, move the decimal point until only one non-zero digit remains on the left. Then count the number of places you moved the decimal point and use that number as a power of ten. For example, the average distance from the Sun to Mars is 227,800,000,000 m. In scientific notation, this distance is 2.278×10^{11} m. Because you moved the decimal point to the left, the number is a positive power of ten.

The mass of an electron is about 0.000 000 000 000 000 000 000 000 000 000 911 kg. Expressed in scientific notation, this mass is 9.11×10^{-31} kg. Because the decimal point was moved to the right, the number is a negative power of ten.

Example

Earth is 149,600,000 km from the Sun. Express this in scientific notation.

Step 1 Move the decimal point until one non-zero digit remains on the left.

1.496 000 00

Step 2 Count the number of decimal places you have moved. In this case, eight.

Step 2 Show that number as a power of ten, 10^8.

Earth is 1.496×10^8 km from the Sun.

Practice Problem 1 How many significant digits are in 149,600,000 km? How many significant digits are in 1.496×10^8 km?

Practice Problem 2 Parts used in a high performance car must be measured to 7×10^{-6} m. Express this number as a decimal.

Practice Problem 3 A CD is spinning at 539 revolutions per minute. Express this number in scientific notation.

Make and Use Graphs

Data in tables can be displayed in a graph—a visual representation of data. Common graph types include line graphs, bar graphs, and circle graphs.

Line Graph A line graph shows a relationship between two variables that change continuously. The independent variable is changed and is plotted on the x-axis. The dependent variable is observed, and is plotted on the y-axis.

Example

Draw a line graph of the data below from a cyclist in a long-distance race.

Table 4 Bicycle Race Data

Time (h)	Distance (km)
0	0
1	8
2	16
3	24
4	32
5	40

Step 1 Determine the x-axis and y-axis variables. Time varies independently of distance and is plotted on the x-axis. Distance is dependent on time and is plotted on the y-axis.

Step 2 Determine the scale of each axis. The x-axis data ranges from 0 to 5. The y-axis data ranges from 0 to 50.

Step 3 Using graph paper, draw and label the axes. Include units in the labels.

Step 4 Draw a point at the intersection of the time value on the x-axis and corresponding distance value on the y-axis. Connect the points and label the graph with a title, as shown in **Figure 8.**

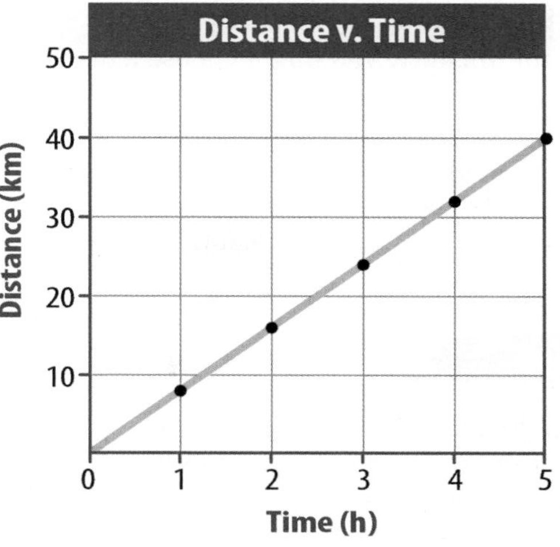

Figure 8 This line graph shows the relationship between distance and time during a bicycle ride.

Practice Problem A puppy's shoulder height is measured during the first year of her life. The following measurements were collected: (3 mo, 52 cm), (6 mo, 72 cm), (9 mo, 83 cm), (12 mo, 86 cm). Graph this data.

Find a Slope The slope of a straight line is the ratio of the vertical change, rise, to the horizontal change, run.

$$\text{Slope} = \frac{\text{vertical change (rise)}}{\text{horizontal change (run)}} = \frac{\text{change in } y}{\text{change in } x}$$

Example

Find the slope of the graph in **Figure 8**.

Step 1 You know that the slope is the change in y divided by the change in x.

$$\text{Slope} = \frac{\text{change in } y}{\text{change in } x}$$

Step 2 Determine the data points you will be using. For a straight line, choose the two sets of points that are the farthest apart.

$$\text{Slope} = \frac{(40 - 0)\text{ km}}{(5 - 0)\text{ h}}$$

Step 3 Find the change in y and x.

$$\text{Slope} = \frac{40\text{ km}}{5\text{ h}}$$

Step 4 Divide the change in y by the change in x.

$$\text{Slope} = \frac{8\text{ km}}{\text{h}}$$

The slope of the graph is 8 km/h.

SCIENCE SKILL HANDBOOK MATH SKILL HANDBOOK FOLDABLES HANDBOOK REFERENCE HANDBOOK GLOSSARY/GLOSARIO INDEX

Bar Graph To compare data that does not change continuously you might choose a bar graph. A bar graph uses bars to show the relationships between variables. The x-axis variable is divided into parts. The parts can be numbers such as years, or a category such as a type of animal. The y-axis is a number and increases continuously along the axis.

Example

A recycling center collects 4.0 kg of aluminum on Monday, 1.0 kg on Wednesday, and 2.0 kg on Friday. Create a bar graph of this data.

Step 1 Select the x-axis and y-axis variables. The measured numbers (the masses of aluminum) should be placed on the y-axis. The variable divided into parts (collection days) is placed on the x-axis.

Step 2 Create a graph grid like you would for a line graph. Include labels and units.

Step 3 For each measured number, draw a vertical bar above the x-axis value up to the y-axis value. For the first data point, draw a vertical bar above Monday up to 4.0 kg.

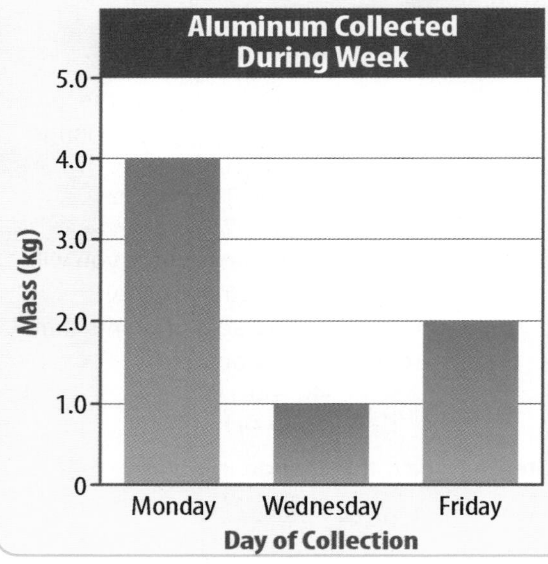

Practice Problem Draw a bar graph of the gases in air: 78% nitrogen, 21% oxygen, 1% other gases.

Circle Graph To display data as parts of a whole, you might use a circle graph. A circle graph is a circle divided into sections that represent the relative size of each piece of data. The entire circle represents 100%, half represents 50%, and so on.

Example

Air is made up of 78% nitrogen, 21% oxygen, and 1% other gases. Display the composition of air in a circle graph.

Step 1 Multiply each percent by 360° and divide by 100 to find the angle of each section in the circle.

$$78\% \times \frac{360°}{100} = 280.8°$$

$$21\% \times \frac{360°}{100} = 75.6°$$

$$1\% \times \frac{360°}{100} = 3.6°$$

Step 2 Use a compass to draw a circle and to mark the center of the circle. Draw a straight line from the center to the edge of the circle.

Step 3 Use a protractor and the angles you calculated to divide the circle into parts. Place the center of the protractor over the center of the circle and line the base of the protractor over the straight line.

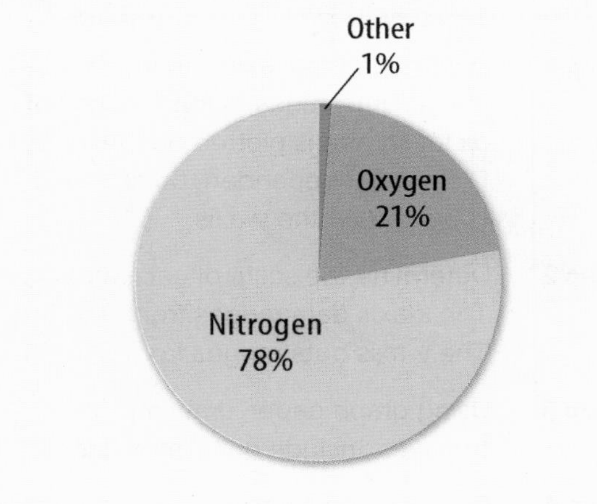

Practice Problem Draw a circle graph to represent the amount of aluminum collected during the week shown in the bar graph to the left.

Student Study Guides & Instructions

By Dinah Zike

1. You will find suggestions for Study Guides, also known as Foldables or books, in each chapter lesson and as a final project. Look at the end of the chapter to determine the project format and glue the Foldables in place as you progress through the chapter lessons.

2. Creating the Foldables or books is simple and easy to do by using copy paper, art paper, and internet printouts. Photocopies of maps, diagrams, or your own illustrations may also be used for some of the Foldables. Notebook paper is the most common source of material for study guides and 83% of all Foldables are created from it. When folded to make books, notebook paper Foldables easily fit into 11" × 17" or 12" × 18" chapter projects with space left over. Foldables made using photocopy paper are slightly larger and they fit into Projects, but snugly. Use the least amount of glue, tape, and staples needed to assemble the Foldables.

3. Seven of the Foldables can be made using either small or large paper. When 11" × 17" or 12" × 18" paper is used, these become projects for housing smaller Foldables. Project format boxes are located within the instructions to remind you of this option.

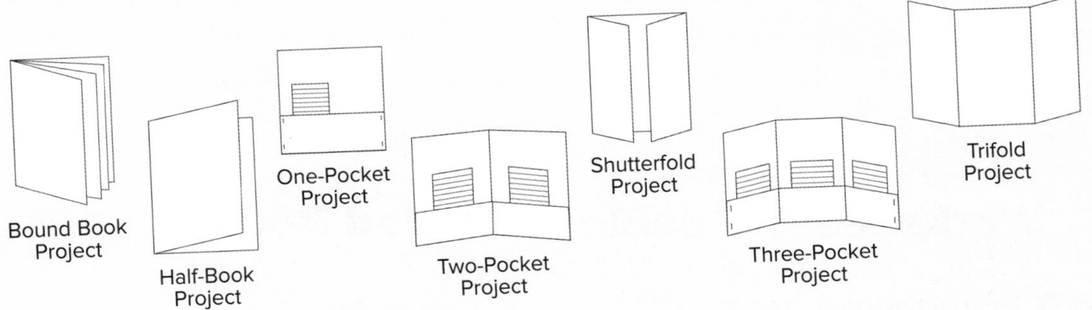

4. Use one-gallon self-locking plastic bags to store your projects. Place strips of two-inch clear tape along the left, long side of the bag and punch holes through the taped edge. Cut the bottom corners off the bag so it will not hold air. Store this Project Portfolio inside a three-hole binder. To store a large collection of project bags, use a giant laundry-soap box. Holes can be punched in some of the Foldable Projects so they can be stored in a three-hole binder without using a plastic bag. Punch holes in the pocket books before gluing or stapling the pocket.

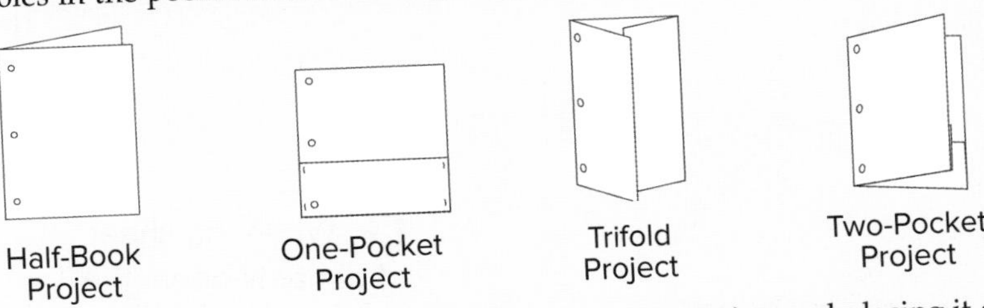

5. Maximize the use of the projects by collecting additional information and placing it on the back of the project and other unused spaces of the large Foldables.

SCIENCE SKILL HANDBOOK | MATH SKILL HANDBOOK | FOLDABLES HANDBOOK | REFERENCE HANDBOOK | GLOSSARY/GLOSARIO | INDEX

Half-Book Foldable® By Dinah Zike

Step 1 Fold a sheet of notebook or copy paper in half.

Label the exterior tab and use the inside space to write information.

PROJECT FORMAT

Use 11" × 17" or 12" × 18" paper on the horizontal axis to make a large project book.

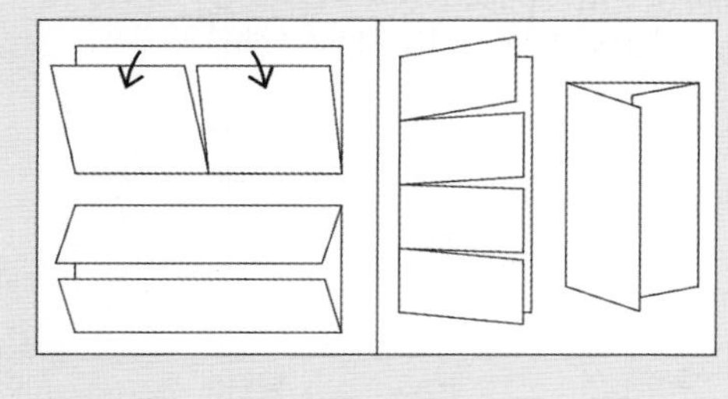

Variations

Paper can be folded horizontally, like a *hamburger* or vertically, like a *hot dog.*

A

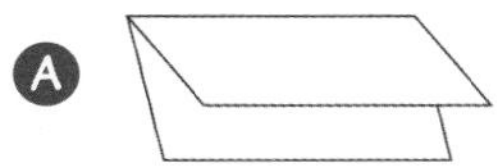

B

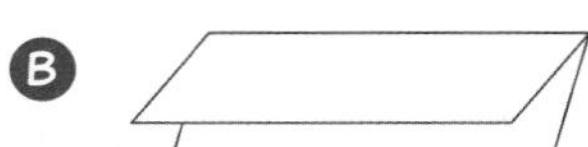

C Half-books can be folded so that one side is ½ inch longer than the other side. A title or question can be written on the extended tab.

Worksheet Foldable or Folded Book® By Dinah Zike

Step 1 Make a half-book (see above) using work sheets, internet printouts, diagrams, or maps.

Step 2 Fold it in half again.

Variations

A This folded sheet as a small book with two pages can be used for comparing and contrasting, cause and effect, or other skills.

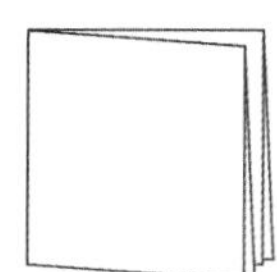

B When the sheet of paper is open, the four sections can be used separately or used collectively to show sequences or steps.

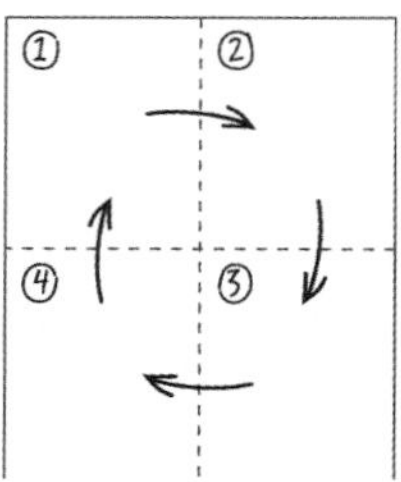

Two-Tab and Concept-Map Foldable® By Dinah Zike

Step 1 Fold a sheet of notebook or copy paper in half vertically or horizontally.

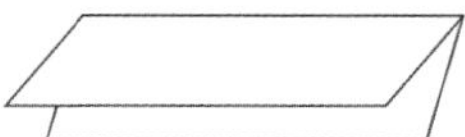

Step 2 Fold it in half again, as shown.

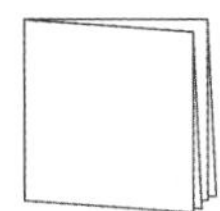

Step 3 Unfold once and cut along the fold line or valley of the top flap to make two flaps.

Variations

A Concept maps can be made by leaving a ½ inch tab at the top when folding the paper in half. Use arrows and labels to relate topics to the primary concept.

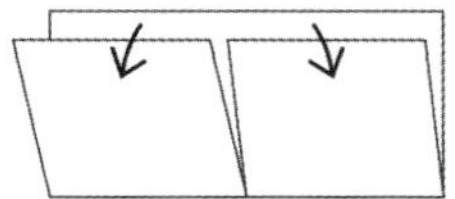

B Use two sheets of paper to make multiple page tab books. Glue or staple books together at the top fold.

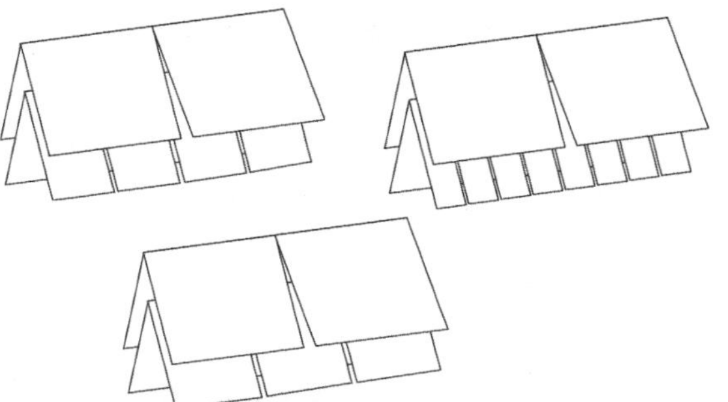

Three-Quarter Foldable® By Dinah Zike

Step 1 Make a two-tab book (see above) and cut the left tab off at the top of the fold line.

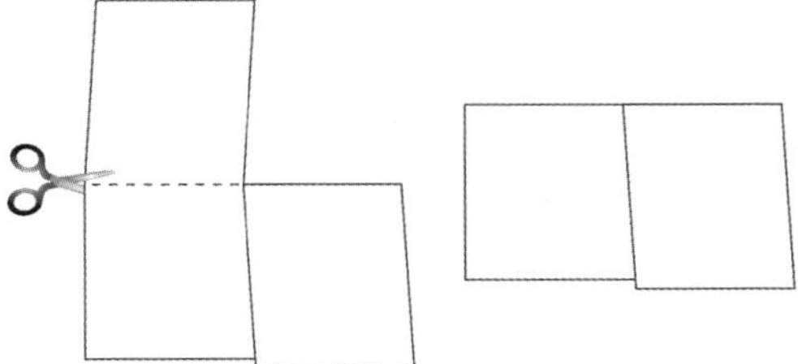

Variations

A Use this book to draw a diagram or a map on the exposed left tab. Write questions about the illustration on the top right tab and provide complete answers on the space under the tab.

B Compose a self-test using multiple choice answers for your questions. Include the correct answer with three wrong responses. The correct answers can be written on the back of the book or upside down on the bottom of the inside page.

Three-Tab Foldable® By Dinah Zike

Step 1 Fold a sheet of paper in half horizontally.

Step 2 Fold into thirds.

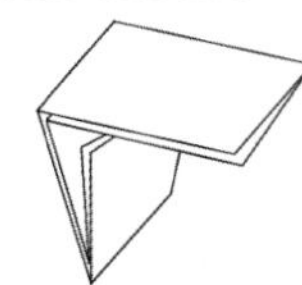

Step 3 Unfold and cut along the folds of the top flap to make three sections.

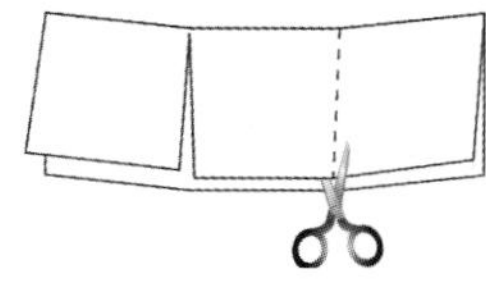

Variations

A Before cutting the three tabs draw a Venn diagram across the front of the book.

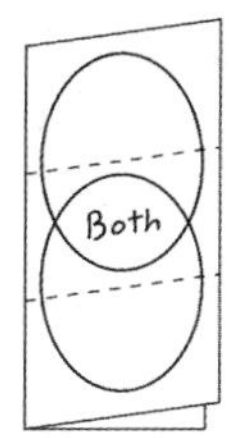

B Make a space to use for titles or concept maps by leaving a ½ inch tab at the top when folding the paper in half.

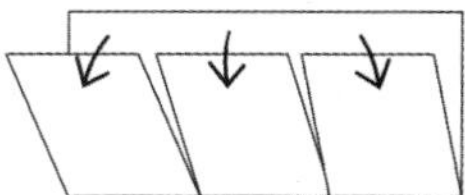

Four-Tab Foldable® By Dinah Zike

Step 1 Fold a sheet of paper in half horizontally.

Step 2 Fold in half and then fold each half as shown below.

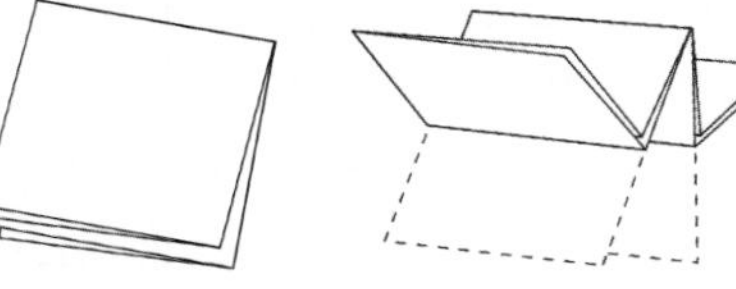

Step 3 Unfold and cut along the fold lines of the top flap to make four tabs.

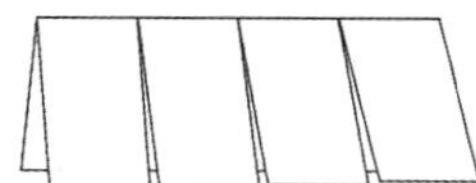

Variations

A Make a space to use for titles or concept maps by leaving a ½ inch tab at the top when folding the paper in half.

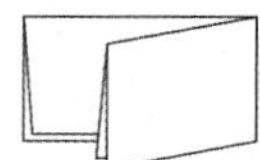

B Use the book on the vertical axis, with or without an extended tab.

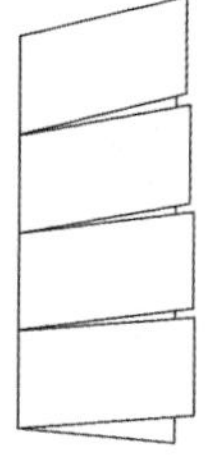

Folding Fifths for a Foldable® By Dinah Zike

Step 1 Fold a sheet of paper in half horizontally.

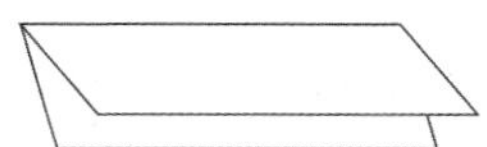

Step 2 Fold again so one-third of the paper is exposed and two-thirds are covered.

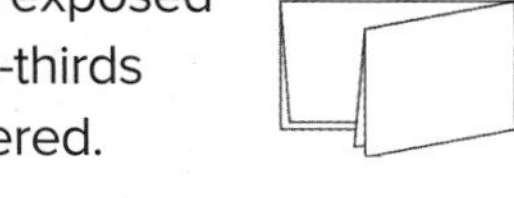

Step 3 Fold the two-thirds section in half.

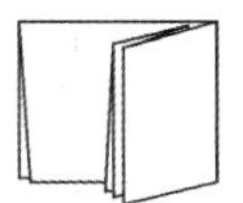

Step 4 Fold the one-third section, a single thickness, backward to make a fold line.

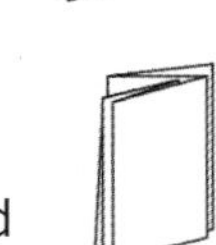

Variations

A Unfold and cut along the fold lines to make five tabs.

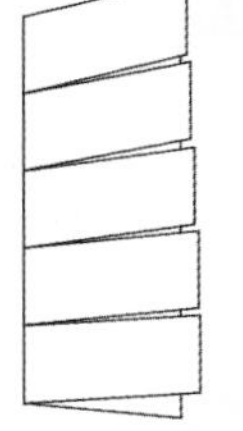

B Make a five-tab book with a ½ inch tab at the top (see two-tab instructions).

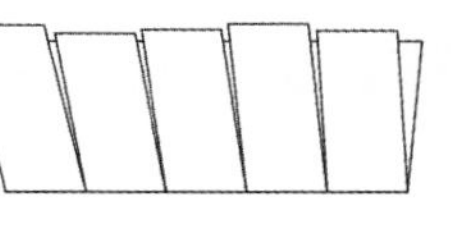

C Use 11” × 17” or 12” × 18” paper and fold into fifths for a five-column and/or row table or chart.

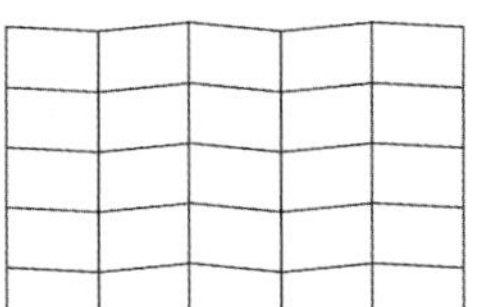

Folded Table or Chart, and Trifold Foldable® By Dinah Zike

Step 1 Fold a sheet of paper in the required number of vertical columns for the table or chart.

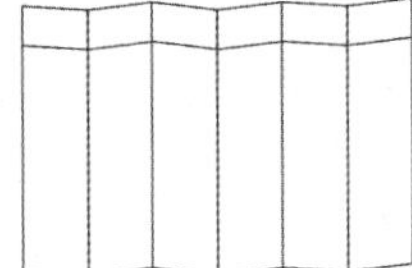

Step 2 Fold the horizontal rows needed for the table or chart.

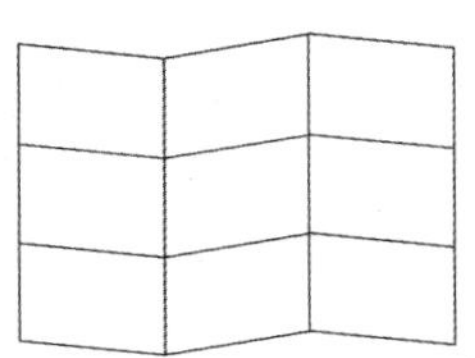

PROJECT FORMAT
Use 11” × 17” or 12” × 18” paper and fold it to make a large trifold project book or larger tables and charts.

Variations

A Make a trifold by folding the paper into thirds vertically or horizontally.

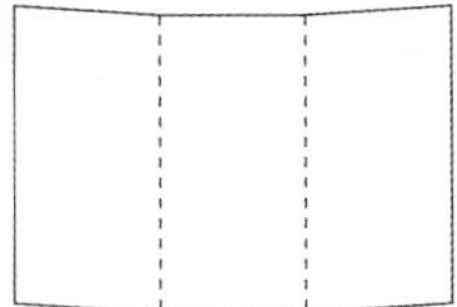

B Make a trifold book. Unfold it and draw a Venn diagram on the inside.

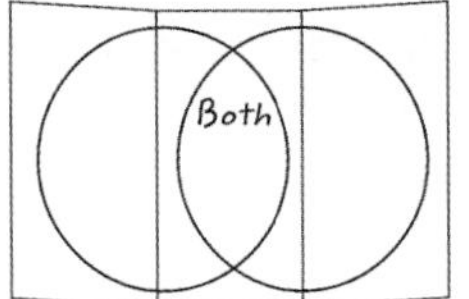

Two or Three-Pockets Foldable® By Dinah Zike

Step 1 Fold up the long side of a horizontal sheet of paper about 5 cm.

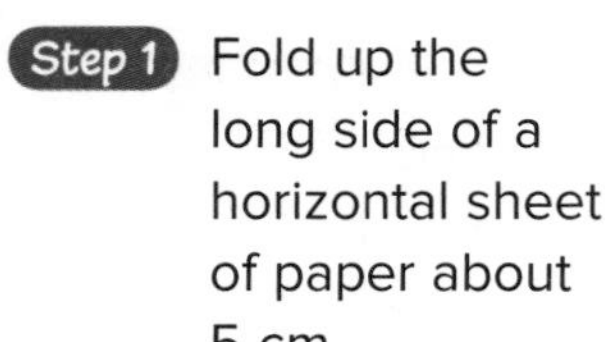

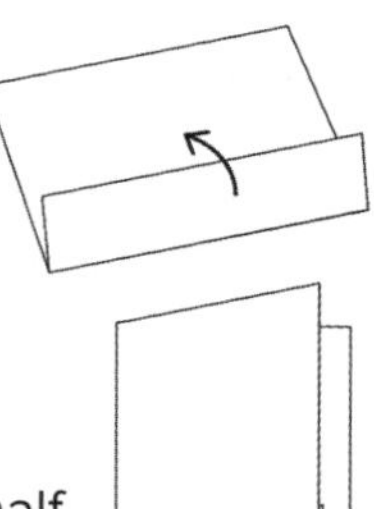

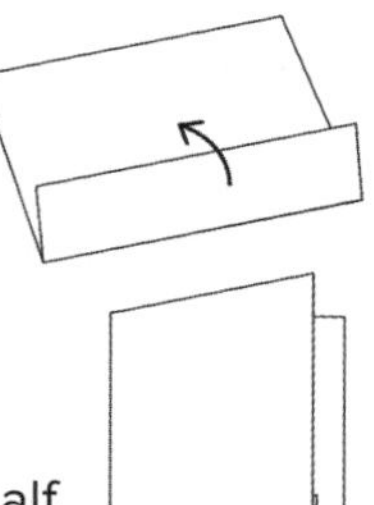

Step 2 Fold the paper in half.

Step 3 Open the paper and glue or staple the outer edges to make two compartments.

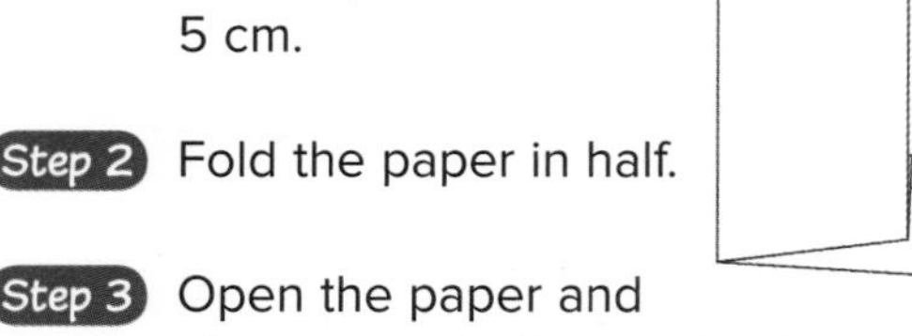

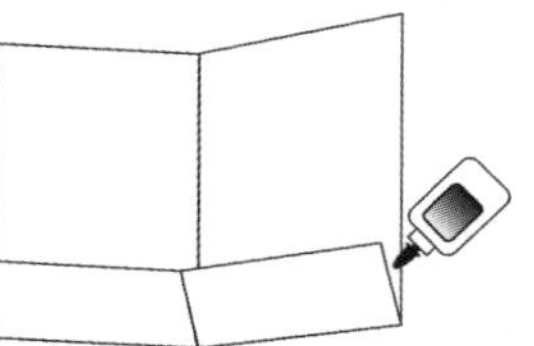

Variations

A Make a multi-page booklet by gluing several pocket books together.

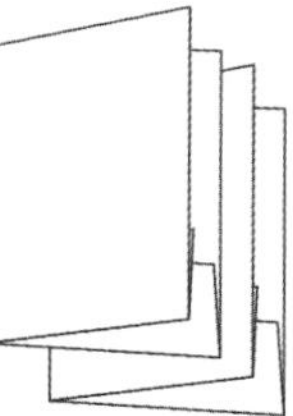

B Make a three-pocket book by using a trifold (see previous instructions).

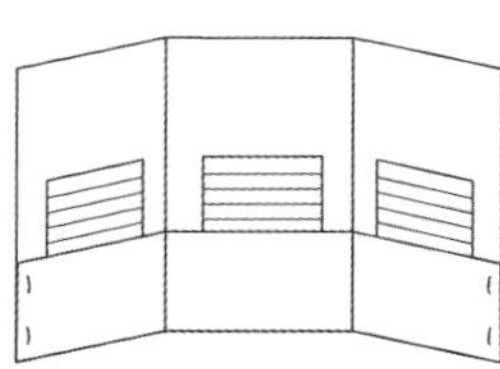

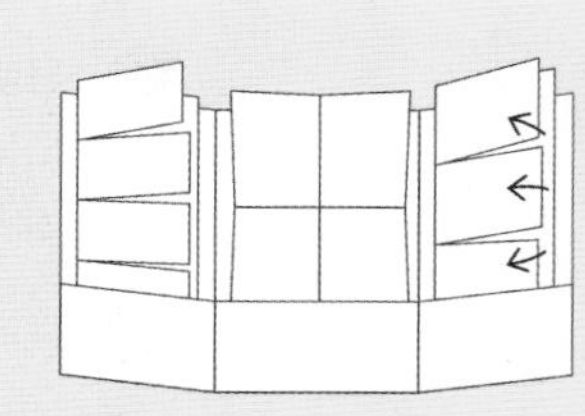

PROJECT FORMAT
Use 11" × 17" or 12" × 18" paper and fold it horizontally to make a large multi-pocket project.

Matchbook Foldable® By Dinah Zike

Step 1 Fold a sheet of paper almost in half and make the back edge about 1–2 cm longer than the front edge.

Step 2 Find the midpoint of the shorter flap.

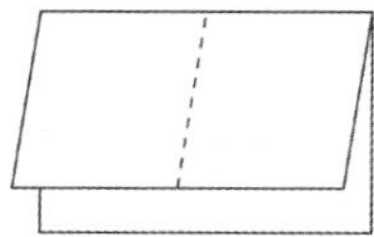

Step 3 Open the paper and cut the short side along the midpoint making two tabs.

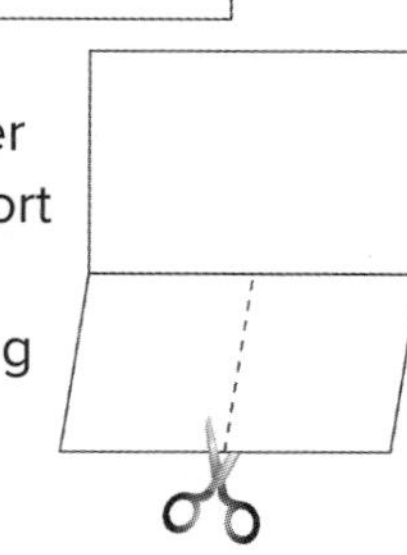

Step 4 Close the book and fold the tab over the short side.

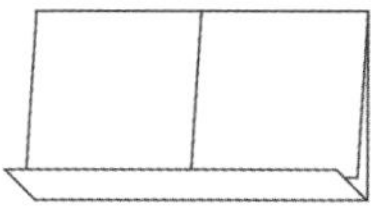

Variations

A Make a single-tab matchbook by skipping Steps 2 and 3.

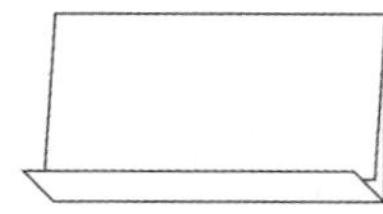

B Make two smaller matchbooks by cutting the single-tab matchbook in half.

Shutterfold Foldable® By Dinah Zike

Step 1 Begin as if you were folding a vertical sheet of paper in half, but instead of creasing the paper, pinch it to show the midpoint.

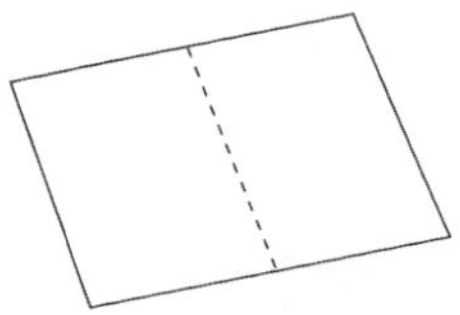

PROJECT FORMAT
Use 11" × 17" or 12" × 18" paper and fold it to make a large shutterfold project.

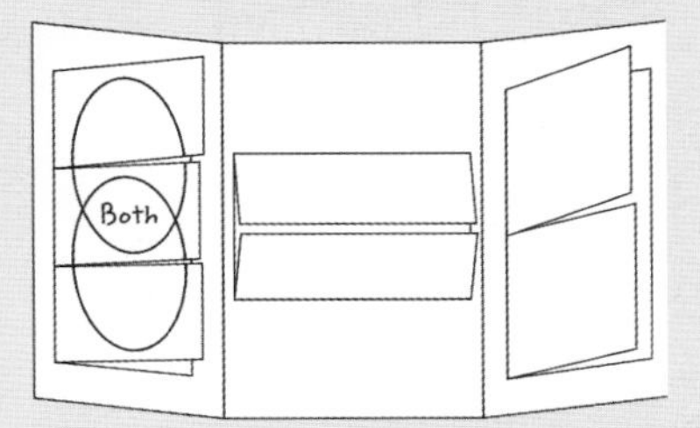

Step 2 Fold the top and bottom to the middle and crease the folds.

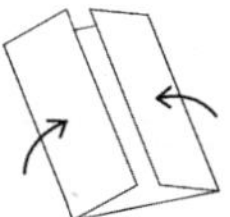

Variations

A Use the shutterfold on the horizontal axis.

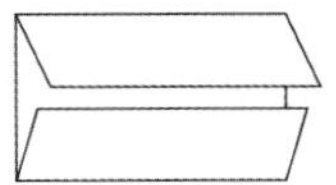

B Create a center tab by leaving .5–2 cm between the flaps in Step 2.

Four-Door Foldable® By Dinah Zike

Step 1 Make a shutterfold (see above).

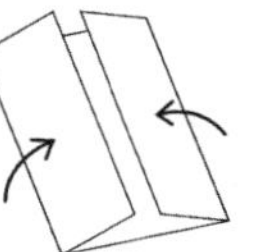

Step 2 Fold the sheet of paper in half.

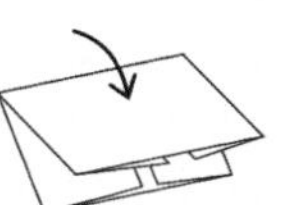

Step 3 Open the last fold and cut along the inside fold lines to make four tabs.

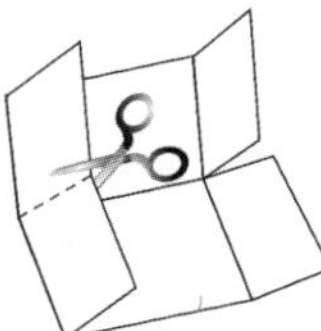

Variations

A Use the four-door book on the opposite axis.

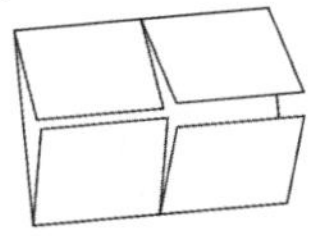

B Create a center tab by leaving .5–2 cm between the flaps in Step 1.

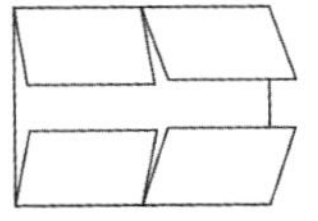

Bound Book Foldable® By Dinah Zike

Step 1 Fold three sheets of paper in half. Place the papers in a stack, leaving about .5 cm between each top fold. Mark all three sheets about 3 cm from the outer edges.

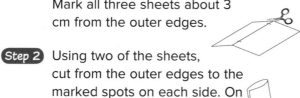

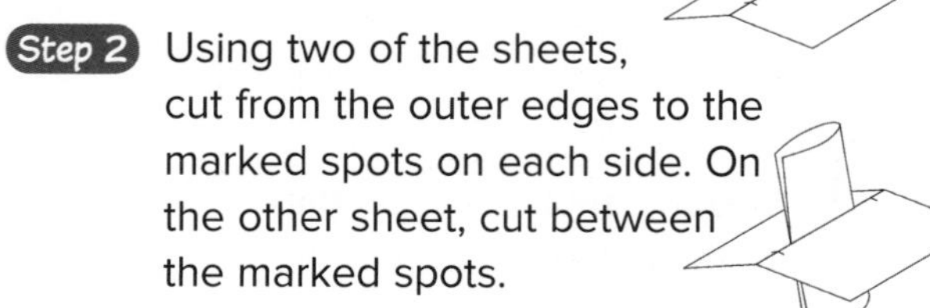

Step 2 Using two of the sheets, cut from the outer edges to the marked spots on each side. On the other sheet, cut between the marked spots.

Step 3 Take the two sheets from Step 1 and slide them through the cut in the third sheet to make a 12-page book.

Step 4 Fold the bound pages in half to form a book.

Variation

A Use two sheets of paper to make an eight-page book, or increase the number of pages by using more than three sheets.

PROJECT FORMAT
Use two or more sheets of 11" × 17" or 12" × 18" paper and fold it to make a large bound book project.

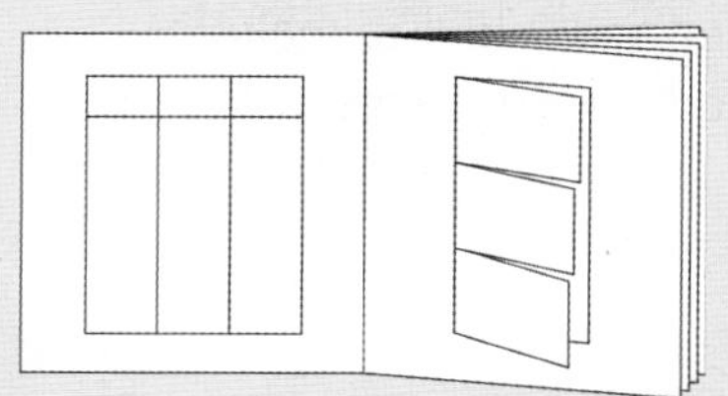

Accordion Foldable® By Dinah Zike

Step 1 Fold the selected paper in half vertically, like a *hamburger*.

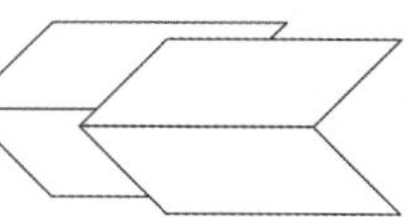

Step 2 Cut each sheet of folded paper in half along the fold lines.

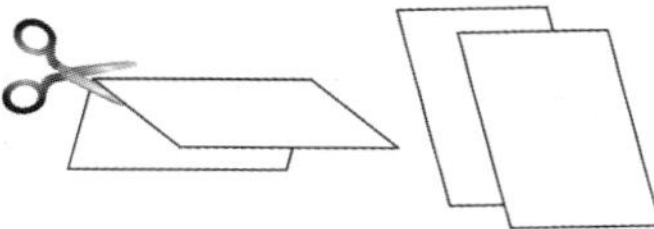

Step 3 Fold each half-sheet almost in half, leaving a 2 cm tab at the top.

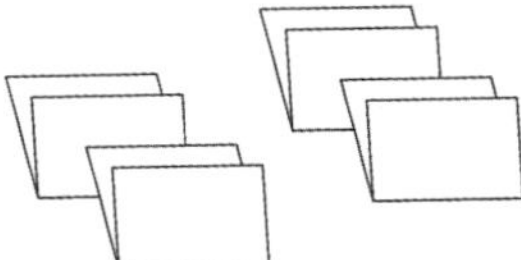

Step 4 Fold the top tab over the short side, then fold it in the opposite direction.

Variations

A Glue the straight edge of one paper inside the tab of another sheet. Leave a tab at the end of the book to add more pages.

B Tape the straight edge of one paper to the tab of another sheet, or just tape the straight edges of nonfolded paper end to end to make an accordion.

C Use whole sheets of paper to make a large accordion.

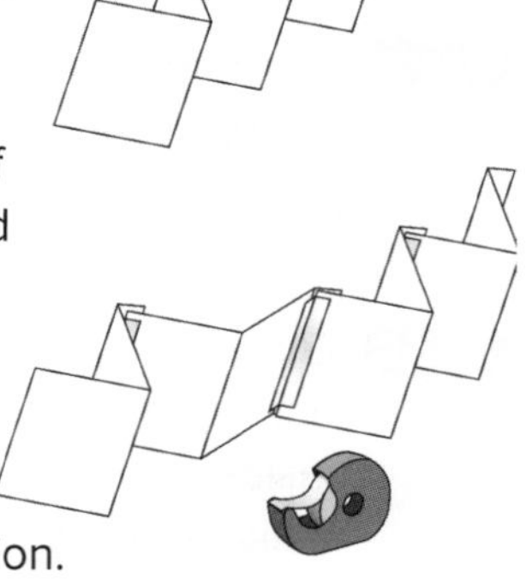

Layered Foldable® By Dinah Zike

Step 1 Stack two sheets of paper about 1–2 cm apart. Keep the right and left edges even.

Step 2 Fold up the bottom edges to form four tabs. Crease the fold to hold the tabs in place.

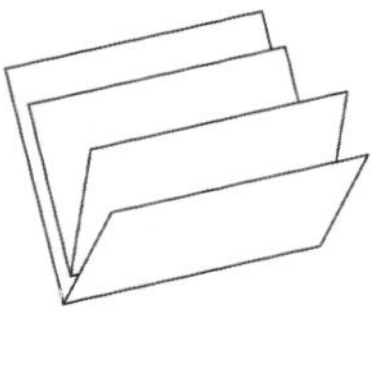

Step 3 Staple along the folded edge, or open and glue the papers together at the fold line.

Variations

A Rotate the book so the fold is at the top or to the side.

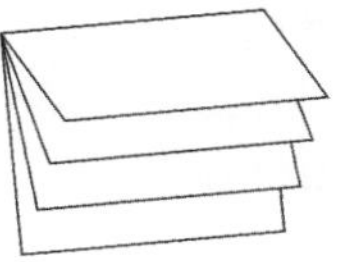

B Extend the book by using more than two sheets of paper.

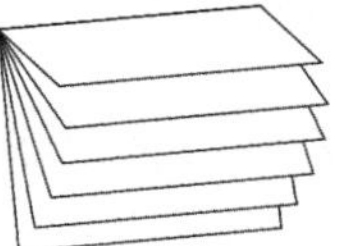

Envelope Foldable® By Dinah Zike

Step 1 Fold a sheet of paper into a *taco*. Cut off the tab at the top.

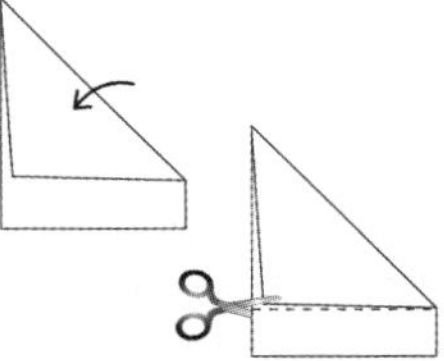

Step 2 Open the *taco* and fold it the opposite way making another *taco* and an X-fold pattern on the sheet of paper.

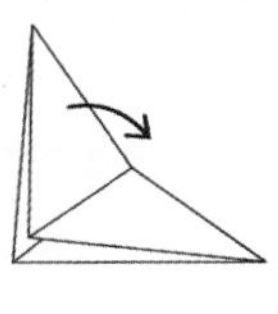

Step 3 Cut a map, illustration, or diagram to fit the inside of the envelope.

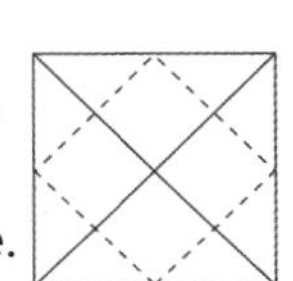

Step 4 Use the outside tabs for labels and inside tabs for writing information.

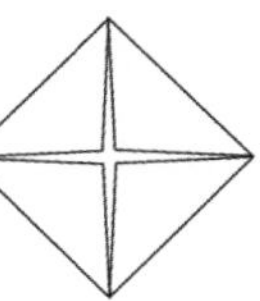

Variations

A Use 11" × 17" or 12" × 18" paper to make a large envelope.

B Cut off the points of the four tabs to make a window in the middle of the book.

Sentence Strip Foldable® By Dinah Zike

Step 1 Fold two sheets of paper in half vertically, like a *hamburger*.

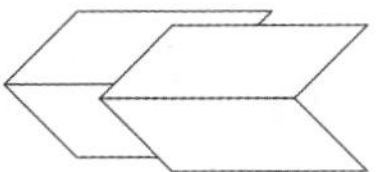

Step 2 Unfold and cut along fold lines making four half sheets.

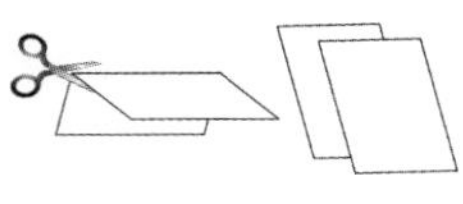

Step 3 Fold each half sheet in half horizontally, like a *hot dog*.

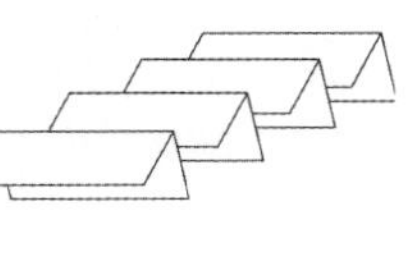

Step 4 Stack folded horizontal sheets evenly and staple together on the left side.

Step 5 Open the top flap of the first sentence strip and make a cut about 2 cm from the stapled edge to the fold line. This forms a flap that can be raised and lowered. Repeat this step for each sentence strip.

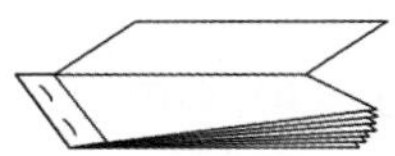

Variations

A Expand this book by using more than two sheets of paper.

B Use whole sheets of paper to make large books.

Pyramid Foldable® By Dinah Zike

Step 1 Fold a sheet of paper into a *taco*. Crease the fold line, but do not cut it off.

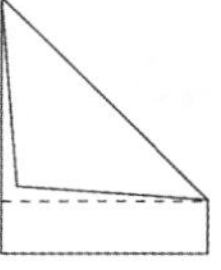

Step 2 Open the folded sheet and refold it like a *taco* in the opposite direction to create an X-fold pattern.

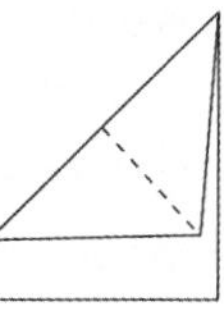

Step 3 Cut one fold line as shown, stopping at the center of the X-fold to make a flap.

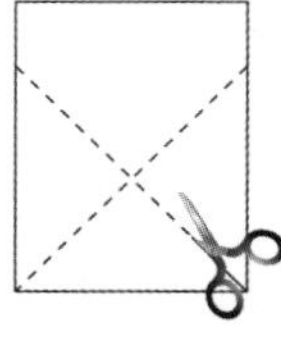

Step 4 Outline the fold lines of the X-fold. Label the three front sections and use the inside spaces for notes. Use the tab for the title.

Step 5 Glue the tab into a project book or notebook. Use the space under the pyramid for other information.

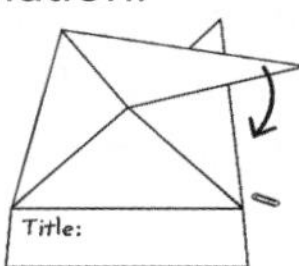

Step 6 To display the pyramid, fold the flap under and secure with a paper clip, if needed.

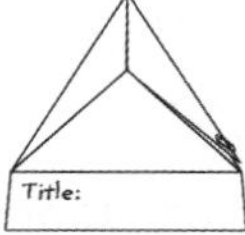

Single-Pocket or One-Pocket Foldable® By Dinah Zike

Step 1 Using a large piece of paper on a vertical axis, fold the bottom edge of the paper upwards, about 5 cm.

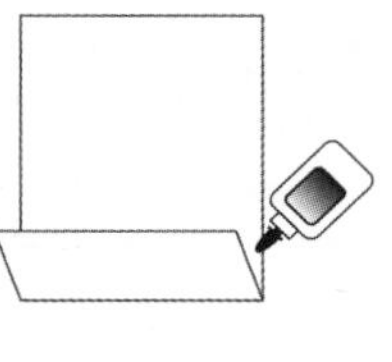

Step 2 Glue or staple the outer edges to make a large pocket.

PROJECT FORMAT
Use 11" × 17" or 12" × 18" paper and fold it vertically or horizontally to make a large pocket project.

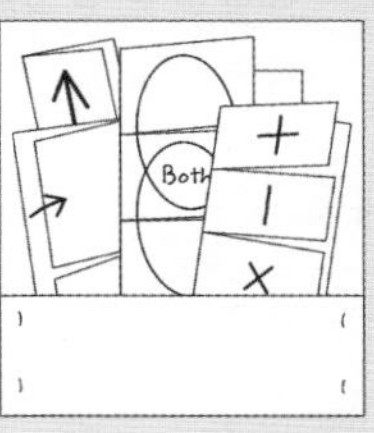

Variations

A Make the one-pocket project using the paper on the horizontal axis.

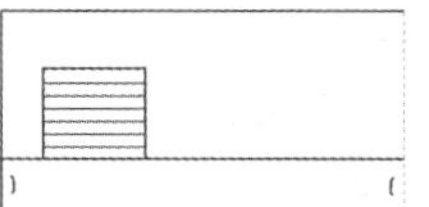

B To store materials securely inside, fold the top of the paper almost to the center, leaving about 2–4 cm between the paper edges. Slip the Foldables through the opening and under the top and bottom pockets.

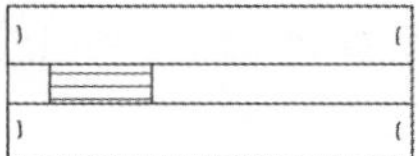

Multi-Tab Foldable® By Dinah Zike

Step 1 Fold a sheet of notebook paper in half like a *hot dog*.

Step 2 Open the paper and on one side cut every third line. This makes ten tabs on wide ruled notebook paper and twelve tabs on college ruled.

Step 3 Label the tabs on the front side and use the inside space for definitions or other information.

Variation

A Make a tab for a title by folding the paper so the holes remain uncovered. This allows the notebook Foldable to be stored in a three-hole binder.

PERIODIC TABLE OF THE ELEMENTS

Element — Hydrogen
Atomic number — 1
Symbol — H
Atomic mass — 1.01
State of matter

Gas
Liquid
Solid
Synthetic

A column in the periodic table is called a **group.**

A row in the periodic table is called a **period.**

Period	1	2	3	4	5	6	7	8	9
1	Hydrogen 1 H 1.01 (Gas)								
2	Lithium 3 Li 6.94 (Solid)	Beryllium 4 Be 9.01 (Solid)							
3	Sodium 11 Na 22.99 (Solid)	Magnesium 12 Mg 24.31 (Solid)							
4	Potassium 19 K 39.10 (Solid)	Calcium 20 Ca 40.08 (Solid)	Scandium 21 Sc 44.96 (Solid)	Titanium 22 Ti 47.87 (Solid)	Vanadium 23 V 50.94 (Solid)	Chromium 24 Cr 52.00 (Solid)	Manganese 25 Mn 54.94 (Solid)	Iron 26 Fe 55.85 (Solid)	Cobalt 27 Co 58.93 (Solid)
5	Rubidium 37 Rb 85.47 (Solid)	Strontium 38 Sr 87.62 (Solid)	Yttrium 39 Y 88.91 (Solid)	Zirconium 40 Zr 91.22 (Solid)	Niobium 41 Nb 92.91 (Solid)	Molybdenum 42 Mo 95.96 (Solid)	Technetium 43 Tc (98) (Synthetic)	Ruthenium 44 Ru 101.07 (Solid)	Rhodium 45 Rh 102.91 (Solid)
6	Cesium 55 Cs 132.91 (Solid)	Barium 56 Ba 137.33 (Solid)	Lanthanum 57 La 138.91 (Solid)	Hafnium 72 Hf 178.49 (Solid)	Tantalum 73 Ta 180.95 (Solid)	Tungsten 74 W 183.84 (Solid)	Rhenium 75 Re 186.21 (Solid)	Osmium 76 Os 190.23 (Solid)	Iridium 77 Ir 192.22 (Solid)
7	Francium 87 Fr (223) (Solid)	Radium 88 Ra (226) (Solid)	Actinium 89 Ac (227) (Solid)	Rutherfordium 104 Rf (267) (Synthetic)	Dubnium 105 Db (268) (Synthetic)	Seaborgium 106 Sg (271) (Synthetic)	Bohrium 107 Bh (272) (Synthetic)	Hassium 108 Hs (270) (Synthetic)	Meitnerium 109 Mt (276) (Synthetic)

The number in parentheses is the mass number of the longest lived isotope for that element.

Lanthanide series	Cerium 58 Ce 140.12 (Solid)	Praseodymium 59 Pr 140.91 (Solid)	Neodymium 60 Nd 144.24 (Solid)	Promethium 61 Pm (145) (Synthetic)	Samarium 62 Sm 150.36 (Solid)	Europium 63 Eu 151.96 (Solid)
Actinide series	Thorium 90 Th 232.04 (Solid)	Protactinium 91 Pa 231.04 (Solid)	Uranium 92 U 238.03 (Solid)	Neptunium 93 Np (237) (Synthetic)	Plutonium 94 Pu (244) (Synthetic)	Americium 95 Am (243) (Synthetic)

- Metal
- Metalloid
- Nonmetal
- Recently discovered

10	11	12	13	14	15	16	17	18
								Helium 2 **He** 4.00
			Boron 5 **B** 10.81	Carbon 6 **C** 12.01	Nitrogen 7 **N** 14.01	Oxygen 8 **O** 16.00	Fluorine 9 **F** 19.00	Neon 10 **Ne** 20.18
			Aluminum 13 **Al** 26.98	Silicon 14 **Si** 28.09	Phosphorus 15 **P** 30.97	Sulfur 16 **S** 32.07	Chlorine 17 **Cl** 35.45	Argon 18 **Ar** 39.95
Nickel 28 **Ni** 58.69	Copper 29 **Cu** 63.55	Zinc 30 **Zn** 65.38	Gallium 31 **Ga** 69.72	Germanium 32 **Ge** 72.64	Arsenic 33 **As** 74.92	Selenium 34 **Se** 78.96	Bromine 35 **Br** 79.90	Krypton 36 **Kr** 83.80
Palladium 46 **Pd** 106.42	Silver 47 **Ag** 107.87	Cadmium 48 **Cd** 112.41	Indium 49 **In** 114.82	Tin 50 **Sn** 118.71	Antimony 51 **Sb** 121.76	Tellurium 52 **Te** 127.60	Iodine 53 **I** 126.90	Xenon 54 **Xe** 131.29
Platinum 78 **Pt** 195.08	Gold 79 **Au** 196.97	Mercury 80 **Hg** 200.59	Thallium 81 **Tl** 204.38	Lead 82 **Pb** 207.20	Bismuth 83 **Bi** 208.98	Polonium 84 **Po** (209)	Astatine 85 **At** (210)	Radon 86 **Rn** (222)
Darmstadtium 110 **Ds** (281)	Roentgenium 111 **Rg** (280)	Copernicium 112 **Cn** (285)	Ununtrium * 113 **Uut** (284)	Flerovium * 114 **Fl** (289)	Ununpentium * 115 **Uup** (288)	Livermorium * 116 **Lv** (293)	Ununseptium * 117 **Uus** (294)	Ununoctium * 118 **Uuo** (294)

* The names and symbols for elements 113, 115, 117, and 118 are temporary. Final names will be approved by IUPAC (International Union of Pure and Applied Chemistry).

Gadolinium 64 **Gd** 157.25	Terbium 65 **Tb** 158.93	Dysprosium 66 **Dy** 162.50	Holmium 67 **Ho** 164.93	Erbium 68 **Er** 167.26	Thulium 69 **Tm** 168.93	Ytterbium 70 **Yb** 173.05	Lutetium 71 **Lu** 174.97
Curium 96 **Cm** (247)	Berkelium 97 **Bk** (247)	Californium 98 **Cf** (251)	Einsteinium 99 **Es** (252)	Fermium 100 **Fm** (257)	Mendelevium 101 **Md** (258)	Nobelium 102 **No** (259)	Lawrencium 103 **Lr** (262)

Topographic Map Symbols

Topographic Map Symbols	
Primary highway, hard surface	Index contour
Secondary highway, hard surface	Supplementary contour
Light-duty road, hard or improved surface	Intermediate contour
Unimproved road	Depression contours
Railroad: single track	
Railroad: multiple track	Boundaries: national
Railroads in juxtaposition	State
	County, parish, municipal
Buildings	Civil township, precinct, town, barrio
cem Schools, church, and cemetery	Incorporated city, village, town, hamlet
Buildings (barn, warehouse, etc.)	Reservation, national or state
Wells other than water (labeled as to type)	Small park, cemetery, airport, etc.
Tanks: oil, water, etc. (labeled only if water)	Land grant
Located or landmark object; windmill	Township or range line, U.S. land survey
Open pit, mine, or quarry; prospect	
	Township or range line, approximate location
Marsh (swamp)	
Wooded marsh	Perennial streams
Woods or brushwood	Elevated aqueduct
Vineyard	Water well and spring
Land subject to controlled inundation	Small rapids
Submerged marsh	Large rapids
Mangrove	Intermittent lake
Orchard	Intermittent stream
Scrub	Aqueduct tunnel
Urban area	Glacier
	Small falls
x7369 Spot elevation	Large falls
670 Water elevation	Dry lake bed

Rocks

Rocks		
Rock Type	**Rock Name**	**Characteristics**
Igneous (intrusive)	Granite	Large mineral grains of quartz, feldspar, hornblende, and mica. Usually light in color.
	Diorite	Large mineral grains of feldspar, hornblende, and mica. Less quartz than granite. Intermediate in color.
	Gabbro	Large mineral grains of feldspar, augite, and olivine. No quartz. Dark in color.
Igneous (extrusive)	Rhyolite	Small mineral grains of quartz, feldspar, hornblende, and mica, or no visible grains. Light in color.
	Andesite	Small mineral grains of feldspar, hornblende, and mica or no visible grains. Intermediate in color.
	Basalt	Small mineral grains of feldspar, augite, and possibly olivine or no visible grains. No quartz. Dark in color.
	Obsidian	Glassy texture. No visible grains. Volcanic glass. Fracture looks like broken glass.
	Pumice	Frothy texture. Floats in water. Usually light in color.
Sedimentary (clastic)	Conglomerate	Coarse grained. Gravel or pebble-size grains.
	Sandstone	Sand-sized grains 1/16 to 2 mm.
	Siltstone	Grains are smaller than sand but larger than clay.
	Shale	Smallest grains. Often dark in color. Usually platy.
Sedimentary (chemical or biochemical)	Limestone	Major mineral is calcite. Usually forms in oceans and lakes. Often contains fossils.
	Coal	Forms in swampy areas. Compacted layers of organic material, mainly plant remains.
Sedimentary (chemical)	Rock Salt	Commonly forms by the evaporation of seawater.
Metamorphic (foliated)	Gneiss	Banding due to alternate layers of different minerals, of different colors. Parent rock often is granite.
	Schist	Parallel arrangement of sheetlike minerals, mainly micas. Forms from different parent rocks.
	Phyllite	Shiny or silky appearance. May look wrinkled. Common parent rocks are shale and slate.
	Slate	Harder, denser, and shinier than shale. Common parent rock is shale.
Metamorphic (nonfoliated)	Marble	Calcite or dolomite. Common parent rock is limestone.
	Soapstone	Mainly of talc. Soft with greasy feel.
	Quartzite	Hard with interlocking quartz crystals. Common parent rock is sandstone.

Minerals

Minerals					
Mineral (formula)	**Color**	**Streak**	**Hardness Pattern**	**Breakage Properties**	**Uses and Other**
Graphite (C)	black to gray	black to gray	1–1.5	basal cleavage (scales)	pencil lead, lubricants for locks, rods to control some small nuclear reactions, battery poles
Galena (PbS)	gray	gray to black	2.5	cubic cleavage perfect	source of lead, used for pipes, shields for X rays, fishing equipment sinkers
Hematite (Fe_2O_3)	black or reddish-brown	reddish-brown	5.5–6.5	irregular fracture	source of iron; converted to pig iron, made into steel
Magnetite (Fe_3O_4)	black	black	6	conchoidal fracture	source of iron, attracts a magnet
Pyrite (FeS_2)	light, brassy, yellow	greenish-black	6–6.5	uneven fracture	fool's gold
Talc ($Mg_3Si_4O_{10}(OH)_2$)	white, greenish	white	1	cleavage in one direction	used for talcum powder, sculptures, paper, and tabletops
Gypsum ($CaSO_4{\cdot}2H_2O$)	colorless, gray, white, brown	white	2	basal cleavage	used in plaster of paris and dry wall for building construction
Sphalerite (ZnS)	brown, reddish-brown, brown, greenish	light to dark brown	3.5–4	cleavage in six directions	main ore of zinc; used in paints, dyes, and medicine
Muscovite ($KAl_3Si_3O_{10}(OH)_2$)	white, light gray, yellow, rose, green	colorless	2–2.5	basal cleavage	occurs in large, flexible plates; used as an insulator in electrical equipment, lubricant
Biotite ($K(Mg,Fe)_3(AlSi_3O_{10})(OH)_2$)	black to dark brown	colorless	2.5–3	basal cleavage	occurs in large, flexible plates
Halite ($NaCl$)	colorless, red, white, blue	colorless	2.5	cubic cleavage	salt; soluble in water; a preservative

Minerals

Minerals					
Mineral (formula)	**Color**	**Streak**	**Hardness**	**Breakage Pattern**	**Uses and Other Properties**
Calcite ($CaCO_3$)	colorless, white, pale blue	colorless, white	3	cleavage in three directions	fizzes when HCl is added; used in cements and other building materials
Dolomite ($CaMg(CO_3)_2$)	colorless, white, pink, green, gray, black	white	3.5–4	cleavage in three directions	concrete and cement; used as an ornamental building stone
Fluorite (CaF_2)	colorless, white, blue, green, red, yellow, purple	colorless	4	cleavage in four directions	used in the manufacture of optical equipment; glows under ultraviolet light
Hornblende ($(CaNa)_{2\text{-}3}(Mg,Al,Fe)_5\text{-}(Al,Si)_2Si_6O_{22}(OH)_2$)	green to black	gray to white	5–6	cleavage in two directions	will transmit light on thin edges; 6-sided cross section
Feldspar ($KAlSi_3O_8$) ($NaAlSi_3O_8$), ($CaAl_2Si_2O_8$)	colorless, white to gray, green	colorless	6	two cleavage planes meet at 90° angle	used in the manufacture of ceramics
Augite ($(Ca,Na)(Mg,Fe,Al)(Al,Si)_2O_6$)	black	colorless	6	cleavage in two directions	square or 8-sided cross section
Olivine ($(Mg,Fe)_2SiO_4$)	olive, green	none	6.5–7	conchoidal fracture	gemstones, refractory sand
Quartz (SiO_2)	colorless, various colors	none	7	conchoidal fracture	used in glass manufacture, electronic equipment, radios, computers, watches, gemstones

Weather Map Symbols

Sample Station Model

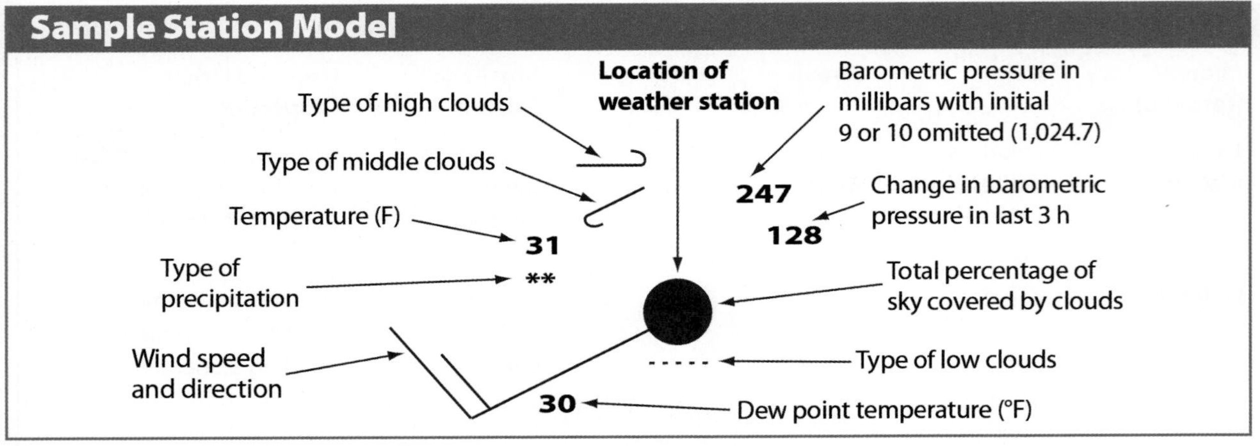

Sample Plotted Report at Each Station

Precipitation	Wind Speed and Direction	Sky Coverage	Some Types of High Clouds
≡ Fog	○ 0 calm	○ No cover	Scattered cirrus
★ Snow	1–2 knots	1/10 or less	Dense cirrus in patches
● Rain	3–7 knots	2/10 to 3/10	Veil of cirrus covering entire sky
Thunderstorm	8–12 knots	4/10	Cirrus not covering entire sky
Drizzle	13–17 knots	–	
▽ Showers	18–22 knots	6/10	
	23–27 knots	7/10	
	48–52 knots	Overcast with openings	
	1 knot = 1.852 km/h	● Completely overcast	

Some Types of Middle Clouds	Some Types of Low Clouds	Fronts and Pressure Systems
Thin altostratus layer	Cumulus of fair weather	(H) or High (L) or Low — Center of high- or low-pressure system
Thick altostratus layer	Stratocumulus	Cold front
Thin altostratus in patches	Fractocumulus of bad weather	Warm front
Thin altostratus in bands	Stratus of fair weather	Occluded front
		Stationary front

Use and Care of a Microscope

Eyepiece Contains magnifying lenses you look through.

Arm Supports the body tube.

Low-power objective Contains the lens with the lowest power magnification.

Stage clips Hold the microscope slide in place.

Coarse adjustment Focuses the image under low power.

Fine adjustment Sharpens the image under high magnification.

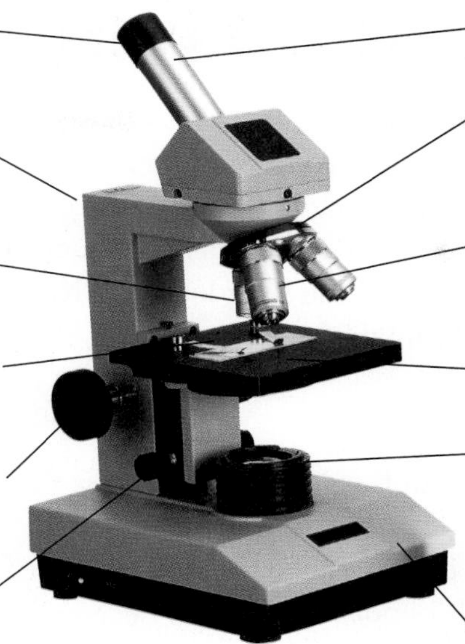

Body tube Connects the eyepiece to the revolving nosepiece.

Revolving nosepiece Holds and turns the objectives into viewing position.

High-power objective Contains the lens with the highest magnification.

Stage Supports the microscope slide.

Light source Provides light that passes upward through the diaphragm, the specimen, and the lenses.

Base Provides support for the microscope.

Caring for a Microscope

1. Always carry the microscope holding the arm with one hand and supporting the base with the other hand.
2. Don't touch the lenses with your fingers.
3. The coarse adjustment knob is used only when looking through the lowest-power objective lens. The fine adjustment knob is used when the high-power objective is in place.
4. Cover the microscope when you store it.

Using a Microscope

1. Place the microscope on a flat surface that is clear of objects. The arm should be toward you.
2. Look through the eyepiece. Adjust the diaphragm so light comes through the opening in the stage.
3. Place a slide on the stage so the specimen is in the field of view. Hold it firmly in place by using the stage clips.
4. Always focus with the coarse adjustment and the low-power objective lens first. After the object is in focus on low power, turn the nosepiece until the high-power objective is in place. Use ONLY the fine adjustment to focus with the high-power objective lens.

Making a Wet-Mount Slide

1. Carefully place the item you want to look at in the center of a clean, glass slide. Make sure the sample is thin enough for light to pass through.
2. Use a dropper to place one or two drops of water on the sample.
3. Hold a clean coverslip by the edges and place it at one edge of the water. Slowly lower the coverslip onto the water until it lies flat.
4. If you have too much water or a lot of air bubbles, touch the edge of a paper towel to the edge of the coverslip to draw off extra water and draw out unwanted air.

Diversity of Life: Classification of Living Organisms

A six-kingdom system of classification of organisms is used today. Two kingdoms—Kingdom Archaebacteria and Kingdom Eubacteria—contain organisms that do not have a nucleus and that lack membrane-bound structures in the cytoplasm of their cells. The members of the other four kingdoms have a cell or cells that contain a nucleus and structures in the cytoplasm, some of which are surrounded by membranes. These kingdoms are Kingdom Protista, Kingdom Fungi, Kingdom Plantae, and Kingdom Animalia.

Kingdom Archaebacteria

one-celled; some absorb food from their surroundings; some are photosynthetic; some are chemosynthetic; many are found in extremely harsh environments including salt ponds, hot springs, swamps, and deep-sea hydrothermal vents

Kingdom Eubacteria

one-celled; most absorb food from their surroundings; some are photosynthetic; some are chemosynthetic; many are parasites; many are round, spiral, or rod-shaped; some form colonies

Kingdom Protista

Phylum Euglenophyta one-celled; photosynthetic or take in food; most have one flagellum; euglenoids

Phylum Bacillariophyta one-celled; photosynthetic; have unique double shells made of silica; diatoms

Phylum Dinoflagellata one-celled; photosynthetic; contain red pigments; have two flagella; dinoflagellates

Phylum Chlorophyta one-celled, many-celled, or colonies; photosynthetic; contain chlorophyll; live on land, in freshwater, or salt water; green algae

Phylum Rhodophyta most are many-celled; photosynthetic; contain red pigments; most live in deep, saltwater environments; red algae

Phylum Phaeophyta most are many-celled; photosynthetic; contain brown pigments; most live in saltwater environments; brown algae

Phylum Rhizopoda one-celled; take in food; are free-living or parasitic; move by means of pseudopods; amoebas

Kingdom Eubacteria
Bacillus anthracis

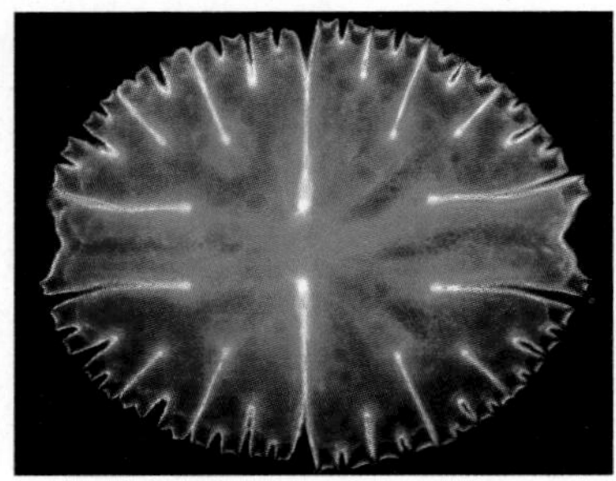

Phylum Chlorophyta
Desmid

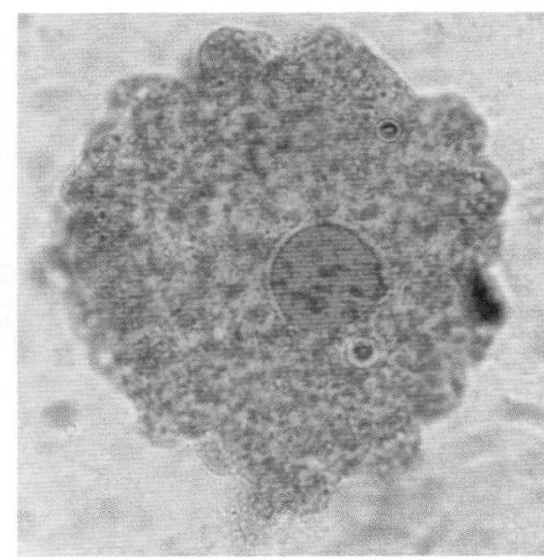

Amoeba

Phylum Zoomastigina one-celled; take in food; free-living or parasitic; have one or more flagella; zoomastigotes

Phylum Ciliophora one-celled; take in food; have large numbers of cilia; ciliates

Phylum Sporozoa one-celled; take in food; have no means of movement; are parasites in animals; sporozoans

Phylum Myxomycota
Slime mold

Phylum Oomycota
Phytophthora infestans

Phyla Myxomycota and Acrasiomycota one- or many-celled; absorb food; change form during life cycle; cellular and plasmodial slime molds

Phylum Oomycota many-celled; are either parasites or decomposers; live in freshwater or salt water; water molds, rusts and downy mildews

Kingdom Fungi

Phylum Zygomycota many-celled; absorb food; spores are produced in sporangia; zygote fungi; bread mold

Phylum Ascomycota one- and many-celled; absorb food; spores produced in asci; sac fungi; yeast

Phylum Basidiomycota many-celled; absorb food; spores produced in basidia; club fungi; mushrooms

Phylum Deuteromycota members with unknown reproductive structures; imperfect fungi; *Penicillium*

Phylum Mycophycota organisms formed by symbiotic relationship between an ascomycote or a basidiomycote and green alga or cyano-bacterium; lichens

Lichens

Kingdom Plantae

Divisions Bryophyta (mosses), **Anthocerophyta** (hornworts), **Hepaticophyta** (liverworts), **Psilophyta** (whisk ferns) many-celled nonvascular plants; reproduce by spores produced in capsules; green; grow in moist, land environments

Division Lycophyta many-celled vascular plants; spores are produced in conelike structures; live on land; are photosynthetic; club mosses

Division Arthrophyta vascular plants; ribbed and jointed stems; scalelike leaves; spores produced in conelike structures; horsetails

Division Pterophyta vascular plants; leaves called fronds; spores produced in clusters of sporangia called sori; live on land or in water; ferns

Division Ginkgophyta deciduous trees; only one living species; have fan-shaped leaves with branching veins and fleshy cones with seeds; ginkgoes

Division Cycadophyta palmlike plants; have large, featherlike leaves; produces seeds in cones; cycads

Division Coniferophyta deciduous or evergreen; trees or shrubs; have needlelike or scalelike leaves; seeds produced in cones; conifers

Division Gnetophyta shrubs or woody vines; seeds are produced in cones; division contains only three genera; gnetum

Division Anthophyta dominant group of plants; flowering plants; have fruits with seeds

Kingdom Animalia

Phylum Porifera aquatic organisms that lack true tissues and organs; are asymmetrical and sessile; sponges

Phylum Cnidaria radially symmetrical organisms; have a digestive cavity with one opening; most have tentacles armed with stinging cells; live in aquatic environments singly or in colonies; includes jellyfish, corals, hydra, and sea anemones

Phylum Platyhelminthes bilaterally symmetrical worms; have flattened bodies; digestive system has one opening; parasitic and free-living species; flatworms

Division Anthophyta
Tomato plant

Division Bryophyta
Liverwort

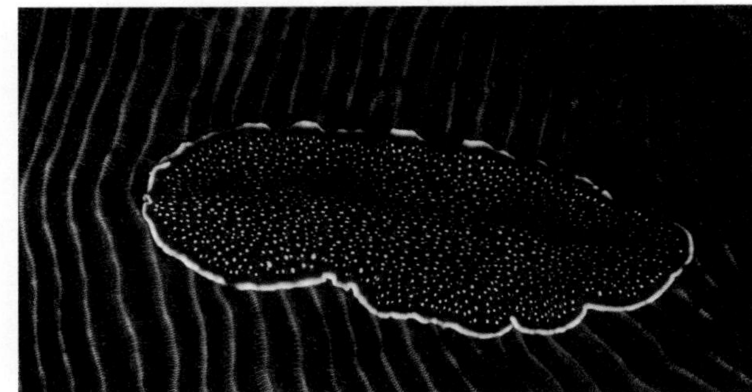

Phylum Platyhelminthes
Flatworm

Phylum Chordata

Phylum Nematoda round, bilaterally symmetrical body; have digestive system with two openings; free-living forms and parasitic forms; roundworms

Phylum Mollusca soft-bodied animals, many with a hard shell and soft foot or footlike appendage; a mantle covers the soft body; aquatic and terrestrial species; includes clams, snails, squid, and octopuses

Phylum Annelida bilaterally symmetrical worms; have round, segmented bodies; terrestrial and aquatic species; includes earthworms, leeches, and marine polychaetes

Phylum Arthropoda largest animal group; have hard exoskeletons, segmented bodies, and pairs of jointed appendages; land and aquatic species; includes insects, crustaceans, and spiders

Phylum Echinodermata marine organisms; have spiny or leathery skin and a water-vascular system with tube feet; are radially symmetrical; includes sea stars, sand dollars, and sea urchins

Phylum Chordata organisms with internal skeletons and specialized body systems; most have paired appendages; all at some time have a notochord, nerve cord, gill slits, and a post-anal tail; include fish, amphibians, reptiles, birds, and mammals

Glossary/Glosario

Multilingual eGlossary

A science multilingual glossary is available on ConnectEd. The glossary includes the following languages:

Arabic	Hmong	Tagalog
Bengali	Korean	Urdu
Chinese	Portuguese	Vietnamese
English	Russian	
Haitian Creole	Spanish	

Cómo usar el glosario en español:

1. Busca el término en inglés que desees encontrar.
2. El término en español, junto con la definición, se encuentran en la columna de la derecha.

Pronunciation Key

Use the following key to help you sound out words in the glossary:

a	back (BAK)	**ew**	f**oo**d (FEWD)
ay	day (DAY)	**yoo**	p**ure** (PYOOR)
ah	**father** (FAH thur)	**yew**	**few** (FYEW)
ow	fl**ow**er (FLOW ur)	**uh**	comm**a** (CAH muh)
ar	**car** (CAR)	**u (+ con)**	r**u**b (RUB)
e	**less** (LES)	**sh**	**sh**elf (SHELF)
ee	**leaf** (LEEF)	**ch**	na**t**ure (NAY chur)
ih	**trip** (TRIHP)	**g**	**g**ift (GIHFT)
i (i + con + e)	**i**dea (i DEE uh)	**j**	**g**em (JEM)
oh	**go** (GOH)	**ing**	s**ing** (SING)
aw	**soft** (SAWFT)	**zh**	vi**s**ion (VIH zhun)
or	**or**bit (OR buht)	**k**	**c**ake (KAYK)
oy	**coi**n (COYN)	**s**	**s**eed, **c**ent (SEED, SENT)
oo	f**oo**t (FOOT)	**z**	**z**one, rai**s**e (ZOHN, RAYZ)

English — A — Español

abiotic factor/angiosperm — **factor abiótico/angiosperma**

abiotic factor (ay bi AH tihk • FAK tuhr): a nonliving thing in an ecosystem. (p. 316)

acid precipitation: precipitation that has a lower pH than that of normal rainwater (pH 5.6). (p. 170)

adaptation: an inherited trait that increases an organism's chance of surviving and reproducing in a particular environment. (pp. 226, 282)

amnion: a protective membrane that surrounds an embryo. (p. 300)

amplitude: the maximum distance a wave varies from its rest position. (p. 452)

angiosperm: a plant that produces flowers and develops fruit. (p. 249)

factor abiótico: componente no vivo de un ecosistema. (pág. 316)

precipitación ácida: precipitación que tiene un pH más bajo que el del agua de la lluvia normal (pH 5.6). (pág. 170)

adaptación: rasgo heredado que aumenta la oportunidad de un organismo de sobrevivir y reproducirse en un medioambiente. (pág. 226, 282)

saco amniótico: membrana que rodea y protege al embrión. (pág. 300)

amplitud: distancia máxima que varía una onda desde su posición de reposo. (pág. 452)

angiosperma: planta que produce flores y desarrolla frutos. (pág. 249)

asteroid: a small, rocky object that orbits the Sun. (p. 53)

asthenosphere (as THE nuh sfir): the partially melted portion of the mantle below the lithosphere. (p. 108)

asymmetry: a body plan in which an organism cannot be divided into any two parts that are nearly mirror images. (p. 280)

atmosphere: a thin layer of gases surrounding Earth. (p. 77)

atom: a small particle that is the building block of matter. (p. 353)

atomic number: the number of protons in an atom of an element. (p. 370)

autotroph (AW tuh trohf): an organism that converts light energy to usable energy. (p. 192)

asteroide: objeto pequeño y rocoso que orbita el Sol. (pág. 53)

astenosfera: porción parcialmente fundida del manto debajo de la litosfera. (pág. 108)

asimetría: plano corporal en el cual un organismo no se puede dividir en dos partes que sean casi imágenes al espejo una de otra. (pág. 280)

atmósfera: capa delgada de gases que rodean la Tierra. (pág. 77)

átomo: partícula pequeña que es el componente básico de la materia. (pág. 353)

número atómico: número de protones en el átomo de un elemento. (pág. 370)

autotrófo: organismo que convierte la energía lumínica en energía útil. (pág. 192)

B

Big Bang theory: the scientific theory that states that the universe began from one point and has been expanding and cooling ever since. (p. 62)

bilateral symmetry: a body plan in which an organism can be divided into two parts that are nearly mirror images of each other. (p. 280)

binomial nomenclature (bi NOH mee ul • NOH mun klay chur): a naming system that gives each organism a two-word scientific name. (p. 194)

biomass energy: energy produced by burning organic matter, such as wood, food scraps, and alcohol. (p. 155)

biosphere: the Earth system that contains all living things. (p. 76)

biotic factor (bi AH tihk • FAK tuhr): a living or once-living thing in an ecosystem. (p. 317)

Teoría del Big Bang: teoría científica que establece que el universo se originó de un punto y se ha ido expandiendo y enfriando desde entonces. (pág. 62)

simetría bilateral: plano corporal en el cual un organismo se puede dividir en dos partes que sean casi imágenes al espejo una de otra. (pág. 280)

nomenclatura binomial: sistema de nombrar que le da a cada organismo un nombre científico de dos palabras. (pág. 194)

energía de biomasa: energía producida por la combustión de materia orgánica, como la madera, las sobras de comida y el alcohol. (pág. 155)

biosfera: el sistema de la Tierra que contiene todo que está viviendo. (pág. 76)

factor biótico: ser vivo o que una vez estuvo vivo en un ecosistema. (pág. 317)

C

camouflage (KAM uh flahj): an adaptation that enables a species to blend in with its environment. (p. 229)

chemical change: a change in matter in which the substances that make up the matter change into other substances with different chemical and physical properties. (p. 400)

camuflage: adaptación que permite a las especies mezclarse con su medioambiente. (pág. 229)

cambio químico: cambio de la materia en el cual las sustancias que componen la materia se transforman en otras sustancias con propi-edades químicas y físicas diferentes. (pág. 400)

chemical property: the ability or inability of a substance to combine with or change into one or more new substances. (p. 391)

propiedad química: capacidad o incapacidad de una sustancia para combinarse con una o más sustancias o transformarse en una o más sustancias. (pág. 391)

chemical weathering: the process that changes the composition of rocks and minerals due to exposure to the environment. (p. 127)

meteorización química: proceso que cambia la composición de las rocas y los minerales debido a la exposición al medioambiente. (pág. 127)

climate: the long-term average weather conditions that occur in a particular region. (p. 91)

clima: promedio a largo plazo de las condiciones del tiempo atmosférico de una región en particular. (pág. 91)

comet: a small, rocky, and icy object that orbits the Sun. (p. 53)

cometa: objeto pequeño, rocoso y helado que orbita el Sol. (pág. 53)

commensalism: a symbiotic relationship that benefits one species but does not harm or benefit the other. (p. 329)

comensalismo: relación simbiótica que beneficia a una especie pero no causa daño ni beneficia a la otra. (pág. 329)

community: all the populations living in an ecosystem at the same time. (p. 319)

comunidad: todas las poblaciones que viven en un ecosistema al mismo tiempo. (pág. 319)

competition: the demand for resources, such as food, water, and shelter, in short supply in a community. (p. 326)

competición: demanda de recursos, como alimento, agua y refugio, cuyo suministro es escaso en una comunidad. (pág. 326)

compound: a substance containing atoms of two or more different elements chemically bonded together. (p. 356)

compuesto: sustancia que contiene átomos de dos o más elementos diferentes unidos químicamente. (pág. 356)

compression: region of a longitudinal wave where the particles of the medium are closest together. (p. 471)

compresión: región de una onda longitudinal donde las partículas del medio están más cerca. (pág. 471)

condensation: the process by which a gas changes to a liquid. (p. 89)

condensación: proceso mediante el cual un gas cambia a líquido. (pág. 89)

consumer: an organism that cannot make its own food and gets energy by eating other organisms. (p. 335)

consumidor: organismo que no elabora su propio alimento y obtiene energía comiendo otros organismos. (pág. 335)

contour interval: the elevation difference between contour lines that are next to each other on a map. (p. 21)

intervalo de contorno: diferencia de elevación entre las líneas de contorno cercanas en un mapa. (pág. 21)

contour line: a line on a topographic map that connects points of equal elevation. (p. 20)

línea de contorno: línea que conecta puntos de igual elevación en un mapa topográfico. (pág. 20)

convection: the circulation of particles within a material caused by differences in thermal energy and density. (p. 111)

convección: circulación de partículas en el interior de un material causada por diferencias en la energía térmica y la densidad. (pág. 111)

convergent boundary: the boundary between two plates that move toward each other. (p. 109)

límite convergente: límite entre dos placas que se acercan una hacia la otra. (pág. 109)

critical thinking: comparing what you already know with information you are given in order to decide whether you agree with it. (p. NOS 10)

pensamiento crítico: comparación que se hace cuando se sabe algo acerca de información nueva, y se decide si se está o no de acuerdo con ella. (pág. NOS 10)

cross section: profile view that shows a vertical slice through rocks below the surface. (p. 23)

sección transversal: vista de perfil que muestra un corte vertical en las rocas bajo la superficie. (pág. 23)

cytoplasm: the liquid part of a cell inside the cell membrane; contains salts and other molecules. (p. 202)

citoplasma: fluido en el interior de una célula que contiene sales y otras moléculas. (pág. 202)

D

decibel: the unit used to measure sound intensity or loudness. (p. 473)

decibel: unidad usada para medir la intensidad o el volumen del sonido. (pág. 473)

deforestation: the removal of large areas of forests for human purposes. (p. 164)

deforestación: eliminación de grandes áreas de bosques con propósitos humanos. (pág. 164)

density: the mass per unit volume of a substance. (p. 389)

densidad: cantidad de masa por unidad de volumen de una sustancia. (pág. 389)

dependent variable: the factor a scientist observes or measures during an experiment. (p. NOS 21)

variable dependiente: factor que el científico observa o mide durante un experimento. (pág. NOS 21)

deposition: the laying down or settling of eroded material. (p. 128)

deposición: establecimiento o asentamiento de material erosionado. (pág. 128)

description: a spoken or written summary of observations. (p. NOS 12)

descripción: resumen oral o escrito de las observaciones. (pág. NOS 12)

divergent boundary: the boundary between two plates that move away from each other. (p. 109)

límite divergente: límite entre dos placas que se alejan una de la otra. (pág. 109)

dormancy: a period of no growth. (p. 255)

latencia: período sin crecimiento. (pág. 255)

dwarf planet: an object that orbits the Sun and is nearly spherical in shape, but shares its orbital path with other objects of similar size. (p. 52)

planeta enano: objeto de forma casi esférica que orbita el Sol y que comparte el recorrido de la órbita con otros objetos de tamaño similar. (pág. 52)

E

earthquake: vibrations caused by the rupture and sudden movement of rocks along a break or a crack in Earth's crust. (p. 115)

terremoto: vibraciones causadas por la ruptura y el movimiento repentino de las rocas en una fractura o grieta en la corteza de la Tierra. (pág. 115)

eclipse: the movement of one solar system object into the shadow of another object. (p. 47)

eclipse: movimiento de un objeto del sistema solar hacia la sombra de otro. (pág. 47)

ecosystem: all the living things and nonliving things in a given area. (p. 315)

ecosistema: todos los seres vivos y los componentes no vivos de un área dada. (pág. 315)

ectotherm: an animal that heats its body from heat in its environment. (p. 300)

electrically charged: the condition of having an unbalanced amount of positive charge or negative charge. (p. 488)

electrically neutral: a particle with equal amounts of positive charge and negative charge. (p. 488)

electric circuit: a closed, or complete, path in which an electric current flows. (p. 497)

electric current: the movement of electrically charged particles. (p. 495)

electric discharge: the process of an unbalanced electric charge becoming balanced. (p. 490)

electric energy: energy carried by an electric current. (p. 425)

electric field: the invisible region around a charged object where an electric force is applied. (p. 491)

electric force: the force that two electrically charged objects apply to each other. (p. 491)

electric resistance: a measure of how difficult it is for an electric current to flow in a material. (p. 499)

electromagnet: a magnet created by wrapping a current-carrying wire around a ferromagnetic core. (p. 512)

electromagnetic wave: a transverse wave that can travel through empty space and through matter. (p. 448)

electron: a negatively charged particle that occupies the space in an atom outside the nucleus. (p. 368)

electron cloud: the region surrounding an atom's nucleus where one or more electrons are most likely to be found. (p. 369)

element: a substance that consists of only one type of atom. (p. 355)

elevation: the height above sea level of any point on Earth's surface. (p. 20)

endoskeleton: the internal rigid framework that supports humans and other animals. (p. 282)

endotherm: an animal that generates its body heat from the inside. (p. 301)

ectotérmico: animal que calienta el cuerpo con el calor del medioambiente. (pág. 300)

cargado eléctricamente: condición de tener una cantidad no balanceada de carga positiva o negativa. (pág. 488)

eléctricamente neutro: partícula con cantidades iguales de carga positiva y negativa. (pág. 488)

circuito eléctrico: trayectoria cerrada, o completa, por la que fluye corriente eléctrica. (pág. 497)

corriente eléctrica: movimiento de partículas cargadas eléctricamente. (pág. 495)

descarga eléctrica: proceso por el cual una carga eléctrica no balanceada se vuelve balanceada. (pág. 490)

energía eléctrica: energía transportada por una corriente eléctrica. (pág. 425)

campo eléctrico: región invisible alrededor de un objeto cargado en donde se aplica una fuerza eléctrica. (pág. 491)

fuerza eléctrica: fuerza que dos objetos cargados eléctricamente se aplican entre sí. (pág. 491)

resistencia eléctrica: medida de qué tan difícil es para una corriente eléctrica fluir en un material. (pág. 499)

electroimán: imán fabricado al enrollar un alambre que transporta corriente alrededor de un núcleo ferromagnético. (pág. 512)

onda electromagnética: onda transversal que puede viajar a través del espacio vacío y de la materia. (pág. 448)

electrón: partícula cargada negativamente que ocupa el espacio por fuera del núcleo de un átomo. (pág. 368)

nube de electrones: región que rodea el núcleo de un átomo en donde es más probable encontrar uno o más electrones. (pág. 369)

elemento: sustancia que consiste de un sólo tipo de átomo. (pág. 355)

elevación: altura sobre el nivel del mar de cualquier punto de la superficie de la Tierra. (pág. 20)

endoesqueleto: armazón interno y rígido que soporta a los seres humanos y a otros animales. (pág. 282)

endotérmico: animal que genera calor corporal de su interior. (pág. 301)

energy: the ability to cause change. (p. 421)

equinox (EE kwuh nahks): when Earth's rotation axis is tilted neither toward nor away from the Sun. (p. 43)

erosion: the moving of weathered material, or sediment, from one location to another. (p. 128)

eukaryotic (yew ker ee AH tihk) cell: a cell that has a nucleus and other membrane-bound organelles. (p. 200)

evaporation: the process of a liquid changing to a gas at the surface of the liquid. (p. 88)

exoskeleton: a thick, hard outer covering; protects and supports an animal's body. (p. 282)

explanation: an interpretation of observations. (p. NOS 12)

energía: capacidad de ocasionar cambio. (pág. 421)

equinoccio: cuando el eje de rotación de la Tierra se inclina sin acercarse ni alejarse del Sol. (pág. 43)

erosión: transporte de material meteorizado, o de sedimento, de un lugar a otro. (pág. 128)

célula eucariótica: célula que tiene un núcleo y otros organelos limitados por una membrana. (pág. 200)

evaporación: proceso mediante el cual un líquido cambia a gas en la superficie del líquido. (pág. 88)

exoesqueleto: cubierta externa, gruesa y dura, que protege y soporta el cuerpo de un animal. (pág. 282)

explicación: interpretación de las observaciones. (pág. NOS 12)

fault: a crack or a fracture in Earth's lithosphere along which movement occurs. (p. 115)

food chain: a model that shows how energy flows in an ecosystem through feeding relationships. (p. 336)

food web: a model of energy transfer that can show how the food chains in a community are interconnected. (p. 336)

frequency: the number of wavelengths that pass by a point each second. (p. 451)

friction: a contact force that resists the sliding motion of two surfaces that are touching. (p. 431)

falla: grieta o fractura en la litosfera de la Tierra en la cual ocurre el movimiento. (pág. 115)

cadena alimentaria: modelo que explica cómo la energía fluye en un ecosistema a través de relaciones alimentarias. (pág. 336)

red alimentaria: modelo de transferencia de energía que explica cómo las cadenas alimentarias están interconectadas en una comunidad. (pág. 336)

frecuencia: número de longitudes de onda que pasan por un punto cada segundo. (pág. 451)

fricción: fuerza que resiste el movimiento de dos superficies que están en contacto. (pág. 431)

G

galaxy: a huge collection of stars, gas, and dust. (p. 61)

gas: matter that has no definite volume and no definite shape. (p. 386)

gene (JEEN): a section of DNA on a chromosome that has genetic information for one trait. (p. 218)

galaxia: conjunto enorme de estrellas, gas, y polvo. (pág. 61)

gas: materia que no tiene volumen ni forma definidos. (pág. 386)

gen: parte del ADN en un cromosoma que contiene información genética para un rasgo. (pág. 218)

generator: a machine that transforms mechanical energy to electric energy. (p. 498)

genotype (JEE nuh tipe): an organism's complete set of genes. (p. 220)

geologic map: a map that shows the surface geology of an area. (p. 23)

geosphere: the solid part of Earth. (p. 81)

geothermal energy: thermal energy from Earth's interior. (p. 155)

gill: an organ that exchanges carbon dioxide for oxygen in the water. (p. 298)

glacier: a large mass of ice formed by snow accumulation on land that moves slowly across Earth's surface. (p. 130)

groundwater: water that is stored in cracks and pores beneath Earth's surface. (p. 80)

gymnosperm: a plant that produces seeds that are not part of a flower. (p. 248)

generador: máquina que transforma energía mecánica en energía eléctrica. (pág. 498)

genotipo: juego completo de genes de un organismo. (pág. 220)

mapa geológico: mapa que muestra la geología de la superficie de un área. (pág. 23)

geosfera: parte sólida de la Tierra. (pág. 81)

energía geotérmica: energía térmica del interior de la Tierra. (pág. 155)

branquia: órgano que intercambia dióxido de carbono por oxígeno en el agua. (pág. 298)

glaciar: masa enorme de hielo formada por la acumulación de nieve en la tierra que se mueve lentamente por la superficie de la Tierra. (pág. 130)

agua subterránea: agua almacenada en grietas y poros debajo de la superficie de la Tierra. (pág. 80)

gimnosperma: planta que produce semillas que no son parte de una flor. (pág. 248)

H

habitat: the place within an ecosystem where an organism lives; provides the biotic and abiotic factors an organism needs to survive and reproduce. (pp. 193, 318)

heterogeneous mixture: a mixture in which substances are not evenly mixed. (p. 359)

heterotroph (HE tuh roh trohf): an organism that obtains energy from other organisms. (p. 192)

homogeneous mixture: a mixture in which two or more substances are evenly mixed but not bonded together. (p. 360)

hydroelectric power: electricity produced by flowing water. (p. 154)

hydrosphere: the system containing all Earth's water. (p. 79)

hydrostatic skeleton: a fluid-filled internal cavity surrounded by muscle tissue. (p. 282)

hypothesis: a possible explanation for an observation that can be tested by scientific investigations. (p. NOS 6)

hábitat: lugar en un ecosistema donde vive un organismo; proporciona los factores bióticos y abióticos de un organismo necesita para sobrevivir y reproducirse. (pág. 193, 318)

mezcla heterogénea: mezcla en la cual las sustancias no están mezcladas de manera uniforme. (pág. 359)

heterótrofo: organismo que obtiene energía de otros organismos. (pág. 192)

mezcla homogénea: mezcla en la cual dos o más sustancias están mezcladas de manera uniforme, pero no están unidas químicamente. (pág. 360)

energía hidroeléctrica: electricidad producida por agua que fluye. (pág. 154)

hidrosfera: sistema que contiene toda el agua de la Tierra. (pág. 79)

esqueleto hidrostático: cavidad interna llena de fluido y rodeada de tejido muscular. (pág. 282)

hipótesis: explicación posible de una observación que se puede probar por medio de investigaciones científicas. (pág. NOS 6)

I

independent variable: the factor that is changed by the investigator to observe how it affects a dependent variable. (p. NOS 21)

variable independiente: factor que el investigador cambia para observar cómo afecta la variable dependiente. (pág. NOS 21)

inference: a logical explanation of an observation that is drawn from prior knowledge or experience. (p. NOS 6)

inferencia: explicación lógica de una observación que se extrae de un conocimiento previo o experiencia. (pág. NOS 6)

infrared wave: an electromagnetic wave that has a wavelength shorter than a microwave but longer than visible light. (p. 460)

onda infrarroja: onda electromagnética que tiene una longitud de onda más corta que la de una microonda, pero más larga que la de la luz visible. (pág. 460)

inheritance: the passing of traits from generation to generation. (p. 217)

herencia: paso de rasgos de generación en generación. (pág. 217)

intensity: the amount of energy that passes through a square meter of space in one second. (p. 464)

intensidad: cantidad de energía que atraviesa un metro cuadrado de espacio en un segundo. (pág. 464)

International Date Line: the line of longitude 180° east or west of the prime meridian. (p. 14)

Línea de Fecha Internacional: línea de 180° de longitud al este u oeste del Meridiano de Greenwich. (pág. 14)

International System of Units (SI): the internationally accepted system of measurement. (p. NOS 12)

Sistema Internacional de Unidades (SI): sistema de medidas aceptado internacionalmente. (pág. NOS 12)

ion (I ahn): an atom that is no longer neutral because it has gained or lost electrons. (p. 371)

ión: átomo que no es neutro porque ha ganado o perdido electrones. (pág. 371)

isotopes (I suh tohps): atoms of the same element that have different numbers of neutrons. (p. 371)

isótopos: átomos del mismo elemento que tienen diferente número de neutrones. (pág. 371)

K

kinetic (kuh NEH tik) energy: energy due to motion. (p. 422)

energía cinética: energía debida al movimiento. (pág. 422)

L

latitude: the distance in degrees north or south of the equator. (p. 12)

latitud: distancia en grados al norte o al sur del Ecuador. (pág. 12)

lava: magma that erupts onto Earth's surface. (p. 118)

lava: magma que sale a la superficie de la Tierra. (pág. 118)

law of conservation of energy: law that states that energy can be transformed from one form to another, but it cannot be created or destroyed. (p. 430)

ley de la conservación de la energía: ley que plantea que la energía puede transformarse de una forma a otra, pero no puede crearse ni destruirse. (pág. 430)

law of conservation of mass: law that states that the total mass of the reactants before a chemical reaction is the same as the total mass of the products after the chemical reaction. (p. 403)

light-year: the distance light travels in one year. (p. 59)

liquid: matter with a definite volume but no definite shape. (p. 386)

lithosphere (LIH thuh sfihr): the rigid outermost layer of Earth that includes the uppermost mantle and crust. (p. 108)

longitude: the distance in degrees east or west of the prime meridian. (p. 12)

longitudinal (lahn juh TEWD nul) wave: a wave in which the disturbance is parallel to the direction the wave travels. (p. 449)

ley de la conservación de la masa: ley que plantea que la masa total de los reactivos antes de una reacción química es la misma que la masa total de los productos después de la reacción química. (pág. 403)

año luz: distancia que recorre la luz en un año. (pág. 59)

líquido: materia con volumen definido y forma indefinida. (pág. 386)

litosfera: capa rígida más externa de la Tierra que incluye el manto superior y la corteza. (pág. 108)

longitud: distancia en grados al este u oeste del Meridiano de Greenwich. (pág. 12)

onda longitudinal: onda en la que la perturbación es paralela a la dirección en que viaja la onda. (pág. 449)

magma: molten rock stored below Earth's surface. (p. 118)

magnet: an object that attracts iron. (p. 506)

magnetic domain: region in a magnetic material in which the magnetic fields of the atoms all point in the same direction. (p. 509)

magnetic force: the force that a magnet applies to another magnet. (p. 507)

magnetic material: any material that a magnet attracts. (p. 506)

mammary gland: special tissue that produces milk for young mammals. (p. 302)

mantle: a thin layer of tissue that covers a mollusk's internal organs. (p. 290)

map legend: a key that lists all the symbols used on a map. (p. 10)

map scale: the relationship between a distance on the map and the actual distance on the ground. (p. 11)

map view: a map drawn as if you were looking down on an area from above Earth's surface. (p. 9)

mass: the amount of matter in an object. (p. 388)

mass wasting: the downhill movement of a large mass of rocks or soil due to gravity. (p. 128)

magma: roca derretida almacenada debajo de la superficie de la Tierra. (pág. 118)

imán: objeto que atrae al hierro. (pág. 506)

dominio magnético: región en un material magnético en el que los campos magnéticos de los átomos apuntan en la misma dirección. (pág. 509)

fuerza magnética: fuerza que un imán lica a otro imán. (pág. 507)

material magnético: cualquier material que un imán atrae. (pág. 506)

glándula mamaria: tejido especial que produce leche para los mamíferos jóvenes. (pág. 302)

manto: capa delgada de tejido que cubre los órganos internos del molusco. (pág. 290)

leyenda del mapa: clave que lista todos los símbolos usados en un mapa. (pág. 10)

escala del mapa: relación entre la distancia en el mapa y la distancia real sobre tierra. (pág. 11)

vista del mapa: mapa trazado como si se estuviera mirando un área hacia abajo, desde arriba de la superficie de la Tierra. (pág. 9)

masa: cantidad de materia en un objeto. (pág. 388)

transporte en masa: movimiento cuesta debajo de gran cantidad de roca o suelo debido a la fuerza de gravedad. (pág. 128)

matter: anything that has mass and takes up space. (p. 353)

mechanical energy: sum of the potential energy and the kinetic energy in a system. (p. 425)

mechanical wave: a wave that can travel only through matter. (p. 448)

metamorphosis (me tuh MOR fuh sihs): a developmental process in which the body form of an animal changes as it grows from an egg to an adult. (p. 291)

meteor: a meteoroid that has entered Earth's atmosphere and produces a streak of light. (p. 53)

meteoroid: a small rocky particle that moves through space. (p. 53)

mid-ocean ridge: long, narrow mountain range on the ocean floor; formed by magma at divergent plate boundaries. (p. 120)

mimicry (MIH mih kree): an adaptation in which one species looks like another species. (p. 229)

mineral: a solid that is naturally occurring, inorganic, and has a crystal structure and definite chemical composition. (p. 81)

mitochondrion (mi tuh KAHN dree ahn): an organelle that breaks down food and releases energy. (p. 203)

mixture: matter that can vary in composition. (p. 358)

molecule (MAH lih kyewl): two or more atoms that are held together by covalent bonds and act as a unit. (p. 355)

molting: a process in which an outer covering, such as an exoskeleton, is shed and replaced. (p. 290)

moon: a natural satellite that orbits an object other than a star. (p. 53)

mutation (myew TAY shun): a permanent change in the sequence of DNA, or the nucleotides, in a gene or a chromosome. (p. 222)

mutualism: a symbiotic relationship in which both organisms benefit. (p. 329)

materia: cualquier cosa que tiene masa y ocupa espacio. (pág. 353)

energía mecánica: suma de la energía potencial y la energía cinética en un sistema. (pág. 425)

onda mecánica: onda que puede viajar sólo a través de la materia. (pág. 448)

metamorfosis: proceso de desarrollo en el cual la forma del cuerpo de un animal cambia a medida que crece de huevo a adulto. (pág. 291)

meteoro: meteorito que ha entrado a la atmósfera de la Tierra y produce un haz de luz. (pág. 53)

meteorito: partícula rocosa pequeña que se mueve por el espacio. (pág. 53)

dorsal oceánica: cordillera larga y angosta en el lecho del océano, formada por magma en los límites de las placas divergentes. (pág. 120)

mimetismo: adaptación en la cual una especie se parece a otra especie. (pág. 229)

mineral: sólido inorganico que se encuentra en la naturaleza, tiene una estructura cristalina y una composición química definida. (pág. 81)

mitocondria: organelo que descompone el alimento y libera energía. (pág. 203)

mezcla: materia que puede variar en composición. (pág. 358)

molécula: dos o más átomos que están unidos mediante enlaces covalentes y actúan como una unidad. (pág. 355)

muda: proceso en el cual una cubierta externa, como un exoesqueleto, se muda y reemplaza. (pág. 290)

luna: satélite natural que orbita un objeto diferente de una estrella. (pág. 53)

mutación: cambio permanente en la secuencia de ADN, de los nucleótidos, en un gen o en un cromosoma. (pág. 222)

mutualismo: relación simbiótica en la cual los dos organismos se benefician. (pág. 329)

SCIENCE SKILL HANDBOOK
MATH SKILL HANDBOOK
FOLDABLES HANDBOOK
REFERENCE HANDBOOK
GLOSSARY/GLOSARIO
INDEX

N

natural selection: the process by which organisms with variations that help them survive in their environment live longer, compete better, and reproduce more than those that do not have the variation. (p. 227)

neutron: a neutral particle in the nucleus of an atom. (p. 368)

niche (NICH): the way a species interacts with abiotic and biotic factors to obtain food, find shelter, and fulfill other needs. (p. 325)

nonrenewable resource: resource that is used faster than it can be replaced by natural processes. (p. 143)

nonvascular plant: a plant that lacks specialized tissues for transporting water and nutrients. (p. 246)

notochord: a flexible, rod-shaped structure that supports the body of a developing chordate. (p. 296)

nuclear energy: energy stored in and released from the nucleus of an atom. (pp. 147, 425)

nucleus: the region in the center of an atom where most of an atom's mass and positive charge is concentrated. (p. 368)

selección natural: proceso por el cual los organismos con variaciones que les ayudan a sobrevivir en sus medioambientes viven más, compiten mejor y se reproducen más que aquellos que no tienen la variación. (pág. 227)

neutrón: partícula neutra en el núcleo de un átomo. (pág. 368)

nicho: forma como una especie interactúa con los factores abióticos y bióticos para obtener alimento, encontrar refugio y satisfacer otras necesidades. (pág. 325)

recurso no renovable: recurso que se usa más rápidamente de lo que se puede reemplazar mediante procesos naturales. (pág. 143)

planta no vascular: planta que carece de tejidos especializados para transportar agua y nutrientes. (pág. 246)

notocordio: estructura flexible con forma de varilla que soporta el cuerpo de un cordado en desarrollo. (pág. 296)

energía nuclear: energía almacenada en y liberada por el núcleo de un átomo. (pág. 147, 425)

núcleo: región en el centro de un átomo donde se concentra la mayor cantidad de masa y las cargas positivas. (pág. 368)

O

observation: the act of using one or more of your senses to gather information and take note of what occurs. (p. NOS 6)

opaque: a material through which light does not pass. (p. 462)

ore: a deposit of minerals that is large enough to be mined for a profit. (p. 163)

overpopulation: condition that occurs when a population becomes so large that it causes damage to the environment. (p. 327)

observación: acción de usar uno o más sentidos para reunir información y tomar notar de lo que ocurre. (pág. NOS 6)

opaco: material por el que no pasa la luz. (pág. 462)

mena: depósito de minerales suficientemente grandes como para ser explotados con un beneficio. (pág. 163)

sobrepoblación: condición que ocurre cuando una población se vuelve tan grande que causa daño al medioambiente. (pág. 327)

P

parasite: an animal that survives by living inside or on another organism, gets food from the organism, and does not help in the organism's survival. (p. 287)

parásito: animal que vive en el interior o encima de otro organismo, y obtiene alimento del organismo sin ayudar a que el organismo sobreviva. (pág. 287)

parasitism: a symbiotic relationship in which one organism benefits and the other is harmed. (p. 329)

pharyngeal (fuh run JEE uhl) pouches: grooves along the side of a developing chordate. (p. 296)

phenotype (FEE nuh tipe): how a trait appears or is expressed. (p. 220)

photochemical smog: air pollution that forms from the interaction between chemicals in the air and sunlight. (p. 170)

physical change: a change in the size, shape, form, or state of matter that does not change the matter's identity. (p. 398)

physical property: a characteristic of matter that you can observe or measure without changing the identity of the matter. (p. 388)

physical weathering: the process of breaking down rocks and minerals without changing their compositions. (p. 126)

pistil: female reproductive organ of a flower. (p. 256)

pitch: the perception of how high or low a sound is; related to the frequency of a sound wave. (p. 471)

planet: an object that orbits the Sun, is large enough to be nearly spherical in shape, and has no other large object in its orbital path. (p. 51)

plate tectonics (tek TAH nihks): theory that Earth's surface is broken into large, rigid pieces that move with respect to each other. (p. 107)

pollination (pah luh NAY shun): the process that occurs when pollen grains land on a female reproductive structure of a plant that is the same species as the pollen grains. (p. 255)

population: all the organisms of the same species that live in the same area at the same time. (p. 319)

population density: the size of a population compared to the amount of space available. (p. 320)

potential (puh TEN chul) energy: stored energy due to the interactions between objects or particles. (p. 422)

parasitismo: relación simbiótica en la cual un organismo se beneficia y el otro se perjudica. (pág. 329)

hendiduras faríngeas: surcos a lo largo del lado de un cordado en desarrollo. (pág. 296)

fenotipo: forma como aparece o se expresa un rasgo. (pág. 220)

smog fotoquímico: polución del aire que se forma de la interacción entre los químicos en el aire y la luz solar. (pág. 170)

cambio físico: cambio en el tamaño, la forma o estado de la materia en el que no cambia la identidad de la materia. (pág. 398)

propiedad física: característica de la materia que puede observarse o medirse sin cambiar la identidad de la materia. (pág. 388)

meteorización física: proceso mediante el cual se rompen las rocas y los minerales, sin cambiar su composición. (pág. 126)

pistilo: órgano reproductor femenino de una flor. (pág. 256)

tono: percepción de qué tan alto o bajo es el sonido; relacionado con la frecuencia de la onda sonora. (pág. 471)

planeta: objeto que orbita el Sol, lo suficientemente grande para tener forma casi esférica, y que no tiene otro objeto grande en el recorrido de su órbita. (pág. 51)

tectónica de placas: teoría que afirma que la superficie de la Tierra está divida en piezas enormes y rígidas que se mueven una con respecto a la otra. (pág. 107)

polinización: proceso que ocurre cuando los granos de polen posan en una estructura reproductora femenina de una planta que es de la misma especie que los granos de polen. (pág. 255)

población: todos los organismos de la misma especie que viven en la misma área al mismo tiempo. (pág. 319)

densidad poblacional: tamaño de una población comparado con la cantidad de espacio disponible. (pág. 320)

energía potencia: energía almacenada debido a las interacciones entre objetos o partículas. (pág. 422)

precipitation: water, in liquid or solid form, that falls from the atmosphere. (p. 89)

predation: the act of one organism, the predator, feeding on another organism, its prey. (p. 328)

prediction: a statement of what will happen next in a sequence of events. (p. NOS 6)

producer: an organism that uses an outside energy source, such as the Sun, and produces its own food. (p. 334)

profile view: a drawing showing a vertical "slice" through the ground. (p. 93)

prokaryotic (pro kayr ee AH tihk) cell: a cell that does not have a nucleus or other membrane-bound organelles. (p. 200)

proton: a positively charged particle in the nucleus of an atom. (p. 368)

precipitación: agua, en forma líquida o sólida, que cae de la atmósfera. (pág. 89)

depredación: acción en la cual un organismo, el depredador, come a otro organismo, la presa. (pág. 328)

predicción: afirmación de lo que ocurrirá después en una secuencia de eventos. (pág. NOS 6)

productor: organismo que usa una fuente de energía externa, como el Sol, para elaborar su propio alimento. (pág. 334)

vista de perfil: dibujo que muestra un "corte" vertical a través de la tierra. (pág. 93)

célula procariota: célula que no tiene núcleo ni otros organelos limitados por una membrana. (pág. 200)

protón: partícula cargada positivamente en el núcleo de un átomo. (pág. 368)

R

radial symmetry: a body plan in which an organism can be divided into two parts that are nearly mirror images of each other anywhere through its central axis. (p. 280)

radiant energy: energy carried by an electromagnetic wave. (p. 425)

radio wave: a low-frequency, low-energy electromagnetic wave that has a wavelength longer than about 30 cm. (p. 459)

rarefaction (rayr uh FAK shun): region of a longitudinal wave where the particles of the medium are farthest apart. (p. 471)

reclamation: a process in which mined land must be recovered with soil and replanted with vegetation. (p. 149)

refraction: the change in direction of a wave as it changes speed in moving from one medium to another. (p. 454)

relief: the difference in elevation between the highest and lowest point in an area. (p. 20)

remote sensing: the process of collecting information about an area without coming into contact with it. (p. 27)

renewable resource: a resource that can be replenished by natural processes at least as quickly as it is used. (p. 143)

simetría radial: plano corporal en el cual un organismo se puede dividir en dos partes que sean casi imágenes al espejo una de la otra, en cualquier parte de su eje axial. (pág. 280)

energía radiante: energía que transporta una onda electromagnética. (pág. 425)

onda de radio: onda electromagnética de baja frecuencia y baja energía que tiene una longitud de onda mayor de más o menos 30 cm. (pág. 459)

rarefacción: region de una onda longitudinal donde las partículas del medio están más alejadas. (pág. 471)

recuperación: proceso por el cual las tierras explotadas se deben recubrir con suelo y se deben replantar con vegetación. (pág. 149)

refracción: cambio en la dirección de una onda a medida que cambia de rapidez al moverse de un medio a otro. (pág. 454)

relieve: diferencia de elevación entre el punto más alto y el más bajo en un área. (pág. 20)

teledetección: proceso de recolectar información sobre un área sin entrar en contacto con ella. (pág. 27)

recurso renovable: recurso natural que se reabastece por procesos naturales al menos tan rápidamente como se usa. (pág. 143)

SCIENCE SKILL HANDBOOK
MATH SKILL HANDBOOK
FOLDABLES HANDBOOK
REFERENCE HANDBOOK
GLOSSARY/GLOSARIO
INDEX

revolution: the orbit of one object around another object. (p. 42)

rhizoid: a structure that anchors a nonvascular plant to a surface. (p. 244)

rock: a naturally occurring solid composed of minerals, rock fragments, and sometimes other materials such as organic matter. (p. 82)

rock cycle: the series of processes that change one type of rock into another type of rock. (p. 92)

rotation: the spin of an object around its axis. (p. 42)

revolución: movimiento de un objeto alrededor de otro objeto. (pág. 42)

rizoide: estructura que sujeta una planta no vascular a una superficie. (pág. 244)

roca: sólido de origen natural compuesto de minerales, acumulación de fragmentos y algunas veces de otros materiales como materia orgánica. (pág. 82)

ciclo geológico: series de procesos que cambian un tipo de roca en otro tipo de roca. (pág. 92)

rotación: movimiento giratorio de un objeto sobre su eje. (pág. 42)

S

science: the investigation and exploration of natural events and of the new information that results from those investigations. (p. NOS 4)

scientific law: a rule that describes a pattern in nature. (p. NOS 9)

scientific theory: an explanation of observations or events that is based on knowledge gained from many observations and investigations. (p. NOS 9)

sediment: rock material that forms when rocks are broken down into smaller pieces or dissolved in water as rocks erode. (p. 125)

selective breeding: the selection and breeding of organisms for desired traits. (p. 228)

significant digits: the number of digits in a measurement that are known with a certain degree of reliability. (p. NOS 14)

slope: a measure of the steepness of the land. (p. 21)

solar energy: energy from the Sun. (p. 153)

solid: matter that has a definite shape and a definite volume. (p. 386)

solstice (SAHL stuhs): when Earth's rotation axis is tilted directly toward or away from the Sun. (p. 43)

solubility (sahl yuh BIH luh tee): the maximum amount of solute that can dissolve in a given amount of solvent at a given temperature and pressure. (p. 390)

ciencia: la investigación y exploración de los eventos naturales y de la información nueva que es el resultado de estas investigaciones. (pág. NOS 4)

ley científica: regla que describe un patrón dado en la naturaleza. (pág. NOS 9)

teoría científica: explicación de observaciones o eventos con base en conocimiento obtenido de muchas observaciones e investigaciones. (pág. NOS 9)

sedimento: material rocoso formado cuando las rocas se rompen en piezas pequeñas o se disuelven en agua al erosionarse. (pág. 125)

cría selectiva: selección y cría de organismos para características deseadas. (pág. 228)

cifras significativas: número de dígitos que se conoce con cierto grado de fiabilidad en una medida. (pág. NOS 14)

pendiente: medida de la inclinación de un terreno. (pág. 21)

energía solar: energía proveniente del Sol. (pág. 153)

sólido: materia con forma y volumen definidos. (pág. 386)

solsticio: cuando el eje de rotación de la Tierra se inclina acercándose o alejándose del Sol. (pág. 43)

solubilidad: cantidad máxima de soluto que puede disolverse en una cantidad dada de solvente a temperatura y presión dadas. (pág. 390)

sound energy: energy carried by sound waves. (p. 425)

stamen: the male reproductive organ of a flower. (p. 256)

star: a large sphere of hydrogen gas, held together by gravity, that is hot enough for nuclear reactions to occur in its core. (p. 59)

stimulus (STIHM yuh lus): a change in an organism's environment that causes a response. (p. 265)

stoma (STOH muh): a small opening in the epidermis, or surface layer, of a leaf. (p. 245)

subduction: the process that occurs when one tectonic plate moves under another tectonic plate. (p. 109)

substance: matter with a composition that is always the same. (p. 354)

symbiosis (sihm bee OH sus): a close, long-term relationship between two species that usually involves an exchange of food or energy. (p. 329)

energía sonora: energía que transportan las ondas sonoras. (pág. 425)

estambre: órgano reproductor masculino de una flor. (pág. 256)

estrella: esfera enorme de gas de hidrógeno, que se mantiene unida por la gravedad, lo suficientemente caliente para producir reacciones nucleares en el núcleo. (pág. 59)

estímulo: cambio en el medioambiente de un organismo que causa una respuesta. (pág. 265)

estoma: abertura pequeña en la epidermis, capa superficial, de una hoja. (pág. 245)

subducción: proceso que ocurre cuando una placa tectónica se mueve debajo de otra placa tectónica. (pág. 109)

sustancia: materia cuya composición es siempre la misma. (pág. 354)

simbiosis: relación estrecha a largo plazo entre dos especies que generalmente involucra intercambio de alimento o energía. (pág. 329)

T

taxon: a group of organisms. (p. 195)

technology: the practical use of scientific knowledge, especially for industrial or commercial use. (p. NOS 8)

thermal energy: the sum of the kinetic energy and the potential energy of the particles that make up an object. (p. 425)

tide: the periodic rise and fall of the ocean's surface caused by the gravitational force between Earth and the Moon, and Earth and the Sun. (p. 46)

time zone: the area on Earth's surface between two meridians where people use the same time. (p. 14)

topographic map: a map showing the detailed shapes of Earth's surface, along with its natural and human-made features. (p. 20)

trait: a distinguishing characteristic of an organism. (p. 217)

transform boundary: the boundary between two plates that slide past each other. (p. 109)

taxón: grupo de organismos. (pág. 195)

tecnología: uso práctico del conocimiento científico, especialmente para uso industrial o comercial. (pág. NOS 8)

energía térmica: suma de la energía cinética y potencial de las partículas que componen un objeto. (pág. 425)

marea: ascenso y descenso periódico de la superficie del océano causados por la fuerza gravitacional entre la Tierra y la Luna, y entre la Tierra y el Sol. (pág. 46)

zona horaria: área en la superficie de la Tierra entre dos meridianos donde la gente maneja la misma hora. (pág. 14)

mapa topográfico: mapa que muestra las formas detalladas de la superficie de la Tierra junto con sus características naturales y artificiales. (pág. 20)

rasgo: característica distintiva de un organismo. (pág. 217)

límite transformante: límite entre dos placas que se deslizan una con respecto a la otra. (pág. 109)

Science Skill Handbook
Math Skill Handbook
Foldables Handbook
Reference Handbook
Glossary/Glosario
Index

translucent: a material that allows most of the light that strikes it to pass through, but through which objects appear blurry. (p. 462)

transparent: a material that allows almost all of the light striking it to pass through, and through which objects can be seen clearly. (p. 462)

transpiration: the process by which plants release water vapor through their leaves. (pp. 88, 264)

transverse wave: a wave in which the disturbance is perpendicular to the direction the wave travels. (p. 449)

tropism (TROH pih zum): plant growth toward or away from an external stimulus. (p. 266)

translúcido: material que permite el paso de la mayor cantidad de luz que lo toca, pero a través del cual los objetos se ven borrosos. (pág. 462)

transparente: material que permite el paso de la mayor cantidad de luz que lo toca, y a través del cual los objetos pueden verse con nitidez. (pág. 462)

transpiración: proceso por el cual las plantas liberan vapor de agua por medio de las hojas. (pág. 88, 264)

onda transversal: onda en la que la perturbación es perpendicular a la dirección en que viaja la onda. (pág. 449)

tropismo: crecimiento de una planta hacia o alejado de un estímulo externo. (pág. 266)

ultraviolet wave: an electromagnetic wave that has a slightly shorter wavelength and higher frequency than visible light. (p. 460)

uplift: the process that moves large bodies of Earth materials to higher elevations. (p. 92)

onda ultravioleta: onda electromagnética que tiene una longitud de onda ligeramente menor y mayor frecuencia que la luz visible. (pág. 460)

levantamiento: proceso por el cual se mueven grandes cuerpos de materiales de la Tierra hacia elevaciones mayores. (pág. 92)

variable: any factor that can have more than one value. (p. NOS 21)

variation (ver ee AY shun): a slight difference in an inherited trait among individual members of a species. (p. 226)

vascular plant: a plant that has specialized tissues, called vascular tissues, that transport water and nutrients throughout the plant. (p. 247)

volcano: a vent in Earth's crust through which molten rock flows. (p. 118)

voltage: the amount of energy used to move one coulomb of electrons through an electric circuit. (p. 501)

volume: the amount of space a sample of matter occupies. (p. 386)

variable: cualquier factor que tenga más de un valor. (pág. NOS 21)

variación: ligera diferencia en un rasgo hereditario entre los miembros individuales de una especie. (pág. 226)

planta vascular: planta que tiene tejidos especializados, llamados tejidos vasculares, que transportan agua y nutrientes por la planta. (pág. 247)

volcán: abertura en la corteza terrestre por donde fluye la roca derretida. (pág. 118)

voltaje: cantidad de energía usada para mover un culombio de electrones por un circuito eléctrico. (pág. 501)

volumen: cantidad de espacio que ocupa la materia. (pág. 386)

W

waning: a phase of the Moon during which less of the lit part of the Moon is visible each night. (p. 45)

water cycle: the series of natural processes by which water continually moves throughout the hydrosphere. (p. 87)

waxing: a phase of the Moon during which more of the lit part of the Moon is visible each night. (p. 45)

weather: the atmospheric conditions, along with short-term changes, of a certain place at a certain time. (p. 90)

weathering: the mechanical and chemical processes that change Earth's surface over time. (p. 125)

wind farm: a group of wind turbines that produce electricity. (p. 154)

work: the amount of energy used as a force that moves an object over a distance. (p. 424)

menguante: fase de la Luna durante la cual se ve menos del lado iluminado de la Luna cada noche. (pág. 45)

ciclo del agua: serie de procesos naturales por los que el cual el agua se mueve continuamente en toda la hidrosfera. (pág. 87)

creciente: fase de la Luna durante la cual se ve más del lado iluminado de la Luna es cada noche. (pág. 45)

tiempo atmosférico: condiciones atmosféricas, junto con cambios a corto plazo, de un lugar determinado a una hora determinada. (pág. 90)

meteorización: procesos mecánicos y químicos que con el paso del tiempo cambian la superficie de la Tierra. (pág.125)

parque eólico: grupo de turbinas de viento que produce electricidad. (pág. 154)

trabajo: cantidad de energía usada como fuerza que mueve un objeto a cierta distancia. (pág. 424)

A

B

C

SCIENCE SKILL HANDBOOK
MATH SKILL HANDBOOK
FOLDABLES HANDBOOK
REFERENCE HANDBOOK
GLOSSARY/GLOSARIO
INDEX

F

SCIENCE SKILL HANDBOOK
MATH SKILL HANDBOOK
FOLDABLES HANDBOOK
REFERENCE HANDBOOK
GLOSSARY/GLOSARIO
INDEX

N

T

V

W

Z

Domain.
Kingdom.
Phylum.
Class.
Order.
family.
genus:
Species.

PERIODIC TABLE OF THE ELEMENTS

Element	Hydrogen
Atomic number	1
Symbol	H
Atomic mass	1.01
State of matter	Gas

Gas
Liquid
Solid
Synthetic

A column in the periodic table is called a **group.**

A row in the periodic table is called a **period.**

Period	1	2	3	4	5	6	7	8	9
1	Hydrogen 1 **H** 1.01								
2	Lithium 3 **Li** 6.94	Beryllium 4 **Be** 9.01							
3	Sodium 11 **Na** 22.99	Magnesium 12 **Mg** 24.31							
4	Potassium 19 **K** 39.10	Calcium 20 **Ca** 40.08	Scandium 21 **Sc** 44.96	Titanium 22 **Ti** 47.87	Vanadium 23 **V** 50.94	Chromium 24 **Cr** 52.00	Manganese 25 **Mn** 54.94	Iron 26 **Fe** 55.85	Cobalt 27 **Co** 58.93
5	Rubidium 37 **Rb** 85.47	Strontium 38 **Sr** 87.62	Yttrium 39 **Y** 88.91	Zirconium 40 **Zr** 91.22	Niobium 41 **Nb** 92.91	Molybdenum 42 **Mo** 95.96	Technetium 43 **Tc** (98)	Ruthenium 44 **Ru** 101.07	Rhodium 45 **Rh** 102.91
6	Cesium 55 **Cs** 132.91	Barium 56 **Ba** 137.33	Lanthanum 57 **La** 138.91	Hafnium 72 **Hf** 178.49	Tantalum 73 **Ta** 180.95	Tungsten 74 **W** 183.84	Rhenium 75 **Re** 186.21	Osmium 76 **Os** 190.23	Iridium 77 **Ir** 192.22
7	Francium 87 **Fr** (223)	Radium 88 **Ra** (226)	Actinium 89 **Ac** (227)	Rutherfordium 104 **Rf** (267)	Dubnium 105 **Db** (268)	Seaborgium 106 **Sg** (271)	Bohrium 107 **Bh** (272)	Hassium 108 **Hs** (270)	Meitnerium 109 **Mt** (276)

The number in parentheses is the mass number of the longest lived isotope for that element.

Lanthanide series	Cerium 58 **Ce** 140.12	Praseodymium 59 **Pr** 140.91	Neodymium 60 **Nd** 144.24	Promethium 61 **Pm** (145)	Samarium 62 **Sm** 150.36	Europium 63 **Eu** 151.96
Actinide series	Thorium 90 **Th** 232.04	Protactinium 91 **Pa** 231.04	Uranium 92 **U** 238.03	Neptunium 93 **Np** (237)	Plutonium 94 **Pu** (244)	Americium 95 **Am** (243)